Elementary Modular
Iwasawa Theory

Series on Number Theory and Its Applications   Vol. 16

Elementary Modular
Iwasawa Theory

Haruzo Hida
University of California, Los Angeles, USA

World Scientific

NEW JERSEY · LONDON · SINGAPORE · BEIJING · SHANGHAI · HONG KONG · TAIPEI · CHENNAI · TOKYO

Published by

World Scientific Publishing Co. Pte. Ltd.

5 Toh Tuck Link, Singapore 596224

USA office: 27 Warren Street, Suite 401-402, Hackensack, NJ 07601

UK office: 57 Shelton Street, Covent Garden, London WC2H 9HE

Library of Congress Cataloging-in-Publication Data
Names: Hida, Haruzo, author.
Title: Elementary modular Iwasawa theory / Haruzo Hida,
    University of California, Los Angeles, USA.
Description: New Jersey : World Scientific Publishing, [2022] |
    Series: Series on number theory and its applications, 1793-3161 ; Vol. 16 |
    Includes bibliographical references and index.
Identifiers: LCCN 2021034577 | ISBN 9789811241369 (hardcover) |
    ISBN 9789811241376 (ebook)
Subjects: LCSH: Iwasawa theory. | Galois theory. | Modules (Algebra)
Classification: LCC QA247.3 .H53 2022 | DDC 512.7/4--dc23
LC record available at https://lccn.loc.gov/2021034577

British Library Cataloguing-in-Publication Data
A catalogue record for this book is available from the British Library.

For any available supplementary material, please visit
https://www.worldscientific.com/worldscibooks/10.1142/12398#t=suppl

Printed in Singapore

Preface

Towards the end of his research career, Kenkichi Iwasawa became somehow more speculative [I79]. Perhaps, he was not totally happy with his achievement up to that point and wanted to make explicit what he had originally aimed to accomplish.

He ventured into Number Theory, particularly into the area dealing with cyclotomic fields, in his mid-career, resurrecting the theory studied in depth by Kummer/Kronecker and Stickelberger (which was a little outdated at the time because of the rapid advance of arithmetic geometry).

Though some theorems he proved on the arithmetic of the p-cyclotomic field $\mathbb{Q}(\mu_{p^\infty})$ (with Galois group over $\mathbb{Q}(\mu_p)$ canonically isomorphic to the p-adic integer ring $\mathbb{Z}_p$) have been extended to general extensions of a number field with Galois group isomorphic (often non-canonically) to $\mathbb{Z}_p$, the cyclotomic $\mathbb{Z}_p$-extensions still exhibit miraculous features incomparable to general cases. Here μ_{p^n} is the group of p^n-th roots of unity. A $\mathbb{Z}_p$-*extension* is a tower $K_\infty = \bigcup_n K_n$ of cyclic extensions K_n/K with $\mathrm{Gal}(K_n/K) \cong \Gamma/\Gamma^{p^n}$ for the multiplicative group $\boldsymbol{\Gamma} = 1+p\mathbb{Z}_p$ isomorphic to the additive group $\mathbb{Z}_p$ by the p-adic logarithm (assuming that the prime p is odd). If $K = \mathbb{Q}(\mu_p)$ and $K_\infty = \mathbb{Q}(\mu_{p^\infty}) = \bigcup_n \mathbb{Q}(\mu_{p^n})$, the isomorphism is given by the p-adic cyclotomic character ν_p and is canonical.

His starting point is the class number formulas Dirichlet and Kummer invented for $K_n := \mathbb{Q}(\mu_{p^{n+1}})$ with $K = \mathbb{Q}(\mu_p)$, which compute the order of the class group of the cyclotomic field in terms of Dirichlet L-values at $s = 1$. The formula has a conjectural generalization to arithmetic cohomology groups in terms of algebro-geometric (i.e., motivic) L-values. Although he started with the order formula of a specific group, his proof [I69] of his main conjecture under the Kummer–Vandiver conjecture reveals that he did actually determine the structure of the p-primary part C_n of the class group Cl_{K_n} of K_n via Galois action. The group algebras $\mathbb{Z}_p[\mathrm{Gal}(K_n/K)] = \mathbb{Z}_p[\Gamma/\Gamma^{p^n}] \cong \mathbb{Z}_p[t]/(t^{p^n} - 1)$ after passing to the limit towards the top of the tower $\mathbb{Z}_p[[\Gamma]] = \varprojlim_n \mathbb{Z}_p[t]/(t^{p^n} - 1)$ is isomorphic to the one variable power series ring $\Lambda := \mathbb{Z}_p[[T]]$ ($t = 1 + T$) called the Iwasawa algebra. Cutting out $C_\infty := \varprojlim_n C_n$ by an odd character $\omega^a : \mathrm{Gal}(K/\mathbb{Q}) \to \mathbb{Z}_p^\times$ (regarding $\mathrm{Gal}(K/\mathbb{Q}) \subset \mathrm{Gal}(K_\infty/\mathbb{Q}) = \Gamma \times \mathrm{Gal}(K/\mathbb{Q})$), the ω^a-eigenspace $C_\infty[\omega^a]$ is a Λ-module. Here we say ω^a is odd if $\omega^a(c) = -1$ for complex conjugation c. What Iwasawa proved under the Kummer–Vandiver conjecture is that $C_\infty[\omega^a] \cong \Lambda/(L_a)$ for the power series L_a of a branch of the p-adic Riemann zeta function constructed by Kubota–Leopoldt and Iwasawa (a cyclicity result of $C_\infty[\omega^a]$ as a Λ-module). We can recover $C_n[\omega^a]$ as $\mathbb{Z}_p[\Gamma/\Gamma^{p^n}]/(\overline{L}_a)$ for the image of $\overline{L}_a$ in the group ring of finite level.

Note here that Λ is a unique factorization domain; so, a square-free power series makes sense. Without assuming the Kummer–Vandiver conjecture, Iwasawa's speculation on this cyclicity question in [179] goes on as follows:

(1) Is $C_\infty[\omega^a] \cong \Lambda/(L_a)$ *up to finite error?*

(2) Is L_a *square-free in* Λ?

(3) Is $L_a = (T - \alpha)u(T)$ *with* $\alpha \in p\mathbb{Z}_p$ *and* $u(T) \in \Lambda^\times$?

Plainly (3) $\Rightarrow$ (2) $\Rightarrow$ (1). Because of the present limit in computation as of the date of publication, all these are numerically verified for primes up to two billion [HHO17]; so, it is rather difficult to find numerically a counter example.

Long ago, when I tried to venture out of the abelian world to explore somehow similar features in non-abelian theory, I wanted to find easy but still systematic examples full of miraculous phenomena (not a grandiose theory, like class field theory). My old paper [H86b] which won the Leroy P. Steele Prize for Seminal Contribution to Research in 2019 was written with this aspiration in mind, and Chapter 4 in this book contains a detailed exposition and a generalization of the results of the paper. See [H20] and an interview of mine (recorded in Chinese) posted on:

https://web.math.sinica.edu.tw/math_media/d442/44201.pdf

for some historical facts and my thoughts at the time. Some of such phenomena might be out of the reach of geometry, rather living in a more arithmetic world. After Grothendieck, we fully exploited algebro-geometric tools to understand number theoretic phenomena, but number theory cannot be just confined to the geometric world only. After the basic theory of deformation of modular forms and Galois representations, we present our point that if the universal deformation ring is generated by (at most) one element over the Iwasawa algebra, we have cyclicity of the (corresponding) adjoint Selmer group over the deformation ring. As we will see, the Iwasawa algebra often appears as a ring carrying the universal deformation. After all, Iwasawa's original example is a 1-dimensional adjoint representation twisted by a Galois character; so, the phenomenon we found is systematic.

Here is an outline of this book, in which plenty of open questions are scattered. Hopefully, some of them would interest some adventurous readers. Iwasawa's elementary proof which we mentioned above is described concisely in Chapter 1 with some extra facts which we will need later. Kubert–Lang found another (modular) elementary example to which Iwasawa's argument apply, which is recalled in Chapter 2. In Chapter 3, we construct the p-adic L-function L_a analytically, and its modular counterpart is also described. In Chapter 4, we start deforming modular forms p-adically. We go back to abelian cases in Chapter 5. We explain why the abelian case is a part of the "adjoint" world. In addition, we show that the Iwasawa algebra is the abelian universal deformation ring. In Chapter 6, we describe the identity of the universal Galois deformation ring R and the universal modular deformation ring $\mathbb{T}$ (i.e., a local ring of a Hecke algebra), first conjectured by Mazur and later proven by Taylor–Wiles and others (this result is called the $R = \mathbb{T}$ theorem). In Chapter 7, we prove many cyclicity results for the adjoint Selmer groups. In Chapter 8, we characterize the generator of the universal ring over the Iwasawa algebra as a generator of the unipotent part of the p-inertia group. Via this fact, we prove that a modular non-induced deformation (of weight ≥ 2) of an induced representation is indecomposable over the p-inertia group (a conjecture of Greenberg made for general modular non-induced Galois representation). In the final Chapter 9, we prove that the order of the (p-adic) adjoint Selmer group is equal to the algebraic part of the adjoint L-value up to p-adic units. In addition, at the end, to conclude the book, we give a sketch of the proof by Taylor–Wiles of the $R = \mathbb{T}$ theorem.

It is a pleasure for me to express my gratitude to Ashay Burungale, who read the entire manuscript and contributed many useful suggestions.

Acknowledgment: While preparing this book, the author is partially supported by the NSF grant: DMS 1464106.

Haruzo Hida
April 2021 in Los Angeles

Notes to the reader

Articles are quoted by abbreviating the author's name with the last two digits of the year of publication, for example, an article by Hida–Tilouine is quoted as [HT93]. If there are many co-authors, the quotation symbol is shortened. The articles by the author in a preprint form is quoted, for example, as [H21] (though their publication year may differ). Books are quoted by abbreviating their title. For example, one of my earlier books with the title: "Geometric Modular Forms and Elliptic Curves" is quoted as [GME]. Our style of reference is unconventional but has been used in my earlier books and the abbreviation is (basically) common to my books. We quote for example, Theorem 3 of §1.2 in [LFE] as [LFE, Theorem 1.2.3].

As for the notation, we describe here some standard ones. For a prime p, the symbol $\mathbb{Z}_p$ denotes the p-adic integer ring inside the field $\mathbb{Q}_p$ of p-adic numbers, and the symbol $\mathbb{Z}_{(p)}$ is used to indicate the valuation ring $\mathbb{Z}_p \cap \mathbb{Q}$. We fix an algebraic closure $\overline{\mathbb{Q}}$ of $\mathbb{Q}$. A subfield E of $\overline{\mathbb{Q}}$ is called a number field (often assuming $[E : \mathbb{Q}] := \dim_{\mathbb{Q}} E < \infty$). For a number field E, unless otherwise indicated, O_E denotes the integer ring of E, $O_{E,p} = O_E \otimes_{\mathbb{Z}} \mathbb{Z}_p \subset E_p = E \otimes_{\mathbb{Q}} \mathbb{Q}_p$ and $O_{E,(p)} = O_E \otimes_{\mathbb{Z}} \mathbb{Z}_{(p)} \subset E$. A quadratic extension M/F is called a CM field if F is totally real and M is totally imaginary.

We fix an algebraic closure $\overline{\mathbb{Q}}_p$ with p-adic absolute value $|\cdot|_p$ with $|p|_p = p^{-1}$ and write $\mathbb{C}_p$ for the completion of $\overline{\mathbb{Q}}_p$ under this absolute value. The symbol W is exclusively used to indicate a complete discrete valuation ring inside $\mathbb{C}_p$ with residual characteristic p, usually unramified over $\mathbb{Z}_p$ but the residue field $\mathbb{F}$ could be an infinite extension of the prime field $\mathbb{F}_p$. Choose an algebraic closure $\overline{\mathbb{Q}}$ of $\mathbb{Q}$. We fix two embeddings $i_\infty : \overline{\mathbb{Q}} \hookrightarrow \mathbb{C}$ and $i_p : \overline{\mathbb{Q}} \hookrightarrow \overline{\mathbb{Q}}_p \subset \mathbb{C}_p$ throughout the paper. For a local ring A, $\mathfrak{m}_A$ denotes its maximal ideal.

We write μ_N for the group of N-th roots of unity. If K is a field, $\mu_N(K)$

is the group of N-th roots of unity in K; so, we could have $\mu_N(K) = \{1\}$. For a field, taking its algebraic closure $\overline{K}$, $K[\mu_N]$ denotes the subfield of $\overline{K}$ generated by all elements of $\mu_N(\overline{K})$. We write $\mu_{Np^\infty} = \bigcup_n \mu_{Np^n}$. Identifying $\mu_p(\overline{K})$ with $\mathbb{Z}/p\mathbb{Z}$, the Galois character giving the action of $\mathrm{Gal}(\overline{K}/K)$ is denoted by $\omega = \omega_p$. Similarly the p-adic *cyclotomic character* giving the Galois action on $\mathbb{Z}_p(1) = \varprojlim_n \mu_{p^n}(\overline{K})$ is denoted by ν_p.

We write $\mathbb{Q}_\infty \subset \mathbb{Q}[\mu_{p^\infty}]$ for the unique $\mathbb{Z}_p$-extension of $\mathbb{Q}$, $\mathbf{\Gamma} = \mathrm{Gal}(\mathbb{Q}_\infty/\mathbb{Q})$ and identify it with $1 + p\mathbb{Z}_p$ for $\mathbf{p} = 4$ and p according to whether $p = 2$ or not. The Iwasawa algebra $\Lambda = W[[\mathbf{\Gamma}]] = \varprojlim_n W[\mathbf{\Gamma}/\mathbf{\Gamma}^{p^n}]$ is often identified with the power series ring $W[[T]]$ of one variable by $\gamma := (1 + \mathbf{p}) \leftrightarrow t = 1 + T$. When $W = \mathbb{Z}_p$, we sometimes write Λ_0 for $\mathbb{Z}_p[[\mathbf{\Gamma}]]$ to emphasize that the coefficient ring is $\mathbb{Z}_p$ (the smallest).

Unless otherwise mentioned, all rings have the identity. Often B denotes a commutative local p-profinite algebra, used as a base ring in the argument. Usually $B = W$ or Λ. Then CL_B is the category of commutative p-profinite local B-algebras with residue field $\mathbb{F} = B/\mathfrak{m}_B$. Morphisms in CL_B are *local B-algebra homomorphisms* (so, $\phi : A \to A'$ pull back $\mathfrak{m}_{A'}$ to $\mathfrak{m}_A$). For a prime ideal P of $A \in CL_B$, A_P denotes the localization at P as in [CRT, §4]. A prime ideal P of A is called a *prime divisor* if every prime ideal $P' \subsetneq P$ is minimal (i.e., $\dim A/P = \dim A - 1$). If A is an integral domain, its field of fraction is denoted by $\mathrm{Frac}(A) = A_{(0)}$ (the localization at (0)).

For a continuous representation $\rho : \mathrm{Gal}(\overline{\mathbb{Q}}/K) \to \mathrm{GL}_n(A)$ with a profinite local ring $A \in CL_B$, we write $F(\rho) := \overline{\mathbb{Q}}^{\mathrm{Ker}(\rho)}$ (the splitting field of ρ). We let $\mathrm{Gal}(\overline{\mathbb{Q}}/\mathbb{Q})$ act on $M_n(A)$ (the matrix algebra of $n \times n$ matrices with coefficients in A) and $\mathfrak{sl}_n(A) := \{x \in M_n(A) | \mathrm{Tr}(x) = 0\}$ by conjugation. We write $ad(\rho) = M_n(A)$ and $Ad(\rho) = \mathfrak{sl}_n(A)$ for these newly-produced Galois modules and representations. These representations are called the *adjoint representations* of ρ. For a number field K, we write $K^{(p)}/K$ for the maximal p-profinite extension unramified outside p. For a given representation $\overline{\rho} : \mathrm{Gal}(\overline{\mathbb{Q}}/\mathbb{Q}) \to \mathrm{GL}_n(\mathbb{F})$, we write $G := \mathrm{Gal}(F^{(p)}(\overline{\rho})/\mathbb{Q})$ $(F^{(p)}(\overline{\rho}) := F(\overline{\rho})^{(p)})$ and we study deformations with coefficients in a ring in CL_B of $\overline{\rho}$ over G. For a profinite group $\mathcal{G}$, we write $\mathcal{G}^{ab}$ (resp. $\mathcal{G}_p^{ab}$) for the maximal profinite (resp. p-profinite) abelian quotient of $\mathcal{G}$.

We write $\log_p$ for the p-adic logarithm defined as usual by the power series $\mathrm{Log}(X) = \sum_{n=1}^\infty (-1)^{n+1} \frac{(X-1)^n}{n}$ in the convergent range of the open unit (p-adic) disk of radius 1. We often extend $\log_p$ to the entire $\mathbb{C}_p^\times$ by putting $\log_p(p) = 0$ and $\log_p(\zeta) = 0$ for any root of unity ζ (e.g., [PAN, Corollary 5.8.3]).

Contents

Chapter 1

Cyclotomic Iwasawa theory

Assuming basic knowledge of algebraic number theory and commutative algebra, we start with topics from the theory of cyclotomic fields. This is the first example of the serious study of the structure of class groups (or arithmetic cohomology groups in modern language) not just the order of the group. In the 1960's to the 1970's, Iwasawa combined his new ideas with older ones conceived by Kummer in the mid 19-th century and made a systematic theory about cyclicity of the groups (though his breakthrough was often limited to cyclotomic fields under well-known but difficult conjectures). Our treatment is elementary. In this chapter, we plan to discuss the following four topics:

(1) Class number formulas,
(2) Basics of cyclotomic fields and Iwasawa's cyclotomic theory,
(3) Stickelberger's theorem,
(4) Cyclicity over the Iwasawa algebra of the cyclotomic Iwasawa module.

For some of the topics, we just give the results without detailed proofs. Our main reference is the text book [ICF, Chapters 4, 5, 6, 7 and 10]. Here is an overview of the goal of this chapter. We write $\overline{\mathbb{Q}}$ for the field of all algebraic numbers in $\mathbb{C}$. Any finite extension of $\mathbb{Q}$ inside $\overline{\mathbb{Q}}$ is called a number field. Write μ_N for the group of N-th roots of unity inside $\overline{\mathbb{Q}}^\times$ and $\mathbb{Q}(\mu_N)$ for the field generated by roots of unity in μ_N, which is called a cyclotomic field.

For any given number field K, the class group Cl_K defined by the quotient of the group of fractional ideals of K modulo principal ideals is a basic invariant of K. It has been a desire of algebraic number theorists to determine the module structure of Cl_K systematically. Or if $K/\mathbb{Q}$ is a Galois extension, $G_{K/\mathbb{Q}} := \mathrm{Gal}(K/\mathbb{Q})$ acts on Cl_K. Thus it might be easier to see the module structure of Cl_K over the group ring $\mathbb{Z}[G_{K/\mathbb{Q}}]$ larger than $\mathbb{Z}$.

1

The first step towards this goal of determining Cl_K for $K = \mathbb{Q}[\mu_N]$ was given in 1839 by Dirichlet as a formula of the order of the class group (his class number formula). The cyclotomic field K has its maximal real subfield K^+ and K/K^+ is a quadratic extension if $N \geq 3$ with G_{K/K^+} generated by complex conjugation c. The norm map gives rise to a homomorphism $Cl_K \to Cl_K^+ := Cl_{K^+}$ whose kernel is written as Cl_K^- (the minus part of Cl_K). By the formula, if N is an odd prime p, the order $|Cl_K^-|$ of the Cl_K^- is given by

$$|Cl_K^-| = 2p \prod_{\chi \bmod p; \chi(-1)=-1} -\frac{1}{p}\left(\sum_{j=1}^{p-1}\chi^{-1}(a)a\right) \quad \text{(Dirichlet/Kummer)},$$

where χ runs over all odd characters of $(\mathbb{Z}/p\mathbb{Z})^\times$ with values in $\mathbb{C}^\times$. Since $G_{K/\mathbb{Q}} \cong (\mathbb{Z}/p\mathbb{Z})^\times$ sending $\sigma_a \in G_{K/\mathbb{Q}}$ with $\sigma_a(\zeta) = \zeta^a$ ($\zeta \in \mu_p$) to $a \in (\mathbb{Z}/p\mathbb{Z})^\times$, we have $\mathbb{Z}[G_{K/\mathbb{Q}}] \cong \mathbb{Z}[(\mathbb{Z}/p\mathbb{Z})^\times]$. Since each character χ of $G_{K/\mathbb{Q}}$ extends to an algebra homomorphism $\chi : \mathbb{Z}[G_{K/\mathbb{Q}}] \to \overline{\mathbb{Q}}$ sending σ_a to $\chi(a)$, Kummer–Stickelberger guessed that

$$\theta_0 := \sum_{a=1}^{p-1}\frac{a}{p}\sigma_a^{-1} \text{ annihilates } Cl_K^- \text{ as } \chi(\theta_0) = \frac{1}{p}\left(\sum_{j=1}^{p-1}\chi^{-1}(a)a\right).$$

This "symbolic" statement means that $\mathfrak{A}^{\beta\theta_0}$ (for any fractional ideal $\mathfrak{A}$ of K) is principal as long as $\beta\theta_0 \in \mathbb{Z}[G_{K/\mathbb{Q}}]$ for $\beta \in \mathbb{Z}[G_{K/\mathbb{Q}}]$. Writing $\mathfrak{a}$ for the $\mathbb{Z}[G_{K/\mathbb{Q}}]$-ideal generated by elements of the form $\beta\theta_0 \in \mathbb{Z}[G_{K/\mathbb{Q}}]$, we might expect:

$$Cl_K^- \cong \mathbb{Z}[G_{K/\mathbb{Q}}]/\mathfrak{a}? \quad \text{(Cyclicity over } \mathbb{Z}[G_{K/\mathbb{Q}}])$$

which is **not** generally true. After summarizing the basics of cyclotomic fields, we will prove in this chapter Stickelberger's theorem:

$$Cl_K^- \otimes_{\mathbb{Z}} \mathbb{Z}_p \cong \mathbb{Z}_p[G_{K/\mathbb{Q}}]^-/(\mathfrak{a} \otimes \mathbb{Z}_p)^- \quad \text{(}p\text{-Cyclicity)}$$

assuming *Kummer–Vandiver conjecture*: $p \nmid |Cl_K^+|$ (see [HHO17] for numerical examples for the conjecture). Here $\mathfrak{A}^- = \{x \in \mathfrak{A} | cx = -x\}$ for complex conjugation c for an ideal $\mathfrak{A}$ of $\mathbb{Z}_p[G_{K/\mathbb{Q}}]$. Set $\Lambda = \mathbb{Z}_p[[T]]$ (one variable power series ring). Then we can easily prove that

$$\varprojlim_n \mathbb{Z}_p[G_{\mathbb{Q}[\mu_{p^n}]/\mathbb{Q}}] \cong \mathbb{Z}_p[\text{Gal}(\mathbb{Q}[\mu_p]/\mathbb{Q})][[T]] \quad (\varprojlim_n \sigma_{1+p} \mapsto t = 1+T),$$

where we identify $\text{Gal}(\mathbb{Q}[\mu_p]/\mathbb{Q}) \cong \mu_{p-1} \subset \mathbb{Z}_p^\times \cong \text{Gal}(\mathbb{Q}[\mu_{p^\infty}]/\mathbb{Q}) = \mathbb{Z}_p[\text{Gal}(\mathbb{Q}[\mu_p]/\mathbb{Q})][[T]]$ and the limit is taken via restriction maps

$G_{\mathbb{Q}[\mu_{p^{n+1}}]/\mathbb{Q}} \ni \sigma \mapsto \sigma|_{\mathbb{Q}[\mu_{p^n}]} \in G_{\mathbb{Q}[\mu_{p^n}]/\mathbb{Q}}$. Then, assuming again Kummer–Vandiver conjecture, we further go on to show Iwasawa's way of proving his main conjecture and cyclicity of the cyclotomic Iwasawa module $X := \varprojlim_n (Cl^-_{\mathbb{Q}[\mu_{p^n}]} \otimes \mathbb{Z}_p)$:

$$X \cong \mathbb{Z}_p[\mathrm{Gal}(\mathbb{Q}[\mu_p]/\mathbb{Q})][[T]]^-/(L_p)$$

for the T-expansion L_p of the Kubota–Leopoldt p-adic L-function (considered as an element in the group algebra $\Lambda[\mathrm{Gal}(\mathbb{Q}[\mu_p]/\mathbb{Q})]^- = \mathbb{Z}_p[\mathrm{Gal}(\mathbb{Q}[\mu_p]/\mathbb{Q})]^-[[T]]$ including all branches). The main conjecture (see §1.9.3) tells us that the characteristic ideal of X is given by (L_p) but not the above isomorphism, and it was first proven in an automorphic way by Mazur–Wiles [MW84] generalizing an idea of Ribet [R76], but we now have elementary proofs by Rubin (see [ICF, §15.7]) and Coates–Sujatha [CFZ]. We fix an odd prime $p > 2$ in this chapter.

1.1 Cyclotomic fields

We recall the basic structure theory of cyclotomic fields. Any finite extension of $\mathbb{Q}$ inside $\overline{\mathbb{Q}}$ is called a number field. In algebraic number theory, the theory of cyclotomic fields occupies a peculiar place as it supplies us a first example where difficult theory can be proven rather explicitly and sometimes can be delved into deeper than the general treatment. To a good extent, cyclotomic theory can be considered to be the origin of the development of algebraic number theory and still inspires us with many miraculous special features.

1.1.1 *Cyclotomic integers*

Fix a prime $p > 2$ and consider a primitive root of unity $\zeta_{p^n} := \exp(\frac{2\pi i}{p^n})$. Then ζ_p satisfies the equation

$$\Phi_1(X) = \frac{X^p - 1}{X - 1} = 1 + X + X^2 + \cdots + X^{p-1} = \prod_{j=1}^{p-1}(X - \zeta_p^j).$$

Note that

$$\Phi_1(X + 1) = \frac{(X+1)^p - 1}{X} = \sum_{j=1}^{p}\binom{p}{j}X^{j-1} = X^{p-1} + pX^{p-2} + \cdots + p$$

is an Eisenstein polynomial. Therefore $\Phi_1(X + 1)$ is irreducible, and hence $\Phi_1(X)$ is irreducible. In the same manner, ζ_{p^n} is a root of $\Phi_n(X) =$

$\frac{X^{p^n}-1}{X^{p^{n-1}}-1} = \Phi_1(X^{p^{n-1}})$ is irreducible. Therefore $\mathbb{Q}(\zeta_{p^n})/\mathbb{Q}$ is a field extension of degree equal to $\deg(\Phi_n) = p^{n-1}(p-1)$. In particular, in $\mathbb{Q}(\zeta_{p^n})/\mathbb{Q}$, p fully ramifies with $\zeta_{p^n} - 1$ giving the unique prime ideal $(\zeta_{p^n} - 1)$ over p; so, $(\zeta_{p^n} - 1)^{p^{n-1}(p-1)} = (p)$ in $\mathbb{Q}[\zeta_{p^n}]$, and $\mathbb{Z}_p[\zeta_{p^n} - 1]$ is the p-adic integer ring of $\mathbb{Q}_p[\zeta_{p^n}]$ (see [PAN, §6.7] for elementary details of these facts). This shows

Theorem 1.1.1. *The ring $\mathbb{Z}[\zeta_{p^n}]$ is the integer ring of $\mathbb{Q}[\zeta_{p^n}]$ and the roots of $\Phi_n(X)$ (i.e., all primitive p^n-th roots) give rise to a basis of $\mathbb{Z}[\zeta_{p^n}]$ over $\mathbb{Z}$, p fully ramifies in $\mathbb{Z}[\zeta_{p^n}]$, and $\zeta_{p^n} - 1$ generates a unique prime ideal of $\mathbb{Z}[\zeta_{p^n}]$ over p.*

Since all Galois conjugate of ζ_{p^n} is again a root of $\Phi_n(X)$, $\mathbb{Q}[\zeta_{p^n}]$ is a Galois extension. See [CRT, §9] for a general theory of integral extension.

For another $l \neq p$, taking a prime $\mathfrak{l}$ of $\mathbb{Q}[\zeta_{p^n}]$ above l, $\overline{\zeta}_{p^n} := (\zeta_{p^n} \bmod \mathfrak{l})$ is a primitive p^n-th root in a finite field of characteristic l; so, $\overline{\zeta}_{p^n}^j$ are all distinct for $j = 1, \ldots, p^{n-1}(p-1)$. Thus $\mathbb{Q}_l[\zeta_{p^n}]$ is unramified at l, and $\mathbb{Z}_l[\zeta_{p^n}]$ is an unramified valuation ring over $\mathbb{Z}_l$; so, it is the l-adic integer ring of $\mathbb{Q}_l[\zeta_{p^n}]$. Conversely, if $W_{/\mathbb{Z}_l}$ is a discrete valuation ring unramified over $\mathbb{Z}_l$, writing q for the order of the residue field $\mathbb{F}$ of W, we have $|(W/p^n W)^\times| = q^n - q^{n-1}$. Thus $x \in W^\times$ satisfies $x^{q^n - q^{n-1}} \equiv 1 \bmod p^n$ (or equivalently $|x^{q^n} - x^{q^{n-1}}|_p \leq p^{-n}$). Therefore $\zeta = \lim_{n \to \infty} x^{q^n}$ satisfies $\zeta^{q-1} = 1$ and $\zeta \equiv x \bmod p$. Thus we can choose x so that ζ generates $\mathbb{F}$; so, $W = \mathbb{Z}_l[\mu_{q-1}]$. In other words, the unramified extension W of $\mathbb{Z}_l$ with a given residue field $\mathbb{F}$ is unique (up to isomorphisms) and generated by a root of unity of order prime to l, which is called *the ring of Witt vectors* with coefficients in $\mathbb{F}$. See [BCM, IX.1] for a general theory of Witt vectors.

1.1.2 *Cyclotomic character*

Since $\sigma \in G_{\mathbb{Q}[\zeta_{p^n}]/\mathbb{Q}}$ is determined by $\zeta_{p^n}^\sigma = \zeta_{p^n}^{\nu_n(\sigma)}$ for $\nu_n(\sigma) \in (\mathbb{Z}/p^n\mathbb{Z})^\times$, we have a character $G_{\mathbb{Q}[\zeta_{p^n}]/\mathbb{Q}} \to (\mathbb{Z}/p^n\mathbb{Z})^\times$ which is injective. Since the two sides have the same order, ν_n is an isomorphism. Writing the residue field of $\mathfrak{l} \nmid p$ as $\mathbb{F} = \mathbb{Z}[\zeta_{p^n}]/\mathfrak{l}$, we have $\mathbb{F} = \mathbb{F}_l[\overline{\zeta}_{p^n}]$. Then the Frobenius element $\mathrm{Frob}_l \in G_{\mathbb{F}/\mathbb{F}_l} = D_l$ sends $\overline{\zeta}_{p^n}$ to $\overline{\zeta}_{p^n}^l$. Thus we get

Lemma 1.1.2. *For a prime $l \neq p$, we have $\nu_n(\mathrm{Frob}_l) = (l \bmod p^n)$.*

It is customary to write $\mu_{p^n} \subset \overline{\mathbb{Q}}^\times$ for the cyclic group generated by ζ_{p^n}. Since $\mathbb{Q}[\zeta_{p^n}]$ contains μ_{p^n}, we write hereafter $\mathbb{Q}[\mu_{p^n}]$ for $\mathbb{Q}[\zeta_{p^n}]$ freeing the notation from a choice of a generator of μ_{p^n}. Write $\mathbb{Q}[\mu_{p^\infty}] := \bigcup_n \mathbb{Q}[\mu_{p^n}]$. Then the restriction map $\mathrm{Res}_{m,n}(\sigma) = \sigma|_{\mathbb{Q}[\mu_{p^n}]}$ for $\sigma \in G_{\mathbb{Q}[\mu_{p^m}]/\mathbb{Q}}$ $(m > n)$ gives the following commutative diagram:

$$
\begin{array}{ccc}
G_{\mathbb{Q}[\mu_{p^m}]/\mathbb{Q}} & \xrightarrow{\ \nu_m\ } & (\mathbb{Z}/p^m\mathbb{Z})^\times \\
{\scriptstyle \mathrm{Res}}\downarrow & & \downarrow{\scriptstyle \bmod p^n} \\
G_{\mathbb{Q}[\mu_{p^n}]/\mathbb{Q}} & \xrightarrow{\ \nu_n\ } & (\mathbb{Z}/p^n\mathbb{Z})^\times .
\end{array}
$$

Passing to the limit, we get

$$
G_{\mathbb{Q}[\mu_{p^\infty}]/\mathbb{Q}} \xrightarrow[\sim]{\nu_p} \mathbb{Z}_p^\times \text{ with } \nu_p(\mathrm{Frob}_l) = l \text{ for } l \neq p. \tag{1.1}
$$

1.1.3 *Decomposition group*

Let $\mathbf{\Gamma} := 1 + p\mathbb{Z}_p \subset \mathbb{Z}_p^\times$. Then $\mathbb{Z}_p^\times/\mathbf{\Gamma} \cong (\mathbb{Z}/p\mathbb{Z})^\times$ which has order $p-1$. Thus $z \mapsto z^p$ is the identity map on $(\mathbb{Z}/p\mathbb{Z})^\times$. On the other hand, for any $z \in \mathbb{Z}_p^\times$, $z^{p^n - p^{n-1}} \equiv 1 \mod p^n$ as $(\mathbb{Z}/p^n\mathbb{Z})^\times$ has order $p^n - p^{n-1}$. Therefore $|z^{p^n} - z^{p^{n-1}}|_p \le \frac{1}{p^n}$, and as before, we have a limit $\omega(z) = \lim_{n \to \infty} z^{p^n}$ which satisfies plainly $\omega(z)^p = \omega(z)$; i.e, $\omega(z) \in \mu_{p-1}$ and $\omega(z) \equiv z \mod p$. Hence $\mu_{p-1} \subset \mathbb{Z}_p^\times$ is a cyclic subgroup of order $p-1$. We thus have $\mathbb{Z}_p^\times = \mathbf{\Gamma} \times \mu_{p-1}$, and we get

Lemma 1.1.3. *Let D_l be the decomposition subgroup of $G_{\mathbb{Q}[\mu_{p^\infty}]/\mathbb{Q}}$ of a prime l. Then if $l \neq p$, D_l is the infinite cyclic subgroup topologically generated by Frob_l isomorphic to $\langle l \rangle^{\mathbb{Z}_p} \times \omega(l)^{\mathbb{Z}}$. If $l = p$, D_p is equal to the inertia subgroup I_p which is the entire group $G_{\mathbb{Q}[\mu_{p^\infty}]/\mathbb{Q}}$. In particular, for each prime, the number of prime ideals in $\mathbb{Z}_p[\mu_{p^\infty}]$ over l is finite and is equal to $[G_{\mathbb{Q}[\mu_{p^\infty}]/\mathbb{Q}} : D_l]$.*

Remark 1.1.4. More generally, for an integer $N > 1$, making prime factorization $N = \prod_l l^{e(l)}$, $\mathbb{Q}[\mu_N]$ is the composite of $\mathbb{Q}[\mu_{l^{e(l)}}]$ with degree $|(\mathbb{Z}/l^{e(l)}/\mathbb{Z})^\times| = \varphi(l^{e(l)}) = l^{e(l)} - l^{e(l)-1}$. In the field $\mathbb{Q}[\mu_{l^{e(l)}}]$, the prime l ramifies fully and all other primes are unramified. Therefore the fields $\{\mathbb{Q}[\mu_{l^{e(l)}}]\}_{l|N}$ are linearly disjoint over $\mathbb{Q}$, and hence

$$
G_{\mathbb{Q}[\mu_N]/\mathbb{Q}} \cong \prod_{l|N} G_{\mathbb{Q}[\mu_{l^{e(l)}}]/\mathbb{Q}} \cong \prod_{l|N} (\mathbb{Z}/l^{e(l)}\mathbb{Z})^\times \cong (\mathbb{Z}/N\mathbb{Z})^\times .
$$

This also tells us that the integer ring of $\mathbb{Q}[\mu_N]$ is given by $\mathbb{Z}[\mu_N] \cong \bigotimes_l \mathbb{Z}[\mu_{l^{e(l)}}]$. The decomposition subgroup of a prime $q \nmid N$ in $G_{\mathbb{Q}[\mu_N]/\mathbb{Q}}$ is isomorphic to the subgroup of $(\mathbb{Z}/N\mathbb{Z})^\times$ generated by the class $(q \bmod N)$. If $l|N$, I_l is isomorphic to $(\mathbb{Z}/l^{e(l)}\mathbb{Z})^\times$ and writing $N^{(l)} = N/l^{e(l)}$, D_l is isomorphic to the product of I_l and the subgroup of $(\mathbb{Z}/N^{(l)}\mathbb{Z})^\times$ generated by the class of l. Since the Galois group $G_{\mathbb{Q}[\mu_N]/\mathbb{Q}}$ is generated by inertia groups of primes $l|N$, for any intermediate extension $\mathbb{Q} \subset E \subset F \subset \mathbb{Q}[\mu_N]$, some prime ramifies. In other words, if F/E is unramified everywhere, we conclude $E = F$.

Exercise 1.1.1. Give a detailed proof of the fact: $\mathbb{Z}[\mu_N] \cong \bigotimes_l \mathbb{Z}[\mu_{l^{e(l)}}]$ and $\mathbb{Z}[\mu_N]$ is the integer ring of $\mathbb{Q}[\mu_N]$.

Exercise 1.1.2. For a finite set $S = \{l_1, \ldots, l_r\}$ of primes, prove that the number of prime ideals over any given prime l in $\mathbb{Z}[\mu_{l_1^\infty}, \ldots, \mu_{l_r^\infty}]$ is finite.

1.2   An outline of class field theory

We give a brief outline of class field theory (a more detailed reference is, for example, [CFN] or [BNT]).

1.2.1   *Ray class groups*

For a given number field F with integer ring O, pick an O-ideal $\mathfrak{n}$, let $I_\mathfrak{n}$ be the group of all fractional ideals prime to $\mathfrak{n}$. We put $O_\mathfrak{n} := \varprojlim_n O/\mathfrak{n}^n$ (the $\mathfrak{n}$-adic completion of O). Write the prime factorization of $\mathfrak{n} = \prod_{\mathfrak{l}|\mathfrak{n}} \mathfrak{l}^{e(\mathfrak{l})}$. Then by Chinese remainder theorem, we have $O_\mathfrak{n} \cong \prod_\mathfrak{l} O/\mathfrak{l}^{e(\mathfrak{l})}$. We write $O_{(\mathfrak{l})}$ for $O_\mathfrak{l} \cap F$ which is the localization

$$O_{(\mathfrak{l})} = \left\{ \frac{\beta}{\alpha} \middle| \alpha, \beta \in O \text{ with } \alpha O + \mathfrak{l} = O \right\}.$$

For a principal ideal $(\alpha) \in I_\mathfrak{n}$, we write $\alpha \equiv 1 (\bmod \mathfrak{n})^\times$ if $\alpha \in 1 + \mathfrak{l}^{e(\mathfrak{l})} O_\mathfrak{l}$ for all prime $\mathfrak{l}|\mathfrak{n}$. More generally, we write $\alpha \equiv \beta (\bmod \mathfrak{n})^\times$ for $(\alpha), (\beta) \in I_\mathfrak{n}$ if $\alpha/\beta \equiv 1 (\bmod \mathfrak{n})^\times$.

Exercise 1.2.1. For $\alpha \in F$, prove that $\alpha \equiv 1 (\bmod \mathfrak{n})^\times$ if and only if $\alpha \in 1 + \mathfrak{l}^{e(\mathfrak{l})} O_\mathfrak{l}$ for all prime $\mathfrak{l}|\mathfrak{n}$.

Define

$$\begin{aligned} P_\mathfrak{n} &:= \{(\alpha) \in I_\mathfrak{n} | \alpha \equiv 1 (\bmod \mathfrak{n})^\times\} \\ P_\mathfrak{n}^+ &:= \{(\alpha) \in P_\mathfrak{n} | \sigma(\alpha) > 0 \text{ for all field embeddings } \sigma : F \to \mathbb{R}\}. \end{aligned} \qquad (1.2)$$

If F has no real embedding (i.e., F is totally imaginary), we have $P_{\mathfrak{n}} = P_{\mathfrak{n}}^+$.

Exercise 1.2.2. Let $F = \mathbb{Q}[\sqrt{5}]$. Is $P_O = P_O^+$ true? How about $\mathbb{Q}[\sqrt{15}]$?

Then $Cl_F(\mathfrak{n}) = Cl_{\mathfrak{n}} := I_{\mathfrak{n}}/P_{\mathfrak{n}}$ (resp. $Cl_F^+(\mathfrak{n}) = Cl_{\mathfrak{n}}^+ = I_{\mathfrak{n}}/P_{\mathfrak{n}}^+$) are called the (resp. *strict*) *ray class group modulo* $\mathfrak{n}$ of F. They are finite groups. We have written Cl_F for $Cl_F(O)$. The order $|Cl_F(O)|$ is called the *class number* of F.

1.2.2   Main theorem of class field theory

Theorem 1.2.1. *For each $\mathfrak{n}$ as above, there is a unique abelian extension $H_{\mathfrak{n}}/F$ (resp. $H_{\mathfrak{n}}^+/F$) such that*

(1) *Prime ideals coprime to $\mathfrak{n}$ is unramified in $H_{\mathfrak{n}}/F$ and $H_{\mathfrak{n}}^+/F$, and every real embedding of F extends to a real embedding of $H_{\mathfrak{n}}$; in particular, if all Galois conjugates of F are in $\mathbb{R}$ (i.e., F is totally real), $H_{\mathfrak{n}}$ is totally real;*

(2) *$Cl_{\mathfrak{n}} \cong G_{H_{\mathfrak{n}}/F}$ and $Cl_{\mathfrak{n}}^+ \cong G_{H_{\mathfrak{n}}^+/F}$ by an isomorphism sending the class of prime ideal $\mathfrak{l}$ coprime to $\mathfrak{n}$ in $Cl_{\mathfrak{n}}$ (resp. $Cl_{\mathfrak{n}}^+$) to the corresponding $\mathrm{Frob}_{\mathfrak{l}} \in G_{H_{\mathfrak{n}}/F}$ (resp. $G_{H_{\mathfrak{n}}^+/F}$), where $\mathrm{Frob}_{\mathfrak{l}}$ is a unique element in $G_{K/F}$ for $K = H_{\mathfrak{n}}, H_{\mathfrak{n}}^+$ such that $\mathrm{Frob}_{\mathfrak{l}}(\mathfrak{l}) = \mathfrak{l}$ and $\mathrm{Frob}_{\mathfrak{l}}(x) \equiv x^{N(\mathfrak{l})}$ mod $\mathfrak{l}$ for $N(\mathfrak{l}) := |O/\mathfrak{l}|$;*

(3) *For any finite abelian extension K/F, there exists an O-ideal $\mathfrak{n}$ such that $K \subset H_{\mathfrak{n}}^+$ (the ideal maximal among $\mathfrak{n}$ with $K \subset H_{\mathfrak{n}}^+$ is called the "conductor" of K with respect to F);*

(4) *For a finite extension F'/F with integer ring O', we write $H_{\mathfrak{n}}/F$ and $H_{\mathfrak{n}}'/F'$ (resp. $Cl_{\mathfrak{n}}^+$ and $Cl_{\mathfrak{n}}'^+$ for the corresponding class fields (resp. the corresponding class groups). Then we have $H_{\mathfrak{n}}' \supset H_{\mathfrak{n}}$ and the following commutative diagram*

$$
\begin{array}{ccc}
G_{H_{\mathfrak{n}}'^+/F'} & \xrightarrow{\ \mathrm{Res}\ } & G_{H_{\mathfrak{n}}^+/F} \\[4pt]
\wr \uparrow & & \wr \uparrow \\[4pt]
Cl_{\mathfrak{n}}'^+ & \xrightarrow{\ N_{F'/F}\ } & Cl_{\mathfrak{n}}^+.
\end{array}
$$

Here $N_{F'/F}$ is induced by the norm map sending a O'-prime ideal $\mathfrak{L}$ prime to $\mathfrak{n}$ to $\mathfrak{l}^f$ ($\mathfrak{l} = \mathfrak{L} \cap O$) with $f = [O'/\mathfrak{L} : O/\mathfrak{l}]$ (note $N_{F'/F}\mathfrak{L} = \prod_{\sigma \in G_{F'/F}} \mathfrak{L}^\sigma \cap O$ if F'/F is a Galois extension).

For a proof, we refer to [BNT] and [CFN].

Example 1.2.1. Suppose $F = \mathbb{Q}$. Consider the map $\{n \in \mathbb{Z}|n\mathbb{Z} + \mathfrak{n} = \mathbb{Z}\} \ni n \mapsto (n) \in Cl_{(N)}^+$ for each integer $n > 0$ prime to N. This induces an injective homomorphism $(\mathbb{Z}/N\mathbb{Z})^\times \to Cl_N^+$. For each $(\alpha) \in I_N$, take an integer n prime to N with $\alpha \equiv n(\bmod N)^\times$ and $\alpha/n > 0$. Then the class of (α) and (n) in Cl_N^+ coincide by definition. Therefore, we get $Cl_N^+ \cong (\mathbb{Z}/N\mathbb{Z})^\times$, and hence $H_N^+ = \mathbb{Q}[\mu_N]$.

Exercise 1.2.3. What is $H_N \subset H_N^+ = \mathbb{Q}[\mu_N]$? Determine $G_{H_N/\mathbb{Q}}$ as a quotient of $(\mathbb{Z}/N\mathbb{Z})^\times$.

Corollary 1.2.2 (Kronecker–Weber–Hilbert). *Any finite abelian extension of $\mathbb{Q}$ is contained in $\mathbb{Q}[\mu_N]$ for some positive integer N.*

1.3   Class number formula and Stickelberger's theorem

Out of the cyclotomic class number formula of Dirichlet–Kummer, we try to understand the module structure of cyclotomic class groups over the group algebra of the Galois group.

1.3.1   *Class number formula*

Let $F_n := \mathbb{Q}[\mu_{p^{n+1}}]$. Complex conjugation c acts non-trivially on F_n as $\nu_n(c) = -1$. Let F_n^+ be the fixed field of c; so, $[F_n^+ : \mathbb{Q}] = \frac{p^n(p-1)}{2}$. We have the following field diagram

$$
\begin{array}{ccc}
F_n & \overset{\hookrightarrow}{\longrightarrow} & H_n \\
\cup\uparrow & & \uparrow\cup \\
F_n^+ & \overset{\hookrightarrow}{\longrightarrow} & H_n^+
\end{array}
$$

for the Hilbert class fields H_n/F_n and H_n^+/F_n^+. Write the class group of F_n (resp. F_n^+) as Cl_n and Cl_n^+. Define then $Cl_n^- := \mathrm{Ker}(N_{F_n/F_n^+} : Cl_n \to Cl_n^+)$. Since H_n^+ is real, $H_n^+ \cap F_n = F_n^+$; so, $G_{H_n^+ F_n/F_n} \cong G_{F_n/F_n^+} \times G_{H_n^+/F_n^+}$. Thus the restriction map Res $: G_{H_n/F_n} \to G_{H_n^+/F_n^+}$ is onto. Therefore $N_{F_n/F_n^+} : Cl_n \to Cl_n^+$ is onto.

Recall Dirichlet's L-function of a character $\chi : (\mathbb{Z}/N\mathbb{Z})^\times \to \mathbb{C}^\times$ defined by an absolutely and locally uniformly converging sum for $s \in \mathbb{C}$ with $\mathrm{Re}(s) > 1$:

$$
L(s, \chi) = \sum_{n=1}^{\infty} \chi(n) n^{-s},
$$

where we use the convention that $\chi(n) = 0$ if n and N are not co-prime. The function $L(s,\chi)$ extends to a meromorphic function on the whole complex plane $\mathbb{C}$ with only possible pole at $s = 1$, and if χ is non-trivial, it is holomorphic everywhere [LFE, §2.2–3]. Here is a formula of $|Cl_n^-|$ (e.g. [ICF, Theorem 4.17] and a shorter formula in [EDM, Theorem 5.7]):

Theorem 1.3.1. *Let $m = n + 1$. Then we have*

$$|Cl_n^-| = 2p^m \prod_{\chi:(\mathbb{Z}/p^m\mathbb{Z})^\times \to \mathbb{C}^\times ; \chi(-1)=-1} \frac{1}{2}L(0,\chi^{-1})$$

with $L(0,\chi^{-1}) = -\frac{1}{p^f}\sum_{a=1}^{p^f}\chi^{-1}(a)a$, where χ is a primitive character modulo p^f with $0 < f \leq m$.

By functional equation for primitive character modulo N (see [LFE, §2.3] or [ICF, Chapter 4]):

$$L(s,\chi) = \begin{cases} \frac{\tau(\chi)(2\pi/N)^s L(1-s,\chi^{-1})}{2\Gamma(s)\cos(\pi s/2)} & \text{if } \chi(-1) = 1, \\ \frac{\tau(\chi)(2\pi/N)^s L(1-s,\chi^{-1})}{2\sqrt{-1}\Gamma(s)\sin(\pi s/2)} & \text{if } \chi(-1) = -1 \end{cases}$$

with the Gauss sum $\tau(\chi) = \sum_{a=1}^{N}\chi(a)\exp(2\pi\sqrt{-1}a/N)$, $L(0,\chi^{-1})$ is almost $L(1,\chi)/2\pi i$ which is directly related to the class number. Because of this, we put χ^{-1} in the formula, though we can replace them by χ for an obvious reason.

Here is another remark. The Gamma function is defined by $s \mapsto \Gamma(s) = \int_0^\infty t^s e^{-t}\frac{dt}{t}$ for the multiplicative Haar measure $\frac{dt}{t}$ and the product of a multiplicative character t^s and an additive character e^{-x} of $\mathbb{R}_+^\times$, and the Gauss sum $\chi \mapsto \tau(\chi)$ is its analogue for the finite groups $(\mathbb{Z}/N\mathbb{Z})^\times$. The two appears in the functional equation naturally (this might be a motivation of Gauss to introduce his sum).

1.3.2 *Galois module structure of Cl_n^-*

We would like to study the group structure of $Cl_n = Cl_{F_n}$ and the module structure of Cl_n over $\mathbb{Z}[G_{F_n/\mathbb{Q}}] \cong \mathbb{Z}[(\mathbb{Z}/p^m\mathbb{Z})^\times]$ ($m = n + 1$). We first describe Kummer–Stickelberger theory to determine the annihilator of Cl_n^- in $\mathbb{Z}[(\mathbb{Z}/p^n\mathbb{Z})^\times]$. Writing $\sigma_a \in G_{F_n/\mathbb{Q}}$ for the element with $\zeta^{\sigma_a} = \zeta^a$ ($\zeta \in \mu_{p^m}$) for $a \in (\mathbb{Z}/p^m\mathbb{Z})^\times$, from the above class number formula, Kummer/Stickelberger guessed that $\theta = \theta_n := \sum_{a=1}^{p^m}\frac{a}{p^m}\sigma_a^{-1} \in \mathbb{Z}[(\mathbb{Z}/p^m\mathbb{Z})^\times]$ would kill Cl_n as $\chi(\theta_n) \doteq L(0,\chi^{-1})$. More generally, let $F/\mathbb{Q}$ be an abelian extension with $F \subset \mathbb{Q}[\mu_N]$ for a minimal $N > 0$. Define $\theta_F :=$

$\sum_{a \in (\mathbb{Z}/N\mathbb{Z})^\times} \left\{ \frac{a}{N} \right\} \sigma_a^{-1}|_F \in \mathbb{Q}[G]$ for $G = G_{F/\mathbb{Q}}$, where $0 \le \{x\} < 1$ is the fractional part of a real number x (i.e., $x - \{x\} \in \mathbb{Z}$) and $\sigma_a \in G_{\mathbb{Q}[\mu_N]/\mathbb{Q}}$ sends every N-th root of unity ζ o ζ^a for $a \in (\mathbb{Z}/N\mathbb{Z})^\times$.

The first Bernoulli polynomial is given by $B_1(x) = x - \frac{1}{2}$. If we ignore the term involving $\sum_a \sigma_a^{-1}$, θ_n is $\sum_{a=1}^{p^m} B_1(\frac{a}{p^m})\sigma_a^{-1}$. Since $\zeta(0, x) = -B_1(x)$ for the *Hurwitz zeta function* $\zeta(s, x) = \sum_{n=0}^{\infty}(n + x)^{-s}$ for $x \in (0, 1]$ [LFE, §2.3], one can generalize the construction of the Stickelberger element to a totally real number field using partial zeta values in place of Hurwitz zeta values (see [CS74], [C77]) or one can use the n-th Bernoulli polynomial $B_n(x)$ in place of $B_1(x)$ (an idea of Kubert–Lang), which we explore in Chapter 2. There is one more different expression going back to Euler found by Shimura in 2007 [EDM, Theorem 4.14]. Is there an analogue of the Stickelberger relation via Shimura's formula?

Theorem 1.3.2 (Stickelberger). *Pick $\beta \in \mathbb{Z}[G]$ such that $\beta\theta_F \in \mathbb{Z}[G]$. Then for any fractional ideal $\mathfrak{a}$ of F, $\mathfrak{a}^{\beta\theta_F}$ is principal.*

This theorem was first proven for $F = \mathbb{Q}[\mu_p]$ by Kummer in 1847 [K47] and by Stickelberger in 1890 in general [S90]. After proving basic properties of Gauss sum, we prove this theorem in a slightly different form (Theorem 1.5.2) in §1.5. Our treatment follows Washington's exposition in [ICF] to a good extent.

1.4   Gauss sum

The idea of proving Stickelberger's theorem is to compute prime factorization of the Gauss sum $G(\chi)$ over a finite field $\mathbb{F} = \mathbb{Z}[\mu_{(q-1)p}]/\mathfrak{P}$ with q elements of characteristic p and to show roughly $(G(\chi)) \doteq \mathfrak{P}^{\theta_F}$. The finding might have been accidental, but for a mathematics addict, it is quite a natural challenge to find an explicit formula of prime factorization of well-behaved numbers arithmetically given.

1.4.1   *Gauss sum over a finite field*

Let us define the *Gauss sum* we study precisely. Take a finite field $\mathbb{F}$ of characteristic p and write $\mathrm{Tr} : \mathbb{F} \to \mathbb{F}_p$ for the trace map. Then for any character $\chi : \mathbb{F}^\times \to \overline{\mathbb{Q}}^\times$, the Gauss sum is defined to be

$$G(\chi) := -\sum_{a\in\mathbb{F}} \chi(a)\zeta_p^{\mathrm{Tr}(a)} \quad \text{for } \zeta_p := \exp(\frac{2\pi i}{p}),$$

where we put $\chi(0) = 0$ as before. We have "$-$" sign in front of the sum defining $G(\chi)$. This is to simplify some of the relation of the Gauss sums (and also Jacobi sums) we discuss later (by this, we can also remove the sign factor of the Hasse–Davenport relation, though we do not use this fact). Therefore $G(\chi) = -\tau(\chi)$ if χ is a character of $\mathbb{F}_p^\times = (\mathbb{Z}/p\mathbb{Z})^\times$. The character $\psi : \mathbb{F} \to \mathbb{C}^\times$ given by $\psi(a) = \exp(\frac{2\pi i}{p})^{\mathrm{Tr}(a)}$ is non-trivial as $\mathbb{F}$ is generated by primitive N-th root of unity for $N = (|\mathbb{F}| - 1)$ whose minimal polynomial is a factor of $X^N + X^{N-1} + \cdots + 1$ (i.e., $\mathrm{Tr}(a) \neq 0$ for some $a \in \mathbb{F}$).

Exercise 1.4.1. Give a detailed proof of the fact: $\mathrm{Tr}(a) \neq 0$ for some $a \in \mathbb{F}$; so, $\mathrm{Tr} : \mathbb{F} \to \mathbb{F}_p$ is onto.

Lemma 1.4.1. *We have*

(1) $-\sum_{a\in\mathbb{F}} \chi(a)\zeta_p^{\mathrm{Tr}(ab)} = \overline{\chi}(b)G(\chi)$ *for* $b \in \mathbb{F}^\times$,
(2) $\overline{G(\chi)} = \chi(-1)G(\overline{\chi})$,
(3) *If* $\chi \neq \mathbf{1}$ *for the identity character* $\mathbf{1}$, $G(\chi)G(\overline{\chi}) = \chi(-1)|\mathbb{F}|$,
(4) *If* $\chi \neq \mathbf{1}$, $|G(\chi)|^2 = |\mathbb{F}|$.

Since $\tau(\chi) = -G(\chi)$ if $\mathbb{F} = \mathbb{F}_p$, we get the corresponding formula for $\tau(\chi)$.

Proof. The assertion (1) holds by the variable change $ab \mapsto a$ combined with $\overline{\chi(b)} = \chi^{-1}(b)$. Then we have

$$\overline{G(\chi)} = -\sum_{a\in\mathbb{F}} \overline{\chi}(a)\zeta_p^{-\mathrm{Tr}(a)} = -\sum_{a\in\mathbb{F}} \overline{\chi}(a)\zeta_p^{\mathrm{Tr}(a(-1))} \overset{(1)}{=} \chi(-1)G(\overline{\chi})$$

proving (2). Note that for $c \neq 1$,

$$\sum_{b\in\mathbb{F}^\times} \zeta_p^{\mathrm{Tr}(b(c-1))} = -1$$

as $1 + \sum_{b\in\mathbb{F}^\times} \zeta_p^{\mathrm{Tr}(b(c-1))} = 0$ (character sum). We then have

$$G(\chi)\overline{G(\chi)} = \sum_{a,b\in\mathbb{F}^\times} \chi(ab^{-1})\zeta_p^{\mathrm{Tr}(a-b)}$$

$$\overset{c=ab^{-1}}{=} \sum_{b,c\in\mathbb{F}^\times} \chi(c)\zeta_p^{\mathrm{Tr}(bc-b)} = \sum_{b\in\mathbb{F}^\times} \chi(1) + \sum_{c\neq 0,1} \chi(c)\sum_{b\in\mathbb{F}^\times} \zeta_p^{\mathrm{Tr}(b(c-1))}$$

$$= (|\mathbb{F}| - 1) + \sum_{c\neq 0,1} \chi(c)(-1) = |\mathbb{F}|.$$

This finishes the proof of (3), and (4) follows from (2) and (3). $\qquad\square$

1.4.2   *Jacobi sum*

For two characters $\varphi, \phi : \mathbb{F}^\times \to \overline{\mathbb{Q}}^\times$, we define the *Jacobi sum* as
$$J(\varphi, \phi) := -\sum_{a \in \mathbb{F}} \varphi(a)\phi(1 - a).$$
Jacobi sum was originally introduced as an analogue of the beta function $B(x, y) = \int_0^1 t^x(1 - t)^{y-1}\frac{dt}{t}$ (similar to Gauss sum which is an analogue of the gamma function as already mentioned), and it satisfies a relation analogous to the relation between the beta function and the gamma function as in Lemma 1.4.2 (4). See [W49] and [W52] for amazing properties of Gauss sum and Jacobi sum which was the origin of Weil's conjecture (Riemann hypothesis for zeta function of algebraic varieties over finite fields), which was solved by P. Deligne. Here are some such properties:

Lemma 1.4.2.

(1) $J(\mathbf{1}, \mathbf{1}) = 2 - |\mathbb{F}|$,
(2) $J(\mathbf{1}, \chi) = J(\chi, \mathbf{1}) = 1$ *if* $\chi \neq \mathbf{1}$,
(3) $J(\chi, \overline{\chi}) = \chi(-1)$ *if* $\chi \neq \mathbf{1}$,
(4) $J(\varphi, \phi) = \frac{G(\varphi)G(\phi)}{G(\varphi\phi)}$ *if* $\varphi \neq \mathbf{1}, \phi \neq \mathbf{1}, \varphi\phi \neq \mathbf{1}$.

Proof.   Since there are $|\mathbb{F}| - 2$ elements in $\mathbb{F} - \{0, 1\}$, we get (1). When one character is $\mathbf{1}$ and the other not, the Jacobi sum is the sum over $\mathbb{F} - \{0, 1\}$, and hence the result (2) follows.

To show (4), we set $\varphi(0) = \phi(0) = 1$ and we compute $G(\varphi)G(\phi)$:

$$G(\varphi)G(\phi) = \sum_{a,b \in \mathbb{F}} \varphi(a)\phi(b)\zeta_p^{\mathrm{Tr}(a+b)}$$

$$\overset{a+b \mapsto c}{=} \sum_{a,c \in \mathbb{F}} \varphi(a)\phi(c-a)\zeta_p^{\mathrm{Tr}(c)} = \sum_{a \in \mathbb{F}, c \in \mathbb{F}^\times} \varphi(a)\phi(c-a)\zeta_p^{\mathrm{Tr}(c)} + \sum_{a \in \mathbb{F}} \varphi(a)\phi(-a).$$

If $\varphi\phi \neq \mathbf{1}$, we have

$$\sum_{a \in \mathbb{F}} \varphi(a)\phi(-a) = \phi(-1)\sum_{a \in \mathbb{F}^\times} \varphi\phi(a) = 0.$$

As for the first sum, without assuming $\varphi\phi \neq \mathbf{1}$, we have

$$\sum_{a \in \mathbb{F}, c \in \mathbb{F}^\times} \varphi(a)\phi(c-a)\zeta_p^{\mathrm{Tr}(c)} \overset{a=bc}{=} \sum_{b \in \mathbb{F}, c \in \mathbb{F}^\times} \varphi(c)\phi(c)\varphi(b)\phi(1-b)\zeta_p^{\mathrm{Tr}(c)}$$

$$= G(\varphi\phi)J(\varphi, \phi).$$

Therefore we get

$$G(\varphi)G(\phi) = G(\varphi\phi)J(\varphi, \phi)$$

as desired if $\varphi\phi \neq \mathbf{1}$.

Suppose now $\varphi\phi = 1$. Then we have

$$\sum_{a\in\mathbb{F}} \varphi(a)\phi(-a) = \phi(-1)\sum_{a\in\mathbb{F}} 1(a) = \phi(-1)(|\mathbb{F}| - 1).$$

Thus

$$\varphi(-1)|\mathbb{F}| = G(\varphi)G(\varphi^{-1}) = \varphi(-1)(|\mathbb{F}| - 1) + G(1)J(\varphi, \overline{\varphi})$$

with $G(1) = -\sum_{a\in\mathbb{F}^\times}\zeta_p^{\mathrm{Tr}(a)} = -(\sum_{a\in\mathbb{F}}\zeta_p^{\mathrm{Tr}(a)} - 1) = 1$. This shows (3). $\square$

Corollary 1.4.3. *Suppose that* $\varphi^N = \phi^N = 1$ *for* $0 < N \in \mathbb{Z}$. *Then* $\frac{G(\varphi)G(\phi)}{G(\varphi\phi)}$ *is an algebraic integer in* $\mathbb{Q}(\mu_N)$.

1.4.3 Basic properties of Gauss sum

Let $0 < N \in \mathbb{Z}$, and suppose $p \nmid N$. Then $\mathbb{Q}(\mu_N)$ and $\mathbb{Q}(\mu_p)$ is linearly disjoint as p is unramified in $\mathbb{Q}(\mu_N)$ while p fully ramifies in $\mathbb{Q}(\mu_p)$. Thus

$$G_{\mathbb{Q}(\mu_{pN})/\mathbb{Q}} \cong G_{\mathbb{Q}(\mu_N)/\mathbb{Q}} \times G_{\mathbb{Q}(\mu_p)/\mathbb{Q}} \cong (\mathbb{Z}/N\mathbb{Z})^\times \times (\mathbb{Z}/p\mathbb{Z})^\times.$$

Let $\sigma_a \in G_{\mathbb{Q}(\mu_{pN})/\mathbb{Q}}$ be the automorphism of $\mathbb{Q}(\mu_{pN})$ corresponding to $(a, 1) \in (\mathbb{Z}/N\mathbb{Z})^\times \times (\mathbb{Z}/p\mathbb{Z})^\times$ for $a \in (\mathbb{Z}/N\mathbb{Z})^\times$.

Lemma 1.4.4. *If* $\chi^N = 1$ *and* a *is an integer prime to* Np, *then*

$$\frac{G(\chi)^a}{G(\chi)^{\sigma_a}} = G(\chi)^{a-\sigma_a} \in \mathbb{Q}(\mu_N)$$

and $G(\chi)^N \in \mathbb{Q}(\mu_N)$.

Proof. Since $\zeta^{\sigma_a} = \zeta^a$ for $\zeta \in \mu_N$ and $\zeta_p^{\sigma_a} = \zeta_p$, we have

$$G(\chi)^{\sigma_a} = (-\sum_{x\in\mathbb{F}}\chi(x)\zeta_p^{\mathrm{Tr}(x)})^{\sigma_a} = -\sum_{x\in\mathbb{F}}\chi^a(x)\zeta_p^{\mathrm{Tr}(x)} = G(\chi^a).$$

Similarly, for $\sigma \in G_{\mathbb{Q}[\mu_{pN}]/\mathbb{Q}[\mu_N]}$, we have some $0 < b \in \mathbb{Z}$ prime to p such that $\zeta_p^\sigma = \zeta_p^b$. Then we have

$$G(\chi)^\sigma = (-\sum_{x\in\mathbb{F}}\chi(x)\zeta_p^{\mathrm{Tr}(x)})^\sigma = -\sum_{x\in\mathbb{F}}\chi(x)\zeta_p^{\mathrm{Tr}(bx)} \overset{bx\mapsto b}{=} \chi^{-1}(b)G(\chi).$$

Replacing χ by χ^a, we get

$$G(\chi^a)^\sigma = \chi^{-a}(b)G(\chi^a).$$

Thus σ fixes $\frac{G(\chi)^a}{G(\chi)^{\sigma_a}}$, and hence we get $G(\chi)^{a-\sigma_a} \in \mathbb{Q}(\mu_N)$. Taking $a := 1 + N$, we get the last assertion. $\square$

Here is the last lemma in this section:

Lemma 1.4.5. *We have* $G(\chi^p) = G(\chi)$.

Proof. The Frobenius automorphism Frob_p of $\mathbb{F}$ acts on $\mathrm{Frob}_p(a) = a^p$. Thus $\mathrm{Tr}(a^p) = \mathrm{Tr}(\mathrm{Frob}_p(a)) = \mathrm{Tr}(a)$. Then we have

$$G(\chi^p) = -\sum_{a\in\mathbb{F}} \chi^p(a)\zeta_p^{\mathrm{Tr}(a)} = -\sum_{a\in\mathbb{F}} \chi(a^p)\zeta_p^{\mathrm{Tr}(a^p)}$$

$$= -\sum_{a\in\mathbb{F}} \chi(\mathrm{Frob}_p(a))\zeta_p^{\mathrm{Tr}(\mathrm{Frob}_p(a))} \overset{\mathrm{Frob}_p(a)\mapsto a}{=} -\sum_{a\in\mathbb{F}} \chi(a)\zeta_p^{\mathrm{Tr}(a)} = G(\chi),$$

as desired. □

1.5 Prime factorization of Gauss sum

As already mentioned, determination of the prime factorization of the Gauss sum is a key to the proof of Stickelberger's theorem. We give the formula and prove the theorem in this section.

1.5.1 *Integrality ideal*

Let $I' = I'_F := \{\beta \in \mathbb{Z}[G] | \beta\theta_F \in \mathbb{Z}[G]\}$ for $G = G_{F/\mathbb{Q}}$ for an abelian extension $F/\mathbb{Q}$. We put $\mathfrak{s}_F := I'\theta_F = (\theta_F) \cap \mathbb{Z}[G]$ which is called the Stickerberger ideal of F. We start with the following lemma.

Lemma 1.5.1. *Suppose* $F = \mathbb{Q}[\mu_N]$ *and put* $G := G_{F/\mathbb{Q}}$. *Then the ideal* I' *is generated by* $c - \sigma_c$ *for* $c \in \mathbb{Z}$ *prime to* N.

Proof. We first show that $I' \supset I'' := (c - \sigma_c)_c$ in $\mathbb{Z}[G]$. We have plainly

$$(c - \sigma_c)\theta = \sum_a \left(c\left\{\frac{a}{N}\right\} - \left\{\frac{ac}{N}\right\} \right) \sigma_a^{-1} \in \mathbb{Z}[G]$$

which shows the claimed result.

Now we show the converse. Suppose that $x = \sum_a x_a\sigma_a \in I'$. Then

$$x\theta = \sum_c \left(\sum_a x_c\left\{\frac{a}{N}\right\} \right) \sigma_{ac^{-1}}^{-1} \overset{ac^{-1}\mapsto b, c\mapsto a}{=} \sum_b \left(\sum_a x_a\left\{\frac{ab}{N}\right\} \right) \sigma_b^{-1}.$$

The coefficient of σ_1^{-1} is given by

$$\sum_a x_a\left\{\frac{a}{N}\right\} \equiv \sum_a \left\{\frac{x_a a}{N}\right\} \equiv \left\{\frac{\sum_a x_a a}{N}\right\} \quad \mod 1.$$

Thus $\sum_a x_a \equiv 0 \mod N$. Since $N = (1 + N) - \sigma_{1+N} \in I''$, we find

$$\sum_a x_a \sigma_a = \sum_a x_a (\sigma_a - a) + \sum_a x_a \in I''.$$

Thus $I' \subset I''$. □

1.5.2 *Reformulation of the Stickelberger theorem*

Theorem 1.5.2 (Stickelberger). *Let $F = \mathbb{Q}[\mu_N]$. Then $\mathfrak{s}_F$ annihilates Cl_F; i.e., for any $\beta \in I'$ and a fractional ideal $\mathfrak{A}$ of F, $\mathfrak{A}^{\beta\theta}$ for $\theta := \theta_F$ is principal. Here for $x = \sum_a x_a \sigma_a$, $\mathfrak{A}^x = \prod_a \mathfrak{A}^{x_a \sigma_a}$.*

This theorem can be extended to a general finite (imaginary) abelian extension $F/\mathbb{Q}$ with some more technicality (see [ICF, Theorem 6.10]), and there is a weaker analogue in the relative abelian case over a totally real field [C77].

We prepare several lemmas before going into the final phase of the proof. We show that a suitable Gauss sum gives a canonical generator of $\mathfrak{p}^{\beta\theta}$ for a prime $\mathfrak{p}$.

1.5.3 *Lemmas and key steps*

Let p be a prime and $\mathbb{F}$ be a finite extension of $\mathbb{F}_p$; so, $|\mathbb{F}| = p^f =: q$. Let $\mathfrak{p}$ be a prime ideal of $\mathbb{Z}[\mu_{q-1}]$ above p. Since $\mathbb{F}^\times$ is a cyclic group of order $q - 1$, we have an isomorphism $\omega = \omega_{\mathfrak{p}} : \mathbb{F}^\times \cong \mu_{q-1} \subset \mathbb{Z}[\mu_{q-1}]^\times$. Since all $q - 1$-th roots of unity are distinct modulo $\mathfrak{p}$, we may assume that $\omega(a) \mod \mathfrak{p} = a \in \mathbb{F}^\times$.

Step 1: *Analysis of the exponent of prime factors of the Gauss sum.* Pick a prime $\mathfrak{P}|\mathfrak{p}$ in $\mathbb{Z}[\mu_{(q-1)p}]$. Write prime factorization of an ideal $\mathfrak{A}$ of $\prod_{\mathfrak{L}} \mathfrak{L}^{v_{\mathfrak{L}}(\mathfrak{A})}$. If $\mathfrak{A} = (\alpha)$ is principal, we simply write $v_{\mathfrak{L}}(\alpha)$ for $v_{\mathfrak{L}}(\mathfrak{A})$. Simply write $v(i)$ for $v_{\mathfrak{P}}(G(\omega^{-i}))$. Here is a lemma on the behavior of the exponent v.

Lemma 1.5.3.

(1) $v(0) = 0$;

(2) $0 \leq v(i + j) \leq v(i) + v(j)$;

(3) $v(i + j) \equiv v(i) + v(j) \mod(p - 1)$;

(4) $v(pi) = v(i)$;

(5) $\sum_{i=1}^{q-2} v(i) = (q - 2)f(p - 1)/2$ *if* $q = p^f$;

(6) $v(i) > 0$ *if* $i \not\equiv 0 \mod (q - 1)$;

(7) $v(1) = 1$, *in particular,* $G(\omega^{-1}) \equiv \pi \mod \mathfrak{P}^2$.

Proof. Since $G(1) = 1$ as $\sum_{a \in \mathbb{F}} \zeta_p^{\mathrm{Tr}(a)} = 0$ (character sum), we get (1). Since $\frac{G(\varphi)G(\phi)}{G(\varphi\phi)}$ is an algebraic integer (Lemma 1.4.2), (2) follows. Moreover, again by Lemma 1.4.2, $\frac{G(\varphi)G(\phi)}{G(\varphi\phi)}$ is an algebraic integer in the smaller field $\mathbb{Q}[\mu_N]$ in which $\mathfrak{p}$ does not ramify, the difference $v(i+j) - (v(i)+v(j))$ is divisible by the ramification index $p-1$ of $\mathfrak{P}/\mathfrak{p}$. Therefore, we get (3). By the existence of Frobenius automorphism on $\mathbb{F}$, we have $G(\chi^p) = G(\chi)$ (Lemma 1.4.5), which shows (4). Since $G(\omega^{-i})G(\omega^i) = G(\overline{\omega}^i)G(\omega^i) = \pm q = \pm p^f$ as long as $\omega^i \neq 1$ (i.e., $1 \leq i \leq p^f - 2$), $v(-i) + v(i) = f(p-1)$ as $p-1$ is the ramification index of $\mathfrak{P}/(p)$, summing these up, we get (5). To show (6), we put $\pi := \zeta_p - 1$ which is a generator of the unique prime in $\mathbb{Z}[\mu_p]$ (Theorem 1.1.1); so, $\pi \in \mathfrak{P}$. Then we see

$$G(\omega^{-i}) = -\sum_a \omega^{-i}(a)\zeta_p^{\mathrm{Tr}(a)} \equiv -\sum_a \omega^{-i}(a) \equiv 0 \mod \mathfrak{P},$$

which shows $v(i) > 0$. Now we prove (7) by showing $G(\omega^{-1}) \equiv \pi \mod \mathfrak{P}^2$. By a computation similar to the case of (6), we see

$$G(\omega^{-1}) = -\sum_a \omega^{-1}(a)\zeta_p^{\mathrm{Tr}(a)} = -\sum_a \omega^{-1}(a)(1+\pi)^{\mathrm{Tr}(a)}$$

$$\equiv -\sum_a \omega^{-1}(a)(1+\pi\,\mathrm{Tr}(a)) \mod \mathfrak{P}^2 \equiv -\pi\sum_a \omega^{-1}(a)\,\mathrm{Tr}(a) \mod \mathfrak{P}^2.$$

Note that $G_{\mathbb{F}/\mathbb{F}_p} = \langle \mathrm{Frob}_p \rangle = \{1, \mathrm{Frob}_p, \mathrm{Frob}_p^2, \ldots, \mathrm{Frob}_p^{f-1}\}$. Thus we have $\mathrm{Tr}(a) = a + a^p + \cdots + a^{p^{f-1}}$. Therefore, we get

$$\sum_a \omega^{-1}(a)\,\mathrm{Tr}(a) \equiv \sum_{0 \neq a \in \mathbb{Z}[\mu_N]/\mathfrak{p}} a^{-1}(a + a^p + \cdots + a^{p^{f-1}}) \mod \mathfrak{p}.$$

Since $a \mapsto a^{p^b-1}$ is a non-trivial character of $\mathbb{F}^\times$ if $0 < b \leq f$, the sum $\sum_{0 \neq a \in \mathbb{Z}[\mu_N]/\mathfrak{p}} a^{p^b-1}$ vanishes modulo $\mathfrak{p}$, and hence the sum reduces to $\sum_{0 \neq a \in \mathbb{Z}[\mu_N]/\mathfrak{p}} 1 = q - 1 \equiv -1 \mod \mathfrak{p}$; therefore, we conclude $G(\omega^{-1}) \equiv \pi \mod \mathfrak{P}^2$ as desired. $\square$

Step 2: *Prime factorization of $G(\chi)$.* To clarify the notation, fix a positive integer N, and choose a prime $p \nmid N$. Write f for the order of the class of p in $(\mathbb{Z}/N\mathbb{Z})^\times$. Thus $N \parallel p^f - 1$. Let $\chi := \omega^{-d}$ for $d = (q-1)/N$; so, χ has values in μ_N and therefore $\chi^N = 1$. Let R be a complete representative set for $(\mathbb{Z}/N\mathbb{Z})^\times/\langle p \rangle$. Let $\mathfrak{p}$ be a prime over p of $\mathbb{Z}[\mu_{q-1}]$ such that $\omega(a) \mod \mathfrak{p} = a \in \mathbb{F}^\times$, and put $\mathfrak{p}_0 = \mathfrak{p} \cap \mathbb{Z}[\mu_N]$ (which is the base prime for the prime factorization of $G(\chi)$). Then by Remark 1.1.4, $\{\mathfrak{p}_0^{\sigma_a^{-1}} \mid a \in R\}$ is the set of all distinct primes above (p) in $\mathbb{Z}[\mu_N]$. Let $\mathfrak{P}_0$ be the unique prime above $\mathfrak{p}_0$ in $\mathbb{Z}[\mu_{pN}]$ as any prime above p fully ramifies in $\mathbb{Z}[\mu_{pN}]/\mathbb{Z}[\mu_N]$. Let $\mathcal{P}_0$ (resp. $\widetilde{\mathcal{P}}_0$) be a prime in $\mathbb{Z}[\mu_{q-1}]$ (resp. $\mathbb{Z}[\mu_{(q-1)p}]$) over $\mathfrak{p}_0$ (resp. $\mathfrak{P}_0$).

Lemma 1.5.4. *Let the notation be as above. Then we have*
$$(G(\chi)) = \mathfrak{P}_0^{\sum_{a \in R} v(ad)\sigma_a^{-1}} = \prod_{a \in R} \mathfrak{P}_0^{\sigma_a^{-1} v(ad)}.$$

Proof. Note that $\mathfrak{P}_0^{\sigma_a^{-1}}$ is the unique prime above $\mathfrak{p}_0^{\sigma_a^{-1}}$. Write $\mathfrak{P} := \mathfrak{P}_0^{\sigma_a^{-1}}$. Then we have
$$v_{\mathfrak{P}}(G(\chi)) = v_{\mathfrak{P}_0}(G(\chi)^{\sigma_a}) = v_{\mathfrak{P}_0}(G(\chi^a)) = v_{\widetilde{\mathcal{P}}_0}(G(\chi^a)) = v(ad).$$
This shows that the exponent of $\mathfrak{P} = \mathfrak{P}_0^{\sigma_a^{-1}}$ in $G(\chi)$ is given by $v(ad)$. $\quad\square$

Step 3: *Determination of $v(i)$ via p-adic expansion.*

Lemma 1.5.5. *Let $0 \leq i < q-1$ and expand i into a standard p-adic expansion $i = a_0(i) + a_1(i)p + \cdots + a_{f-1}(i)p^{f-1}$ with $0 \leq a_j(i) \leq p-1$. Then $v(i) = a_0(i) + a_1(i) + \cdots + a_{f-1}(i)$.*

Proof. Since $v(i) = v(\overbrace{1 + 1 + \cdots + 1}^{i}) \leq i \cdot v(1) = i$ as $v(i+j) \leq v(i) + v(j)$ and and $v(1) = 1$ by Lemma 1.5.3 (2) and (7). Since $v \bmod (p-1)$ is linear by Lemma 1.5.3 (3), we find $v(i) \equiv i \bmod (p-1)$. Thus if $0 \leq i < p-1$, we get $v(i) = i$. Now assume that $i \geq p$. Since $v(pi) = v(i)$ by Lemma 1.5.3 (3), we can sharpen $v(i) \leq i$ to $v(i) \leq a_0(i) + a_1(i) + \cdots + a_{f-1}(i)$. Then we have
$$\sum_{i=0}^{q-1}(a_0(i) + a_1(i) + \cdots + a_{f-1}(i))$$
$$= (1 + 2 + \cdots + p-1)fp^{f-1} = \frac{p(p-1)}{2}fp^{f-1} = \frac{p-1}{2}fq$$
as each $a_j(i)$ takes value 0 to $p-1$ exactly p^{f-1} times when i varies from 0 to p^{f-1}. Removing $i = q-1 = (p-1) + (p-1)p + \cdots + (p-1)p^{f-1}$, we get
$$\sum_{i=0}^{q-2} v(i) \leq \sum_{i=0}^{q-2}(a_0(i) + a_1(i) + \cdots + a_{f-1}(i))$$
$$= \frac{p-1}{2}fq - (p-1)f \stackrel{\text{Lemma 1.5.3 (5)}}{=} \sum_{i=0}^{q-2} v(i).$$
Thus each term $v(i)$ must be equal to $a_0(i) + a_1(i) + \cdots + a_{f-1}(i)$. $\quad\square$

We now make more explicit the value $v(i)$:

Lemma 1.5.6. *If* $0 \le a < q - 1$, *then*

$$v(a) = (p-1)\sum_{j=0}^{f-1}\left\{\frac{p^j a}{q-1}\right\} \quad \text{and} \quad v(ad) = (p-1)\sum_{j=0}^{f-1}\left\{\frac{p^j a}{N}\right\}.$$

Proof. Expand $a = a_0(a) + a_1(a)p + \cdots + a_{f-1}(a)p^{f-1}$. Then we have

$$p^j a = a_0(a)p^j + a_1(a)p^{j+1} + \cdots + a_{f-1}(a)p^{j+f-1}.$$

Since $p^f \equiv 1 \mod (q-1)$, once the exponent $j+k$ of p in $a_k(a)p^{j+k}$ exceeds f, we can remove p^f modulo $q - 1$. Thus we have

$$p^j a \equiv \sum_{k=j}^{f-1} a_{k-j}(a)p^k + \sum_{\ell=0}^{j-1} a_{f+\ell-j}(a)p^\ell \mod (q-1).$$

Here p^k, p^ℓ runs through $\{1, p, \ldots, p^{f-1}\}$ once for each, and $a_j(a)$ for $j = 0, 1 \ldots, f - 1$ appears once for each.

Since the right-hand-side is less than $q - 1$, we have

$$\left\{\frac{p^j a}{q-1}\right\} = \frac{\sum_{k=j}^{f-1} a_{k-j}(a)p^k + \sum_{\ell=0}^{j-1} a_{f+\ell-j}(a)p^\ell \mod (q-1)}{q-1}.$$

Now we sum up over j. By moving j from 0 to $f - 1$, modulo $q-1$, for each p^k, the term $a_i(a)p^k$ shows up once for each i with $0 \le i \le f - 1$. Then each term involving $a_k(a)$ is given by

$$a_k(a)(1 + p + \cdots + p^{f-1}) = a_k(a)\frac{p^f - 1}{p - 1} = a_k(a)\frac{q - 1}{p - 1}.$$

Thus we conclude

$$\sum_{j=0}^{f-1}\left\{\frac{p^j a}{q-1}\right\} = \frac{1}{q-1}\frac{q-1}{p-1}\sum_k a_k(a) = \frac{1}{p-1}\sum_k a_k(a) = \frac{v(a)}{p-1}.$$

Then we see for $d = \frac{q-1}{N}$,

$$v(ad) = (p-1)\sum_{j=0}^{f-1}\left\{\frac{p^j ad}{q-1}\right\} = (p-1)\sum_{j=0}^{f-1}\left\{\frac{p^j a}{N}\right\}$$

as desired.   □

Corollary 1.5.7. *We have* $(G(\chi)^N) = \mathfrak{p}_0^{N\theta_F}$ *in* $\mathbb{Z}[\mu_N]$.

Proof. Since $G(\chi)^\tau = \overline{\chi}(a)G(\chi)$ for $G_{\mathbb{Q}[\mu_{pN}]/F}$ with $\zeta_p^\tau = \zeta_p^a$ (Lemma 1.4.1 (1)), we have $G(\chi)^N \in \mathbb{Z}[\mu_N]$. Since $\mathbb{Q}[\mu_{pN}]/F$ fully ramifies at $\mathfrak{p}_0$ with ramification index $p-1$, we find $\mathfrak{P}_0^{(p-1)} = \mathfrak{p}_0$. We have

$$\sum_{a\in(\mathbb{Z}/N\mathbb{Z})^\times/\langle p\rangle} v(ad)\sigma_a^{-1} = (p-1)\sum_{j=0}^{f-1}\sum_{a\in(\mathbb{Z}/N\mathbb{Z})^\times/\langle p\rangle}\left\{\frac{p^j a}{N}\right\}\sigma_a^{-1}.$$

Note that $\mathfrak{p}_0^{\sigma_p} = \mathfrak{p}_0$ as σ_p is the generator of the decomposition group of p; so, the effect of σ_{ap^j} on $\mathfrak{p}_0$ is the same as the effect of σ_a on $\mathfrak{p}_0$. Thus we conclude

$$\mathfrak{p}_0^{\sum_{j=0}^{f-1}\sum_{a\in(\mathbb{Z}/N\mathbb{Z})^\times/\langle p\rangle}\{\frac{p^j a}{N}\}\sigma_a^{-1}}$$

$$= \mathfrak{p}_0^{\sum_{j=0}^{f-1}\sum_{a\in(\mathbb{Z}/N\mathbb{Z})^\times/\langle p\rangle}\{\frac{p^j a}{N}\}\sigma_{ap^j}^{-1}} \overset{(*)}{=} \mathfrak{p}_0^{\sum_{a\in(\mathbb{Z}/N\mathbb{Z})^\times}\{\frac{a}{N}\}\sigma_a^{-1}}.$$

The last equality $(*)$ follows from the bijection $(\mathbb{Z}/N\mathbb{Z})^\times \leftrightarrow \langle p\rangle \times (\mathbb{Z}/N\mathbb{Z})^\times/\langle p\rangle$ Then by Lemma 1.5.6,

$$(G(\chi)^N) = \mathfrak{P}_0^{N(p-1)\theta_F} = \mathfrak{p}_0^{N\theta_F}$$

as desired. $\qquad\square$

1.5.4 *Proof of Stickelberger's theorem*

We now finish the proof of Stickelberger's theorem. Though the theorem itself is suggested by the class number formula, the proof does not use the validity of class number formula (so, to some extent, the argument of Stickelberger is an algebraic proof of the formula).

Let $\mathfrak{A}$ be an ideal of $\mathbb{Q}[\mu_N]$ prime to (N). Without losing generality, $\mathfrak{A}$ can be chosen so that $\mathfrak{A} = \prod_i \mathfrak{p}_i$ (here $\mathfrak{p}_i$'s may overlap). Let χ_i be the character of $(\mathbb{Z}[\mu_N]/\mathfrak{p}_i)^\times = \mathbb{F}^\times$ given by $\omega_{\mathfrak{p}_i}^d$ for $d = \frac{q-1}{N}$ for $q = N(\mathfrak{p}_i) = |\mathbb{F}|$. By Corollary 1.5.7 applied to each $\mathfrak{p}_i$, we have

$$\mathfrak{A}^{N\theta_F} = \prod_i \mathfrak{p}_i^{N\theta_F} = (\prod_i G(\chi_i)^N).$$

Write $\gamma := \prod_i G(\chi_i) \in \mathbb{Q}[\mu_{PN}]$ for $P = \prod_i p_i$. If $\beta\theta_F \in \mathbb{Z}[G]$ $(G = G_{F/\mathbb{Q}})$, then

$$\mathfrak{A}^{N\beta\theta_F} = (\gamma^{\beta N}).$$

Since $\gamma^{N\beta} \in F$ as $G(\chi)^{b-\sigma_b} \in F$ (Lemma 1.4.4), $F[\gamma^\beta]/F$ is a Kummer extension adding N-th root γ^β of $\gamma^{N\beta} \in F$. Now we claim that

$$F(\gamma^\beta)/F \text{ can ramify only at prime factors of } N. \qquad (\text{Ur})$$

Here is the proof of (Ur): By adding N-th root, only ramified primes in the extension are factors of N and prime factors of $\gamma^{N\beta}$. Since $(\gamma^{N\beta})$ is N-th power of ideal $\mathfrak{A}$, for a prime factor $\mathfrak{l}$ of $\gamma^{\beta N}$, $F_{\mathfrak{l}}[\gamma^{\beta}] = F_{\mathfrak{l}}[\sqrt[N]{u}]$ for a unit u of the $\mathfrak{l}$-adic completion $F_{\mathfrak{l}}$. Since $\mathfrak{l} \nmid N$, $F_{\mathfrak{l}}[\sqrt[N]{u}]/F_{\mathfrak{l}}$ is unramified.

By definition,

$$F \subset F[\gamma^{\beta}] \subset \mathbb{Q}[\mu_{NP}] = F[\mu_P].$$

The only primes ramifying in $F[\mu_P]/F$ are factors of P which is prime to N. Therefore by (Ur), $F[\gamma^{\beta}]$ is unramified everywhere over F and is abelian over $\mathbb{Q}$, which is impossible by Remark 1.1.4 unless $F = F[\gamma^{\beta}]$. Therefore, $\mathfrak{A}^{\beta\theta} = (\gamma^{\beta})$ with $\gamma^{\beta} \in F$.

1.6   A consequence of the Kummer–Vandiver conjecture

Our next goal is to prove cyclicity of the relative class group over the group algebra, finding an explicit plausible condition to guarantee the fact (as proving cyclicity without a concrete input seems out of reach).

1.6.1   *Conjecture*

Recall $F_n = \mathbb{Q}[\mu_{p^{n+1}}]$ with $m = n + 1$. Let $h_n^+ = |Cl_{F_n^+}| = |Cl_n^+|$. The following conjecture is well known but we do not have theoretical evidence except for numerical evidences verified for primes up to 2 billion [HHO17]:

Conjecture 1.6.1 (Kummer–Vandiver). $p \nmid h_0^+$ *(so, $Cl_0 \otimes_{\mathbb{Z}} \mathbb{Z}_p = 0$).*

Suppose hereafter Conjecture 1.6.1. Let $A_n^{\pm} := Cl_n^{\pm} \otimes_{\mathbb{Z}} \mathbb{Z}_p$ (the p-Sylow part of $Cl_n^{\pm}$, and put $R_n := \mathbb{Z}_p[G_{F_n/\mathbb{Q}}]$. Then $A_n^{\pm}$ is a module over R_n. Define $\mathfrak{s}_n := R_n \cap \theta_{F_n} R_n$. Since $R_n = \mathbb{Z}[G_{F_n/\mathbb{Q}}] \otimes_{\mathbb{Z}} \mathbb{Z}_p$, we have $\mathfrak{s}_n = \mathfrak{s}_{F_n} \otimes_{\mathbb{Z}} \mathbb{Z}_p$, and hence by Stickelberger's theorem (Theorem 1.3.2), we have $\mathfrak{s}_n A_n^- = 0$ without assuming Conjecture 1.6.1. Here is Iwasawa's cyclicity theorem which is a goal of this chapter:

Theorem 1.6.2 (Iwasawa). *Suppose $p \nmid h_0^+$. Then we have an isomorphism $A_n^- \cong R_n^-/\mathfrak{s}_n^-$ as R_n-modules, where $X^- = (1 - c)X$ (the "$-$" eigenspace of complex conjugation c).*

We start preparing several facts necessary for the proof of the theorem. The proof ends in §1.8.

1.6.2 *Index calculation*

We first compute the index $[R_n^- : \mathfrak{s}_n^-]$.

Lemma 1.6.3. *We have* $[R_n^- : \mathfrak{s}_n^-] = |A_n^-|$.

Proof. Let $G = G_{F_n/\mathbb{Q}}$ and $\theta = \theta_{F_n}$. Let $\mathrm{Tr} := \sum_{\sigma \in G} \sigma \in R_n$. Since $\{z\} + \{-z\} = 1$ and $c\sigma_a = \sigma_{-a}$, $(1+c)\theta$ is equal to

$$
\sum_{a \in (\mathbb{Z}/p^m\mathbb{Z})^\times} \left\{\frac{a}{p^m}\right\} \sigma_a^{-1} + \sum_{a \in (\mathbb{Z}/p^m\mathbb{Z})^\times} \left\{\frac{a}{p^m}\right\} \sigma_{-a}^{-1}
$$

$$
= \sum_{a \in (\mathbb{Z}/p^m\mathbb{Z})^\times} \left(\left\{\frac{a}{p^m}\right\} + \left\{\frac{-a}{p^m}\right\}\right) \sigma_a^{-1} = \mathrm{Tr}.
$$

Let $\theta^\pm := \frac{1 \pm c}{2}\theta = \theta - \frac{1 + c}{2}\theta = \theta - \frac{\mathrm{Tr}}{2} \in \frac{1-c}{2}R_n = R_n^-$. Thus we have

$$
\theta^+ := \frac{1+c}{2}\theta = \frac{1}{2}\mathrm{Tr}. \tag{1.3}
$$

Since $\mathfrak{s}_n = R_n \cap \theta R_n$, we have $\mathfrak{s}_n^- = \frac{1-c}{2}R_n \cap \frac{1-c}{2}\theta R_n = R_n^- \cap \theta^- R_n$.

We now compute the index $[R_n\theta^- : \mathfrak{s}_n^-] = [R_n\theta^- : R_n^- \cap \theta^- R_n]$. Take $x = \sum_{b \in (\mathbb{Z}/p^m\mathbb{Z})^\times} x_b\sigma_b \in R_n$. Since $\theta^+ = \frac{1}{2}\mathrm{Tr} \in R_n$, we have $x\theta^- \in R_n^- \Leftrightarrow x\theta \in R_n$. Then

$$
x\theta = \frac{1}{p^m}\sum_b\sum_a ax_b\sigma_a^{-1}\sigma_b = \frac{1}{p^m}\sum_b\sum_a ax_b\sigma_{ba^{-1}} \overset{a^{-1}b \mapsto b}{=} \frac{1}{p^m}\sum_b\sum_a ax_{ab}\sigma_b.
$$

This shows $x\theta^- \in R_n^- \Leftrightarrow x\theta \in R_n \Leftrightarrow \sum_a ax_{ab} \equiv 0 \mod p^m$ for all b prime to p. But

$$
\sum_a ax_{ab} \equiv b^{-1}\sum_a(ab)x_{ab} \equiv b^{-1}\sum_a ax_a \mod p^m.
$$

Therefore

$$
x\theta^- \in R_n^- \Leftrightarrow x\theta \in R_n \Leftrightarrow \sum_a ax_a \equiv 0 \mod p^m.
$$

Thus $R_n\theta^- \cap R_n^- = \{x\theta^- \in R_n | \sum_a ax_a \equiv 0 \mod p^m\}$. Then $R_n\theta^-/(R_n\theta^- \cap R_n^-) \hookrightarrow \mathbb{Z}/p^m\mathbb{Z}$ by sending $x\theta^-$ to $\sum_s x_a a \mod p^m$. This map is surjective taking $x := \sigma_1$ as $\sum_s x_a a = 1$. Therefore we conclude $[R_n\theta^- : R_n\theta^- \cap R_n^-] = p^m$.

Consider the linear map $T : R_n^- \to R_n^-$ given by the multiplication by $p^m\theta^-$. Then

$$
[R_n^- : p^mR_n^-\theta] \overset{(*)}{=} |\det(T)|_p^{-1} = |p^{m[F_n^+:\mathbb{Q}]}\prod_\chi L(0,\chi)|_p^{-1}
$$

$$
= p^{m|G|/2 - m}|A_n^-| = [R_n^-\theta^- : p^mR_n^-\theta^-]p^{-m}|A_n^-|
$$

by the class number formula (Theorem 1.3.1). Here the first equality $(*)$ uses the fact that $\det(T)$ is up to units a product of the order of cyclic factors of $R_n^-/T(R_n^-)$ (i.e., the product of elementary divisors of T), and therefore $|\det(T)|_p^{-1} = [R_n^- : T(R_n^-)]$. Thus

$$|A_n^-| = \frac{[R_n^- : p^m R_n^- \theta][R_n \theta^- : R_n \theta^- \cap R_n^-]}{[R_n^- \theta^- : p^m R_n^- \theta^-]}.$$

Since $R_n^- \supset \mathfrak{s}_n^- \supset p^m R_n^- \theta$ and $R_n^- \theta \supset \mathfrak{s}_n^- \supset p^m R_n^- \theta$, we have

$$\frac{[R_n^- : p^m R_n^- \theta]}{[R_n^- \theta^- : p^m R_n^- \theta^-]} = \frac{[R_n^- : \mathfrak{s}_n^-]}{[R_n^- \theta^- : \mathfrak{s}_n^-]}.$$

From this we see

$$|A_n^-| = \frac{[R_n^- : \mathfrak{s}_n^-]][R_n \theta^- : R_n \theta^- \cap R_n^-]}{[R_n^- \theta^- : \mathfrak{s}_n^-]} = \frac{[R_n^- : \mathfrak{s}_n^-][R_n^- \theta^- : \mathfrak{s}_n^-]}{[R_n^- \theta^- : \mathfrak{s}_n^-]} = [R_n^- : \mathfrak{s}_n^-].$$

This shows the desired index formula. $\square$

1.7   Kummer theory

As we will see repeatedly, somehow, in the known cases of cyclicity for Selmer groups, the use of Kummer theory of units in the integer ring is essential in the proof. This is the first appearance of the use of the theory.

1.7.1   *Kummer pairing*

Let F/F_0 be a finite extension inside $\overline{\mathbb{Q}}$. If $\alpha \in F^\times$ is not a p-power in $F^\times$, $F[\sqrt[p]{\alpha}]/F$ is a Galois extension of degree p, as a complete set of conjugates of $\sqrt[p]{\alpha}$ over F is given by $\{\zeta \sqrt[p]{\alpha} | \zeta \in \mu_p\}$. Thus $G_{F[\sqrt[p]{\alpha}]/F} \cong \mathbb{Z}/p\mathbb{Z}$ by sending $\sigma_a \in G_{F[\sqrt[p]{\alpha}]/F}$ with $\sigma_a(\sqrt[p]{\alpha}) = \zeta_p^a \sqrt[p]{\alpha}$ to $a \in \mathrm{Aut}(\mu_p) = \mathbb{Z}/p\mathbb{Z}$. The extension $F[\sqrt[p]{\alpha}]/F$ only depends on $\alpha \mod (F^\times)^p$. Thus we simply write $F[\sqrt[p]{\alpha}]$ for the extension corresponding to $\alpha \in F^\times/(F^\times)^p$. For a subset B of $F^\times/(F^\times)^p$, we put $F[\sqrt[p]{B}] := F[\sqrt[p]{\alpha}]_{\alpha \in B}$, which is a $(p, p, \ldots, p)$-Galois extension of F.

Conversely, we take a p-cyclic extension K/F with a fixed isomorphism $G_{K/F} \cong \mathbb{Z}/p\mathbb{Z}$ with $\sigma \in G_{K/F}$ corresponding to $1 \in \mathbb{Z}/p\mathbb{Z}$. By the normal basis theorem in Galois theory, K is free of rank 1 over the group algebra $F[G_{K/F}]$. Thus for any character $\xi : \mathbb{Z}/p\mathbb{Z} \to \mu_p$, the ξ-eigenspace $F[\xi] = \{x \in K | \sigma(x) = \xi(\sigma)x \text{ for all } \sigma \in G_{K/F}\}$ is one dimensional over F. Suppose that $\xi(\sigma_a) = \zeta_p^a$. Pick $\beta \in F[\xi]$. Then $\alpha := \beta^p \in F$ as it is invariant under $G_{K/F}$. Since $K \supset F[\sqrt[p]{\alpha}]$ and $F[\sqrt[p]{\alpha}]$ has degree p over F, we conclude $K = F[\sqrt[p]{\alpha}]$. Thus every $(p, p, \ldots, p)$-extension of F is of the form $F[\sqrt[p]{B}]$. Since $F[\sqrt[p]{\alpha^a}] \subset F[\sqrt[p]{\alpha}]$ for any $a \in \mathbb{F}_p$, replacing B by the $\mathbb{F}_p$-span of B in $F^\times \otimes_\mathbb{Z} \mathbb{F}_p$, we may assume that B is an $\mathbb{F}_p$-vector subspace of $F^\times \otimes_\mathbb{Z} \mathbb{F}_p = F^\times/(F^\times)^p$.

Lemma 1.7.1. *Let the notation be as above. Suppose* $\dim_{\mathbb{F}_p} B < \infty$. *We have a non-degenerate pairing* $\langle \cdot, \cdot \rangle : G_{F[\sqrt[p]{B}]/F} \times B \to \mu_p$ *given by* $\langle \tau, \beta \rangle := \tau(\sqrt[p]{\beta})/\sqrt[p]{\beta}$.

Proof. If $\langle \tau, \beta \rangle = 1$ for all $\tau \in G_{F[\sqrt[p]{B}]/F}$, plainly $\sqrt[p]{\beta} \in F^\times$; so, $\beta = 0$ in $F^\times/(F^\times)^p$. Thus the pairing is non-degenerate on the B side. Thus $\dim_{\mathbb{F}_p} G_{F[\sqrt[p]{B}]/F} \geq \dim_{\mathbb{F}_p} B$.

Since $F[\sqrt[p]{\beta}] \supset F[\sqrt[p]{\beta^a}]$ for $a \in \mathbb{F}_p$ and $F[\sqrt[p]{\alpha\beta}] \subset F[\sqrt[p]{\alpha}, \sqrt[p]{\beta}]$, we find that $F[\sqrt[p]{B}] = F[\sqrt[p]{\beta_1}, \ldots, \sqrt[p]{\beta_d}]$ for a basis $\{\beta_1, \ldots, \beta_d\}$ of B over $\mathbb{F}_p$; so, $p^{\dim_{\mathbb{F}_p} B} = p^d \geq [F[\sqrt[p]{B}] : F] = |G_{F[\sqrt[p]{\alpha}]/F}|$, which shows $\dim_{\mathbb{F}_p} G_{F[\sqrt[p]{B}]/F} = d$, and this finishes the proof. $\square$

1.7.2 *Kummer theory for p-units*

Lemma 1.7.2. *Suppose* $F \supset F_0 = \mathbb{Q}[\mu_p]$. *Let* K/F *be the maximal* $(p, p, \ldots, p)$-*extension unramified outside* p. *Then for* $B := O[\frac{1}{p}]^\times \otimes_\mathbb{Z} \mathbb{F}_p$, *we have* $K = F[\sqrt[p]{B}]$ *and* $\dim_{\mathbb{F}_p} B < \infty$.

Proof. Let $L := F[\sqrt[p]{\alpha}]$. By multiplying p-power of a non-zero integer, we may assume that $\alpha \in O - \{0\}$. Write prime factorization of $(\alpha) = \prod_\mathfrak{l} \mathfrak{l}^{e(\mathfrak{l})}$. Plainly if $\mathfrak{l} \nmid p$ and $p \nmid e(\mathfrak{l})$, $\mathfrak{l}$ ramifies in $L_{/F}$. As for p, residually, any p-th root does not give any non-trivial extension as Frob_p just raises p-power. Thus p can ramify independent of $e(\mathfrak{l})$ for $\mathfrak{l}|p$. This shows the result. By Dirichlet's unit theorem in §7.2.8 (see [LFE, Theorem 1.2.3] for a proof), $\mathrm{rank}_\mathbb{Z} O^\times < [F : \mathbb{Q}]$; so, writing r for the distinct prime factors of p in O,

$$\mathrm{rank}_\mathbb{Z} O[\frac{1}{p}]^\times \leq [F : \mathbb{Q}] + r.$$

Thus $\dim_{\mathbb{F}_p} B \leq [F : \mathbb{Q}] + r < \infty$. $\square$

Exercise 1.7.1. Compute $\dim_{\mathbb{F}_p} B$ in Lemma 1.7.2.

Corollary 1.7.3. *Let the notation be as in Lemma 1.7.2. Assume* $F = F_n$. *Then* $K = F[\sqrt[p]{B}]$ *for* B *generated by* $O^\times$ *and* $1 - \zeta_{p^m}$ $(m = n + 1)$.

This is because the unique prime ideal $\mathfrak{p}$ over p in F_n is principal generated by $1 - \zeta_{p^{n+1}}$ by Theorem 1.1.1.

1.8   Proof of cyclicity theorem

We start the proof of Iwasawa's theorem Theorem 1.6.2. The first step is
to deal with $F_0 = \mathbb{Q}[\mu_p]$, and then we reduce the general case to F_0 by
Nakayama's lemma.

1.8.1   *Cyclicity for F_0*

Let $F = F_n$ and $L = L_{n/F_n}$ be the maximal p-elementary abelian extension
unramified everywhere. Here elementary means that $\mathrm{Gal}(L_n/F_n)$ is killed
by p. Write A_n for the maximal p-abelian quotient $Cl_n \otimes_{\mathbb{Z}} \mathbb{Z}_p$ of Cl_n. Since
L/F is elementary p-abelian, $\mathrm{Gal}(L/F)$ is an $\mathbb{F}_p$-vector space. By class
field theory (and Galois theory), $\mathrm{Gal}(L/F) \cong Cl_n/pCl_n = A_n/pA_n$. By
Kummer theory, $L = F[\sqrt[p]{B}]$ for an $\mathbb{F}_p$-vector subspace B of $F^{\times} \otimes_{\mathbb{Z}} \mathbb{F}_p =$
$F^{\times}/(F^{\times})^p$. Since $F[\sqrt[p]{b}]_{/F}$ is everywhere unramified for $b \in F^{\times}$ with $\bar{b} =$
$(b \mod (F^{\times})^p) \in B$, the principal ideal (b) is a p-power $\mathfrak{a}^p$ for an O-ideal
$\mathfrak{a}$. The class of $\mathfrak{a}$ in A_n only depends on the class of b modulo $(F^{\times})^p$ as
$F[\sqrt[p]{a^p b}] = F[\sqrt[p]{b}]$. Thus sending $\bar{b} := b \mod (F^{\times})^p$ to the class of $\mathfrak{a}$, we get
a homomorphism $\phi : B \to A_n[p] = \{x \in A_n | px = 0\}$, which is obviously
$\mathbb{Z}_p[G]$-linear for $G := G_{F/\mathbb{Q}}$. Assume $\bar{b} \in \mathrm{Ker}(\phi)$. Then $(b) = (a)^p$; so,
$b = a^p \varepsilon$ with $\varepsilon \in O^{\times}$. In other words, $F[\sqrt[p]{b}] = F[\sqrt[p]{\varepsilon}]$. Thus we conclude
that $\mathrm{Ker}(\phi) \subset O^{\times}/(O^{\times})^p = O^{\times} \otimes_{\mathbb{Z}} \mathbb{F}_p$ as $\mathbb{Z}_p[G]$-modules.

Now we assume that $F = F_0$. Then $\hat{G} = \mathrm{Hom}(G, \mathbb{Z}_p^{\times})$ is generated by
Teichmüller character ω of order $p - 1$. Then $\mathbb{Z}_p[G] = \bigoplus_{i \mod p-1} \mathbb{Z}_p e_i$ for
the idempotent $e_i = \frac{1}{|G|} \sum_{\sigma} \omega^{-i}(\sigma)\sigma$. For a finite p-abelian group H, we de-
fine p-rank$(H) = \dim_{\mathbb{F}_p} H \otimes_{\mathbb{Z}} \mathbb{F}_p = \dim_{\mathbb{F}_p} H[p]$ for $H[p] = \{x \in H | px = 0\}$
(which is the minimal number of generators of H by the fundamental theo-
rem of finite abelian groups); thus, H is cyclic if and only if p-rank$(H) = 1$.
We first prove

Theorem 1.8.1. *Let A be the p-Sylow subgroup of Cl_0 and put $A =$*
$\bigoplus_i e_i A$. *If i is even and j is odd with $i + j \equiv 1 \mod (p - 1)$, then*

$$p\text{-rank}(e_i A) \leq p\text{-rank}(e_j A) \leq 1 + p\text{-rank}(e_i A).$$

This implies a famous result of Kummer (when he proved FLT for regular
primes): $p | |Cl_0^+| \Rightarrow p | |Cl_0^-|$. There is a more general version than the above
theorem called Leopoldt's reflection theorem (Spiegelungssatz; [CNF, XI.4]
or [ANT, Theorem 2.116]), and its non-abelian adjoint version (in §7.2.3)
is used in Wiles' proof of Fermat's last theorem.

Proof. One guesses that the case $i = 0$ could be slightly more involved as it corresponds to the trivial character (and the Riemann zeta function has a pole at $s = 1$); so, we need to argue this case a bit differently.

By class field theory, $A/pA \cong \mathrm{Gal}(L/F)$ for the maximal p-abelian elementary extension unramified everywhere. Then we have the perfect Kummer pairing as in Lemma 1.7.1 $\langle \cdot, \cdot \rangle : A/pA \times B \to \mu_p$. Note that $\sigma_a \mathfrak{a} = \omega^i(a)\mathfrak{a}$ for all $a \in (\mathbb{Z}/p\mathbb{Z})^\times$ if $\mathfrak{a} \in A_i := e_i A$. Since $\langle \mathfrak{a}, b \rangle^{\omega(a)} = \langle \mathfrak{a}, b \rangle^{\sigma_a} = \langle \mathfrak{a}^{\sigma_a}, b^{\sigma_a} \rangle = \langle \omega^i(a)\mathfrak{a}, \omega^k(a)b \rangle = \langle \mathfrak{a}, b \rangle^{\omega^{i+k}(a)}$ for $b \in B_k = e_k B$, $\langle \mathfrak{a}, b \rangle = 1$ unless $i + k \equiv 1 \mod (p-1)$. This shows that $\langle \cdot, \cdot \rangle$ indices a perfect pairing on $A_i \times B_j$. Thus $\dim_{\mathbb{F}_p} A_i = \dim_{\mathbb{F}_p} B_j$.

Now $\phi : B_j \to A_i[p] = e_i(A[p])$ is $\mathbb{Z}_p[G]$-linear. Since $\mathrm{Ker}(\phi) \cap B_j \subset e_j(O^\times \otimes_\mathbb{Z} \mathbb{F}_p)$ and

$$E_k := e_k(O^\times \otimes_\mathbb{Z} \mathbb{F}_p) \cong \begin{cases} \mathbb{F}_p & \text{if } k \text{ is even, } k \not\equiv 0 \mod(p-1), \\ \mathbb{F}_p & \text{if } k \equiv 1 \mod(p-1), \\ 0 & \text{otherwise} \end{cases}$$

by Dirichlet's unit theorem [LFE, Theorem 1.2.3], we have

$$p\text{-rank}(A_l) = \dim B_k \leq \dim E_k + \dim A_k[p] \text{ if } l + k \equiv 1 \mod (p-1). \tag{1.4}$$

If i is even (and j is odd), taking $k = i$ (and $l = j$), we have

$$p\text{-rank}(A_j) = \dim B_i \leq \dim E_i + \dim A_i[p] = 1 + p\text{-rank}(A_i).$$

If j is odd and $j \not\equiv 1 \mod (p-1)$, we have

$$p\text{-rank}(A_i) \leq \dim A_j[p] = p\text{-rank}(A_j).$$

Thus if $j \not\equiv 1 \mod (p-1)$, the result follows.

Suppose that $j \equiv 1 \mod (p-1)$. Then $i = 0$. Let L_0 be the subfield of L with $G_{L_0/F} = A_0$. From the exact sequence:

$$1 \to G_{L_0/F} \to G_{L_0/\mathbb{Q}} \to G \to 1$$

with G acting on the normal subgroup $G_{L_0/F}$ by conjugation, $G_{L_0/\mathbb{Q}}$ is abelian. Then for the inertia subgroup I at p in $G_{L_0/\mathbb{Q}}$ is isomorphic to $G_{F/\mathbb{Q}}$; so, L_0^I is everywhere unramified extension of $\mathbb{Q}$; so, $A_0 \cong G_{L_0^I/\mathbb{Q}}$ is trivial.

Consider $c = 1 + p$. Then $\omega(c - \sigma_c) = 1 + p - 1 = p$. For the Stickelberger element $\theta = \frac{1}{p} \sum_{a=1}^{p-1} \sigma_a^{-1} a$, $\omega((c - \sigma_c)\theta_F) = \sum_a a\omega^{-1}(a) \equiv 1(p-1) \mod p$ kills A_1; so, $A_1 = 0$. This shows the result when $j \equiv 1 \mod (p-1)$. $\square$

Corollary 1.8.2. *Assume the Kummer–Vandiver conjecture. Then we have $A_0^- \cong R_0^-/\mathfrak{s}_0^-$ as R_0-modules for $R_0 := \mathbb{Z}_p[G_{F_0/\mathbb{Q}}]$.*

Proof. By Kummer–Vandiver, p-rank$(e_i A) = 0$ for i even. Then by the above theorem (Theorem 1.8.1), p-rank$(A_j) \leq 1$; so, A_j is cyclic. Thus we have a surjective module homomorphism $e_j R_0 \twoheadrightarrow A_j$. We write the image of e_j in A_j as $\bar{e}_j$. $A_0^- = \bigoplus_{j:odd} e_j A = \sum_{j:odd} R_0 \bar{e}_j = R_0(\sum_{j:odd} \bar{e}_j)$; so, A_0^- is cyclic. Since $\mathfrak{s}_0$ kills A_0^- and $|A_0^-| = [R_0^- : \mathfrak{s}_0^-]$ by Lemma 1.6.3, we conclude $A_0^- \cong R_0^-/\mathfrak{s}_0^-$. $\qquad\square$

1.8.2　Proof in general

We first recall Nakayama's lemma: Let R be a local ring with a unique maximal ideal $\mathfrak{m}_R$ and M be a finitely generated R-module.

Lemma 1.8.3 (NAK). *If $M = \mathfrak{m}_R M$ ($\Leftrightarrow M \otimes_R R/\mathfrak{m}_R = 0$), then $M = 0$.*

See [CRT, p. 8] for a proof of this lemma. This can be used to determine the number of generators. Take a basis $\bar{m}_1, \ldots, \bar{m}_r$ of $M/\mathfrak{m}_R M$ over $R/\mathfrak{m}_R$ and lift it to $m_j \in M$ so that $(m_j \mod \mathfrak{m}_R) = \bar{m}_j$.

Corollary 1.8.4. *The elements $m_1, \ldots, m_r$ generate M over R, and r is the minimal number of generators.*

Proof. For the R-linear map $\pi : A^r \to M$ given by $(a_1, \ldots, a_r) \mapsto \sum_j a_j m_j$, $\mathrm{Coker}(\pi) \otimes_R R/\mathfrak{m}_R = 0$ as $\bar{m}_j$ generate $M \otimes_R R/\mathfrak{m}_R$. Thus $\mathrm{Coker}(\pi) = 0$ by NAK; so, $m_1, \ldots, m_r$ generate M. $\qquad\square$

We recall the cyclicity theorem:

Theorem 1.8.5 (Iwasawa). *Suppose $p \nmid h_0^+$. Then we have an isomorphism $A_n^- \cong R_n^-/\mathfrak{s}_n^-$ as R_n-modules and $A_n^+ = 0$.*

Proof. We have already proven the result when $n = 0$. Note that $R_0 = \prod_{i=0}^{p-1} e_i R_0$ and $e_i R_0 \cong \mathbb{Z}_p$ as a ring. The restriction map $G_{F_{n'}/\mathbb{Q}} \ni \sigma \mapsto \sigma|_{F_n} \in G_{F_n/\mathbb{Q}}$ induces a surjective ring homomorphism $\pi_{n'}^n : R_{n'} \to R_n$ for $n' > n$. Take $\epsilon_i \in R_n$ projecting down to e_i. Since $e_i^2 = e_i$ (so, $e_i^j = e_i$), $\epsilon_i^2 \equiv \epsilon_i$. Since $(\pi_n^0)^{-1}(e_i R_0)$ is a p-profinite ring, $\lim_{j \to \infty} \epsilon_j^{p^j}$ converges to an idempotent, lifting e_j. We again wrote this lift as e_j; so, $R_n^- = \prod_{j=0, j:odd}^{p-1} e_j R_n$ as a ring direct product.

Note that $G_{F_n/\mathbb{Q}} \cong (\mathbb{Z}/p^m\mathbb{Z})^\times = \mu_{p-1} \times \Gamma/\Gamma^{p^n}$ for $\Gamma = 1 + p\mathbb{Z}_p$ as $\mathbb{Z}_p^\times = \mu_{p-1} \times \Gamma$. Let us write $\Lambda_n := \mathbb{Z}_p[\Gamma/\Gamma^{p^n}]$. Thus $R_n^- = R_0^- \otimes_{\mathbb{Z}_p} \Lambda_n = \prod_{j=0, j:odd}^{p-1} e_j R_0 \otimes_{\mathbb{Z}_p} \Lambda_n$; so, $e_j R_n \cong \Lambda_n$. Since Γ/Γ^{p^n} is a p-group, Λ_n is a local ring with $\Lambda_n/(\gamma - 1) \cong \mathbb{Z}_p$ as rings for the generator $\gamma = 1 + \mathbf{p}$ of Γ.

To show cyclicity, we need to show that $M := e_i A_n^-$ is generated by one element over Λ_n. This is equivalent to $M/\mathfrak{m}_n M \cong \mathbb{F}_p$ for the maximal ideal $\mathfrak{m}_n = (p, \gamma - 1) \subset \Lambda_n$ by Nakayama's lemma. Let L_n/F_n be the maximal p-abelian extension unramified everywhere. Let $X_n = G_{L_n/F_n}$. We have the following field diagram

$$
\begin{array}{ccccc}
F_n & \longrightarrow & L_0 F_n & \longrightarrow & L_n \\
\uparrow & & \uparrow & & \\
F_0 & \longrightarrow & L_0. & &
\end{array}
$$

Since F_n/F_0 is fully ramified at p, F_n and L_0 is linearly disjoint. Thus

$$G_{L_n/F_0} = G_{F_n/F_0} \ltimes X_n$$

identifying G_{F_n/F_0} with the inertia subgroup at p of G_{L_n/F_0}. Therefore, $G_{L_0 F_n/F_n} \cong G_{L_0/F_0}$, and we have an exact sequence

$$1 \to G_{L_n/L_0 F_n} \to X_n \to X_0 \to 1.$$

Since X_0 is the maximal abelian quotient of G_{L_n/F_0}, $G_{L_n/L_0 F_0}$ is the commutator subgroup of G_{L_n/F_0}. Since $G_{L_n/F_0} = G_{F_n/F_0} \ltimes X_n$ and G_{F_n/F_0} is generated by γ, any element in G_{L_n/F_0} is of the form $\gamma^j x$ for $x \in X_n$ uniquely. Then the commutator subgroup is generated by $(\gamma, x) = x^{\gamma-1} = (\gamma - 1)x$ for $x \in X_n$ since X_n is abelian. In other words, $G_{L_n/F_0} = (\gamma - 1)X_n$ written additively as Λ_n-module. This implies $e_i X_n/(\gamma - 1)e_i X_n \cong e_i X_0$ and hence $e_i X_n/\mathfrak{m}_n e_i X_n = e_i X_0/\mathfrak{m}_0 e_i X_0 \cong \mathbb{F}_p$. Thus by Nakayama's lemma, $e_i X_n$ is cyclic over Λ_n, and hence X_n is cyclic over R_n. In particular, if i is even, by Kummer–Vandiver, $e_i X_0 = 0$, and hence $e_i X_n = 0$. This implies $A_n^+ = 0$. Therefore $A_n^- \cong X_n = R_n^-/\mathfrak{a}_n$ for an ideal $\mathfrak{a}_n \supset \mathfrak{s}_n^-$.

We now prove $\mathfrak{a}_n = \mathfrak{s}_n$. By Lemma 1.6.3, we have $[R_n^- : \mathfrak{s}_n^-] = |A_n^-| = |X_n| = [R_n : \mathfrak{a}_n]$, we conclude $\mathfrak{a}_n = \mathfrak{s}_n$. $\qquad\square$

Remark 1.8.6. There is another proof of this theorem via the class number formula of F_n^+ which we will describe in §1.11.2 with more input from Kummer theory. For the proof, see [ICF, §10.3].

1.9 Iwasawa theoretic interpretation

Once we pass to the limit climbing up the n-th layers cyclotomic tower moving n towards infinity, the Stickelberger element converges to a Kubota–Leopoldt p-adic L-function. We describe this limit process.

1.9.1 Limit of Stickelberger's elements

Let $R_\infty = \varprojlim_n R_n = \Lambda \otimes_{\mathbb{Z}_p} \mathbb{Z}_p[\mu_{p-1}]$ for $\Lambda := \varprojlim_j \Lambda_j = \varprojlim_j \mathbb{Z}_p[\Gamma/\Gamma^{p^j}]$.
Since $\Lambda_j := \mathbb{Z}_p[\Gamma/\Gamma^{p^j}]$ is a local ring with maximal ideal $\mathfrak{m}_{\Lambda_j} = (p, \gamma - 1)$,
Λ is a local ring with unique maximal ideal $\mathfrak{m}_\Lambda = (p, \gamma - 1)$. The local ring
Λ is called the Iwasawa algebra. Set $X = \varprojlim_n X_n$ for $X_n := G_{L_n/F_n} \cong A_n$
(taking the restriction $G_{L_m/F_m} \ni \sigma \mapsto \sigma|_{L_n} \in G_{L_n/F_n}$ for $m > n$ as
projection maps), which is a R_∞-module.

Lemma 1.9.1. *Under the projection* $G_{F_{n'}/\mathbb{Q}} \to G_{F_n/\mathbb{Q}}$ *sending* $\sigma \in G_{F_{n'}/\mathbb{Q}}$
to $\sigma|_{G_{F_n/\mathbb{Q}}}$, $\theta_{n'}^- \in \mathbb{Q}[G_{F_{n'}/\mathbb{Q}}]$ *for* $n' > n$ *projects down to* $\theta_n^- \in \mathbb{Q}[G_{F_n/\mathbb{Q}}]$.

This is the so-called distribution relation, which we will study in more
details in §2.4.3.

Proof. Define Hurwitz zeta function by

$$\zeta(s, x) = \sum_{n=0}^\infty \frac{1}{(x+n)^s} \ (\mathrm{Re}(s) > 1, 0 < x \le 1).$$

This function can be analytically continued to $s \in \mathbb{C}$ and holomorphic
outside $s = 1$, and it is known that $\zeta(1 - n, x) = -\frac{B_n(x)}{n}$ $(0 < n \in \mathbb{Z})$ for
the Bernoulli polynomial $B_n(x)$ (cf. [LFE, §2.3]). By definition,

$$f^{-s}\zeta(s, \frac{a}{f}) = \sum_{\substack{n \equiv a \mod f, n > 0}}^\infty \frac{1}{n^s}.$$

Thus

$$\mathbb{C}[(\mathbb{Z}/p^{m'}\mathbb{Z})^\times] \ni \sum_{a \in (\mathbb{Z}/p^{m'}\mathbb{Z})^\times} p^{-m's}\zeta(s, \frac{a}{p^{m'}})\sigma_a^{-1}$$

$$\mapsto \sum_{a \in (\mathbb{Z}/p^m\mathbb{Z})^\times} p^{-ms}\zeta(s, \frac{a}{p^m})\sigma_a^{-1} \in \mathbb{C}[(\mathbb{Z}/p^m\mathbb{Z})^\times]$$

under the reduction map modulo p^m (here $m = n+1$ and $m' = n'+1$). Note
that $B_1(x) = x - \frac{1}{2}$, and hence, taking $s = 0$, $\sum_{a=1,(a,p)=1}^{p^m} B_1(a/p^m)\sigma_a^{-1} = \theta_n^-$ by (1.3). $\qquad\square$

1.9.2 Cyclicity over Iwasawa algebra

Since $a - \sigma_a \in \mathbb{Z}_p[G_{F_n/\mathbb{Q}}] = R_n$ for $a \in \mathbb{Z}_p^\times$ also gives a compatible system
with respect to the projective system $R_\infty = \varprojlim_n R_n$, we have $(a - \sigma_a)\theta_\infty^- :=$
$\varprojlim_n (a - \sigma_a)\theta_n^-$. Then we get an idempotent $e_j \in \mathbb{Z}_p[\mu_{p-1}] \subset R_\infty$. Since

we can choose a as above such that $a - \omega^j(a) \in \mathbb{Z}_p^\times$ if $j \neq 1$ $(0 < j < p-1)$, $e_j(a - \sigma_a) \in (e_j R_\infty)^\times$. Thus we have $L_j := e_j \theta_\infty^- \in \Lambda$ for odd j, and hence $e_j \mathfrak{s}_n^-$ is generated by $e_j \theta_n^-$ for all odd $1 < j < p-1$, and hence $\mathfrak{s}_\infty^{(j)} = e_j \mathfrak{s}_\infty^- = \varprojlim_n e_j \mathfrak{s}_n^- = (L_j) \subset \Lambda$. Recall $X = \varprojlim_n X_n$ for $X_n := G_{L_n/F_n} \cong A_n$, and put $X^{(j)} = e_j X$. Since we know that $e_1 A_n^- = 0$ for all n, we find $X^{(1)} = 0$. The following is the consequence of the cyclicity theorem (see [I69]).

Theorem 1.9.2 (Iwasawa). *Let* $3 \leq j < p - 1$ *be an odd integer. If* $p \nmid |Cl_n^+|$, *then*

$$X^{(j)} \cong \Lambda/\mathfrak{s}_\infty^{(j)} = \Lambda/(L_j)$$

as Λ-*modules. In particular* $X_n^{(j)} = e_j X_n = X^{(j)}/(\gamma^{p^n} - 1)X^{(j)} \cong \Lambda/(L_j, \gamma^{p^n} - 1)$.

Recall $\gamma := \sigma_{1+p} \in G_{F_\infty/F_0}$. Since $\mathbb{Z}_p$-module Γ satisfies $\Gamma/\Gamma^p \cong \mathbb{F}_p$, by Nakayama's lemma, Γ is generated by γ over $\mathbb{Z}_p$. Thus $\Gamma = \{(1+p)^s = \sum_{n=0}^\infty \binom{n}{s} X^n | s \in \mathbb{Z}_p\}$, and γ is a topological generator of the group Γ.

1.9.3 *p-adic L-function*

Lemma 1.9.3. *We have* $\Lambda \cong \mathbb{Z}_p[[T]]$ *by sending* γ *to* $t := 1 + T$, *where* $\mathbb{Z}_p[[T]]$ *is the one variable power series ring. More generally, for a discrete valuation ring* W *free of finite rank over* $\mathbb{Z}_p$, *writing* Λ_W *for* $W[[\Gamma]] = \varprojlim_n W[\Gamma/\Gamma^{p^n}]$, *we have* $\Lambda_W \cong W[[T]]$ *in the same way.*

Proof. We have $\Lambda = \varprojlim_n \mathbb{Z}_p[G_{F_n/F_0}] \cong \varprojlim_n \mathbb{Z}_p[t]/(t^{p^n} - 1)$ as $G_{F_n/F_0} = \Gamma/\Gamma^{p^n}$ is a cyclic group of order p^n generated by $\gamma = \sigma_{1+p}$.

Inside $\mathbb{Z}_p[[T]]$, $t^{p^n} - 1 = \prod_{j=0}^n \Phi_j(t)$ for the minimal polynomial $\Phi_j(t)$ of p^j-th roots of unity. Since $|\alpha_n|_p < 1$ for $\alpha_n := \zeta_{p^n} - 1$, any power series $f(T) \in \mathbb{Z}_p[[T]]$ converges at α_n; so, $f(t) \mapsto f(\alpha_n)$ gives a onto algebra homomorphism $\mathbb{Z}_p[[T]] \to \mathbb{Z}_p[\mu_{p^n}]$ whose kernel is generated by $\Phi_n(t)$.

Thus $(\Phi_n(t)) \subset \mathfrak{m}_\Lambda$; so,

$$\bigcap_n (t^{p^n} - 1) = \bigcap_n (\Phi_1 \cdots \Phi_n) \subset \bigcap_n \mathfrak{m}_\Lambda^n = (0).$$

This shows the desired result for $\mathbb{Z}_p$. We leave the proof for general W to the attentive reader. $\square$

For each character $\chi : G_{F_\infty/\mathbb{Q}} \cong \mathbb{Z}_p^\times \to \overline{\mathbb{Q}}_p^\times$ with $\chi|_{\mu_{p-1}} = \omega^{-j}$ for $\mu_{p-1} \subset \mathbb{Z}_p^\times$ factoring through $G_{F_n/\mathbb{Q}}$, we have $\chi(L_j) = \chi(e_j \theta_n) = L(0, \chi)$.

Since $\chi(T) = \chi(t) - 1 = \chi(\gamma) - 1$ with $|\chi(\gamma) - 1|_p < 1$ as $\chi(\gamma) \in \mu_{p^\infty} = \bigcup_j \mu_{p^j}$, we find $\chi(L_j) = L_j(\chi(\gamma)-1)$ regarding L_j as a power series $L_j(T) \in \mathbb{Z}_p[[T]]$. Actually we can show that $L_j((1+p)^k - 1) - (1 - p^{k-1})\zeta(1-k)$ for all integers $k = j + 1 \mod (p-1)$ (see [ICF, Theorem 5.11] or [LFE, §3.5 and §4.4]). The p-adic analytic function $\mathbb{Z}_p \ni s \mapsto L_p(s, \omega^{-j}) := L_j((1+p)^s - 1) \in \mathbb{Z}_p$ is called the Kubota–Leopoldt p-adic L-function. We will prove evaluation of $L(s, \omega^{-j})$ at non-positive integers s later in §3.1.6.

Here is a general theory of Λ-modules (see [ICF, §13.2]), claiming the theory of elementary divisors is valid for Λ up to finite error. If M is a finitely generated Λ-module, then there exists finitely many non-zero elements $f_j \in \Lambda$ ($j = 1, 2, \ldots, t$), a non-negative integer r and a Λ-linear map $i : M \to \Lambda^r \oplus \bigoplus_{j=1}^t \Lambda/(f_j)$ such that $|\ker(i)| < \infty$ and $|\mathrm{Coker}(i)| < \infty$ (i.e., i is a pseudo isomorphism). Plainly $r = \dim_{\mathrm{Frac}(\Lambda)} M \otimes_\Lambda \mathrm{Frac}(\Lambda)$ which is called the Λ-rank of M (and written as $r = \mathrm{rank}_\Lambda M$). Moreover the set of ideals $(f_1), \ldots, (f_t)$ is independent of the choice of i, and the ideal $\mathrm{char}_\Lambda(M) := (\prod_j f_j)$ is called the *characteristic ideal* of M if $r = 0$. In §6.2.6, we give a notion finer than the characteristic ideal $\mathrm{char}(M)$ for Λ.

It is easy to see that $X^{(j)}$ is a torsion Λ-module of finite type as $X_1^{(j)} = X^{(j)}/(\gamma - 1)X^{(j)} \subset Cl_1^-$ is finite. Iwasawa conjectured that $\mathrm{char}(X^{(j)}) = (L_j)$ in general for odd j, and it was first proven by Mazur–Wiles in 1984 [MW84] and there are more elementary proofs by Rubin [ICF, §15.7] and Coates–Sujatha [CFZ]. But the above theorem tells more: that $X^{(j)}$ for odd j is cyclic over Λ, and Iwasawa conjectured an existence of a Λ-linear map: $X^{(j)} \to \Lambda/(L_j)$ with finite kernel and cokernel (the *pseudo-cyclicity conjecture*), which is not known yet. Iwasawa himself seems to have had a belief of an isomorphism without finite error (see [I79, C.1]).

1.10   Iwasawa's formula for $|A_n^-|$

Since the size $|A_n^-|$ is closely related to the p-adic power series $L_j(T)$ modulo $(1+T)^{p^n} - 1$, one expects to get a formula of $|A_n^-|$ up to finite error.

1.10.1   *Iwasawa's formula*

We would like to prove the following theorem of Iwasawa:

Theorem 1.10.1. *There exist integer constants λ, μ, ν such that*
$$|A_n^-| = p^{\lambda n + \mu p^n + \nu}$$
for all n sufficiently large.

These constants are called Iwasawa's λ-*invariant*, μ-*invariant* and ν-*invariant* for the cyclotomic $\mathbb{Z}_p$-extension. There is another theorem by Ferrero–Washington [FW79] (see also [S84]) which was conjectured by Iwasawa when he proved the above theorem:

Theorem 1.10.2. *We have $\mu = 0$.*

In this section, we prove Theorem 1.10.1 for the cyclotomic $\mathbb{Z}_p$-extension under the Kummer–Vandiver conjecture. See [ICF, Chapter 13] for the proof for general $\mathbb{Z}_p$-extension.

1.10.2 *Weierstrass preparation theorem*

Let V be a complete discrete valuation ring with residue field F and prime element ϖ. Write ord_V for the valuation of V with $\mathrm{ord}_V(\varpi) = 1$ and $|\cdot|_V$ for the absolute value of V. Choose an algebraic closure $\overline{K}$ of $K := \mathrm{Frac}(V)$ and extend $|\cdot|_V$ to $\overline{K}$ [BCM, VI.8]. A polynomial $P(T)$ in $V[T]$ is called *distinguished* if $P(T) = T^n + a_{n-1}T^{n-1} + \cdots + a_0$ with $\varpi | a_i$ for all i. We first quote Weierstrass' preparation theorem in our setting:

Theorem 1.10.3 (Weierstrass preparation theorem). *Let $f(T) = \sum_{i=0}^{\infty} a_i T^i \in V[[T]]$, and suppose $(f(T) \mod \varpi V[[T]]) = \sum_{i \geq n > 0} \overline{a}_i T^i \in F[[T]]$ for $\overline{a}_n \neq 0$ with $n > 0$. Then there exists a distinguished polynomial $P(T)$ of degree n and a unit $U(T) \in V[[T]]$ such that $f(T) = P(T)U(T)$. More generally, for any non-zero $f(T) \in V[[T]]$, we can find an integer $\mu \geq 0$ and a distinguished polynomial $P(T)$ and a unit such that $f(T) = \varpi^\mu P(T)U(T)$. The triple $P(T), \mu, U(T)$ is uniquely determined by $f(T)$.*

Once the first assertion is proven, $P(T) = \prod_\alpha (T - \alpha)$ for zeros $\alpha \in \overline{K}$ with $|\alpha|_V < 0$; so, the decomposition is unique. In other words, $\mathrm{Spec}(V[[T]])(\overline{K}) \cong \{\alpha \in \overline{K} : |\alpha|_V < 1\}$ (the open unit disk) which contains a "natural" V-line pV. Before proving this theorem, we state a division algorithm for $\Lambda_V := V[[T]]$:

Proposition 1.10.4 (Euclidean algorithm). *Let $f(T) = \sum_{i=0}^{\infty} a_i T^i \in \Lambda_V = V[[T]]$ for a discrete valuation ring V free of finite rank over $\mathbb{Z}_p$. Suppose $(f(T) \mod \varpi V[[T]]) = \sum_{i \geq n} \overline{a}_i T^i \in F[[T]]$ with $\overline{a}_n \neq 0$. Suppose that the index n is the minimal with $\overline{a}_n \neq 0$. For each $g(T) \in \Lambda_V$, there exists a pair $(q(T), r(T))$ with $q(T) \in \Lambda_V$ and $r(T) \in V[T]$ (a polynomial) such that $\deg(r(T)) < n$ and $g(T) = q(T)f(T) + r(T)$.*

Proof. Supposing existence of the pair (q, r), we show its uniqueness. If $g = 0$, we have $qf + r = 0$. Since $\bar{f}$ has leading term $\bar{a}_n T^n$, we find $\bar{r} = 0$. Thus $\bar{q}\bar{f} = 0$; so, $\bar{q} = 0$. Dividing by p and repeating this argument, we find $q \equiv r \equiv 0 \mod p^j$ for all $j > 0$; so, $q = r = 0$. This shows the uniqueness of (q, r) for $g = 0$. If $g = qf + r = q'f + r'$ for $g \neq 0$, we find $0 = (q - q')f + (r - r')$. By the above argument, we find $q - q' = r - r' = 0$. This finishes the proof of uniqueness.

Define $R : \Lambda_V \to \Lambda_V$ removing first n terms and dividing by T^n; so, $R(\sum_{j=0}^{\infty} a_j T^j) = \sum_{j=n}^{\infty} a_j T^{j-n}$. We put $A := \mathrm{Id} - T^n R$; so, A projects a power series to the first n-term up to degree $n - 1$. Thus we have

(1) $R(T^n h(T)) = h(T)$;
(2) $R(h) = 0 \Leftrightarrow h$ is a polynomial of degree $< n$.

Since $R(f)^{-1} = a_n^{-1}(1 + T\phi(T))^{-1} = a_n^{-1} \sum_{j=0}^{\infty}(-1)^j T^j \phi(T)^j \in \Lambda_V$ for $\phi(T) \in \mathbb{Z}_p[[T]]$, we have $R(f) \in \Lambda_V^{\times}$.

We would like to solve $g = qf + r$. This is to solve $R(g) = R(qf)$ by (2) above. Note that $f = A(f) + T^n R(f)$; so, we need to solve

$$R(g) = R(qA(f)) + R(qT^n R(f)) = R(qA(f)) + qR(f)$$

by (1) above. Write $X = qR(f)$. Then the above equation becomes

$$R(g) = R(X\frac{A(f)}{R(f)}) + X = (\mathrm{id} + R \circ \frac{A(f)}{R(f)})(X).$$

We need to solve this equation of X. As a linear map, $R \circ \frac{A(f)}{R(f)}$ has values in $\mathfrak{m}_{\Lambda_V}$ as $A(f)$ is divisible by p. Thus the map $\mathrm{id} + R \circ \frac{A(f)}{R(f)} : X \mapsto X + R(X\frac{A(f)}{R(f)})$ is invertible with inverse given by

$$(\mathrm{id} + R \circ \frac{A(f)}{R(f)})^{-1} = \sum_{j=0}^{\infty}(-R \circ \frac{A(f)}{R(f)})^j.$$

In particular, $X = (\mathrm{id} + R \circ \frac{A(f)}{R(f)})^{-1}(R(g))$, and hence $q = XR(f)^{-1}$ and $r = g - qf$. □

Once an Euclidean algorithm is known, it is standard to have a unique factorization theorem from the time of Euclid:

Corollary 1.10.5. *The ring Λ_V is a unique factorization domain.*

Proof of Weierstrass theorem. Dividing $f(T)$ by p^μ for $\mu = \min_j \mathrm{ord}_V(a_j)$, we may assume that $\bar{a}_n \neq 0$ for n as above. Therefore we only prove the first part. Apply Euclidean algorithm above to $g = T^n$, we get $T^n = qf + r$ with a polynomial r of degree $< n$. Writing $q(T) = \sum_{j=0}^\infty c_n T^n$ and comparing the coefficient of T^n, we get $1 \equiv c_0 a_n$ mod $\mathfrak{m}_{\Lambda_V}$; so, $c_0 \in V^\times$; so, $f = q^{-1}(T^n - r)$. Thus $U(T) := q^{-1}$ and $P(T) := T^n - r$ satisfies the property of the preparation theorem. $\square$

1.10.3 Proof of Iwasawa's formula

We now prove Iwasawa's formula, applying Theorem 1.10.3 to $V = \mathbb{Z}_p$. We start with two preliminary estimates:

Lemma 1.10.6. *If $E = \Lambda/p^\mu\Lambda$, then $|E/(\gamma^{p^n} - 1)E| = p^{\mu p^n}$.*

Proof. Since $E = \Lambda/p^\mu\Lambda = (\mathbb{Z}/p^\mu\mathbb{Z})[[T]]$, from
$$(\mathbb{Z}/p^\mu\mathbb{Z})[[T]]/(\gamma^{p^n} - 1)(\mathbb{Z}/p^\mu\mathbb{Z})[[T]] \cong (\mathbb{Z}/p^\mu\mathbb{Z})[[T]][\Gamma/\Gamma^{p^n}],$$
we conclude the desired assertion. $\square$

Lemma 1.10.7. *If $E = \Lambda/g(T)\Lambda$ for a distinguished polynomial of degree λ with $g(\zeta - 1) \neq 0$ for all $\zeta \in \mu_{p^\infty}$, then there exists an integer $n_0 > 0$ and a constant $\nu \in \mathbb{Z}$ such that $|E/(\gamma^{p^n} - 1)E| = p^{\lambda n + \nu}$ for all $n > n_0$.*

Proof. A monic polynomial $f(T) \in \mathbb{Z}_p[T]$ is distinguished if and only if $f(T) \equiv T^\lambda \bmod p$; so, a product and a factor of distinguished polynomials are distinguished.

We put $N_{n,n'} := \frac{\gamma^{p^n}-1}{\gamma^{p^{n'}}-1} = \sum_{j=0}^{p^n-p^{n'}} \gamma^{jp^{n'}}$ for $n > n'$. Writing $g(T) = T^\lambda - pQ(T)$ with $Q(T) \in \mathbb{Z}_p[T]$, we have $T^\lambda \equiv pQ(T) \bmod g$; so, $T^k \equiv pQ_k(T) \bmod g$ for all $k \geq \lambda$ with some polynomial $Q_k(T) \in \mathbb{Z}_p[T]$. Therefore if $p^n > \lambda$,
$$\gamma^{p^n} = (1+T)^{p^n} = 1 + pR(T) + T^{p^n} \equiv 1 + pS_n(T) \bmod g(T)$$
for $R(T), S_n(T) \in \mathbb{Z}_p[T]$. Thus
$$(1+T)^{p^{n+1}} \equiv (1 + pS_n(T))^p \equiv 1 + p^2 S_n'(T) \bmod g(T).$$
Thus we find
$$\gamma^{p^{n+2}} - 1 = (\gamma^{p^{n+1}})^p - 1 = (\gamma^{p^{n+1}} - 1)(1 + \gamma^{p^{n+1}} + \cdots + \gamma^{(p-1)p^{n+1}})$$
$$\equiv (\underbrace{1 + \cdots + 1}_{p} + p^2 P(T))(\gamma^{p^{n+1}} - 1) \bmod g(T) = p(1 + pP(T))(\gamma^{p^{n+1}} - 1),$$

where $P(T) \in \mathbb{Z}_p[T]$. Since $(1 + pP(T))^{-1} = \sum_{j=0}^{\infty} (-pP(T))^j \in \Lambda$, $1 + pP(T)$ is a unit in Λ. Therefore multiplication by $N_{n+2,n+1} = \frac{\gamma^{p^{n+2}} - 1}{\gamma^{p^{n+1}} - 1}$ is equal to multiplication by p on E as long as E is $\mathbb{Z}_p$-free of rank λ; so, we find $|E/pE| = p^\lambda$. Thus under the conditions: $p^{n_0} > \lambda$ and $n \geq n_0$, we get the desired formula. $\qquad\square$

Proof of Theorem 1.10.1. Since $X^{(j)} = \Lambda/(L_j)$ for j odd with $j > 1$, by the above two lemmas, we get Theorem 1.10.1 for $\lambda = \sum_{1<j<p-1,\text{odd}} \lambda_j$ and $\mu = \sum_{1<j<p-1,j:\text{odd}} \mu_j$, where $L_j = p^{\mu_j} D_j(T) U_j(T)$ with distinguished polynomial D_j of degree λ_j and units $U_j(T)$. $\qquad\square$

Remark 1.10.8. Actually for any $\mathbb{Z}_p$-extension $K_\infty/K = \bigcup_n K_{n/K}$ (i.e., $G_{K_\infty/K} \cong \mathbb{Z}_p$ and $G_{K_\infty/K_n} \cong p^n\mathbb{Z}_p$) for a number field K, writing the p-primary part of the class number of $K_n = K_\infty^{p^n\mathbb{Z}_p}$ as p^{e_n}, it is known that $e_n = ln + mp^n + c$ for constants l, m, c if n is sufficiently large (see [ICF, Theorem 13.3]).

1.10.4   *Iwasawa's heuristic*

If $\mathcal{X}$ is a smooth projective curve over the finite field $\mathbb{F}_p$, we can think of the extension $\overline{\mathbb{F}}_p(\mathcal{X})/\mathbb{F}_p(\mathcal{X})$ of the function fields, where $\overline{\mathbb{F}}_p$ is a fixed algebraic closure of $\mathbb{F}_p$. Then $\mathrm{Gal}(\overline{\mathbb{F}}_p(\mathcal{X})/\mathbb{F}_p(\mathcal{X})) \cong \widehat{\mathbb{Z}} = \mathrm{Frob}_p^{\widehat{\mathbb{Z}}}$. Put $X = \mathrm{Gal}(L/\overline{\mathbb{F}}_p(\mathcal{X}))$ for the maximal abelian extension unramified everywhere $L/\overline{\mathbb{F}}_p(\mathcal{X})$. Then $\mathrm{Gal}(\overline{\mathbb{F}}_p(\mathcal{X})/\mathbb{F}_p(\mathcal{X}))$ acts on X by conjugation. Let $\mathcal{J}_{/\mathbb{F}_p}$ be the Jacobian variety of $\mathcal{X}$ (i.e., degree 0 divisors modulo principal divisors). Then $X = \prod_l T_l \mathcal{J}$ for the Tate module $T_l \mathcal{J} = \varprojlim_n \mathcal{J}[l^n]$ for primes l. In particular, $T_l \mathcal{J} \cong \mathbb{Z}_l^{2g}$ for the genus g of $\mathcal{X}$ if $l \neq p$. If $l = p$, then $T_p \mathcal{J} \cong \mathbb{Z}_p^r$ for $0 \leq r \leq g$ is called the p-rank of $\mathcal{X}$. The Frobenius has reciprocal characteristic polynomial $\det(1_{2g} - \mathrm{Frob}_p|_{T_l \mathcal{J}} x) = \phi(x) \in \mathbb{Z}[x]$ independent of $l \neq p$. Then (the main part of) the zeta function of $\mathcal{X}$ is given by $L(s, \mathcal{X}) = \phi(p^{-s})$. For an eigenvalue α of Frob_p on $T_l \mathcal{J}$, Weil proved that $|\alpha| = \sqrt{p}$. Thus by definition, $L(s, \mathcal{X}) = 0 \leftrightarrow p^{-s} = \alpha$ for an eigenvalue α. This implies $\mathrm{Re}(s) = \frac{1}{2}$ (Riemann hypothesis for $\mathcal{X}$).

In Iwasawa's case, for the maximal p-abelian extension L/F_∞ unramified everywhere, $X := \mathrm{Gal}(L/F_\infty)$ is a module over $\mathbf{\Gamma} := \mathrm{Gal}(F_\infty/F_0) = \gamma^{\mathbb{Z}_p}$. By $\mu = 0$, X is $\mathbb{Z}_p$-free (under the Kummer–Vandiver conjecture). We have an isomorphism of Λ-modules $X \cong \bigoplus_{0<j<p-1,j\neq 1, j:\text{odd}} \Lambda/(L_j)$. Write $X_j := e_j X = \Lambda/(L_j)$ and $L_j(T) = P_j(T) U_j(T)$ for $U_j \in \Lambda^\times$ and a

distinguished polynomial $P_j(T)$. Regard $P_j(T)$ as a polynomial of $t = 1 + T$ and write $P_j(t)$. Since $X_j \cong \Lambda/(P_j(t))$, the action of γ on X_j satisfies $P_j(\gamma) = 0$; so, $P_j(t) = \det(t - \gamma|_{X_j})$. Perhaps an analogue of the Riemann hypothesis is to believe that $P_j(t)$ factors into a product of linear polynomials in $\mathbb{Z}_p[T]$ as $\{z \in \mathbb{Z}_p : |z|_p < 1\}$ is a line in $D = \{z \in \overline{\mathbb{Q}}_p : |z|_p < 1\}$. This is something which Iwasawa seems to have believed to be true (see [I79, C.6]).

1.11 Class number formula for F_n^+

We now want to give a formula of $|Cl_n^+|$ as an index of the cyclotomic units in the entire units in O_n^+. This is a base of the proof of the cyclicity theorem in [ICF, §10.3], a bit different from our proof in §1.8. For a number field, if $F \otimes_{\mathbb{Q}} \mathbb{R} \cong \mathbb{R}^r \times \mathbb{C}^s$, by Dirichlet's unit theorem, we know $\operatorname{rank}_{\mathbb{Z}} O^\times = \dim_{\mathbb{Q}} O^\times \otimes_{\mathbb{Z}} \mathbb{Q} = r + s - 1$.

1.11.1 *Cyclotomic units*

Since $F_n \otimes_{\mathbb{Q}} \mathbb{R} = \mathbb{C}^{(p^{n+1}-p^n)/2}$, we find $\operatorname{rank} O_n^\times = (p^{n+1} - p^n)/2 - 1$ for the integer ring O_n of F_n. Let V_n be the multiplicative group generated by $\mu_{p^{n+1}} \cup \{1 - \zeta_{p^{n+1}}^a | 1 < a \le p^{n+1} - 1\}$. Put $C_n := O_n^\times \cap V_n$. A unit in C_n is called a *cyclotomic unit*.

Lemma 1.11.1.

(1) $C_n^+ := C_n \cap F_n^+$ *is generated by* -1 *and the units*

$$\xi_a := \zeta_{p^{n+1}}^{(1-a)/2} \frac{1 - \zeta_{p^{n+1}}^a}{1 - \zeta_{p^{n+1}}} \quad \text{with } 1 < a < p^{n+1}/2 \text{ and } (a,p) = 1;$$

(2) *we have* $C_n = C_n^+ \mu_{p^{n+1}}$.

Proof. Let $m = n + 1$, and write $\zeta := \zeta_{p^m}$. Note that

$$\xi_a = \zeta^{(1-a)/2} \frac{1 - \zeta^a}{1 - \zeta} = \frac{\zeta^{-a/2} - \zeta^{a/2}}{\zeta^{-1/2} - \zeta^{1/2}} = \pm \frac{\sin(\pi a/p^m)}{\sin(\pi/p^m)}.$$

Since $\zeta^{1/2} = -\zeta_{p^m}$ as p is odd, $\xi_a \in F_m^+$. Note that $(1 - \zeta) = \mathfrak{p} = \mathfrak{p}^{\sigma_a} = (1 - \zeta^a)$ for the unique prime ideal $\mathfrak{p}$ of $\mathbb{Z}[\mu_{p^m}]$ above p, we find $(\xi_a) = O_m$, and hence $\xi_a \in (O_m^+)^\times$. Thus the assertion (2) implies (1).

We now prove (2). Note that $\zeta^{p^{m-k}}$ generates μ_{p^k}; so, we have for $0 \leq k < m$

$$1 - X^{p^k} = \prod_{j=0}^{p^k - 1} (1 - \zeta^{jp^{m-k}} X).$$

Thus making $X = \zeta^b$ for $0 < b \in \mathbb{Z}$ prime to p, we have

$$1 - \zeta^{bp^k} = \prod_{j=0}^{p^k - 1} (1 - \zeta^{b+jp^{m-k}}).$$

Since $(1 - \zeta^{-a}) = -\zeta^{-a}(1 - \zeta^a)$, to show (2), we only need to consider $(1 - \zeta^a)$ for $1 \leq a < p^m/2$ prime to p.

Suppose that $\xi := \pm\zeta^d \prod_{1<a<p^m/2,(a,p)=1}(1 - \zeta^a)^{e_a} \in O_m^\times$. Since $\mathfrak{p} = (1 - \zeta) = \mathfrak{p}^{\sigma_a} = (1 - \zeta^a)$, we have $(\xi) = \mathfrak{p}^{\sum_a e_a}$; therefore $\sum_a e_a = 0$, which implies $\prod_a (1 - \zeta^a)^{e_a} = 1$, and hence

$$\xi = \pm\zeta^d \prod_a \frac{(1 - \zeta^a)^{e_a}}{(1 - \zeta)^{e_a}} = \pm\zeta^f \prod_a \xi_a$$

with $f = d + \sum_a e_a(a - 1)/2$. This shows (2). $\qquad\qquad\square$

1.11.2   *Class number as residue of Dedekind zeta function*

Generally, take a number field F with integer ring O. Identify $F \otimes_\mathbb{Q} \mathbb{R} = \mathbb{R}^r \times \mathbb{C}^{r'}$ as semi-simple algebras, and let $\sigma_1, \ldots, \sigma_{r+r'}$ be the projection of F into each simple factor of $F \otimes_\mathbb{Q} \mathbb{R}$ so that $\sigma_1, \ldots, \sigma_r$ have values in $\mathbb{R}$. We write the corresponding simple factor as F_{σ_i}. By Dirichlet's unit theorem, $O^\times$ has rank $R := r + r' - 1$; so, it has $r + r' - 1$ independent units $\varepsilon_1, \ldots, \varepsilon_R$. By Dirichlet's unit theorem, $R_F(\varepsilon_1, \ldots, \varepsilon_R) := \det(\log |\varepsilon_i^{\sigma_j}|^{d_i})_{1 \leq i,j \leq R}$ is a non-zero real number called the regulator of $\{\varepsilon_1, \ldots, \varepsilon_R\}$. Here $d_i = \dim_\mathbb{R} F_{\sigma_i}$ (so, $d_i = 1, 2$ according to whether F_{σ_i} is real or complex embedding). If $\{\varepsilon_1, \ldots, \varepsilon_R\}$ span the maximal free quotient of $O^\times$, $R_F(\varepsilon_1, \ldots, \varepsilon_R)$ is independent of the choice of the basis $R_F(\varepsilon_1, \ldots, \varepsilon_R)$ and is just written as R_F (and is called the *regulator* o F).

Let $\zeta_F(s) := \sum_\mathfrak{n} N(\mathfrak{n})^{-s} = \prod_\mathfrak{l}(1 - N(\mathfrak{l})^{-s})^{-1}$ be the Dedekind zeta function of F, where $\mathfrak{n}$ (resp. $\mathfrak{l}$) runs over all non-zero (resp. prime) O-ideals and $N(\mathfrak{n}) = |O/\mathfrak{n}|$. The sum converges absolutely and locally uniformly if $\mathrm{Re}(s) > 1$. By Hecke, this zeta function is continued meromorphically to the whole complex plane having an only simple pole at $s = 1$ (e.g., [LFE, §2.7] and [CFN, V.2]). So $\lim_{s\to+1}(s - 1)\zeta_F(s)$ exists which was proven by Dedekind earlier than Hecke. Here is his limit formula (e.g., [CFN, V.2.2]):

Theorem 1.11.2 (Dedekind). *We have*

$$\text{Res}_{s=1}\, \zeta_F(s) = \lim_{s \to +1} (s-1)\zeta_F(s) = \frac{2^r (2\pi)^{r'} |Cl_F| R_F}{w\sqrt{|D_F|}},$$

where w is the number of roots of unity in F and D_F is the discriminant of F.

The following theorem is due to Dirichlet if $F = \mathbb{Q}$, and Hecke in general.

Theorem 1.11.3. *The L-functions $L(s, \chi)$ and $\zeta_F(s)$ can be continued analytically to $\mathbb{C} - \{1\}$ and satisfies*

$$\Lambda_F(s)\zeta_F(s) = \Lambda_F(1-s)\zeta_F(s) \quad and \quad \Lambda_\chi(s)L(s,\chi) = \varepsilon_\chi \Lambda_\chi(1-s)L(1-s, \chi^{-1}),$$

where $\Lambda_F(s) = A^r \Gamma(\frac{s}{2})^r \Gamma(s)^{r'}$ with $A = 2^{-s}\pi^{-[F:\mathbb{Q}]/2}\sqrt{|D_F|}$, $\varepsilon_\chi = \frac{-G(\chi)}{\sqrt{\chi(-1)f}}$ for the conductor f of χ and $\Lambda_\chi(s) = (f/\pi)^{s/2}\Gamma(\frac{s+\delta}{2})$ with $\delta = \frac{1-\chi(-1)}{2}$. If $\chi \neq 1$, $L(s,\chi)$ is holomorphic everywhere on $\mathbb{C}$.

Since $\text{Res}_{s=1}\, \zeta(s) = 1$, we get

Corollary 1.11.4. *We have*

$$\lim_{s \to +1} \frac{\zeta_F(s)}{\zeta(s)} = \frac{2^r (2\pi)^{r'} |Cl_F| R_F}{w\sqrt{|D_F|}}.$$

If $F_{/\mathbb{Q}}$ is a Galois extension, by the existence of a normal basis, the Galois module F is isomorphic to the group algebra $\mathbb{Q}[\text{Gal};(F/\mathbb{Q}) = \text{Ind}_F^{\mathbb{Q}}\, 1$, which is the direct sum of the trivial character $\mathbf{1} : \text{Gal}(F/\mathbb{Q}) \to \{1\}$ and a complementary Artin representation $\rho_F : \text{Gal}(F/\mathbb{Q}) \to \text{GL}_{d-1}(\mathbb{Q})$ for $d = [F : \mathbb{Q}]$. For each prime l, writing V for the representation space of ρ_F, on the subspace $V_l := V^{I_l}$ fixed by the inertia subgroup $I_l \subset \text{Gal}(F/\mathbb{Q})$, the Frobenius element Frob_l is well defined. Write $\rho_F(\text{Frob}_l)|_{V_l}$ for this action. Then we have the Euler factor $\det(1 - \rho_F(\text{Frob}_l)|_{V_l}l^{-s})^{-1}$. In this way, we have the Euler product expansion of the Artin L-function $L(s, \rho_F) := \prod_l \det(1 - \rho_F(\text{Frob}_l)|_{V_l}l^{-s})^{-1}$. Since $\zeta_F(s) = \zeta(s)L(s, \rho_F)$, we have

$$\lim_{s \to +1} \frac{\zeta_F(s)}{\zeta(s)} = L(1, \rho_F).$$

Thus the above corollary is equivalent to the Dedekind class number formula.

Lemma 1.11.5. *Let $\{\eta_1, \ldots, \eta_R\}$ be a basis of a subgroup E of $O^\times$ modulo torsion. Then we have*

$$\frac{R_F}{R_F(\eta_1, \ldots, \eta_R)} = [O^\times/torsion : E].$$

This is because $R_F(\eta_1, \ldots, \eta_R)$ is the volume of the lattice spanned by $\mathrm{Log}(\eta) = (\log |\eta^{\sigma_j}|^{d_i})_i$ in the subspace $\{x \in F | \mathrm{Tr}_{F/\mathbb{Q}}(x) = 0\} \otimes_{\mathbb{Q}} \mathbb{R}$ of the real vector space $F \otimes_{\mathbb{Q}} \mathbb{R}$. See [ICF, Lemma 4.15] for more details.

1.11.3   *Specialization to F_n^+*

Suppose now that $F = F_n^+$ with $m = n + 1$ for $n \geq 0$. We now quote from [ICF, Theorem 4.9]:

Theorem 1.11.6 (Dirichlet–Kummer). *For a primitive character* $\chi :$ $(\mathbb{Z}/p^m\mathbb{Z})^{\times} \to \overline{\mathbb{Q}}^{\times}$ *with* $\chi(-1) = 1$, *we have*

$$L(1, \chi) = -\frac{\tau(\chi)}{p^m} \sum_{a=1}^{p^m} \overline{\chi}(a) \log |1 - \zeta_{p^m}^a|.$$

The Galois group $G_{F/\mathbb{Q}} \cong (\mathbb{Z}/p^m\mathbb{Z})^{\times}/\{\pm 1\}$ acts on F, and hence $F = \mathbb{Q} \oplus \mathrm{Ker}(\mathrm{Tr}_{F/\mathbb{Q}})$ as a Galois module. Let ρ be the representation of $G_{F/\mathbb{Q}}$ on $\mathrm{Ker}(\mathrm{Tr}_{F/\mathbb{Q}})$. By a normal basis theorem of Galois theory, we have $F \cong \mathbb{Q}[G_{F/\mathbb{Q}}]$ as Galois modules; so, $F \otimes_{\mathbb{Q}} \overline{\mathbb{Q}} \cong \bigoplus_{\chi} \chi$, where χ runs over all characters of $(\mathbb{Z}/p^m\mathbb{Z})^{\times}$ with $\chi(-1) = 1$. This shows $\rho \cong \bigoplus_{\chi \neq 1} \chi$. Then

Lemma 1.11.7. *We have*

$$\zeta_{F_n^+}(s) = \zeta(s) L(s, \rho) = \zeta(s) \prod_{\chi \neq 1, \chi(-1)=1} L(s, \chi),$$

where χ *runs over all characters of* $(\mathbb{Z}/p^m\mathbb{Z})^{\times}/\{\pm 1\}$ *and* $L(s, \chi)$ *is the Dirichlet L-function of a character* $\chi : (\mathbb{Z}/p^m\mathbb{Z})^{\times}/\{\pm 1\} \to \overline{\mathbb{Q}}^{\times}$.

Proof. Let $0 < l \neq p$ be a prime. Then $(l) = \prod_{j=1}^{g} \mathfrak{l}_j$ for primes $\mathfrak{l}_j$ in F, where g is the index of the subgroup D_l of $(\mathbb{Z}/p^m\mathbb{Z})^{\times}$ generated by l (see Lemma 1.1.3). Then writing $f = |D_l|$, we have $gf = [F : \mathbb{Q}]$. Note that

$$(1 - N(\mathfrak{l}_j)^{-s}) = (1 - l^{-fs}) = \prod_{\zeta \in \mu_f} (1 - \zeta l^{-s}) = \prod_{\chi} (1 - \chi(l) l^{-s}),$$

where χ runs over all characters of D_l. The restriction map $\mathrm{Hom}((\mathbb{Z}/p^m\mathbb{Z})^{\times}, \mathbb{Q}^{\times}) \to \mathrm{Hom}(D_l, \overline{\mathbb{Q}}^{\times})$ has fiber containing g elements, we get the desired formula by Euler factorization of ζ_F. $\square$

Corollary 1.11.8. *For* $F = F_n^+$, *we have*

$$\prod_{\chi \neq 1} L(1, \chi) = \frac{2^{[F:\mathbb{Q}]} |Cl_F| R_F}{w \sqrt{|D_F|}},$$

where χ runs over all non-trivial characters of $(\mathbb{Z}/p^{n+1}\mathbb{Z})^\times$ with $\chi(-1) = 1$. In addition, we have $\prod_{\chi \neq 1} \varepsilon_\chi = 1$ and $\prod_{\chi \neq 1}(-G(\chi)) = \sqrt{|D_F|}$.

This follows from Lemma 1.11.7 and the functional equation of Dirichlet L-function (see [LFE, Theorem 2.3.2] and (4.5) in the text).

Exercise 1.11.1. For a general field F not necessarily abelian over $\mathbb{Q}$, we write the Galois representation on $V := \mathrm{Ker}(\mathrm{Tr}_{F/\mathbb{Q}}) \subset F$ as ρ and define

$$L(s, \rho) = \prod_l \det(1 - \rho|_{V^{I_l}}(\mathrm{Frob}_l)l^{-s})^{-1} \quad \text{(Artin L-function of } \rho).$$

Prove that $\zeta_F(s) = \zeta(s)L(s, \rho)$.

1.11.4 *Class number as a unit index*

Theorem 1.11.9. *Let $F = F_n^+$. Then we have*
$$|Cl_{F_n}^+| = [(O_n^+)^\times : C_n^+],$$
and $\{\xi_a\}_{1 \leq a < p^m/2, p \nmid a}$ $(m = n + 1)$ is a set of independent units giving a basis of $(O_n^+)^\times \otimes_{\mathbb{Z}} \mathbb{Q}$.

Iwasawa seems to have believed $A_0^+ \cong ((O_0^+)^\times/C_0^+) \otimes_{\mathbb{Z}} \mathbb{Z}_p$ as Galois modules (this follows from $A_0^+ \cong R_0^-/\mathfrak{s}_0^-$; see the statement below [I79, C.1]).

Proof. Recall $\zeta = \zeta_{p^m}$. Let $\{\xi_a\}_{1 < a < p^m/2, (a,p)=1}$ be the generators of C_n^+ in Lemma 1.11.1. Note that

$$\xi_a = \frac{(\zeta^{-1/2}(1 - \zeta))^{\sigma_a}}{\zeta^{-1/2}(1 - \zeta)}$$

and $|\{\xi_a\}| = [F : \mathbb{Q}] - 1$ for $F = F_n^+$. Thus we can think of the regulator $R_C := R_F(\{\xi_a\})$. Write $l(\sigma) = \log|(\zeta^{-1/2}(1 - \zeta))^\sigma| = \log|1 - \zeta^\sigma|$ for $\sigma \in G = G_{F/\mathbb{Q}}$. We have for $G = G_{F/\mathbb{Q}}$

$$R_C = \pm \det(\log|\xi_a^\tau|)_{a,\tau \in G - \{1\}}$$

$$= \pm \det(l(\sigma\tau) - l(\tau))_{\sigma, \tau \neq 1}$$

$$\overset{\text{Lemma 1.11.10 (2)}}{=} \pm \prod_{\chi \neq 1} \sum_{\sigma \in G} \chi(\sigma)l(\sigma)$$

$$= \pm \prod_{\chi \neq 1} \sum_{1 \leq a < p^m/2} \chi(a) \log|(1 - \zeta)^{\sigma_a}|$$

$$= \pm \prod_{\chi \neq 1} \sum_{1 \leq a < p^m/2} \chi(a) \log|1 - \zeta^a|$$

$$= \pm \prod_{\chi \neq 1} \frac{1}{2} \sum_{a=1}^{p^m} \chi(\sigma) \log|1 - \zeta^a|$$

as $|1 - \zeta^a| = |-\zeta^a(1 - \zeta^{-a})| = |1 - \zeta^{-a}|$. Since

$$\prod_{1 < a < p^m, a \equiv b \mod p^k} (1 - \zeta^a) = 1 - \zeta^b_{p^k}$$

for $1 \leq k \leq m$, we get for χ primitive modulo p^k

$$\sum_{a=1}^{p^m} \chi(\sigma) \log |1 - \zeta^a| = \sum_{b=1}^{p^k} \chi(\sigma) \log |1 - \zeta^b|$$

$$= -\frac{p^k}{\tau(\chi^{-1})} L(1, \chi^{-1}) = -\tau(\chi)L(1, \chi^{-1}).$$

Therefore we conclude from the formula in Corollary 1.11.8

$$R_C = \pm \prod_{\chi \neq 1} -\frac{1}{2} L(1, \chi) = |Cl_F| R_F.$$

Since $R_C/R_F = [(O_n^+)^\times : C_n^+]$ by Lemma 1.11.5, the theorem follows. $\qquad\square$

1.11.5   *Key lemma*

We now prove Lemma 1.11.10 quoted below.

Lemma 1.11.10. *Let G be a finite abelian group, and let $f : G \to \mathbb{C}$ be a function. Then*

$$\det(f(\tau\sigma^{-1}))_{\sigma, \tau \in G} = \det(f(\sigma\tau))_{\sigma, \tau \in G} = \prod_{\chi \in \mathrm{Hom}(G, \mathbb{C}^\times)} \sum_{\sigma \in G} \chi(\sigma)f(\sigma),$$

$$\det(f(\tau\sigma^{-1}) - f(\tau))_{\sigma, \tau \neq 1} = \det(f(\sigma\tau) - f(\tau))_{\sigma, \tau \neq 1} = \prod_{\chi \neq 1} \sum_{\sigma \in G} \chi(\sigma)f(\sigma).$$

Proof. The proof is representation theoretic, and left regular representation and right regular representation of G is isomorphic, we get $\det(f(\tau\sigma^{-1}))_{\sigma, \tau \in G} = \det(f(\sigma\tau))_{\sigma, \tau \in G}$ and $\det(f(\tau\sigma^{-1}) - f(\tau))_{\sigma, \tau \neq 1} = \det(f(\sigma\tau) - f(\tau))_{\sigma, \tau \neq 1}$. Thus we prove

(1) $\det(f(\tau\sigma^{-1}))_{\sigma, \tau \in G} = \prod_{\chi \in \mathrm{Hom}(G, \mathbb{C}^\times)} \sum_{\sigma \in G} \chi(\sigma)f(\sigma)$;
(2) $\det(f(\tau\sigma^{-1}) - f(\tau))_{\sigma, \tau \neq 1} = \prod_{\chi \neq 1} \sum_{\sigma \in G} \chi(\sigma)f(\sigma)$.

Let V be the complex vector space of $\mathbb{C}$-valued functions on G, and let $\sigma \in G$ act on V by inner left multiplication. Consider the linear transformation $T : V \to V$ given by $\phi \mapsto \sum_\sigma f(\sigma)\phi(\sigma x)$. Let ϕ_τ be the characteristic function of $\{\tau\}$; i.e, $\phi_\tau(\sigma) = \delta_{\sigma, \tau}$ for the Kronecker symbol δ. Then $\beta := \{\phi_\tau\}_\tau$ is a basis of V. Since

$$\phi_\tau(\sigma x) = 1 \Leftrightarrow \sigma x = \tau \Leftrightarrow x = \sigma^{-1}\tau \Leftrightarrow \phi_{\sigma^{-1}\tau}(x) = 1,$$

we have

$$T\phi_\tau(x) = \sum_\sigma f(\sigma)\phi_\tau(\sigma x) = \sum_\sigma f(\sigma)\phi_{\sigma^{-1}\tau}(x) \overset{\sigma^{-1}\tau \mapsto \sigma}{=} \sum_\sigma f(\tau\sigma^{-1})\phi_\sigma(x).$$

Thus the matrix expression of T with respect to the basis β is given by $(f(\tau\sigma^{-1}))_{\sigma,\tau \in G}$. Since $\{\chi\}_{\chi \in \mathrm{Hom}(G,\mathbb{C}^\times)}$ also forms a T-eigen basis of V with eigenvalue $\sum_\sigma \chi(\sigma)f(\sigma)$, this shows (1).

To show (2), let $V_0 = \{h \in V \mid \sum_s h(\sigma) = 0\}$ which is stable under the action of G. Set $\psi_\sigma = \phi_\sigma - \frac{1}{|G|}$ which is in V_0, and $\beta_0 := \{\psi_\sigma\}_{\sigma \neq 1}$ forms a basis of V_0. Since $\psi_1 + \sum_{\sigma \neq 1} \psi_\sigma = 0$, we have $\psi_1 = -\sum_{\tau \neq 1} \psi_\tau$. Since

$$\psi_\tau(\sigma x) = 1 - (1/|G|) \Leftrightarrow \tau x = \tau \Leftrightarrow x = \sigma^{-1}\tau \Leftrightarrow \phi_{\sigma^{-1}\tau}(x),$$

we have

$$T\psi_\tau(x) = \sum_{\sigma \in G} f(\tau\sigma^{-1})\phi_\sigma(x) = \sum_{\sigma \neq 1}(f(\tau\sigma^{-1}) - f(\tau))\phi_\sigma(x).$$

Then $T|_{V_0}$ has the matrix expression $(f(\tau\sigma^{-1}) - f(\tau))_{\sigma,\tau \neq 1}$. This shows (2). $\qquad\square$

1.12 Local Iwasawa theory

For our later use, we insert here a local Iwasawa theory related to the cyclicity problem. The result in this subsection will not be used until Chapter 8. Indeed, we would like to prove that for the composite $\mathcal{K}_\infty$ of the unramified $\mathbb{Z}_p$-extension and the cyclotomic $\mathbb{Z}_p$-extension over a p-adic field k, the p-abelian inertia group is generated by one element over the two variable Iwasawa algebra $\mathbb{Z}_p[[\mathrm{Gal}(\mathcal{K}_\infty/k)]]$.

1.12.1 *A theorem of Iwasawa*

Let $k/\mathbb{Q}_p$ (inside $\overline{\mathbb{Q}}_p$) be a finite Galois extension of degree d. Write K_∞/k for a $\mathbb{Z}_p$-extension inside $\overline{\mathbb{Q}}_p$ (i.e., $\mathrm{Gal}(K_\infty/k) \cong \mathbb{Z}_p$). Let $\Gamma := \mathrm{Gal}(K_\infty/k) = \gamma^{\mathbb{Z}_p}$ and put $\Gamma_n = \Gamma^{p^n}$. Set $K_n := K_\infty^{\Gamma_n}$ with p-adic integer ring $\mathfrak{O}_n$. Pick a generator ϖ_n of $\mathfrak{m}_{\mathfrak{O}_n}$. Let L (resp. L_n) be the maximal abelian p-extension of K_∞ (resp. K_n); so, $L_n \supset K_\infty$. Write $X_n := \mathrm{Gal}(L_n/K_\infty)$ and $X := \mathrm{Gal}(L/K_\infty)$. In his celebrated paper [I73, Theorem 25], Iwasawa proved the following structure theorem of X:

Theorem 1.12.1. *Set* $\Lambda_0 := \mathbb{Z}_p[[\mathbf{\Gamma}]]$.

(1) *Suppose that* $K_\infty \neq k[\mu_{p^\infty}]$ *(i.e.,* $\mu_{p^\infty}(K_\infty)$ *is finite). Then* X *has a* Λ_0*-linear embedding into* Λ_0^d *such that* $\Lambda_0^d/X \cong \mu_{p^\infty}(K_\infty)$ *as* Λ_0*-modules.*

(2) *Suppose* $K_\infty = k[\mu_{p^\infty}(\overline{\mathbb{Q}}_p)]$. *Then* $X \cong \mathbb{Z}_p(1) \oplus \Lambda_0^d$ *as* Λ_0*-modules, where* $\mathbb{Z}_p(1) := \varprojlim_n \mu_{p^n}(\overline{\mathbb{Q}}_p)$ *as* $\mathbf{\Gamma}$*-modules.*

This theorem holds also for $p = 2$ as is clear from the following proof.

Proof. Lifting the generator $\gamma \in \mathbf{\Gamma}$ to $\mathrm{Gal}(L/k)$, we can split $\mathrm{Gal}(L/k) \xrightarrow{\tau \mapsto \tau|_{K_\infty}} \mathbf{\Gamma}$, and we get an isomorphism $\mathrm{Gal}(L/k) = \mathbf{\Gamma} \ltimes X$. Then writing additively the $\mathbf{\Gamma}$-module X, the commutator subgroup of $\mathrm{Gal}(L/K_n)$ is given by $(\gamma^{p^n} - 1)X$. Thus

$$X_n = \mathrm{Gal}(L_n/K_\infty) = X/(\gamma^{p^n} - 1)X,$$

and $X = \varprojlim_n X_n$. For a multiplicative module M, let $\widehat{M} := \varprojlim_m M/M^{p^m}$. By local class field theory

$$\mathrm{Gal}(L_n/k_n) \cong \widehat{K}_n^\times \cong \mu_{p^\infty}(K_n) \times (\widehat{\mathfrak{O}}_n^\times/\mu_{p^\infty}(K_n)) \times \varpi_n^{\mathbb{Z}_p} \cong \mu_{p^\infty}(K_n) \times \mathbb{Z}_p^{d_n+1}$$

for $d_n = [K_n : \mathbb{Q}_p] = p^n d$, since $\widehat{\mathfrak{O}}_n^\times/\mu_{p^\infty}(K_n) \cong \mathbb{Z}_p^{d_n}$ by the p-adic logarithm. We have an exact sequence

$$1 \to X_n \to \mathrm{Gal}(L_n/K_n) \xrightarrow{\tau \mapsto \tau|_{K_\infty}} \mathbf{\Gamma}_n \to 1.$$

Since $\mathbf{\Gamma}_n \cong \mathbb{Z}_p$, we find

$$X_n \cong \mu_{p^\infty}(K_n) \times \mathbb{Z}_p^{dp^n}. \tag{1.5}$$

By Nakayama's lemma applied to $(X, (\gamma - 1) \subset \mathfrak{m}_{\Lambda_0})$, the module X is a Λ_0-module of finite type. Then, as explained in §1.9.3, we have a Λ_0-linear map with finite kernel and cokernel

$$i : X \to \Lambda_0^r \oplus \bigoplus_{j=1}^{t} \Lambda_0/(f_j)$$

for finitely many non-zero power series $f_j \in \Lambda_0$. For a $\mathbb{Z}_p$-module M, we write $\mathrm{rank}_{\mathbb{Z}_p} M := \dim_{\mathbb{Q}_p} M \otimes_{\mathbb{Z}_p} \mathbb{Q}_p$ as long as it is finite. We call $\mathrm{rank}_{\mathbb{Z}_p} M$ the $\mathbb{Z}_p$-*rank* of M. Since $\Lambda/(f_j)$ has finite $\mathbb{Z}_p$-rank by Theorem 1.10.3 and

$$\mathrm{rank}_{\mathbb{Z}_p} \Lambda_0/(\gamma^{p^n} - 1) = \mathrm{rank}_{\mathbb{Z}_p} \mathbb{Z}_p[\mathbf{\Gamma}/\mathbf{\Gamma}^{p^n}] = p^n,$$

we have

$$|\mathrm{rank}_{\mathbb{Z}_p} X_n - rp^n| = |\mathrm{rank}_{\mathbb{Z}_p} X/(\gamma^{p^n} - 1)X - rp^n| \leq B \tag{1.6}$$

for a bound $B > 0$ independent of n. From this combined with (1.5), we conclude $r = d$.

Write $f = \prod_j f_j$ and decompose $f(T) = p^\mu D(T)U(T)$ for a distinguished polynomial $D(T)$ of degree λ by Theorem 1.10.3. If $(\gamma^{p^n} - 1)$ and $D(X)$ has a common factor, X_n has $\mathbb{Z}_p$-component produced by the factor in addition to $(\Lambda_0/(\gamma^{p^n} - 1))^r$, and therefore $\mathrm{rank}_{\mathbb{Z}_p} X_n > dp^n$ which is not the case, therefore f is prime to $(\gamma^{p^n} - 1)$ for all n. Let Z be the maximal Λ-torsion submodule of X. Then we have a Λ-linear morphism $Z \to \bigoplus_j \Lambda/(f_j)$ with finite kernel and cokernel. Writing $Z_n := Z/(\gamma^{p^n} - 1)Z$, we find

$$\left| |\mu_{p^\infty}(K_n)| - |Z_n| \right| < B_1 \tag{1.7}$$

for a bound $B_1 > 0$ independent of n. By the proof of Theorem 1.10.1, we find a constant ν such that

$$|Z_n| = p^{\mu p^n + \lambda n + \nu}$$

for all $n \gg 0$. This combined with (1.7) shows

$$\mu = 0 \quad \text{and} \quad \lambda = \begin{cases} 1 & \text{if } K_\infty = k[\mu_{p^\infty}], \\ 0 & \text{otherwise.} \end{cases} \tag{1.8}$$

Now assume that $K_\infty \neq k[\mu_{p^\infty}(\overline{\mathbb{Q}}_p)]$. Then $\mu_{p^\infty}(K_\infty) = \mu_{p^N}(\overline{\mathbb{Q}}_p)$ for some $N \geq 0$. Therefore for $m \geq n \geq N$, N_{K_m/K_n} sends $\zeta \in \mu_{p^\infty}(K_\infty)$ to $\zeta^{p^{m-n}}$. Passing to the limit, we find that X is p-torsion-free. Thus $i : X \hookrightarrow \Lambda_0^d$ is an embedding with finite cokernel.

Let $X^{(n)} \subset X$ be a Λ_0-submodule such that $X^{(n)} \supset (\gamma^{p^n} - 1)X$ with $X^{(n)}/(\gamma^{p^n} - 1)X = \mu_{p^\infty}(K_n)$. Since Λ_0^d/X is finite, it is killed by $\gamma^{p^n} - 1$ for $n \gg 0$. Then $(\gamma^{p^n} - 1)\Lambda_0^d \subset X$; so, $X/(\gamma^{p^n} - 1)\Lambda_0^d \hookrightarrow \Lambda_0^d/(\gamma^{p^n} - 1)\Lambda_0^d$ is $\mathbb{Z}_p$-free of rank dp^n, while $(\gamma^{p^n} - 1)\Lambda_0^d/(\gamma^{p^n} - 1)X$ is the image of Λ_0^d/X which is finite. This shows $X^{(n)} = (\gamma^{p^n} - 1)\Lambda_0^d$, and hence

$$\Lambda_0^d/X \cong X^{(n)}/(\gamma^{p^n} - 1)X \cong \mu_{p^\infty}(K_\infty).$$

This shows the assertion (1).

Finally we assume that $K_\infty = k[\mu_{p^\infty}]$. In this case, for $m \geq n \geq 0$, the norm map $N_{m,n} = N_{K_m/K_n} : \mu_{p^\infty}(K_m) \to \mu_{p^\infty}(K_n)$ is surjective and hence $X^{(n)} = X^{(m)} + (\gamma^{p^n} - 1)X$. Let $Y = \bigcap_n X^{(n)}$. Since $X^{(m)} \subset X^{(n)}$ if $m \geq n$, we have $X^{(n)} = Y + (\gamma^{p^n} - 1)X$. Thus

$$Y = \varprojlim_n Y/((\gamma^{p^n} - 1)X \cap Y) \cong \varprojlim_n X^{(n)}/(\gamma^{p^n} - 1)X \cong \varprojlim_n \mu_{p^\infty}(K_n) = \mathbb{Z}_p(1).$$

Putting $X' := X/Y$, we have

$$X'/(\gamma^{p^n} - 1)X' \cong X/(\gamma^{p^n} - 1)X \cong \mathbb{Z}_p^{dp^n} \quad \text{for } n \geq 0.$$

In the same manner as in the case where $K_\infty \neq k[\mu_{p^\infty}]$, we conclude $X' \cong \Lambda_0^d$, which implies $X \cong \mathbb{Z}_p(1) \oplus \Lambda_0^d$. $\square$

Exercise 1.12.1. Give a detailed proof of (1.6) and (1.7).

Remark 1.12.2. In [I73, Theorem 25], Iwasawa treated also a $\mathbb{Z}_p$-extension of an l-adic field. Let us describe his result without proof: for a prime $l \neq p$ and a $\mathbb{Z}_p$-extension K_∞ of a finite extension $k/\mathbb{Q}_l$, writing $X = \mathrm{Gal}(L/K_\infty)$ for the maximal p-abelian extension L/K_∞,

$$X \cong \begin{cases} X = \{1\} & \text{if } K_\infty \neq k[\mu_{p^\infty}], \\ \mathbb{Z}_p(1) & \text{if } K_\infty = k[\mu_{p^\infty}]. \end{cases}$$

We return to the setting of Theorem 1.12.1. We now assume that $k/\mathbb{Q}_p$ is a finite Galois extension such that $p \nmid [k : \mathbb{Q}_p]$. Then $L/\mathbb{Q}_p$ is a Galois extension and $\mathrm{Gal}(L/\mathbb{Q}_p) = (\mathbf{\Gamma} \times \mathrm{Gal}(k/\mathbb{Q}_p)) \ltimes X$. Thus X is naturally a $\Lambda_0[\mathrm{Gal}(k/\mathbb{Q})]$-module.

Corollary 1.12.3. *Suppose $p > 2$, $K_\infty \subset k[\mu_{p^\infty}]$ and $p \nmid [k : \mathbb{Q}_p]$. Then we have a canonical decomposition*

$$X \cong \begin{cases} \mathbb{Z}_p[[\mathrm{Gal}(K_\infty/\mathbb{Q}_p)]] & \text{if } \mu_p(k) = \{1\}, \\ \mathbb{Z}_p[[\mathrm{Gal}(K_\infty/\mathbb{Q}_p)]] \oplus \mathbb{Z}_p(1) & \text{if } \mu_p(k) = \mu_p(\overline{\mathbb{Q}}_p) \end{cases}$$

as $\Lambda_0[\mathrm{Gal}(k/\mathbb{Q}_p)]$-modules. Thus for each character η of $\mathrm{Gal}(k/\mathbb{Q}_p)$ with values in the discrete valuation ring $W = \mathbb{Z}_p[\eta]$ inside $\overline{\mathbb{Q}}_p$ generated by the values of η, writing $X[\eta]$ for the maximal η-isotypical quotient of X (i.e., $X[\eta] = X \otimes_{\mathbb{Z}_p[\mathrm{Gal}(k/\mathbb{Q}_p)]} \eta$ regarding η as a rank 1 W-module on which $\mathrm{Gal}(k/\mathbb{Q}_p)$ acts by η), we have

$$X[\eta] \cong \begin{cases} W[[\mathbf{\Gamma}]] & \text{if } \eta \neq \omega, \\ \mathbb{Z}_p[[\mathbf{\Gamma}]] \oplus \mathbb{Z}_p(1) & \text{if } \eta = \omega \end{cases}$$

as $\Lambda_0[\mathrm{Gal}(k/\mathbb{Q}_p)]$-modules.

By the assumptions $p > 2$ and $K_\infty \subset k[\mu_{p^\infty}]$, we have convenient equivalences: $\mu_p(k) = \{1\} \Leftrightarrow \mu_{p^\infty}(K_\infty) = \{1\}$ and $\mu_p(k) \neq \{1\} \Leftrightarrow \mu_{p^\infty}(K_\infty) = \mu_{p^\infty}(\overline{\mathbb{Q}}_p)$, and plainly the assertion fails for $p = 2$ unless $\mu_{p^\infty}(K_\infty) = \mu_{p^\infty}(\overline{\mathbb{Q}}_p)$.

Proof. We first prove the assertion (1). By Theorem 1.12.1, we have either $X \cong \mathbb{Z}_p[[\Gamma]]^{[k:\mathbb{Q}_p]}$ or $\mathbb{Z}_p[[\Gamma]]^{[k:\mathbb{Q}_p]} \oplus \mathbb{Z}_p(1)$ as $\mathbb{Z}_p[[\Gamma]]$-modules. Write Y for the maximal $\mathbb{Z}_p[[\Gamma]]$-free quotient of X. Since $\mathrm{Gal}(k/\mathbb{Q}_p)$ has order prime to p, $\mathrm{Gal}(K/\mathbb{Q}_p) \cong \mathrm{Gal}(k/\mathbb{Q}_p) \ltimes \Gamma$, and its action on Y is determined by its action on $Y_0 = Y/(\gamma - 1)Y$. We need to show $Y_0 \cong \mathbb{Z}_p[\mathrm{Gal}(k/\mathbb{Q}_p)]$ as $\mathrm{Gal}(k/\mathbb{Q}_p)$-modules (which implies $Y \cong Y_0[[\Gamma]] \cong \mathbb{Z}_p[[\mathrm{Gal}(K/\mathbb{Q}_p)]]$). Let $\mathbb{Q}_{p,\infty} \subset \mathbb{Q}_p[\mu_{p^\infty}]$ be the cyclotomic $\mathbb{Z}_p$-extension of $\mathbb{Q}_p$. By class field theory, $\mathrm{Gal}(L_0 K/K)$ fits into the following commutative diagram with exact rows and surjective vertical maps:

$$
\begin{array}{ccccc}
\mathrm{Gal}(L_0 K/K) & \xhookrightarrow{\quad} & \widehat{k^\times} & \xrightarrow{\ N_{k/\mathbb{Q}_p}\ } & \widehat{\mathbb{Q}_p^\times} \\
\Big\| & & \Big\downarrow {\scriptstyle \text{Artin rec.}} & & \Big\downarrow {\scriptstyle a} \\
\mathrm{Gal}(L_0 K/K) & \xhookrightarrow{\quad} & \mathrm{Gal}(L_0/k) & \xrightarrow[\text{Res}]{\ \twoheadrightarrow\ } & \mathrm{Gal}(\mathbb{Q}_{p,\infty}/\mathbb{Q}_p),
\end{array}
$$

where the composite $a \circ N_{k/\mathbb{Q}_p}$ for the norm map $N_{k/\mathbb{Q}_p}$ has image $\mathrm{Gal}(\mathbb{Q}_{p,\infty}/\mathbb{Q}_p) \cong 1 + p\mathbb{Z}_p \cong \Gamma$.

First suppose that $\mu_p(k) = \{1\}$. Then $\widehat{k^\times}$ is torsion-free. Since $p \nmid |\mathrm{Gal}(k/\mathbb{Q}_p)|$, the isomorphism class of a torsion-free $\mathbb{Z}_p[\mathrm{Gal}(k/\mathbb{Q}_p)]$-module M of finite rank over $\mathbb{Z}_p$ is determined by the $\mathbb{Q}_p[\mathrm{Gal}(k/\mathbb{Q}_p)]$-module $M \otimes_{\mathbb{Z}_p} \mathbb{Q}_p$. Since $\mathbb{Q}_p[\mathrm{Gal}(k/\mathbb{Q}_p)]$ is semi-simple, we conclude $\widehat{k^\times} \cong \mathbb{Z}_p[\mathrm{Gal}(k/\mathbb{Q}_p)] \oplus \Gamma$ with $\mathrm{Gal}(k/\mathbb{Q}_p)$ acting on Γ trivially. Thus we conclude $Y_0 \cong \mathbb{Z}_p[\mathrm{Gal}(k/\mathbb{Q}_p)]$ in which the η-isotypical component has rank $\mathrm{rank}_{\mathbb{Z}_p} \mathbb{Z}_p[\eta]$ over $\mathbb{Z}_p$.

Now assume that $\mu_p(k)$ is non-trivial. Since $p \nmid [k : \mathbb{Q}_p]$, $\mu_{p^\infty}(k) = \mu_p(k)$; so, the torsion part of $\widehat{k^\times}$ is cyclic of order p. Let $\widehat{k_f^\times}$ be the maximal torsion-free quotient of $\widehat{k^\times}$. Then by the same argument as in the case where $\mu_p(k) = \{1\}$, we find $\widehat{k_f^\times} \cong \mathbb{Z}_p[\mathrm{Gal}(k/\mathbb{Q}_p)] \oplus \Gamma$ as $\mathbb{Z}_p[\mathrm{Gal}(k/\mathbb{Q}_p)]$-modules. By Iwasawa's expression, $X/(\gamma - 1)X \cong \mathbb{Z}_p^{[k:\mathbb{Q}_p]} \oplus \mu_p(k)$ in which $\mu_p(k)$ is identified with $\mathbb{Z}_p(1)/(\gamma - 1)\mathbb{Z}_p(1)$. Again we have $(X/(\gamma - 1)X)/\mu_p(k) \cong \mathbb{Z}_p[\mathrm{Gal}(k/\mathbb{Q}_p)]$ as $\mathbb{Z}_p[\mathrm{Gal}(k/\mathbb{Q}_p)]$-modules. We have a commutative diagram with exact row

$$
\begin{array}{ccccc}
\mathbb{Z}_p(1)/(\gamma - 1)\mathbb{Z}_p(1) & \xhookrightarrow{\quad} & X/(\gamma - 1)X & \xrightarrow{\ \twoheadrightarrow\ } & Y/(\gamma - 1)Y \\
\Big\downarrow {\scriptstyle \wr} & & \Big\| & & \Big\downarrow \\
\mu_p(k) & \xhookrightarrow{\quad} & X/(\gamma - 1)X & \xrightarrow{\ \twoheadrightarrow\ } & \mathbb{Z}_p[\mathrm{Gal}(k/\mathbb{Q}_p)]
\end{array}
$$

of $\mathbb{Z}_p[\mathrm{Gal}(k/\mathbb{Q}_p)]$-modules. This shows $Y_0 = Y/(\gamma - 1)Y \cong \mathbb{Z}_p[\mathrm{Gal}(k/\mathbb{Q}_p)]$ as $\mathbb{Z}_p[\mathrm{Gal}(k/\mathbb{Q}_p)]$-modules, and hence $Y \cong \mathbb{Z}_p[[\mathrm{Gal}(F/\mathbb{Q}_p)]]$. Therefore the surjective $\mathbb{Z}_p[[\mathrm{Gal}(F/\mathbb{Q}_p)]]$-morphism $X \twoheadrightarrow Y$ splits, and hence $X \cong \mathbb{Z}_p(1) \oplus \mathbb{Z}_p[[\mathrm{Gal}(K/\mathbb{Q}_p)]]$ as desired. $\qquad\qquad\square$

Exercise 1.12.2. Determine the $\Lambda_0[\mathrm{Gal}(k/\mathbb{Q})]$-module structure of X if K_∞/k is a general $\mathbb{Z}_p$-extension with $\mu_{p^\infty}(K_\infty) = \{1\}$.

1.12.2 A version for a $\mathbb{Z}_p^2$-extension

Suppose $K_\infty \subset k[\mu_{p^\infty}]$. Let k_∞/k be the unramified $\mathbb{Z}_p$-extension inside $\overline{\mathbb{Q}}_p$ with its n-th layer k_n, and put $\mathcal{K}_n = K_\infty k_n$. Such an unramified k_∞/k is unique as there is only one $\mathbb{Z}_p$-extension over a finite field. Write $\Upsilon := \mathrm{Gal}(\mathcal{K}_\infty/K_\infty)$. Let $\mathcal{L}$ (resp. $\mathcal{L}_n$) be the maximal abelian p-extension of $\mathcal{K}_\infty$ (resp. $\mathcal{K}_n$). Set $\mathcal{X} := \mathrm{Gal}(\mathcal{L}/\mathcal{K}_\infty)$ and $\mathcal{X}_n := \mathrm{Gal}(\mathcal{L}_n/\mathcal{K}_n)$. Pick a lift $\phi \in \mathrm{Gal}(\mathcal{L}/k)$ of the Frobenius element $[p, \mathbb{Q}_p]^f$ (for the residual degree f of $k/\mathbb{Q}_p$) with $\phi|_{\mathcal{K}_\infty}$ generating Υ and a lift $\widetilde{\gamma} \in \mathrm{Gal}(\mathcal{L}/k)$ of the generator γ of $\mathrm{Gal}(k\mathbb{Q}_{p,\infty}/k) = \Gamma$ (note $k\mathbb{Q}_{p,\infty} = K_\infty$). The commutator $\tau := [\phi, \widetilde{\gamma}]$ acts on $\mathcal{X}$ by conjugation, and $(\tau - 1)x := [\tau, x] = \tau x \tau^{-1} x^{-1}$ for $x \in \mathcal{X}$ is uniquely determined independent of the choice of $\widetilde{\gamma}$ and ϕ. Define $L' \subset \mathcal{L}$ and $L_n' \subset \mathcal{L}_n$ by the fixed field of $(\tau - 1)\mathcal{X}$ (i.e., the fixed field of τ), which is independent of the choice of $\widetilde{\gamma}$ and ϕ. Let $X' = \mathrm{Gal}(L'/\mathcal{K}_\infty)$ and $X_n' = \mathrm{Gal}(L_n'/\mathcal{K}_n)$. Recall that $\mathbb{Z}_p[[\mathrm{Gal}(K_\infty/\mathbb{Q}_p)]]$ is a direct factor of X by Corollary 1.12.3.

Theorem 1.12.4. *Let the notation be as above, and assume $K_\infty \subset k[\mu_{p^\infty}]$ and $p > 2$. Then we have the following assertion:*

(1) *The restriction map $X' \to X$ induces an isomorphism of $X'/(\phi - 1)X'$ onto the augmentation ideal of $\mathbb{Z}_p[[\mathrm{Gal}(K_\infty/\mathbb{Q})]] \subset X$.*

(2) *For the character $\eta : \mathrm{Gal}(k/\mathbb{Q}_p) \to W^\times$ in Corollary 1.12.3, the factor $X'[\eta]$ is a cyclic $W[[\Gamma \times \Upsilon]]$-module (i.e., it is generated topologically over $W[[\Gamma \times \Upsilon]]$ by one element).*

Proof. First suppose $\mu_p(k) = \mu_p(\overline{\mathbb{Q}}_p)$. Since $K_\infty \subset k[\mu_{p^\infty}]$, we have $\mu_{p^\infty}(K_\infty) = \mu_{p^\infty}(\overline{\mathbb{Q}}_p)$. By Corollary 1.12.3 applied to the $\mathbb{Z}_p$-extension $\mathcal{K}_n/k_n$, we have $\mathcal{X}_n \cong \mathcal{Y}_n \oplus \mathbb{Z}_p(1)$ as $\mathbb{Z}_p[[\mathrm{Gal}(k_n/\mathbb{Q}_p) \times \Gamma]]$-modules for a unique direct summand $\mathcal{Y}_n$. On $\mathbb{Z}_p(1)$, ϕ acts trivially (as $\nu_p([p, \mathbb{Q}_p]) = 1$ for the p-adic cyclotomic character ν_p); so, $[\widetilde{\gamma}, \phi]$ acts trivially on the factor $\mathbb{Z}_p(1)$. Hence we still have the decomposition $X_n' = Y_n' \oplus \mathbb{Z}_p(1)$. The restriction from $X_m' \to X_n'$ for $m > n$ induces on $\mathbb{Z}_p(1)$ multiplication by p^{m-n} as $\phi = [p, \mathbb{Q}_p]^f$ acts trivially on $\mu_{p^\infty}(\overline{\mathbb{Q}}_p)$. Passing to the limit, the factor $\mathbb{Z}_p(1)$ disappears. Therefore we have

$$\mathrm{Coker}(X' \to X'_0) = \mathrm{Gal}(\mathcal{K}_\infty) \oplus \mathbb{Z}_p(1) \cong \mathbb{Z}_p \oplus \mathbb{Z}_p(1).$$

The factor $\mathbb{Z}_p$ corresponds to the Galois group of K_∞/k inside L'_0 (as $p \nmid [k : \mathbb{Q}_p]$). This shows, for $Y' = \varprojlim_n Y'_n$, $\mathrm{Coker}(Y' \to Y'_0) = \mathbb{Z}_p$. Since $Y'_0 = \mathbb{Z}_p[[\mathrm{Gal}(K_\infty/\mathbb{Q}_p)]]$ by Corollary 1.12.3, the image of Y' in Y_0 is the augmentation ideal of $\mathbb{Z}_p[[\mathrm{Gal}(K_\infty/\mathbb{Q}_p)]]$. Since $\mathrm{Ker}(X' \to X)$ is plainly $(\phi - 1)X'$, we find that $X'/(\phi - 1)X'$ is isomorphic to the augmentation ideal of $\mathbb{Z}_p[[\mathrm{Gal}(K_\infty/\mathbb{Q}_p)]]$ if $\mu_p(k) = \mu_p(\overline{\mathbb{Q}}_p)$.

Suppose $\mu_p(k) \neq \mu_p(\overline{\mathbb{Q}}_p)$. Since $p > 2$, we find $\mu_{p^\infty}(K_\infty) = \mu_p(k) = \{1\}$. Replacing Y' by X' and Y'_n by X'_n, the same argument works well when $\mu_p(k) = \{1\}$. In this case, the argument is easier as the factor $\mathbb{Z}_p(1)$ does not show up.

We now prove (2). Note that $\mathbb{Z}_p[[\mathrm{Gal}(K_\infty/\mathbb{Q}_p)]] = \bigoplus_\chi \mathbb{Z}_p[\chi][[\Gamma]]$ for χ running over all characters of $\mathrm{Gal}(k/\mathbb{Q})$ (up to Galois conjugation over $\mathbb{Q}_p$). We have $\mathbb{Z}_p[\chi] = W(\kappa_\chi)$ for the finite field κ_χ generated by the values of $\chi \bmod p$ over $\mathbb{F}_p$. Then its augmentation ideal is given by $(\gamma - 1)\mathbb{Z}_p[[\Gamma]] \oplus \bigoplus_{\chi \neq 1} W(\kappa_\chi)[[\Gamma]]$. Thus $X'[\eta]/(\phi - 1)X'[\eta] \cong W(\kappa_\eta)[[\Gamma]]$ as $W(\kappa_\eta)[[\Gamma]]$-modules by the assertion (1). This is clear if η is non-trivial. If $\eta = 1$, we note that $(\gamma - 1)\mathbb{Z}_p[[\Gamma]] \cong \mathbb{Z}_p[[\Gamma]]$ as $\mathbb{Z}_p[[\Gamma]]$-modules. So $X'[\eta]/(\phi - 1)X'[\eta]$ is cyclic over $W[[\Gamma]]$. By Nakayama's lemma, we get the desired cyclicity of $X'[\eta]$ over the two variable Iwasawa algebra $W[[\Gamma \times \Upsilon]]$. $\square$

Chapter 2

Cuspidal Iwasawa theory

We describe in this chapter one more example of Stickelberger type theorems (founded by Kubert–Lang) for the divisor class group generated by cusps of modular curves. Since this is good for introducing an explicit theory of modular curves and elliptic functions, we insert this chapter in this book. Though we follow the book of Kubert–Lang [MUN] in our exposition, our treatment is concise and coherent, and the original idea of Stickelberger is somehow more outstanding. We quote results in [MUN] adding chapter number at the top (so, [MUN, Theorem 5.1.1] means Theorem 1.1 in Chapter 5). If the reader prefers to go right into the study (via the theory of Hecke algebras) of a more classical Selmer group of arithmetic Galois representation (rather than the geometric one treated here), he/she may skip the material after §2.3.6 and jump into the following chapter.

In this chapter, we discuss the following three topics:

(1) Explicit construction of modular forms and modular functions;
(2) Units in the elliptic modular function fields (modular units);
(3) The cuspidal class group of modular curves including a proof of the cuspidal class number formula.

A modular curve is an affine plane curve (i.e., an open Riemann surface) classifying elliptic curves with certain additional structure (called level structure) naturally defined over $\mathbb{Q}$. As a Riemann surface, it is a quotient of the upper half complex plane by $\mathrm{SL}_2(\mathbb{Z})$ (and its subgroups). Adding finite number of points (called cusps), we can complete the curve into a projective curve (i.e., a compact Riemann surface). Most of recent progress in number theory and arithmetic geometry is based on the study of modular curves and modular forms defined on them; e.g., proof of Iwasawa main conjectures (Mazur–Wiles) and Fermat's last theorem (Wiles).

49

Modular units are the units in the ring of holomorphic functions of the affine modular curve. Divisors supported on cusps modulo principal divisors (divisors of modular units) give the cuspidal class group (which is the torsion subgroup of rational points on the Jacobian of the modular curve). We can determine explicitly the group of modular units via classical results of Weierstrass and Siegel (like the determination of cyclotomic units in the field generated by roots of unity by Dirichlet–Kummer).

Striking points are that the class group is finite and that we have an explicit class number formula of Kubert–Lang in terms of the Dirichlet L-values at $s = 2$ (while the classical class number formula of Dirichlet–Kummer of the cyclotomic field is in terms of the values at $s = 1$). This is done by generalizing Stickelberger's theory of cyclotomic class groups to the setting of modular curves. To a good extent, the proof of the Iwasawa main conjecture by Mazur–Wiles [MW84] is based upon this theory in the following sense: The cusps over ∞ on $X_1(p^n)$ are rational, and hence $\mathrm{Gal}(\overline{\mathbb{Q}}/\mathbb{Q})$ acts trivially on the p-cuspidal class group in the Jacobian. By the self-duality of the Jacobian, on the dual of the p-primary part of the p-cuspidal class group, the Galois action factors through the cyclotomic Galois group $\mathrm{Gal}(\mathbb{Q}[\mu_{p^\infty}]/\mathbb{Q})$. Then elaborating on the idea going back to Ribet's proof [R76] of the converse of the Herbrand theorem, they produce non-trivial unramified extensions of μ_{p^ε} by $\mathbb{Z}/p^\varepsilon\mathbb{Z}$, which after passing to the limit to $X_1(p^\infty)$, give rise to a sufficiently large unramified p-extension of $\mathbb{Q}[\mu_{p^\infty}]$ as predicted by Iwasawa.

For each prime $p > 2$, $\{\sin(\pi a/p)/\sin(\pi/p)\}_{0<a<p/2}$ gives independent units in the integer ring of the field of p-th root of unity for each prime p (the cyclotomic units; see §1.11.1). Analogously, the value of modular units at a point on the upper half complex plane belonging to an imaginary quadratic field K gives independent units in the (certain) Hilbert ring class field over K, and this is a base of the generalization of the cyclotomic Iwasawa theory to the elliptic Iwasawa theory of imaginary quadratic fields [UNE] (see also [MUN, Chapters 10, 13]).

We start with a sketch of the theory of affine plane curves and how to compactify them in a projective space. Then we turn towards analytic aspects and construct rational functions over modular curves in analytic means. This explicit construction helps us to compute cuspidal class groups inside the Jacobian of the projective modular curves.

2.1 Curves over a field

Any algebraic curve over an algebraically closed field can be embedded into the 3-dimensional projective space $\mathbf{P}^3$ (e.g., [ALG, IV.3.6]) and any closed curve in $\mathbf{P}^3$ is birationally isomorphic to a curve inside $\mathbf{P}^2$ (a plane curve; see [ALG, IV.3.10]), we give some details of the theory of plane curves defined over a field $k \subset \mathbb{C}$ in this section to accustom with the theory of curves. We also write $\overline{k}$ for the algebraic closure of k inside $\mathbb{C}$.

2.1.1 *Plane curves*

We start with a plane curve (i.e., a curve inside the 2-dimensional affine space) defined over the field k. The curve is assumed to be purely of dimension 1 (i.e., it does not have an irreducible component of dimension 0). Since $k[X,Y]$ is factorial, the set of geometric points has the defining ideal $\mathfrak{a} \subset k[X,Y]$ which is principal. We thus have prime factorization $\mathfrak{a} = \prod_{\mathfrak{p}} \mathfrak{p}^{e(\mathfrak{p})}$ with principal primes $\mathfrak{p}$. We call $\mathfrak{a}$ square free if $0 \leq e(\mathfrak{p}) \leq 1$ for all principal primes $\mathfrak{p}$. Fix a square-free $\mathfrak{a}$. The set of A-rational points for any k-algebra A of a *plane curve* is given by the zero set

$$V_{\mathfrak{a}}(A) = \left\{ (x,y) \in A^2 \middle| f(x,y) = 0 \text{ for all } f(X,Y) \in \mathfrak{a} \right\}.$$

Obviously, for a generator $f(X,Y)$ of $\mathfrak{a}$, we could have defined

$$V_f(A) = \left\{ (x,y) \in A^2 \middle| f(\mathbf{x}) = 0 \right\},$$

but V_f does not depend on the choice of generators and depends only on the ideal $\mathfrak{a}$; so, it is more appropriate to write it as $V_{\mathfrak{a}}$. As an exceptional case, we note $V_{(0)}(A) = A^2$. Geometrically, we think of $V_{\mathfrak{a}}(\mathbb{C})$ as a curve in $\mathbb{C}^2 = V_{(0)}(\mathbb{C})$ (the 2-dimensional "plane"). This view point is more geometric. For any algebraically closed field K over k, a point $x \in V_{\mathfrak{a}}(K)$ is called a geometric point with coefficients in K, and $V_{(f)}(K)$ is called the geometric curve in $V_{(0)}(K) = K^2$ defined by the equation $f(X,Y) = 0$.

By Hilbert's zero theorem (Nullstellensatz; e.g., [CRT, Theorem 5.4]), writing $\overline{\mathfrak{a}}$ the principal ideal of $\overline{k}[X,Y]$ generated by $\mathfrak{a}$, we have

$$\overline{\mathfrak{a}} = \left\{ f(X,Y) \in \overline{k}[X,Y] \middle| f(x,y) = 0 \text{ for all } (x,y) \in V_{\mathfrak{a}}(\overline{k}) \right\}. \qquad (2.1)$$

Thus we have a bijection

$$\{\text{square-free ideals of } \overline{k}[X,Y]\} \leftrightarrow \{\text{plane curves } V_{\mathfrak{a}}(\overline{k}) \subset V_{(0)}(\overline{k})\}.$$

The association $V_{\mathfrak{a}} : A \mapsto V_{\mathfrak{a}}(A)$ is a covariant functor from the category of k-algebras to the category of sets (denoted by $SETS$). Indeed, for any

k-algebra homomorphism $\sigma : A \to A'$, $V_{\mathfrak{a}}(A) \ni (x,y) \mapsto (\sigma(x),\sigma(y)) \in V_{\mathfrak{a}}(A')$ as $0 = \sigma(0) = \sigma(f(x,y)) = f(\sigma(x),\sigma(y))$. Thus $\mathfrak{a} = \overline{\mathfrak{a}} \cap k[X,Y]$ is determined uniquely by this functor, but the value $V_{\mathfrak{a}}(A)$ for an individual A may not determine $\mathfrak{a}$.

From number theoretic view point, studying $V_{\mathfrak{a}}(A)$ for a small field is important. Thus it would be better to regard $V_{\mathfrak{a}}$ as a functor in some number theoretic setting.

If $\mathfrak{a} = \prod_{\mathfrak{p}} \mathfrak{p}$ for principal prime ideals $\mathfrak{p}$, by definition, we have

$$V_{\mathfrak{a}} = \bigcup_{\mathfrak{p}} V_{\mathfrak{p}}.$$

The plane curve $V_{\mathfrak{p}}$ (for each prime $\mathfrak{p}|\mathfrak{a}$) is called an *irreducible* component of $V_{\mathfrak{a}}$. Since $\mathfrak{p}$ is a principal prime, we cannot further have non-trivial decomposition $V_{\mathfrak{p}} = V \cup W$ with plane curves V and W. A prime ideal $\mathfrak{p} \subset k[X,Y]$ may decompose into a product of primes in $\overline{k}[X,Y]$. If $\mathfrak{p}$ remains prime in $\overline{k}[X,Y]$, we call $V_{\mathfrak{p}}$ *geometrically irreducible*.

Suppose that we have a map $F_A = F(\phi)_A : V_{\mathfrak{a}}(A) \to V_{\mathfrak{b}}(A)$ given by two polynomials $\phi_X(X,Y), \phi_Y(X,Y) \in k[X,Y]$ (independent of A) such that $F_A(x,y) = (\phi_X(x), \phi_Y(y))$ for all $(x,y) \in V_{\mathfrak{a}}(A)$ and all k-algebras A. Such a map is called *a regular k-map* or *a k-morphism* from a plane k-curve $V_{\mathfrak{a}}$ into $V_{\mathfrak{b}}$. Here $V_{\mathfrak{a}}$ and $V_{\mathfrak{b}}$ are plane curve defined over k. If $\mathbb{A}^1 = V_{\mathfrak{b}}$ is the affine line, i.e., $V_{\mathfrak{b}}(A) \cong A$ for all A (taking for example $\mathfrak{b} = (y)$), a regular k-map $V_{\mathfrak{a}} \to \mathbb{A}^1$ is called a *regular k-function*. Regular k-functions are just functions induced by the polynomials in $k[x,y]$ on $V_{\mathfrak{a}}$; so, $R_{\mathfrak{a}}$ is the ring of regular k-functions of $V_{\mathfrak{a}}$ defined over k.

We write $\mathrm{Hom}_{k\text{-curves}}(V_{\mathfrak{a}}, V_{\mathfrak{b}})$ for the set of regular k-maps from $V_{\mathfrak{a}}$ into $V_{\mathfrak{b}}$. Obviously, only $\phi_?$ mod $\mathfrak{a}$ can possibly be unique. We have a commutative diagram for any k-algebra homomorphism $\sigma : A \to A'$:

$$
\begin{array}{ccc}
V_{\mathfrak{a}}(A) & \xrightarrow{\ F_A\ } & V_{\mathfrak{b}}(A) \\
{\scriptstyle \sigma}\downarrow & & \downarrow{\scriptstyle \sigma} \\
V_{\mathfrak{a}}(A') & \xrightarrow[\ F_{A'}\ ]{} & V_{\mathfrak{b}}(A').
\end{array}
$$

Indeed,

$$\sigma(F_A((x,y))) = (\sigma(\phi_X(x,y)), \sigma(\phi_Y(x,y)))$$
$$= (\phi_X(\sigma(x), \sigma(y)), \phi_Y(\sigma(x), \sigma(y))) = F_{A'}(\sigma(x), \sigma(y)).$$

Thus the k-morphism is a *natural transformation of functors* (or a *morphism of functors*) from $V_{\mathfrak{a}}$ into $V_{\mathfrak{b}}$. We write $\mathrm{Hom}_{COF}(V_{\mathfrak{a}}, V_{\mathfrak{b}})$ for the set of natural transformations (we will see later that $\mathrm{Hom}_{COF}(V_{\mathfrak{a}}, V_{\mathfrak{b}})$ is a set).

The polynomials (ϕ_X, ϕ_Y) induce a k-algebra homomorphism $\underline{F} : k[X, Y] \to k[X, Y]$ by pull-back, that is, $\underline{F}(\Phi(X, Y)) = \Phi(\phi_X(X, Y), \phi_Y(X, Y))$. Take a class $[\Phi]_\mathfrak{b} = \Phi + \mathfrak{b}$ in $B = k[X, Y]/\mathfrak{b}$. Then look at $\underline{F}(\Phi) \in k[X, Y]$ for $\Phi \in \mathfrak{b}$. Since $(\phi_X(x), \phi_Y(y)) \in V_\mathfrak{b}(\overline{k})$ for all $(x, y) \in V_\mathfrak{a}(\overline{k})$, $\Phi(\phi_X(x, y), \phi_Y(x, y)) = 0$ for all $(x, y) \in V_\mathfrak{a}(\overline{k})$. By Nullstellensatz, $\underline{F}(\Phi) \in \overline{\mathfrak{a}} \cap k[X, Y] = \mathfrak{a}$. Thus $\underline{F}(\mathfrak{b}) \subset \mathfrak{a}$, and $\underline{F}$ induces a (reverse) k-algebra homomorphism

$$\underline{F} : k[X, Y]/\mathfrak{b} \to k[X, Y]/\mathfrak{a}$$

making the following diagram commutative:

$$
\begin{array}{ccc}
k[X, Y] & \xrightarrow{\;F\;} & k[X, Y] \\
\downarrow & & \downarrow \\
k[X, Y]/\mathfrak{b} & \xrightarrow[\;F\;]{} & k[X, Y]/\mathfrak{a}.
\end{array}
$$

We write $R_\mathfrak{a} = k[X, Y]/\mathfrak{a}$ and call it the affine ring of $V_\mathfrak{a}$. Here is a useful (but tautological) lemma which is a special case of Yoneda's lemma:

Lemma 2.1.1. *We have a canonical isomorphism:*

$$\mathrm{Hom}_{COF}(V_\mathfrak{a}, V_\mathfrak{b}) \cong \mathrm{Hom}_{k\text{-}curves}(V_\mathfrak{a}, V_\mathfrak{b}) \cong \mathrm{Hom}_{k\text{-}alg}(R_\mathfrak{b}, R_\mathfrak{a}).$$

The first association is covariant and the second is contravariant.

Since $R_\mathfrak{a} = \mathrm{Hom}_{k\text{-alg}}(k[X] = R_{(Y)}, R_\mathfrak{a}) = \mathrm{Hom}_{k\text{-curves}}(V_\mathfrak{a}, \mathbb{A}) = \mathrm{Hom}_{COF}(V_\mathfrak{a}, \mathbb{A})$, we can recover the ring $R_\mathfrak{a}$ as a collection of morphisms from $V_\mathfrak{a}$ into the affine line $\mathbb{A}$. Here is a sketch of the proof.

Proof. First we note $V_\mathfrak{a}(A) \cong \mathrm{Hom}_{ALG_{/k}}(R_\mathfrak{a}, A)$ via $(a, b) \leftrightarrow (\Phi(X, Y) \mapsto \Phi(a, b))$. Thus as functors, we have $V_\mathfrak{a}(?) \cong \mathrm{Hom}_{ALG_{/k}}(R_\mathfrak{a}, ?)$. We identify the two functors $A \mapsto V_\mathfrak{a}(A)$ and $A \mapsto \mathrm{Hom}(R_\mathfrak{a}, A)$ in this way, and in this sense, we write the functor corresponding to $V_\mathfrak{a}$ as $\mathrm{Spec}(R_\mathfrak{a})$. Then the main point of the proof of the lemma is to construct from a given natural transformation $F \in \mathrm{Hom}_{COF}(V_\mathfrak{a}, V_\mathfrak{b})$ a k-algebra homomorphism $\underline{F} : R_\mathfrak{b} \to R_\mathfrak{a}$ giving F by $V_\mathfrak{a}(A) = \mathrm{Hom}_{ALG_{/k}}(R_\mathfrak{a}, A) \ni \phi \overset{F_A}{\mapsto} \phi \circ \underline{F} \in \mathrm{Hom}_{ALG_{/k}}(R_\mathfrak{b}, A) = V_\mathfrak{b}(A)$. Then the following exercise finishes the proof, as plainly if we start with $\underline{F}$, the above association gives rise to F. $\qquad\square$

Exercise 2.1.1. Let $\underline{F} = F_{R_\mathfrak{a}}(\mathrm{id}_{R_\mathfrak{a}}) \in V_{R_\mathfrak{b}}(R_\mathfrak{a}) = \mathrm{Hom}_{ALG_{/k}}(R_\mathfrak{b}, R_\mathfrak{a})$, where $\mathrm{id}_{R_\mathfrak{a}} \in V_\mathfrak{a}(R_\mathfrak{a}) = \mathrm{Hom}_{ALG_{/k}}(R_\mathfrak{a}, R_\mathfrak{a})$ is the identity map. Then prove that $\underline{F}$ does the job.

We call $V_{\mathfrak{a}}$ *irreducible* (resp. *geometrically irreducible*) if $\mathfrak{a}$ is a prime ideal (resp. $\bar{\mathfrak{a}} = \mathfrak{a}\bar{k}[X,Y]$ is a prime ideal in $\bar{k}[X,Y]$). For a general noetherian k-algebra $\mathcal{A}$, we define $V_{\mathcal{A}} = \mathrm{Spec}(\mathcal{A})$ to be the functor $A \mapsto \mathrm{Hom}_{k\text{-alg}}(\mathcal{A}, A) = \mathrm{Spec}(\mathcal{A})(A)$, and call $\mathrm{Spec}(\mathcal{A})$ an *affine scheme* associated to the ring $\mathcal{A}$. In the same way as above, $\mathcal{A} \cong \mathrm{Hom}_{COF}(V_{\mathcal{A}}, \mathbb{A})$. The functor $V_{\mathcal{A}}$ is also called an affine scheme with affine ring $\mathcal{A}$. We give many geometric notions for plane curves in two ways, one is an intuitive definition particular to the plane curve and another a ring theoretic interpretation of the notion which is valid for general $V_{\mathcal{A}}$. For example, $\mathrm{Hom}_{k\text{-curves}}(V_{\mathcal{A}}, V_{\mathcal{B}}) = \mathrm{Hom}_{COF}(V_{\mathcal{A}}, V_{\mathcal{B}}) = \mathrm{Hom}_{k\text{-alg}}(\mathcal{B}, \mathcal{A})$.

An element in the total quotient ring of $R_{\mathfrak{a}}$ (resp. $\mathcal{A}$) is called a *rational k-function* on $V_{\mathfrak{a}}$ (resp. $V_{\mathcal{A}}$). If $V_{\mathfrak{a}}$ is irreducible, then rational k-functions form a field. This field is called the *rational function field of $V_{\mathfrak{a}}$ over k*.

2.1.2 Tangent space and local rings

Suppose $\mathfrak{a} = (f(X,Y))$. Write $V = V_{\mathfrak{a}}$ and $R = R_{\mathfrak{a}}$. Let $P = (a,b) \in V_{\mathfrak{a}}(K)$. We consider partial derivatives

$$\frac{\partial f}{\partial X}(P) := \frac{\partial f}{\partial X}(a,b) \quad \text{and} \quad \frac{\partial f}{\partial Y}(P) := \frac{\partial f}{\partial Y}(a,b).$$

Then the line tangent to $V_{\mathfrak{a}}$ at (a,b) has equation

$$\frac{\partial f}{\partial X}(a,b)(X-a) + \frac{\partial f}{\partial Y}(a,b)(Y-b) = 0.$$

We write the line as $T_P = V_{\mathfrak{b}}$ for the principal ideal $\mathfrak{b}$ generated by $\frac{\partial f}{\partial X}(a,b)(X-a) + \frac{\partial f}{\partial Y}(a,b)(Y-b)$. We say $V_{\mathfrak{a}}$ is *non-singular* or *smooth* at $P = (a,b) \in V_{\mathfrak{a}}(K)$ for a subfield $K \subset \mathbb{C}$ if this T_P is really a line; in other words, if $(\frac{\partial f}{\partial X}(P), \frac{\partial f}{\partial Y}(P)) \neq (0,0)$. We call $V_{\mathfrak{a}}$ non-singular or smooth if $V_{\mathfrak{a}}$ is non-singular at every $P \in V_{\mathfrak{a}}(\bar{k})$.

Example 2.1.1. Let $\mathfrak{a} = (f)$ for $f(X,Y) = Y^2 - X^3$. Then

$$\frac{\partial f}{\partial X}(a,b)(X-a) + \frac{\partial f}{\partial Y}(a,b)(Y-b) = -3a^2(X-a) + 2b(Y-b)$$

with $b^2 = a^3$. Thus this curve is singular only at $(0,0)$.

Example 2.1.2. Suppose that k has characteristic different from 2. Let $\mathfrak{a} = (Y^2 - g(X))$ for a cubic polynomial $g(X) = X^3 + aX + b$. Then the tangent line at (x_0, y_0) is given by $2y_0(X - x_0) - g'(x_0)(Y - y_0)$. This equation vanishes if $0 = y_0^2 = g(x_0)$ and $g'(x_0) = 0$; so, singular at only $(x_0, 0)$ for a multiple root x_0 of $g(X)$. Thus $V_{\mathfrak{a}}$ is a non-singular curve if and only if $g(X)$ is separable ($\Leftrightarrow \Delta := 4a^3 - 27b^2 \neq 0$ for the discriminant Δ of $g(X)$).

Suppose that K/k is an algebraic field extension. Then $K[X,Y]/\mathfrak{a}K[X,Y]$ contains $R_\mathfrak{a}$ as a subring. The maximal ideal $(X - a, Y - b) \subset K[X,Y]/\mathfrak{a}K[X,Y]$ induces a maximal ideal $P = (X - a, Y - b) \cap R_\mathfrak{a}$ of $R_\mathfrak{a}$. The *local ring* $\mathcal{O}_{V_\mathfrak{a},P}$ at P is the localization

$$\mathcal{O}_{V_\mathfrak{a},P} = \left\{ \frac{a}{b} \middle| b \in R_\mathfrak{a},\ b \in R_\mathfrak{a}P \right\},$$

where $\frac{a}{b} = \frac{a'}{b'}$ if there exists $s \in R_\mathfrak{a} \setminus P$ such that $s(ab' - a'b) = 0$. Write the maximal ideal of $\mathcal{O}_{V_\mathfrak{a},P}$ as $\mathfrak{m}_P$. Then $\mathfrak{m}_P \cap R = P$.

Lemma 2.1.2. *The linear vector space $T_P(K)$ is the dual vector space of $P/P^2 = \mathfrak{m}_P/\mathfrak{m}_P^2$.*

In general, for a maximal ideal P of $\mathcal{A}$ with residue field K, we define the *tangent space* $T_P := \mathrm{Hom}_K(P/P^2, K)$.

Proof. Write $\mathfrak{a} = (f)$. Replacing $k[X,Y]/(f)$ by $K[X,Y]/(f)$, we may assume that $K = k$. A K-derivation $\partial : \mathcal{O}_{V,P} \to K$ (at P) is a K-linear map with $\partial(\phi\varphi) = \varphi(P)\partial(\phi) + \phi(P)\partial(\varphi)$. Write $D_{V,P}$ for the space of K-derivations at P, which is a K-vector space. Plainly for $\mathbb{A} := V_{(0)}$, $D_{\mathbb{A},P}$ is a 2-dimensional vector space generated by $\partial_X : \phi \mapsto \frac{\partial\phi}{\partial X}(P)$ and $\partial_Y : \phi \mapsto \frac{\partial\phi}{\partial Y}(P)$. We have a natural injection $i : D_{V,P} \to D_{\mathbb{A},P}$ given by $i(\partial)(\phi) = \partial(\phi|_V)$. Note that $\Omega_{(a,b)} = (X - a, X - b)/(X - a, X - b)^2$ is a 2-dimensional vector space over K generated by $X - a$ and $Y - b$. Thus $D_{\mathbb{A},P}$ and $\Omega_{(a,b)}$ are dual to each other under the pairing $(\alpha(X-a)+\beta(Y-b), \partial) = \partial(\alpha(X - a) + \beta(Y - b))$. The projection $k[X,Y] \twoheadrightarrow R$ induces a surjection

$$\Omega_{(a,b)} \to P/P^2 =: \Omega_{V,P},$$

whose kernel is spanned by $f \mod (X - a, Y - b)^2 = \frac{\partial f}{\partial X}(a,b)(X - a) + \frac{\partial f}{\partial Y}(a,b)(Y-b)$ if $\mathfrak{a} = (f)$, since $\phi(X,Y) \equiv \frac{\partial\phi}{\partial X}(a,b)(X-a)+\frac{\partial\phi}{\partial Y}(a,b)(Y-b)$ $\mod (X - a, Y - b)^2$. Thus the above duality between $\Omega_{(a,b)}$ and $D_{\mathbb{A},(a,b)}$ induces the duality $\Omega_{V,P} = P/P^2$ and $T_P(K)$ given by $(\omega, t) = t(\omega)$, where we regard t as a derivation $\mathcal{O}_{V,P} \to K$. $\square$

We call T_P the *tangent space* at P and $\Omega_P = \Omega_{V,P}$ the *cotangent space* at P of V. More generally, a k-derivation $\partial : R_\mathfrak{a} \to R_\mathfrak{a}$ is a k-linear map satisfying the Leibniz condition $\partial(\phi\varphi) = \phi\partial(\varphi) + \varphi\partial(\phi)$ and $\partial(k) = 0$. For a k-derivation as above, $f\partial : \varphi \mapsto f \cdot \partial(\varphi)$ for $f \in R_\mathfrak{a}$ is again a k-derivation. The totality of k-derivations $Der_{V_\mathfrak{a}/k}$ is therefore an $R_\mathfrak{a}$-module.

First take $\mathfrak{a} = (0)$; so, $V_\mathfrak{a} = \mathbb{A}^2$. By the Leibniz relation, $\partial(X^n) = nX^{n-1}\partial X$, $\partial(Y^m) = mY^{m-1}\partial Y$ and $\partial(X^nY^m) = nX^{n-1}Y^m\partial X + mX^nY^{m-1}\partial Y$ for $\partial \in Der_{\mathbb{A}^2/k}$; so, ∂ is determined by its value $\partial(X)$ and $\partial(Y)$. Note that $(\partial X)\frac{\partial}{\partial X} + (\partial Y)\frac{\partial}{\partial Y}$ lies in $Der_{\mathbb{A}^2/k}$ and the original ∂ has the same value at X and Y; so, we have

$$\partial = (\partial X)\frac{\partial}{\partial X} + (\partial Y)\frac{\partial}{\partial Y}.$$

Thus $\{\frac{\partial}{\partial X}, \frac{\partial}{\partial Y}\}$ gives a basis of $Der_{\mathbb{A}^2/k}$.

Assuming $V_\mathfrak{a}$ is non-singular (including $\mathbb{A}^2 = V_{(0)}$), we write the $R_\mathfrak{a}$-dual as $\Omega_{V_\mathfrak{a}/k} := \mathrm{Hom}(Der_{V_\mathfrak{a}/k}, R_\mathfrak{a})$ (the *space of k-differentials*) with the pairing $(\cdot, \cdot) : \Omega_{V_\mathfrak{a}/k} \times Der_{V_\mathfrak{a}/k} \to R_\mathfrak{a}$. We have a natural map $d : R_\mathfrak{a} \to \Omega_{V_\mathfrak{a}/k}$ given by $\phi \mapsto (d\phi : \partial \mapsto \partial(\phi)) \in Der_{V_\mathfrak{a}/k}$. Note

$$(d(\phi\varphi), \partial) = \partial(\phi\varphi) = \phi\partial(\varphi) + \varphi\partial(\phi) = (\phi d\varphi + \varphi d\phi, \partial)$$

for all $\partial \in Der_{V_\mathfrak{a}/k}$. Thus we have $d(\phi\varphi) = \phi d\varphi + \varphi d\phi$, and d is a k-linear derivation with values in $\Omega_{V_\mathfrak{a}/k}$.

Again let us first look into $\Omega_{\mathbb{A}^2/k}$. Then by definition $(dX, \partial) = \partial X$ and $(dY, \partial) = \partial Y$; so, $\{dX, dY\}$ is the dual basis of $\{\frac{\partial}{\partial X}, \frac{\partial}{\partial Y}\}$. We have $d\Phi = \frac{\partial\Phi}{\partial X}dX + \frac{\partial\Phi}{\partial Y}dY$ as we can check easily that the left-hand-side and right-hand-side have the same value on any $\partial \in Der_{\mathbb{A}^2/k}$.

If $\partial : R_\mathfrak{a} = k[X,Y]/(f) \to R_\mathfrak{a}$ is a k-derivation, we can apply it to any polynomial $\Phi(X,Y) \in k[X,Y]$ and hence regard it as $\partial : k[X,Y] \to R_\mathfrak{a}$. By the above argument, $Der_k(k[X,Y], R_\mathfrak{a})$ has a basis $\{\frac{\partial}{\partial X}, \frac{\partial}{\partial Y}\}$ now over $R_\mathfrak{a}$. Since ∂ factor through the quotient $k[X,Y]/(f)$, it satisfies $\partial(f(X,Y)) = (df, \partial) = 0$. Thus we have

Lemma 2.1.3. *We have an inclusion* $Der_{V_\mathfrak{a}/k} \hookrightarrow (R_\mathfrak{a}\frac{\partial}{\partial X} \oplus R_\mathfrak{a}\frac{\partial}{\partial Y})$ *whose image is given by* $\{\partial \in Der_k(k[X,Y], R_\mathfrak{a}) | \partial f = 0\}$. *This implies* $\Omega_{V_\mathfrak{a}/k} = (R_\mathfrak{a}dX \oplus R_\mathfrak{a}dY)/R_\mathfrak{a}df$ *for* $df = \frac{\partial f}{\partial X}dX + \frac{\partial f}{\partial Y}dY$ *by duality.*

Remark 2.1.4. If $V_\mathfrak{a}$ is an irreducible curve; so, $R_\mathfrak{a}$ is an integral domain, for its quotient field $k(V_\mathfrak{a})$, $k(V_\mathfrak{a})\Omega_{V_\mathfrak{a}/k} = (k(V_\mathfrak{a})dX \oplus k(V_\mathfrak{a})dY)/k(V_\mathfrak{a})df$ is 1 dimensional, as $df \neq 0$ in $\Omega_{\mathbb{A}^2/k}$. In particular, if we pick $\psi \in R_\mathfrak{a}$ with $d\psi \neq 0$ (i.e., a non-constant), any differential $\omega \in \Omega_{V_\mathfrak{a}/k}$ can be uniquely written as $\omega = \phi d\psi$ for $\phi \in k(V_\mathfrak{a})$.

Lemma 2.1.5. *The following four conditions are equivalent:*

(1) *A point P of $V(\overline{k})$ is a smooth point.*
(2) $\mathcal{O}_{V,P}$ *is a local principal ideal domain, not a field.*

(3) $\mathcal{O}_{V,P}$ *is a discrete valuation ring with residue field $\overline{k}$.*
(4) $\varprojlim_n \mathcal{O}_{V,P}/\mathfrak{m}_P^2 \cong \overline{k}[[T]]$ *(a formal power series ring).*

Proof. Let $K = \overline{k}$. By the above lemma, T_P is a line if and only if $\dim T_P(K) = 1$ if and only if $\dim P/P^2 = 1$. Thus by Nakayama's lemma, P is principal. Any prime ideal of $k[X,Y]$ is either minimal or maximal (i.e., the ring $k[X,Y]$ has Krull dimension 2). Thus any prime ideal of R and $\mathcal{O}_{V,P}$ is maximal. Thus (1) and (2) are equivalent. The equivalence of (2) and (3) follows from general ring theory (see [CRT, Theorem 11.2]). We leave the equivalence (3) $\Leftrightarrow$ (4) as an exercise. $\qquad\square$

Write x, y for the image of $X, Y \in k[X,Y]$ in $R_\mathfrak{a}$. Any $\omega \in \Omega_{V_\mathfrak{a}/k}$ can be written as $\phi dx + \varphi dy$. Suppose that $V_\mathfrak{a}$ is non-singular. Since $\mathcal{O}_{V_\mathfrak{a},P} \hookrightarrow k[[T]]$ (for $P \in V_\mathfrak{a}(k)$) for a local parameter T as above, ϕ, φ, x, y have the "Taylor expansion" as an element of $k[[T]]$, for example, $x(T) = \sum_{n\geq 0} a_n(x)T^n$ with $a_n(x) \in k$. Thus dx, dy also have a well defined expansion, say, $dx = d(\sum_{n\geq 0} a_n(x)T^n) = \sum_{n\geq 1} a_n(x)T^{n-1}dT$. Thus we may expand $\omega = \phi dx + \varphi dy = \sum_{n\geq 0} a_n(\omega)T^n dT$ once we choose a parameter T at P. This expansion is unique independent of the expression $\phi dx + \varphi dy$. Indeed, if we allow meromorphic functions Φ as coefficients, as we remarked already, we can uniquely write $\omega = \Phi dx$ and the above expansion coincides with the Taylor expansion of Φdx.

Exercise 2.1.2. Let $P \in V_\mathfrak{a}(K)$ for a finite field extension K/k, and pull back P to a maximal ideal $(X - a, Y - b) \subset K[X,Y]$. Project the prime ideal $(X - a, Y - b) \cap k[X,Y]$ down to a maximal ideal $p \subset R_\mathfrak{a} = k[X,Y]/\mathfrak{a}$. Write $\mathcal{O}_{V_\mathfrak{a},p}$ for the localization of $R_\mathfrak{a}$ at p. Prove the following facts:

(1) p is a maximal ideal and its residue field is isomorphic to the field $k(a, b)$ generated by a and b over k.
(2) $(p/p^2) \otimes_{k(a,b)} K \cong P/P^2$ as K-vector space.
(3) Any maximal ideal of $R_\mathfrak{a}$ is the restriction of $P \in V_\mathfrak{a}(K)$ for a suitable finite field extension K/k.
(4) $\mathcal{O}_{V_\mathfrak{a},p}$ is a discrete valuation ring if and only if $\mathcal{O}_{V_\mathfrak{a},P}$ is a discrete valuation ring.

Write $Max(R_\mathfrak{a})$ for the set of maximal ideals of $R_\mathfrak{a}$. Then plainly, we have a natural inclusion $V_\mathfrak{a}(k) \hookrightarrow Max(R_\mathfrak{a})$ sending (a, b) to $(x - a, y - b)$ for the image x, y in $R_\mathfrak{a}$ of $X, Y \in k[X,Y]$. For $P \in Max(R_\mathfrak{a})$, we say P is *smooth* on $V_\mathfrak{a}$ if $\mathcal{O}_{V,P}$ is a discrete valuation ring. By the above exercise, this is consistent with the earlier definition (no more and no less).

For any given affine plane irreducible curve $V_{\mathfrak{a}}$, we say $V_{\mathfrak{a}}$ is *normal* if $R_{\mathfrak{a}}$ is integrally closed in its field of fractions.

Corollary 2.1.6. *Any normal irreducible affine plane curve is smooth everywhere.*

Proof. By ring theory, any localization of a normal domain is normal. Thus $\mathcal{O}_{V,P}$ is a normal domain. By the exercise below, we may assume that $P \cap k[X, Y] \neq (0)$. Then P is a maximal ideal, and hence $K = k[X, Y]/P$ is an algebraic extension of k. In this case, $\mathcal{O}_{V,P}$ is a normal local domain with principal maximal ideal, which is a discrete valuation ring (cf. [CRT, Theorem 11.1]). $\qquad\square$

Exercise 2.1.3.

(1) For a number field $k \subset \mathbb{C}$, let $P = k[X, Y] \cap (X - a, Y - b)$ for $(a, b) \in V_{\mathfrak{a}}(\mathbb{C})$, where $(X - a, Y - b)$ is the ideal of $\mathbb{C}[X, Y]$. Is it possible to have $P = (0) \subset k[X, Y]$ for a point $(a, b) \in V_{\mathfrak{a}}(\Omega)$?

(2) If $\mathfrak{a} = (XY)$, is the ring $A := \mathcal{O}_{V,0}$ for $0 = (0, 0)$ an integral domain? What is $\dim_k \mathfrak{m}_A/\mathfrak{m}_A^2$?

(3) For a point $P \in V_{\mathfrak{a}}(\mathbb{C})$, if $R_{\mathfrak{a}} \cap P = (0)$, prove that V is smooth at P.

(4) If A is a discrete valuation ring containing a field $k \subset A$ which is naturally isomorphic to the residue field of A, prove $\widehat{A} = \varprojlim_n A/\mathfrak{m}_A^2 \cong k[[T]]$, where $\mathfrak{m}_A$ is the maximal ideal of A.

2.1.3 *Projective space*

Let A be a commutative ring. Write A_P for the localization at a prime ideal P of A. Thus

$$A_P = \left\{ \frac{b}{s} \Big| s \in A \setminus P \right\} / \sim,$$

where $\frac{b}{s} \sim \frac{b'}{s'}$ if there exists $s'' \in A \setminus P$ such that $s''(s'b - sb') = 0$. An A-module M is called *locally free* at P if

$$M_P = \{ \frac{m}{s} | s \in A \setminus P\} / \sim\, = A_P \otimes_A M$$

is free over A_P. We call M locally free if it is free at all prime ideals of A. If $\mathrm{rank}_{A_P} M_P$ is constant r independent of P, we write $\mathrm{rank}_A M$ for r.

Write $ALG_{/k}$ for the category of k-algebras; so, $\mathrm{Hom}_{ALG_{/k}}(A, A')$ is made up of k-algebra homomorphisms from A into A' sending the identity 1_A to the identity $1_{A'}$. Here k is a general base ring, and we write ALG for

$ALG_{/\mathbb{Z}}$ (as ALG is the category of all commutative rings with identity). We consider a covariant functor $\mathbf{P}^n = \mathbf{P}^n_{/k} : ALG_{/k} \to SETS$ given by

$$\mathbf{P}^n(A) = \left\{ L \subset A^{n+1} \,\middle|\, L \text{ and } A^{n+1}/L \text{ are locally } A\text{-free with rank } L = 1 \right\}.$$

This is a covariant functor. Indeed, if $\sigma : A \to A'$ is a k-algebra homomorphism, letting it act on A^{n+1} component-wise, $L \mapsto \sigma(L) \otimes_A A'$ induces a map $\mathbf{P}^2(A) \to \mathbf{P}^2(A')$. If A is a field K, then $L \in \mathbf{P}^n(K)$ has to be free of dimension 1 generated by a non-zero vector $x = (x_0, x_1, \dots, x_n)$. The vector x is unique up to multiplication by non-zero elements of K. Thus we have proven (for a field) the following fact:

Lemma 2.1.7. *Suppose that K is a field. Then we have*

$$\mathbf{P}^n(K) \cong \left\{ \underline{x} = (x_0, x_1, \dots, x_n) \in K^{n+1} \,\middle|\, x \neq (0, \dots, 0) \right\} / K^\times.$$

Moreover, writing $D_i : ALG_{/k} \to SETS$ for the subfunctor $D_i(A) \subset \mathbf{P}^n(A)$ made up of the classes L whose projection to the i-th coordinate is surjective onto A (i.e., $x_i \neq 0$), we have $\mathbf{P}^n(K) = \bigcup_i D_i(K)$ and $D_i \cong \mathbb{A}^n$ canonically for all k-algebras A. The isomorphism: $D_i \cong \mathbb{A}^n$ is given by sending $(x_0, \dots, x_n)$ to $(\frac{x_0}{x_i}, \dots, \frac{x_n}{x_i}) \in \mathbb{A}^n$ removing the i-th coordinate.

Proof. If $L \in D_i(A)$, we have the following commutative diagram

$$
\begin{array}{ccc}
L & \overset{\hookrightarrow}{\longrightarrow} & A^{n+1} \\
\| \big\downarrow & & \big\downarrow {\scriptstyle i\text{-th proj}} \\
L & \overset{\sim}{\longrightarrow} & A
\end{array}
$$

Thus L is free of rank 1 over A; so, it has a generator $(x_0, \dots, x_n)$ with $x_i \in A^\times$. Then $(x_0, \dots, x_n) \mapsto (\frac{x_0}{x_i}, \dots, \frac{x_n}{x_i}) \in A^n$ gives rise to a natural transformation of D_i onto $\mathbb{A}^n$ (which is an isomorphism of functors). $\square$

If K is a field, we write $(x_0 : x_1 : \cdots : x_n)$ for the point of $\mathbf{P}^n(K)$ represented by $(x_0, \dots, x_n)$ as only the ratio matters. We assume that K is a field for a while. When $n = 1$, we see $\mathbf{P}^1(K) = K^\times \sqcup \{\infty\}$ by $(x : y) \mapsto \frac{x}{y} \in K \sqcup \{\infty\}$. Thus $\mathbf{P}^1(\mathbb{R})$ is isomorphic to a circle and $\mathbf{P}^1(\mathbb{C})$ is a Riemann sphere.

We now assume that $n = 2$. Writing $L = \{(x : y : 0) \in \mathbf{P}^2(K)\}$. Then $\mathbf{P}^1 \cong L$ by $(x : y) \mapsto (x : y : 0)$; so, L is isomorphic to the projective line. We have $\mathbf{P}^2(K) = D(K) \sqcup L$ for fields K, where $D = D_2$. Thus geometrically (i.e., over fields), $\mathbf{P}^2$ is the union of the affine plane added L. We call $L = L_\infty$ (the line at ∞).

2.1.4   *Projective plane curve*

For a plane curve defined by $\mathfrak{a} = (f(x,y))$ for $f(x,y)$ of degree m, $F(X,Y,Z) = Z^m f(\frac{X}{Z}, \frac{Y}{Z})$ is a (square-free) homogeneous polynomial of degree m in $k[X,Y,Z]$. If $L \in \mathbf{P}^2(A)$, we can think of $F(\ell)$ for $\ell \in L$. We write $F(L) = 0$ if $F(\ell) = 0$ for all $\ell \in L$. Thus for any k-algebra A, we define the functor $\overline{V}_\mathfrak{a} : ALG_{/k} \to SETS$ by

$$\overline{V}_\mathfrak{a}(A) = \left\{ L \in \mathbf{P}^2(A) | F(L) = 0 \right\}.$$

If A is a field K, we sent $L \in \mathbf{P}^2(K)$ to its generator $(a : b : c) \in L$ when we identified $\mathbf{P}^2(K)$ with the (classical) projective space with homogeneous coordinates. Since $F(L) = 0$ if and only if $F(a : b : c) = 0$ in this circumstance, we have

$$\overline{V}_\mathfrak{a}(K) = \left\{ (a : b : c) \in \mathbf{P}^2(K) | F(a, b, c) = 0 \right\}$$

which is called a *projective plane k-curve*. Since $D_2 \cong \mathbb{A}^2$ canonically via $(x : y : 1) \mapsto (x, y)$ (and this coordinate is well defined even over A which is not a field), we have $\overline{V}_\mathfrak{a}(A) \cap D_2(A) = V_\mathfrak{a}(A)$. In this sense, we can think of $\overline{V}_\mathfrak{a}$ as a completion of $V_\mathfrak{a}$ adding the boundary $\overline{V}_\mathfrak{a} \cap L_\infty$. Since in $D_j \cong \mathbb{A}^2$ $(j = 0, 1)$, $\overline{V}_\mathfrak{a} \cap D_j$ is a plane affine curve (for example, $\overline{V}_\mathfrak{a} \cap D_0$ is defined by $F(1, y, z) = 0$), $(L_\infty \cap \overline{V}_\mathfrak{a})(\overline{k})$ is a finite set. Thus $\overline{V}_\mathfrak{a}$ is a sort of completion/compactification of the (open) affine curve $V_\mathfrak{a}$ (we sort out this point more rigorously later). Of course, we can start with a homogeneous polynomial $F(X,Y,Z)$ (or a homogeneous ideal of $k[X,Y,Z]$ generated by $F(X,Y,Z)$) to define a *projective plane curve*. Following Lemma 2.1.1, we define $\mathrm{Hom}_{\text{proj } k\text{-curves}}(\overline{V}_\mathfrak{a}, \overline{V}_\mathfrak{b}) := \mathrm{Hom}_{COF}(\overline{V}_\mathfrak{a}, \overline{V}_\mathfrak{b})$.

Example 2.1.3. Suppose $\mathfrak{a} = (y^2 - f(x))$ for a cubic $f(x) = x^3 + ax + b$. Then $F(X,Y,Z) = Y^2 Z - X^3 - aXZ^2 - bZ^3$. Since L_∞ is defined by $Z = 0$, we find $L_\infty \cap \overline{V}_\mathfrak{a} = \{(0 : 1 : 0)\}$ made of a single point (with multiplicity 3). This point we call the origin $\mathbf{0}$ of $V_\mathfrak{a}$.

A projective plane curve $\overline{V}_\mathfrak{a}$ is non-singular (or smooth) if $\overline{V}_\mathfrak{a} \cap D_j$ is a non-singular plane curve for all $j = 0, 1, 2$. The tangent space at $P \in \overline{V}_\mathfrak{a}(K)$ is defined as before since P is in one of $D_j \cap V_\mathfrak{a}$.

Exercise 2.1.4. Suppose $\overline{V}_\mathfrak{a}$ is defined by $F(X,Y,Z) = 0$. Let $f(x,y) = F(x,y,1)$ and $g(y,z) = F(1,y,z)$. Then the projective plane curve $\overline{V}_\mathfrak{a}$ for $\mathfrak{a} = (f(x,y))$ satisfies $\overline{V}_\mathfrak{a} \cap D_0 = V_{(g)}$. Show that $\mathcal{O}_{V_\mathfrak{a}, P} \cong \mathcal{O}_{V_{(g)}, P}$ canonically if $P \in \overline{V}_\mathfrak{a} \cap D_0 \cap D_2$.

By the above exercise, the tangent space (the dual of $\mathfrak{m}_P/\mathfrak{m}_P^2$) at $P \in \overline{V}_\mathfrak{a}(K)$ does not depend on the choice of j with $P \in \overline{V}_\mathfrak{a} \cap D_j$. If a projective plane curve C is irreducible, the rational function field over k is the field of fractions of $\mathcal{O}_{C,P}$ for any $P \in C(\overline{k})$; so, independent of $C \cap D_j$.

Lemma 2.1.8. *Take a nonzero $f \in k(C)$. Then there exist homogeneous polynomials $G(X,Y,Z), H(X,Y,Z) \in k[X,Y,Z]$ with $\deg(G) = \deg(H)$ such that $f(x:y:z) = \frac{H(x,y,z)}{G(x,y,z)}$ for all $(x:y:z) \in C(\overline{k})$.*

Proof. We may write $f(x,y,1) = \frac{h(x,y)}{g(x,y)}$ on $C \cap D_2$. If $m = \deg(h) = \deg(g)$, we just define $H(X,Y,Z) = h(\frac{X}{Z}, \frac{Y}{Z})Z^m$ and $G(X,Y,Z) = g(\frac{X}{Z}, \frac{Y}{Z})Z^m$. If $\deg(h) > \deg(g)$, we define $H(X,Y,Z) = h(\frac{X}{Z}, \frac{Y}{Z})Z^{\deg(h)}$ and $G(X,Y,Z) = g(\frac{X}{Z}, \frac{Y}{Z})Z^{\deg(h)}$. If $\deg(h) < \deg(g)$, we define $H(X,Y,Z) = h(\frac{X}{Z}, \frac{Y}{Z})Z^{\deg(g)}$ and $G(X,Y,Z) = g(\frac{X}{Z}, \frac{Y}{Z})Z^{\deg(g)}$. Multiplying h or g by a power of Z does not change the above identity $f(x,y,1) = \frac{h(x,y)}{g(x,y)}$, because $Z = 1$ on $C \cap D_2$. Thus by adjusting in this way, we get G and H. $\qquad\square$

Example 2.1.4. Consider the function $\phi = cx + dy$ in $k(C)$ for $C = \overline{V}_\mathfrak{a}$ with $\mathfrak{a} = (y^2 - x^3 - ax - b)$. Then C is defined by $Y^2Z - X^3 - aXZ^2 - bZ^3 = 0$, and

$$\phi(X : Y : Z) = c\frac{X}{Z} + d\frac{Y}{Z} = \frac{cX + dY}{Z}.$$

So ϕ has pole of order 3 at $Z = 0$ (as the infinity on C has multiplicity 3) and three zeros at the intersection of $L := \{cx + dy = 0\}$ and $C \cap D_2 \cap L$.

Take a projective non-singular plane k-curve $C_{/k}$. Put $C_i = C \cap D_i$ which is an affine non-singular plane curve. Then we have well defined global differentials $Der_{C_i/k}$. Since $\partial : Der_{C_i/k}$ induces $\partial_P : \mathcal{O}_{C_i,P} \to K$ for any $P \in C_i(K)$ by $f \mapsto \partial(f)(P)$, we have $\partial_P \in T_P$. If $\partial_i \in Der_{C_i/k}$ given for each $i = 0, 1, 2$ satisfies $\partial_{i,P} = \partial_{j,P}$ for all (i,j) and all $P \in (D_i \cap D_j)(\overline{k})$, we call $\partial = \{\partial_i\}_i$ a global tangent vector defined on C. Plainly the totality $T_{C/k}$ of global tangent vectors is a k-vector space. The k-dual of $T_{C/k}$ is called the space of k-differentials over k and written as $\Omega_{C/k}$. It is known that $\Omega_{C/k}$ is finite dimensional over k.

Corollary 2.1.9. *Suppose that C is non-singular. Each $\phi \in k(C)$ induces $\phi \in \mathrm{Hom}_{proj\ k\text{-}curves}(C, \mathbf{P}^1)$. Indeed, we have $k(C) \sqcup \{\infty\} \cong \mathrm{Hom}_{proj\ k\text{-}curves}(C, \mathbf{P}^1)$, where ∞ stands for the constant function sending all $P \in C(A)$ to the image of $\infty \in \mathbf{P}^1(k)$ in $\mathbf{P}^1(A)$.*

Here we recall that "$\mathrm{Hom}_{\mathrm{proj}\ k\text{-curves}}$" is defined just above Example 2.1.3.

Proof. We prove only the first assertion. Suppose $k = \overline{k}$. Write $\phi(x : y : z) = \frac{h(x,y,z)}{g(x,y,z)}$ as a reduced fraction by the above lemma. For $L \in C(A) \subset \mathbf{P}^2(A)$, we consider the sub A-module $\phi(L)$ of A^2 generated by $\{(h(\ell), g(\ell)) \in A^2 | \ell \in L\}$. We now show that $\phi(L) \in \mathbf{P}^1(A)$; so, we will show that the map $C(A) \ni L \mapsto \phi(L) \in \mathbf{P}^1(A)$ induces the natural transformation of C into $\mathbf{P}^1$. If A is local, by Lemma 2.1.7, L is generated by (a, b, c) with at least one unit coordinate. Then any $\ell \in L$ is of the form $\lambda(a, b, c)$ and therefore $\phi(\ell) = \lambda^{\deg(h)}\phi(a, b, c)$. Thus $\phi(L) = A \cdot \phi(a, b, c)$. Since A is a k-algebra, k is naturally a subalgebra of the residue field $A/\mathfrak{m}$ of A. Since $\phi(P)$ for all $P \in C(k)$ is either a constant in k or ∞, we may assume that $(h(P), g(P)) \neq (0, 0)$ for all $P \in C(k)$. Since $(a, b, c) \not\equiv 0 \mod \mathfrak{m}$ as (a, b, c) generates a direct summand of A^3. Thus $(h(a, b, c), g(a, b, c)) \not\equiv (0, 0)$ mod $\mathfrak{m}$. After tensoring $A/\mathfrak{m}$ over A, $(A/\mathfrak{m})^2/(\phi(L)/\mathfrak{m}\phi(L))$ is one dimensional. Thus by Nakayama's lemma (e.g., [CRT, Theorem 2.2–3]), $A/\phi(L)$ is generated by a single element and has to be a free module of rank 1 as $\phi(L)$ is a free A-module of rank 1. Thus $\phi(L) \in \mathbf{P}^1(A)$. If k is not algebraically closed, replacing A by $\overline{A} = A \otimes_k \overline{k}$, we find $\phi(L) \otimes_k \overline{k} \in \mathbf{P}^2(\overline{k})$ and hence $\phi(L) \otimes_A A/\mathfrak{m} \in \mathbf{P}^2(k)$, which implies $\phi(L) \in \mathbf{P}^2(A)$.

If A is not necessarily local, applying the above argument to the local ring A_P for any prime ideal P of A, we find that $\phi(L)_P = \phi(L_P)$ and $A_P^2/\phi(L_P)$ are free of rank 1; so, $\phi(L)$ and $A^2/\phi(L)$ are locally free of rank 1; therefore, $\phi(L) \in \mathbf{P}^2(A)$.

Now it is plain that $L \mapsto \phi(L)$ induces a natural transformation of functors.   □

2.1.5 *Divisors*

The divisor group $\mathrm{Div}(C)$ of a non-singular projective geometrically irreducible plane curve C is a formal free $\mathbb{Z}$-module generated by points $P \in C(\overline{k})$. When we consider a point P as a divisor, we write it as $[P]$. For each divisor $D = \sum_P m_P[P]$, we define $\deg(D) = \sum_P m_P$. Since C is non-singular, for any point $P \in C(\overline{k})$, $\mathcal{O}_{C,P}$ is a discrete valuation ring, and the rational function field $\overline{k}(C)$ is the quotient field of $\mathcal{O}_{C,P}$ (regarding C as defined over $\overline{k}$). Thus if we write the valuation $v_P : \overline{k}(C) \twoheadrightarrow \mathbb{Z} \cup \{\infty\}$ for the additive valuation of $\mathcal{O}_{C,P}$, we have a well defined $v_P(f) \in \mathbb{Z}$ for any non-zero rational $\overline{k}$-function $f \in \overline{k}(C)$. Since $\mathfrak{m}_P = (t_P)$ and $t_P^{v_P(f)} \parallel f$ in $\mathcal{O}_{C,P}$, f has a zero of order $v_p(f)$ at P if $v_P(f) > 0$ and a pole of order $|v_p(f)|$ if

$v_P(f) < 0$. In other words, if the Taylor expansion of f at P is given by $\sum_n a_n(f)t_P^n$, we have $v_p(f) = \min(n : a_n(f) \neq 0)$. For a global differential $\omega \in \Omega_{C/\overline{k}}$, we have its Taylor expansion $\sum_n a_n(f)t_P^n dt_P$ at each $P \in C(\overline{k})$; so, we may also define $v_P(\omega) := \min(n : a_n(\omega) \neq 0)$. We extend this definition for meromorphic differentials $k(C) \cdot \Omega_{C/k} = \{f \cdot \omega | f \in k(C), \omega \in \Omega_{C/k}\}$. Here we quote Bézout's theorem:

Theorem 2.1.10. *Let C and C' be two plane projective k-curves inside $\mathbf{P}^2$ defined by relatively prime homogeneous equations $F(X, Y, Z) = 0$ and $G(X, Y, Z) = 0$ of degree m and n respectively. Then counting with multiplicity (as defined below), $|C(\overline{k}) \cap C'(\overline{k})| = m \cdot n$.*

If C is smooth at $P \in C \cap C'$ in $C \cap D_2$, $\phi = \frac{G(X,Y,Z)}{Z^2}$ is a function vanishing at P. The multiplicity of P in $C \cap C'$ is just $v_P(\phi)$. More generally, if $P = (a, b)$ is not necessarily a smooth point, writing $C \cap D_2 = V_{\mathfrak{a}}$ and $C' \cap D_2 = V_{\mathfrak{b}}$ for principal ideals $\mathfrak{a}, \mathfrak{b}$ in $\overline{k}[X, Y]$ and regarding P as an ideal $(X - a, Y - b) \subset \overline{k}[X, Y]$, the multiplicity is given by the dimension of the localization $(\overline{k}[x, y]/\mathfrak{a} + \mathfrak{b})_P$ over $\overline{k}$. The same definition works well for any points in $C \cap D_0$ and $C \cap D_1$. One can find the proof of this theorem with (possibly more sophisticated) definition of multiplicity in a text of algebraic geometry (e.g. [ALG, Theorem I.7.7]).

Since there are only finitely many poles and zeros of f, we can define the divisors $\operatorname{div}(f) = \sum_{P \in C(k)} v_P(f)[P]$, $\operatorname{div}_0(f) = \sum_{P \in C(k), v_P(f) > 0} v_P(f)[P]$ and $\operatorname{div}_\infty(f) = \sum_{P \in C(k), v_P(f) < 0} v_P(f)[P]$ of f. Similarly, for meromorphic differential ω, we define again $\operatorname{div}(\omega) = \sum_P v_P(\omega)[P]$. By Lemma 2.1.8, $f(x : y : z) = \frac{h(x:y:z)}{g(x:y:z)}$ for a homogeneous polynomial h, g in $\overline{k}[x, y, z]$ of the same degree. If the degree of equation defining C is m and C' is defined by $h(X, Y, Z) = 0$, $\deg_0(\operatorname{div}(f)) = |C(\overline{k}) \cap C'(\overline{k})| = m \deg(h) = m \deg(g) = \deg_\infty(\operatorname{div}(f))$. This shows $\deg(\operatorname{div}(f)) = 0$ as $\sum_{P, v_P(f) > 0} m_P = m \deg(h)$ and $-\sum_{P, v_P(f) < 0} m_P = m \deg(g)$.

Lemma 2.1.11. *Let C be a non-singular projective plane curve. For any $f \in \overline{k}(C)$, $\deg(\operatorname{div}(f)) = 0$, and if $f \in \overline{k}(C)$ is regular at every $P \in C$, f is a constant in $\overline{k}$.*

Lemma 2.1.12. *If $f \in k(C)$ satisfies $\deg(\operatorname{div}_0(f)) = \deg(\operatorname{div}_\infty(f)) = 1$, $f : C \to \mathbf{P}^1$ induces an isomorphism of projective plane curve over k.*

Proof. By the proof of Corollary 2.1.9, $\deg(\operatorname{div}_0(f))$ is the number of points over 0 (counting with multiplicity) of the regular map $f : C \to \mathbf{P}^1$. Taking off a constant $\alpha \in k \subset \mathbf{P}^1$ from f, we conclude $\deg(\operatorname{div}_0(f - \alpha)) =$

$1 = \deg(\mathrm{div}_\infty(f - \alpha))$ and $|f^{-1}(\alpha)| = \deg(\mathrm{div}_0(f - \alpha)) = 1$. Therefore f is 1-1 onto, and f is an isomorphism.   □

Write $\mathrm{Div}^0(C) := \{D \in \mathrm{Div}(C_{/\overline{k}})| \deg(D) = 0\}$. Inside $\mathrm{Div}^0(C)$, we have the subgroup $\{\mathrm{div}(f)|f \in \overline{k}(C)^\times\}$. We call two divisors D, D' *linearly equivalent* if $D = \mathrm{div}(f) + D'$ for $f \in \overline{k}(C)$. We say that D and D' are *algebraically equivalent* if $\deg(D) = \deg(D')$. The quotient groups $J(C) = \mathrm{Div}^0(C)/\{\mathrm{div}(f)|f \in k(C)^\times\}$ and $\mathrm{Pic}(C) = \mathrm{Div}(C)/\{\mathrm{div}(f)|f \in k(C)^\times\}$ are called the Jacobian and the Picard group of C, respectively. Sometimes, $J(C)$ is written as $\mathrm{Pic}^0(C)$ (the degree 0 Picard group).

2.1.6   Theorem of Riemann–Roch

We write $D = \sum_P m_P[P] \geq 0$ (resp. $D > 0$) for a divisor D on C if $m_P \geq 0$ for all P (resp. $D \geq 0$ and $D \neq 0$). For a divisor D on $C_{\overline{k}}$

$$L(D) = \{f \in \overline{k}(C)| \mathrm{div}(f) + D \geq 0\} \cup \{0\}.$$

Plainly, $L(D)$ is a vector space over $\overline{k}$. It is known that $\ell(D) := \dim_{\overline{k}} L(D) < \infty$. For $\phi \in k(C)^\times$, $L(D) \ni f \mapsto f\phi \in L(D - \mathrm{div}(\phi))$ is an isomorphism. Thus $\ell(D)$ only depends on the class of D in $\mathrm{Pic}(C)$.

Example 2.1.5. Let $C = \mathbf{P}^1$. For a positive divisor $D = \sum_{a \in \overline{k}} m_a[a]$ with $m_a \geq 0$ and $m_a > 0$ for some a, regard $a \in \overline{k}$ as a point $[a] \in \mathbf{P}^1(\overline{k}) = \overline{k} \sqcup \{\infty\}$. On $\mathbb{A}^1(\overline{k}) = \overline{k}$, forgetting about the infinity, $\mathrm{div}(f) + D \geq 0$ if $f = \frac{g(x)}{\prod_a(x-a)^{m_a}}$ for a polynomial $g(x)$. If $\deg(D) \geq \deg(g(x))$, the function f does not have pole at ∞. Thus $L(D) = \{g(x)| \deg(g(x)) \leq \deg(D)\}$ and we have $\ell(D) = 1 + \deg(D)$ if $D > 0$. If C is a plane projective curve, we can write $f = \frac{h(X,Y,Z)}{g(X,Y,Z)}$ as a reduced fraction by Lemma 2.1.8. Write $D = \sum_P m_P[P]$, and put $|D| = \{P|D = \sum_P m_P[P]$ with $m_P \neq 0\}$. If $|D|$ is inside $D_2 \cap C \subset \mathbb{A}^2$ and $D > 0$, we may assume that $V_{(g(X,Y,1))} \cap C$ contains $|D|$. Then not to have pole at $C \setminus D_2$, $\deg(h)$ has to be bounded; so, $\ell(D) < \infty$. Since $L(D) \subset L(D_+)$ in general, writing $D = D_+ + D_-$ so that $D_+ \geq 0$ and $-D_- \geq 0$, this shows $\ell(D) < \infty$.

Theorem 2.1.13 (Riemann–Roch). *Let $C = \overline{V}_a$ be a non-singular projective curve defined over a field k. Then for $g = g(C) = \dim_{\overline{k}} \Omega_{C/\overline{k}}$ and a divisor K of degree $2g - 2$ of the form $\mathrm{div}(\omega)$ for a meromorphic differential ω on C such that $\ell(D) = 1 - g + \deg(D) + \ell(K - D)$ for all divisor D on $C(\overline{k})$. If $g = 1$, K is linearly equivalent to 0.*

The number $g(C)$ above is called the *genus* of the curve C. The divisor K is called a *canonical divisor* K (whose linear equivalence class is unique). Note that

$$L(K) = \{f \in \overline{k}(C)|\, \mathrm{div}(f\omega) = \mathrm{div}(f) + \mathrm{div}(\omega) \geq 0\} \cong \Omega_{C/\overline{k}}$$

by $f \mapsto f\omega \in \Omega_{C/k}$. Then by the above theorem,

$$g(C) = \dim \Omega_{C/\overline{k}} = \ell(K) = 1 - g + \deg(K) + \ell(0) = 2 + \deg(K) - g(C),$$

and from this, we conclude $\deg(K) = 2g(C) - 2$. One can find a proof of this theorem in any algebraic geometry book (e.g., [ALG, IV.1] or [GME, Theorem 2.1.3]).

Corollary 2.1.14. *If $g(C) = 1$, then $\ell(D) = \deg(D)$ if $\deg(D) > 0$.*

Proof. For a non-constant $f \in \overline{k}(E)$, $\deg(\mathrm{div}(f)) = 0$ implies that f has a pole somewhere. If $D > 0$, $f \in L(-D)$ does not have pole; so, constant. Since $D > 0$, f vanishes at $P \subset D$. Thus $f = 0$. More generally, if $\deg(D) > 0$ and $\phi \in L(-D)$, then $0 > \deg(-D) = \deg(\phi) - \deg(D) \geq 0$; so, $\phi = 0$. Thus if $\deg(D) > 0$, then $\ell(-D) = 0$. Since $K = 0$, we have by the Riemann–Roch theorem that $\ell(D) = \deg(D) + \ell(0 - D) = \deg(D)$ if $\deg(D) > 0$. $\square$

Since $\deg(\mathrm{div}(f)) = 0$, if $D \gg 0$, we get $\ell(-D) = 0$. Thus in particular $\ell(K - D) = 0$ if $D \gg 0$. Thus the above theorem implies what Riemann originally proved:

Corollary 2.1.15 (Riemann). *Let $C = \overline{V}_a$ be a non-singular projective curve defined over a field k. Then there exists a non-negative integer $g = g(C)$ such that $\ell(D) \geq 1 - g + \deg(D)$ for all divisors D on $C(\overline{k})$ and the equality holds for sufficiently positive divisors D.*

By Example 2.1.5, we conclude $g(\mathbf{P}^1) = 0$ from the corollary.

Exercise 2.1.5. Prove $\Omega_{\mathbf{P}^1/\overline{k}} = 0$.

2.1.7 Regular maps from a curve into projective space

Take a divisor D on a non-singular projective plane curve C. Suppose $\ell(D) = n > 0$. Take a basis $(f_1, f_2, \ldots, f_n)$ of $L(D)$. Thus we can write $f_j = \frac{h_j}{g_j}$ with homogeneous polynomials g_j, h_j having $\deg(g_j) = \deg(h_j)$. Replacing (g_j, h_j) by $(g'_0 := g_1 g_2 \cdots g_n, h'_j := h_j g^{(j)})$ for $g^{(j)} = \prod_{i \neq j} g_i$, we may assume $\deg(g'_j) = \deg(h'_j)$ for all j, and further dividing them by the GCD of $(h'_1, \ldots, h'_n, g'_0)$, we may assume that $f_j = \frac{h_j}{g_0}$ with $\deg(h_j) = \deg(g_0)$ for all j and $(g_0, h_1, \ldots, h_n)$ do not have a nontrivial common divisor.

Lemma 2.1.16. *Let the assumptions on $(g_0, h_1, \ldots, h_n)$ be as above. Suppose that $(g_0(P), h_1(P), \ldots, h_n(P)) \neq (0, 0, \ldots, 0)$ for all $P \in C(\overline{k})$. Define $L \in C(A) \subset \mathbf{P}^n(A)$, $\phi_A(L)$ for an A-submodule of A^{n+1} generated by $\phi(\ell) = (g_0(\ell), h_1(\ell), \ldots, h_n(\ell)) \in A^{n+1}$ for all $\ell \in L$. Then $\phi = \{\phi_A\}_A : C \to \mathbf{P}^n$ is a k-morphism of the projective plane k-curve C into $\mathbf{P}^n_{/k}$.*

We can prove the above lemma similarly to Corollary 2.1.9.

2.2   Elliptic curves

An *elliptic curve* $E_{/k}$ is a non-singular projective geometrically irreducible plane curve having $g(E) = 1$ with a point $\mathbf{0}_E \in E(k)$ specified. Here we define the genus $g(E)$, regarding E as defined over $\overline{k}$. We study elliptic curves in more details.

2.2.1   *Abel's theorem*

When we regard $P \in E(k)$ as a divisor, we just write $[P]$. So $3[P]$ is a divisor supported on P with multiplicity 3. We prove

Theorem 2.2.1 (Abel). *Let $E_{/k}$ be an elliptic curve with origin $\mathbf{0}_E$. The correspondence $P \mapsto [P] - [\mathbf{0}_E]$ induces a bijection $E(\overline{k}) \cong J(E)$. In particular, $E(\overline{k})$ is an ableian group.*

Proof.   Injectivity: if $[P] - [Q] = [P] - [\mathbf{0}_E] - ([Q] - [\mathbf{0}_E]) = \mathrm{div}(f)$ with $P \neq Q$ in $E(\overline{k})$, by Lemma 2.1.12, f is an isomorphism. This is wrong as $g(\mathbf{P}^1) = 0$ while $g(E) = 1$. Thus $P = Q$.

Surjectivity: Pick $D \in \mathrm{Div}^0(E)$. Then $D + [\mathbf{0}_E]$ has degree 1; so, $\ell(D + [\mathbf{0}_E]) = 1$ by Corollary 2.1.14, and we have $\phi \in L(D + [\mathbf{0}_E])$. Then $\mathrm{div}(\phi) + D + [\mathbf{0}_E] \geq 0$ and has degree 1. Any non-negative divisor with degree 1 is a single point $[P]$. Thus $D + [\mathbf{0}_E]$ is linearly equivalent to $[P]$; so, the map is surjective.   □

Corollary 2.2.2. *If $0 \neq \omega \in \Omega_{E/\overline{k}}$, then $\mathrm{div}(\omega) = 0$.*

Proof. Since $E(\overline{k})$ is a group, for each $P \in E(\overline{k})$, $T_P : Q \mapsto Q + P$ gives an automorphism of E. Thus $\omega \circ T_P$ is another element in $\Omega_{E/\overline{k}}$. Since $\dim \Omega_{E/\overline{k}} = 1$, we find $\omega \circ T_P = \lambda(P)\omega$ for $\lambda \in \overline{k}$. Since $\omega \neq 0$, at some point $P \in E(\overline{k})$, $v_P(\omega) = 0$. Since $v_Q(\omega \circ T_P) = v_{P+Q}(\omega)$ and we can bring any point to P by translation, we have $v_P(\omega) = 0$ everywhere. Thus $\operatorname{div}(\omega) = 0$. $\qquad\square$

We can show easily $\lambda(P) = 1$ for all P (see [GME, §2.2.3]). The nonzero differentials ω in $\Omega_{E/k}$ are called *nowhere vanishing differentials* as $\operatorname{div}(\omega) = 0$. They are unique up to constant multiple.

Exercise 2.2.1. Take a line L defined by $aX + bY + cZ$ in $\mathbf{P}^2$ and suppose its intersection with an elliptic curve $E \subset \mathbf{P}^2$ to be $\{P, Q, R\}$. Prove that $[P] + [Q] + [R] \sim 3[\mathbf{0}_E]$.

A field k is called a *perfect* field if any finite field extension of k is separable (i.e., generated by θ over k whose minimal equation over k does not have multiple roots). Fields of characteristic 0 and finite fields are perfect.

Remark 2.2.3. If k is perfect, $\overline{k}/k$ is possibly an infinite Galois extension; so, by Galois theory, we have a bijection between open subgroups G of $\operatorname{Gal}(\overline{k}/k)$ and finite extensions K/k inside $\overline{k}$ by $G \mapsto \overline{k}^G = \{x \in \overline{k} | \sigma(x) = x \text{ for all } \sigma \in G\}$ and $K \mapsto \operatorname{Gal}(\overline{k}/K)$. Since the isomorphism $E(\overline{k}) \cong J(C)$ is Galois equivariant, we have

$$E(K) \cong J(E)^{\operatorname{Gal}(\overline{k}/K)} = \{D \in J(E) | \sigma(D) = D \text{ for all } \sigma \in G\},$$

where $\sigma \in \operatorname{Gal}(\overline{k}/k)$ acts on $D = \sum_P m_P[P]$ by $\sigma(D) = \sum_P m_P[\sigma(P)]$. Basically by definition, we have

$$J(E)(K) := J(E)^{\operatorname{Gal}(\overline{k}/K)} = \frac{\{D \in \operatorname{Pic}^0(E) | \sigma(D) = D\}}{\{\operatorname{div}(f) | f \in K(E)^\times\}}.$$

Since any subfield $K \subset \overline{k}$ is a union of finite extensions, the identity $E(K) \cong J(E)(K)$ is also true for an infinite extension K/k inside $\overline{K}$. Actually we have a good definition of $\operatorname{Pic}(E)(A)$ for any k-algebra A, and we can generalize the identity $E(K) \cong J(E)(K)$ to all k-algebras A in place of fields K inside $\overline{k}$.

2.2.2 Weierstrass equations of elliptic curves

We now embed $E_{/k}$ into the two-dimensional projective space $\mathbf{P}^2_{/k}$ using a basis of $L(3[0])$ and determine the equation of the image in $\mathbf{P}^2_{/k}$. Choose a parameter $T = t_0$ at the origin $\mathbf{0} = \mathbf{0}_E$. We first consider $L(n[\mathbf{0}])$ which has dimension n if $n > 0$. We have $L([\mathbf{0}]) = k$ and $L(2[\mathbf{0}]) = k1 + kx$. Since x has to have a pole of order 2 at $\mathbf{0}$, we may normalize x so that $x = T^{-2}(1+\text{higher terms})$ in $k[[T]]$. Here x is unique up to translation: $x \mapsto x+a$ with $a \in k$. Then $L(3[\mathbf{0}]) = k1+kx+ky$. We may then normalize y so that $y = -T^{-3}(1 + \text{higher terms})$ (following the tradition, we later rewrite y for $2y$; thus, the normalization will be $y = -2T^{-3}(1 + \text{higher terms})$ at the end). Then y is unique up to the affine transformation: $y \mapsto y + ax + b$ $(a, b \in k)$.

Proposition 2.2.4. *Suppose that the characteristic of the base field k is different from 2 and 3. Then for a given pair (E, ω) of an elliptic curve E and a nowhere-vanishing differential ω both defined over k, we can find a unique base $(1, x, y)$ of $L(3[\mathbf{0}])$ such that E is embedded into $\mathbf{P}^2_{/k}$ by $(1, x, y)$ whose image is defined by the affine equation*

$$y^2 = 4x^3 - g_2 x - g_3 \quad \text{with} \quad g_2, g_3 \in k, \tag{2.2}$$

and ω on the image is given by $\frac{dx}{y}$. Conversely, a projective algebraic curve defined by the above equation is an elliptic curve with a specific nowhere-vanishing differential $\frac{dx}{y}$ if and only if the discriminant

$$\Delta(E, \omega) = g_2^3 - 27 g_3^2$$

of $4X^3 - g_2 X - g_3$ does not vanish.

An equation of an elliptic curve E as in (2.2) is called a *Weierstrass equation* of E, which is determined by the pair (E, ω).

Proof. By the dimension formulas, counting the order of poles at $\mathbf{0}$ of monomials of x and y, we have

$$L(4[\mathbf{0}]) = k + kx + ky + kx^2,$$
$$L(5[\mathbf{0}]) = k + kx + ky + kx^2 + kxy \quad \text{and}$$
$$L(6[\mathbf{0}]) = k + kx + ky + kx^2 + kxy + kx^3$$
$$= k + kx + ky + kx^2 + kxy + ky^2,$$

from which the following relation results,

$$y^2 + a_1 xy + a_3 y = x^3 + a_2 x^2 + a_4 x + a_6 \quad \text{with } a_j \in k, \tag{2.3}$$

because the poles of order 6 of y^2 and x^3 have to be cancelled. We homogenize the equation (2.3) by putting $x = \frac{X}{Z}$ and $y = \frac{Y}{Z}$ (and multiplying by Z^3). Write C for the projective plane k-curve in $\mathbf{P}^2$ defined by the (homogenized) equation. Thus we have a k-regular map: $\phi : E \to C \subset \mathbf{P}^2$ given by $P \mapsto (x(P) : y(P) : 1)$. Thus the function field $k(E)$ contains the function field $k(C)$ by the pull back of ϕ. By definition, $k(C) = k(x, y)$. Since $\mathrm{div}_\infty(x) = 2[\mathbf{0}_E]$ for $x = \frac{X}{Z} : E \to \mathbf{P}^1$, this gives a covering of degree 2; so, $[k(E) : k(x)] = 2$. Similarly $[k(E) : k(y)] = 3$. Since $[k(E) : k(C)]$ is a common factor of $[k(E) : k(x)] = 2$ and $[k(E) : k(y)] = 3$, we get $k(E) = k(C)$. Thus if C is smooth, $E \cong C$ by ϕ as a smooth geometrically irreducible curve is determined by its function field. Therefore, assuming C is smooth, $E_{/k}$ can be embedded into $\mathbf{P}^2_{/k}$ via $P \mapsto (x(P), y(P))$. The image is defined by the equation (2.3).

Let T be a local parameter at $\mathbf{0}_E$ normalized so that

$$\omega = (1 + \text{higher degree terms})dT.$$

Anyway $\omega = (a + \text{higher degree terms})dT$ for $a \in k^\times$, and by replacing T by aT, we achieve this normalization. The parameter T normalized as above is called a parameter adapted to ω. Then we may normalize x so that $x = T^{-2} + \text{higher degree terms}$. We now suppose that 2 is invertible in k. Then we may further normalize y so that $y = -2T^{-3} + \text{higher degree terms}$ (which we will do soon but not yet; so, for the moment, we still assume $y = T^{-3} + \text{higher degree terms}$).

The above normalization is not affected by variable change of the form $y \mapsto y + ax + b$ and $x \mapsto x + a'$. Now we make a variable change $y \mapsto y + ax + b$ in order to remove the terms of xy and y (i.e., we are going to make $a_1 = a_3 = 0$):

$$(y + ax + b)^2 + a_1 x(y + ax + b) + a_3(y + ax + b)$$
$$= y^2 + (2a + a_1)xy + (2b + a_3)y + \text{polynomial in } x.$$

Assuming that 2 is invertible in k, we take $a = -\frac{a_1}{2}$ and $b = -\frac{a_3}{2}$. The resulting equation is of the form $y^2 = x^3 + b_2 x^2 + b_4 x + b_6$. We now make the change of variable $x \mapsto x + a'$ to make $b_2 = 0$:

$$y^2 = (x + a')^3 + b_2(x + a')^2 + b_4(x + a') + b_6 = x^3 + (3a' + b_2)x^2 + \cdots.$$

Assuming that 3 is invertible in k, we take $a' = -\frac{b_2}{3}$. We can rewrite the equation as in (2.2) (making a variable change $-2y \mapsto y$). By the variable change as above, we have $y = -2T^{-3}(1 + \text{higher terms})$, and from this, we conclude $\omega = \frac{dx}{y}$. The numbers g_2 and g_3 are determined by T adapted to a given nowhere-vanishing differential form ω.

If the discriminant $\Delta(E, \omega)$ of $g(x) = 4x^3 - g_2 x - g_3$ vanishes, C has only singularity at $(x_0 : 0 : 1)$ for a multiple root x_0 of $g(x) = 0$. If $g(x)$ has a double zero, C is isomorphic over $\overline{k}$ to the curve defined by $y^2 = x^2(x-a)$ for $a \neq 0$. Let $t = \frac{x}{y}$. Then for $P \in E(\overline{k})$ mapping to $(0,0)$, $v_P(y) = v_P(x)$; so, P is neither a zero nor a pole of t. The function t never vanish outside $\mathbf{0}_E$ (having a pole at $(a,0)$). It has a simple zero at $\mathbf{0}_E$ by the normalization of x and y. Thus $\deg(\mathrm{div}_0(t)) = 1$, and $\overline{k}(C) = \overline{k}(t)$, which is impossible as $k(C) = k(E)$ and $g(E) = 1$. The case of triple zero can be excluded similarly. Thus we conclude $\Delta(E, \omega) \neq 0$ ($\Leftrightarrow C$ is smooth: Example 2.1.3), and we have $E \cong C$ by ϕ.

Conversely, we have seen that any curve defined by equation (2.2) is smooth in Example 2.1.3 if the cubic polynomial $F(X) = 4X^3 - g_2X - g_3$ has three distinct roots in k. In other words, if the discriminant $\Delta(E, \omega)$ of $F(X)$ does not vanish, E is smooth.

For a given equation, $Y^2 = F(X)$, the algebraic curve E defined by the homogeneous equation $Y^2Z = 4X^3 - g_2XZ^2 - g_3Z^3$ in $\mathbf{P}^2_{/k}$ has a rational point $\mathbf{0} = (0, 1, 0) \in E(k)$, which is ∞ in $\mathbf{P}^2$. Thus E is smooth over k if and only if $\Delta(E, \omega) \neq 0$ (an exercise following this proof).

We show that there is a canonical nowhere-vanishing differential $\omega \in \Omega_{E/k}$ if E is defined by (2.2). If such an ω exists, all other holomorphic differentials ω' are of the form $f\omega$ with $\mathrm{div}(f) \geq 0$, which implies $f \in k$; so, $g = \dim_k \Omega_{E/k} = 1$, and $E_{/k}$ is an elliptic curve. It is an easy exercise to show that $y^{-1}dx$ does not vanish on E (an exercise following this proof).

We summarize what we have seen. Returning to the starting elliptic curve $E_{/k}$, for the parameter T at the origin, we see by definition

$$x = T^{-2}(1+\text{higher degree terms}) \text{ and } y = -2T^{-3}(1+\text{higher degree terms}).$$

This shows

$$\frac{dx}{y} = \frac{-2T^{-3}(1 + \cdots)}{-2T^{-3}(1 + \cdots)}dT = (1 + \text{higher degree terms})dT = \omega.$$

Thus the nowhere-vanishing differential form ω to which T is adapted is given by $\frac{dx}{y}$. Conversely, if $\Delta \neq 0$, the curve defined by $y^2 = 4x^3 - g_2x - g_3$ is an elliptic curve over k with origin $\mathbf{0} = \infty$ and a standard nowhere-vanishing differential form $\omega = \frac{dx}{y}$. This finishes the proof. $\square$

Exercise 2.2.2.

(1) If C is defined by $y^2 = x^3$, prove $k(C) = k(t)$ for $t = \frac{x}{y}$.

(2) Compute $v_P(dx/y)$ explicitly at any point P on $E(\overline{k})$.

(3) Show that if $\Delta \neq 0$, the curve defined by $y^2 = 4x^3 - g_2x - g_3$ is also smooth at $\mathbf{0} = \infty$.

2.2.3 *Moduli of Weierstrass type*

We continue to assume that the characteristic of k is different from 2 and 3. Suppose that we are given two elliptic curves $(E, \omega)_{/k}$ and $(E', \omega')_{/k}$ with nowhere-vanishing differential forms ω and ω'. We call two pairs (E, ω) and (E', ω') isomorphic if we have an isomorphism $\varphi : E \to E'$ with $\varphi^* \omega' = \omega$. Here for $\omega' = f dg$, $\varphi^* \omega' = (f \circ \varphi) d(g \circ \varphi)$; in other words, if $\sigma : k(E') \to k(E)$ is the isomorphism of the function fields associated with φ, $\varphi^* \omega' = \sigma(f) d(\sigma(g))$. Let T' be the parameter at the origin $\mathbf{0}$ of E' adapted to ω'. If $\varphi : (E, \omega) \cong (E', \omega')$, then the parameter $T = \varphi^* T'$ mod T^2 is adapted to ω (because $\varphi^* \omega' = \omega$). We choose coordinates (x, y) for E and (x', y') for E' relative to T and T' as above. By the uniqueness of the choice of (x, y) and (x', y'), we know $\varphi^* x' = x$ and $\varphi^* y' = y$. Thus the Weierstrass equations of (E, ω) and (E', ω') coincide. We write $g_2(E, \omega)$ and $g_3(E, \omega)$ for the g_2 and g_3 of the coefficients of the Weierstrass equation of (E, ω). If a field K has characteristic different from 2 and 3, we have

$$\mathcal{P}(K) := \big[(E, \omega)_{/K}\big] \cong \big\{(g_2, g_3) \in K^2 \big| \Delta(E, \omega) \neq 0\big\} \cong \mathrm{Spec}(\mathcal{R})(K), \tag{2.4}$$

where $\mathcal{R} := \mathbb{Q}[g_2, g_3, \frac{1}{g_2^3 - 27g_3^2}]$ (the polynomial ring of variables g_j with $g_2^3 - 27g_3^2$ inverted) and $[\cdot]$ indicates the set of isomorphism classes of the objects inside the bracket and $\mathrm{Spec}(R)(K)$ for a ring R is the set of all algebra homomorphisms: $R \to K$. The last isomorphism sends (g_2, g_3) to the algebra homomorphism ϕ with $\phi(X) = g_2$ and $\phi(Y) = g_3$.

There is an elliptic curve $\mathbb{E}$ defined by $Y^2 Z = 4X^3 - g_2 X Z^2 - g_3 Z^3$ over $\mathcal{R}$. This is a universal curve in the sense that for any pair $(E, \omega)_{/A}$ defined by $Y^2 Z = 4X^3 - a_2 X Z^2 - a_3 Z^3$ with $\omega = \frac{dX}{Y}$ over A, we have a unique morphism $\mathcal{R} \xrightarrow{\varphi} A$ such that $\varphi(g_j) = a_j$ induces the pair (E, ω). In other words, (2.4) means that for each $\phi \in \mathrm{Hom}_{ALG}(\mathcal{R}, K)$, there is a unique object $(E, \omega)_{/K}$ defined by the equation $Y^2 Z = 4X^3 - \phi(g_2) X Z^2 - \phi(g_3) Z^3$ with $\omega = \frac{dX}{Y}$ (not just an isomorphism class of $(E, \omega)_{/K}$); so, for such representability, it is absolutely necessary that $\mathrm{Aut}(E, \omega)_{/K} = \{\mathrm{id}\}$ as otherwise, we would have several choices of ϕ for the single isomorphism class of $(E, \omega)_{/K}$.

We now classify elliptic curves E eliminating the contribution of the differential from the pair (E, ω). If $\varphi : E \cong E'$ for (E, ω) and (E', ω'), we have $\varphi^* \omega' = \lambda \omega$ with $\lambda \in K^\times$, because $\varphi^* \omega'$ is another nowhere-vanishing differential. Therefore we study $K^\times$-orbit: (E, ω) mod $K^\times$ under the action of $\lambda \in K^\times$ given by $(E, \omega)_{/K} \longmapsto (E, \lambda \omega)_{/K}$, computing the dependence of $g_j(E, \lambda \omega)$ $(j = 2, 3)$ on λ for a given pair $(E, \omega)_{/K}$. Let T be the parameter adapted to ω. Then λT is adapted to $\lambda\omega$. We see

$$x(E,\omega) = \frac{(1+T\phi(T))}{T^2} \Rightarrow x(E,\lambda\omega) = \frac{(1+\text{higher terms})}{(\lambda T)^2} = \lambda^{-2}x(E,\omega),$$

$$y(E,\omega) = \frac{(-2+T\psi(T))}{T^3} \Rightarrow y(E,\lambda\omega) = \frac{(-2+\text{higher terms})}{(\lambda T)^3} = \lambda^{-3}y(E,\omega).$$

Since $y^2 = 4x^3 - g_2(E,\omega)x - g_3(E,\omega)$, we have

$$(\lambda^{-3}y)^2 = 4\lambda^{-6}x^3 - g_2(E,\omega)\lambda^{-6}x - \lambda^{-6}g_3(E,\omega)$$
$$= 4(\lambda^{-2}x)^3 - \lambda^{-4}g_2(E,\omega)(\lambda^{-2}x) - \lambda^{-6}g_3(E,\omega).$$

This shows

$$g_2(E,\lambda\omega) = \lambda^{-4}g_2(E,\omega) \quad \text{and} \quad g_3(E,\lambda\omega) = \lambda^{-6}g_3(E,\omega). \qquad (2.5)$$

Thus we have

Theorem 2.2.5. *If two elliptic curves $E_{/K}$ and $E'_{/K}$ are isomorphic, then choosing nowhere-vanishing differentials $\omega_{/E}$ and $\omega'_{/E'}$, we have $g_j(E',\omega') = \lambda^{-2j}g_j(E,\omega)$ for $\lambda \in K^\times$. The constant λ is given by $\varphi^*\omega' = \lambda\omega$.*

We define the J-invariant of E by $J(E) = \frac{(12g_2(E,\omega))^3}{\Delta(E,\omega)}$. Then J only depends on E (not the chosen differential ω). If $J(E) = J(E')$, then we have

$$\frac{(12g_2(E,\omega))^3}{\Delta(E,\omega)} = \frac{(12g_2(E',\omega'))^3}{\Delta(E',\omega')} \iff g_j(E',\omega') = \lambda^{-2j}g_j(E,\omega)$$

for a twelfth root λ of $\Delta(E,\omega)/\Delta(E',\omega')$. Note that the twelfth root λ may not be in K if K is not algebraically closed.

Conversely, for a given $j \notin \{0,1\}$, the elliptic curve defined by $y^2 = 4x^3 - gx - g$ for $g = \frac{27j}{j-1}$ has J-invariant 12^3j. If $j = 0$ or 1, we can take the following elliptic curve with $J = 0$ or 12^3. If $J = 0$, then $y^2 = 4x^3 - 1$ and if $J = 12^3$, then $y^2 = 4x^3 - 4x$. Thus we have

Corollary 2.2.6. *If K is algebraically closed, then $J(E) = J(E') \Leftrightarrow E \cong E'$ for two elliptic curves over K. Moreover, for any field K, there exists an elliptic curve E with a given $J(E) \in K$.*

Exercise 2.2.3.

(1) Prove that $g_j(E',\omega') = \lambda^{-2j}g_j(E,\omega)$ for suitable ω and ω' and a suitable twelfth root λ of $\Delta(E,\omega)/\Delta(E',\omega')$ if $J(E) = J(E')$.

(2) Explain what happens if $J(E) = J(E')$ but $E \not\cong E'$ over a field K not necessarily algebraically closed.

Note that $\mathcal{R}$ is a graded ring such that g_2 is of degree 4 and g_3 is of degree 6. Then the degree 0 subring $\mathcal{R}_0 = \mathbb{Q}[J]$. Note that $\mathrm{Spec}(\mathbb{Q}[J])$ is the 1 dimensional affine space $\mathbb{A}^1$, as $\mathrm{Spec}(\mathbb{Q}[J])(A) = \mathrm{Hom}_{alg}(\mathbb{Q}[J], A) \cong A$ by $\phi \mapsto \phi(J) \in A$.

Consider functors

$$\mathcal{P}_{1,N}(A) := \left[(E, \omega, \phi)_{/A} | \phi : N^{-1}\mathbb{Z}/\mathbb{Z} \hookrightarrow E[N] := \mathrm{Ker}(E \xrightarrow{N} E) \right],$$

$$\mathcal{E}_{1,N}(A) := \left[(E, \phi)_{/A} | \phi : N^{-1}\mathbb{Z}/\mathbb{Z} \hookrightarrow E[N] := \mathrm{Ker}(E \xrightarrow{N} E) \right] \tag{2.6}$$

for a positive integer N. The functor has natural transformation $\mathcal{P}_{1,N} \to \mathcal{P}$ sending $(E, \omega, \phi)_{/A}$ to $(E, \omega)_{/A}$. This is represented by

$$\mathcal{Y}_1(N) := \mathbb{E}[N] - \bigcup_{0 < d | N, D \neq N} \mathbb{E}[d] \tag{2.7}$$

which is affine in the sense that $\mathcal{Y}_1(N) = \mathrm{Spec}(\mathcal{R}_{1,N})$ for the ring $\mathcal{R}_{1,N} := \mathrm{Hom}_{COF}(\mathcal{Y}_1(N), \mathbb{A})$, and $\mathcal{R}_{1,N}$ is finite locally free over $\mathcal{R}$. By the action $\omega \mapsto \lambda \omega$ on $\mathcal{P}_{1,N}$, the multiplicative group $\mathbf{G}_m$ given by $A \mapsto A^\times$ acts on $\mathcal{R}_{1,N}$. The subring $\mathcal{A}_{1,N} := H^0(\mathbf{G}_m, \mathcal{R}_{1,N})$ fixed by this action represents $\mathcal{E}_{1,N}$, i.e.,

$$\mathcal{E}_{1,N}(A) = \mathrm{Hom}_{\mathbb{Q}\text{-alg}}(\mathcal{A}_{1,N}, A).$$

The corresponding plane curve $Y_1(N)_{/\mathbb{Q}} := \mathrm{Spec}(\mathcal{A}_{1,N})$ (resp. $Y_0(N)_{/\mathbb{Q}} := \mathrm{Spec}(\mathcal{A}_{1,N}^{\Gamma_0(N)})$) is the modular curve of level $\Gamma_1(N)$ (resp. $\Gamma_0(N)$). In other words, for a triple $(E, \omega, \phi : \mathbb{Z}/N\mathbb{Z} \hookrightarrow E[N])_{/A}$, the value $\phi(1)$ gives rise to a unique point of $\mathcal{Y}_1(N)(A)$ over the point corresponding $(E, \omega)_{/A} \in \mathcal{P}_{1,N}(A)$.

Similarly,

$$\mathcal{P}_{1,N}(A) := \left[(E, \omega, \phi)_{/A} | \phi : (N^{-1}\mathbb{Z}/\mathbb{Z})^2 \cong E[N] := \mathrm{Ker}(E \xrightarrow{N} E) \right],$$

$$\mathcal{E}_{1,N}(A) := \left[(E, \phi)_{/A} | \phi : (N^{-1}\mathbb{Z}/\mathbb{Z})^2 \cong E[N] := \mathrm{Ker}(E \xrightarrow{N} E) \right] \tag{2.8}$$

is represented by

$$\mathcal{Y}(N) = \{ (x, y) \in \mathbb{E}[N] \times_y \mathbb{E}[N] | x \wedge y \in \mathbb{E}[N] \wedge \mathbb{E}[N] - \bigcup_{0 < d | N, D \neq N} \mathbb{E}[d] \wedge \mathbb{E}[d] \}$$

and $Y(N) = \mathrm{Spec}(\mathcal{A}_N)$ with $\mathcal{A}_N = H^0(\mathbf{G}_m, \mathcal{R}_N)$, respectively, where $\mathcal{Y}(N) = \mathrm{Spec}(\mathcal{R}_N)$ for $\mathcal{R}_N := \mathrm{Hom}_{COF}(\mathcal{Y}(N), \mathbb{A})$.

Remark 2.2.7. We will see later that the curve $Y(N)$ is irreducible over $\mathbb{Q}$ but becomes reducible over $\mathbb{Q}[\mu_N]$ (the cyclotomic field of N-th root of unity), looking into the q-expansion of Weierstrass $\wp$-functions. The function field $\mathbb{Q}(Y(N))$ thus contains $\mathbb{Q}[\mu_N]$ as algebraic closure of $\mathbb{Q}$ in it.

2.3   Modular forms and functions

We give an algebraic definition of modular forms and then relate it to classical theory (due to Weierstrass, Klein, Flicke).

2.3.1   *Geometric modular forms*

Let A be an algebra over $\mathbb{Q}$. We restrict the functor $\mathcal{P}$ to $ALG_{/A}$ and write the restriction $\mathcal{P}_{/A}$. Then by (2.4), for $\mathcal{R}_A := A[g_2, g_3, \frac{1}{\Delta}]$,

$$\mathcal{P}_{/A}(?) = \mathrm{Hom}_{ALG_{/A}}(\mathcal{R}_A, ?).$$

A morphism of functors $\phi : \mathcal{P}_{/A} \to \mathbb{A}^1_{/A}$ is by definition given by maps $\phi_R : \mathcal{P}_{/A}(R) \to \mathbb{A}^1(R) = R$ indexed by $R \in ALG_{/A}$ such that for any $\sigma : R \to R'$ in $\mathrm{Hom}_{ALG_{/A}}(R, R')$, $\phi_{R'}((E, \omega) \otimes_R R') = \sigma(f((E, \omega)_{/R}))$. Note that $\mathbb{A}^1_{/A}(?) = \mathrm{Hom}_{ALG_{/A}}(A[X], ?)$ by $R \ni a \leftrightarrow (\varphi : A[X] \to R) \in \mathrm{Hom}_{ALG_{/A}}(A[X], ?)$ with $\varphi(X) = a$. Thus in particular,

$$\phi_{\mathcal{R}_A} : \mathcal{P}(\mathcal{R}_A) = \mathrm{Hom}_{ALG_{/A}}(\mathcal{R}_A, \mathcal{R}_A) \to \mathbb{A}^1(A[X], \mathcal{R}_A) = \mathcal{R}_A.$$

Thus $\phi_{\mathcal{R}_A}(\mathrm{id}_{\mathcal{R}_A}) \in \mathcal{R}_A$; so, write $\phi_{\mathcal{R}_A}(\mathrm{id}_{\mathcal{R}_A}) = \Phi(g_2, g_3)$ for a two variable rational function $\Phi(x, y) \in A[x, y, \frac{1}{x^3 - 27y^2}]$. Let $\mathbf{E}_{/\mathcal{R}_A}$ be the universal elliptic curve over $\mathcal{R}_A$ defined by $Y^2 Z = 4X^3 - g_2 X Z^2 - g_3 Z^3$ with the universal differential $\boldsymbol{\omega} = \frac{dX}{Y}$. For each $(E, \omega)_{/R}$, we have a unique A-algebra homomorphism $\sigma : \mathcal{R}_A \to R$ given by $\sigma(g_j) = g_j(E, \omega)$; in other words, $(E, \omega)_{/R} \cong (\mathbf{E}, \omega)_{\mathcal{R}_A} \otimes_{\mathcal{R}_A} R$. Thus

$$\phi_R(E, \omega) = \phi_R((\mathbf{E}, \omega) \otimes_{\mathcal{R}_A} R) = \sigma(\phi_{\mathcal{R}_A}(\mathbf{E}, \omega))$$
$$= \sigma(\phi_{\mathcal{R}_A}(\mathrm{id}_{\mathcal{R}_A})) = \Phi(\sigma(g_2), \sigma(g_3)) = \Phi(g_2(E, \omega), g_3(E, \omega)).$$

Theorem 2.3.1. *Any functor morphism* $\phi : \mathcal{P}_{/A} \to \mathbb{A}^1_{/A}$ *is given by a rational function* $\Phi \in \mathcal{R}_A$ *of* g_2 *and* g_3 *so that* $\phi(E, \omega) = \Phi(g_2(E, \omega), g_2(E, \omega))$ *for every elliptic curve* (E, ω) *over an* A-*algebra.*

Define a weight function $w : A[g_2, g_3] \to \mathbb{Z}$ by $w(g_2^a g_3^b) = 4a + 6b$, and for a polynomial $\Phi = \sum_{a,b} c_{a,b} g_2^a g_3^b$, we put $w(\Phi) = \max(w(g_2^a g_3^b) | c_{a,b} \neq 0)$. A polynomial $\Phi = \sum_{a,b} c_{a,b} g_2^a g_3^b$ of g_2 and g_3 is called *isobaric* if $c_{a,b} \neq 0 \Rightarrow 4a + 6b = w$.

A weight w (geometric) *modular form* defined over A is a morphism of functors $\mathcal{P}_{/A} \to \mathbb{A}^1_{/A}$ given by an isobaric polynomial of g_2 and g_3 of weight w with coefficients in A. Write $G_w(A)$ for the A-module of modular forms of weight w. Then $f \in G_w(A)$ is a functorial rule assigning each isomorphism class of $(E, \omega)_{/R}$ for an A-algebra R an element $f(E, \omega) \in R$ satisfying the following properties:

(g0) $f \in A[g_2, g_3]$,

(g1) If (E, ω) is defined over an A-algebra R, we have $f(E, \omega) \in R$, which depends only on the isomorphism class of (E, ω) over R,

(g2) $f((E, \omega) \otimes_R R') = \sigma(f(E, \omega))$ for A-algebra homomorphism $\sigma : R \to R'$,

(g3) $f((E, \lambda\omega)_{/R}) = \lambda^{-w} f(E, \omega)$ for any $\lambda \in R^\times$.

Remark 2.3.2. We defined the space $G_w(A)$ of geometric modular forms over any k-algebras A purely algebro-geometrically, and hence we use the symbol "G_w" for the space. Later we define analytically and cohomologically the space $M_k(A)$ of analytic modular forms with A-integral q-expansions in §3.2.1 and §4.1.5. The two spaces are isomorphic canonically, although the definitions are quite different in appearance. To emphasize the geometric nature, we use different symbols for these spaces.

Exercise 2.3.1. For a field $K \supset \mathbb{Q}$, prove for $0 < w \in 2\mathbb{Z}$,

$$\dim_K G_w(K) = \begin{cases} \left[\frac{w}{12}\right] & \text{if } w \equiv 2 \mod 12, \\ \left[\frac{w}{12}\right] + 1 & \text{otherwise.} \end{cases}$$

We can define modular forms of level N rational over A replacing the functor $\mathcal{P}$ in the previous section by $\mathcal{P}_{1,N}$ or $\mathcal{P}_N$. We list the functorial properties:

(G_N0) f is integral over $\in A[g_2, g_3]$,

(G_N1) If (E, ω, ϕ) is defined over an A-algebra R, we have $f(E, \omega, \omega) \in R$, which depends only on the isomorphism class of (E, ω) over R,

(G_N2) $f((E, \omega, \phi) \otimes_R R') = \sigma(f(E, \omega, \phi))$ for A-algebra homomorphism $\sigma : R \to R'$,

(G_N3) $f((E, \lambda\omega, \phi)_{/R}) = \lambda^{-w} f(E, \omega, \phi)$ for any $\lambda \in R^\times$.

Here ϕ is the level structure depending on our choice of $\mathcal{P}_{1,N}$ and $\mathcal{P}_N$.

We then define $G_k(\Gamma_1(N); A)$ (resp. $G_k(\Gamma(N); A)$) for the space of modular forms of weight k defined over A for $\mathcal{P}_{1,N}$ (resp. $\mathcal{P}_N$).

Remark 2.3.3. As we remarked in Remark 2.3.2, we will have the analytic incarnation $M_k(\Gamma_1(N); A)$ of $G_k(\Gamma_1(N); A)$ in Chapters 3 and 4. They are equivalent under the q-expansion principle [GME, Corollary 3.2.11].

2.3.2 Topological fundamental groups

In the following four subsections, we would like to give a sketch of Weierstrass' theory of elliptic curves defined over the complex field $\mathbb{C}$. By means of Weierstrass $\wp$-functions, we can identify $E(\mathbb{C})$ (for each elliptic curve

$E_{/\mathbb{C}}$) with a quotient of $\mathbb{C}$ by a lattice L. In this way, we can identify $[(E, \omega)_{/\mathbb{C}}]$ with the space of lattices in $\mathbb{C}$. This method is analytic.

We can deduce from the analytic parameterization (combining with geometric technique of Weil–Shimura) many results on the moduli space of elliptic curves, like, the exact field of definition of the moduli, determination of the field of moduli (of each member), and so on (e.g., [IAT, Chapter 6]). We have come here in a reverse way: starting algebraically, mainly by the Riemann–Roch theorem, we have determined a unique Weierstrass equation over A for a given pair $(E, \omega)_{/A}$, and therefore, we know the exact shape of the moduli space before setting out to study the analytic method. After studying analytic theory over $\mathbb{C}$, combining these techniques, we start studying modular units.

Let $(E, \omega)_{/\mathbb{C}}$ be an elliptic curve over $\mathbb{C}$. Then

$$E(\mathbb{C}) = E(g_2, g_3)(\mathbb{C}) = \{(x : y : z) \in \mathbf{P}^2(\mathbb{C}) | y^2 z - 4x^3 + g_2 z^2 x + g_3 z^3 = 0\},$$

and $E(\mathbb{C})$ is a compact Riemann surface of genus 1. A path $\gamma : y \to x$ on $E(\mathbb{C})$ is a piecewise smooth continuous map γ from the interval $[0, 1]$ into $E(\mathbb{C})$ (under the Euclidean topology on $E(\mathbb{C})$) such that $\gamma(0) = y$ and $\gamma(1) = x$. Two paths $\gamma, \gamma' : x \to x$ are homotopy equivalent (for which we write $\gamma \approx \gamma'$) if there is a bi-continuous map $\varphi : [0, 1] \times [0, 1] \to E(\mathbb{C})$ such that $\varphi(0, t) = \gamma(t)$ and $\varphi(1, t) = \gamma'(t)$. Let $\mathcal{Z}$ be the set of all equivalence classes of paths emanating from $\mathbf{0}$.

More generally, for each complex manifold X, we can think of the space $\mathcal{Z} = \mathcal{Z}(X)$ of homotopy classes of paths emanating from a fixed point $x \in X$ [AFC, §3.5]. An open neighborhood U of x is called *simply connected* if $\mathcal{Z}(U) \cong U$ by projecting $(\gamma : x \to y)$ down to y. For example, if U is diffeomorphic to an open disk with center x, it is simply connected (that is, every loop is equivalent to x). If $\gamma : x \to y$ and $\gamma' : y \to z$ are two paths, we define their product path $\gamma\gamma' : x \to z$ by

$$\gamma\gamma'(t) = \begin{cases} \gamma(2t) & \text{if } 0 \le t \le 1/2 \\ \gamma'(2t - 1) & \text{if } 1/2 \le t \le 1. \end{cases}$$

By this multiplication, $\pi_1(X) = \pi(X, x) = \{\gamma \in \mathcal{Z}(X) | \gamma : x \to x\}/\approx$ becomes a group called *the topological fundamental group* of X. Taking a fundamental system of neighborhoods $\mathcal{U}_y$ of $y \in X$ made of simply connected open neighborhoods of y, we define a topology on $\mathcal{Z}(X)$ so that a fundamental system of neighborhoods of $\gamma : x \to y$ is given by $\{\gamma U | U \in \mathcal{U}_x\}$. Then $\pi_1(X)$ acts on $\mathcal{Z}(X)$ freely without fixed points. By definition, we have a continuous map $\pi : \pi_1(X) \backslash \mathcal{Z}(X) \to X$ given by $\pi(\gamma : x \to y) = y$, which is a local isomorphism. Since $\pi^{-1}(x) = \{x\}$, $\pi : \pi_1(X)\backslash\mathcal{Z}(X) \cong X$ is a homeomorphism. Since $\pi : \mathcal{Z}(X) \to X$ is local isomorphism, we can regard $\mathcal{Z} = \mathcal{Z}(X)$ as a complex manifold. This space $\mathcal{Z}(X)$ is called a *universal covering* space of X.

2.3.3 *Fundamental group of an elliptic curve*

We now return to the original setting: $\mathcal{Z} = \mathcal{Z}(E(\mathbb{C}))$, and write $\Pi = \pi(E, \mathbf{0})$. Since $E(\mathbb{C})$ is a commutative group, writing its group multiplication additively, we define the sum $\gamma + \gamma'$ on $\mathcal{Z}$ by, noting that γ and γ' originate at the origin $\mathbf{0}$,

$$(\gamma + \gamma')(t) = \begin{cases} \gamma(2t) & \text{if } 0 \le t \le 1/2 \\ \gamma(1) + \gamma'(2t - 1) & \text{if } 1/2 \le t \le 1. \end{cases}$$

Then $(\gamma+\gamma')(1) = \gamma(1)+\gamma'(1)$, and we claim that $\gamma+\gamma' \approx \gamma'+\gamma$. In fact, on the square $[0, 1] \times [0, 1]$, we consider the path α on the boundary connecting the origin $(0, 0)$ and $(1, 1)$ passing $(0, 1)$, and write β the opposite path from $(0, 0)$ to $(1, 1)$ passing $(1, 0)$. They are visibly homotopy equivalent. Thus we have a continuous map $\phi : [0, 1] \times [0, 1] \to [0, 1] \times [0, 1]$ such that $\phi(0, t) = \alpha(t)$ and $\phi(1, t) = \beta(t)$. Define

$$f : [0, 1] \times [0, 1] \to E(\mathbb{C}) \text{ by } f(t, t') = \gamma(t) + \gamma'(t').$$

Then it is easy to see $f \circ \phi(0, t) = (\gamma' + \gamma)(t)$ and $f \circ \phi(1, t) = (\gamma + \gamma')(t)$.

By the above addition, $\mathcal{Z}$ is an additive complex Lie group. Since $\gamma+\gamma' = \gamma\gamma'$ if $\gamma \in \Pi$ and $\gamma' \in \mathcal{Z}$ by definition, Π is an additive subgroup of $\mathcal{Z}$ and $\Pi\backslash\mathcal{Z} \cong E(\mathbb{C})$, where the quotient is made through the group action.

Now we define, choosing a C^∞-path $[\gamma]$ in each class of $\gamma \in \mathcal{Z}$ and a nowhere vanishing differential form ω on E, a map $I : \mathcal{Z} \to \mathbb{C}$ by $\gamma \mapsto \int_{[\gamma]} \omega \in \mathbb{C}$. Since ω is holomorphic on $\mathcal{Z}$, the value of I is independent of the choice of the representative $[\gamma]$ by Cauchy's integration theorem. Since ω is translation invariant on $E(\mathbb{C})$, it is translation invariant on $\mathcal{Z}$ and $I(\gamma+\gamma') = I(\gamma)+I(\gamma')$. In particular, I is a local homeomorphism because $E(\mathbb{C})$ is one dimensional and for simply connected U, $\mathcal{Z}(U) \cong I(U)$. The pair $(E(\mathbb{C}), \omega)$ is isomorphic locally to the pair of the additive group $\mathbb{C}$ and du for the coordinate u on $\mathbb{C}$, because du is the unique translation invariant differential (up to constant multiple). Since $I^{-1}([\mathbf{0}]) = \{\mathbf{0}\}$, I is a linear isomorphism into $\mathbb{C}$. For an open neighborhood U of $\mathbf{0}$ with $U \cong \mathcal{Z}(U) \ni \gamma \mapsto I(\gamma) = \int_\gamma \omega \in \mathbb{C}$ giving an isomorphism onto a small open disk D in $\mathbb{C}$ centered at 0, we have two $\gamma_1, \gamma_2 \in U$ giving rise to a two $\mathbb{R}$-linearly independent $I(\gamma_j)$ $(j = 1, 2)$. Then $I(m\gamma_1+n\gamma_2) = mI(\gamma_1)+nI(\gamma_2)$

for all $m, n \in \mathbb{Z}$. Replacing γ_j by $\frac{1}{a}\gamma_j \in \mathcal{Z}(U)$ such that $I(\frac{1}{a}\gamma_j) = \frac{I(\gamma_j)}{a}$ for any positive integer a, by the same argument, we find $I(m\gamma_1 + n\gamma_2) = mI(\gamma_1) + nI(\gamma_2)$ for all $m, n \in \mathbb{Q}$; so, I is an isomorphism.

This also shows that if $\alpha : E \to E$ is an endomorphism of E with $\alpha(\mathbf{0}_E) = \mathbf{0}_E$, α lifts an endomorphism of $\mathcal{Z}$ sending a path γ from $\mathbf{0}_E$ to $z \in \mathbb{C}$ to a path $\alpha(\gamma)$ from $\alpha(\mathbf{0}_E) = \mathbf{0}_E$ to $\alpha(z)$. In particular, $\alpha(\gamma + \gamma') = \alpha(\gamma) + \alpha(\gamma')$. Thus α induces a linear map from $\mathbb{C} = \mathcal{Z}$ to $\mathbb{C}$. Since α is holomorphic (as it is a polynomial map of the coordinates of $\mathbf{P}^2_{/\mathbb{C}}$), α is a $\mathbb{C}$-linear map. We thus get a natural inclusion:

$$\text{End}(E_{/\mathbb{C}}) \hookrightarrow \mathbb{C}. \tag{2.9}$$

Writing $L = L_E$ for $I(\Pi)$, we can find a base w_1, w_2 of L over $\mathbb{Z}$. Thus we have a map

$$\wp(\mathbb{C}) \ni (E, \omega) \longmapsto L_E \in \{L|L : \text{lattice in } \mathbb{C}\} =: Lat,$$

and we have $(E(\mathbb{C}), \omega) \cong (\mathbb{C}/L_E, du)$. Therefore the map: $\wp(\mathbb{C}) \to Lat$ is injective. We show its surjectivity in the next subsection.

By the above fact combined with (2.9), we get

Proposition 2.3.4. *We have a ring embedding* $\text{End}(E_{/\mathbb{C}}) \hookrightarrow \{u \in \mathbb{C}|u \cdot L_E \subset L_E\}$, *and hence* $\text{End}(E_{/\mathbb{C}})$ *is either* $\mathbb{Z}$ *or an order of an imaginary quadratic field.*

Proof. The first assertion follows from (2.9). Pick $\alpha \in \text{End}(E_{/\mathbb{C}})$ corresponding $u \in \mathbb{C}$ as above. Note that $L_E = \mathbb{Z}w_1 + \mathbb{Z}w_2$. Then $uw_1 = aw_1 + bw_2$ and $uw_2 = cw_1 + dw_2$ for integers a, b, c, d. In short, writing $w = \binom{w_1}{w_2}$ and $\rho(\alpha) = \left(\begin{smallmatrix} a & b \\ c & d \end{smallmatrix}\right)$, we get $uw = \rho(\alpha)w$; so, $\rho : \text{End}(E_{/\mathbb{C}}) \to M_2(\mathbb{Z})$ is a ring homomorphism. By the first assertion, the image has to be an order of imaginary quadratic field or just $\mathbb{Z}$. $\square$

When $\text{End}(E_{/\mathbb{C}}) \neq \mathbb{Z}$, E is said to have *complex multiplication*.

2.3.4 *Classical Weierstrass $\wp$-function*

For a given $L \in Lat$, we define the Weierstrass $\wp$–functions by

$$x_L(u) = \wp(u) = \frac{1}{u^2} + \sum_{\ell \in L - \{0\}} \left\{ \frac{1}{(u - \ell)^2} - \frac{1}{\ell^2} \right\} = \frac{1}{u^2} + \frac{g_2}{20}u^2 + \frac{g_3}{28}u^4 + \cdots$$

$$y_L(u) = \wp'(u) = -\frac{2}{u^3} - 2 \sum_{\ell \in L - \{0\}} \frac{1}{(u - \ell)^3} = -2u^{-3} + \cdots ,$$

where

$$g_2 = g_2(L) = 60 \sum_{\ell \in L-\{0\}} \frac{1}{\ell^4} \quad \text{and} \quad g_3 = g_3(L) = 140 \sum_{\ell \in L-\{0\}} \frac{1}{\ell^6}.$$

Then $\varphi = y_L^2 - 4x_L^3 + g_2 x_L + g_3$ is holomorphic everywhere. Since these functions factors through the compact space $\mathbb{C}/L$, φ has to be constant, because any non-constant holomorphic function is an open map (the existence of power series expansion and the implicit function theorem). Since x_L and y_L do not have constant terms, we conclude $\varphi = 0$.

We have obtained a holomorphic map $(x_L, y_L) : \mathbb{C}/L - \{0\} \to \mathbf{A}_{/\mathbb{C}}^2$. Looking at the order of poles at $\mathbf{0}$, we know the above map is of degree 1, that is, an isomorphism onto its image and extends to

$$\Phi = (x_L : y_L : 1) = (u^3 x_L : u^3 y_L : u^3) : \mathbb{C}/L \to \mathbf{P}_{/\mathbb{C}}^2.$$

Thus we have an elliptic curve $E_L = \Phi(\mathbb{C}/L) = E(g_2(L), g_3(L))$. We then have

$$\omega_L = \frac{dx_L}{y_L} = du.$$

This shows

Theorem 2.3.5 (Weierstrass). *We have* $[(E, \omega)_{/\mathbb{C}}] \cong Lat$.

We would like to make the space *Lat* a little more explicit. We see easily that $w_1, w_2 \in (\mathbb{C}^\times)^2$ span a lattice if and only if $\mathrm{Im}(w_1/w_2) \neq 0$. Let $\mathfrak{H} = \{z \in \mathbb{C} | \mathrm{Im}(z) > 0\}$. By changing the order of w_1 and w_2 without affecting the lattice, we may assume that $\mathrm{Im}(w_1/w_2) > 0$. Thus we have a natural isomorphism of complex manifolds:

$$\mathcal{B} = \left\{ v = \left(\begin{smallmatrix} w_1 \\ w_2 \end{smallmatrix}\right) \in (\mathbb{C}^\times)^2 \,\Big|\, \mathrm{Im}(w_1/w_2) > 0 \right\} \cong \mathbb{C}^\times \times \mathfrak{H}$$

via $\left(\begin{smallmatrix} w_1 \\ w_2 \end{smallmatrix}\right) \mapsto (w_2, w_1/w_2)$. Since v and v' span the same lattice L if and only if $v' = \alpha v$ for $\alpha \in SL_2(\mathbb{Z})$,

$$Lat \cong SL_2(\mathbb{Z})\backslash\mathcal{B}.$$

This action of $\alpha = \left(\begin{smallmatrix} a & b \\ c & d \end{smallmatrix}\right) \in SL_2(\mathbb{Z})$ on $\mathcal{B}$ can be interpreted on $\mathbb{C}^\times \times \mathfrak{H}$ as follows:

$$\alpha(u, z) = (cu + d, \alpha(z)) \quad \text{for} \quad \alpha(z) = \frac{az + b}{cz + d}.$$

By definition, $\wp(u)$ is an even function. Let $L = \mathbb{Z}w_1 + \mathbb{Z}w_2$ and put $w_3 = w_1 + w_2$. Put $e_j := \wp(\frac{w_j}{2})$. Then $\wp(u) - e_j$ has zero at $\frac{w_j}{2}$. Since $\wp$ is even, the order of zero is even; so, $\wp'(u)$ has zero at $\frac{w_j}{2}$. Therefore, e_j are roots of $4x^3 - g_2 x - g_3$. Comparing the zeros and poles of $\wp'^2$ and $4(\wp - e_1)(\wp - e_2)(\wp - e_3)$, we get

$$\wp'^2 = 4(\wp - e_1)(\wp - e_2)(\wp - e_3), \quad \Delta = 16[(e_1 - e_2)(e_2 - e_3)(e_3 - e_1)]^2. \quad (2.10)$$

Since $E(\mathbb{C})$ is smooth $\Leftrightarrow \Delta \neq 0$, Δ never vanishes over $\mathfrak{H}$.

2.3.5 *Complex modular forms*

We want to write down definitions of modular forms over $\mathbb{C}$. We consider $f \in G_w(\mathbb{C})$. Writing $L(v) = L(w_1, w_2)$ for the lattice spanned by $v \in \mathcal{B}$, we can regard f as a holomorphic function on $\mathcal{B}$ by $f(v) = f(E_{L(v)}, \omega_{L(v)})$. Then the conditions (g0–3) can be interpreted as

(g0) $\qquad\qquad\qquad f \in \mathbb{C}[g_2(v), g_3(v)];$
(g1) $\qquad\qquad\qquad f(\alpha v) = f(v)$ for all $\alpha \in SL_2(\mathbb{Z});$
(g2) $\qquad\qquad\qquad f \in \mathbb{C}[g_2(v), g_3(v), \Delta(v)^{-1}];$
(g3) $\qquad\qquad\qquad f(\lambda v) = \lambda^{-w} f(v) \ (\lambda \in \mathbb{C}^\times).$

We may also regard $f \in G_w(\mathbb{C})$ as a function on $\mathfrak{H}$ by $f(z) = f(v(z))$ for $v(z) = 2\pi i \left(\begin{smallmatrix} z \\ 1 \end{smallmatrix}\right)$ $(z \in \mathfrak{H})$. Here multiplying $\left(\begin{smallmatrix} z \\ 1 \end{smallmatrix}\right)$ by $2\pi i$ is to adjust the rationality coming from q-expansion to the rationality coming from the universal ring $\mathbb{Q}[g_2, g_3]$, as we will see later $(2\pi i)^{-j} g(\left(\begin{smallmatrix} z \\ 1 \end{smallmatrix}\right))$ has Fourier expansion in $\mathbb{Q}[[q]]$ for $q = \exp(2\pi i z)$. Then we have the following interpretation:

(g0) $\qquad\qquad\qquad f \in \mathbb{C}[g_2(z), g_3(z)];$
(g1,3) $\qquad f(\alpha(z)) = f(z)(cz + d)^w$ for all $\alpha = \left(\begin{smallmatrix} a & b \\ c & d \end{smallmatrix}\right) \in SL_2(\mathbb{Z});$
(g2) $\qquad\qquad\qquad f \in \mathbb{C}[g_2(z), g_3(z), \Delta(z)^{-1}].$

Since $\left(\begin{smallmatrix} 1 & 1 \\ 0 & 1 \end{smallmatrix}\right)(z) = z + 1$, any $f \in \mathbb{C}[g_2(z), g_3(z), \Delta^{-1}(z)]$ is translation invariant. Defining $\mathbf{e}(z) := \exp(2\pi i z)$ for $i = \sqrt{-1}$, the function $\mathbf{e} : \mathbb{C} \to \mathbb{C}^\times$ induces an analytic isomorphism: $\mathbb{C}/\mathbb{Z} \cong \mathbb{C}^\times$. Let $q = \mathbf{e}(z)$ be the variable on $\mathbb{C}^\times$. Since f is translation invariant, f can be considered as a function of q. Thus it has a Laurent expansion $f(q) = \sum_{n \gg -\infty} a(n, f) q^n$. We have the following examples (see the following section and [LFE, Chapter 5]):

$$12 g_2 = 1 + 240 \sum_{n=1}^{\infty} \left\{ \sum_{0 < d | n} d^3 \right\} q^n \in \mathbb{Z}[[q]]^\times,$$

$$-6^3 g_3 = 1 - 504 \sum_{n=1}^{\infty} \left\{ \sum_{0 < d | n} d^5 \right\} q^n \in \mathbb{Z}[[q]]^\times, \qquad (2.11)$$

$$\Delta = q \prod_{n=1}^{\infty} (1 - q^n)^{24} \in q(\mathbb{Z}[[q]]^\times).$$

This shows that

$$J = \frac{(12 g_2)^3}{\Delta} = q^{-1} + \cdots \in q^{-1}(1 + \mathbb{Z}[[q]]).$$

In particular, we may regard g_2 and g_3 as elements of $\mathbb{Q}[[q]]$.

We consider a projective plane curve $E_{\infty/\mathbb{Z}[[q]]}$ called the Tate curve defined over the power series ring $\mathbb{Z}[[q]]$ by the equation $Y^2 Z = 4X^3 - g_2(q) X Z^2 - g_3(q) Z^3$ and define $\omega_\infty = \frac{dX}{Y}$. Since Δ is a unit in $\mathbb{Z}[1/6]((q)) := \mathbb{Z}[1/6][[q]][\frac{1}{q}]$, we see that $(E_\infty, \omega_\infty)$ gives an elliptic curve over $\mathbb{Z}[1/6]((q))$ with nowhere vanishing differential ω_∞. For any $f \in G_w(A)$, $f(q) = f((E_\infty, \omega_\infty) \otimes_{\mathbb{Q}((q))} A((q))) \in A[[q]]$ is called the q-*expansion* of f. In particular, if $f \in G_w(\mathbb{C})$, the q-expansion $f(q)$ coincides with the analytic Fourier expansion via $q = \mathbf{e}(z)$, because f is an isobaric polynomial in g_2 and g_3 and by definition $g_2(q)$ and $g_3(q)$ are their analytic expansions.

Write $\mathbf{P}^1(J)_{/\mathbb{Q}}$ for the projective line over $\mathbb{Z}[\frac{1}{6}]$ whose coordinate is given by J (in other words, $\mathbf{P}^1(J) = D_0 \cup D_1$ over local rings with $D_1 = \mathbb{A}^1$ defined by the affine ring $\mathbb{Z}[\frac{1}{6}][J]$). Since the coordinate at ∞ of $\mathbf{P}^1(J)$ can be given by J^{-1} ($J^{-1} \in q(1 + q\mathbb{Z}[[q]])$), we know that $\mathbb{Z}[[q]] = \mathbb{Z}[[J^{-1}]]$ and

$$\widehat{\mathcal{O}}_{\mathbf{P}^1(J),\infty} \cong \mathbb{Z}[1/6][[q]] \text{ via } q\text{-expansion,} \qquad (2.12)$$

where $\widehat{\mathcal{O}}_{\mathbf{P}^1(J),\infty}$ is the (q)-adic completion of the local ring $\mathcal{O}_{\mathbf{P}^1(J),\infty}$ at ∞. Since we have

$$M_1(\mathbb{C}) = Lat/\mathbb{C}^\times = \mathfrak{H} \times \mathbb{C}^\times / (SL_2(\mathbb{Z}) \times \mathbb{C}^\times) \cong SL_2(\mathbb{Z}) \backslash \mathfrak{H},$$

which is isomorphic to $\mathbf{P}^1(J) - \{\infty\}$ by J. Thus we see that (g0) over $\mathbb{C}$ is equivalent to

(g0′) f is a holomorphic function on $\mathfrak{H}$ satisfying the automorphic property (g1,3), and its analytic q-expansion $f(q)$ is contained in $\mathbb{C}[[q]]$.

More generally, for modular forms $f \in G_w(A)$, we can interpret (g0) as

(g0″) $f : \mathcal{P}_{/A} \to \mathbb{A}^1_{/A}$ is a morphism of functors satisfying the automorphic property (g3) in §2.3.1, and its algebraic q-expansion $f(E_\infty, \omega_\infty)$ is contained in $A[[q]]$.

2.3.6   *Weierstrass ζ and σ functions*

Pick a lattice L of $\mathbb{C}$; so, for $E(\mathbb{C}) = \mathbb{C}/L$, $L = \pi_1(E(\mathbb{C}))$, and put $L' := \{\ell \in L | \ell \neq 0\}$. We define Weierstrass σ-function by the following infinite product:

$$\sigma(u) = \sigma(u; L) = u \prod_{\ell \in L'} \left(1 - \frac{u}{\ell}\right) \exp((u/\ell) + \frac{1}{2}(u/\ell)^2). \qquad (\sigma)$$

Plainly $\sigma(\lambda u; \lambda L) = \lambda \sigma(u; L)$ (homogeneous of degree 1). Taking the logarithmic derivative of σ formally, we get Weierstrass ζ-function given by the following infinite sum:

$$\zeta(u) = \zeta_W(u) = \frac{1}{u} + \sum_{\ell \in L'} \left[\frac{1}{u-\ell} + \frac{1}{\ell} + \frac{u}{\ell^2} \right], \qquad (\zeta)$$

which converges absolutely and locally uniformly outside L as the denominator of the term is of degree 2 in ℓ. Therefore σ also converges. By definition,

$$\zeta'(u) = \frac{d\zeta}{du}(u) = -\wp(u) \quad \text{and} \quad \zeta(\lambda u; \lambda l) = \frac{1}{\lambda}\zeta(u; L).$$

Since $\wp$-function is periodic, we find $\zeta(u+\ell) = \zeta(u) + \eta(\ell)$ for a linear map $\eta = \eta_W : L \to \mathbb{C}$. If $L = \mathbb{Z}w_1 + \mathbb{Z}w_2$ with $\text{Im}(w_1/w_2) > 0$, we define $\eta_j := \eta(w_j)$ and extend η to a $\mathbb{R}$-linear map from $\mathbb{C}$ into $\mathbb{C}$ by $\eta(a_1 w_1 + a_2 w_2) := a_1\eta_1 + a_2\eta_2$. Also we can easily check

$$\sigma(-u) = -\sigma(u) \quad \text{and} \quad \zeta(-u) = -\zeta(u).$$

Proposition 2.3.6 (Legendre relation). *We have* $\eta_2 w_1 - \eta_1 w_2 = 2\pi i$.

Proof. Take the fundamental parallelogram P with 4 vertices α, $\alpha + w_j$ and $\alpha + w_1 + w_2$ so that $L \cap P = \{0\}$. Write the path connecting $\alpha, \alpha + w_1$ as γ. Then $\gamma + w_2$ connects $\alpha + w_2$ and $\alpha + w_1 + w_2$. Similarly, writing δ for the path connecting α to $\alpha + w_2$, then $\delta + w_1$ connects $\alpha + w_1$ and $\alpha + w_1 + w_2$. On the one hand, we have

$$\int_{\partial P} \zeta(u)du = \int_\gamma \zeta(u)du - \int_\gamma \zeta(u+w_2)du + \int_\delta \zeta(u+w_1)du - \int_\delta \zeta(u)du$$

$$= \int_\gamma \zeta(u)du - \int_\gamma \zeta(u)du + \eta_2 \int_\gamma du + \int_\delta \zeta(u)du + \eta_1 \int_\delta du - \int_\delta \zeta(u)du$$

$$= \eta_2 w_1 - \eta_1 w_2.$$

On the other hand, $\zeta(u)$ has pole of residue 1 at 0 and no other pole in P; so,

$$\int_{\partial P} \zeta(u)du = 2\pi i \cdot \text{Res}_{u=0}\,\zeta = 2\pi i$$

as desired. $\qquad\qquad\qquad\qquad\qquad\qquad\qquad\qquad\qquad\qquad\qquad\qquad\qquad\square$

Theorem 2.3.7. *For* $a \in \mathbb{C}$ *not in* L, *we have*

$$\wp(u) - \wp(a) = -\frac{\sigma(u+a)\sigma(u-a)}{\sigma^2(u)\sigma^2(a)}.$$

Proof. We may assume that a is in the parallelogram P. The function $\wp(u) - \wp(a)$ has zeros at a and $-a$ (as $\wp$ is an even function) and has a double pole at 0. The product expansion of σ tells us that the same holds for $\frac{\sigma(u+a)\sigma(u-a)}{\sigma^2(u)}$; so,

$$\wp(u) - \wp(a) = C\frac{\sigma(u+a)\sigma(u-a)}{\sigma^2(u)}$$

for a constant C. We see easily

$$\lim_{u\to 0} \sigma^2(u)/u^2 = 1 \quad \text{and} \quad \lim_{u\to 0} u^2\wp(u) = 1.$$

So $C = -1/\sigma^2(a)$. □

Recall $\mathbf{e}(x) = \exp(2\pi i x)$. A theta function for E is an entire function $\theta : \mathbb{C} \to \mathbb{C}$ satisfying the following functional equation:

$$\theta(u + \ell) = \theta(z)\mathbf{e}(l(u, \ell) + c(\ell)), \qquad (\theta)$$

where l is $\mathbb{C}$-linear in u and $\mathbb{R}$-linear in ℓ and a function $c : L \to \mathbb{C}$. There is a trivial theta function $\phi(u) = \exp(au^2 + bu)$ as

$$\phi(u + \ell) = \exp(a(u + \ell)^2 + b(u + \ell)) = \phi(u)\exp(2au\ell + c(\ell))$$

for $c(\ell) := a\ell^2 + b\ell$. Writing $\mathcal{H} := \{f : \mathbb{C} \to \mathbb{C} | f \text{ is holomorphic}\}$, we may view $\mathcal{H}^\times$ as an L-module by $\ell f(u) = f(u + \ell)$. Then plainly $\ell \mapsto \theta(u + \ell)/\theta(u)$ is a 1-cocycle of L with values in $\mathcal{H}^\times$.

Since $E(\mathbb{C}) = \mathbb{C}/L$ is embedded into $\mathbf{P}^2$, it is projective; so, any non-constant function $f \in \mathbb{C}(E)$ has poles and zeros. Therefore it appears to be difficult to write a section s of a line bundle as a function on E, though if we could write s as a function we possibly understand it better. Once we pull it back to its universal covering $\mathbb{C}$, for a simply-connected open set $U \subset \mathbb{C}$ overlaps with $\ell + U$ with $0 \neq \ell \in L$, s is a function s_U on U and $s_{\ell+U}$ on $\ell + U$. Take $U = \mathbb{C}$. Then $C : \ell \mapsto s_{\ell+U}/s_U$ is a cocycle of L with values in the L-module $\mathcal{H}^\times$. Since $C(\ell) \neq 0$, we can write $C(\ell) = \mathbf{e}(\phi(l))$. We can normalize ϕ as in the definition (θ) [TTF, II.2–3]. This is a great idea of Jacobi (and Riemann) to make sections of a line bundle visible to us as a holomorphic function on the universal cover. This way of studying a section of a line bundle via cocycle appears repeatedly after Jacobi; e.g., in the definition of modular forms (g1,3) in §2.3.5 as $\mathfrak{H}$ is simply connected on which $\mathrm{SL}_2(\mathbb{Z})$ acts (here $\mathrm{SL}_2(\mathbb{Z}) \ni \left(\begin{smallmatrix} a & b \\ c & d \end{smallmatrix}\right) \mapsto j(\gamma, z)^w = (cz + d)^w$ is a 1-cocycle). See [PAF, §1.1] for a philosophical background on the series of ideas started by Jacobi.

Theorem 2.3.8. *The σ-function is a theta function; i.e.,*

$$\sigma(u + \ell) = \psi(\ell)\exp(\eta(\ell)(u + \ell/2))\sigma(u),$$

where $\psi(\ell) = \begin{cases} 1 & \text{if } \ell \in 2L, \\ -1 & \text{if } \ell \notin 2L. \end{cases}$

Proof. We have

$$\frac{d}{du}\log\frac{\sigma(u+\ell)}{\sigma(u)} = \eta(\ell)$$

by definition. By the fundamental theorem of Calculus, we get

$$\log\frac{\sigma(u+\ell)}{\sigma(u)} = \eta(\ell)u + c(\ell).$$

Exponentiating, we get

$$\sigma(u + \ell) = \sigma(u)\exp(\eta(\ell)u + c(\ell)).$$

Here $c \bmod (2\pi i)$ is well defined.

Define $\psi(\ell) := \exp(\eta(\ell)u + c(\ell))/\exp(\eta(\ell)(u+\ell/2))$, and we compute ψ. Suppose $\ell \notin 2L$. Set $u = -\ell/2$. Then $\sigma(\ell/2) = \psi(\ell)\sigma(-\ell/2)$; so, $\psi(\ell) = -1$ as σ is an odd function and $\sigma(\ell/2) \neq 0$ by $\ell \notin 2L$.

We see

$$\psi(2\ell)\exp(\eta(2\ell)(u + \ell)) = \frac{\sigma(u+2\ell)}{\sigma(u)} = \frac{\sigma(u+2\ell)}{\sigma(u+\ell)}\frac{\sigma(u+\ell)}{\sigma(u)}$$

$$= \psi(\ell)^2\exp(\eta(\ell)(u + \frac{3}{2}\ell) + \eta(\ell)(u + \frac{1}{2}\ell)).$$

In other words, we have $\psi(\ell) = \psi(\ell/2)^2$. Iterating this to reach $\ell/2^n \notin L$ first time, we get $\psi(\ell) = (-1)^{2n} = 1$. $\qquad\square$

2.3.7 *Product q-expansion*

Hereafter we take $L = \mathbb{Z} + \mathbb{Z}z$ for $z \in \mathfrak{H}$. We then write $(u; L)$ as (u, z). Define $\varphi(z) = \varphi(u, z) := \exp(-\frac{1}{2}\eta_2 u^2)q_u^{1/2}\sigma(u; z)$, where $\eta_2 = \eta(1)$ for $\eta = \eta_W : L \to \mathbb{C}$ given by $\eta(\ell) = \zeta(u+\ell) - \zeta(u)$ and $q_u = \exp(2\pi i u)$. Then we see

$$\varphi(u + 1) = \varphi(u) \text{ and } \varphi(u + z) = -\frac{1}{q_u}\varphi(u). \tag{2.13}$$

By Theorem 2.3.8, we find

$$\varphi(u + 1) = \exp(-\frac{1}{2}\eta_2(u+1)^2 + \pi i(u+1))\sigma(u+1)$$

$$= -\exp(-\frac{1}{2}\eta_2(u+1)^2 + \pi i(u+1) + \eta_2(u + \frac{1}{2}))\sigma(u) = \varphi(u).$$

Similarly for $\eta_1 = \eta(z)$

$$\varphi(u+z) = \exp(-\frac{1}{2}\eta_2(u+z)^2) + \pi i(u+z))\sigma(u+\tau)$$

$$= -\exp(-\frac{1}{2}\eta_2(u+z)^2) + \pi i(u+z) + \eta_1(u+z/2))\sigma(u).$$

By Legendre's relation: $\eta_2 z - \eta_1 = 2\pi i$, we can eliminate η_1 and get (2.13). Indeed, we have

$$-\frac{1}{2}\eta_2(u+z)^2) + \pi i(u+z) + \eta_1(u+z/2)$$

$$= -\frac{1}{2}\eta_2 u^2 - \eta_2 uz - \frac{1}{2}\eta_2 z^2 + \pi i(u+z) + (\eta_2 z - 2\pi i)u + \frac{(\eta_2 z - 2\pi i)z}{2}$$

$$= -\frac{1}{2}\eta_2 u^2 + \pi i(u+z) - 2\pi iu - \pi iz = -\frac{1}{2}\eta_2 u^2 + \pi iu - 2\pi iu$$

as desired.

Theorem 2.3.9. *Let $q_w = \exp(2\pi i w)$ for $w = u, z$. Then we have*

$$\sigma(u,z) = (2\pi i)^{-1}\exp(\frac{1}{2}\eta_2 z^2)(q_u^{1/2} - q_u^{-1/2})\prod_{n=1}^{\infty}\frac{(1-q_z^n q_u)(1-q_z^n/q_u)}{(1-q_z^n)^2}.$$

Proof. We prove the equivalent form of the product formula of $\varphi(u,z) = g(u)$:

$$g(u) := (2\pi i)^{-1}(q_u - 1)\prod_{n=1}^{\infty}\frac{(1-q_z^n q_u)(1-q_z^n/q_u)}{(1-q_z^n)^2}.$$

By definition, it is easy to see $g(u+1) = g(u)$ and $g(u+z) = -\frac{1}{q_u}g(u)$ and that g has exactly the same zeros of order 1 at $u = 0$ as σ (and hence as φ). Thus φ/g is an entire function on $E(\mathbb{C}) = \mathbb{C}/L$, and hence $\varphi = C \cdot g$ for a constant C. Then we compute C by $C = \lim_{u \to 0}\varphi(u)/g(u) = 1$. $\square$

Corollary 2.3.10. *We have*

$$\Delta(z) = (2\pi i)^{12}q_z\prod_{n=1}^{\infty}(1-q_z^n)^{24}.$$

This can be proven by the above theorem because

$$\Delta = 16[(e_1 - e_2)(e_2 - e_3)(e_3 - e_1)]^2$$

by (2.10) with $e_j = \wp(\frac{w_j}{2})$ and invoking Theorem 2.3.7

$$e_i - e_j = \wp(w_i/2) - \wp(w_j/2) = -\frac{\sigma((w_i + w_j)/2)\sigma((w_i - w_j)/2)}{\sigma^2(w_i/2)\sigma^2(w_j/2)}.$$

Define the Dedekind η-function (as a 24-th root of Δ) by

$$\eta(z) = \eta_D(z) = q_z^{1/24}\prod_{n=1}^{\infty}(1-q_z^n). \tag{η}$$

Theorem 2.3.11 (Dedekind). *We have* $\eta_D(z + 1) = \eta_D(z)$ *and* $\eta_D(-1/z) = \sqrt{-iz}\,\eta_D(z)$, *where the square root* $\sqrt{-iz}$ *has positive value on the imaginary axis in* $\mathfrak{H}$.

Proof. The first formula follows from the definition. Since Δ is on $\mathrm{SL}_2(\mathbb{Z})$ of weight 12, $\frac{\eta_D(-1/z)}{\sqrt{z}\eta_D(z)}$ is holomorphic on $\mathfrak{H}$ with $\left|\frac{\eta_D(-1/z)}{\sqrt{z}\eta_D(z)}\right| = 1$. By Maximum modulus principle, it must be a constant C. Putting $z = i$, we have $1 = C\sqrt{i}$; so, $C = \sqrt{-i}$. □

2.3.8 Klein forms

Let $L = L(v) = \mathbb{Z}w_1 + \mathbb{Z}w_2$ for $v = \binom{w_1}{w_2}$. Then for any of Weierstrass functions $f(u; L) = \sigma(u; L), \zeta(u; L)$ and $\wp(u; L)$, if one cuts it down to one variable function on Lat at an order N torsion point $\frac{a}{N}w_1 + \frac{b}{N}w_2$ in $E(\mathbb{C}) = \mathbb{C}/L$, we expect from the construction of the moduli $Y_1(N)$ in (2.7) that $f(\frac{a}{N}w_1 + \frac{b}{N}w_2; L)$ would be a modular form of level close to N. If one wants a unit in $\mathcal{R}_{1,N}$, the choice is to use $\sigma(u; L)$ as the σ-function having a product expansion does not vanish on Lat. This was proposed by Klein with an astute modification of the cut-value by the Legendre period relation. We now state the definition of the so-called Klein form which is a modular form of weight negative -1 made out of the σ-function through this heuristic. Let $\eta = \eta_W$ be Weierstrass eta function given by $\eta_W(\ell; L) = \zeta(u + \ell; L) - \zeta(\ell; L)$ for $\ell \in L \in Lat$ (not the Dedekind eta function), and extend $\eta_W : L \to \mathbb{C}$ linearly to the $\mathbb{Q}$ span $\mathbb{Q} \cdot L$.

Definition 2.3.12. For $a = (a_1, a_2) \in \mathbb{Q}^2 - \mathbb{Z}^2$ (a row vector), define $\mathfrak{k}_a(v) = \mathfrak{k}_a(L(v)) := \exp(-\eta_W(a \cdot v; L(v))a \cdot v/2)\sigma(a \cdot v; L(v))$, where $a \cdot v = a_1 w_1 + a_2 w_2$ (matrix product).

Since $\sigma(\lambda u, \lambda L) = \lambda\sigma(u, L)$ (homogeneous of weight -1) and $\zeta(\lambda u, \lambda L) = \lambda^{-1}\zeta(u, L)$ (homogeneous of weight 1), we have

$$\mathfrak{k}_a(\lambda v) = \lambda\mathfrak{k}_a(v). \tag{2.14}$$

Theorem 2.3.13. *Let* $a \in \frac{1}{N}\mathbb{Z}^2$ *but* $a \notin \mathbb{Z}^2$. *Then* $\mathfrak{k}_a$ *is a meromorphic modular form of weight* -1 *of level* $\Gamma(2N^2)$ *(whose poles and zeros are limited to cusps), and* $\mathfrak{k}_a^{2N}$ *is of level* $\Gamma(N)$ *and if* N *is odd,* $\mathfrak{k}_a^N$ *is on* $\Gamma(N)$.

Proof. Since $\mathfrak{k}_a(v) = \mathfrak{k}_a(L(v)) := \exp(-\eta_W(a \cdot v; L(v))a \cdot v/2)\sigma(a \cdot v; L(v))$, for $\alpha \in \mathrm{SL}_2(\mathbb{Z})$, we have

$$\mathfrak{k}_{a\alpha}(v) = \exp(-\eta_W(a\alpha \cdot v; L(v))a\alpha \cdot v/2)\sigma(a\alpha \cdot v; L(v))$$
$$= \exp(-\eta_W(a \cdot \alpha v; L(v))a \cdot \alpha v/2)\sigma(a \cdot \alpha v; L(v)).$$

In short,

$$\mathfrak{k}_{a\alpha}(v) = \mathfrak{k}_a(\alpha v). \tag{2.15}$$

We are going to show for $b = (b_1, b_2) \in \mathbb{Z}^2$

$$\mathfrak{k}_{a+b}(v) = \varepsilon(a,b)\mathfrak{k}_a(v) \text{ for } \varepsilon(a,b) = (-1)^{b_1 b_2 + b_1 + b_2}\mathbf{e}((a_1 b_2 - a_2 b_1)/2), \tag{2.16}$$

where $\mathbf{e}(z) = \exp(2\pi i z)$. By (2.15) combined with (2.16), as long as $\varepsilon(a, m\mathbb{Z}^2)^j = 1$ for $0 < m \in \mathbb{Z}$, we find that $\mathfrak{k}_a^j$ is on $\Gamma(mN)$ as $\Gamma(mN)$ induces an identity on $(N^{-1}\mathbb{Z}/m\mathbb{Z})^2$.

By Theorem 2.3.8, we know

$$\sigma(u + \ell) = \psi(\ell)\exp(\eta(\ell)(u + \ell/2))\sigma(u)$$

with $\psi(\ell) = \begin{cases} 1 & \text{if } \ell \in 2L, \\ -1 & \text{if } \ell \notin 2L. \end{cases}$

We can then write $\psi(b_1 w_1 + b_2 w_2) = (-1)^{b_1 b_2 + b_1 + b_2}$ for $b = (b_1, b_2) \in \mathbb{Z}^2$. Take $\ell = b \cdot v \in L(v)$. Then

$$\mathfrak{k}_{a+\ell}(v) = \exp(-\eta((a+b) \cdot v)(a+b) \cdot v/2)\sigma((a+\ell) \cdot v)$$
$$= \psi(\ell)\exp(-\eta((a+b) \cdot v)(a+b) \cdot v/2)\exp(\eta(b \cdot v)(a \cdot v + b \cdot v/2))\sigma(a \cdot v)$$
$$= \psi(\ell)\exp(-\eta((a+b) \cdot v)(a+b) \cdot v/2 + \eta(b \cdot v)(2a+b) \cdot v/2 + \eta(a \cdot v)a \cdot v/2)\mathfrak{k}_a(v).$$

The inside of the exponential function is

$$-\eta((a+b) \cdot v)(a+b) \cdot v/2 + \eta(b \cdot v)(2a+b) \cdot v/2 + \eta(a \cdot v)a \cdot v/2$$
$$= \frac{1}{2}(-a_2 b_1(\eta_2 w_1 - \eta_1 w_2) + b_2 a_1(\eta_2 w_1 - \eta_1 w_2)) \overset{(*)}{=} \pi i(a_1 b_2 - a_2 b_1).$$

Here the identity at $(*)$ is by Legendre's relation $\eta_2 w_1 - \eta_1 w_2 = 2\pi i$. Thus (2.16) follows.

Thus if $m = N^2$, we have $\mathfrak{k}_{a+\ell}(v) = \psi(\ell)\mathfrak{k}_a(v)$ and hence, if $m = 2N^2$, $\mathfrak{k}_a$ depends only on $a \in N^{-1}\mathbb{Z}^2/2N\mathbb{Z}^2$ as desired. Since $\varepsilon(a,b)$ is a $2N$-th root of unity, $\mathfrak{k}_a^{2N}$ only depends on $a \in (N^{-1}\mathbb{Z}/\mathbb{Z})^2$; so, it is on $\Gamma(N)$ in the stabilizer of $a \in (N^{-1}\mathbb{Z}/\mathbb{Z})^2$.

To deal with $\ell_{a'}^N$ for odd N, we need to be more careful. Let $\alpha = \begin{pmatrix} a & b \\ c & d \end{pmatrix} \in \Gamma(N)$. Write $a' = (r, s)/N$. Then $a'\alpha = (\frac{r}{N} + (\frac{a-1}{N}r + \frac{c}{N}s), \frac{s}{N} + (\frac{b}{N}b + \frac{d-1}{N}s))$. Therefore

$$\ell_{a'}(\alpha v) = \ell_{a'\alpha}(v) = \ell_{a'+b'}(v) = \varepsilon_{a'}(\alpha)\ell_{a'}(v), \qquad (2.17)$$

where from (2.16), for $Na' = (r, s)$,

$$\varepsilon_{a'}(\alpha) = -(-1)^{(\frac{a-1}{N}r + \frac{c}{N}s+1)(\frac{b}{N}r + \frac{d-1}{N}s+1)} \mathbf{e}((br^2 + (d-a)rs - cs^2)/2N^2)$$

because $b' = (\frac{a-1}{N}r + \frac{c}{N}s, \frac{b}{N}b + \frac{d-1}{N}s)$.

We now need to show

$$\varepsilon_{a'}(\alpha)^N = 1 \quad \text{if } N \text{ is odd.} \qquad (2.18)$$

Equivalently, as $N \equiv 1 \mod 2$,

$$((a-1)r + cs + 1)(br + (d-1)s + 1) + (br^2 + (d-a)rs - cs^2) \equiv 1 \mod 2. \qquad (2.19)$$

If $r, s \in 2\mathbb{Z}$, it is plain. Suppose $(r, s) \equiv (1, 0) \mod 2$. Then (2.19) is equivalent to

$$a(b + 1) + b \equiv 1 \mod 2.$$

If b is odd, this is plain. By $ad - bc = 1$, if b is even, then a is odd; so, the result follows. In the same way, we can settle the case where $(r, s) \equiv (0, 1) \mod 2$.

Now suppose $(r, s) \equiv (1, 1) \mod 2$. Then (2.19) is equivalent to

$$(a + c)(b + d) + a + b + c + d \equiv 1 \mod 2.$$

Again if b is even, then a and d are odd, and hence

$$(a + c)(b + d) + a + b + c + d \equiv (1 + c) + (1 + c) + 1 \equiv 1 \mod 2.$$

We can settle the case where c is even in the same way. If $bc \equiv 1 \mod 2$, then a or d is even. Supposing that a is even, again we have

$$(a + c)(b + d) + a + b + c + d \equiv 1 + d + 1 + 1 + d \equiv 1 \mod 2.$$

We settle the case where d is even in the same way. $\qquad \square$

Let $p \geq 5$ be a prime and put $N = p^m$ for $m > 0$. For $m : (N^{-1}\mathbb{Z}/\mathbb{Z})^2 - \{(0,0)\} \to \mathbb{Z}$, we say m satisfies (Q_N) if the following sum over $0 \neq a \in (N^{-1}\mathbb{Z}/\mathbb{Z})^2$ vanishes modulo N (the so-called quadratic relation modulo N):

$$\sum_{a \neq 0} m(a)a_1^2 \equiv \sum_{a \neq 0} m(a)a_2^2 \equiv \sum_{a \neq 0} m(a)a_1 a_2 \equiv 0 \mod N^{-1}\mathbb{Z}, \qquad (Q_N)$$

which is equivalent to, writing $a_1 = \frac{r}{N}$, $a_2 = \frac{s}{N}$ and $m(a) = m(r, s)$,

$$\sum_{(r,s)\neq 0} m(r,s)r^2 \equiv \sum_{(r,s)\neq 0} m(r,s)s^2 \equiv \sum_{(r,s)\neq 0} m(r,s)rs \equiv 0 \mod N. \quad (Q'_N)$$

Remark 2.3.14. If $m(a) = 1$ for all a, then

$$\sum_a m(a)a_1^2 = N^{-2} \sum_{r=0}^{N-1}\sum_{s=0}^{N-1} r^2 = N^{-1}\sum_{r=0}^{N-1} r^2 = \frac{(N-1)(2N-1)}{6} \in \mathbb{Z}.$$

Similarly

$$\sum_a m(a)a_1 a_2 = N^{-2}\sum_{r=0}^{N-1}\sum_{s=0}^{N-1} rs = N^{-2}(\sum_{r=0}^{N-1} r)^2 = N^{-2}\frac{N^2(N-1)^2}{4} \in \mathbb{Z}.$$

Thus a constant function m satisfies the quadratic relation (Q_N).

2.3.9 *Kubert's theorem*

We would like to prove

Theorem 2.3.15 (Kubert). *Let m be as above. Then $\mathfrak{k}^m = \prod_{a\neq 0}\mathfrak{k}_a^{m(a)}$ is on $\Gamma(N)$ if and only if m satisfies (Q_N).*

We give a sketch of the proof given by Kubert.

Proof. Recall $\mathbf{e}(z) = \exp(2\pi i z)$ and define $\varepsilon_a(\alpha)$ by $\mathfrak{k}_{a\alpha}(v) = \varepsilon_a(\alpha)\mathfrak{k}_a$ for $\alpha = \left(\begin{smallmatrix} a & b \\ c & d \end{smallmatrix}\right) \in \Gamma(N)$. By (2.16), we have $\varepsilon(a,b) = (-1)^{b_1 b_2 + b_1 + b_2}\exp(\pi i(a_1 b_2 - a_2 b_1))$. Replacing b by $(\frac{a-1}{N}r + \frac{c}{N}s, \frac{b}{N}b + \frac{d-1}{N}s)$ as in (2.17) and writing $a = (\frac{r}{N}, \frac{s}{N})$,

$$\varepsilon_a(\alpha) = -(-1)^{(\frac{a-1}{N}r + \frac{c}{N}s + 1)(\frac{b}{N}r + \frac{d-1}{N}s + 1)}\mathbf{e}((br^2 + (d-a)rs - cs^2)/2N^2).$$

Then we need to compute $\varepsilon^m(\alpha) = \prod_a \varepsilon_a(\alpha)^{m(a)}$. Define, writing $m(r,s) = m(\frac{r}{N}, \frac{s}{N})$,[1]

$$E(r,s;a,b,c,d) := m(r,s)\Big[\frac{ab}{N^2}r^2 + \frac{c(d-1)-c}{N^2}s^2 + \Big(\frac{b}{N} + \frac{a-1}{N}\Big)r$$
$$+ \Big(\frac{d-1}{N} + \frac{c}{N}\Big)s + \Big(\frac{bc}{N^2} + \frac{(a-1)(d-1)}{N^2}\Big)rs + \frac{d-a}{N^2}rs\Big]. \quad (2.20)$$

Then we have

$$\varepsilon^m(\alpha) = \exp(\pi i \sum_{(r,s)\in Z} E(r,s;a,b,c,d)),$$

---

[1]In the definition of the formula (2.20) in [MUN, page 69], the term $\frac{d-a}{N^2}rs$ is incorrectly written as $\frac{d-a}{N}rs$.

where $Z := \{(r, s) \in \mathbb{Z}^2 \cap [0, N)^2 | (r, s) \neq (0, 0)\}$. From this, taking $\alpha = \left(\begin{smallmatrix} 1 & N \\ 0 & 1 \end{smallmatrix}\right)$, we have $E(r, s; 1, N, 0, 1) = m(r, s)(\frac{1}{N}r^2 + r)$ and hence

$$\sum_{(r,s)} m(r, s)(rN + r^2) \equiv 0 \mod 2N.$$

Similarly, taking $\alpha = \left(\begin{smallmatrix} 1 & 0 \\ N & 1 \end{smallmatrix}\right)$, we conclude

$$\sum_{(r,s)} m(r, s)(sN + s^2) \equiv 0 \mod 2N.$$

However from (2.18), we can replace the modulus $2N$ by N. This implies the first two identities of (Q_N).

As for the third term involving rs, again we only need to compute modulo N in place of $2N$. We take $\alpha = \left(\begin{smallmatrix} 1-N & N \\ -N & 1+N \end{smallmatrix}\right)$. Since we know the first two identities of (Q_N), we can ignore the square terms in r, s and also terms only having denominator N of (2.20). Then

$$E(r, s; 1-N, N, -N, 1+N) \equiv m(r, s)\left(\frac{bc}{N^2} + \frac{(a-1)(d-1)}{N^2} + \frac{d-a}{N^2}\right)rs$$

$$\equiv m(r, s)\frac{2}{N}rs \mod \mathbb{Z}.$$

Thus the last identity of (Q_N) holds if $\mathfrak{t}^m$ is on $\Gamma(N)$.

Conversely, we already know from (2.18) that $\mathfrak{t}_a$ is on $\Gamma(N^2)$ if $N = p^m$ with $p \geq 5$. Writing $\gamma \in \Gamma(N)$ as $\gamma = 1 + N\gamma_1$ and sending γ_1 to $(\gamma_1 \mod N) \in M_2(\mathbb{Z}/N\mathbb{Z})$, we can easily verify that $i : \Gamma(N)/\Gamma(N^2) \cong \mathfrak{sl}_2(\mathbb{Z}/N\mathbb{Z}) = \{\gamma_1 \in M_2(\mathbb{Z}/N\mathbb{Z}) | \text{Tr}(\gamma_1) = 0\}$. This is an isomorphism of groups regarding $\mathfrak{sl}_2(\mathbb{Z}/N\mathbb{Z})$ as an additive group. Then it is easy to see that $i\left(\begin{smallmatrix} 1 & N \\ 0 & 1 \end{smallmatrix}\right)$, $i\left(\begin{smallmatrix} 1 & 0 \\ N & 1 \end{smallmatrix}\right)$ and $i\left(\begin{smallmatrix} 1-N & N \\ -N & 1+N \end{smallmatrix}\right)$ generate the right-hand-side. Therefore the relation (Q_N) is sufficient for $\mathfrak{t}^m$ is on $\Gamma(N)$. $\square$

2.4   Modular Stickelberger theory

The units $\mathcal{A}_N^\times$ and $\mathcal{A}_{1,N}^\times$ of affine ring of the modular curves $Y_1(N)$ and $Y(N)$ are called *modular units*. They are holomorphic and do not vanish on $\mathfrak{H}$, and indeed often they have product expansion similar to the Δ-function. The compactification $\mathbf{P}^1(J)$ of $\mathcal{A}_1 = \mathbb{Q}[J]$ has one cusp ∞. Since $\mathcal{A} = \mathcal{A}_{1,N}$ and $\mathcal{A}_N$ are finite flat over $Y(1) = \text{Spec}(\mathcal{A}_1)$, the normalization of $\mathbf{P}^1(J)$ in $\mathcal{A}$ gives a projective curve X finite flat over $\mathbf{P}^1(J)$. The curve X for $\mathcal{A}_{1,N}$ (resp. $\mathcal{A}_{1,N}^{\Gamma_0(N)}$, $\mathcal{A}_N$) is denoted by $X_1(N)$ (resp. $X_0(N)$, $X(N)$).

Points above ∞ are called cusps of X. The cuspidal divisor group of X is defined to be

$$Cl_X := \frac{\{D \in \bigoplus_{P:cusps} \mathbb{Z}[P] | \deg(D) = 0\}}{\{\operatorname{div}(u) | u: \text{ modular units}\}}.$$

We study this cuspidal divisor group and prove its finiteness in this section and a formula for its order in the following section. Moreover we see it is a cyclic module over a suitable group algebra.

2.4.1 *Siegel units*

We now introduce a typical modular unit called a Siegel unit. Modular units on an open modular curve often extend to the compactified curve as a rational function with possible poles at cusps. Siegel units have order at the infinity cusp given essentially by the rational values of the second Bernoulli polynomial $B_2(x) = x^2 - x + \frac{1}{6}$ (so, the values of Hurwitz zeta functions at $s = -1$).

Definition 2.4.1. We define the Siegel unit $g_a(z)$ by

$$g_a(z) = \mathfrak{k}_a(z)\Delta(z)^{1/12} \quad \text{for } a = (a_1, a_2) \in \mathbb{Q}^2 \cap [0,1)^2,$$

where $\Delta^{1/12}(z) = 2\pi i q^{1/12} \prod_{n=1}^{\infty}(1 - q^n)^2$ for $q = \exp(2\pi i z)$. For a general $a \in \mathbb{Q}^2$, taking the fraction part $\langle a \rangle := (\langle a_1 \rangle, \langle a_2 \rangle)$ we define $g_a := g_{\langle a \rangle}$. Here $x - \langle x \rangle \in \mathbb{Z}$ with $\langle x \rangle \in [0,1)$ for $x \in \mathbb{R}$.

Note that $\Delta^{1/12}(z)$ is the square of the Dedekind η-function η_D up to a constant.

Proposition 2.4.2. *We have*

$$g_a(z) = -q^{B_2(a_1)/2}\mathbf{e}(a_2(a_1 - 1)/2)(1 - \mathbf{e}(a_2)q^{a_1})$$

$$\times \prod_{n=1}^{\infty}[(1 - \mathbf{e}(a_2)q^{n+a_1})(1 - \mathbf{e}(-a_2)q^{n-a_1})], \quad (2.21)$$

where $q = \exp(2\pi i z)$.

Even if $Na = N(a_1, a_2) \in \mathbb{Z}^2$ for $0 < N \in \mathbb{Z}$, the term $q^{B_2(a_1)/2}$ only can have the exponent with denominator N^2. This is an important feature of Siegel units.

Proof. This is just a computation out of the product expansion of the σ-function (combined with definition of $\mathfrak{k}_a$) and $\Delta^{1/12}$. An important point later is the leading term $q^{B_2(a_1)/2}$; so, we describe the computation for the leading term. Here is the contribution of each of σ, $\mathfrak{k}_a$ and $\Delta^{-1/12}$ to the leading term:

(σ) $\exp(\frac{1}{2}\eta_2(a_1z + a_2)^2)$;
($\mathfrak{k}$) $\exp(-(\eta_1 a_1 + \eta_2 a_2)(a_1 z + a_2)/2)$;
(Δ) $q^{1/12}$.

The product of the terms (σ), ($\mathfrak{k}$) and (Δ) has inside exp the following:

$$\frac{1}{2}\eta_2(a_1z + a_2)^2 - (\eta_1 a_1 + \eta_2 a_2)(a_1 z + a_2)/2 + \frac{2\pi i z}{12}$$

$$= \frac{1}{2}(\eta_1 a_1^2 z - \eta_1 a_1 a_2 - \eta_2 a_2 a_1 z - \eta_2 a_2^2 + \eta_2 a_1^2 z^2 + 2\eta_2 a_1 a_2 z + \eta_2 a_2^2) + \frac{2\pi i z}{12}$$

$$= \frac{1}{2}a_1^2(-\eta_1 + \eta_2 z) + \frac{a_1 a_2}{2}(-\eta_1 + \eta_2 z) + \frac{2\pi i z}{12} \overset{(*)}{=} 2\pi i(\frac{a_1^2}{2} + \frac{a_1 a_2}{2}) + \frac{2\pi i z}{12}$$

$$= 2\pi i(B_2(a_1)/2 + a_2(a_1 - 1)/2).$$

The identity $(*)$ follows from Legendre's relation: $\eta_2 z - \eta_1 = 2\pi i$, as we brought the modulus (w_1, w_2) of Proposition 2.3.6 into the inhomogeneous form $(z, 1)$.   □

By the analysis of Klein forms and the fact that $\Delta \in S_{12}(\mathrm{SL}_2(\mathbb{Z}))$, we get

Theorem 2.4.3. *Suppose that* $0 \neq a \in (N^{-1}\mathbb{Z}/\mathbb{Z})^2$. *Then* g_a^{12N} *is a modular function on* $\Gamma(N)$, *and it does not have poles and zeros on* $\mathfrak{H}$. *Further for* $\alpha \in \mathrm{SL}_2(\mathbb{Z})$, *we have* $g_{a\alpha}^{12N}(z) = g_a^{12N}(\alpha(z))$; *therefore,* $g_{a\alpha}(z) = \zeta_{a,\alpha}g_a(\alpha(z))$ *for* $\zeta_{a,\alpha} \in \mu_{12N}(\overline{\mathbb{Q}})$.

First g_a^{12N} is a modular function of level N by Theorem 2.3.13 as $\Delta^{1/12}$ is of weight 1 while $\mathfrak{k}_a$ is of weight -1. Since $(g_{a\alpha}/g \circ \alpha)^{12N}$ has order 0 at all cusps, it is holomorphic everywhere over $X(N)(\mathbb{C})$ with value 1 at ∞; so, it is a non-zero constant. Thus the ratio $g_{a\alpha}/g \circ \alpha$ is a $12N$-th root of unity.

Remark 2.4.4. We see that g_a^{12N} generate the function field $\mathbb{Q}(Y(N))$ over $\mathbb{Q}$ as it is generated by $\Delta^{-1/12}\wp(a \cdot v; L(v))$ for a running over $(N^{-1}\mathbb{Z}/\mathbb{Z})^2$ (essentially logarithmic derivative of g_a) basically by definition. Since its q-expansion involves N-th roots of unity, $\mathbb{Q}(Y(N))$ becomes reducible over $\mathbb{Q}[\mu_N]$ (see [IAT, Chapter 6] to see Shimura's proof of the fact that each geometrically irreducible component of $Y(N)$ is defined over $\mathbb{Q}[\mu_N]$).

Proposition 2.4.5. *We have* $\prod_{Na=0, a\neq 0} g_a^{12N} \in \mathbb{Q}^\times$, *where a runs over* $N^{-1}\mathbb{Z}^2 \cap [0,1)^2$.

Proof. Since the product $\varepsilon := \prod_{Na=0, a\neq 0} g_a^{12N}$ is invariant under $\mathrm{Gal}(X(N)/\mathbf{P}^1(J)) = \mathrm{Gal}(\mathbb{Q}(Y(N))/\mathbb{Q}(J))$, we find that $\varepsilon \in \mathbb{Q}[J]^\times$; so, ε is a constant in $\mathbb{Q}^\times$. $\qquad\square$

Remark 2.4.6. It is known that $\prod_{Na=0, a\neq 0} g_a^{12N} = N^{12N}$ (see [MUN, 2.4, p. 45]).

Theorem 2.4.7. *Let* $Z_N := \{a \in N^{-1}\mathbb{Z}^2 \cap [0,1)^2 | a \neq (0,0)\}$ *for* $N = p^m$ *(with a prime* $p \geq 5$*), and put* $g^m := \prod_{a\in Z_N} g_a^{m(a)}$ *for a function* $m : Z_N \to \mathbb{Z}$*. Then* $g^m \in \mathcal{A}_N^\times$ *if and only if* $\sum_{a\in Z_N} m(a) \equiv 0 \mod 12$ *and* (Q_N) *is satisfied.*

Proof. By Theorem 2.3.15, $\mathfrak{k}^m$ is on $\Gamma(N)$ if and only if (Q_N) is satisfied. Let $M := \sum_{a\in Z_N} m(a)$. Thus we need to show $\Delta^{M/12}$ is on $\Gamma(N)$ if and only if $M \equiv 0 \mod 12$. Since $p \geq 5$, we need to prove that $f := \Delta^{1/2}$ is on $\Gamma(6)$ but not on $\mathrm{SL}_2(\mathbb{Z})$ (as $\dim S_6(\mathrm{SL}_2(\mathbb{Z})) = 0$). For $\alpha = \left(\begin{smallmatrix} a & b \\ c & d \end{smallmatrix}\right) \in \mathrm{SL}_2(\mathbb{Z})$, we have $f|\alpha(z) = f(\alpha(z))(cz+d)^{-6} = \chi(\alpha)f(z)$. Indeed, by q-expansion, we have $\chi\left(\begin{smallmatrix} 1 & 1 \\ 0 & 1 \end{smallmatrix}\right) = -1$ and by Theorem 2.3.11, $\chi\left(\begin{smallmatrix} 0 & 1 \\ -1 & 0 \end{smallmatrix}\right) = -1$. Since Δ is on $\mathrm{SL}_2(\mathbb{Z})$ and the above two elements generate $\mathrm{SL}_2(\mathbb{Z})$, χ is a character of $\mathrm{SL}_2(\mathbb{Z})$ of order 2. By the strong approximation theorem (e.g., [LFE, §6.1]), the abelian quotient of $\mathrm{SL}_2(\mathbb{Z})$ is isomorphic to the abelian quotient of $\mathrm{SL}_2(\mathbb{Z}/6\mathbb{Z})$. Therefore f is on $\Gamma(6)$ as desired. $\qquad\square$

Actually, for $\eta = \eta_D$, we know $\eta^2 = \Delta^{1/12}$ is on $\Gamma(12)$, η^4 is on $\Gamma(3)$, $\Delta^{1/2}$ is on $\Gamma(2)$ (see [MUN, Lemma 3.5.1]).

2.4.2 *Distribution on p-divisible groups*

We seek to prove that $\mathcal{A}_N^\times$ for the affine ring $\mathcal{A}_N$ of $X = Y(N)$ is generated by Siegel units (up to constants). Since $J(C)$ is the quotient of the degree 0 divisor group by linear equivalence, the cuspidal divisor group is the quotient of the degree 0 divisor group generated by cusps by the subgroup of divisors of rational functions whose poles and zeros are limited to cusps. Such functions do not have zero on $\mathfrak{H}$ (or equivalently on $Y(N)$); i.e, modular units in $\mathcal{A}_N^\times$. Thus this is equivalent to showing

$$Cl_X := \frac{\{D \in \bigoplus_{P:cusps} \mathbb{Z}P | \deg(D) = 0\}}{\langle \mathrm{div}(g_a) | a \in (N^{-1}\mathbb{Z}/\mathbb{Z})^2\rangle},$$

where the denominator is the $\mathbb{Z}$-span of $\mathrm{div}(g_a)$.

Hereafter we assume that $N = p^m$ for a prime $p \geq 5$ (in [MUN], $p \geq 5$ is often assumed). To show linear independence of $\{\mathrm{div}(g_a)\}_{a\in(N^{-1}\mathbb{Z}/\mathbb{Z})^2}$, we now recall a theory of distributions.

Let W be the unique discrete valuation ring unramified over $\mathbb{Z}_p$ of rank 2 (the Witt vector ring with coefficients in $\mathbb{F}_{p^2}$). Then the field K of fractions of W is the unique unramified quadratic extension of $\mathbb{Q}_p$ generated by $(p^2 - 1)$-st roots of unity. Let $V = \mathbb{Z}_p$ or W and put $Q = V \otimes_{\mathbb{Z}_p} \mathbb{Q}_p$ (the field of fractions of V). We say $\mu : Q/V = \bigcup_n p^{-n} V/V \to A$ for an abelian group A with identity is a *distribution* of weight $k \geq 0$ if $\mu : Q/V \to A$ is a function satisfying

$$p^{k(m-n)} \sum_{p^{m-n} y = x} \mu(y) = \mu(x)$$

for all $m > n \geq 0$. If we modify μ into $\mu_{p^n}(a) = p^{kn} \mu(\frac{a}{p^n})$ for $a \in V/p^n V$, this is equivalent to the usual distribution relation $\sum_{a \equiv b \pmod{p^n}} \mu_{p^m}(a) = \mu_{p^n}(b)$ for all $a \in V/p^m V$ and $b \in V/p^n V$.

We restrict μ to $p^{-m} V/V$ and write it as $\mu|_m$ if we need to indicate its order p^m. For a distribution μ, $\mathrm{rank}(\mu|_m) = \dim_{\mathbb{Q}} \mathrm{Im}(\mu|_m) \otimes_{\mathbb{Z}} \mathbb{Q}$. A distribution $\mathcal{M} : p^{-m} V/V \to \mathcal{A}$ is called *universal* if for any distribution $\mu : p^{-m} V/V \to A$, there exists a unique homomorphism $\phi : \mathcal{A} \to A$ such that $\mu = \phi \circ \mathcal{M}$. Universal distribution is unique up to isomorphism if it exists. Indeed $\mathcal{A} = \bigoplus_{a \in p^{-m} V/V} \mathbb{Z}(a)/R$ (the free abelian group generated by symbol (a) indexed by $a \in p^{-n} V/V$ and R is a submodule generated by $(x) - p^{k(m-n)} \sum_{p^{m-n} y = x} (y)$ for all possible $x \in p^{-n} V/V$). Plainly $\mathcal{M}(x) = (x)$ is the universal distribution.

2.4.3　*Stickelberger distribution*

Let $C_m := (V/p^m V)^{\times}$ and consider its group ring $A[C_m]$ (assuming A is a ring). For $a \in C_m$, we write σ_a for the group element in $\mathbb{Q}[C_m]$ corresponding to $a \in C_m$. We define $St_\mu = St_{\mu|_m} : p^{-m} V/V \to A[C_m]$ by $St_\mu(x) := \sum_{a \in C_m} \mu(ax) \sigma_a^{-1}$, which is obviously a distribution. Let $B_n(x)$ be the n-th Bernoulli polynomial; so,

$$\frac{t e^{tx}}{e^t - 1} = \sum_{n=1}^{\infty} B_n(x) \frac{x^n}{n!}.$$

By definition, $B_n(1 - x) = (-1)^n B_n(x)$. For example, $B_1(x) = x - \frac{1}{2}$ and $B_2(x) = x^2 - x + \frac{1}{6}, \dots$. Fix $0 < n \in \mathbb{Z}$. Let $V = \mathbb{Z}_p$ and define $\beta(x) = p^{n-1} B_n(\langle x \rangle)$ for $x \in p^{-m} \mathbb{Z}/\mathbb{Z}$ with representative $0 \leq \langle x \rangle < 1$. Here is an obvious way of creating a distribution:

Lemma 2.4.8. *Let $V = \mathbb{Z}_p$, and fix $0 < n \in \mathbb{Z}$. Then β is a distribution with values in $\mathbb{Q}$.*

Proof. Recall Hurwitz zeta function given by

$$\zeta(s,x) = \sum_{n=0}^{\infty} \frac{1}{(x+n)^s} \quad (\mathrm{Re}(s) > 1, 0 < x \leq 1).$$

This function can be analytically continued to $s \in \mathbb{C}$ and holomorphic outside $s = 1$, and it is known that $\zeta(1-n,x) = -\frac{B_n(x)}{n}$ $(0 < n \in \mathbb{Z})$ for the Bernoulli polynomial $B_n(x)$ (cf. [LFE, §2.3]). By definition,

$$f^{-s}\zeta(s, \frac{a}{f}) = \sum_{n \equiv a \bmod f, n > 0}^{\infty} \frac{1}{n^s}.$$

Thus

$$\sum_{a \in (p^{-m'}\mathbb{Z}/\mathbb{Z})^{\times}, p^{m'-m}a = b} p^{-m's}\zeta(s, \frac{a}{p^{m'}}) = p^{-ms}\zeta(s, \frac{b}{p^m}).$$

Thus β is a distribution of weight $n - 1$. □

For a function $f : \mathbb{Z}/p^m\mathbb{Z} \to \mathbb{C}$, we define

$$\zeta(s,f) = \sum_{n=1}^{\infty} f(n)n^{-s} = \sum_{a \in p^{-m}\mathbb{Z}/\mathbb{Z}} f(p^m a)p^{-ms}\zeta(s, \langle a \rangle).$$

We have the following functional equation of Hurwitz zeta function (e.g. [LFE, §2.3] and [EDM, §3]):

$$\zeta(s,f) = \frac{1}{2\pi i}\left(\frac{2\pi}{p^m}\right)^s \Gamma(1-s)[\zeta(1-s,\widehat{f}^-)e^{\pi i s/2} - \zeta(1-s,\widehat{f})e^{-\pi i s/2}], \quad (2.22)$$

where $\widehat{f}^-(x) := \widehat{f}(-x)$ and $\widehat{f}$ is the Fourier transform given by

$$\widehat{f}(a) = \sum_{b \in \mathbb{Z}/p^m\mathbb{Z}} f(b)\mathbf{e}(-\langle p^{-m}ab \rangle).$$

We consider the trace map $\mathrm{Tr} : W \to \mathbb{Z}_p$, which induces a surjection $\mathrm{Tr} : K/W \to \mathbb{Q}_p/\mathbb{Z}_p$, where $K = W \otimes_{\mathbb{Z}_p} \mathbb{Q}_p$. Then $\beta \circ \mathrm{Tr}$ is a distribution defined on K/W. We then have Stickelberger distribution $St_{\beta|_m}$ and $St_{\beta \circ \mathrm{Tr}|_m}$ with values in $\mathbb{Q}[C_m]$.

2.4.4   *Rank of distribution*

For a distribution $\mu|_m : C_m \to A$, we write $\langle \mu|_m \rangle$ for the submodule of A generated by the values of $\mu|_m$. Then we define rank $\mu|_m = \dim_{\mathbb{Q}}\langle \mu|_m \rangle \otimes_{\mathbb{Z}} \mathbb{Q}$. By distribution relation, the value at an element x of order $p^n \neq 0$ is the sum of values at y with $py = x$. Thus we can modify the value at 0 to be zero taking off $\mu(0)\sum_a \sigma_a$ from St_μ. This does not affect distribution relation and the new modified distribution will be denoted as St_μ^0. This distribution satisfies $\deg(St_{\mu|_m}^0) = 0$ for the degree map of the group algebra $A[C_m]$. By definition, $B_n(1-x) = (-1)^n B_n(x)$; so, $B_n(\langle x \rangle) = (-1)^n B_n(\langle x \rangle)$. Thus $St_{\beta \circ \mathrm{Tr}}^0$ factors through $C_m/\{\pm 1\}$ if n is even.

Theorem 2.4.9. *Let k be the weight of β. For a prime p, if $p^m > 3$, we have*

$$
\mathrm{rank}\, St_{\beta \circ \mathrm{Tr}\,|_m}^0 = \begin{cases} |(W/p^m W)^\times/\{\pm 1\}| - 1 & \text{if } (-1)^{k+1} = 1, \\ |(W/p^m W)^\times/\{\pm 1\}| & \text{if } (-1)^{k+1} = -1. \end{cases}
$$

Proof. Write $St_m = St_{\beta \circ \mathrm{Tr}\,|_m}$ simply. Since

$$
St_m(x) = \sum_{a \in C_m} \beta(\mathrm{Tr}(ax))\sigma_a^{-1},
$$

we have $St(x) \cdot \sigma_b = \sum_{a \in C_m} \beta(\mathrm{Tr}(ax))\sigma_{ab^{-1}}^{-1} = \sum_{a \in C_m} \beta(\mathrm{Tr}(abx))\sigma_a^{-1} = St_m(bx)$. Thus the module $\langle St_m \rangle$ is stable under the multiplication by σ_b; so, to prove that $\mathrm{rank}\, St_m = |C_m| - 1$ or $|C_m|$, we need to show the χ-eigenspace of $\langle St_m \rangle \otimes_{\mathbb{Z}} \overline{\mathbb{Q}}$ is non-trivial for all characters $\chi \neq 1$ of C_m with $\chi(-1) = (-1)^n$. We therefore compute

$$
\sum_{a \in p^{-m} W/W} \beta(\langle \mathrm{Tr}(a) \rangle)\chi(p^m a) = \sum_{a \in C_m} \beta(\langle \mathrm{Tr}(a/p^m) \rangle)\chi(a).
$$

By distribution relation, $\langle St_m \rangle \supset \langle St_{m'} \rangle$ for all $m' < m$, we only need to do this for primitive characters modulo p^m. Then

$$
\sum_{a \in C_m} \beta(\langle \mathrm{Tr}(a/p^m) \rangle)\chi(\mathrm{Tr}(a))
$$

$$
= \sum_{a \in C_m} p^{-ms}\chi(\mathrm{Tr}(a))\zeta(s, \langle \mathrm{Tr}(a/p^m) \rangle))|_{s=1-n} = \zeta(1-n, f_\chi),
$$

where $f_\chi(a) = \sum_{x \in C_m, \mathrm{Tr}(x)=a} \chi(x)$ for $a \in \mathbb{Z}/p^m\mathbb{Z}$. We compute the Fourier transform of f_χ:

$$
\widehat{f_\chi}(a) = \sum_{b=1}^N f_\chi(b)\mathbf{e}(-b\langle a/p^m \rangle) = \sum_{b=1}^N \sum_{x \in C_m, \mathrm{Tr}(x)=b} \chi(x)\mathbf{e}(-\langle ba/p^m \rangle)
$$

$$
= \sum_{x \in C_m} \chi(x)\mathbf{e}(-\langle \mathrm{Tr}(xa/p^m) \rangle)) = G(\chi)\overline{\chi}(-a).
$$

for the Gauss sum $G(\chi) = \sum_{x \in C_m} \chi(x) \mathbf{e}(\langle \mathrm{Tr}(xp^{-m}) \rangle)) \neq 0$ (as χ is primitive). Thus by the functional equation (2.22), we get from $-nN^{n-1}\zeta(1 - n, \langle a \rangle) = B_n(\langle a \rangle)$

$$\sum_{a \in C_m} \beta(\langle \mathrm{Tr}(a/p^m) \rangle)) \chi(\mathrm{Tr}(a))$$

$$= \frac{-nN^{n-1}}{2\pi i} \left(\frac{2\pi}{p^m} \right)^s \Gamma(1-s)[\zeta(1-s, \widehat{f_\chi}) e^{\pi i s/2} - \zeta(1-s, \widehat{f_\chi}) e^{-\pi i s/2}]|_{s=1-n}$$

$$= \frac{-nN^{n-1}}{2\pi i} \left(\frac{2\pi}{p^m} \right)^{1-n} \Gamma(n)[e^{\pi i(1-n)/2} - \chi(-1) e^{-\pi i(1-n)/2}] G(\chi) L(n, \overline{\chi}|_{\mathbb{Z}})$$

$$= \frac{-nN^{n-1}}{2\pi i} \left(\frac{2\pi}{p^m} \right)^{1-n} \Gamma(n)[i^{1-n} - \chi(-1)(-i)^{1-n}] G(\chi) L(n, \overline{\chi}|_{\mathbb{Z}})$$

$$= \frac{-nN^{n-1}}{2\pi i} \left(\frac{2\pi}{p^m} \right)^{1-n} \Gamma(n) i^{n-1} [(-1)^{n-1} - \chi(-1)] G(\chi) L(n, \overline{\chi}|_{\mathbb{Z}}) \quad (2.23)$$

which does not vanish if and only if $\chi(-1) = (-1)^n$. □

2.4.5 Cusps of $X(N)$

We prove

Theorem 2.4.10. *Let* $C_m = (W/p^m W)^\times$ *and embed* C_m *into* $\mathrm{GL}_2(\mathbb{Z}/p^m\mathbb{Z})$ *by its action on* $W/p^m W \cong (\mathbb{Z}/p^m\mathbb{Z})^2$. *Then* $C_m/\{\pm 1\}$ *acts on the cusp of* $X(p^m)$ *freely and transitively.*

Proof. Let $N = p^m$. The set S_N of cusps is one to one onto correspondence with $\Gamma(N)\backslash \mathbf{P}^1(\mathbb{Q})/\{\pm 1\}$. Since $Y(N)$ classifies elliptic curves E with level structure $\phi : (\mathbb{Z}/N\mathbb{Z})^2 \cong E[N]$, $\mathrm{GL}_2(\mathbb{Z}/N\mathbb{Z})$ acts on $Y(N)$ and $X(N)$ and $X(N)/\mathrm{GL}_2(\mathbb{Z}/N\mathbb{Z}) = \mathbf{P}^1(J)$. Since $\{\pm 1\} \subset \mathrm{Aut}(E)$, the action factors through $\mathrm{GL}_2(\mathbb{Z}/N\mathbb{Z})/\{\pm 1\}$. The action sends the Siegel unit g_a^{12N} to $g_{a\alpha}^{12N}$ as in Theorem 2.4.3, and hence as is shown by Shimura [IAT, (6.6.1–2)], $\overline{\alpha} \in \mathrm{GL}_2(\mathbb{Z}/N\mathbb{Z})$ in the image $\overline{\alpha}$ of $\alpha \in \mathrm{SL}_2(\mathbb{Z})$ acts on $\mathbb{Q}(X(N))$ by $f \mapsto f \circ \alpha$ and on $\mathbb{Q}(X(N)) \cap \overline{\mathbb{Q}} = \mathbb{Q}(\mu_N)$, $g \in \mathrm{GL}_2(\mathbb{Z}/N\mathbb{Z})$ acts by $\zeta_N \mapsto \zeta_N^{\det(g)}$. Also $\left(\begin{smallmatrix} 1 & 0 \\ 0 & d \end{smallmatrix} \right)$ acts on q-expansion of $\mathbb{Q}(X(N))$ through its coefficients by $\sigma_a : \zeta_N \mapsto \zeta_N^d$. This follows from the fact that $\mathbb{Q}(X(N)) = \mathbb{Q}(g_a^{12N})$ for g_a in Theorem 2.4.3.

Writing one geometrically irreducible component of $X(N)$ as $X^\circ(N)$, we therefore have $\mathbb{Q}(X(N)) = \mathbb{Q}(\mu_N)(X^\circ(N))$ and $X^\circ(N)$ is geometrically irreducible over $\mathbb{Q}(\mu_N)$. Thus $\mathrm{Gal}(X(N)/\mathbf{P}^1(J)) \cong \mathrm{GL}_2(\mathbb{Z}/N\mathbb{Z})/\{\pm 1\}$

canonically. Since $\mathbf{P}^1(J)$ has one cusp ∞, the cusps of $\Gamma(N)\backslash\mathfrak{H}$ are one to one onto $\mathrm{SL}_2(\mathbb{Z})(\infty) \cong \mathrm{SL}_2(\mathbb{Z})\backslash\mathbf{P}^1(\mathbb{Q}) \cong \Gamma(N)\backslash\mathrm{SL}_2(\mathbb{Z})/\Gamma_\infty$ for

$$\Gamma_\infty = \left\{ \pm \left(\begin{smallmatrix} 1 & m \\ 0 & 1 \end{smallmatrix}\right) \,\middle|\, m \in \mathbb{Z} \right\}.$$

Since $\Gamma(N)\backslash\mathfrak{H}$ gives rise to a geometrically irreducible component of $X(N)$ indexed by $\mathrm{Gal}(\mathbb{Q}(\mu_N)/\mathbb{Q})$. Thus the stabilizer of ∞ in $\mathrm{Gal}(X(N)/\mathbf{P}^1(J)) = \mathrm{GL}_2(\mathbb{Z}/N\mathbb{Z})/\{\pm 1\}$ is given by the image $M(\mathbb{Z}/N\mathbb{Z})$ of $\left\{ \left(\begin{smallmatrix} 1 & u \\ 0 & d \end{smallmatrix}\right) \,\middle|\, u \in \mathbb{Z}/N\mathbb{Z}, d \in (\mathbb{Z}/N\mathbb{Z})^\times \right\}.$ We have

$$S_N \cong \mathrm{GL}_2(\mathbb{Z}/N\mathbb{Z})/M(\mathbb{Z}/N\mathbb{Z})$$
$$\cong \left\{ v := \left(\begin{smallmatrix} a \\ c \end{smallmatrix}\right) \in (\mathbb{Z}/N\mathbb{Z})^2 \,\middle|\, \text{order of } v = N \right\}/\{\pm 1\}.$$

Taking a basis $\tau, 1$ of W over $\mathbb{Z}_p$, we have for $w = a + c\tau$ with $a, c \in \mathbb{Z}_p$,

$$w(1, \tau) = (w, w\tau) = (1, \tau)\left(\begin{smallmatrix} a & * \\ c & * \end{smallmatrix}\right).$$

This shows that $C_m \cdot M(\mathbb{Z}/N\mathbb{Z}) = \mathrm{GL}_2(\mathbb{Z}/N\mathbb{Z})/\{\pm 1\}$ (the Iwasawa decomposition) and $S_N \cong C_m/\{\pm 1\}$, and the action of C_m on the left is transitive, as desired. $\square$

2.4.6   *Finiteness of* $Cl_{X(N)}$

Note that the parameter at the cusp $\infty \in X(N)$ is $q^{1/N}$ (as the parameter at the cusp of $\mathbf{P}^1(J)$ is q). We take a basis of W of the form $(\frac{1}{2}, \tau)$ with $\mathrm{Tr}(\tau) = 0$. Then $\mathrm{Tr}(a_1\frac{1}{2} + a_2\tau) = a_1$. In this way, we identify $W \cong \mathbb{Z}_p^2$ and $p^{-m}W/W$ with $(p^{-m}\mathbb{Z}/\mathbb{Z})^2$. Then we consider Siegel units g_a for $a \in p^{-m}W/W$ with $\mathrm{Tr}(a) = a_1$. Identify S_{p^m} with $C_m/\{\pm 1\}$. Thus by Proposition 2.4.2, noting that the parameter at ∞ of $X(N)$ is given by $q^{1/N}$, we find

$$\mathrm{div}(g_a) = N \sum_{b \in C_m/\{\pm 1\}} \frac{1}{2} B_2(\langle \mathrm{Tr}(ab) \rangle) \sigma_b^{-1}(\infty). \tag{2.24}$$

Though this formula is defined for $0 \neq a \in p^{-m}W/W$, we can put $g_0 := 1$ and then the $a \mapsto \mathrm{div}(g_a) \in \mathrm{Div}^0(X(N))$ is a distribution proportional to $St := St_{\beta \circ \mathrm{Tr}\,|_m}^0$, where $\mathrm{Div}^0(X(N))$ is the degree 0 divisor group of $X(N)$. Write $\mathrm{Div}^0(S_N) \subset \mathrm{Div}^0(X(N))$ for the subgroup generated by cusps of $X(N)$. Thus $Cl_{X(N)}$ is a surjective image of $\mathrm{Div}^0(S_N)/\langle \mathrm{div}(g_a^{12p^m}) \rangle_{a \in p^{-m}W/W}$. By Theorem 2.4.9, $\mathrm{Div}^0(S_N)/\langle \mathrm{div}(g_a^{12p^m}) \rangle_{a \in p^{-m}W/W}$ is a finite group. Therefore we obtain

Theorem 2.4.11. *If* $N = p^m > 3$ *for a prime* p, *the class group* $Cl_{X(N)}$ *is finite, and the unit group* $\mathcal{A}_N^\times$ *is non-trivial; in other words,* $\mathrm{rank}_{\mathbb{Z}}\,\mathcal{A}_N^\times/\mathbb{Q}[\mu_N]^\times > 0$.

The non-triviality of $\mathcal{A}_N^\times/\mathbb{Q}[\mu_N]^\times$ follows since $X(N)$ $(N > 0)$ has more than one cusp (i.e., $|S_N| = |\,\mathrm{PSL}_2(\mathbb{Z}/N\mathbb{Z})|/N$ [IAT, Proposition 1.40]). Indeed, we have $\mathrm{rank}_\mathbb{Z}\,\mathcal{A}_N^\times/\mathbb{Q}[\mu_N]^\times = \mathrm{rank}_\mathbb{Z}\,\mathrm{Div}^0(S_N) = |S_N| - 1 > 0$ by the finiteness of $Cl_{X(N)}$.

By the genus formula [IAT, (1.6.4)], the genus g of $X(N)$ $(N > 1)$ is given by $g = 1 + |\,\mathrm{PSL}_2(\mathbb{Z}/N\mathbb{Z})|(N-6)/12N$. So $g = 0 \Leftrightarrow N \le 5$. Under this circumstance, $X(N) \cong \mathbf{P}^1 - S_N$, and hence $Y := X(N) - \{\infty\} \cong \mathbb{A}^1$; so, the affine ring of Y is given by $\mathbb{Q}[\mu_N](x)$. Thus writing the x-coordinate of $s \in S_N$ as $\alpha_s \in \mathbb{Q}[\mu_N]$, $\mathcal{A}_N$ is given by the localization $\mathbb{Q}[\mu_N](x)[\frac{1}{x-\alpha_s}]_{s\in S_N-\{\infty\}}$. Thus $Cl_{X(N)} = \{1\}$ if $N \le 5$. The formula $\mathrm{rank}_\mathbb{Z}\,\mathcal{A}_N^\times/\mathbb{Q}[\mu_N]^\times = |S_N| - 1$ remains true for all N. In particular, if $N = 1$, $X(1) = \mathbf{P}^1(J) - \{\infty\}$; so, $\mathcal{A}_1 = \mathbb{Q}[\mu_N][J]$ for the modular J-invariant J. This implies $\mathcal{A}_1 = \mathbb{Q}[\mu_N]^\times$.

2.4.7 *Siegel units generate* $\mathcal{A}_{p^m}^\times$

We state the following theorem due to Shimura which follows from a solution of the moduli problem for the functor $\mathcal{P}$ we discussed in §2.2.3:

Theorem 2.4.12. *If $f \in \mathbb{Q}(X)$ for $X = X_0(N)$, $X_1(N)$ and $X(N)$, for every $\sigma \in \mathrm{GL}_2(\mathbb{Z}/N\mathbb{Z}) = \mathrm{Gal}(X(N)/\mathbf{P}^1(J))$, the q-expansion of f^σ has bounded denominator; in other words, for some $0 < M \in \mathbb{Z}$ independent of σ, $Mf^\sigma(q)$ has integral q-expansion coefficients.*

For a function $m : \mathbb{Q}^2/\mathbb{Z}^2 \to \mathbb{Z}$ with finite support, we define $g(m) := \prod_a g_{(a)}^{m(a)}$. Here $\langle a \rangle = (\langle a_1 \rangle, \langle a_2 \rangle)$. Since $g_{a+b} = cg_a$ with $c \in \mu_{2N}$ for $b \in \mathbb{Z}^2$ as long as $Na \in \mathbb{Z}^2$ (see (2.16)), up to constant, for any choice of representatives $[a]$ of $a \in \mathbb{Q}^2/\mathbb{Z}^2$, $g(m)/\prod_a g_{[a]}^{m(a)} \in \mu_{2N}(\mathbb{C})$. Note that $\mathrm{End}(\mathbb{Q}^2/\mathbb{Z}^2) = M_2(\widehat{\mathbb{Z}})$ for $\widehat{\mathbb{Z}} = \varprojlim_N \mathbb{Z}/N\mathbb{Z} = \prod_l \mathbb{Z}_l$. By Theorems 2.3.13 and 2.4.3, g_a modulo scalars only depends on $a \mod \mathbb{Z}^2$; so, $m\sigma(a) := m(a\sigma^{-1})$ (modulo scalars). Then $\sigma \in \mathrm{GL}_2(\widehat{\mathbb{Z}}) = \varprojlim_N \mathrm{GL}_2(\mathbb{Z}/N\mathbb{Z})$ acts (modulo scalars) on $g(m)$ by $g(m) \mapsto \sigma g(m) := g(m\sigma)$.

We write $g(m) = \sum_{n=n_0}^\infty a_n q^{n/N}$ with $a_{n_0} \ne 0$ and put $g^*(m) := a_{n_0}^{-1} q^{-n_0} g(m)$ which is independent of scaling of $g(m)$. Since $g_a = q^{(1/2)B_2(\langle a_1 \rangle)}(1 - \mathbf{e}(a_2)q^{a_1})\prod_{n=1}^\infty (1 - \mathbf{e}(a_2)q^{n+a_1})(1 - \mathbf{e}(-a_2)q^{n-a_1})$, we see, writing $\mathbf{e}(x) = \exp(2\pi ix)$,

$$g_a^* = \begin{cases} (1 - \mathbf{e}(a_2)q^{a_1})\prod_{n=1}^\infty (1 - \mathbf{e}(a_2)q^{n+a_1})(1 - \mathbf{e}(-a_2)q^{n-a_1}) & \text{if } \langle a_1 \rangle \ne 0, \\ \prod_{n=1}^\infty (1 - \mathbf{e}(a_2)q^{n+a_1})(1 - \mathbf{e}(-a_2)q^{n-a_1}) & \text{if } \langle a_1 \rangle = 0. \end{cases}$$

$$(2.25)$$

An immediate feature from this definition is

Lemma 2.4.13. *We have* $g_a^* \circ \alpha = g_{a\alpha}^*$ *for* $\alpha \in \mathrm{GL}_2(\widehat{\mathbb{Z}})$ *and* $g_a^* \in \mathbb{Z}[\mu_N][[q_N]]$.

Since $\mathfrak{k}_a$ only depends on $a \in N^{-1}\mathbb{Z}^2/2N\mathbb{Z}^2 \cong \mathbb{Z}/2N^2\mathbb{Z}$, g_a only depends on $a \in N^{-1}\mathbb{Z}/12N\mathbb{Z}^2$ (basically by (2.16)), $\mathrm{GL}_2(\widehat{\mathbb{Z}})$ ($\widehat{\mathbb{Z}} = \prod_l \mathbb{Z}_l$) acts on g_a via its quotient $\mathrm{GL}_2(\mathbb{Z}/12N^2)$ as $\mathrm{Aut}(N^{-1}\mathbb{Z}/12N\mathbb{Z}^2) = \mathrm{GL}_2(\mathbb{Z}/12N^2\mathbb{Z})$.

Proof. By (2.16), $\mathfrak{k}_{a+b} = \varepsilon\mathfrak{k}_a$ and $\Delta^{1/12} \circ \alpha = \epsilon\Delta^{1/12}$ for roots of unity ε, ϵ. Therefore $g_a = \mathfrak{k}_a\Delta^{1/12}$ also satisfies $g_{a\alpha} = \zeta(\alpha)g_a$ for a root of unity $\zeta(\alpha)$ (dependent on α). Thus $(g_a \circ \alpha)^* = (\zeta(\alpha)g_{a\alpha})^* = g_{a\alpha}^*$. The integrality of g_a^* is obvious from the product expansion. $\square$

Here by (2.16), $g_a \circ \alpha = \varepsilon g_a$ for a root of unity ε. Therefore, for $q_N = q^{1/N}$ if $Na \in \mathbb{Z}^2$, we have $g_a^* = 1 + q_N f(q_N)$ for $f(q_N) \in \mathbb{Q}[\mu_N][[q_N]]$. Thus as a formal power series, for any $1 < M \in \mathbb{Z}$, we have a well defined

$$(g_a^*)^{1/M} = \sum_{n=0}^{\infty} \binom{1/M}{n} q_N^n f(q_N)^n \in \mathbb{Q}[\mu_N][[q_N]]$$

with

$$\binom{1/M}{n} = \frac{(1/M)((1/M) - 1)((1/M) - 2) \cdots ((1/M) - n + 1)}{n!}$$
$$= \frac{(1 - M)(1 - 2M) \cdots (1 - (n-1)M)}{M^n n!}.$$

However the coefficients of $g_a^{*1/M}$ and hence of $g^*(m)^{1/M}$ would have growing denominator caused by the denominator of $1/M$; in other words, if $g^*(m)^{1/M}$ remains integral, m is forced to be divisible by M. We prove this divisibility. By the above formula, we remark that for a prime l,

$$g^*(m)^{1/l} \text{ remains in } \mathbb{Q}[\mu_N][[q^{1/N}]] \text{ if } g^*(m) \in \mathbb{Q}[\mu_N][[q^{1/N}]]. \qquad (2.26)$$

Fix a prime p and consider a primitive root of unity $\zeta_{p^n} := \exp(\frac{2\pi i}{p^n})$. We need the following fact which follows from Theorem 1.1.1.

Lemma 2.4.14. *Decompose* $\mu_{p^n} = \bigsqcup_{j=1}^{p^{n-1}} \zeta_j \mu_p$. *Then* $S = \mu_{p^n} - \{\zeta_j\}_j$ *gives a basis of* $\mathbb{Z}[\zeta_{p^n}]$ *over* $\mathbb{Z}$.

We shall give a short proof.

Proof. Since $\sum_{\zeta \in \mu_p} \zeta = 0$, we get $\sum_{\zeta \in \mu_p} \zeta_j \zeta = 0$. Since $\mathbb{Z}[\mu_{p^n}] = \sum_{\zeta \in \mu_{p^n}} \mathbb{Z}\zeta$, by $\sum_{\zeta \in \mu_p} \zeta_j \zeta = 0$, we find $\mathbb{Z}[\mu_{p^n}] = \sum_{\zeta \in S} \mathbb{Z}\zeta$. Since $|S| = \deg \Phi_n(X) = [\mathbb{Q}[\mu_{p^n}] : \mathbb{Q}]$, the set S must be a basis of $\mathbb{Z}[\mu_{p^n}]$. $\square$

Corollary 2.4.15. *For a prime l, if $l| \sum_{\zeta \in \mu_{p^n}} a_\zeta \zeta$ in $\mathbb{Z}[\mu_{p^n}]$, for any pair ζ, ζ' with $\zeta' \zeta^{-1} \in \mu_p$, $l|(a_\zeta - a_{\zeta'})$.*

Proof. By Lemma 2.4.14, $\sum_{j=1}^{p^{n-1}} \sum_{\zeta \in \mu_p \zeta_j - \{\zeta_j\}} b_\zeta \zeta$ is divisible by l if and only if $l|b_\zeta$ for all $\zeta \in \mu_{p^n} - \{\zeta_j\}_j$. Since $-\zeta_j = \sum_{\zeta \in \mu_p \zeta_j - \{\zeta_j\}} \zeta$, we have $\sum_{\zeta \in \mu_{p^n}} a_\zeta \zeta = \sum_j \sum_{\mu_p \zeta_j - \{\zeta_j\}} (a_\zeta - a_{\zeta_j}) \zeta$, which shows the result. $\square$

We say

- a function $m : \mathbb{Q}^2/\mathbb{Z}^2 \to \mathbb{Z}$ has *prime period* if for any a with $m(a) \neq 0$, there exists a prime p such that $pa \in \mathbb{Z}^2$,
- $0 < n \in \mathbb{Z}$ *occurs as a denominator* of m if there exists $a \in \mathbb{Q}^2/\mathbb{Z}^2$ with order n such that $m(a) \neq 0$.

Note that $g^*(m\sigma/l)$ has l-integral coefficients for all $\sigma \in \mathrm{GL}_2(\widehat{\mathbb{Z}})$ by Lemma 2.4.13.

Lemma 2.4.16. *Let l be a prime number. Assume that m has prime period. Then there exists an expression $g^*(m) = cg^*(m')$ with a constant c such that m' has values in $l\mathbb{Z}$ and every denominator occurring in m' also occurs in m.*

Proof. For each prime p, we put $g_p^*(m/l) = \prod_{pa=0} (g_a^*)^{m(a)/l}$. Then by (2.26), $g_p^*(m/l)$ lives in $\mathbb{Q}[\mu_p][[q^{1/p}]]$ with the identity $g_p^*(m/l)(0) = 1$ of its leading term. Therefore, the coefficient of $q^{1/p}$ of $g_p^*(m/l)$ and $g^*(m/l)$ are equal. In the same way, the coefficient of $q^{1/p}$ of $g_p^*(m\sigma/l)$ and $g^*(m\sigma/l)$ are equal for any $\sigma \in \mathrm{GL}_2(\widehat{\mathbb{Z}})$.

We claim

$$\text{if } a, b \text{ in } \mathbb{Q}^2/\mathbb{Z}^2 \text{ has order } p, \text{ then } l|m(a) - m(b).$$

Writing $\langle a \rangle$ for a subgroup generated by a, to prove the claim we may assume $\langle a \rangle \cap \langle b \rangle = \{0\}$ as we can write $m(a) - m(b) = m(a) - m(a') + m(a') - m(b)$ for a choice $a' \in (p^{-1}\mathbb{Z}/\mathbb{Z})^2$ with $\langle a' \rangle \cap \langle a \rangle = \langle a' \rangle \cap \langle b \rangle = \{0\}$. Thus a, b is a basis over $\mathbb{F}_p$ of $(p^{-1}\mathbb{Z}/\mathbb{Z})^2$, and we can find $\sigma \in \mathrm{GL}_2(\mathbb{F}_p)$ so that $a\sigma = (\frac{1}{p}, 0)$ and $b\sigma = (\frac{1}{p}, \frac{1}{p})$. Then the coefficient in $q^{1/p}$ of $g_p^*(m\sigma/l)$ is equal to the coefficient in $q^{1/p}$ of the following product:

$$\prod_c \left[(1 - \mathbf{e}(c_2)q^{c_1}) \prod_{n=1}^{\infty} (1 - \mathbf{e}(c_2)q^{c_1}q^n)(1 - \mathbf{e}(-c_2)q^{-c_1}q^n) \right]^{m(c)/l},$$

where c runs over elements of the following form:

$$c = a\sigma + j(b\sigma - a\sigma) \ (j = 0, \ldots, p-1).$$

This is because the coefficient of $q^{1/p}$ comes from a with $a_1 = \frac{1}{p}$. Therefore the coefficient is equal to

$$-\frac{1}{l} \sum_c m(c)\zeta_c \quad \text{for } \zeta_c = \mathbf{e}(\frac{j}{p}).$$

Thus by Corollary 2.4.15, l-integrality of this sum implies $l|m(c) - m(c')$ for any two $c \neq c'$. We take $j = 0, 1$ (i.e., $c = a\sigma$ and $c' = a\sigma + b\sigma - a\sigma = b\sigma$), and we get $l|m(a\sigma) - m(b\sigma)$ as claimed.

We now finish the proof. Take a_0 of order p. Then we have

$$\prod_{pa=0, a\neq 0} = \prod_{pa=0, a\neq 0} g_a^{m(a)-m(a_0)} \prod_{pa=0, a\neq 0} g_a^{m(a_0)}.$$

By Proposition 2.4.5, we know $c := \prod_{pa=0, a\neq 0} g_a^{m(a_0)} \in \mathbb{Q}^\times$, and by the above claim $m' = m - m(a_0)$ has values in $l\mathbb{Z}$. $\square$

We now prove

Theorem 2.4.17. *Let l, p be prime numbers (which can be equal). Assume that $g^*(m\sigma/l) := \prod_a (g_{a\sigma-1}^*)^{m(a)/l}$ has l-integral coefficients for all $\sigma \in \mathrm{GL}_2(\widehat{\mathbb{Z}})$ and that m is supported on $(p^{-r}\mathbb{Z}/\mathbb{Z})^2$ for some $r > 0$. Then there exists an even function $m' : \mathbb{Q}^2/\mathbb{Z}^2 \to l\mathbb{Z}$ supported on $(p^{-r'}\mathbb{Z}/\mathbb{Z})^2$ for some $r' > 0$ such that $\prod_a g_a^{m(a)} = c \prod_a g_a^{m'(a)}$ for a constant $c \in \mathbb{Q}[\mu_{p^r}]^\times$.*

The left-hand-side resides in $\mathcal{A}_N^\times$ and the right-hand-side is stable under $\mathrm{GL}_2(\mathbb{Z}/N\mathbb{Z})$; so, $c \in \mathbb{Q}[\mu_{p^r}]^\times$.

Proof. The case of $r = 1$ follows from Lemma 2.4.16. We proceed by induction on r; so, we may assume that $m(a) \neq 0$ for some a of order p^r. We claim

$$\text{if } p^r a = p^r b = 0 \text{ in } \mathbb{Q}^2/\mathbb{Z}^2 \text{ with } p(a - b) = 0, \text{ then } l|m(a) - m(b).$$

If $\langle a \rangle = \langle b \rangle$, choose a' of order p^r so that $\langle a' \rangle \neq \langle a \rangle$. Then again $m(a) - m(b) = m(a) - m(a') + m(a') - m(b)$, and hence, we may assume that $\langle a \rangle \neq \langle b \rangle$ (so, in particular, $a - b \neq 0$). Choose c so that $p^{r-1}c = b - a$. As $p(b - a) = 0$ and (we may assume) $b - a \neq 0$, c has order p^r. Since $\langle a \rangle \neq \langle b \rangle$, $\langle a \rangle + \langle b \rangle \supset (p^{-1}\mathbb{Z}/\mathbb{Z})^2$; so, $p^{r-1}a$ and $a - b = -p^r c$ form a basis of $(p^{-1}\mathbb{Z}/p\mathbb{Z})^2$. By (the divisible version of) Nakayama's lemma, a and c form a basis of $(p^{-r}\mathbb{Z}/\mathbb{Z})^2$. Through base-change via multiplication by σ, we may assume that $a = \left(\frac{1}{p^r}, 0\right)$ and $c = \left(0, \frac{1}{p^r}\right)$. Define $g_{p^r}^*(m/l) :=$

$\prod_{p^r a=0, p^{r-1} a \neq 0} (g_a^*)^{m(a)/l}$. As in the proof of Lemma 2.4.16, we look into the coefficient of q^{1/p^r} of $g_{p^r}^*(m/l)$ and $g^*(m/l)$ which are equal. It is

$$-\frac{1}{l} \sum_{j=0}^{p^r-1} m\left(\frac{1}{p^r}, \frac{j}{p^r}\right) \zeta_{p^r}^j \quad (\zeta_{p^r} = \mathbf{e}(\frac{1}{p^r})).$$

Again by Corollary 2.4.15, we find $l|m(a) - m(b)$ as claimed.

Given a coset C of $(p^{-1}\mathbb{Z}/\mathbb{Z})^2$ in $(p^{-r}\mathbb{Z}/\mathbb{Z})^2$, we see $p \cdot C = \{c\}$ for a single $c \in (p^{1-r}\mathbb{Z}/\mathbb{Z})^2$, and the distribution relation in Proposition 2.4.5 tells us $\prod_{x \in C} g_x = \lambda g_c$ for a constant $\lambda \in \mathbb{Q}[\mu_{12p^r}]^\times$. Taking $c_0 \in C$, we see

$$\prod_{x \in C} g_x^{m(x)} = \prod_{x \in C} g_x^{m(x)-m(c_0)} \prod_{x \in C} g_x^{m(c_0)} = \lambda^{m(c_0)} g_c^{m(c_0)} \prod_{x \in C} g_x^{m(x)-m(c_0)}.$$

Since c has order $\leq p^{r-1}$, for $\prod_c g_c^{m(c_0)}$, induction assumption applies. As for the remaining $\prod_{x \in C} g_x^{m(x)-m(c_0)}$, we have proven $l|m(x) - m(c_0)$. This finishes the proof. $\square$

We already know that Siegel units $\{g_a^{12N}\}_{a \in p^{-m}W/W}$ (for $N = p^m$) generate a subgroup of finite index in $\mathcal{A}_N^\times$. So for any $g \in \mathcal{A}_N^\times$, we find $0 < M \in \mathbb{Z}$ such that $g^M = g(m)$ for some $m : (p^{-m}\mathbb{Z}/\mathbb{Z})^2 - \{(0,0)\} \to \mathbb{Z}$. Then by Theorem 2.4.17, we can remove prime-factor by prime-factor from M replacing the exponent m; so, we obtain

Corollary 2.4.18. *Suppose $p \geq 5$ and that N is a p-power. Then for $g \in \mathcal{A}_N^\times$, there exists $m : (N^{-1}\mathbb{Z}/\mathbb{Z})^2 - \{(0,0)\} \to \mathbb{Z}$ such that $g = g(m)$ up to constants.*

The next step is to make explicit the set

$$\{m : (N^{-1}\mathbb{Z}/\mathbb{Z})^2 - \{(0,0)\} \to \mathbb{Z} | g(m) \in \mathcal{A}_N^\times\}.$$

2.4.8 *Fricke–Wohlfahrt theorem*

We need a theorem due to Fricke–Klein (1890) and its generalization by Wohlfahrt (1964). Let $\Delta \subset \mathrm{SL}_2(\mathbb{Z})$ be a subgroup of finite index. Such a group Δ is called a congruence subgroup if Δ contains

$$\Gamma(N) = \{\alpha \in \mathrm{SL}_2(\mathbb{Z}) | \alpha \equiv 1 \mod N \cdot M_2(\mathbb{Z})\}$$

for a positive integer N. If Δ is a congruence subgroup, the smallest N such that $\Delta \supset \Gamma(N)$ is called the arithmetic level of Δ. For each cusp s

of general Δ, let $\Delta_s = \{\alpha \in \Delta | \alpha(s) = s\}$. Taking $\gamma = \gamma_s \in \mathrm{SL}_2(\mathbb{Z})$ with $\gamma(s) = \infty$, we find

$$\gamma\{\pm 1\}\Delta_s \gamma^{-1} \subset \{\pm \left(\begin{smallmatrix} 1 & m \\ 0 & 1 \end{smallmatrix}\right) | m \in \mathbb{Z}\} =: \Gamma_\infty$$

as a subgroup of finite index N_s. Since the coset $\Gamma_\infty \gamma_s$ is determined uniquely by s, N_s depends only on s. We define the geometric level of Δ to be the least common multiple of N_s for s running over all cusps of Δ.

Theorem 2.4.19. *If Δ is a congruence subgroup of $\mathrm{SL}_2(\mathbb{Z})$, its geometric level and arithmetic level coincide. In other words, writing N for the geometric level of Δ, if $\Delta \supset \Gamma(N')$ for some $N' > 0$ and $E(N)$ is the minimal normal subgroup of $\mathrm{SL}_2(\mathbb{Z})$ generated by $\left(\begin{smallmatrix} 1 & N \\ 0 & 1 \end{smallmatrix}\right)$, then $\Gamma(N) = \Gamma(N')E(N)$.*

Since the congruence subgroup problem fails for $\mathrm{SL}_2(\mathbb{Z})$ (see [S70, §1.4]), the condition $\Delta \supset \Gamma(N')$ is perhaps necessary for this theorem, and also this might be a reason for not (perhaps) having a geometric proof of this fact. We give a computational proof due to Wohlfahrt. Let $\Gamma := \Gamma(N')E(N)$. We define M to be the order of $T = \left(\begin{smallmatrix} 1 & 1 \\ 0 & 1 \end{smallmatrix}\right)$ in $\mathrm{SL}_2(\mathbb{Z})/\Gamma$. Since $U = jTj^{-1} = \left(\begin{smallmatrix} 1 & 0 \\ 1 & 1 \end{smallmatrix}\right)$ also has order M. Plainly $M|N$. Thus we need to prove that $\Gamma(M) \subset \Gamma(N)$, which will show $M = N$.

Proof. We define subsets E_1, E_2, and E_3 of $\Gamma(M)$ whose union equals $\Gamma(M)$ and show that each E_i is a subset of $\Gamma(N)$. To this end, let $A = \left(\begin{smallmatrix} a & b \\ c & d \end{smallmatrix}\right) \in \Gamma(M)$. First, let E_1 be the collection of elements whose upper right and lower left entries are divisible by N and suppose $A \in E_1$. If we let

$$A_0 = \begin{pmatrix} a & ad-1 \\ 1-ad & d(2-ad) \end{pmatrix}$$

then $\det(A_0) = 1$, and since $ad - bc = 1$, we have $ad \equiv 1 \mod N$ and so $A \equiv A_0 \mod N$. Equivalently, $AA_0^{-1} - 1 \in \Gamma(N)$. It suffices to show $A_0 \in \Gamma(N)$. This follows from writing $V = TU^{-1}T^3U^{-1}T$ and computing $A_0 = (ST^{d-1}S)^{-1}(VT^{a-1}V^{-1})T^{d-1}$ with $S = TU^{-1}T$ as an element of $E(N)$ since $a \equiv d \equiv 1 \mod M$.

Now let E_2 be the set of matrices whose diagonal entries are relatively prime to N. If $A \in E_2$ then $\gcd(a, N) = 1$ so a is a unit modulo N say with inverse a'. We can write $c = M\widetilde{c}$ and let $k = -a'\widetilde{c}$. Then

$$(S'T^mS)^{-k}A = \left(\begin{smallmatrix} 1 & 0 \\ mk & 1 \end{smallmatrix}\right)\left(\begin{smallmatrix} a & b \\ c & d \end{smallmatrix}\right) = \left(\begin{smallmatrix} a & b \\ c+amk & d+bmk \end{smallmatrix}\right)$$

and $c + amk = m(\widetilde{c} - a'a\widetilde{c})$ is congruent to 0 modulo N. Let $\widetilde{d} = d + bmk$. Taking the determinant of $(ST^mS)^{-k}A$, we see that $\widetilde{d}$ is relatively prime to N. Therefore we can also choose ℓ so that $b + \widetilde{d}m\ell \equiv 0 \mod N$, and

$$T^{m\ell}(ST^mS)^{-k}A = \left(\begin{smallmatrix} 1 & m\ell \\ 0 & 1 \end{smallmatrix}\right)\left(\begin{smallmatrix} a & b \\ c+amk & \widetilde{d} \end{smallmatrix}\right) = \left(\begin{smallmatrix} * & b+\widetilde{d}m\ell \\ c+amk & \widetilde{d} \end{smallmatrix}\right)$$

and this is an element of E_1. This says A is an element of $E(M)E_1$ and is therefore in $\Gamma(N)$. Finally, define $E_3 = \Gamma(M)\backslash(E_1 \cup E_2)$. If $A \in E_3$, then $\gcd(a, n) \neq 1$ or $\gcd(d, n) \neq 1$. Suppose first that $\gcd(a, n) \neq 1$. Since $ad - bc = 1$, we know $\gcd(a, b) = 1$. Then since $a \equiv 1 \mod M$, we also have $\gcd(a, bM) = 1$. Recall Dirichlet's theorem on primes in arithmetic progressions: Suppose r and s are relatively prime integers. Then $\{r + st | t \in Z\}$ contains infinitely many prime numbers. In particular, taking $r = a$ and $s = bm$, we can choose an integer k so that $a + bmk$ is a prime number larger than N. Then

$$A(ST^m S)^{-k} = \begin{pmatrix} a & b \\ c & d \end{pmatrix} \begin{pmatrix} 1 & 0 \\ mk & 1 \end{pmatrix} = \begin{pmatrix} a+bmk & b \\ c+dmk & d \end{pmatrix}$$

has upper left entry relatively prime to N. If $\gcd(d, n) = 1$, then $A(ST^m S)^{-k} \in E_2$ and we are done. Otherwise, let $\tilde{a} = a + bmk$, let $\tilde{c} = c + dmk$ and note that $\tilde{a}d - b\tilde{c} = 1$ so $\gcd(\tilde{c}, d) = 1$. As $\gcd(d, m) = 1$, we can choose ℓ so that $d + \tilde{c}m\ell$ is a prime number larger than N. Then

$$A(ST^m S)^{-k} T^{m\ell} = \begin{pmatrix} \tilde{a} & b \\ \tilde{c} & d \end{pmatrix} \begin{pmatrix} 1 & m\ell \\ 0 & 1 \end{pmatrix} = \begin{pmatrix} \tilde{a} & * \\ \tilde{c} & d+\tilde{c}m\ell \end{pmatrix}$$

and $A(ST^m S)^{-k} T^{m\ell}$ must be in E_2. Thus A is an element of $E_2 E(M)$ and the theorem follows. $\qquad\qquad\square$

2.4.9 *Siegel units and Stickelberger's ideal*

We put

$$\mathfrak{D} = \mathfrak{D}_N = \bigoplus_{P:\text{cusp of } X(N)} \mathbb{Z}[P],$$

$$\mathfrak{D}^0 = \mathfrak{D}_N^0 = \{D \in \bigoplus_{P:\text{cusp of } X(N)} \mathbb{Z}[P] | \deg(D) = 0\},$$

$$\mathfrak{F} = \mathfrak{F}_N = \{\text{div}(u) \in \bigoplus_{P:\text{cusp of } X(N)} \mathbb{Z}[P] | u \in \mathcal{A}_N^\times\}, \qquad (2.27)$$

$$\mathfrak{S}_N = \{\text{div}(g^m) | g^m = \prod_{a \in N^{-1}\mathbb{Z}^2/\mathbb{Z}^2 - \{0\}} g_a^{m(a)} \in \mathcal{A}_N\},$$

$$\mathfrak{S} = \{\text{div}(g^m) | g^m = \prod_{a \in \mathbb{Q}^2/\mathbb{Z}^2 - \{0\}} g_a^{m(a)} \in \mathcal{A}_N\}.$$

Then we have $\mathfrak{D}^0 \supset \mathfrak{F} \supset \mathfrak{S}$ and $Cl_{X(N)} = \mathfrak{D}^0/\mathfrak{F}$, which is a finite abelian group by Theorem 2.4.11.

Theorem 2.4.20 (Kubert–Lang). *We have* $\mathfrak{S} = \mathfrak{S}_N = \mathfrak{F}_N$ *if* $N = p^r$ *for a prime* $p \geq 5$.

When N is not a prime power as in the theorem, then $[\mathfrak{F} : \mathfrak{S}]$ is a 2-power (see [MUN, Theorem 5.1.1]).

Proof. We know that the q-expansion coefficients of g^m lies in $\mathbb{Q}[\mu_N]$ if m is supported on $N^{-1}\mathbb{Z}^2 - \mathbb{Z}^2$. Since $g_a \circ \alpha = g_{a\alpha}$ for all $\alpha \in \mathrm{SL}_2(\mathbb{Z})$ up to a non-zero constant by Theorem 2.4.3, if an integer which is not a factor of N occurs as a divisor of m, $g^m \circ \alpha \notin \mathbb{C}((q^{1/N}))$ by looking into the leading term of g^m. Therefore to have $g^m \in \mathcal{A}_N$, it is necessary to have m supported on $N^{-1}\mathbb{Z}^2 - \mathbb{Z}^2$. Thus $\mathfrak{S} \subset X_N := \{\mathrm{div}(g^m)|m : N^{-1}\mathbb{Z}^2/\mathbb{Z}^2 - \{0\} \to \mathbb{Z}\}$. Suppose $\mathrm{div}(g^m) \in X_N$ for $m : N^{-1}\mathbb{Z}^2/\mathbb{Z}^2 - \{0\} \to \mathbb{Z}$. Then the leading term of g^m is an integer power of q_N, and by the product q-expansion of Siegel units in (2.21), we conclude $g^m \in \mathbb{Q}[\mu_N]((q_N))$. Then by Theorem 2.3.13 and the remark after Theorem 2.4.7, g^m is invariant under $\Gamma(N')$ for $N' = 12N^2$. Since $g^m \circ \alpha \in \mathbb{C}((q_N))$ for all $\alpha \in \mathrm{SL}_2(\mathbb{Z})$, g^m has geometric level N. Then by Fricke–Wohlfahrt theorem Theorem 2.4.19, $g^m \in \mathbb{C}(X(N))$ and hence $\mathfrak{S} = X_N$, as desired. $\qquad\square$

By Theorem 2.4.10, we identify $R = R_N = \mathbb{Z}[C_m/\{\pm 1\}]$ with $\mathfrak{D}$ by sending $\alpha \in C_m/\{\pm 1\}$ to $[\alpha(\infty)]$. Recall (2.24):

$$\mathrm{div}(g_a) = N \sum_{b \in C_m/\{\pm 1\}} \frac{1}{2} B_2(\langle \mathrm{Tr}(ab)\rangle)\sigma_b^{-1}.$$

We then put

$$\theta = N \sum_{b \in C_m/\{\pm 1\}} \frac{1}{2} B_2(\langle \mathrm{Tr}(b/N)\rangle)\sigma_b^{-1} \in \mathbb{Q} \cdot R_N = \mathbb{Q}[C_m/\{\pm 1\}]. \quad (2.28)$$

Taking $a = \mathrm{Tr}(1/N)$, we find $\mathrm{div}(g_a) = \theta$; so,

$$\deg(\theta) = \deg(\mathrm{div}(g_a)) = 0. \quad (2.29)$$

Since $\mathrm{div}(g_a \circ \alpha) = \mathrm{div}(g_{a\alpha})$ for $\alpha \in \mathrm{GL}_2(\mathbb{Z})$, we find $\mathfrak{S} = R \cap R\theta$ by Fricke–Wohlfahrt theorem, where $R\theta$ is the fractional ideal inside $\mathbb{Q}[C_m/\{\pm 1\}]$ generated by θ. As in §2.4.6, we identify $W_m = W/p^m W$ with $(\mathbb{Z}/p^m\mathbb{Z})^2$ by sending $\frac{1}{2}a_1 + a_2\tau$ to $a = (a_1, a_2)$. Note that $C_m = W_m - pW_m$; so, if $a, b \in W_m$ are generators of $\mathbb{Z}/p^m\mathbb{Z}$-module W_m, $a, b, a + b$ and $a - b$ belong to C_m, where the sum $a + b$ and difference $a - b$ is computed in the additive group W_m. Let I be the ideal of R_N generated by parallelograms:

$$\pi(a, b) = (a + b) + (a - b) - 2(a) - 2(b) \in R_N,$$

where we regard $a, b \in C_m \subset (\mathbb{Z}/p^m\mathbb{Z})^2$ and $(a + b), (a - b), (a)$ and (b) is the image of $a + b, a - b, a$ and b in $C_m/\{\pm 1\}$.

Lemma 2.4.21. *Let $N = p^m$ with $m \geq 1$ and B_N be the set of pairs (a, b) in $\mathbb{Z}^2/N\mathbb{Z}^2$ such that a, b, $a + b$ and $a - b$ all have order N. Define I by the ideal of R_N generated by $\{\pi(a, b)\}_{(a,b) \in B_N}$. Then I contains $\sum_{a \in C_m} m(a/N)[a]$ if and only if $m : N^{-1}\mathbb{Z}^2/\mathbb{Z}^2 - \{(0, 0)\} \to \mathbb{Z}$ satisfies (Q_N).*

If $m(a) \neq 0$ for a of order d less than N with $N = dM$, by the distribution relation: $g_a = c \prod_{Mb=a} g_b$ for b's of order N, replacing m by m' such that $m'(b) = m(a)$ for all b with $Mb = a$ and $m'(a) = 0$ without changing any other values, we may assume that $m(a) \neq 0$ implies that a has order N (i.e., m is supported on C_m).

Proof. Multiplying N, we prove the equivalence between

$$\sum_{(r,s) \neq 0} m(r/N, s/N)r^2 \equiv \sum_{(r,s) \neq 0} m(r/N, s/N)s^2$$
$$\equiv \sum_{(r,s) \neq 0} m(r/N, s/N)rs \equiv 0 \mod N, \quad (Q'_N)$$

and $\sum_{(r,s)} m(\frac{r}{N}, \frac{s}{N})[(r, s)] \in I$. Without taking congruence modulo N of r and s, it is a plain computation to check that $\pi(a, b)$ for $a = (r, s)$ and $b = (r', s')$ satisfies the following condition:

$$\sum_{(r,s) \neq 0} m(r/N, s/N)r^2 = \sum_{(r,s) \neq 0} m(r/N, s/N)s^2$$
$$= \sum_{(r,s) \neq 0} m(r/N, s/N)rs = 0. \quad (Q)$$

For example,

left-hand-side of (Q) for $\pi(a, b) = (r + r')^2 + (r - r')^2 - 2r^2 - 2r'^2 = 0$.

Since r and s are actually classes modulo N, the equality in (Q) becomes the identity (Q'_N) modulo N.

Let $\deg(m) = \sum_{(r,s)} m(r, s)$. To forget about congruence classes, we argue in the group ring $\mathbb{Z}[\mathbb{Z}^2]$ regarding $m = \sum_{(r,s)} m(r/N, s/N)(r, s) \in \mathbb{Z}[\mathbb{Z}^2]$. In other words, assuming (Q) for $m \in \mathbb{Z}[\mathbb{Z}^2]$ and prove that m is a linear combination of $\pi(a, b)$ for finitely many (a, b) as long as $\deg(m)$ is even. Since we take modulo N at the end and N is odd, by multiplying 2 (which is a unit in $\mathbb{Z}/N\mathbb{Z}$), we may assume that $\deg(m)$ is even. Let $\mathcal{I}$ be the $\mathbb{Z}$-linear span in $\mathbb{Z}[\mathbb{Z}^2]$ generated by $\pi(a, b)$ with $a + b, a - b, a, b$ all having order N in $(\mathbb{Z}/N\mathbb{Z})^2$. Since the image of $\mathcal{I}$ in R_N and NR_N

span the ideal of elements satisfying (Q'_N), we are going to prove that any $m : \mathbb{Z}^2 \to \mathbb{Z}$ satisfying (Q) is in $\mathcal{I}$.

Put $h(r,s) = |r| + |s|$ and define $h(m) = \max_{m(r,s) \neq 0} h(r,s)$. Suppose $m(r,s) \neq 0$ with maximal $h(r,s) \geq 3$. As we remarked, $a := (r,s)$ has order N in $(\mathbb{Z}/N\mathbb{Z})^2$. Since the argument is the same for each quadrant by rotation of ± 90 and 180 degrees we may assume $r \geq 0$ and $s > 0$ with $r+s \geq 3$. We now show that we are able to choose $(0,0) \neq (i,j) \in [0, r/2] \times [0, s/2]$ such that $x = (r-i, s-j)$ and $y = (i,j)$ have both order N in $(\mathbb{Z}/N\mathbb{Z})^2$. Then $\pi(x,y) = (x+y)+(x-y)-2(x)-2(y) = (a)+(r-2i, r-2j)-2(x)-2(y)$, and $m' := m - m(a)\pi(x,y)$ removes a from its support and $h(m') \leq h(m)$, and repeating this process, eventually we can bring m modulo $\mathcal{I}$ to another function supported in $P := \{(c,d) : |\pm c \pm d| \leq 3\}$.

We separate our argument in the following four cases:

(1) $(r,s) \bmod p \in (\mathbb{Z}/p\mathbb{Z})^2 - \{0,1\}^2$,
(2) $(r,s) \equiv (0,1) \mod p$,
(3) $(r,s) \equiv (1,0) \mod p$,
(4) $(r,s) \equiv (1,1) \mod p$.

First suppose $r \geq 2$ and $s \geq 2$, we may choose in Case (1) and (2), $y = (1,0)$; Case (3) $y = (0,1)$ and Case (4) $y = (2,0)$ (as $r > 2$ and $s > 2$ in Case (4) by $N \geq 5$). If $r = 1$ (then $s \geq 2$), we are in Case (3), and $y = (0,1)$ still works. In the same way, if $s = 1$, $y = (1,0)$ works. If $r = 0$, $s \geq 3$, we can choose $y = (0,j)$ with $j \in \{1,2\}$ so that $s - 2j \not\equiv 0$ mod p. This is possible as $p \geq 5$. If $s = 0$, we take $y = (j,0)$ with $j \in \{1,2\}$ so that $r - 2j \not\equiv 0 \mod p$. Thus we are able to choose x, y as desired, and by induction, we may now assume that m is supported on $P_0 := \{(\pm 2, 0), (0, \pm 2), (\pm 1, 0), (0, \pm 1), (\pm 1, \pm 1)\}$. We check that $m : P_0 \to \mathbb{Z}$ satisfying (Q'_N) is a linear combination of $\pi(a,b)$'s.

To pin down, we follow [MUN, §3.1]. Note that $(1,1) + (1,-1) \equiv 2(1,0) + 2(0,1) \mod \mathcal{I}$ and $(2,0) + (0,2) \equiv 2(1,1) + 2(1,-1) \mod \mathcal{I}$. By this, we are reduced to a linear combination of $P_1 := \{(1,0), (0,1), (1,1), (2,0)\}$. Note that $8(1,0) - 2(2,0) = \alpha - \beta - 2\gamma \in \mathcal{I}$ with $\alpha = \pi((2,0), (3,0)) - \pi((1,0), (4,0))$, $\beta = \pi((1,0), (2,0))$ and $\gamma = \pi((1,0), (3,0))$. Suppose $m : P_1 \to \mathbb{Z}$ satisfies (Q'_N), and write $x = m(1,0)$, $y = m(0,1)$, $z = m(2,0)$ and $u = m(1,1)$. Then, we have

$$x + 4z + u = 0, y + u = 0, u = 0 \Rightarrow y = u = 0.$$

Since $\deg(m) = x + z$ is even, z is even. Writing $z = 2a$ and $x = -8a$. Since $8(1,0) - 2(2,0) \in \mathcal{I}$, we are done. $\qquad \square$

By Theorem 2.4.7, to have $g^m \in \mathcal{A}_N^\times$ is equivalent to having m satisfy $\deg(m) \equiv 0 \mod 12$ and (Q_N). This lemma therefore shows

Theorem 2.4.22. *Let* $I_{12} := \{\alpha \in I \mid \deg(\alpha) \equiv 0 \mod 12\}$. *Then we have* $\mathfrak{S}_N = R_N \theta \cap R_N = I_{12}\theta$.

2.4.10 Cuspidal class number formula

Let $R^0 = R_N^0 = \{D \in R_N \mid \deg(D) = 0\}$. Then we have

Theorem 2.4.23. *Suppose* $N = p^m$ *(*$m > 0$*) with a prime* $p \geq 5$. *Then*

$$|Cl_{X(N)}| = [R_N^0 : \mathfrak{S}_N] = \frac{6N^3}{|C_m|} \prod_{\chi \neq 1} \frac{N}{2} B_{2,\chi}$$

for the generalized Bernoulli number $B_{2,\chi} = \sum_{a \in C_m/\{\pm 1\}} B_2(\langle \frac{\mathrm{Tr}(a)}{N} \rangle)$, *where* χ *runs over all even character of* C_m.

The above $B_{2,\chi}$ is a non-zero multiple of $L(-1, \chi)$ (see (2.23)). The assumption $p \geq 5$ is enforced as we need to separate the denominator of the order of poles and vanishing coming from the factor $\Delta^{1/12}$ of the Siegel unit and the denominator coming from the Klein forms (i.e., a power of p).

Let $s := \sum_{\sigma \in C_m/\{\pm 1\}} \sigma \in R_N$ and $R^d := \{D \in R \mid \deg(D) \in d\mathbb{Z}\}$. Then for $D \in R_N$, $D \cdot s = \deg(D)s$, and $\mathrm{div}(\Delta) = N \cdot s$. Put $\theta' := \theta - \frac{N}{12}s$. First we prove a sequence of inclusions for $d = \frac{N|G|}{6}$:

$$R \supset R^d \supset I\theta' + RN \cdot s \supset I\theta',$$

and compute indices of adjacent terms (and related terms) one by one by a series of lemmas (to reach the formula).

Lemma 2.4.24. *We have* $R_N^0 \cap (I\theta' + R\,\mathrm{div}(\Delta)) = I_{12}\theta$.

Proof. Since $\xi s = \deg(\xi)s$ and $\mathrm{div}(\Delta) = Ns$, we find $R\,\mathrm{div}(\Delta) = \mathbb{Z}\,\mathrm{div}(\Delta)$. Write $G := C_m/\{\pm 1\}$. Recall $\deg(\theta) = 0$ by (2.29); so,

$$\deg(\theta') = -\frac{N}{12}|G| = -\frac{N}{12}\deg(s).$$

Thus, for $\xi \in I$, $\xi\theta = \xi\theta' + \frac{1}{12}\deg(\xi)Ns = \xi\theta' + \frac{1}{12}\deg(\xi)\,\mathrm{div}(\Delta) \in R^0$. This implies

$$R_N^0 \cap (I\theta' + R\,\mathrm{div}(\Delta)) = R_N^0 \cap (I\theta' + \mathbb{Z}\,\mathrm{div}(\Delta)) \subset I_{12}\theta.$$

The reverse inclusion follows from Theorem 2.4.22. $\qquad\square$

Lemma 2.4.25. *We have*

$$\deg(R^0 + I\theta' + RNs) = \deg(I\theta' + RNs) = \frac{N|G|}{6}\mathbb{Z};$$

so, $R^d = R^0 + I\theta' + RNs$ for $d = \frac{N|G|}{6}$.

Proof.   Since θ has degree 0 and $\mathbb{Q} \cdot R^0$ is the augmentation ideal of $\mathbb{Q}[G]$, $\deg(I\theta) = 0$. Thus we conclude

$$\deg(I\theta') = \deg(I(\frac{N}{12}s)) = 2\frac{N|G|}{12}\mathbb{Z} = \frac{N|G|}{6}\mathbb{Z}.$$

Here we used the fact that $\deg(\pi(a,b)) = -2$ and hence $\deg(I) = 2\mathbb{Z}$. Since $|G| = (p^2 - 1)/2 \equiv 0 \mod 6$ as $p \geq 5$, $\deg(I\theta') \subset \mathbb{Z}$. Therefore

$$\deg(\deg(I\theta' + RNs)) = \frac{N|G|}{6}\mathbb{Z} + N|G|\mathbb{Z} = \frac{N|G|}{6}\mathbb{Z}$$

as desired.   □

Thus for $d = \frac{N|G|}{6}$, we have inclusions:

$$R \supset R^d \supset I\theta' + RN \cdot s \supset I\theta'.$$

Lemma 2.4.26. *We have $[R : R^d] = d$ and $[I\theta' + RN \cdot s : I\theta'] = \frac{|G|}{12}$.*

Proof.   By the surjectivity of $\deg : R \to \mathbb{Z}$, we see

$$[R : R^d] = [\deg(R) : \deg(R^d)] = [\mathbb{Z} : d\mathbb{Z}] = d.$$

By the second isomorphism theorem, we have $[I\theta' + RN \cdot s : I\theta'] = [RN \cdot s : RN \cdot s \cap I\theta']$. We argue via modular forms. The module $I\theta'$ is the module generated by divisors of $\mathfrak{k}^m$ for $m : N^{-1}\mathbb{Z}^2/\mathbb{Z}^2 - \{0\} \to \mathbb{Z}$ satisfying (Q_N). Thus $D \in I\theta' \cap RN \cdot s = I\theta' \cap R(\mathrm{div}(\Delta))$ means that $D = \mathrm{div}(\mathfrak{k}^m) = \nu\,\mathrm{div}(\Delta)$ with $\nu \in \mathbb{Z}$. Thus $\mathfrak{k}^m = C \cdot \Delta^\nu$ with $0 \neq C \in \mathbb{C}$. Comparing the weight, we get

$$\nu = -\frac{1}{12}\sum_a m(a).$$

Since $g_a = \mathfrak{k}_a \Delta^{1/12}$, we find $g^m = C$. This implies that m is a constant function as $g^m \circ \alpha = g^{m \circ \alpha}$ for $\alpha = \mathrm{GL}_2(\mathbb{Z}/N\mathbb{Z})/\{\pm 1\} = \mathrm{Gal}(X(N)/\mathbf{P}^1(J))$ (and g_a for $a \in N^{-1}\mathbb{Z}^2/\mathbb{Z}^2 - \{0\}$ are independent; see Proposition 2.4.5). Since a constant function m satisfies (Q_N) (see Remark 2.3.14), m is arbitrary, and $\xi Ns \mapsto \xi\,\mathrm{div}(\Delta) = Ns\deg(\xi) \in \mathbb{Z}Ns$ sends RNs (resp. $RNs \cap I\theta'$) isomorphically to $\mathbb{Z}Ns$ (resp. $\frac{|G|}{12}\mathbb{Z}Ns$). Thus we conclude $[RN \cdot s : RN \cdot s \cap I\theta'] = \frac{|G|}{12}$.   □

For any two lattices L and L' in a finite dimensional $\mathbb{Q}$-vector space, we define $[L : L'] = \frac{[L:L \cap L']}{[L':L \cap L']} \in \mathbb{Q}$, which behaves just like an index. For example, we have $[L : L'][L' : L''] = [L : L'']$ and the value equals the index $[L : L']$ if $L \supset L'$. Moreover for a linear transformation T taking L onto L', we have $[L : L'] = |\det(T)|$. Since $[R^0 : R^0\theta] \neq 0$ (by Theorem 2.4.11) and $\deg(\theta') = -\frac{N|G|}{12} \neq 0$, we find $[R : R\theta'] \neq 0$.

Lemma 2.4.27. *We have* $[R_N : R_N\theta'] = \deg(\theta') \prod_{\chi \neq 1} \chi(\theta)$.

Proof. The index is just the determinant of the multiplication by θ' on R. The determinant can be calculated in $\mathbb{C}[G]$ which is the product of the eigenvalues on each eigenspace. The χ eigenspace is of dimension 1, and hence we get $\chi(\theta)$ as its eigenvalue if $\chi \neq 1$ (since $\theta \in R^0$). For the trivial eigenspace, R/R^0, the eigenvalue is $\deg(\theta') = \frac{N|G|}{12}$. $\qquad\square$

Lemma 2.4.28. *We have* $[R : I] = [R\theta' : I\theta'] = N^3$.

Proof. We may identify C_m with $P := \{(r,s) \in (\mathbb{Z}/N\mathbb{Z})^2 | (r) + (s) = \mathbb{Z}/N\mathbb{Z}\}$ (having order N). Then we claim that R/I is free of rank 3 with basis $(0,1),(1,0),(1,1)$. Indeed, in the proof of Lemma 2.4.21, we have shown that R/I is generated by $P_1 := \{(1,0),(0,1),(1,1),(2,0)\}$ and $8(1,1) \equiv 2(2,0) \mod I$; so, we can remove $(2,0)$.

We need to show that for any given $(r,s) \in P$, we have a unique triple $(x,y,z) \in (\mathbb{Z}/N\mathbb{Z})^3$ such that $(r,s) + x(1,0) + y(0,1) + z(1,1)$ such that this sum satisfies (Q'_N). The condition (Q'_N) is equivalent to

$$r^2 + x + z = 0, s^2 + y + z = 0 \text{ and } rs + z = 0,$$

which can be solved: $x = -rs - r^2, y = -rs - s^2$ and $z = -rs$. $\qquad\square$

Proof of Theorem 2.4.23. Recall $d = \frac{N|G|}{6} = [R : R^d]$. We have

$$|Cl_{X(N)}| = [R^0 : I_{12}\theta] \overset{\text{Lemma } 2.4.24}{=} [R^0 : R^0 \cap (I\theta' + RN \cdot s)]$$

$$\overset{\text{Lemma } 2.4.25}{=} [R^d : (I\theta' + RN \cdot s)] = \frac{[R : I\theta']}{[(I\theta' + RN \cdot s) : I\theta'][R : R^d]}$$

$$\overset{\text{Lemma } 2.4.26}{=} \frac{12 \cdot 6}{N|G|^2}[R : R\theta'][R\theta' : I\theta'] \overset{\text{Lemma } 2.4.28}{=} \frac{12 \cdot 6N^2}{|G|^2}[R : R\theta']$$

$$\overset{\text{Lemma } 2.4.27}{=} \frac{12 \cdot 6N^2}{|G|^2}\deg(\theta') \prod_{\chi \neq 1} \chi(\theta) = \frac{6N^3}{|G|} \prod_{\chi \neq 1} \chi(\theta),$$

where the last identity follows from $\deg(\theta') = \frac{N|G|}{12}$. This finishes the proof as $\chi(\theta) = \frac{N}{2}B_{2,\chi}$ by definition. $\qquad\square$

2.4.11   *Cuspidal class number formula for $X_1(N)$*

We give a description of the cuspidal class number formula for $X_1(N)$ without proof. Let $N = p^r$ for a prime p. We assume that $p \geq 5$ for simplicity. Let $X_0(p) = \Gamma_0(p) \backslash (\mathfrak{H} \sqcup \mathbf{P}^1(\mathbb{Q}))$ for $\Gamma_0(N) := \left\{ \left(\begin{smallmatrix} a & b \\ c & d \end{smallmatrix} \right) \big| c \equiv 0 \mod N \right\}$. It is easy to see $\mathrm{SL}_2(\mathbb{Z}) = \Gamma_0(p) \sqcup \bigsqcup_{j=1}^{p} \Gamma_0(p) \delta \left(\begin{smallmatrix} 1 & j \\ 0 & 1 \end{smallmatrix} \right)$ for $\delta = \left(\begin{smallmatrix} 0 & -1 \\ 1 & 0 \end{smallmatrix} \right)$. Thus the set of cusps $S_0(N)$ of $X_0(N)$ can be explicitly given by

$$S_0(p) = \Gamma_0(p) \backslash \mathrm{SL}_2(\mathbb{Z}) / \Gamma_\infty = \{0, \infty\} = \{\delta(\infty), \infty\},$$

where $\Gamma_\infty := \{\gamma \in \mathrm{SL}_2(\mathbb{Z}) | \gamma(\infty) = \infty)\} = \left\{ \pm \left(\begin{smallmatrix} 1 & m \\ 0 & 1 \end{smallmatrix} \right) \big| m \in \mathbb{Z} \right\}$. Thus one can classify the set $S_1(p^r)$ of cusps of $X_1(p^r)$ into three classes:

$$S_1(p^r) = S^\infty \sqcup S_m \sqcup S^0.$$

The classification goes as follows: Since $\left(\begin{smallmatrix} x & b \\ y & d \end{smallmatrix} \right) \left(\begin{smallmatrix} 1 & m \\ 0 & 1 \end{smallmatrix} \right) = \left(\begin{smallmatrix} x & xm+b \\ y & ym+d \end{smallmatrix} \right)$ for $\left(\begin{smallmatrix} x & b \\ y & d \end{smallmatrix} \right) \in \mathrm{SL}_2(\mathbb{Z})$,

$$\Gamma_1(p^r) \backslash \mathrm{SL}_2(\mathbb{Z}) / \Gamma_\infty \cong \left\{ \mathbf{x} := \left(\begin{smallmatrix} x \\ y \end{smallmatrix} \right) \in (\mathbb{Z}/N\mathbb{Z})^2 \big| \mathbf{x} \text{ has order } N \right\} / \sim,$$

where $\left(\begin{smallmatrix} x \\ y \end{smallmatrix} \right) \sim \left(\begin{smallmatrix} x+my \\ y \end{smallmatrix} \right)$ for $m \in \mathbb{Z}$. Thus

$$S^0 \cong \left\{ \left(\begin{smallmatrix} 0 \\ y \end{smallmatrix} \right) \big| y \in (\mathbb{Z}/N\mathbb{Z})^\times / \{\pm 1\} \right\} \quad \text{and} \quad S^\infty \cong \left\{ \left(\begin{smallmatrix} x \\ 0 \end{smallmatrix} \right) \big| x \in (\mathbb{Z}/N\mathbb{Z})^\times / \{\pm 1\} \right\}.$$

The set S^∞ is made up of cusps unramified over $\infty \in X_0(p)$. The action of $W := \left(\begin{smallmatrix} 0 & -1 \\ p^r & 0 \end{smallmatrix} \right)$ induces $S^0 \cong S^\infty$. On $S^?$, the group $G = G_r = (\mathbb{Z}/p^r\mathbb{Z})^\times / \{\pm\}$ acts transitively and freely, and hence, writing $\mathfrak{D}^? := \bigoplus_{s \in S^?} \mathbb{Z}[s]$ and $\mathfrak{D}_0^? := \{D \in \mathfrak{D}^? | \deg(D) = 0\}$, we have $\mathbb{Z}[G] \cong \mathfrak{D}^?$ by sending $\sum_g a_g g$ to $\sum_g a_g g(?)$ for $? = 0, \infty$.

The Riemann surface $\Gamma_1(p^r) \backslash (\mathfrak{H} \sqcup \mathbf{P}^1(\mathbb{Q}))$ regarded as a projective curve $X_1(p^r)_{/\mathbb{C}}$ has a canonical rational model. Indeed, for each $z \in \mathfrak{H}$, taking the corresponding lattice $L_z = \mathbb{Z} + \mathbb{Z}z$, the pair $(\mathbb{C}/L_z, \frac{1}{p^r})$ is stabilized by $\Gamma_1(p^r)$ embedded into $\mathrm{Aut}(L_z)$ with respect to the basis $(1, z)$. Thus $Y_1(N) = X_1(N) - \{\text{cusps}\}$ classifies pairs $(E_z, \mu_{p^r} \hookrightarrow E_z)$ (with $E_z(\mathbb{C}) = \mathbb{C}/L_z$) of elliptic curves $E_{z/\mathbb{C}}$ with an embedding $\mu_{p^n} \ni \mathbf{e}(1/p^r) \mapsto 1/p^r \in E(\mathbb{C})$. By applying $\mathbf{e} : \mathbb{C}/\mathbb{Z} \cong \mathbb{C}^\times$, $\mathbb{C}/L_z \cong \mathbb{C}^\times / q^\mathbb{Z}$ for $q = \mathbf{e}(z)$; so, the embedding $\mu_{p^r} \hookrightarrow E_z$ is actually induced by the inclusion $\mu_{p^r}(C) \hookrightarrow \mathbb{C}^\times$. This classification problem is well defined over $\mathbb{Q}$ and gives rise to a model $X_1(N)_{/\mathbb{Q}}$ with one to one identification $X_1(K)$ with isomorphism classes of $(E, \mu_{p^r} \hookrightarrow E)_{/K}$ defined over any field K of characteristic 0. Similarly we have its dual model $X_1^*(p^r)$ classifying $(E, \mathbb{Z}/p^r \hookrightarrow E)$.

Write $X_1(p^r) - \{\text{cusps}\} = \mathrm{Spec}(\mathcal{A}_{1,N})$ and $X_1^*(p^r) - \{\text{cusps}\} = \mathrm{Spec}(\mathcal{A}_{1,N}^*)$. Define

$$\mathfrak{F}^\infty := \{\mathrm{div}(u) | u \in \mathcal{A}_{1,N}^\infty\} \quad \text{with} \quad \mathcal{A}_{1,N}^\infty := \{u \in \mathcal{A}_{1,N}^\times | \mathrm{Supp}(\mathrm{div}(u)) \subset S^\infty\}$$
$$\mathfrak{F}^0 := \{\mathrm{div}(u) | u \in \mathcal{A}_{1,N}^{*,0}\} \quad \text{with} \quad \mathcal{A}_{1,N}^{*,0} := \{u \in (\mathcal{A}_{1,N}^*)^\times | \mathrm{Supp}(\mathrm{div}(u)) \subset S^0\}.$$
$$(2.30)$$

The partial cuspidal group is defined as follows:

$$Cl_{X_1(N)}^\infty := \mathfrak{D}_0^\infty / \mathfrak{F}^\infty \quad \text{and} \quad Cl_{X_1^*(N)}^0 := \mathfrak{D}_0^0 / \mathfrak{F}^0.$$

By the action of the involution W, we have $Cl_{X_1(N)}^\infty \cong Cl_{X_1^*(N)}^0$. In this case, the Stickelberger element is the traditional one:

$$\theta = N \sum_{b \in G} \frac{1}{2} B_2(\langle b/N \rangle) \sigma_b^{-1},$$

where $\sigma_b \in \mathrm{Gal}(X_1(p^r)/X_0(p^r))$ is associated to $b \in G$. Let I be the ideal of $R = \mathbb{Z}[G]$ generated by $\sigma_c - c^2$ and $I_0 := \{\xi \in I | \deg(\xi) = 0\}$. Then the following lemma is easier than Theorem 2.4.22 (see [MUN, Lemma 6.1.2]:

Lemma 2.4.29. *For* $\theta' = \theta - \frac{p^r}{12} s$ *with* $s = \sum_{\sigma \in G} \sigma$, $R\theta' \cap R = I\theta'$ *and* $R_0\theta' \cap R = I_0\theta' = I_0\theta$.

Consider functions $m : (p^{-r}\mathbb{Z}/\mathbb{Z} \to p^{1-r}\mathbb{Z}/\mathbb{Z})/\{\pm 1\} \to \mathbb{Z}$ satisfying

(1) For every coset C of $p^{-1}\mathbb{Z}/\mathbb{Z}$, $\sum_{a \in C} m(a) = 0$,
(2) $\sum_a m(a)(p^r a)^2 \equiv 0 \mod p^r$.

Let $g_a := g_{(0,a)}$ and $_a g := g_{(a,0)}$ for $a \in \mathbb{Q} - \mathbb{Z}$ with $Na \in \mathbb{Z}$, and put $g^m = \prod_a g_a^{m(a)}$ and $^m g := \prod_a {}_a g^{m(a)}$. It is shown in [MUN, Theorem 6.3.1] that the units $g^m \in \mathfrak{F}^0$ and $^m g \in \mathfrak{F}^\infty$ satisfying the above two conditions generate $\mathcal{A}_{1,N}^{*,0}$ up to scalars. Note that $\mathrm{div}(_{1/N}g) = \theta/N$ and $\theta = \mathrm{div}(g_{1/N})$ identifying $R[G] = \mathfrak{D}^0 = \mathfrak{D}^\infty$ by $(\mathbb{Z}/N\mathbb{Z})^\times \ni 0 \in S^0$ and $(\mathbb{Z}/N\mathbb{Z})^\times \ni \infty \in S^\infty$. This enable us to get

Theorem 2.4.30 (Kubert–Lang). *Assume* $N = p^r$ *for* $p \geq 5$. *Then*

$$|Cl_{X_1(N)}^\infty| = |Cl_{X_1^*(N)}^0| = \frac{p^{p^{r-1}}}{p^{2r-2}} \prod_{\chi : G \to \overline{\mathbb{Q}}^\times, \chi \neq 1} \frac{1}{2} B_{2,\chi}.$$

Though we do not know the cyclicity of $Cl_{X_1(N)}$ over $\mathbb{Z}[G]$ except for the case where $N = p$ [KL78, Theorem 3.4], the limit $\varprojlim_r Cl_{X_1(p^r)}^j \otimes_{\mathbb{Z}} \mathbb{Z}_p$ $(j = 0, \infty)$ is likely to be isomorphic (after a suitable modification) to the

Tate twist of the Iwasawa module we studied in the previous chapter; so, one would expect the cyclicity for the p-part (at least under the Kummer–Vandiver conjecture). The proof by Mazur–Wiles of the Iwasawa main conjecture [MW84, Chapter 4] and the following study by Ohta [O99] and Sharifi [S11] of Eisenstein ideals may shed some light upon this question.

Chapter 3

Cohomological modular forms and p-adic L-functions

We will study p-adically cohomological modular forms on $GL(1)$ and $GL(2)$ limiting ourselves to weight 2 modular forms and constant coefficients. A more general theory including the case of higher weight and locally constant sheaves will be discussed in the next Chapter 4; so, this chapter is intended to be a warm-up for more advanced and technical studies.

We discuss the following four topics in this chapter:

(1) Isomorphism of Eichler–Shimura type connecting modular forms of weight 2 and cohomology groups with constant coefficients (higher weight forms will be treated in Chapter 4),
(2) Rationality and integrality of L-values,
(3) p-Adic measure theory,
(4) Construction of analytic p-adic L-functions for GL(1) and GL(2).

Along with these main topics, we will give a brief description of different cohomology theories we will use.

3.1 The multiplicative group G_m and Dirichlet L-function

In this first section, without going into technical details, we describe a prototypical example of the cohomology groups we deal with and construction of p-adic L-functions. To justify our construction, we shall give a brief description of cohomology theory. The example we describe is from [LFE, Chapter 4].

3.1.1   Betti cohomology groups

We consider the multiplicative group $\mathbf{G}_m$ as an algebraic group. Thus as a scheme, $\mathbf{G}_m = \mathrm{Spec}(\mathbb{Z}[t, t^{-1}])$ whose A-points $\mathbf{G}_m(A)$ are given by $\mathrm{Hom}_{\mathrm{alg}}(\mathbb{Z}[t, t^{-1}], A) \cong A^\times$ by $\phi \mapsto \phi(t)$ for each commutative ring A.

Exercise 3.1.1. Prove the following assertion:

(1) $\mathrm{Hom}_{\mathrm{alg}}(\mathbb{Z}[t, t^{-1}], A) \cong A^\times$ as sets for commutative algebras A, where $\mathrm{Hom}_{\mathrm{alg}}$ denotes the set of algebra homomorphisms.
(2) The "map" assigning each commutative algebra A the group $\mathbf{G}_m(A)$ gives rise to a covariant functor from the category of commutative algebras with identity into the category of abelian groups.

The space $\mathbf{G}_m(\mathbb{C})$ has nontrivial homology group of positive degree. Indeed, we have $H_1(\mathbf{G}_m(\mathbb{C}), A) \cong A$. Here intuitively, for any given commutative ring A, a C^∞ n-chain in a C^∞ manifold X is a formal A-linear combination of C^∞ maps from $\Delta^n = [0, 1]^n$ for the closed interval $[0, 1]$ into X. Then the totality of n-chains forms an A-free module $C_n(X; A)$ generated by $\phi : \Delta^n \to X$. Since Δ^n has natural boundaries

$$[0, 1] \times \cdots \times [0, 1] \times \overset{i}{x} \times [0, 1] \times \cdots \times [0, 1] = \Delta_x^{n-1}$$

with orientation, identifying $\Delta_{i,x}^{n-1}$ for $x = 0, 1$, we can think of the boundary $\partial \phi = \sum_{i,x} (-1)^{i+1+x} \phi|_{\Delta_{i,x}^{n-1}}$ which is a $n-1$ chain, identifying $\Delta_{i,x}^{n-1}$ with Δ^n (removing the i-th coordinate). We set Δ^0 to be the origin 0 (one point). By linearity, we can extend ∂ to an A-linear map $\partial : C_n(X; A) \to C_{n-1}(X; A)$ if $n \geq 1$. Thus we have a chain complex

$$\cdots \overset{\partial}{\to} C_n(X; A) \overset{\partial}{\to} C_{n-1}(X; A) \overset{\partial}{\to} \cdots \overset{\partial}{\to} C_0(X; A) \overset{\deg}{\longrightarrow} A \to 0,$$

where $\deg(\sum_\phi a_\phi \phi) = \sum_\phi a_\phi$. An n-chain ϕ with $\partial \phi = 0$ is called an n cycle.

Exercise 3.1.2. Prove $\partial \circ \partial = 0$.

By taking A-dual, we put $C^n(X; A) = \mathrm{Hom}(C_n(X; A), A)$. Then we have a reversed chain complex

$$0 \to A \overset{\deg^*}{\longrightarrow} C^0(X; A) \overset{\partial^*}{\longrightarrow} \cdots \overset{\partial^*}{\longrightarrow} C^n(X; A) \overset{\partial^*}{\longrightarrow} \cdots,$$

where $\partial^* \phi = \phi \circ \partial$. An n-cochain $\phi \in C^n(X; A)$ with $\partial^* \phi = 0$ is called an n-cocycle. Similarly, an n-cocycle (resp. an n-cycle) which is in the image of ∂^* (resp. ∂) is called an n-coboundary (resp. an n-boundary). Here is a definition of the singular (or Betti) cohomology group of degree $n \geq 1$:

$$H_n(X, A) = \frac{\text{Ker}(\partial : C_n(X; A) \to C_{n-1}(X; A))}{\text{Im}(\partial : C_{n+1}(X; A) \to C_n(X; A))},$$

$$H^n(X, A) = \frac{\text{Ker}(\partial^* : C^n(X; A) \to C^{n+1}(X; A))}{\text{Im}(\partial^* : C^{n-1}(X; A) \to C^n(X; A))}. \tag{3.1}$$

When $n = 0$, we put

$$H_0(X, A) = C_0(X; A) / \text{Im}(\partial),$$

$$H^0(X, A) = \text{Ker}(\partial : C^0(X; A) \to C^1(X; A)).$$

Exercise 3.1.3. Prove that $H_0(X, A) \cong H^0(X, A) \cong A$ if X is connected.

If X is a real C^∞-manifold of dimension d, $H^n(X, A) = H_n(X, A) = 0$ for $n > d$.

By definition, taking a cohomology class $[c]$ represented by an n-cocycle c and a homology class $[\gamma]$ represented by a cycle γ. Then $c \in \text{Hom}(C_n(X; A), A)$; so, we have the value $c(\gamma)$. If we replace γ by $\gamma + \partial b$, we have $c(\gamma) = c(\gamma) + c(\partial b) = c(\gamma)$, because $c(\partial b) = (\partial^* c)(b) = 0$. Similarly, $(c + \partial^* \beta)(\gamma) = c(\gamma)$, and hence $c(\gamma) \in A$ only depends on the classes $[c]$ and $[\gamma]$. Thus we have the Poincaré duality pairing:

$$\langle \cdot, \cdot \rangle : H^n(X, A) \times H_n(X, A) \to A \tag{3.2}$$

given by $\langle [c], [\gamma] \rangle = c(\gamma)$.

Here is an example.

Example 3.1.1. Let S^1 be the unit circle. Then 1-chain $\phi : [0, 1] \to S^1$ has $\partial \phi = \phi(0) - \phi(1)$. Thus $\phi \in \text{Ker}(\partial) \Leftrightarrow \phi(0) = \phi(1)$, and therefore, if one moves x from 0 to 1, $\phi(x)$ circles m times forward in counterclockwise or backward. Assigning ϕ to this number $m = [\phi]$ when $\phi(x)$ moves forward and $-m = [\phi]$ when $\phi(x)$ moves backwards, we have a linear map $H_1(S^1, \mathbb{Z}) \to \mathbb{Z}$, and actually, we have $H_1(S^1, \mathbb{Z}) \cong \mathbb{Z}$.

Exercise 3.1.4. Prove that $H_1(S^1, \mathbb{Z}) \cong \mathbb{Z}$. (Hint: if $[\phi] = [\varphi] = 1$ and $\phi(0) = \varphi(0) = 0 \in S^1 = \mathbb{R}/\mathbb{Z}$, then defining $\Delta^2 \xrightarrow{\Phi} S^1$ by $\Phi(x, y) = \phi(x)y + \varphi(x)(1 - y)$ identifying $S^1 - \{0\}$ with $(0, 1)$ naturally, we have $\partial \Phi = \phi - \varphi$; so, the class of ϕ and φ are equal in $H_1(S^1, \mathbb{Z})$.)

Example 3.1.2. We now consider $X = \mathbb{R}$. Then it is well known that $H_1(\mathbb{R}, A) = H^1(\mathbb{R}, A) = 0$. By the Künneth formula (see [CGP, I.0.8]),

$$H^n(X \times Y, A) \cong \bigoplus_{i+j=n} H^i(X, A) \otimes_A H^j(Y, A)$$

as long as A is a principal ideal domain (or Dedekind domain) and $H^j(X, A)$ and $H^j(Y, A)$ are A-free for all $j = 0, 1, \ldots, n$. In particular,

$$H^1(X \times Y, A) \cong (H^1(X, A) \otimes_A H^0(Y, A)) \oplus (H^0(X, A) \otimes_A H^1(Y, A))$$

if $H^1(X, A)$ and $H^1(Y, A)$ are A-free. By polar coordinates, we have $\mathbb{C}^\times \cong (0, \infty) \times S^1 \cong \mathbb{R} \times S^1$. Thus $H^1(\mathbb{C}^\times, \mathbb{Z}) \cong \mathbb{Z}$ by regarding a generator $\phi \in H_1(S^1, \mathbb{Z})$ as a cycle having values in $\mathbb{C}^\times$. We may identify $\mathbb{R} \times S^1$ with $T = \mathbb{C}/\mathbb{Z}$, and the isomorphism $T \cong \mathbb{C}^\times$ can be given by $z \mapsto \mathbf{e}(z) = \exp(2\pi i z)$. Then $H_1(T, \mathbb{Z})$ is generated by $\phi : [0, 1] \hookrightarrow \mathbb{R}/\mathbb{Z} \subset T$, which is the real circle in T. In this case, we have $H_1(T, A) = H_1(T, \mathbb{Z}) \otimes_\mathbb{Z} A$, and the above fact holds for all commutative rings A.

There is another way of computing a cohomology group over $\mathbb{C}$ [PAG, Chapter 0, §3]. We consider a C^∞ class differential form ω on X of degree n. Pick a point $x \in X$ and a coordinate neighborhood U of x with coordinate $t_1, \ldots, t_d$ with x giving the origin of U, ω has the following form

$$\omega = \sum_{i_1 < \cdots < i_n} f_{i_1 i_2 \ldots i_n}(t) dt_{i_1} \wedge \cdots \wedge dt_{i_n} \tag{3.3}$$

with C^∞ functions $f_{i_1 i_2 \ldots i_n}(t)$. If one changes the coordinates system, the expression of ω changes according to the chain-rule. For example, if $n = 1$, and $z_1, \ldots, z_d$ is another set of coordinates, $dt_i = \sum_j \frac{\partial t_i}{\partial z_j} dz_j$ and the expression of $\omega = \sum_i f_i dt_i$ with respect to z_j is given by

$$\omega = \sum_i f_i \sum_j \frac{\partial t_i}{\partial z_j} dz_j = \sum_j \left(\sum_i f_i \frac{\partial t_i}{\partial z_j} \right) dz_j.$$

If (U, t_i) and (V, z_j) are coordinate neighborhoods with $U \cap V \neq \emptyset$, the expression of ω on U and V is related by the above chain-rule. Write $\Omega^n(X, \mathbb{C})$ for the $\mathbb{C}$-vector space of differential forms as above of degree n. When $n = 0$, $\Omega^0(X, \mathbb{C})$ is just the vector space of C^∞-class functions on X. We can define the exterior derivative $d\omega \in \Omega^{n+1}(X, \mathbb{C})$ as follows. For $f \in \Omega^0(X, \mathbb{C})$, we just put $df = \sum_i \frac{\partial f}{\partial t_i} dt_i$. For ω as in (3.3), we define

$$d\omega = \sum_{i_1 < \cdots < i_n} df_{i_1 i_2 \ldots i_n}(t) \wedge dt_{i_1} \wedge \cdots \wedge dt_{i_n}.$$

Exercise 3.1.5.

(1) Check that $d\omega$ is a well defined differential form of degree $n + 1$ if ω is of degree n. Here you need to verify that the chain-rule is satisfied by $d\omega$ if one changes coordinates.

(2) Prove $d \circ d = 0$.

We simply put $\Omega^n(X, \mathbb{C}) = 0$ if $n > d = \dim X$. By the above exercise, we have a complex:

$$0 \to \mathbb{C} \to \Omega^0(X, \mathbb{C}) \xrightarrow{d} \Omega^1(X, \mathbb{C}) \xrightarrow{d} \cdots \xrightarrow{d} \Omega^d(X, \mathbb{C}) \to 0.$$

We define the de Rham cohomology group $H^n_{DR}(X, \mathbb{C})$ (in differential geometry) by

$$H^n_{DR}(X, \mathbb{C}) = \frac{\mathrm{Ker}(d : \Omega^n(X, \mathbb{C}) \to \Omega^{n+1}(X, \mathbb{C}))}{\mathrm{Im}(d : \Omega^{n-1}(X, \mathbb{C}) \to \Omega^n(X, \mathbb{C}))}$$

if $n \geq 1$. When $n = 0$, we simply put $H^0_{DR}(X, \mathbb{C}) = \mathrm{Ker}(d : \Omega^0(X, \mathbb{C}) \to \Omega^1(X, \mathbb{C}))$, which is isomorphic to the space of constant functions in $\Omega^0(X, \mathbb{C})$ (and is isomorphic to $\mathbb{C}$ if X is connected) by the fundamental theorem of Calculus.

Theorem 3.1.1. *If X is a compact differentiable manifold of dimension d, we have a canonical isomorphism $H^n(X, \mathbb{C}) \cong H^n_{DR}(X, \mathbb{C})$.*

Here is a brief sketch of the proof. By the Poincaré duality, we have $H^n(X, \mathbb{C}) \cong \mathrm{Hom}_{\mathbb{C}}(H_n(X, \mathbb{C}), \mathbb{C})$. We have a pairing $H^n_{DR}(X, \mathbb{C}) \times H_n(X, \mathbb{C}) \to \mathbb{C}$ given by $(\omega, \gamma) = \int_\gamma \omega$. By Stokes' theorem, $\int_{\partial\gamma} \omega = \int_\gamma d\omega$, and hence the pairing is well defined. G. de Rham proved the non-degeneracy of the above pairing, which implies the theorem. See [PAG, Chapter 0, §3] for the details.

3.1.2   Cohomology of $\mathbf{G}_m(\mathbb{C})$ and Dirichlet L-values

Consider a Dirichlet character $\chi : \mathbb{Z}/N\mathbb{Z} \to \mathbb{C}^\times$; so, $\chi(m)\chi(n) = \chi(mn)$ and $\chi(m) = 0$ if m is not prime to N. The Dirichlet L-function $L(s, \chi)$ is defined by $L(s, \chi) = \sum_{n=1}^\infty \chi(n)n^{-s}$ which is absolutely and locally uniformly convergent if $\mathrm{Re}(s) > 1$. The holomorphic function $L(s, \chi)$ can be continued analytically to the whole complex plane if χ is non-trivial (see [LFE, Chapter 2]). When $N = 1$ and $\chi = 1$, we define $L(s, \chi)$ by the Riemann zeta function $\zeta(s)$. For simplicity, we suppose that χ is nontrivial.

Identify $\mathbf{G}_m(\mathbb{C})$ with $T = \mathbb{C}/\mathbb{Z}$. Let z be the complex coordinate of T, and put $q = \exp(2\pi i z)$ which is the coordinate of $\mathbf{G}_m(\mathbb{C})$. We consider $\theta_\chi(z) = \sum_{n=1}^\infty \chi(n)q^n$. Then we see

$$\theta_\chi(z) = \sum_{a=1}^N \chi(a) \sum_{n=0}^\infty q^{a+nN} = \sum_a \chi(a)\frac{q^a}{1-q^N} = \frac{\sum_a \chi(a)q^a}{1-q^N}.$$

Thus θ_χ is a meromorphic function on T. Since $\sum_a \chi(a) = 0$, $q = 1$ (that is, $z = 0$ is not a pole of θ_χ). Thus θ_χ can have a pole at nontrivial N-th roots of unity in $\mathbb{C}^\times$ (or at $\frac{a}{N} \in T$ for a prime to N).

Lemma 3.1.2. *If χ is a primitive character modulo N, θ_χ has a pole of order 1 only at primitive N-th roots of unity. The residue of θ_χ at $q = \zeta$ for a primitive N-th root ζ of unity is given by $\mathrm{Res}_{q=\zeta} \theta_\chi = -N^{-1}\zeta G(\chi,\zeta)$ for the Gauss sum $G(\chi,\zeta) = \sum_{a \bmod N} \chi(a)\zeta^a$.*

Proof. If ζ is a M-th root of unity for a proper divisor M of N with $MM' = N$. Then we have

$$\sum_{a=1}^{N} \chi(a)\zeta^a = \sum_{a \in \mathbb{Z}/M\mathbb{Z}} \zeta^a \sum_{b \equiv a \bmod M, b \in \mathbb{Z}/N\mathbb{Z}} \chi(b)$$

$$= \sum_{a \in \mathbb{Z}/M\mathbb{Z}} \zeta^a \chi(a) \sum_{b \equiv 1 \bmod M, b \in \mathbb{Z}/N\mathbb{Z}} \chi(b) = 0,$$

because χ restricted to the subgroup $\{b \in (\mathbb{Z}/N\mathbb{Z})^\times | b \equiv 1 \bmod M\}$ is a nontrivial character. This shows that the zero at $q = \zeta$ of $q^N - 1$ is cancelled by the zero at the same location of the numerator of θ_χ.

Take a primitive N-th root ζ of unity. The value of $\frac{(q^N-1)}{(q-\zeta)}$ at ζ is given by $\frac{d(q^N-1)}{dq}|_{q=\zeta} = Nq^{N-1}|_{q=\zeta} = N\zeta^{-1}$. The value at $q = \zeta$ of the numerator at $q = \zeta$ is given by the Gauss sum $G(\chi,\zeta) = \sum_{a \bmod N} \chi(a)\zeta^a$ which is non-zero (e.g., Exercise 3.1.9). Thus $\mathrm{Res}_{q=\zeta} \theta_\chi = -N^{-1}\zeta^{-1}G(\chi,\zeta)$. $\quad\square$

Exercise 3.1.6. Prove the following facts:

(1) the expansion defining θ_χ is absolutely and locally uniformly convergent on the upper half plane $\mathfrak{H} = \{z \in \mathbb{C} | \mathrm{Im}(z) > 0\}$,
(2) the expansion of θ_χ convergent on the lower half plane $\overline{\mathfrak{H}}$ is given by

$$- \sum_{n=-1}^{-\infty} \chi(n)q^n.$$

Write $z = x + iy$ for $i = \sqrt{-1}$. An important fact is

$$\int_0^\infty \theta_\chi(y\sqrt{-1})dy = \sum_{n=1}^\infty \chi(n) \int_0^\infty \exp(-2\pi y)y^{s-1}dy|_{s=1}$$

$$\overset{(*)}{=} (2\pi)^{-s}\Gamma(s) \sum_{n=1}^\infty \chi(n)n^{-s}|_{s=1} = (2\pi)^{-1}L(1,\chi). \quad (3.4)$$

Here at $(*)$, a familiar formula $\Gamma(s) = \int_0^\infty e^{-t}t^{s-1}dt$ is applied.

Exercise 3.1.7.

(1) Justify the interchange of the summation and the integral in (3.4) when Re(s) > 1,
(2) Prove the middle identity (*), using Euler's formula $\int_0^\infty e^{-y}y^{s-1}dy = \Gamma(s)$ if Re(s) > 1,
(3) Suppose that χ is non-trivial. Justify the formula (3.4) even if Re(s) < 1. You may use the fact that $\Gamma(s)$ is a well defined meromorphic function on the whole complex plane $\mathbb{C}$.

By a similar calculation using the expansion of θ_χ over the lower half plane, we have

$$\int_{-\infty}^0 \theta_\chi(y\sqrt{-1})dy = -\sum_{n=1}^\infty \chi(n)\int_{-\infty}^0 \exp(2\pi y)|y|^{s-1}dy|_{s=1}$$

$$= -\chi(-1)(2\pi)^{-s}\Gamma(s)\sum_{n=1}^\infty \chi(n)n^{-s}|_{s=1} = -\chi(-1)(2\pi)^{-1}L(1,\chi). \quad (3.5)$$

Exercise 3.1.8. Justify the above formula (3.5).

Thus we have

$$2(2\pi i)^{-1}L(1,\chi) = \int_{-\infty}^\infty \theta_\chi dz = \int_\gamma \theta_\chi dz,$$

because $dy = \sqrt{-1}dz$ on γ_0. Here $\gamma = \gamma_0$ is the vertical line on T passing through 0.

Our idea is to prove $L(0,\chi) = -2NG(\chi^{-1})^{-1}\frac{L(1,\chi^{-1})}{2\pi i} \in \mathbb{Q}(\chi)$ (if $\chi(-1) = -1$) by using cohomology theory. Here $G(\chi) = G(\chi, \mathbf{e}(\frac{1}{N}))$ is the standard *Gauss sum* $G(\chi) = \sum_{a=1}^N \chi(a)\mathbf{e}(\frac{a}{N})$. Here $\mathbb{Q}(\chi)$ is a finite extension of $\mathbb{Q}$ generated by the values $\chi(n)$ for $n = 1, 2, \ldots$. Since θ_χ for primitive χ has a pole at $\frac{a}{N}$, we need to consider $T_N = T - \{\frac{a}{N}|a\mathbb{Z} + N\mathbb{Z} = \mathbb{Z}\}$.

Exercise 3.1.9. Prove $G(\chi^{-1})G(\chi) = \chi(-1)N$ if χ is primitive modulo N.

Then $\omega_\chi := G(\chi)^{-1}\theta_\chi(z)dz$ gives rise to a cohomology class in $H^1_{DR}(T_N, \mathbb{C})$, since $d(\theta_\chi dz) = 0$. If we can prove that $\omega_\chi \in H^1(T_N, \mathbb{Q}(\chi))$, we will have $\int_\gamma \omega_\chi \in \mathbb{Q}(\chi)$.

A Dirichlet character χ is called even or odd according to whether $\chi(-1) = 1$ or $\chi(-1) = -1$. Here is a theorem of Euler (and Hurwitz):

Theorem 3.1.3. *If m is a positive integer whose parity is given by the parity of χ, we have $L(1 - m, \chi) \in \mathbb{Q}(\chi)$.*

We prove this in Subsection 3.1.4 when χ is odd and $m - 1$. See [LFE, Chapter 4] for the general case.

3.1.3   Relative cohomology

Since the vertical line γ_x passing through $x \in S^1 = \mathbb{R}/\mathbb{Z}$ is not a 1-chain, we need to generalize cohomology theory to manifolds with boundary.

Exercise 3.1.10. For $x \notin \{\frac{a}{N} | a\mathbb{Z} + N\mathbb{Z} = \mathbb{Z}\}$, explain why γ_x is not a 1-chain on T_N.

We add S^1 as the two boundaries of T_N at $y = \pm\infty$, and write the resulting space as $\overline{T}_N$. Thus the boundary $\partial \overline{T}_N$ is the disjoint union of S^1 located at $y = \pm\infty$.

More generally we take a complex manifold X of dimension 1; i.e., compact or open Riemann surfaces. Pick a finite set of points $S \subset X$. Let $Y = X - S$ (the punctured open Riemann surface at S). The exponential function $\mathbf{e} : \mathbb{C}/\mathbb{Z} \to \mathbb{C}^\times$ given by $\mathbf{e}(z) = \exp(2\pi\sqrt{-1}z)$ induces a diffeomorphism between a tube $S^1 \times (0, \epsilon)$ and an open disc of radius $\delta = \exp(2\pi\epsilon)$ punctured at origin. Therefore, on the circular neighborhood of $s \in S$, Y is isomorphic to $S^1 \times (0, \epsilon)$. This type of neighborhood of S^1 at $s \in S$ is called a tubular neighborhood of S^1 at $s \in S$. Thus adding at s, the boundary S^1 for each $s \in S$, we get a manifold $\overline{Y}$ with boundary $\partial \overline{Y} \cong \bigsqcup_{s \in S} S^1$. If X is compact, $\overline{Y}$ is another compactification of Y non-isomorphic to X. If we start with $X = \mathbf{P}^1(\mathbb{C})$ (Riemann sphere) punctured at $\mu_N = \{\zeta^N = 1\}$, 0 and ∞ which is isomorphic to T_N, taking $S = \{0, \infty\}$, we get the manifold with boundary $\overline{Y}$ diffeomorphic to $\overline{T}_N$.

We have the subspace of n-chains having values in $\partial\overline{Y}$ as $C_n(\partial\overline{Y}; A) \subset C_n(\overline{Y}; A)$ ($0 \le n \le 2$). Then we define, noting $C_2(\partial\overline{Y}; A) = 0$,

$$C_n(\overline{Y}, \partial\overline{Y}; A) = C_n(\overline{Y}; A)/C_n(\partial\overline{Y}; A).$$

We still have a chain simplicial complex

$$0 \to C_2(\overline{Y}, \partial\overline{Y}; A) \xrightarrow{\partial} C_1(\overline{Y}, \partial\overline{Y}; A) \xrightarrow{\partial} C_0(\overline{Y}, \partial\overline{Y}; A) \xrightarrow{\deg} A \to 0,$$

because the boundary map preserves $C_n(\partial \overline{Y}; A)$. The A-linear dual complex of the above complex is written as $C^n(\overline{Y}, \partial \overline{Y}; A)$. Then we define

$$H_n(\overline{Y}, \partial \overline{Y}, A) = \frac{\mathrm{Ker}(\partial : C_n(\overline{Y}, \partial \overline{Y}; A) \to C_{n-1}(\overline{Y}, \partial \overline{Y}; A))}{\mathrm{Im}(\partial : C_{n+1}(\overline{Y}, \partial \overline{Y}; A) \to C_n(\overline{Y}, \partial \overline{Y}; A))},$$

$$H^n(\overline{Y}, \partial \overline{Y}, A) = \frac{\mathrm{Ker}(\partial^* : C^n(\overline{Y}, \partial \overline{Y}; A) \to C^{n+1}(\overline{Y}, \partial \overline{Y}; A))}{\mathrm{Im}(\partial^* : C^{n-1}(\overline{Y}, \partial \overline{Y}; A) \to C^n(\overline{Y}, \partial \overline{Y}; A))}. \tag{3.6}$$

We have the Poincaré duality pairing

$$\langle \cdot, \cdot \rangle : H_n(\overline{Y}, \partial \overline{Y}, A) \times H^n(\overline{Y}, \partial \overline{Y}, A) \to A$$

similarly defined as in the case without boundary.

We consider the space of differential forms $\Omega^1(\overline{T}_N, \partial \overline{T}_N; \mathbb{C})$ given by C^∞ differential forms $\omega = f(z)dx + g(z)dy$ such that $\lim_{y \to \pm \infty} y^M g^{(m)}(z) = 0$ for m-th derivatives $g^{(m)}$ of g for all $m \geq 0$ and all $M \in \mathbb{Z}$. Similarly, we write $\Omega^2(\overline{T}_N, \partial \overline{T}_N; \mathbb{C})$ for the space spanned by $g(z)dx \wedge dy$ with $\lim_{y \to \pm \infty} g^{(m)}(z) = 0$ for m-th derivatives $g^{(m)}$ of g for all $m \geq 0$. The C^∞ function satisfying the above limit property is called rapidly decreasing towards $\pm \infty$. The space $\Omega^0(\overline{T}_N, \partial \overline{T}_N; \mathbb{C})$ is made up of functions rapidly decreasing towards $\pm \infty$. We then define

$$H^1_{DR}(\overline{T}_N, \partial \overline{T}_N; \mathbb{C}) = \frac{\mathrm{Ker}(d : \Omega^1(\overline{T}_N, \partial \overline{T}_N; \mathbb{C}) \to \Omega^2(\overline{T}_N, \partial \overline{T}_N; \mathbb{C}))}{\mathrm{Im}(\Omega^0(\overline{T}_N, \partial \overline{T}_N; \mathbb{C}) \to \Omega^1(\overline{T}_N, \partial \overline{T}_N; \mathbb{C}))}.$$

Exercise 3.1.11. Prove that for $\omega \in \Omega^1(\overline{T}_N, \partial \overline{T}_N; \mathbb{C})$, the integral $\int_{\gamma_x} \omega$ converges absolutely.

We thus have a pairing $H^1_{DR}(\overline{T}_N, \partial \overline{T}_N; \mathbb{C}) \times H_1(\overline{T}_N, \partial \overline{T}_N; \mathbb{C}) \to \mathbb{C}$ given by $([\omega], \gamma) = \int_\gamma \omega$. Thus we get (see [D52] and [STH, V.16])

Lemma 3.1.4. *We have a canonical isomorphism:*

$$H^1_{DR}(\overline{T}_N, \partial \overline{T}_N; \mathbb{C}) \cong H^1(\overline{T}_N, \partial \overline{T}_N; \mathbb{C}).$$

Remark 3.1.5. Though we have given the de Rham isomorphism specifically for $\overline{T}_N$ in Lemma 3.1.4, plainly in a tubular neighborhood at $s \in S$ isomorphic to $(x, y) \in S^1 \times (0, \epsilon)$ we can think of the space of differential forms $\Omega^1(\overline{Y}, \partial \overline{Y}; \mathbb{C})$ given by C^∞ differential forms ω whose restriction to each tubular neighborhood has the form $f(z)dx + g(z)dy$ such that $\lim_{y \to \pm \infty} y^M g^{(m)}(z) = 0$ for m-th derivatives $g^{(m)}$ of g for all $m \geq 0$ and all $M \in \mathbb{Z}$. Similarly, we can think of rapidly decreasing differential 2-forms $\Omega^2(\overline{Y}, \partial \overline{Y}; \mathbb{C})$. Then the de Rham isomorphism as in Lemma 3.1.4 is valid for $\overline{Y}$ and $\partial \overline{Y}$ as verified in [D52].

The vertical line γ_x for $x \in S^1 - \{\frac{a}{N} | a\mathbb{Z} + N\mathbb{Z} = \mathbb{Z}\}$ is a well defined 1-cycle. Let $(N^{-1}\mathbb{Z}/\mathbb{Z})^\times$ be the set of $\frac{a}{N}$ with $a\mathbb{Z} + N\mathbb{Z} = \mathbb{Z}$ in $S^1 = \mathbb{R}/\mathbb{Z}$, and write c_a with $a \in (N^{-1}\mathbb{Z}/\mathbb{Z})^\times$ for a small circle centered at a. Then c_a gives a homology class of $H^1(\overline{T}_N, \partial\overline{T}_N, A)$. We have the following theorem given in [LFE, Section 4.1 (1a)]:

Theorem 3.1.6. *Let $\{c_a | a \in (N^{-1}\mathbb{Z}/\mathbb{Z})^\times\}$ and γ_0 gives a basis of $H_1(\overline{T}_N, \partial\overline{T}_N, A)$ over A for any commutative ring A.*

Here is a sketch of a proof.

Proof. Taking a point $i\infty$ at the boundary S^1_∞ at ∞ on γ_0. Draw a straight line from ∞ to c_a and the boundary $S^1_{-\infty}$ at $-\infty$. We take γ_0 for the line from ∞ to $-\infty$ on $S^1_{-\infty}$. Cut $\overline{T}_N$ along this line. We get a simply connected polygon encircled by these lines and c_a, γ_0 and $S^1_{\pm\infty}$. The sum of all these cycles is zero. Modulo $S^1_{\pm\infty}$, we find that $H^1(\overline{T}_N, \partial\overline{T}_N, A)$ is generated by c_a and γ_0. We have the natural A-linear surjection

$$\pi : H^1(\overline{T}_N, \partial\overline{T}_N, A) \to H^1(\overline{T}, \partial\overline{T}, A),$$

regarding cycles in $\overline{T}_N$ as cycles in $\overline{T} = T \sqcup \partial\overline{T}_N$. Since $H^1(\overline{T}, \partial\overline{T}, A)$ is the dual of $H^1(T, A) = A \cdot S^1$ by the intersection product, we have $H^1(\overline{T}, \partial\overline{T}, A) = A\gamma_0$. The kernel of π is generated by c_as. The cycles c_as are independent by construction. $\square$

Exercise 3.1.12. Give details of the proof of the above theorem.

Corollary 3.1.7. *For the cohomology class of $\omega_\chi = G(\chi)\theta_\chi dz$ in $H^1_{DR}(\overline{T}_N, \partial\overline{T}_N, \mathbb{C})$, the integral $\int_{c_a} \omega_\chi$ is nonzero and belongs to $\mathbb{Q}(\chi)$ for all $a \in (N^{-1}\mathbb{Z}/\mathbb{Z})^\times$.*

Proof. Identify $\mathrm{Gal}(\mathbb{Q}(\mu_N)/\mathbb{Q})$ with $(\mathbb{Z}/N\mathbb{Z})^\times$ $\sigma \leftrightarrow m$ if $\zeta^\sigma = \zeta^m$ for a primitive N-th root ζ of unity. Then $G(\chi, \zeta)^\sigma = \chi(\sigma)^{-1}G(\chi, \zeta)$ for $\sigma \in \mathrm{Gal}(\mathbb{Q}(\chi)[\mu_N]/\mathbb{Q}(\chi))$. Since $2\pi\sqrt{-1}dz = q^{-1}dq$, by Lemma 3.1.2, we find that $\int_{c_a} \omega_\chi = -N^{-1}G(\chi)^{-1}G(\chi, \zeta)$. Since $G(\chi, \zeta)^\sigma = \chi(\sigma)^{-1}G(\chi, \zeta)$, $\frac{G(\chi, \zeta)}{G(\chi)}$ is invariant under any $\sigma \in \mathrm{Gal}(\mathbb{Q}(\chi)[\mu_N]/\mathbb{Q}(\chi))$, we conclude the assertion. $\square$

3.1.4   GL(1) *Hecke operators*

We consider multiplication $[n] : \mathbf{G}_m(\mathbb{C}) \to \mathbf{G}_m(\mathbb{C})$ given by $x \mapsto x^n$ for a positive integer n prime to N. This map is $z \mapsto nz$ on $\overline{T}_N$. Thus

$[n]$ preserves $(N^{-1}\mathbb{Z}/\mathbb{Z})^\times$, $[n]$ acts on cycles and cocycles, by pulling back and pushing forward. We write this action on $H_1(\overline{T}_N, \partial \overline{T}_N, A)$ and $H^1(\overline{T}_N, \partial \overline{T}_N, A)$ as $T(n)$. For a differential form $\omega = f(z)dz$, $\omega | T(n) = [n]^*\omega$.

Exercise 3.1.13. Prove $[\omega]|T(n) = \frac{1}{n}[\sum_{a \bmod n} f(\frac{z+a}{n})dz]$ for $\omega = f(z)dz$.

By the above formula, we have

$$\theta_\chi(z)|T(m) = \frac{1}{m}\sum_{n=1}^{\infty} \chi(n)\mathbf{e}(nz)|T(m) = \frac{1}{m}\sum_{n=1}^{\infty}\chi(n)\mathbf{e}(\frac{nz}{m}) \sum_{a \bmod m}\mathbf{e}(\frac{na}{m})$$

$$= \frac{1}{m}\sum_{n=1}^{\infty}\chi(mn)m\mathbf{e}(\frac{nz}{m}) = \chi(m)\theta_\chi(z),$$

because $\sum_{a \bmod m}\mathbf{e}(\frac{na}{m}) = \begin{cases} 0 & \text{if } m \nmid n \\ m & \text{if } m | n. \end{cases}$ Thus we have

Lemma 3.1.8. *For each primitive character χ modulo N, we have $[\omega_\chi]|T(n) = \chi(n)[\omega_\chi]$.*

Exercise 3.1.14. Give a detailed proof of Lemma 3.1.8.

Theorem 3.1.9. *Write $H^1(\overline{T}_N, \partial \overline{T}_N, \mathbb{Q}(\chi))[\chi]$ for the subspace of $H^1(\overline{T}_N, \partial \overline{T}_N, \mathbb{Q}(\chi))$ on which $T(n)$ for all n prime to N acts by multiplication by $\chi(n)$. Then we have $\dim_{\mathbb{Q}(\chi)} H^1(\overline{T}_N, \partial \overline{T}_N, \mathbb{Q}(\chi))[\chi] = 1$ for a nontrivial primitive character χ modulo N.*

Proof. By definition, $T(n)(c_a) = \sum_{u \bmod n} c_{\frac{a}{n}+\frac{u}{n}}$ for $a \in (N^{-1}\mathbb{Z}/\mathbb{Z})^\times$. If $\frac{a}{n} + \frac{u}{n}$ has reduced numerator bigger than N, the homology class of $c_{\frac{a}{n}+\frac{u}{n}}$ vanishes in $H_1(\overline{T}_N, \partial \overline{T}_N, \mathbb{Z})$. The fraction $\frac{a}{n} + \frac{u}{n}$ has exact numerator N if and only if $aN + nu \equiv 0 \bmod n$. Such u is unique (because n is prime to N). In other words, $aN + nu = mn$; so, $\frac{aN+nu}{n} = m$ and $m \equiv [n]^{-1}aN$ mod N, writing $[n]$ for the multiplication by n mod N on $(N^{-1}\mathbb{Z}/\mathbb{Z})^\times$. We thus have $T(n)(c_a) = c_{[n]^{-1}a}$.

The morphism

$$\pi : H_1(\overline{T}_N, \partial \overline{T}_N, \mathbb{Q}(\chi)) \to H_1(\overline{T}, \partial \overline{T}, \mathbb{Q}(\chi)) = \mathbb{Q}(\chi)\gamma_0$$

is equivariant under $T(n)$, while the section

$$H_1(\overline{T}, \partial \overline{T}, \mathbb{Q}(\chi))\gamma_0 \hookrightarrow H_1(\overline{T}_N, \partial \overline{T}_N, \mathbb{Q}(\chi))$$

sending $\gamma_0 \in H_1(\overline{T}, \partial \overline{T}, \mathbb{Q}(\chi))$ to $\gamma_0 \in H_1(\overline{T}_N, \partial \overline{T}_N, \mathbb{Q}(\chi))$ is not equivariant. The action of $T(n)$ on $H_1(\overline{T}, \partial \overline{T}, \mathbb{Q}(\chi))$ is multiplication by n. Since

c_a and γ_0 span $H_1(\overline{T}_N, \partial \overline{T}_N, \mathbb{Q}(\chi))$ and $T(n)$ permutes c_a, the eigenvalues of $T(n)$ on $\mathrm{Ker}(\pi)$ is given by $\{\chi(n)\}_\chi$ where χ runs over all characters of $\mathbb{Z}/N\mathbb{Z}$. Thus the commutative algebra $\mathcal{H}_N(\mathbb{Q}(\chi))$ generated over $\mathbb{Q}(\chi)$ by $T(n)$s in $\mathrm{End}(H^1(\overline{T}_N, \partial \overline{T}_N, \mathbb{Q}(\chi)))$ acts on $H^1(\overline{T}_N, \partial \overline{T}_N, \mathbb{Q}(\chi))$ semi-simply with multiplicity free, because $\dim_{\mathbb{Q}(\chi)} H^1(\overline{T}_N, \partial \overline{T}_N, \mathbb{Q}(\chi))$ is given by $1 + |(\mathbb{Z}/N\mathbb{Z})^\times|$, which is the number of distinct eigenvalues realized on $H^1(\overline{T}_N, \partial \overline{T}_N, \mathbb{Q}(\chi))$. Thus the result follows.   □

Exercise 3.1.15. Compute the action of $T(n)$ on $\gamma_0 \in H_1(\overline{T}_N, \partial \overline{T}_N, \mathbb{Q}(\chi))$ (for n prime to N).

Corollary 3.1.10. *The cohomology class of ω_χ in $H^1_{DR}(\overline{T}_N, \partial \overline{T}_N, \mathbb{C})$ is rational over $\mathbb{Q}(\chi)$.*

Proof.   The class $[\omega_\chi]$ generates $H^1_{DR}(\overline{T}_N, \partial \overline{T}_N, \mathbb{C})[\chi]$, which is one dimensional. If we find a 1-cycle c in $H_1(\overline{T}_N, \partial \overline{T}_N, \mathbb{Q}(\chi))$ such that $0 \neq \int_c \omega_\chi \in \mathbb{Q}(\chi)$, we may conclude that ω_χ is rational over $\mathbb{Q}(\chi)$. The cycle c_a as in Corollary 3.1.7 does the job.   □

By this corollary, we have finished the proof of Theorem 3.1.3 for $m = 1$.

Exercise 3.1.16. Show the following assertions:

(1) For an integer a prime to N, the operator $T(a) - a$ sends the cohomology group $H_1(\overline{T}_N, \partial \overline{T}_N, \mathbb{Z}[\chi])$ into

$$\mathrm{Ker}(\pi : H_1(\overline{T}_N, \partial \overline{T}_N, \mathbb{Z}[\chi]) \to H_1(\overline{T}, \partial \overline{T}, \mathbb{Z}[\chi])),$$

where $\mathbb{Z}[\chi]$ is the subalgebra of $\mathbb{Q}(\chi)$ generated by all the values of χ.
(2) $(a - \chi^{-1}(a))L(0, \chi) \in \mathbb{Z}[\chi]$ for any integer a prime to N, where χ is a nontrivial primitive character.

3.1.5 *p-adic measure*

Let p be a prime. Though what we present is valid for an arbitrary almost p-profinite group G with a system of open normal subgroups $\{G_i \supset G_{i+1}\}_{0 \leq i \in \mathbb{Z}}$ satisfying $\bigcap_i G_i = \{e\}$ for the neutral element $e \in G$, for simplicity, we assume that G has the form $G = \mu \times \mathbb{Z}_p^r$ for a finite group μ. Most arguments are valid without modification for general groups.

We put $G_i = (p^i \mathbb{Z}_p)^r$. We fix a finite extension $K/\mathbb{Q}_p$ and write A for its p-adic integer ring. We equip K a normalized p adic norm $|\cdot|_p$ such that $|p|_p = \frac{1}{p}$. For each topological space X, we write $LC(G; X)$ for the space of locally constant functions on G with values in X. Thus a function

$\phi : G \to X$ is in $LC(G; X)$ if and only if for any point $g \in G$, there exists an open neighborhood V_g of g in G such that the restriction of ϕ to V_g is a constant function. By definition, it is plain that for any locally constant function ϕ and for any subset S of X, $\phi^{-1}(S) = \bigcup_{g \in \phi^{-1}(S)} V_g$ is open; in particular, ϕ is continuous. Since G is compact, $G = \bigcup_{g \in G} V_g$ implies that we can find finitely many points $g_1, \ldots, g_s$ on G such that $G = \bigcup_{j=1}^{s} V_{g_j}$. By the definition of the topology of G, a basis of open sets of G is given by $\{g + G_i | g \in G, i = 0, 1, \ldots\}$. Thus for large i, $V_{g_j} \supset g_j + G_i$ for all j, that is, ϕ induces a function

$$\phi_i : G/G_i \to X \quad \text{and} \quad \phi = \phi_i \circ \pi_i$$

for the projection $\pi_i : G \to G/G_i$. The space $C(G/G_i; X)$ of continuous functions of G/G_i into X is made of all functions on the finite group G/G_i with values in X and is isomorphic to the set $X[G/G_i]$ of formal linear combinations $\sum_{g \in G/G_i} x_g g$ with $x_g \in X$ via $\phi \mapsto \sum_g \phi(g)g$. Thus we see

$$LC(G; X) = \varinjlim_i C(G/G_i; X) = \varinjlim_i X[G/G_i]. \tag{3.7}$$

For a topological ring R, we define the space of distributions $Dist(G; R)$ by

$$Dist(G; R) = \mathrm{Hom}_R(LC(G; R), R). \tag{3.8}$$

If $\varphi \in Dist(G; R)$ and if χ_S is the characteristic function of an open set S of G, we write $\varphi(S)$ for $\varphi(\chi_S)$. Since $\chi_{h+G_i} = \sum_{g \in G_i/G_j} \chi_{h+g+G_j}$ for $j \geq i$, we have the following distribution relation:

$$\varphi(h + G_i) = \sum_{g \in G_i/G_j} \varphi(h + g + G_j) \quad \text{for all } h \in G \text{ and } j \geq i. \tag{3.9}$$

On the other hand, if we are given a system φ assigning a value $\varphi(g + G_i) \in R$ for all $g \in G/G_i$ and for all i sufficiently large satisfying (3.9), we can extend φ to a distribution as follows. For a given $\phi \in LC(G; R)$, taking sufficiently large i so that $\varphi(g + G_i)$ is well defined and $\phi = \phi_i \circ \pi_i$ with $\phi_i : G/G_i \to R$, we define $\varphi(\phi) = \sum_{g \in G/G_i} \phi_i(g)\varphi(g + G_i)$.

Exercise 3.1.17. Prove that $\varphi(\phi)$ is well defined independent of the choice i of the index if i is sufficiently large.

Thus we have

Proposition 3.1.11. *Let R be a topological ring. Then a function*

$$\varphi : \{g + G_i | i \geq M, \text{ and } g \in G\} \to R$$

is induced from a distribution if and only if φ satisfies (3.9) for all $j \geq i \geq M$.

Let R be a closed subring of K. Let $C(G; R)$ be the space of continuous functions of G into R. Define a norm on $C(G; R)$ by $|\phi|_p = \operatorname{Sup}_{x \in G} |\phi(x)|_p$. A measure φ is a R-linear functional $\varphi \in \operatorname{Hom}_R(C(G; R), R)$ such that $|\varphi(\phi)|_p \le C|\phi|_p$ for a constant $C > 0$ independent of any continuous function $\phi \in C(G; R)$. We often write $\varphi(\phi)$ as $\int_G \phi d\varphi$ following the tradition from the time of Leibniz.

Exercise 3.1.18. Prove that $|\phi|_p$ gives a well defined norm on $C(G; R)$. Is $LC(G; R)$ (resp. $C(G; R)$) a Banach R-module under $|\cdot|_p$?

Write $Meas(G; R)$ for the space of R-valued measures on G. For any measure $\varphi \in Meas(G; R)$, φ induces a distribution, again denoted by φ, by $\varphi(S) = \int \chi_S d\varphi$. Then $|\varphi(\phi)|_p \le C|\phi|_p$ for all $\phi \in C(G; R)$. Thus $|\varphi|_p = \operatorname{Sup}_{0 \ne \phi \in LC(G;R)} |\varphi(\phi)|_p / |\phi|_p$ is finite. Now we want to show the converse. For any continuous function $\phi : G \to R$, we can find for each positive $\varepsilon > 0$ and $g \in G$ a small open neighborhood V_g of g such that $|\phi(h) - \phi(h')|_p < \varepsilon$ for all h and h' in V_g. Cover G by such V_g: $G = \bigcup_{g \in G} V_g$. Since G is compact, we can choose finitely many $g_1, \ldots, g_s \in G$ such that $\bigcup_{j=1}^s V_{g_j}$ and find an index i large such that $V_{g_j} \supset g + G_i$ for all $g \in V_{g_j}$. Choosing a complete representative set Ξ_i for G/G_i and defining $\phi_\varepsilon : G/G_i \to R$ by $\phi_\varepsilon(h) = \phi(g)$ if $h \in (g + G_i) \cap \Xi_i$, we see that $\phi_\varepsilon \in LC(G; R)$ and $|\phi_\varepsilon - \phi|_p < \varepsilon$. Thus $LC(G; R)$ is dense in $C(G; R)$ and $|\phi_{\varepsilon'} - \phi_\varepsilon|_p < |(\phi_{\varepsilon'} - \phi) + (\phi - \phi_\varepsilon)|_p \le \max(|\phi_{\varepsilon'} - \phi|_p, |\phi - \phi_\varepsilon|_p) \le \max(\varepsilon, \varepsilon')$.

$$\tag{3.10}$$

Let φ be a distribution with bounded norm $|\varphi|_p$. This is equivalent to saying that $|\varphi(g + G_i)|_p$ is bounded by $|\varphi|_p$ for all $i \ge M$ and all $g \in G$. Then (3.10) implies

$$\varphi(\phi_\varepsilon) - \varphi(\phi_{\varepsilon'})_p \le |\varphi|_p |\phi_{\varepsilon'} - \phi_\varepsilon|_p \le |\varphi|_p \max(\varepsilon, \varepsilon')$$

and $\{\varphi(\phi_{1/n})\}$ is a Cauchy sequence in R. We then define

$$\int_G \phi d\varphi = \lim_{n \to \infty} \varphi(\phi_{1/n}) \in R.$$

Then it is easy to verify that $\varphi \in Meas(G; R)$. Thus we have

Proposition 3.1.12. *For any closed subring R of K, $LC(G; R)$ is dense in $C(G; R)$. Any bounded distribution on G with values in R can be uniquely extended to a bounded measure with values in R. In particular, $Meas(G; A) \cong Dist(G; A)$ via the restriction to $LC(G; A)$ for the p-adic integer ring A of K.*

Exercise 3.1.19. If $\varphi \in Meas(G; K)$, prove
$$\operatorname{Sup}_{0 \ne \phi \in C(G;K)} |\varphi(\phi)|_p / |\phi|_p = \operatorname{Sup}_{0 \ne \phi \in LC(G;K)} |\varphi(\phi)|_p / |\phi|_p.$$

3.1.6 p-adic L-function of Kubota–Leopoldt

We construct a p-adic measure which interpolates the values of Dirichlet L-functions via cohomology theory. This type of formalism (the formalism of relating p-adic measure and modular symbols) was invented by Mazur and Swinnerton-Dyer, who applied it to L-functions of elliptic modular forms.

Let $N > 1$ be a positive integer prime to the fixed prime p. Let $\overline{T}_N = \mathbb{C}/\mathbb{Z} - (N^{-1}\mathbb{Z}/\mathbb{Z})^\times$ be as above. We have an A-linear map: $H^1(\overline{T}_N, \partial\overline{T}_N, A) \to A$ given by $\omega \mapsto \int_{\gamma_x} \omega$. Then we consider a map

$$c : p^{-\infty}\mathbb{Z} = \bigcup_{i=1}^{\infty} p^{-i}\mathbb{Z} \to \operatorname{Hom}_A(H^1(\overline{T}_N, \partial\overline{T}_N, A), K) \qquad (3.11)$$

given by $c(x)(\omega) = \int_{\gamma_x} \omega$.

For $\omega \in H^1(\overline{T}_N, \partial\overline{T}_N, A)$, we write $c_\omega(r) = \int_{\gamma_r} \omega$. Then $c_\omega(r+1) = c_\omega(r)$ by definition, and c_ω factors through $\mathbb{Q}_p/\mathbb{Z}_p = p^{-\infty}\mathbb{Z}/\mathbb{Z}$. Supposing $\omega|T(p) = a\omega$ with $|a|_p = 1$, we define a distribution φ_ω on $\mathbb{Z}_p$ by

$$\varphi_\omega(z + p^m\mathbb{Z}_p) = a^{-m}c_\omega(\frac{z}{p^m}) \text{ for } z = 1, 2, \ldots \text{ prime to } p. \qquad (3.12)$$

This is well defined because $c_\omega(r+1) = c_\omega(r)$. We take the multiplicative group $G = \mathbb{Z}_p^\times$ and fix an isomorphism $G \cong \mu \times \mathbb{Z}_p$ for a finite group μ, where $\mathbb{Z}_p$ in the right-hand-side is an additive group.

Exercise 3.1.20. Prove that $\mu = \{\zeta \in \mathbb{Z}_p^\times | \zeta^M = 1\}$, where $M = p - 1$ or 2 according to whether p is odd or even.

Then the multiplicative subgroup $G_i = 1 + p^i\mathbb{Z}_p$ corresponds to the additive group $p^i\mathbb{Z}_p$. To show that φ_ω actually gives a distribution, we need to check the distribution relation (3.9). We compute

$$\sum_{j=1}^{p} c_\omega(\frac{x+j}{p}) = \sum_{j} c(\frac{x+j}{p})(\omega) = c(x)(\omega|T(p)) = a \cdot c_\omega(x).$$

This shows

$$\sum_{j=1}^{p} \varphi_\omega(x + jp^m + p^{m+1}\mathbb{Z}_p) = \varphi_\omega(x + p^m\mathbb{Z}_p).$$

The general distribution relation (3.9) then follows from the iteration of this relation. By a similar argument, we see that

$$|\varphi_\omega(z + p^m\mathbb{Z}_p)|_p = |a^{-m}c_\omega(\frac{z}{p^m})|_p = |c(\frac{z}{p^m})(\omega)|_p \leq |\omega|_p, \qquad (3.13)$$

where $|\omega|_p = \text{Sup}_x |c(x)(\omega)|_p$ with x running over $p^{-\infty}\mathbb{Z}$. Thus φ_ω is bounded and, by Proposition 3.1.12, we have a unique measure φ_ω extending the distribution φ_ω. Now we compute $\int_G \phi d\varphi_\omega(x)$. To do this, we may assume that $|\omega|_p \leq 1$ by multiplying by a constant if necessary. For $\phi \in C(G/G_m; A)$, we have

$$\int_G \phi d\varphi_\omega = a^{-m} \sum_{z=1}^{p^m} \phi(z) c_\omega(\frac{z}{p^m}).$$

Let $N > 1$ be a positive integer prime to p. We take $\omega = \omega_{\chi^{-1}}$ for each primitive character χ modulo N (ω may not be p-integral but is bounded because $(a - \chi^{-1}(a))\omega$ is p-integral as seen in Exercise 3.1.16). Then we write φ_ω as $\varphi = \varphi_\chi$ and compute for any primitive character ϕ of $(\mathbb{Z}/p^r\mathbb{Z})^\times$ the integral $\int_G \phi d\varphi_\chi$. Note that $\omega|T(p) = \chi(a)^{-1}\omega$. We then write $\alpha_x : \overline{T}_N \to \overline{T}_N$ to be the translation $\alpha_x(z) = z + x$ for $x \in \mathbb{R}$. We see that, if $\phi \neq 1$, then $\phi\chi$ is primitive modulo Np^r and

$$\begin{aligned}
\int_G \phi d\varphi_\chi &= \chi(p)^r \sum_{x \in (\mathbb{Z}/p^r\mathbb{Z})^\times} \phi(x) c(\frac{x}{p^r})(\omega) \\
&= \chi(p)^r \int_{\gamma_0} \sum_{x \in (\mathbb{Z}/p^r\mathbb{Z})^\times} \phi(x) \alpha^*_{x/p^r}\omega \\
&= \chi(p)^r G(\chi)G(\phi) \int_{\gamma_0} \theta_{\phi^{-1}\chi^{-1}}(z)dz \\
&= -\chi(p)^r G(\chi)G(\phi)(2pi)^{-1}(1 - \chi\phi(-1))L(1, (\chi\phi)^{-1}) \\
&= -\chi(p)^r G(\chi)G(\phi)G(\chi\phi)^{-1}L(0, \chi\phi).
\end{aligned}$$

(3.14)

Here we have used the following formulas

$$L(0, \chi^{-1}) = (1 - \chi(-1))G(\chi^{-1})(2\pi i)^{-1}L(1, \chi)$$

by the functional equation of $L(s, \chi)$ and

$$G(\chi)G(\phi) = \chi^{-1}(p^r)\phi^{-1}(N)G(\chi\phi). \tag{3.15}$$

Exercise 3.1.21. Prove (3.15).

We have basically proved the following theorem of Kubota–Leopoldt–Iwasawa when $j = 1$:

Theorem 3.1.13. *Let p be a prime and N be a positive integer prime to p. For each primitive Dirichlet character $\chi \neq 1$ modulo N, we have a unique* *p*-adic measure φ_χ on $\mathbb{Z}_p^\times$ such that for all finite order characters ϕ of $\mathbb{Z}_p^\times$ and $1 \le j \in \mathbb{Z}$, we have

$$\int \phi(z)z^j d\varphi_\chi = -\phi(N)^{-1}N^{-j}(1 - \chi\phi(p)p^j)L(1-j, \chi\phi).$$

As for the identity character, fixing a prime q prime to p, we have a unique *p*-adic measure φ_q on $\mathbb{Z}_p^\times$ such that for all finite order characters ϕ of $\mathbb{Z}_p^\times$ and $1 \le j \in \mathbb{Z}$, we have

$$\int \phi(z)z^j d\varphi_q = -(1 - \phi^{-1}(q)q^{-1-j})(1 - \phi(p)p^j)L(1-j, \phi).$$

We now define the *p*-adic Dirichlet L-function for each primitive character χ modulo Np^r (with values in K), writing χ_N (resp. χ_p) for the restriction of χ to $(\mathbb{Z}/N\mathbb{Z})^\times$ (resp. $(\mathbb{Z}/p^r\mathbb{Z})^\times$), by

$$L_p(s, \chi) = c(s) \int_{\mathbb{Z}_p^\times} \chi_p\omega^{-1}(x)x_\Gamma^{-s} d\varphi_{\chi_N}(x),$$

where

$$c(s) = \begin{cases} -\chi_p\omega^{-1}(N)N_\Gamma^{-s} & \text{if } \chi_N \ne 1, \\ -(1 - \chi_p\omega^{-1}(\gamma)\gamma_\Gamma^{s-1})^{-1} & \text{if } \chi_N = 1. \end{cases}$$

Here $x_\Gamma \in 1 + p\mathbb{Z}_p$ is given by $x\omega(x)^{-1}$ for the Teichmüller character $\omega : \mathbb{Z}_p^\times \to \mu$ which is given by $\omega(x) = \lim_{n \to \infty} x^{p^n}$ if p is odd, and if $p = 2$, $\omega(x) = \pm 1$ according to whether $x \equiv \pm 1 \mod 4$.

Exercise 3.1.22. Prove that the limit $\omega(z) = \lim_{n \to \infty} z^{p^n}$ exists in $\mathbb{Z}_p$ and that it gives rise to a Dirichlet character modulo p.

We get the following result:

Theorem 3.1.14. *Let χ be a primitive Dirichlet character modulo Np^r with $\chi(-1) = 1$. Then there exists a p-adic analytic function $L_p(s, \chi)$ on $\mathbb{Z}_p$ if $\chi_N \ne 1$ and on $\mathbb{Z}_p - \{1\}$ if $\chi = 1$ such that*

$$L_p(-m, \chi) = (1 - \chi\omega^{-m-1}(p)p^m)L(-m, \chi\omega^{-m-1})$$

for all non-negative integers m.

See [LFE, Chapter 4] (or [ICF, Theorem 5.11]) for more details.

3.2 Modular *p*-adic L-functions

In this section, we will do the same construction of *p*-adic L-functions for elliptic Hecke eigenforms in place of rational functions on $\mathbf{G}_m$.

3.2.1   Elliptic modular forms

Let $\Gamma_0(N) = \{\left(\begin{smallmatrix} a & b \\ c & d \end{smallmatrix}\right) \in SL_2(\mathbb{Z}) | c \equiv 0 \mod N\}$. This a subgroup of finite index in $SL_2(\mathbb{Z})$.

Exercise 3.2.1. Let $\mathbf{P}^1(A)$ be the projective space of dimension 1 over a ring A. Prove $[SL_2(\mathbb{Z}) : \Gamma_0(N)] = |\mathbf{P}^1(\mathbb{Z}/N\mathbb{Z})| = N \prod_{\ell | N}(1 + \frac{1}{\ell})$ if N is square-free, where ℓ runs over all prime factors of N. *Hint: Let* $\left(\begin{smallmatrix} a & b \\ c & d \end{smallmatrix}\right) \in SL_2(\mathbb{Z})$ *act on* $\mathbf{P}^1(A)$ *by* $z \mapsto \frac{az+b}{cz+d}$ *and show that this is a transitive action if* $A = \mathbb{Z}/N\mathbb{Z}$ *and the stabilizer of* ∞ *is* $\Gamma_0(N)$.

We let $\left(\begin{smallmatrix} a & b \\ c & d \end{smallmatrix}\right) \in GL_2(\mathbb{C})$ act on $\mathbf{P}^1(\mathbb{C}) = \mathbb{C} \cup \{\infty\}$ by $z \mapsto \frac{az+b}{cz+d}$ (by linear fractional transformation).

Exercise 3.2.2. Prove the following facts:

(1) there are two orbits of the action of $GL_2(\mathbb{R})$ on $\mathbf{P}^1(\mathbb{C})$: $\mathbf{P}^1(\mathbb{R})$ and $\mathfrak{H} \sqcup \overline{\mathfrak{H}}$, where $\mathfrak{H} = \{z \in \mathbb{C} | \operatorname{Im}(z) > 0\}$ and $\overline{\mathfrak{H}} = \{z \in \mathbb{C} | \operatorname{Im}(z) < 0\}$.

(2) the stabilizer of $i = \sqrt{-1}$ is the center times $SO_2(\mathbb{R}) = \left\{ \left(\begin{smallmatrix} \cos(\theta) & \sin(\theta) \\ -\sin(\theta) & \cos(\theta) \end{smallmatrix}\right) | \theta \in \mathbb{R} \right\}$,

(3) $\gamma \in GL_2(\mathbb{R})$ with $\det(\gamma) < 0$ interchanges the upper half complex plane $\mathfrak{H}$ and lower half complex plane $\overline{\mathfrak{H}}$,

(4) the upper half complex plane is isomorphic to $SL_2(\mathbb{R})/SO_2(\mathbb{R})$ by $SL_2(\mathbb{R}) \ni g \mapsto g(\sqrt{-1}) \in \mathfrak{H}$.

Then $Y_0(N) = \Gamma_0(N) \backslash \mathfrak{H}$ is an open Riemann surface with a hole at each cusp. In other words, $X_0(N) = \Gamma_0(N) \backslash (\mathfrak{H} \sqcup \mathbf{P}^1(\mathbb{Q}))$ is a compact Riemann surface with the hole filled by a disk centered at each cusp.

Exercise 3.2.3. Show that $SL_2(K)$ acts transitively on $\mathbf{P}^1(K)$ for any field K by linear fractional transformation. *Hint:* $\left(\begin{smallmatrix} 1 & a \\ 0 & 1 \end{smallmatrix}\right)(0) = a$.

Let $f : \mathfrak{H} \to \mathbb{C}$ be a holomorphic functions with $f(z + 1) = f(z)$. Since $\mathfrak{H}/\mathbb{Z} \cong D = \{z \in \mathbb{C}^{\times} | |z| < 1\}$ by $z \mapsto q = \mathbf{e}(z) = \exp(2\pi i z)$, we may regard f as a function of q undefined at $q = 0 \Leftrightarrow z = i\infty$. Then the Laurent expansion of f gives

$$f(z) = \sum_n a(n, f)q^n = \sum_n a(n, f)\exp(2\pi i n z).$$

In particular, we may assume that q is the coordinate of $X_0(N)$ around the infinity cusp ∞. We call f is *finite* (resp. *vanishing*) at ∞ if $a(n, f) = 0$ if $n < 0$ (resp. if $n \leq 0$). By Exercise 3.2.3, we can bring any point $c \in \mathbf{P}^1(\mathbb{Q})$

to ∞; so, the coordinate around the cusp c is given by $q \circ \alpha$ for $\alpha \in SL_2(\mathbb{Q})$ with $\alpha(c) = \infty$.

Exercise 3.2.4. Show that the above α can be taken in $SL_2(\mathbb{Z})$. *Hint: write $c = \frac{a}{b}$ as a reduced fraction; then, we can find $x, y \in \mathbb{Z}$ such that $ax - by = 1$.*

We consider the space of holomorphic functions $f : \mathfrak{H} \to \mathbb{C}$ satisfying the following conditions:

$$f(\frac{az + b}{cz + d}) = f(z)(cz + d)^2 \text{ for all } \left(\begin{smallmatrix} a & b \\ c & d \end{smallmatrix} \right) \in \Gamma_0(N). \tag{M1}$$

If f satisfies the above conditions, we find that $f(z + 1) = f(z)$ because $\left(\begin{smallmatrix} 1 & 1 \\ 0 & 1 \end{smallmatrix} \right)(z) = z + 1$; so, we can say that f is finite or not.

Exercise 3.2.5. Define $f| \left(\begin{smallmatrix} a & b \\ c & d \end{smallmatrix} \right)(z) = f(\frac{az+b}{cz+d})(cz + d)^{-2}$. Prove:

(1) $(f|\alpha)|\beta = f|(\alpha\beta)$ for $\alpha \in SL_2(\mathbb{R})$,
(2) if f satisfies (M1), $f|\alpha$ satisfies (M1) replacing $\Gamma_0(N)$ by $\Gamma = \alpha^{-1}\Gamma_0(N)\alpha$,
(3) Let $\Gamma(N) = \{\gamma \in SL_2(\mathbb{Z})|\gamma - 1 \in NM_2(\mathbb{Z})\}$. If $\alpha \in SL_2(\mathbb{Z})$, show that $\alpha^{-1}\Gamma_0(N)\alpha$ contains $\Gamma(N)$.

By (3) of the above exercise, for $\alpha \in SL_2(\mathbb{Z})$, we find $f|\alpha(z + N) = f|\alpha(z)$; thus, $f|\alpha$ has expansion $f|\alpha = \sum_n a(n, f|\alpha)q^{Nn}$. We call f is finite (resp. vanishing) at the cusp $\alpha^{-1}(\infty)$ if $f|\alpha$ is finite (resp. vanishing) at ∞.

$$f \text{ is finite at all cusps of } X_0(N). \tag{M2}$$

We write $M_2(\Gamma_0(N))$ for the space of functions satisfying (M1–2). Replace (M2) by

$$f \text{ is vanishing at all cusps of } X_0(N), \tag{S}$$

we define subspace $S_2(\Gamma_0(N)) \subset M_2(\Gamma_0(N))$ by imposing (S). An element in $S_2(\Gamma_0(N))$ is called a holomorphic cusp form on $\Gamma_0(N)$ of weight 2.

Remark 3.2.1. As mentioned in Remark 2.3.2 and Remark 2.3.3, we have $G_k(\Gamma_0(1)) \cong M_k(\Gamma_0(1))$ by q-expansion.

3.2.2 Modular cohomology group

Take a holomorphic differential ω on $X_0(N)$. Then we pull back ω to $\mathfrak{H}$ and still write ω. We can write $\omega = f(z)dz$ on $\mathfrak{H}$ because $\mathfrak{H}$ is simply connected.

Exercise 3.2.6. Prove $\alpha^* dz = d(\frac{az+b}{cz+d}) = (cz+d)^{-2}dz$ for $\alpha = \left(\begin{smallmatrix} a & b \\ c & d \end{smallmatrix}\right) \in GL_2(\mathbb{R})$ with $\det\left(\begin{smallmatrix} a & b \\ c & d \end{smallmatrix}\right) = 1$. Write down the formula if $\det\left(\begin{smallmatrix} a & b \\ c & d \end{smallmatrix}\right)$ is non-trivial and positive.

Since $\gamma^*\omega = \omega$ for all $\gamma \in \Gamma_0(N)$, we find

$$f(z)dz = \omega = \gamma^*\omega = f(\gamma(z))\gamma^* dz = f(\gamma(z))(cz+d)^{-2}dz$$

if $\gamma = \left(\begin{smallmatrix} a & b \\ c & d \end{smallmatrix}\right)$. Thus f has to satisfy (M1). At infinity, since $dz = 2\pi i\frac{dq}{q}$, ω with respect to the coordinate q is finite at ∞, and hence $f(z)dz = 2\pi i\frac{dq}{q}\sum_n a(n,f)q^n$ is finite at $q = 0$. This implies f has to be vanishing at ∞. Writing $H^0(X_0(N), \Omega_{X_0(N)/\mathbb{C}})$ for the space of holomorphic 1-forms on $X_0(N)$, we thus find

Proposition 3.2.2. *We have a canonical isomorphism* $S_2(\Gamma_0(N)) \cong H^0(X_0(N), \Omega_{X_0(N)/\mathbb{C}})$ *sending f to $f(z)dz$.*

Let C be the divisor on $X_0(N)$ which is the formal sum of all cusps. If we write $H^0(X_0(N), \Omega_{X_0(N)/\mathbb{C}}(-C))$ for the space of meromorphic 1-forms on $X_0(N)$ with at most simple poles at cusps, by the same argument, we have

Corollary 3.2.3. *We have an isomorphism*

$$M_2(\Gamma_0(N)) \cong H^0(X_0(N), \Omega_{X_0(N)/\mathbb{C}}(-C))$$

sending f to $f(z)dz$.

For any compact Riemann surface X, general theory of Riemann surface tells us $H^1(X, \mathbb{C}) \cong H^0(X, \Omega_{X/\mathbb{C}}) \oplus H^0(X, \overline{\Omega}_{X/\mathbb{C}})$ (the Hodge decomposition, where $H^0(X, \Omega_{X/\mathbb{C}})$ is the space of holomorphic 1-forms on X and $H^0(X, \overline{\Omega}_{X/\mathbb{C}})$ is the space of antiholomorphic 1-forms on X. Since $H^0(X, \overline{\Omega}_{X/\mathbb{C}})$ is the complex conjugate of $H^0(X, \Omega_{X/\mathbb{C}})$, we get

Proposition 3.2.4. *We have a canonical isomorphism*

$$H^1(X_0(N), \mathbb{C}) \cong S_2(\Gamma_0(N)) \oplus \overline{S}_2(\Gamma_0(N)),$$

where $\overline{S}_2(\Gamma_0(N))$ is made up of complex conjugate $\overline{f}$ for $f \in S_2(\Gamma_0(N))$. In particular, $S_2(\Gamma_0(N))$ is finite dimensional, and its dimension is given by the genus of $X_0(N)$.

We add a small circle at each cusp of $Y_0(N)$ and get a different compactification $\overline{Y}_0(N)$ of $Y_0(N)$ from $X_0(N)$. Take the circle S around the cusp c. Then $\int_S \omega$ is essentially the residue at c of ω and if we write $\omega = f(z)dz$ it is given by $a(0, f|\alpha)$ for $\alpha \in SL_2(\mathbb{Z})$ taking the cusp to ∞. Thus we get

Corollary 3.2.5. *If we write $g_0(N)$ for the genus of $X_0(N)$ and $c_0(N)$ for the number of cusps of $X_0(N)$, the dimension of the space $M_2(\Gamma_0(N))$ is bounded by $g_0(N) + c_0(N)$. In fact, it is equal to $g_0(N) + c_0(N) - 1$.*

The fact that the dimension is one less than $g_0(N) + c_0(N)$ follows from the fact that $M_2(SL_2(\mathbb{Z})) = 0$ (or $H^1(X_0(1), \Omega_{X_0(1)/\mathbb{C}}(-C)) = 0$), because a punctured sphere is still simply connected.

By the de Rham theorem, we have the following duality given by integration:

Proposition 3.2.6. *The cohomology groups*

$$H_1(X_0(N), \mathbb{C}) \quad and \quad H_1(\overline{Y}_0(N), \partial \overline{Y}_0(N); \mathbb{C})$$

are dual to $H^1(X_0(N), \mathbb{C}) \cong S_2(\Gamma_0(N)) \oplus \overline{S}_2(\Gamma_0(N))$, and $H_1(\overline{Y}_0(N), \mathbb{C})$ is dual to $M_2(\Gamma_0(N)) \oplus \overline{S}_2(\Gamma_0(N))$.

3.2.3 GL(2) *Hecke operators*

Let $GL_2^+(\mathbb{R}) = \{\alpha \in GL_2(\mathbb{R}) | \det(\alpha) > 0\}$ and put $GL_2^+(A) = GL_2^+(\mathbb{R}) \cap GL_2(A)$ for $A \subset \mathbb{R}$. For $\alpha = \left(\begin{smallmatrix} a & b \\ c & d \end{smallmatrix}\right) \in GL_2^+(\mathbb{R})$ and a function $f : \mathfrak{H} \to \mathbb{C}$, we define $f|\alpha(z) = \det(\alpha)f(\alpha(z))(cz + d)^{-2}$.

Exercise 3.2.7. Prove $(f|\alpha)|\beta = f|(\alpha\beta)$ for $\alpha, \beta \in GL_2^+(\mathbb{R})$.

Then $f \in S_2(\Gamma_0(N))$ (resp. $f \in M_2(\Gamma_0(N))$) if and only if f vanishes (resp. finite) at all cusps of $X_0(N)$ and $f|\gamma = f$ for all $\gamma \in \Gamma_0(N)$. Let $\Gamma = \Gamma_0(N)$. For $\alpha \in GL_2(\mathbb{R})$ with $\det(\alpha) > 0$, if $\Gamma\alpha\Gamma$ can be decomposed into a disjoint union of finite left cosets $\Gamma\alpha\Gamma = \bigsqcup_{j=1}^{h} \Gamma\alpha_j$, we can think of the finite sum $g = \sum_j f|\alpha_j$. If $\gamma \in \Gamma$, then $\alpha_j\gamma \in \Gamma\alpha_{\sigma(j)}$ for a unique index $1 \le \sigma(j) \le h$ and σ is a permutation of $1, 2, \ldots, h$. If $f|\gamma = f$ for all $\gamma \in \Gamma$, we have

$$g|\gamma = \sum_j f|\alpha_j\gamma = \sum_j f|\gamma_j\alpha_{\sigma(j)} = \sum_j (f|\gamma_j)|\alpha_{\sigma(j)} = \sum_j f|\alpha_{\sigma(j)} = g.$$

Thus under the condition that $f|\gamma = f$ for all $\gamma \in \Gamma$, $f \mapsto g$ is a linear operator only dependent on the double coset $\Gamma\alpha\Gamma$; so, we write $g = f|[\Gamma\alpha\Gamma]$. More generally, if we have a set $T \subset GL_2^+(\mathbb{R})$ such that $\Gamma T\Gamma = T$ with finite $|\Gamma\backslash T|$, we can define the operator $[T]$ by $f \mapsto \sum_j f|t_j$ if $T = \bigsqcup_j \Gamma t_j$. Put

$$\Delta_0(N) = \left\{ \left(\begin{smallmatrix} a & b \\ c & d \end{smallmatrix}\right) \in M_2(\mathbb{Z}) \cap GL_2^+(\mathbb{R}) | c \equiv 0 \mod N, \ a\mathbb{Z} + N\mathbb{Z} = \mathbb{Z} \right\},$$

which is a semi-group under matrix multiplication.

Exercise 3.2.8. Prove that $\Gamma\Delta_0(N)\Gamma = \Delta_0(N)$ for $\Gamma = \Gamma_0(N)$.

Lemma 3.2.7. *Let* $\Gamma = \Gamma_0(N)$.

(1) *If* $\alpha \in M_2(\mathbb{Z})$ *with positive determinant,* $|\Gamma \backslash (\Gamma \alpha \Gamma)| < \infty$;

(2) *If* p *is a prime,*

$$\Gamma \begin{pmatrix} 1 & 0 \\ 0 & p \end{pmatrix} \Gamma = \{\alpha \in \Delta_0(N) | \det(\alpha) = p\}$$

$$= \begin{cases} \Gamma \begin{pmatrix} p & 0 \\ 0 & 1 \end{pmatrix} \sqcup \bigsqcup_{j=0}^{p-1} \Gamma \begin{pmatrix} 1 & j \\ 0 & p \end{pmatrix} & \text{if } p \nmid N, \\ \bigsqcup_{j=0}^{p-1} \Gamma \begin{pmatrix} 1 & j \\ 0 & p \end{pmatrix} & \text{if } p | N. \end{cases}$$

(3) *for an integer* $n > 0$,

$$T_n := \{\alpha \in \Delta_0(N) | \det(\alpha) = n\}$$

$$= \bigsqcup_a \bigsqcup_{b=0}^{d-1} \Gamma_0(N) \begin{pmatrix} a & b \\ 0 & d \end{pmatrix} \quad (a > 0, \ ad = n, \ (a, N) = 1, \ a, b, d \in \mathbb{Z}),$$

(4) *Write* $T(n)$ *for the operator corresponding to* T_n. *Then we get the following identity of Hecke operators for* $f \in M_2(\Gamma_0(N))$:

$$a(m, f|T(n)) = \sum_{0 < d | (m,n), (d,N)=1} d \cdot a(\frac{mn}{d^2}, f).$$

(5) $T(m)T(n) = T(n)T(m)$ *for all integers* m *and* n.

For primes l dividing the level N, we often write $U(l)$ for $T(l)$.

Proof. Note that (1) and (2) are particular cases of (3). We only prove (2), (4) when $n = p$ for a prime p and (5), leaving the other cases as an exercise (see [IAT, 3.36, (3.5.10)] for a detailed proof of (3) and (4)).

We first deal with (2). Since the argument in each case is essentially the same, we only deal with the case where $p \nmid N$ and $\Gamma = \Gamma_0(N)$. Take any $\gamma = \begin{pmatrix} a & b \\ c & d \end{pmatrix} \in M_2(\mathbb{Z})$ and $ad - bc = p$. If c is divisible by p, then ad is divisible by p; so, one of a and d has a factor p. We then have

$$\gamma = \begin{pmatrix} a & b \\ c & d \end{pmatrix} = \begin{pmatrix} a/p & b \\ c/p & d \end{pmatrix} \begin{pmatrix} p & 0 \\ 0 & 1 \end{pmatrix} \in \Gamma_0(N) \begin{pmatrix} p & 0 \\ 0 & 1 \end{pmatrix}$$

if a is divisible by p. If d is divisible by p and a is prime to p, choosing an integer j with $0 \le j \le p - 1$ with $ja \equiv b \mod p$, we have $\gamma \begin{pmatrix} 1 & j \\ 0 & p \end{pmatrix}^{-1} \in GL_2(\mathbb{Z})$. If c is not divisible by p but a is divisible by p, we can interchange a and c via multiplication by $\begin{pmatrix} 0 & -1 \\ 1 & 0 \end{pmatrix}$ from the left-side. If a and c are not divisible by p, choosing an integer j so that $ja \equiv -c \mod p$, we find that the lower left corner of $\begin{pmatrix} 1 & 0 \\ j & 1 \end{pmatrix} \gamma$ is equal to $ja + c$ and is divisible by p. This finishes the proof of (2).

We now deal with (4) assuming $n = p$. By (2), we have

$$f|T(p)(z) = \begin{cases} p \cdot f(pz) + \sum_{j=0}^{p-1} f\left(\frac{z+j}{p}\right) & \text{if } p \nmid N, \\ \sum_{j=0}^{p-1} f\left(\frac{z+j}{p}\right) & \text{if } p|N. \end{cases} \tag{3.16}$$

Writing $f = \sum_{n=1}^{\infty} a(n, f)q^n$ for $q = \mathbf{e}(z)$, we find

$$a(m, f|T(p)) = a(mp, f) + p \cdot a(\frac{m}{p}, f).$$

Here we put $a(r, f) = 0$ unless r is a non-negative integer.

The formula of Lemma 3.2.7 (4) is symmetric with respect to m and n; so, we conclude $T(m)T(n) = T(n)T(m)$. This proves (5). $\quad\square$

Exercise 3.2.9. Give a detailed proof of the above lemma.

The following exercise is more difficult:

Exercise 3.2.10. Let $\Gamma = SL_2(\mathbb{Z})$. Prove that $|\Gamma\backslash(\Gamma\alpha\Gamma)| < \infty$ for $\alpha \in GL_2(\mathbb{R})$ if and only if $\alpha \in M_2(\mathbb{Q})$ modulo real scalar matrices.

Write $\pi : \mathfrak{H} \to Y_0(N) = \Gamma\backslash\mathfrak{H}$ for the quotient map.

Lemma 3.2.8. *If* $\Gamma\alpha\Gamma = \bigsqcup_{j=1}^{h} \Gamma\alpha_j$ *for* $\Gamma = \Gamma_0(N)$ *and* $\alpha \in GL_2^+(\mathbb{R})$, *then for a chain* $c \in C_1(\mathfrak{H}; A)$ *with* $\partial(\pi(c)) = \pi(\partial c) = 0$, $\partial(\pi(\sum_{j=1}^{h} \alpha_j(c))) = 0$.

Proof. If $\pi(\partial c)$ in $Y_0(N)$ vanishes, writing $\partial c = \sum_z a_z[z]$ for points z in $\mathfrak{H}$, we may assume that $a_z + a_{\gamma(z)} = 0$ for some $\gamma \in \Gamma$. If $\Gamma\alpha\Gamma = \bigsqcup_{j=1}^{h} \Gamma\alpha_j$, then $\alpha_j\gamma = \gamma_j\alpha_{\sigma(j)}$ and $\sum_j(a_{\alpha_j(z)} + a_{\alpha_j\gamma(z)}) = \sum_j(a_{\alpha_j(z)} + a_{\gamma_j\alpha_{\sigma(j)}(z)}) = 0$. This shows that $\pi(\partial(\sum_j \alpha_j(c))) = 0$, which finishes the proof. $\quad\square$

Obviously, for any 2-chain c, $\pi(\sum_j \partial\alpha_j(c)) = \pi(\sum_j \alpha_j(\partial(c)))$, and therefore, the operator $c \mapsto \sum_j \alpha_j(c)$ preserves boundaries and cycles. In this way, the Hecke operator $[\Gamma\alpha\Gamma]$ acts on $H_1(Y_0(N), A)$, and hence on $H^1(Y_0(N), A)$ by the definition of cohomology group. On $H_{DR}^1(Y_0(N), \mathbb{C})$, the action of $[\Gamma\alpha\Gamma]$ is given by $[\omega] \mapsto [\sum_{j=1}^{h} \alpha_j^* \omega]$. Since $z \mapsto \alpha_j(z)$ takes cusps to cusps, we can show similarly that Hecke operators act on $H_1(X_0(N), A)$ and $H_1(\overline{Y}_0(N), \partial\overline{Y}_0(N), A)$. Since we can verify $\alpha^* f(z)dz = (f|\alpha)dz$ by the chain rule, the Eichler–Shimura isomorphism

$$S_2(\Gamma_0(N)) \oplus \overline{S}_2(\Gamma_0(N)) \cong H^1(X_0(N), \mathbb{C})$$

is equivariant under Hecke operators.

3.2.4 Duality

Let N be a fixed positive integer and R be a commutative ring with identity. We suppose the following axiom for an R-module E and a character ϕ of $(\mathbb{Z}/N\mathbb{Z})^\times \times \mathbb{Z}p^\times$ with values in $\mathrm{Aut}_R(E)$:

(du1) The R-module E is of finite type and has R-linear operators $T(n)$ $(n = 1, 2, \dots)$ acting on E, $T(1)$ giving the identity operator;

(du2) We have an R-linear embedding $E \hookrightarrow qR[[q]]$ for a power series ring $R[[q]]$ given by $E \ni f \mapsto \sum_{n=1}^\infty a(n, f)q^n \in qR[[q]]$;

(du3) We have $a(m, f|T(n)) = \sum_{0 < d | (m,n), (d,Np)=1} a(\frac{mn}{d^2}, f|\phi(d))$ for all positive integer m, n.

If R is a $\mathbb{Z}_p$-algebra, the operator $\phi(d) \in \mathrm{Aut}(E)$ can be given by $f|\phi(d) = \chi(z)z_p^{k-1}f$ for characters $\chi : (\mathbb{Z}/N\mathbb{Z})^\times \times \mathbb{Z}_p^\times \to R^\times$ and $z \mapsto z_p^{k-1}$ with the projection z_p of $z \in \mathbb{Z}_p^\times \times (\mathbb{Z}/\mathbb{Z})^\times$ to $\mathbb{Z}_p^\times$. By (du2), E is R-torsion-free if R is a domain.

Exercise 3.2.11. Prove that $T(m)T(n) = T(n)T(m)$ for all $0 < m, n \in \mathbb{Z}$.

Let $\mathcal{H}(E) = \mathcal{H}_R(E)$ be the subring of the R-linear endomorphism algebra $\mathrm{End}_R(E)$ generated over R by $T(n)$ $(n = 1, 2, \dots)$. By (du3) (and Exercise 3.2.11), $\mathcal{H}(E)$ is a commutative algebra. We define a pairing $\langle \cdot, \cdot \rangle : E \times \mathcal{H}(E) \to R$ by $\langle f, h \rangle = a(1, f|h)$. By (du3), we have $\langle f, T(n) \rangle = a(n, f)$. Then by (du2), $\langle f, T(n) \rangle = 0$ for all n implies $f = 0$. On the other hand, if we assume that $\langle f, h \rangle = 0$ for all $f \in E$, we have

$$0 = \langle f|T(n), h \rangle = \langle f, hT(n) \rangle = a(1, f|hT(n)) = \langle f|h, T(n) \rangle = a(n, f|h).$$

This shows that $f|h = 0$ for all $f \in E$, and by definition $h = 0$; so, the pairing is non-degenerate.

Proposition 3.2.9. *If R is a noetherian integral domain and E is a flat R-module of finite type, we have $\mathrm{Hom}_R(\mathcal{H}_R(E), R) \cong E$ and $\mathcal{H}_R(E) = \mathrm{Hom}_R(E, R)$ under the above pairing. If $\lambda \in \mathrm{Hom}_R(\mathcal{H}_R(E), R)$, the isomorphism: $\mathrm{Hom}_R(\mathcal{H}(E), R) \cong E$ sends λ to $\sum_{n=1}^\infty \lambda(T(n))q^n \in E$.*

Proof. For a finite set $\{x_1, \dots, x_n\}$ of R-generators of E, $\mathrm{End}_R(E) \ni \phi \mapsto (\phi(x_1), \dots, \phi(x_n)) \in E^n$ is an R-linear embedding. Since R is noetherian, $\mathrm{End}_R(E)$ is an R-module of finite type. Thus $\mathcal{H}(E) \subset \mathrm{End}_R(E)$ is an R-module of finite type and hence a noetherian ring.

First suppose that R is a field. In this case, E and $\mathcal{H}(E)$ are both finite dimensional vector space. Then non-degeneracy of the pairing implies perfectness as described in the lemma.

We now assume that R is local. Any R-flat module of finite type is free of finite rank if R is local [CRT, Theorem 7.10]; so, E is free of finite rank over R. Let $\mathfrak{m}$ be the maximal ideal of R with $F = R/\mathfrak{m}$. By definition, $\mathcal{H}(E)/\mathfrak{m}\mathcal{H}(E)$ surjects down to $\mathcal{H}_F(E/\mathfrak{m}E)$ as the two algebras are generated by $T(n)$. This shows the morphism: $i_R : \mathcal{H}(E) \to \mathrm{Hom}_R(E, R)$ induced by the pairing gives rise to

$$\mathcal{H}(E)/\mathfrak{m}\mathcal{H}(E) \xrightarrow{i_F} \mathrm{Hom}_R(E, R) \otimes_R R/\mathfrak{m} \overset{(*)}{\cong} \mathrm{Hom}_F(E/\mathfrak{m}E, F).$$

The last identity $(*)$ follows as the R-module E is R-free of finite rank. Since $\mathrm{Hom}_F(E/\mathfrak{m}E, F) \cong \mathcal{H}(E/\mathfrak{m}E)$ by the non-degeneracy over the field F, i_F factors through $\mathcal{H}(E/\mathfrak{m}E)$. Then by Nakayama's lemma, $i_R : \mathcal{H}(E) \to \mathrm{Hom}_R(E, R)$ is surjective. Tensoring the quotient field Q of R, we have

$$i_Q = i \otimes 1 : \mathcal{H}_Q(E \otimes_R Q) = \mathcal{H}(E) \otimes_R Q \to E \otimes_R Q.$$

Again by the result over now the field Q, i_Q is an isomorphism. This shows that i_R is injective, and hence $i_R : \mathcal{H}_R(E) \cong \mathrm{Hom}_R(E, R)$.

Since $\mathcal{H}(E)$ is the R-dual of the R-free module E, $\mathcal{H}(E)$ is R-free. Then by applying $\mathrm{Hom}_R(?, R)$ to $\mathcal{H}(E) \cong \mathrm{Hom}_R(E, R)$, we recover

$$i_R^* : E = \mathrm{Hom}_R(\mathrm{Hom}_R(E, R), R) \cong \mathrm{Hom}_R(\mathcal{H}(E), R),$$

as desired.

Now we treat the general case where R is a noetherian integral domain. We have a natural map $i_R : \mathcal{H}_R(E) \cong \mathrm{Hom}_R(E, R)$ given by the pairing. Localizing at maximal ideals $\mathfrak{m}$ of R, i_R induces we have

$$i_\mathfrak{m} = i_R \otimes 1 : \mathcal{H}_R(E) \otimes_R R_\mathfrak{m} = \mathcal{H}_{R_\mathfrak{m}}(E_\mathfrak{m})$$
$$\to \mathrm{Hom}_{R_\mathfrak{m}}(E_\mathfrak{m}, R_\mathfrak{m}) = \mathrm{Hom}_R(E, R) \otimes_R R_\mathfrak{m}$$

is an isomorphism; so,

$$i_R \otimes 1 : \mathcal{H}_R(E) \otimes_R \Omega = \bigoplus_\mathfrak{m} \mathcal{H}_{R_\mathfrak{m}}(E_\mathfrak{m})$$
$$\cong \bigoplus_\mathfrak{m} \mathrm{Hom}_{R_\mathfrak{m}}(E_\mathfrak{m}, R_\mathfrak{m}) = \mathrm{Hom}_R(E, R) \otimes_R \Omega$$

for $\Omega := \bigoplus_\mathfrak{m} R_\mathfrak{m}$ with $\mathfrak{m}$ running over all maximal ideals of R. Since Ω is faithful flat over R, i_R has to be an isomorphism.

Starting from R-linear map: $i_R^* : E \to \mathrm{Hom}_R(\mathcal{H}(E), R)$, we have an isomorphism

$$i_R^* \otimes 1 : E \otimes_R \Omega = \bigoplus_\mathfrak{m} E_\mathfrak{m}$$
$$\cong \bigoplus_\mathfrak{m} \mathrm{Hom}_{R_\mathfrak{m}}(\mathcal{H}(E_\mathfrak{m}), R_\mathfrak{m}) = \mathrm{Hom}_R(\mathcal{H}(E), R) \otimes_R \Omega.$$

Then by faithfully flat descent, the original i_R^* has to be an isomorphism.

As we computed before stating the proposition, $i_R^*(f) = \lambda \Leftrightarrow \lambda(T(n)) = \langle f, T(n) \rangle = a(n, f)$ for all n; so, $f = \sum_{n=0}^{\infty} \lambda(T(n))q^n$. □

3.2.5   Duality between Hecke algebra and cusp forms

Let $A \subset \mathbb{C}$ be a subring, and define

$$S_2(\Gamma_0(N), A) = \left\{ f \in S_2(\Gamma_0(N)) \big| a(n, f) \in A \right\}.$$

By definition, $S_2(\Gamma_0(N), \mathbb{C}) = S_2(\Gamma_0(N))$. By Lemma 3.2.7, $T(n)$ preserves the A-submodule $S_2(\Gamma_0(N), A)$ of $S_2(\Gamma_0(N))$. Define

$$h(N, A) = \mathbf{h}_2(N; A) := A[T(n)|n = 1, 2, \dots] \subset \operatorname{End}_A(S_2(\Gamma_0(N), A)),$$
$$H(N, A) = \mathbf{H}_2(N; A) := A[T(n)|n = 1, 2, \dots] \subset \operatorname{End}_A(M_2(\Gamma_0(N), A))$$
$$(3.17)$$

and call $h(N, A)$ the Hecke algebra on $\Gamma_0(N)$. By Lemma 3.2.7 (5), $h(N, A)$ is a commutative A-algebra.

After Chapter 4, we deal with the Hecke algebra of general weight k (not necessarily 2), and to emphasize the dependence on the weight k, we use the symbol $\mathbf{h}_k$ and $\mathbf{H}_k$ to denote Hecke algebras of weight k. Since in this chapter, we limit ourselves to $k = 2$, we alleviate our notation and use, for example, $h(N; A)$ in place of $\mathbf{h}_2(N; A)$.

We define an A–bilinear pairing

$$\langle \, , \, \rangle : h(N, A) \times S_2(\Gamma_0(N), A) \to A \quad \text{by } \langle h, f \rangle = a(1, f|h).$$

Proposition 3.2.10. *Assume that $A \subset \mathbb{C}$ is a subring.*

(1) *We have the following canonical isomorphisms:*

$$\operatorname{Hom}_A(S_2(\Gamma_0(N), A), A) \cong h(N, A)$$
$$\text{and } \operatorname{Hom}_A(h(N, A), A) \cong S_2(\Gamma_0(N), A),$$

and the latter is given by sending an A–linear form $\phi : h(N, A) \to A$ to the q-expansion $\sum_{n=1}^{\infty} \phi(T(n))q^n$.

(2) *(Shimura) We have*

$$S_2(\Gamma_0(N), A) = S_2(\Gamma_0(N), \mathbb{Z}) \otimes_{\mathbb{Z}} A \quad \text{and } h(N, A) = h(N, \mathbb{Z}) \otimes_{\mathbb{Z}} A.$$

Proof.   If A is a subfield of $\mathbb{C}$, the result (1) follows applying Proposition 3.2.9 to $h(N, A) = \mathcal{H}(S_2(N, A))$ and $E = S_2(N, A)$.

We have the Poincaré duality pairing

$$(\cdot,\cdot) : H^1(X_0(N), A) \times H^1(X_0(N), A) \to A$$

which is a perfect pairing [STH, V.9]. Define

$$\theta : H^1(X_0(N), A) \otimes_A H^1(X_0(N), A) \to S_2(\Gamma_0(N), A)$$

by $\theta(\xi \otimes \eta) = \sum_{n=1}^{\infty}(\xi, \eta|T(n))q^n$. Indeed, $h \mapsto (\xi, \eta|h)$ is an A-linear form on $h(N, A)$, and by the result already proven, we have $\sum_{n=1}^{\infty}(\xi, \eta|T(n))q^n \in S_2(\Gamma_0(N), A)$ if A is a field. By Proposition 3.2.4, θ is surjective if $A = \mathbb{C}$. Indeed, by the self-duality of $H^1(X_0(N), \mathbb{C})$, the projection $H^1(X_0(N), \mathbb{C}) \twoheadrightarrow S_2(\Gamma_0(N))$ induces a $T(n)$-equivariant inclusion $h(N, \mathbb{C}) = \mathrm{Hom}_{\mathbb{C}}(S_2(\Gamma_0(N)), \mathbb{C}) \hookrightarrow H^1(X_0(N), \mathbb{C})$, and thus any linear form on $h(N, \mathbb{C})$ is a linear combination of $h \mapsto \langle \xi|h, \eta \rangle$. This is equivalent to the surjectivity of θ over $\mathbb{C}$.

Since $H^1(X_0(N), \mathbb{Z}) \otimes A = H^1(X_0(N), A)$, the image under θ of $H^1(X_0(N), \mathbb{Z}) \otimes H^1(X_0(N), \mathbb{Z})$ spans $S_2(\Gamma_0(N), \mathbb{C})$. Thus $S_2(\Gamma_0(N), \mathbb{Z})$ span $S_2(\Gamma_0(N), \mathbb{C})$. This shows $S_2(\Gamma_0(N), \mathbb{Z}) \otimes_{\mathbb{Z}} \mathbb{C} = S_2(\Gamma_0(N), \mathbb{C})$, and therefore $S_2(\Gamma_0(N), \mathbb{Z}) \otimes_{\mathbb{Z}} A = S_2(\Gamma_0(N), A)$ for any ring A. In particular, $h(N, A)$ is a subalgebra of $\mathrm{End}_{\mathbb{C}}(S_2(\Gamma_0(N)))$ generated over A by $T(n)$ for all n; so, $h(N, A) = h(N, \mathbb{Z}) \otimes_{\mathbb{Z}} A$ by definition for any subring $A \subset \mathbb{C}$.

As for $A = \mathbb{Z}$, we only need to show that $\phi \mapsto \sum_{n=1}^{\infty}\phi(T(n))q^n$ is well defined and is surjective onto $S_2(\Gamma_0(N), \mathbb{Z})$ from $h(N, \mathbb{Z})$, because this is the case if we extend scalar to $A = \mathbb{Q}$. The cusp form $f \in S_2(\Gamma_0(N), A)$ corresponding to ϕ satisfies $\langle h, f \rangle = \phi(h)$; so, $a(n, f) = \langle T(n), f \rangle = \phi(T(n))$. Thus $f = \sum_{n=1}^{\infty}\phi(T(n))q^n \in S_2(\Gamma_0(N), A)$. However

$$f \in S_2(\Gamma_0(N), \mathbb{Z}) \iff \phi \in \mathrm{Hom}(h(N, \mathbb{Z}), \mathbb{Z}),$$

because $h(N, \mathbb{Z})$ is generated by $T(n)$ over $\mathbb{Z}$. This is enough to conclude surjectivity. Since $h(N, A) = h(N, \mathbb{Z}) \otimes A$ and $S_2(\Gamma_0(N), \mathbb{Z}) \otimes_{\mathbb{Z}} A = S_2(\Gamma_0(N), A)$, the duality over $\mathbb{Z}$ implies that over A. $\square$

Corollary 3.2.11. *We have the following assertions.*

(1) *For any $\mathbb{C}$-algebra homomorphism $\lambda : h(N, \mathbb{C}) \to \mathbb{C}$, $\lambda(T(n))$ for all n generates an algebraic number field $\mathbb{Q}(\lambda)$ over $\mathbb{Q}$ and $\lambda(T(n))$ is an algebraic integer.*

(2) *For any algebra homomorphism $\lambda : h(N, \mathbb{Z}) \to \mathbb{Q}(\lambda)$,*

$$S_2(\Gamma_0(N), \mathbb{Q}(\lambda))[\lambda]$$
$$= \{f \in S_2(\Gamma_0(N), \mathbb{Q}(\lambda) \big| f|T(n) = \lambda(T(n))f \text{ for all } n\}$$

is one dimensional and is generated by $\sum_{n=1}^{\infty} \lambda(T(n))q^n$.

(3) *For any algebra homomorphism* $\lambda : h(N, \mathbb{Z}) \to \mathbb{Q}(\lambda)$,

$$H^1(X_0(N), \mathbb{Q}(\lambda))[\lambda]$$
$$= \{c \in H^1(X_0(N), \mathbb{Q}(\lambda) | c|T(n) = \lambda(T(n))c \text{ for all } n\}$$
is two dimensional, and is isomorphic to
$$S_2(\Gamma_0(N), \mathbb{Q}(\lambda))[\lambda] \oplus \overline{S_2(\Gamma_0(N), \mathbb{Q}(\overline{\lambda}))[\overline{\lambda}]}.$$

Proof. Since $h(N, \mathbb{Z})$ is of finite rank over $\mathbb{Z}$, $A = \lambda(h(N, \mathbb{Z}))$ has finite rank d over $\mathbb{Z}$. Then the characteristic polynomial $P(X)$ of multiplication by $r \in A$ (regarding $A \cong \mathbb{Z}^d$) is satisfied by r, that is, $P(r) = 0$. Since $P(X) \in \mathbb{Z}[X]$, r is an algebraic integer. Then $A \otimes_{\mathbb{Z}} \mathbb{Q}$ is a finite extension $\mathbb{Q}(\lambda)$ of degree d over $\mathbb{Q}$.

Let K be a field. For any finite dimensional commutative K-algebra A, a K-algebra homomorphism $\lambda : A \to K$ gives rise to a generator of λ-eigenspace of the linear dual $\text{Hom}_K(A, K)$. Applying this fact to $\text{Hom}_K(h(N, K), K) = S_2(\Gamma_0(N), K)$ for $K = \mathbb{Q}(\lambda)$, we get the second assertion.

The third assertion then essentially follows from Proposition 3.2.4. Let us give some more details. Let $L := H^1(X_0(N), \mathbb{Q}(\lambda))[\lambda]$ and $R := S_2(\Gamma_0(N), \mathbb{Q}(\lambda))[\lambda] \oplus \overline{S_2(\Gamma_0(N), \mathbb{Q}(\overline{\lambda}))[\overline{\lambda}]}$. Then $L \otimes_{\mathbb{Q}(\lambda)} \mathbb{C} \cong R \otimes_{\mathbb{Q}(\lambda)} \mathbb{C}$ as modules over $\mathbb{Q}(\lambda)$ by Proposition 3.2.4. This isomorphism is canonical. By flat descent, we have $L \cong R$ as $\mathbb{Q}(\lambda)$-vector spaces. This isomorphism is not canonical as it is just by the descent theory of $\mathbb{Q}(\lambda)$-vector spaces. Since the $h(N; \mathbb{Q}(\lambda))$-module structures on R and L factors through $\mathbb{Q}(\lambda)$-vector space structure of R and L, they are non-canonically isomorphic as $h(N; \mathbb{Q}(\lambda))$-modules. $\qquad\square$

Definition 3.2.12. For a Hecke eigenform $f = \sum_{n=1}^{\infty} \lambda(T(n))q^n$ ($\lambda \in \text{Hom}_{\text{alg}}(h(N; \mathbb{Z}), \mathbb{C})$), define $\mathbb{Q}(f) = \mathbb{Q}(\lambda)$ by the number field generated by $\lambda(T(n))$ for all n, and write $\mathbb{Z}[f] = \mathbb{Z}[\lambda]$ for the integer ring of $\mathbb{Q}(f)$.

Note that $\lambda(h(N; \mathbb{Z})) \subset \mathbb{Z}[f]$ is a subring of finite index but may not be equal [J79].

Let $\varepsilon = \begin{pmatrix} -1 & 0 \\ 0 & 1 \end{pmatrix}$. Define $\varepsilon(z) = -\overline{z}$. Then ε takes $\mathfrak{H}$ to $\mathfrak{H}$. Define the action of $GL_2^+(\mathbb{R})\varepsilon$ on $\mathfrak{H}$ by $\gamma\varepsilon(z) = \gamma(-\overline{z})$. Since $GL_2(\mathbb{R}) = GL_2^+(\mathbb{R}) \sqcup \varepsilon GL_2^+(\mathbb{R})$, we have well defined map $\gamma : \mathfrak{H} \to \mathfrak{H}$ for all $\gamma \in GL_2(\mathbb{R})$.

Exercise 3.2.12. Prove the following facts.

(1) $\alpha(\beta(z)) = (\alpha\beta)(z)$ for $\alpha, \beta \in GL_2(\mathbb{R})$.
(2) We have $\mathfrak{H} \cong GL_2(\mathbb{R})/\mathbb{R}^{\times}O_2(\mathbb{R})$ by $g \mapsto g(\sqrt{-1})$.

Since $\varepsilon^2 = 1$ and $\varepsilon T_n \varepsilon^{-1} = T_n$, the action of ε commutes with Hecke operators. Thus if A is a ring in which 2 is invertible, we have a decomposition $H_1(X_0(N), A)$ and $H^1(X_0(N), A)$ into the direct sum of ± 1 eigenspaces of ε. We write $H_1^{\pm}$ or $H_{\pm}^1$ for the ± 1 eigenspace.

Proposition 3.2.13. *Let $A \subset \mathbb{C}$ be a principal ideal domain. Then for an algebra homomorphism $\lambda : h(N, \mathbb{Z}) \to A$, the three A-modules $H_1^{\pm}(X_0(N), A)[\lambda]$, $H_{\pm}^1(X_0(N), A)[\lambda]$ and $H_1^{\pm}(\overline{Y}_0(N), \partial \overline{Y}_0(N), A)[\lambda]$ are free of rank 1 over A.*

Proof. Since $\varepsilon^* f = f(\varepsilon(z)) \in \overline{S}_2(\Gamma_0(N), \mathbb{C})$ if $f \in S_2(\Gamma_0(N), \mathbb{C})$, $\pi_{\pm}(f) = \frac{f \pm \varepsilon^* f}{2} \neq 0$ if $f \neq 0$. Moreover $\pi_{\pm}$ induces an isomorphism of $S_2(\Gamma_0(N), \mathbb{C})[\lambda]$ onto $(S_2(\Gamma_0(N), \mathbb{C}) \oplus \overline{S}_2(\Gamma_0(N), \mathbb{C}))^{\pm}[\lambda]$, where the superscript "$\pm$" indicates the $\pm$ eigenspace for ε. Since $S_2(\Gamma_0(N), \mathbb{C})[\lambda]$ is one dimensional, we conclude that $(S_2(\Gamma_0(N), \mathbb{C}) \oplus \overline{S}_2(\Gamma_0(N), \mathbb{C}))^{\pm}[\lambda]$ is one dimensional. By Proposition 3.2.4, we have

$$(S_2(\Gamma_0(N), \mathbb{C}) \oplus \overline{S}_2(\Gamma_0(N), \mathbb{C}))^{\pm}[\lambda] \cong H_{\pm}^1(X_0(N), \mathbb{C})[\lambda],$$

which is therefore one dimensional. Then by the Poincaré duality (3.2), $H_1^{\pm}(X_0(N), \mathbb{C})[\lambda]$ is one dimensional. The same argument tells us that $H_1^{\pm}(\overline{Y}_0(N), \partial \overline{Y}_0(N), \mathbb{C})[\lambda]$ is one dimensional by Proposition 3.2.6.

As long as A contains the eigenvalues $\lambda(T(n))$, we have

$$H_1^{\pm}(X_0(N), A)[\lambda] \otimes_A \mathbb{C} \cong H_1^{\pm}(X_0(N), \mathbb{C})[\lambda].$$

Since A is a principal ideal domain, $H_1^{\pm}(X_0(N), A)[\lambda]$ is A-free. By the above identity, we conclude $\operatorname{rank}_A H_1^{\pm}(X_0(N), A)[\lambda] = 1$. The same argument proves the assertion for other homology and cohomology groups. $\square$

3.2.6 Modular Hecke L-functions

Let $\lambda : h(N, \mathbb{Z}) \to \mathbb{C}$ be an algebra homomorphism. Then we define $L(s, \lambda) = \sum_{n=1}^{\infty} \lambda(T(n)) n^{-s}$, which is the modular Hecke L-function of λ. We now give key points of the analytic continuation of this Dirichlet to the whole complex plane. We start with

Lemma 3.2.14. *If $f \in S_2(\Gamma_0(N))$, then $|f(x+iy)| \leq Cy^{-1}$ for a constant C independent of x and y.*

Proof. For $\alpha = \begin{pmatrix} a & b \\ c & d \end{pmatrix} \in GL_2^+(\mathbb{R})$, $\alpha \begin{pmatrix} z & \bar{z} \\ 1 & 1 \end{pmatrix} = \begin{pmatrix} \alpha(z) & \alpha(\bar{z}) \\ 1 & 1 \end{pmatrix} \begin{pmatrix} cz+d & 0 \\ 0 & c\bar{z}+d \end{pmatrix}$.
Taking the determinant of this, we get $\det(\alpha)\,\mathrm{Im}(z) = \mathrm{Im}(\alpha(z))|cz + d|^2$.
Thus $g(z) = |f(z)\,\mathrm{Im}(z)|$ factors through $X_0(N)$. Since $f(z)$ vanishes at
cusps, $g(z)$ also; so, $g(z)$ is a continuous function on the compact space
$X_0(N)$. Thus the positive function $g(z)$ is bounded by a positive constant
C: $|g(z)| \le C$, which proves the lemma. $\square$

Lemma 3.2.15. *For a constant $B > 0$, $|\lambda(T(n))| \le B \cdot n$ for $0 < n \in \mathbb{Z}$.*

Proof. Since $f = f_\lambda = \sum_{n=1}^\infty \lambda(T(n))q^n$ is a cusp form in $S_2(\Gamma_0(N))$
with $f_\lambda | T(n) = \lambda(T(n))f_\lambda$, picking any $f \in S_2(\Gamma_0(N))$, we need to prove
$|a(n, f)| \le Bn$ for all n. Since $a(n, f) = (2\pi i)^{-1} \int_{|q|=r} f(q)q^{-n-1}dq =$
$\int_0^1 f(z)\exp(-2\pi i n z)dz$ for any $r > 0$ by the residue formula (and it is
independent of $y = \mathrm{Im}(z)$), taking $r = \exp(-1/n)$ ($\Leftrightarrow \mathrm{Im}(z) = \frac{1}{2\pi n}$, by
Lemma 3.2.14, we get $|a(n, f)| \le 2Ce\pi n$. Thus $B = 2Ce\pi$. $\square$

Exercise 3.2.13. Prove that $L(s, \lambda)$ converges absolutely if $\mathrm{Re}(s) > 2$.

Exercise 3.2.14. Let $\tau = \begin{pmatrix} 0 & -1 \\ N & 0 \end{pmatrix}$. Prove $\tau\Gamma_0(N)\tau^{-1} = \Gamma_0(N)$ and $f|\tau \in$
$S_2(\Gamma_0(N))$ if $f \in S_2(\Gamma_0(N))$.

Lemma 3.2.16. *If $f \in S_2(\Gamma_0(N))$, then $L(s, f) = \sum_{n=1}^\infty a(n, f)n^{-s}$ con-*
verges absolutely if $\mathrm{Re}(s) > 2$ and can be continuated analytically to a
holomorphic function on the whole complex plane.

Proof. By the same computation as in (3.4), if $\mathrm{Re}(s) > 2$,

$$\int_0^\infty f(iy)y^{s-1}dy = (2\pi)^{-s}\Gamma(s)L(s, f).$$

The integral $\int_0^\infty f(iy)y^{s-1}dy$ is convergent for all $s \in \mathbb{C}$, because for any
power y^s, $\lim_{y\to\infty} f(iy)y^{s-1} = \lim_{y\to 0} f(iy)y^{s-1} = 0$ (by the q-expansion).
Thus $L(s, f)$ is analytically continued to the whole complex plane. $\square$

By the above expression, we have

$$L(1, \lambda) = -\int_0^\infty (2\pi i)f_\lambda(z)dz \overset{(*)}{=} -\int_0^\infty \varepsilon^*((2\pi i)f_\lambda(z)dz). \qquad (3.18)$$

The identity $(*)$ follows since $\varepsilon^*(f(z)dz) - -f(-\bar{z})d\bar{z}$.

3.2.7 *Rationality of Hecke L-values*

Let $A \subset \mathbb{C}$ be a principal ideal domain. We need this assumption as we
compare two A-integral structure of a one dimensional $\mathbb{C}$-vector space V.

In other words, $V = R \otimes_A \mathbb{C} = L \otimes_A \mathbb{C}$ for two A-modules R and L. By the PID assumption, $R = \Omega \cdot L$ for a complex number $0 \neq \Omega \in \mathbb{C}$ unique up to A-unit multiple. In this way, we define our "period".

Let $\lambda : h(N, \mathbb{Z}) \to A$ be an algebra homomorphism. Define $\omega_\pm(\lambda) = \frac{1}{2}((2\pi i)f_\lambda(z)dz \pm \varepsilon^*((2\pi i)f_\lambda(z)dz))$ and $\omega_\pm(f) = \frac{1}{2}((2\pi i)f(z)dz \pm \varepsilon^*((2\pi i)f(z)dz))$ for $f \in S_2(\Gamma_0(N))$. Let $\delta_\pm(\lambda)$ be a generator of $H^1_\pm(X_0(N), A)[\lambda]$ over A; so, $H^1_\pm(X_0(N), A)[\lambda] = A\delta_\pm(\lambda)$. Then by Proposition 3.2.13, we have $[\omega_\pm(\lambda)] = \Omega_\pm(\lambda; A)\delta_\pm(\lambda)$. We call $\Omega_\pm(\lambda; A)$ the $\pm$ period of λ. Let $\gamma_a \in H_1(\overline{Y}_0(N), \partial\overline{Y}_0(N), \mathbb{Z})$ for $a \in \mathbb{Q}$ be the relative 1-cycle represented by vertical line in $\mathfrak{H}$ passing through $a \in \mathbb{Q}$.

Lemma 3.2.17. *We have $\frac{L(1,\lambda)}{\Omega_+(\lambda;A)} \in M^{-1}A$ for a positive integer M only dependent on N. Moreover $\int_{\gamma_a} \delta_\pm(\lambda) \in M^{-1}A$ for all $a \in \mathbb{Q}$.*

Clear from the proof below, this constant M is a factor of the least common multiple D of the denominator of the 1-cycles γ_a in $H_1(X_0(N), \mathbb{Q})$. The integer D therefore kills the cuspidal divisor class group of $X_0(N)$ we studied in Chapter 2. See [M72], [D73] and [M77] for an estimate of D.

Proof. The natural map $\iota : H_1(X_0(N), A) \to H_1(\overline{Y}_0(N), \partial\overline{Y}_0(N), A)$, after tensoring $\mathbb{C}$, becomes an isomorphism by Proposition 3.2.6, $\mathrm{Coker}(\iota)$ is finite of order M. Since the vertical line γ_a passing through $a \in \mathbb{Q}$ is a 1-cycle in $H_1(\overline{Y}_0(N), \partial\overline{Y}_0(N), A)$, we have $M\gamma_a \in H_1(X_0(N), A)$. Thus by (3.18), we get

$$\frac{M\int_{\gamma_0}\omega_+(\lambda)}{\Omega_+(\lambda; A)} = \int_{M\gamma_0} \delta_+(\lambda) \in A.$$

The same argument also applies to γ_a. This finishes the proof. $\square$

Since $\left(\begin{smallmatrix} 1 & 1 \\ 0 & 1 \end{smallmatrix}\right)(z) = z + 1$, $\gamma_a = \gamma_{a+1}$. Thus the cycle γ_a only depends on $a \in \mathbb{Q}/\mathbb{Z}$. Let $\chi : (\mathbb{Z}/m\mathbb{Z}) \to A$ be a primitive Dirichlet character. Consider $\gamma(\chi) = \sum_{u \bmod m} \chi^{-1}(u)\gamma_{\frac{u}{m}} \in H_1(\overline{Y}_0(N), \partial\overline{Y}_0(N), A)$.

Lemma 3.2.18. *We have $\varepsilon(\gamma(\chi)) = \chi(-1)\gamma(\chi)$.*

Proof. Note that $\varepsilon(\gamma_a) = \gamma_{-a}$, and from this, we have

$$\varepsilon(\gamma(\chi)) = \sum_{u \bmod m} \chi^{-1}(u)\varepsilon(\gamma_{u/m})$$

$$= \sum_{u \bmod m} \chi^{-1}(u)(\gamma_{-u/m}) \overset{u \mapsto -u}{=} \chi(-1)\gamma(\chi).$$

$\square$

We consider $f|R_\chi(z) = \sum_{u \bmod m} \chi^{-1}(u)f(z + \frac{u}{m})$. The following exercise is a bit difficult.

Exercise 3.2.15. Let N' be the LCM of N and m^2. Prove $(f|R_\chi)|\gamma = \chi^2(u)f|R_\chi$ for $\gamma = \left(\begin{smallmatrix} a & b \\ c & d \end{smallmatrix}\right) \in \Gamma_0(N')$.

Note that $\int_{\gamma(\chi)} \omega_\pm(\lambda) = \int_{\gamma_0} \omega_\pm(f|R_\chi)$. Then we have, for $\zeta_m = \exp(2\pi/m)$

$$f|R_\chi(z) = \sum_{u \bmod m} \chi^{-1}(u)f(z + \frac{u}{m})$$

$$= \sum_{n=1}^\infty a(n,f)q^n \sum_u \chi^{-1}(u)\zeta_m^{nu} = G(\chi^{-1}) \sum_{n=1}^\infty a(n,f)\chi(n)q^n. \quad (3.19)$$

Then by the same argument proving Lemma 3.2.17 applied to $\gamma(\chi)$, for the constant M in the lemma, we get

Proposition 3.2.19. *Let χ be a primitive Dirichlet character modulo m with values in a principal ideal domain $A \subset \mathbb{C}$. Then there exists a constant M only depending on N such that $\frac{G(\chi^{-1})L(1,\lambda\otimes\chi)}{\Omega_{\chi(-1)}(\lambda;A)} \in M^{-1}A$, where*

$$L(s, \lambda \otimes \chi) = \sum_{n=1}^\infty \chi(n)\lambda(T(n))n^{-s}.$$

3.2.8 *p-old and p-new forms*

Let N be a positive integer prime to p.

Exercise 3.2.16. If $f \in S_2(\Gamma_0(N))$, prove that $f(pz) \in S_2(\Gamma_0(Np))$.

We consider an algebra homomorphism $\lambda : h(N, \mathbb{Z}) \to \overline{\mathbb{Q}}$. Then we have a Hecke eigenform $f = \sum_{n=1}^\infty \lambda(T(n))q^n \in S_2(\Gamma_0(N))$ with $f|T(n) = \lambda(T(n))f$. The L-function for a Dirichlet character χ modulo M $L(s, \lambda \otimes \chi) = \sum_{n=1}^\infty \lambda(T(n))\chi(n)n^{-s}$ has the following Euler product:

$$\prod_l \frac{1}{(1 - \lambda(T(l))\chi(l)l^{-s} + \chi(l)^2 l^{1-2s})} = \prod_l \left[(1 - \frac{\alpha_l\chi(l)}{l^s})(1 - \frac{\beta_l\chi(l)}{l^s})\right]^{-1},$$

where α_l and β_l are two roots of $X^2 - \lambda(T(l))X + l = 0$ if $l \nmid N$. If $l|N$, α_l and β_l are integers in $\mathbb{Q}(\lambda)$ which could be 0. See §9.1.1 for an exact description of the factor. For the compatible system ρ of Galois representations attached to f we introduce later in §6.2, the Euler factor is given by $\det(1 - \rho(\mathrm{Frob}_l)|_{V^{I_l}}l^{-s})^{-1}$, where V is the representation space of the q-adic member ρ of the system (q is chosen to be prime to l).

Exercise 3.2.17.

(1) Prove $T(m)T(n) = \sum_{0<d|(m,n),(d,N)=1} d \cdot T(mn/d^2)$ by Lemma 3.2.7.

(2) Prove the above Euler factorization of $L(s, \lambda \otimes \chi)$.

Lemma 3.2.20. *Let $\alpha = \alpha_p$ and $\beta = \beta_p$, and put $f_\alpha(z) = f(z) - \beta f(pz)$. Then we have $f_\alpha \in S_2(\Gamma_0(Np))$ and $f_\alpha|U(p) = \alpha f_\alpha$ and $f_\alpha|T(n) = \lambda(T(n))f_\alpha$ for all $n > 0$ prime to p, where $U(p)$ is the Hecke operator $T(p)$ acting on $S_2(\Gamma_0(Np))$.*

The form f_α is an α-*stabilization* of f. If a p-adic absolute value $|\cdot|_p$ is chosen, f_α is called the ordinary (resp. critical) stabilization of f if $|\alpha|_p = 1$ (resp. $|\alpha|_p < 1$).

Proof. By Lemma 3.2.7 (4), we have

$$a(m, f|T(n)) = \sum_{0<d|(m,n),\ (d,N)=1} d \cdot a(\frac{mn}{d^2}, f).$$

From this, it is easy to see that $T(m)T(n) = T(mn)$ if $m\mathbb{Z} + n\mathbb{Z} = \mathbb{Z}$. Note that $a(n, f) = \lambda(T(n))$ and hence $a(mn, f) = a(m, f)a(n, f)$ if $m\mathbb{Z} + n\mathbb{Z} = \mathbb{Z}$. Since $f_\alpha = \sum_{m=1}^\infty a(m, f)q^m - \beta \sum_{m=1}^\infty a(m, f)q^{mp}$, we see $a(m, f_\alpha) = a(m, f)$ if $p \nmid m$. In particular, if $p \nmid m$ and $p \nmid n$, we have

$$a(m, f_\alpha|T(n)) = \sum_{0<d|(m,n),\ (d,Np)=1} d \cdot a(\frac{mn}{d^2}, f_\alpha)$$

$$= \sum_{0<d|(m,n),\ (d,N)=1} d \cdot a(\frac{mn}{d^2}, f) = \lambda(T(n))a(m, f) = \lambda(T(n))a(m, f_\alpha).$$

If $m = m_0 p$, we have $a(m, f_\alpha) = a(m, f) - \beta \cdot a(m_0, f)$. Thus if $p|m$ and $p \nmid n$, we have

$$a(m, f_\alpha|T(n)) = \sum_{0<d|(m,n),\ (d,Np)=1} d \cdot a(\frac{mn}{d^2}, f_\alpha)$$

$$= \sum_{0<d|(m,n),\ (d,N)=1} d \cdot (a(\frac{mn}{d^2}, f) - \beta \cdot a(\frac{m_0 n}{d^2}, f))$$

$$= \lambda(T(n))(a(m, f) - \beta \cdot a(m_0, f)) = \lambda(T(n))a(m, f_\alpha).$$

This shows that $f_\alpha|T(n) = \lambda(T(n))f_\alpha$ for n prime to p.

Now we have $a(m, f_\alpha|U(p)) = a(mp, f_\alpha) = a(mp, f) - \beta \cdot a(m, f)$. On the other hand,

$$(\alpha + \beta)a(m, f) = a(m, f|T(p)) = p \cdot a(\frac{m}{p}, f) + a(mp, f)$$

$$= (\alpha\beta) \cdot a(\frac{m}{p}, f) + a(mp, f).$$

This shows

$$a(m, f_\alpha | U(p)) = (\alpha + \beta)a(m, f) - (\alpha\beta) \cdot a(\frac{m}{p}, f)$$

$$- \beta \cdot a(m, f) = \alpha \cdot a(m, f_\alpha).$$

This finishes the proof. □

Corollary 3.2.21. *Let the notation be as in* Lemma 3.2.20. *If* $p \nmid N$ *and* $\lambda : h(N, \mathbb{Z}) \to \overline{\mathbb{Q}}$ *is an algebra homomorphism, we have an algebra homomorphism* $\lambda_\alpha : h(pN, \mathbb{Z}) \to \overline{\mathbb{Q}}$ *such that* $\lambda_\alpha(U(p)) = \alpha$ *and* $\lambda_\alpha(T(n)) = \lambda(T(n))$ *if* $p \nmid n$. *Moreover, we have* $L(s, \lambda_\alpha \otimes \chi) = (1 - \beta\chi(p)p^{-s})L(s, \lambda \otimes \chi)$.

Exercise 3.2.18. Give a detailed proof of the above corollary.

3.2.9 *Elliptic modular p-adic measure*

Take an algebra homomorphism of the Hecke algebra $\lambda : h(pN, \mathbb{Z}) \to \overline{\mathbb{Q}}$. Then we have $f_\lambda = \sum_{n=1}^\infty \in S_2(\Gamma_0(Np))$ with $f_\lambda | T(n) = \lambda(T(n))f_\lambda$. We write the Hecke operator $T(p)$ on $S_2(\Gamma_0(Np))$ as $U(p)$; so, we have

$$f | U(p)(z) = \frac{1}{p} \sum_{j=0}^{p-1} f(\frac{z+j}{p}).$$

The action of $U(p)$ is the same as in the case of $\mathbf{G}_m$. We suppose $a = \alpha_p = \lambda(U(p))$ is a p-adic unit in $\mathbb{Q}_p(\lambda)$. Such λ and f_λ are called p-*ordinary*.

We have an A-linear map: $H^1(\overline{Y}_0(Np), \partial\overline{Y}_0(Np), A) \to A$ given by $\omega \mapsto \int_{\gamma_x} \omega$. Then we consider a map

$$c : p^{-\infty}\mathbb{Z} = \bigcup_{i=1}^\infty p^{-i}\mathbb{Z} \to \operatorname{Hom}_A(H^1(\overline{Y}_0(Np), \partial\overline{Y}_0(Np), A), K) \qquad (3.20)$$

given by $c(x)(\omega) = \int_{\gamma_x} \omega$.

For $\omega \in H^1(\overline{Y}_0(Np), \partial\overline{Y}_0(Np), A)$, we write $c_\omega(r) = \int_{\gamma_r} \omega$. Then $c_\omega(r + 1) = c_\omega(r)$, and c_ω factors through $\mathbb{Q}_p/\mathbb{Z}_p = p^{-\infty}\mathbb{Z}/\mathbb{Z}$. Supposing $\omega | U(p) = a\omega$ with $|a|_p = 1$, we define a distribution φ_ω on $\mathbb{Z}_p$ by

$$\varphi_\omega(z + p^m\mathbb{Z}_p) = a^{-m}c_\omega(\frac{z}{p^m}) \quad \text{for } z = 1, 2, \dots \text{ prime to } p. \qquad (3.21)$$

This is well defined because $c_\omega(r + 1) = c_\omega(r)$. We take the multiplicative group $G = \mathbb{Z}_p^\times$ and fix an isomorphism $G \cong \mu \times \mathbb{Z}_p$ for a finite group μ, where $\mathbb{Z}_p$ in the right-hand-side is an additive group. Then the multiplicative subgroup $G_i = 1 + p^i\mathbb{Z}_p$ corresponds to the additive group $p^i\mathbb{Z}_p$. To show that φ_ω actually gives a distribution, we need to check the distribution relation (3.9). We compute

$$\sum_{j=1}^{p} c_\omega(\frac{x+j}{p}) = \sum_j c(\frac{x+j}{p})(\omega) = c(x)(\omega|U(p)) = a \cdot c_\omega(x).$$

This shows

$$\sum_{j=1}^{p} \varphi_\omega(x + jp^m + p^{m+1}\mathbb{Z}_p) = \varphi_\omega(x + p^m\mathbb{Z}_p).$$

The general distribution relation (3.9) then follows from the iteration of this relation. By a similar argument, we see that

$$|\varphi_\omega(z + p^m\mathbb{Z}_p)|_p = |a^{-m}c_\omega(\frac{z}{p^m})|_p = |c(\frac{z}{p^m})(\omega)|_p \le |\omega|_p, \qquad (3.22)$$

where $|\omega|_p = \text{Sup}_x |c(x)(\omega)|_p$ with x running over $p^{-\infty}\mathbb{Z}$. Thus φ_ω is bounded (by the proof of Lemma 3.2.17) and, by Proposition 3.1.12, we have a unique measure φ_ω extending the distribution φ_ω. Now we compute $\int_G \phi d\varphi_\omega(x)$. To do this, we may assume that $|\omega|_p \le 1$ by multiplying by a constant if necessary. For $\phi \in C(G/G_m; A)$, we have

$$\int_G \phi d\varphi_\omega = a^{-m} \sum_{z=1}^{p^m} \phi(z)c_\omega(\frac{z}{p^m}).$$

Let $N > 1$ be a positive integer prime to p. We take $\omega = \delta_+(\lambda)$ for each algebra homomorphism $\lambda : h(Np, \mathbb{Z}) \to \overline{\mathbb{Q}}$. Then we write φ_ω as $\varphi = \varphi_\lambda$ and compute for any primitive character ϕ of $(\mathbb{Z}/p^r\mathbb{Z})^\times$ the integral $\int_G \phi d\varphi_\lambda$. Note that $\omega|U(p) = a\omega$. We then write $\alpha_x : \overline{Y}_0(Np) \to \overline{Y}_0(Np)$ to be the translation $\alpha_x(z) = z + x$ for $x \in \mathbb{R}$. We see that, if $\phi \ne 1$,

$$\int_G \phi d\varphi_\lambda = a^{-r} \sum_{x \in (\mathbb{Z}/p^r\mathbb{Z})^\times} \phi(x)c(\frac{x}{p^r})(\omega) = a^{-r} \int_{\gamma_0} \sum_{x \in (\mathbb{Z}/p^r\mathbb{Z})^\times} \phi(x)\alpha^*_{x/p^r}\omega$$

$$= a^{-r} \int_{\gamma_0} \omega|R_{\phi^{-1}} = a^{-r} \frac{G(\phi)L(1, \lambda \otimes \phi^{-1})}{\Omega_{\phi(-1)}(\lambda; A)}.$$

$$(3.23)$$

We have basically proved the following theorem of Mazur:

Theorem 3.2.22. *Let p be a prime and N be a positive integer prime to p. Let $A = \{x \in \mathbb{Q}(\lambda)\,|\,|x|_p \le 1\}$ be the discrete valuation ring in $\mathbb{Q}(\lambda)$ for a p-adic valuation $|\cdot|_p$ of $\mathbb{Q}(\lambda)$. For an algebra homomorphism $\lambda : h(Np, \mathbb{Z}) \to A$ with $|\lambda(U(p))|_p = 1$, we have a unique p-adic measure φ_λ on $\mathbb{Z}_p^\times$ such that for all finite order characters ϕ of $\mathbb{Z}_p^\times$ and $1 \le j \in \mathbb{Z}$, we have*

$$\int_{\mathbb{Z}_p^\times} \phi(z)d\varphi_\lambda = \lambda(U(p))^{-r}\frac{G(\phi)L(1, \lambda \otimes \phi^{-1})}{\Omega_{\phi(-1)}(\lambda; A)}.$$

When ϕ is the identity character $\mathbf{1}$, we need to explain what $G(\mathbf{1})$ means. We have

$$f|R_1(z) = \sum_{x \in (\mathbb{Z}/\|\mathbb{Z})^\times} f(z + \frac{x}{p}) = \sum_{n=1}^\infty a(n,f)q^n \left(\sum_{x \in (\mathbb{Z}/\|\mathbb{Z})^\times} \exp(\frac{2\pi n x}{p}) \right)$$

$$= (p-1)\sum_{n=1}^\infty a(np,f)q^{np} - \sum_{n=1, p\nmid n}^\infty a(n,f)q^n = -f(z) + p\sum_{n=1}^\infty a(np,f)q^{np}.$$

Thus we get $\int_0^\infty f_\lambda|R_1 dy = (-2\pi)^{-1}(1 - \alpha_p)L(1,\lambda)$ for $a = \alpha_p = \lambda(U(p))$. Since we have

$$\omega_+(f) = 2^{-1}(2\pi i)(f(z)dz + \varepsilon^*(f(z)dz)) = 2^{-1}(2\pi i)(f(z)dz - (f(-\bar{z})d\bar{z}))$$

whose restriction to γ_0 is $(-2\pi)f(iy)dy$, and replacing f by $f|R_1$, we get

$$\int_{\mathbb{Z}_p^\times} d\varphi_\lambda = \lambda(U(p))^{-1}\frac{(1 - \lambda(U(p)))L(1,\lambda)}{\Omega_+(\lambda; A)}.$$

Thus essentially $G(\mathbf{1}) = 1$. We leave you to formulate the corresponding p-adic L-functions. See [LFE, Chapter 6] for more details of these facts.

Chapter 4

p-adic families of modular forms

In this chapter, we present the basics of *p*-adic deformation theory of elliptic modular forms of weight $k \geq 2$. We discuss the following three topics:

(1) *p*-adic analytic family of modular forms,
(2) *p*-adic measure theory via polynomial functions,
(3) Eichler–Shimura isomorphisms of higher weight.

In §4.1, we recall the method of Wiles for constructing *p*-adic analytic family described in [LFE, Chapter 7] (which is perhaps the most elementary). In the following §4.2, our approach is cohomological but still elementary, developing the ideas first conceived in [H86b]. There is a more sophisticated algebro geometric method (proving equivalent results) of the twin paper [H86a] which has been exposed already in [GME].

Along with these main topics, we will give a brief description of different cohomology theories we will use. As before, all rings are supposed to have the identity, and we fix a prime p. When $p = 2$, we write $\mathbf{p} = 4$ and we just set $\mathbf{p} = p$ if $p > 2$. Often we state the results including $p = 2$, but for simplicity, from time to time we assume $p > 2$. We denote by W the base ring which is a discrete valuation ring free of finite rank over $\mathbb{Z}_p$. We recall the $\mathbf{\Gamma} = 1 + \mathbf{p}\mathbb{Z}_p$ (which is canonically isomorphic to $\mathrm{Gal}(\mathbb{Q}_\infty/\mathbb{Q})$) and the Iwasawa algebra $\Lambda = \Lambda_W = W[[\mathbf{\Gamma}]] := \varprojlim_r W[\mathbf{\Gamma}/\mathbf{\Gamma}^{p^r}]$. Hereafter, often a diagonal matrix with diagonal entries $a_1, \ldots, a_n$ from the top to the bottom is written as $\mathrm{diag}[a_1, \ldots, a_n]$.

4.1 *p*-adic family and slope

First, without going into technical details, we describe a prototypical example of *p*-adic analytic families of modular forms. In later sections, we justify our construction cohomologically. The examples we describe are from [S73] and [LFE, Chapter 7].

4.1.1 *p-adic L-functions as a power series*

We start with a general fact on the Kubota–Leopoldt p-adic L-functions. We consider the binomial formula:

$$(1 + T)^s = \sum_{n=0}^{\infty} \binom{s}{n} T^n. \tag{4.1}$$

Since $s \mapsto \binom{s}{n} = \frac{s(s-1)\cdots(s-(n-1))}{n!}$ is a polynomial in s and has integer value over natural numbers, it is a polynomial on $\mathbb{Z}_p$ with values in $\mathbb{Z}_p$. Thus if $\gamma \equiv 1 \mod p$, we have the p-adic power $\gamma^s = \sum_{n=0}^{\infty} \binom{s}{n}(\gamma - 1)^n \in \mathbb{Z}_p$ well defined (i.e., convergent p-adically) for all $s \in \mathbb{Z}_p$.

Let K be a finite extension of $\mathbb{Q}_p$ with the p-adic integer ring $W = \{w \in K \mid |w|_p \le 1\}$. Let φ be a p-adic measure on $\mathbb{Z}_p$ with values in W (so it is a bounded measure). Since the power series ring $W[[T]]$ is a Banach algebra under the norm $\|\sum_{n=1}^{\infty} a_n T^n\| = \text{Sup}_n |a_n|_p$, we can integrate any continuous function $\phi : \mathbb{Z}_p \to W[[T]]$ under $d\varphi$. In other words, we approximate ϕ by step functions $\phi_n : \mathbb{Z}_p \to W[[T]]$ factoring through $(\mathbb{Z}/p^n\mathbb{Z})$ so that $\lim_{n\to\infty} \phi_n = \phi$ under the norm $\| \cdot \|$ and define

$$\int_{\mathbb{Z}_p^\times} \phi \, d\varphi = \lim_{n\to\infty} \int_{\mathbb{Z}_p^\times} \phi_n \, d\varphi \in W[[T]].$$

Exercise 4.1.1. Prove that

$$\int_{\mathbb{Z}_p^\times} (1 + T)^s d\varphi(s) = \sum_{\mathbb{Z}_p^\times} \binom{s}{n} d\varphi(s) T^n =: \Phi_\varphi(T).$$

Let $\boldsymbol{\Gamma} = 1 + \mathbf{p}\mathbb{Z}_p$ and $z \mapsto z_\Gamma = \omega(z)^{-1}z$ be the projection of $\mathbb{Z}_p^\times$ onto $\boldsymbol{\Gamma}$, where ω is the Teichmüller character defined above Theorem 3.1.14 if $p > 2$ and the Legendre symbol $\left(\frac{-4}{\cdot}\right)$ if $p = 2$. By the existence of a primitive root, for an odd prime p, the multiplicative group $(\mathbb{Z}/p^n\mathbb{Z})^\times$ is a cyclic group, and hence its subgroup $\{x \in (\mathbb{Z}/p^n\mathbb{Z})^\times \mid x \equiv 1 \mod p\}$ is cyclic generated by $\gamma = 1 + p$. If $p = 2$, though $(\mathbb{Z}/2^n\mathbb{Z})^\times$ is not cyclic if $n \ge 2$, the subgroup $\{x \in (\mathbb{Z}/p^n\mathbb{Z})^\times \mid x \equiv 1 \mod \mathbf{p}\}$ is cyclic. Hereafter we write $\gamma := 1 + \mathbf{p}$ which is a topological generator of the multiplicative group $\boldsymbol{\Gamma}$.

Exercise 4.1.2. Let $\Gamma^n = \{u^n | u \in \Gamma\} \subset \Gamma$. Prove the following facts

(1) $\Gamma^n = \Gamma$ if $p \nmid n$.

(2) $\Gamma/\Gamma^{p^{n-1}} \cong \{x \in (\mathbb{Z}/p^n\mathbb{Z})^\times | x \equiv 1 \mod \mathbf{p}\}$ by sending $u\Gamma^{p^{n-1}}$ to u mod $p^{n-1}\mathbf{p}$. In particular, $[\Gamma : \Gamma^{p^n}] = p^n$.

(3) As topological groups, $\Gamma \cong \mathbb{Z}_p$ by $\gamma^s \mapsto s \in \mathbb{Z}_p$ for $\gamma = 1 + \mathbf{p}$.

We have a projection $\mathbb{Z}_p^\times \ni z \mapsto z_\Gamma \in \Gamma$. Thus we can define a bounded measure φ_Γ on Γ by $\int_\Gamma \phi d\varphi_\Gamma = \int_{\mathbb{Z}_p^\times} \phi(z_\Gamma) d\varphi$. Identifying Γ with $\mathbb{Z}_p$ by $\gamma^s \leftrightarrow s \in \mathbb{Z}_p$, consider $\Phi_{\varphi_\Gamma}(T) \in W[[T]]$.

Lemma 4.1.1. *We have* $\int_\Gamma u^s d\varphi_\Gamma(u) = \Phi_{\varphi_\Gamma}(\gamma^s - 1)$.

Proof. For the isomorphism $\iota : \Gamma \cong \mathbb{Z}_p$ with $\iota(\gamma^z) = z$, we can define a measure φ_+ on $\mathbb{Z}_p$ by $\int_{\mathbb{Z}_p} \phi d\varphi_+ = \int_\Gamma \phi \circ \iota d\varphi_\Gamma$. Then we have $\Phi_{\varphi_\Gamma} = \Phi_{\varphi_+}$, and $\Phi_{\varphi_+}(T) = \int_{\mathbb{Z}_p}(1 + T)^z d\varphi_+(z)$. Replacing T by $\gamma^s - 1$ and writing $u = \gamma^z$, we get

$$\Phi_{\varphi_+}(\gamma^s - 1) = \int_{\mathbb{Z}_p} \gamma^{sz} d\varphi_+(z) = \int_\Gamma u^s d\varphi_\Gamma(u),$$

which shows the assertion. $\qquad\qquad\square$

Exercise 4.1.3. Define a Dirac measure δ_z for $z \in \mathbb{Z}_p$ by $\int_{\mathbb{Z}_p} \phi d\delta_z = \phi(z)$. Prove that $\Phi_{\delta_z}(T) = (1 + T)^z$.

Let N be a positive integer prime to p. We defined in Theorem 3.1.14 the p-adic Dirichlet L-function for each primitive character ψ modulo Np^r (with values in K), writing ψ_N (resp. ψ_p) for the restriction of ψ to $(\mathbb{Z}/N\mathbb{Z})^\times$ (resp. $(\mathbb{Z}/p^r\mathbb{Z})^\times$), by

$$L_p(s, \psi) = c(s) \int_{\mathbb{Z}_p^\times} \psi_p \omega^{-1}(z) z_\Gamma^{-s} d\varphi_{\psi_N}(z),$$

where

$$c(s) = \begin{cases} -\psi_p \omega^{-1}(N) N_\Gamma^{-s} & \text{if } \psi_N \neq 1, \\ -(1 - \psi_p \omega^{-1}(z) z_\Gamma^{s-1})^{-1} & \text{if } \psi_N = 1. \end{cases}$$

By the above lemma, we thus get

Theorem 4.1.2. *Let N be a positive integer prime to p and ψ with $\psi(-1) = 1$ be a Dirichlet character modulo Np. Suppose that ψ_N is primitive modulo N. Then there exists a power series $\Phi_\psi(T) \in W[[T]]$ such that $L_p(s, \psi) = \Phi_\psi(\gamma^s - 1)$ if $\psi_N \neq 1$ and $L_p(s, 1) = \frac{\Phi_1(\gamma^s - 1)}{\gamma^{s-1} - 1}$.*

Exercise 4.1.4. Give a detailed proof of the above theorem.

A p-adic analytic function on $\mathbb{Z}_p$ of the form $s \mapsto \Phi(\gamma^s - 1)$ for a power series $\Phi(T) \in W[[T]]$ is called an Iwasawa function. Iwasawa functions form a special subclass of p-adic analytic functions on $\mathbb{Z}_p$.

4.1.2 Eisenstein series

Let $\psi : (\mathbb{Z}/N\mathbb{Z})^\times \to \overline{\mathbb{Q}}^\times$ be a primitive Dirichlet character. We consider the Eisenstein series of weight $0 < k \in \mathbb{Z}$

$$E'_k(z,s) = \sum_{(m,n) \in \mathbb{Z}^2 - \{(0,0)\}} \psi^{-1}(n)(mNz + n)^{-k}|mNz + n|^{-2s},$$

where $z \in \mathfrak{H}$ and $s \in \mathbb{C}$. When $N = 1$, ψ is the trivial character $\mathbf{1}$. For the following exercise, see [MFM, Section 2.6 and Chapter 7].

Exercise 4.1.5. Prove

(1) $E'_{k,\psi}(z,s)$ converges absolutely and locally uniformly with respect to $(z,s) \in \mathfrak{H} \times \mathbb{C}$ if $\mathrm{Re}(s) > 1 - (k/2)$;
(2) $E'_{k,\psi}(z,s) = 0$ if $\psi(-1) \neq (-1)^k$ (assuming convergence);
(3) $E'_{k,\psi}(z) = E'_{k,\psi}(z,0)$ is a holomorphic function of z if $k > 2$ (this fact is actually true if $k = 2$ and $\psi \neq \mathbf{1}$ for the limit $E'_{k,\psi}(z) = \lim_{s \to +0} E'_{k,\psi}(z,s)$; see [MFM, Corollary 7.2.10] what happens in the exceptional case);
(4) $E'_{k,\psi}(\gamma(z)) = \psi(d)(cz + d)^k E'_{k,\psi}(z)$ for $\gamma = \left(\begin{smallmatrix} a & b \\ c & d \end{smallmatrix}\right) \in \Gamma_0(N)$.

A holomorphic function $f : \mathfrak{H} \to \mathbb{C}$ is called a modular form on $\Gamma_0(N)$ of weight k with character ψ if f satisfies the following conditions:

(M1) $f(\frac{az+b}{cz+d}) = \psi(d)f(z)(cz+d)^k$ for all $\left(\begin{smallmatrix} a & b \\ c & d \end{smallmatrix}\right) \in \Gamma_0(N)$;
(M2) f is finite at all cusps of $X_0(N)$; in other words, for all $\alpha = \left(\begin{smallmatrix} a & b \\ c & d \end{smallmatrix}\right) \in SL_2(\mathbb{Z})$, $f|_k\alpha(z) = f(\alpha(z))(cz+d)^{-k}$ has Fourier expansion of the form

$$\sum_{0 \leq n \in N^{-1}\mathbb{Z}} a(n, f|_k\alpha)q^n \quad \text{for} \quad q = \exp(2\pi iz) \quad (\text{with } a(n, f|_k\alpha) \in \mathbb{C}).$$

The above condition means that the function f is finite at the cusp $\alpha(\infty)$ of $X_0(N)$ (whose value at the cusp is $a(0, f|_k\alpha)$. We write $M_{k,\psi}(\Gamma_0(N))$ for the space of functions satisfying (M1–2). Replacing (M2) by

(S) f is vanishing at all cusps of $X_0(N)$ (that is, $a(n, f|_k\alpha) = 0$ for all $\alpha \in SL_2(\mathbb{Z})$ and $n \leq 0$),

we define subspace $S_{k,\psi}(\Gamma_0(N)) \subset M_{k,\psi}(\Gamma_0(N))$ by imposing (S). Functions in the space $S_{k,\psi}(\Gamma_0(N))$ are called holomorphic cusp forms on $\Gamma_0(N)$ of weight k with character ψ.

Exercise 4.1.6. Prove that $M_{0,\psi}(\Gamma_0(N))$ is either $\mathbb{C}$ (constants) or 0 according to whether $\psi = \mathbf{1}$ or not.

Exercise 4.1.7. Prove that $M_{k,\psi}(\Gamma_0(N)) = 0$ if $\psi(-1) \neq (-1)^k$.

Proposition 4.1.3. *Let ψ be a primitive Dirichlet character modulo N. The Eisenstein series $E'_{k,\psi}(z,s)$ for $0 < k \in \mathbb{Z}$ can be meromorphically continued as a function of s for a fixed z giving a real analytic function of z if $E'_{k,\psi}(z,s)$ is finite at $s \in \mathbb{C}$. If $\psi \neq 1$ or $k \neq 2$, $E'_{k,\psi}(z) = E'_{k,\psi}(z,0)$ is an element in $M_{k,\psi}(\Gamma_0(N))$.*

We only prove the last assertion for $k > 2$, since the proof of the other assertions require more preparation from real analysis. See [LFE, Chapter 9] (or [MFM, Chapter 7]) for a proof of these assertions not proven here.

Proof. Suppose $k > 2$. Then $E'_{k,\psi}$ is absolutely and locally uniformly convergent by the exercise above, and hence $E'_{k,\psi}$ is a holomorphic function in $z \in \mathfrak{H}$. Thus we need to compute its Fourier expansion. Since the computation is basically the same for all cusps, we only do the computation at the cusp ∞. Recall $\mathbf{e}(z) = \exp(2\pi\sqrt{-1}z)$. We use the following partial fraction expansion of cotangent function (which can be found in any advanced Calculus text or [LFE, (2.1.5-6)]):

$$
\begin{aligned}
\pi \cot(\pi z) &= \pi i \frac{\mathbf{e}(z)+1}{\mathbf{e}(z)-1} = \frac{1}{z} + \sum_{n=1}^{\infty}\left(\frac{1}{z+n} + \frac{1}{z-n}\right) \\
\pi \cot(\pi z) &= \pi i \frac{\mathbf{e}(z)+1}{\mathbf{e}(z)-1} = \pi i\left(-1 - 2\sum_{n=1}^{\infty} q^n\right), \quad q = \mathbf{e}(z).
\end{aligned}
\tag{4.2}
$$

The two series converge locally uniformly on $\mathfrak{H}$ and periodic on $\mathbb{C}$ by definition. Applying the differential operator $(2\pi i)^{-1}\frac{\partial}{\partial z}$ to the formulas in (4.2) term by term, we get

$$
S_k(z) = \sum_{n\in\mathbb{Z}} \frac{1}{(z+n)^k} = \frac{(-2\pi i)^k}{(k-1)!}\sum_{n=1}^{\infty} n^{k-1}q^n. \tag{4.3}
$$

Form this, assuming $\psi(-1) = (-1)^k$, we have

$$
E'_{k,\psi}(z) = 2\sum_{n=1}^{\infty}\psi(n)^{-1}n^{-k} + 2\sum_{r=1}^{N}\psi^{-1}(r)\sum_{m=1}^{\infty}\sum_{n\in\mathbb{Z}} N^{-k}\left(mz + \frac{r}{N} + n\right)^{-k}
$$

$$
= 2L(k,\psi^{-1}) + 2\sum_{r=1}^{N}\psi^{-1}(r)\sum_{m=1}^{\infty} N^{-k}S_k\left(mz + \frac{r}{N}\right)
$$

$$
\stackrel{(4.3)}{=} 2L(k,\psi^{-1}) + 2N^{-k}\frac{(-2\pi i)^k}{(k-1)!}\sum_{m=1}^{\infty}\sum_{n=1}^{\infty} n^{k-1}q^m \sum_{r=1}^{N}\psi^{-1}(r)q^{nr/N}.
$$

$$
\tag{4.4}
$$

By the functional equation (see [LFE, Theorem 2.3.2]), we have, if $\psi(-1) = (-1)^k$,

$$L(k, \psi^{-1}) = G(\psi^{-1}) \frac{(-2\pi i)^k}{N^k(k-1)!} L(1-k, \psi), \qquad (4.5)$$

where $G(\chi)$ for a primitive character χ modulo C is the Gauss sum $\sum_{r=1}^{C} \chi(r) \mathbf{e}(\frac{u}{C})$. We have

$$\sum_{r=1}^{N} \psi^{-1}(r) \mathbf{e}\left(\frac{nr}{N}\right) = \begin{cases} \psi(n)G(\psi^{-1}) & \text{if } n \text{ is prime to } N, \\ 0 & \text{otherwise,} \end{cases}$$

and we get the formula

$$E'_{k,\psi}(z) = G(\psi^{-1}) \frac{2(-2\pi i)^k}{N^k(k-1)!} E_{k,\psi}(z) \qquad (4.6)$$

for

$$E_{k,\psi}(z) = 2^{-1} L(1-k, \psi) + \sum_{n=1}^{\infty} \sigma_{k-1,\psi}(n) q^n$$

for $\sigma_{k-1,\psi}(n) = \sum_{0<d|n} \psi(d) d^{k-1}$. Here we used the fact that $E_{k,\psi}(z) = 0$ if $\psi(-1) \neq (-1)^k$. $\qquad \square$

Exercise 4.1.8. Give a proof of

$$\sum_{r=1}^{N} \psi^{-1}(r) \mathbf{e}\left(\frac{nr}{N}\right) = \begin{cases} \psi(n)G(\psi^{-1}) & \text{if } n \text{ is prime to } N, \\ 0 & \text{otherwise.} \end{cases}$$

Exercise 4.1.9. Let p be a prime, and write $\mathbf{1}_p$ for the imprimitive identity character of $(\mathbb{Z}/p\mathbb{Z})^\times$. Prove that

$$E_{k,\mathbf{1}}(z) - p^{k-1} E_{k,\mathbf{1}}(pz) = 2^{-1}(1-p^{k-1})\zeta(1-k) + \sum_{n=1}^{\infty} \sigma_{k-1,\mathbf{1}}^{(p)}(n) q^n$$

for $\sigma_{k-1,\mathbf{1}}^{(p)}(n) = \sum_{0<d|n, p\nmid n} d^{k-1}$. More generally, if N is prime to p, prove that

$$E_{k,\psi}(z) - \psi(p) p^{k-1} E_{k,\psi}(pz) = 2^{-1}(1-\psi(p)p^{k-1}) L(1-k, \psi) + \sum_{n=1}^{\infty} \sigma_{k-1,\psi}^{(p)}(n) q^n$$

for $\sigma_{k-1,\psi}^{(p)}(n) = \sum_{0<d|n, p\nmid n} \psi(d) d^{k-1}$.

4.1.3 *Eisenstein family*

We hereafter fix a *level* which is a positive integer N prime to p and a Dirichlet character ψ modulo Np with $\psi(-1) = (-1)^k$. We know by a work of Shimura that $M_{k,\psi}(\Gamma_0(Np^r); A) \otimes_A \mathbb{C} = M_{k,\psi}(\Gamma_0(Np^r))$ for any algebra $A \subset \mathbb{C}$ containing the values of ψ (see Proposition 3.2.10 (2) and [GME, Proposition 3.1.1]), where

$$M_{k,\psi}(\Gamma_0(Np^r); A) = \left\{ f \in M_{k,\psi}(\Gamma_0(Np^r)) \middle| a(n, f) \in A \text{ for all } n \geq 0 \right\}.$$

Here we write the q-expansion of f as $f = \sum_{n=0}^{\infty} a(n, f)q^n$. Then we take $A = W \cap \overline{\mathbb{Q}}$ and define $M_{k,\psi}(\Gamma_0(Np^r); W) = M_{k,\psi}(\Gamma_0(Np^r); A) \otimes_A W$ and $M_{k,\psi}(\Gamma_0(Np^r); K) = M_{k,\psi}(\Gamma_0(Np^r); W) \otimes_W K = M_{k,\psi}(\Gamma_0(Np^r); A) \otimes_A K$. By definition, $M_{k,\psi}(\Gamma_0(Np^r); A) \hookrightarrow A[[q]]$ via q-expansion.

Definition 4.1.4. A p-adic analytic family of modular forms of character ψ (modulo Np) with coefficients in $\Lambda = W[[T]]$ is a formal q-expansion $F(T) = \sum_{n=0}^{\infty} a(n, F)(T)q^n \in \Lambda[[q]]$ such that for all sufficiently large integers $k \gg 0$, $F(\gamma^k - 1)$ is the q-expansion of an element in $M_{k,\psi\omega^{-k}}(\Gamma_0(Np); W)$ for the Teichmüller character ω as in Theorem 3.1.14 (which factors through $\mathbb{Z}/p\mathbb{Z}$).

Exercise 4.1.10. Prove that

(1) $\log_p(z) = \sum_{n=1}^{\infty} (-1)^{n+1} \frac{(z-1)^n}{n}$ converges p-adically for $z \in \Gamma$ to an element in $p\mathbb{Z}_p$;

(2) $\log_p(zw) = \log_p(z) + \log_p(w)$;

(3) $\gamma^{\log_p(n_\Gamma)/\log_p(\gamma)} = n_\Gamma$ for all integer n prime to p (cf. [LFE, §1.3]).

Similarly, prove that $\exp_p(z) = \sum_{n=0}^{\infty} \frac{z^n}{n!}$ converges to an element in Γ p-adically over $p\mathbb{Z}_p$ and show that $\exp_p \circ \log_p$ and $\log_p \circ \exp_p$ are the identity maps.

Since one expects $\log_p(\zeta) = 0$, we hereafter define $\log_p(z) := \log_p(z_\Gamma)$ for any $z \in \mathbb{Z}_p^\times$ as $z = \zeta \cdot z_\Gamma$ for a root of unity.

Define $\Phi_\psi(T) \in W[[T]]$ by

$$\Phi_\psi(\gamma^s - 1) = \begin{cases} 2^{-1}L_p(1-s, \psi) & \text{if } \psi \neq 1 \\ 2^{-1}(\gamma^s - 1)L_p(1-s, 1) & \text{otherwise} \end{cases}$$

for $s \in \mathbb{Z}_p$ and

$$a(n, \mathcal{E}_\psi)(T) = \begin{cases} \sum_{0 < d|n, p \nmid d} \psi(d)(1+T)^{\log_p(d)/\log_p(\gamma)} & \text{if } \psi \neq 1, \\ T \sum_{0 < d|n, p \nmid d} (1+T)^{\log_p(d)/\log_p(\gamma)} & \text{if } \psi = 1. \end{cases}$$

Exercise 4.1.11. Using Theorem 4.1.2, prove the existence and uniqueness of $\Phi_\psi(T) \in W[[T]]$ if ψ_N is primitive modulo N.

Theorem 4.1.5. *Let ψ be an even Dirichlet character modulo $N\mathbf{p}$ with primitive ψ_N. Then the q-expansion $\mathcal{E}_\psi = \Phi_\psi(T) + \sum_{n=1}^\infty a(n, \mathcal{E}_\psi)q^n$ gives a p-adic analytic family of modular form with character ψ. Moreover*

$$\mathcal{E}_\psi(\gamma^k - 1) \in M_{k, \psi\omega^{-k}}(\Gamma_0(N\mathbf{p}); W)$$

for $k \geq 1$ except for the case where $\psi = 1$. When $N = 1$ and $\psi = 1$, $\mathcal{E}_1(\gamma^k - 1) \in M_{k, \omega^{-k}}(\Gamma_0(N\mathbf{p}); W)$ if $k \geq 0$, and $\mathcal{E}_1(0) \in \mathbb{Z}_p^\times$.

Proof. First we assume that $\psi \neq 1$. By computation, we have

$$a(n, \mathcal{E}_\psi)(\gamma^k - 1) = \sum_{0 < d \mid n, p \nmid n} \psi(d) \gamma^{k \log_p(d) / \log_p(\gamma)}$$

$$= \sum_{0 < d \mid n, p \nmid n} \psi(d) \exp_p(\log_p(\gamma))^{k \log_p(d)/\log_p(\gamma)} = \sum_{0 < d \mid n, p \nmid n} \psi(d) n_\Gamma^k$$

$$= \sum_{0 < d \mid n, p \nmid n} \psi\omega^{-k}(d) d^k = \sigma_{k, \psi\omega^{-k}}^{(p)}(n).$$

Similarly by definition, $\Phi_\psi(\gamma^k - 1) = 2^{-1}L_p(1 - k, \psi\omega^{-k})$. Thus we have, from Exercise 4.1.9, $\mathcal{E}_\psi(\gamma^k - 1) = E_{k, \psi\omega^{-k}}$ if $\psi\omega^{-k}$ is primitive modulo $N\mathbf{p}$, and otherwise $E_{k, \psi_N} - \psi_N(p)p^{k-1}E_{k, \psi_N}|[p]$ for $f|[p](z) = f(pz)$. This finishes the proof in the case when $\psi \neq 1$.

Now suppose $\psi = 1$ and $N = 1$. By the same computation as above, for any $k \in \mathbb{Z}$ we get for $n > 0$

$$a(n, \mathcal{E}_1)(\gamma^k - 1) = (\gamma^k - 1) \sum_{0 < d \mid n, p \nmid n} \psi\omega^{-k}(d) d^k = \sigma_{k, \psi\omega^{-k}}^{(p)}(n) \qquad (4.7)$$

and for $0 \neq k \in \mathbb{Z}$

$$a(0, \mathcal{E}_1)(\gamma^k - 1) = 2^{-1}(\gamma^k - 1)L_p(1 - k, 1)$$

$$= \begin{cases} a(0, E_{k, \psi\omega^{-k}}) & \text{if } k \not\equiv 0 \mod \varphi(\mathbf{p}), \\ a(0, E_{k,1} - p^{k-1}E_{k,1}|[p]) & \text{if } k \equiv 0 \mod \varphi(\mathbf{p}), \end{cases}$$

for the Euler function $\varphi(n) = |(\mathbb{Z}/n\mathbb{Z})^\times|$. This shows the result in the case where $k \geq 1$.

If $k = 0$, by (4.7), we find $a(n, \mathcal{E}_1)(0) = 0$. By Von Staudt–Clausen theorem, $\Phi_1(\gamma^{p-1} - 1) \in \mathbb{Z}_p^\times$. Since $\Phi_1(0) \equiv \Phi_1(\gamma^{p-1} - 1) \mod p$, we conclude $\mathcal{E}_1(0) = \Phi_1(0) \in \mathbb{Z}_p^\times$. $\qquad \square$

Exercise 4.1.12. Compute the value $\Phi_\psi(0)$ from the values $L_p(1, \psi)$ ($\psi \neq$ 1) and $\mathrm{Res}_{s=1} L_p(s, 1)$ given in [LFE, Theorem 3.5.2].

Definition 4.1.6. The collection of all p-adic analytic families of modular forms with character ψ form a Λ-module $M_\psi(N; \Lambda)$. If $F \in M_\psi(N; \Lambda)$ specializes to a cusp form $F(\gamma^k - 1) \in S_{k,\psi\omega^{-k}}(\Gamma_0(N\mathbf{p}); W)$ for all sufficiently large $k \gg 0$, F is called a p-adic analytic cuspidal family. The collection of all cuspidal families is written as $S_\psi(N; \Lambda)$.

For a given modular form $f \in M_{\ell,\chi}(\Gamma_0(N\mathbf{p}); W)$, we can define a convoluted product $f * \mathcal{E}_\psi$ by

$$f * \mathcal{E}_\psi(T) = f\mathcal{E}_\psi(\gamma^{-\ell}(1 + T) - 1)).$$

Then $f * \mathcal{E}_k \in \Lambda[[q]]$, and we have $f * \mathcal{E}_\psi(\gamma^k - 1) = f \cdot \mathcal{E}(\gamma^{k-\ell} - 1)$ by computation. Since $\mathcal{E}(\gamma^{k-\ell} - 1) \in M_{k-\ell,\psi\omega^{-k}}(\Gamma_0(N\mathbf{p}); W)$, we find $f \cdot \mathcal{E}(\gamma^{k-\ell} - 1) \in M_{k,\psi\chi\omega^{-k}}(\Gamma_0(N\mathbf{p}); W)$ if $k \geq \ell + 2$. This shows

Corollary 4.1.7. *We have*

$$f * \mathcal{E}_\psi \in M_{\psi\chi\omega^\ell}(N; \Lambda)$$

if $f \in M_{\ell,\chi}(\Gamma_0(N\mathbf{p}); W)$. If $f \in S_{\ell,\chi}(\Gamma_0(N\mathbf{p}); W)$, we have $f * \mathcal{E}_\psi \in S_{\psi\chi\omega^\ell}(N; \Lambda)$.

In this way, we can produce a lot of p-adic analytic families.

Exercise 4.1.13. Prove that $f * \mathcal{E}_\psi \in S_{\psi\chi\omega^\ell}(N; \Lambda)$ if $f \in S_{\ell,\chi}(\Gamma_0(N\mathbf{p}); W)$.

4.1.4 *Hecke operator*

Recall

$$\Delta_0(p^r N) = \left\{ \left(\begin{smallmatrix} a & b \\ c & d \end{smallmatrix}\right) \in M_2(\mathbb{Z}) \big| c \equiv 0 \mod p^r N, a\mathbb{Z} + Np\mathbb{Z} = \mathbb{Z}, ad - bc > 0 \right\}.$$

We define a character ψ_Δ of $\Delta_0(N\mathbf{p})$ by $\psi_\Delta \left(\begin{smallmatrix} a & b \\ c & d \end{smallmatrix}\right) = \psi^{-1}(a)$.

Exercise 4.1.14. Prove that $\psi_\Delta \left(\begin{smallmatrix} a & b \\ c & d \end{smallmatrix}\right) = \psi(d)$ if $\left(\begin{smallmatrix} a & b \\ c & d \end{smallmatrix}\right) \in \Gamma_0(N\mathbf{p})$.

Define for $\alpha = \left(\begin{smallmatrix} a & b \\ c & d \end{smallmatrix}\right) \in \Delta_0(p^r N)$ and a function $f : \mathfrak{H} \to \mathbb{C}$, a new function $f|_{k,\psi}\alpha(z)$ by $f|_{k,\psi}\alpha(z) = \det(\alpha)^{k-1} f(\alpha(z)) \psi_\Delta^{-1}(\alpha)(cz + d)^{-k}$. Splitting $T_n = \{\alpha \in \Delta_0(pN)| \det(\alpha) = n\}$ into a disjoint union $T_n = \bigsqcup_\alpha \Gamma_0(pN)\alpha$, we define $f|T(n) = \sum_\alpha f|_{k,\psi}\alpha$. Then the same proof of Lemma 3.2.7 gives the following fact

Lemma 4.1.8.

(1) *Write $T(n)$ for the operator corresponding to T_n. Then $T(n)$ gives a linear endomorphism of $M_{k,\psi}(\Gamma_0(pN))$.*
(2) *We get the following identity of Hecke operators for $f \in M_{k,\psi}(\Gamma_0(pN))$:*

$$a(m, f|T(n)) = \sum_{0<d|(m,n),(d,pN)=1} \psi(d)d^{k-1} \cdot a(\frac{mn}{d^2}, f).$$

(3) $T(m)T(n) = T(n)T(m)$ *for all integers m and n.*

When $m|pN$, we often write $U(m)$ for $T(m)$.

Exercise 4.1.15. Give a detailed proof of the above lemma.

Corollary 4.1.9. *If $k \geq 1$ and A contains the values of ψ, the Hecke operators $T(n)$ preserves $M_{k,\psi}(\Gamma_0(N\mathbf{p}), \psi; A)$.*

Definition 4.1.10. We consider the operator $T_\Lambda(n)$ acting on $F = \sum_{n=0}^{\infty} a(n, F)(T)q^n \in \Lambda[[q]]$ defined by

$$a(m, F|T_\Lambda(n)) = \sum_{0<d|(m,n),(d,pN)=1} \psi(d)d^{-1}(1+T)^{\log_p(d)/\log_p(\gamma)} \cdot a(\frac{mn}{d^2}, F).$$

Since $(1 + T)^{\log_p(d)/\log_p(\gamma)}|_{T=\gamma^k-1} = d_\Gamma^k = \omega^{-k}(d)d^k$, after specializing $T = \gamma^k - 1$, we find $(F|T_\Lambda(n))(\gamma^k - 1) = (F(\gamma^k - 1)|T(n)$. Thus $T_\Lambda(n)$ preserves $M_\psi(N; \Lambda)$ and $S_\psi(N; \Lambda)$.

Proposition 4.1.11. *We have a linear operator $T_\Lambda(n)$ defined by Definition 4.1.10 acting on $M_\psi(N; \Lambda)$ which preserves $S_\psi(N; \Lambda)$. In particular, $(F|T_\Lambda(n))(\gamma^k - 1) = F(\gamma^k - 1)|T(n)$ for all $F \in M_k(N; \Lambda)$ and all $k \gg 0$ and $T_\Lambda(m)T_\Lambda(n) = T_\Lambda(n)T_\Lambda(m)$ and $T_\Lambda(m)T_\Lambda(n) = T_\Lambda(mn)$ if $m\mathbb{Z} + n\mathbb{Z} = \mathbb{Z}$.*

There are a lot of questions we can ask for p-adic analytic families; for example,

- Is the Λ-module $M_\psi(N; \Lambda)$ finitely generated?
- Is the module $M_\psi(N; \Lambda)$ spanned by Hecke eigenforms (at least topologically if it is infinite rank)?
- What is $F(\zeta - 1)$ for a general $\zeta \in \overline{\mathbb{Q}}_p$ with $|\zeta - 1|_p < 1$?

We try to answer some of these questions.

4.1.5 *Modular forms of level N*

We generalize a bit the notion of modular forms. Let

$$\Gamma_1(N) = \left\{ \left(\begin{smallmatrix} a & b \\ c & d \end{smallmatrix} \right) \in \Gamma_0(N) \middle| a \equiv d \equiv 1 \mod N \right\}.$$

A modular form $f \in M_k(\Gamma_1(N))$ is a holomorphic function on $\mathfrak{H}$ satisfying the conditions (M1–2) in §4.1.2 for $\Gamma_1(N)$ in place of $\Gamma_0(N)$. Since $d \equiv 1$ mod N and ψ is a character modulo N in (M1), this space is independent of the choice of ψ, and hence the subscript ψ is dropped from the notation. Similarly we define the subspace of cusp forms $S_k(\Gamma_1(N))$ inside $M_k(\Gamma_1(N))$ by imposing (S) in addition to (M1–2). Then we define first for a ring $A \subset \mathbb{C}$

$$M_k(\Gamma_1(N); A) = \left\{ f \in M_k(\Gamma_1(N)) \middle| a(n, f) \in A \ \text{for all } n \geq 0 \right\}$$

and put $S_k(\Gamma_1(N); A) = M_k(\Gamma_1(N); A) \cap S_k(\Gamma_1(N))$. Then again it is known by Shimura (cf. [IAT, Theorem 3.52] and [GME, Theorem 3.1.1]) that

$$M_k(\Gamma_1(N); \mathbb{Z}) \otimes_{\mathbb{Z}} A = M_k(\Gamma_1(N); A), \ M_k(\Gamma_1(N); A) \otimes_A \mathbb{C} = M_k(\Gamma_1(N))$$
$$S_k(\Gamma_1(N); \mathbb{Z}) \otimes_{\mathbb{Z}} A = S_k(\Gamma_1(N); A), \ S_k(\Gamma_1(N); A) \otimes_A \mathbb{C} = S_k(\Gamma_1(N)).$$

$$(4.8)$$

A proof of this fact can be given in the same way as in the proof of Proposition 3.2.10 (which gives the result for $k = 2$) and is based on

* The Eichler–Shimura isomorphism in Theorem 4.2.8;
* the self duality of the Eichler–Shimura cohomology group in §4.2.6.

We will give a sketch of Shimura's proof of this fact after proving Theorem 4.2.8.

By (4.8), for an algebra X with $W \subset X \subset \overline{\mathbb{Q}}_p$, taking $A = X \cap \overline{\mathbb{Q}}$, we may define

$$M_k(\Gamma_1(N); X) = M_k(\Gamma_1(N); A) \otimes_A X$$
$$S_k(\Gamma_1(N); X) = S_k(\Gamma_1(N); A) \otimes_A X. \tag{4.9}$$

As easily seen, these spaces can be embedded into $X[[q]]$ by q-expansion. We can also define for $W/\mathfrak{m}_W^n$,

$$M_k(\Gamma_1(N); W/\mathfrak{m}_W^n) = M_k(\Gamma_1(N); W) \otimes_W W/\mathfrak{m}_W^n$$
$$S_k(\Gamma_1(N); W) = S_k(\Gamma_1(N); W) \otimes_W W/\mathfrak{m}_W^n. \tag{4.10}$$

Note that if $pf(q) \in \mathrm{Im}(S_k(\Gamma_1(N); W) \hookrightarrow qW[[q]])$ for $f(q) \in qW[[q]]$, $f(q) \in \mathrm{Im}(S_k(\Gamma_1(N); W) \hookrightarrow qW[[q]])$, since $f \in \mathrm{Im}(M_k(\Gamma_1(N); K) \hookrightarrow qK[[q]])$ with $f(q) \in qW[[q]]$. Thus $C := \mathrm{Coker}(M_k(\Gamma_1(N); W) \hookrightarrow qW[[q]])$

is p-torsion free [BCM, I.2.4]; so, C is W-flat. Therefore, tensoring $W/\mathfrak{m}_W^n$, the map $S_k(\Gamma_1(N), W/\mathfrak{m}_W^n) \to qW/\mathfrak{m}_W^n[[q]]$ is injective [BCM, I.2.5]. In the same way, $M_k(\Gamma_1(N), W/\mathfrak{m}_W^n) \to W/\mathfrak{m}_W^n[[q]]$ is injective. For a given algebraic extension $\mathbb{F}$ of $\mathbb{F}_p$, choosing $W = W(\mathbb{F})$ (the Witt vector ring with coefficients in $\mathbb{F}$), we also know that $M_k(\Gamma_1(N), \mathbb{F}) \to \mathbb{F}[[q]]$ is injective. The injectivity we have shown also holds for S_k.

Since $\Gamma_0(N)/\Gamma_1(N) \cong (\mathbb{Z}/N\mathbb{Z})^\times$ by $\left(\begin{smallmatrix} a & b \\ c & d \end{smallmatrix}\right) \mapsto (d \mod N)$, the finite group $(\mathbb{Z}/N\mathbb{Z})^\times$ acts on $M_k(\Gamma_1(N))$. Then by definition, for a W-algebra A, the ψ-eigenspace of $M_k(\Gamma_1(N); A)$ is the space $M_{k,\psi}(\Gamma_0(N); A) = M_k(\Gamma_0(N), \psi; A)$:

$$M_{k,\psi}(\Gamma_0(N); A) = M_k(\Gamma_0(N), \psi; A)$$
$$= \left\{ f \in M_k(\Gamma_1(N); A) \big| f|\langle d \rangle_k = \psi(d) f \text{ for all } d \in (\mathbb{Z}/N\mathbb{Z})^\times \right\}, \quad (4.11)$$

where $f|\langle d \rangle_k = f|_k \left(\begin{smallmatrix} a & b \\ c & d \end{smallmatrix}\right)$ for $\left(\begin{smallmatrix} a & b \\ c & d \end{smallmatrix}\right) \in \Gamma_0(N)$. We then put

$$S_{k,\psi}(\Gamma_0(N); A) = S_k(\Gamma_0(N), \psi; A) = M_{k,\psi}(\Gamma_0(N); A) \cap S_k(\Gamma_1(N), \psi; A).$$

When $A = \mathbb{C}$, we often omit A from the notation (e.g., $S_k(\Gamma_1(N), \psi) = S_k(\Gamma_1(N), \psi; \mathbb{C})$). Thus we get

Lemma 4.1.12. *We have*
$$M_k(\Gamma_1(N)) = \bigoplus_\psi M_{k,\psi}(\Gamma_0(N))$$
$$S_k(\Gamma_1(N)) = \bigoplus_\psi S_{k,\psi}(\Gamma_0(N)),$$

where ψ runs over all Dirichlet characters modulo N.

The Hecke operator $T(n)$ on each $M_{k,\psi}(\Gamma_0(N))$ gives rise to a Hecke operator on the sum $M_k(\Gamma_1(N))$ over ψ. In other words, writing $f \in M_k(\Gamma_1(N))$ as $f = \oplus_\psi f_\psi$ with $f_\psi \in M_{k,\psi}(\Gamma_0(N))$, we have $f|T(n) = \oplus_\psi (f_\psi|T(n))$.

Exercise 4.1.16. Prove that for $f \in M_k(\Gamma_1(N))$
$$a(m, f|T(n)) = \sum_{0 < d|(m,n), (d,N)=1} d^{k-1} a\left(\frac{mn}{d^2}, f|\langle d \rangle_k\right)$$

and $T(\ell)^2 - T(\ell^2) = \ell^{k-1}\langle \ell \rangle_k$ if ℓ is a prime outside N.

Let
$$\Delta_1(N) = \left\{ \left(\begin{smallmatrix} a & b \\ c & d \end{smallmatrix}\right) \in M_2(\mathbb{Z}) \big| c \equiv 0 \mod N, \ a \equiv 1 \mod N, \ ad - bc > 0 \right\}.$$

Exercise 4.1.17. Splitting $T_n = \{\alpha \in \Delta_1(N)| \det(\alpha) = n\}$ into a disjoint union $T_n = \bigsqcup_\alpha \Gamma_0(pN)\alpha$, prove that $f|T(n) = \sum_\alpha f|_{k,1}\alpha$.

The modular curve $X_1(N)(\mathbb{C}) = \Gamma_1(N)\backslash(\mathfrak{H}\cup\mathbf{P}^1(\mathbb{Q}))$ has a regular model $X_1(N)$ over $\mathbb{Z}$ (a regular model $X_1(N)_{/\mathbb{Z}}$ is a scheme flat over $\mathbb{Z}$ whose local rings are all regular rings such that $X_1(N) \otimes_{\mathbb{Z}} \mathbb{C} \cong X_1(N)_{/\mathbb{C}}$; see [GME, §2.8]). We admit the following nontrivial facts from algebro-geometric definition of modular forms:

Theorem 4.1.13. *For a subgroup Γ with $\Gamma_1(N) \subset \Gamma \subset \Gamma_0(N)$, the space $M_k(\Gamma; A)$ and $S_k(\Gamma; A)$ are stable under the action of $(\mathbb{Z}/N\mathbb{Z})^\times$ and hence under $T(n)$ for all n for any algebra A.*

Here is a very brief sketch of a proof for $\Gamma = \Gamma_1(N)$.

Proof. The (geometric) diamond operator $\langle d \rangle = \langle d \rangle_g := d^k \langle d \rangle_k$ corresponds to the action $(E, \omega, \phi) \mapsto (E, \omega, d \cdot \phi)$ on test objects in $\mathcal{P}_{1,N}(A)$ of §2.3.1, and hence identifying $M_k(\Gamma_1(N); A)$ with $G_k(\Gamma_1(N); A)$, we get the stability of the space under the geometric diamond operators $\langle d \rangle_g$. Strictly speaking, we have defined G_k for A over a field in §2.3.1, and therefore, we need to extend the theory to any commutative ring A with identity. Taking mutually prime d, d' with $d \equiv d' \mod N$ (so, $\langle d \rangle_k = \langle d' \rangle_k$), for $f \in M_k(\Gamma; A)$, $d^k f|\langle d \rangle_k = f|\langle d \rangle_g$ and $d'^k f|\langle d \rangle_k = f|\langle d' \rangle_g$ both belong to $M_k(\Gamma; A)$. Since d and d' are coprime, choosing $a, b \in \mathbb{Z}$ with $ad^k + bd'^k = 1$, we find $f|\langle d \rangle_k = a \cdot f|\langle d \rangle_g + b \cdot f|\langle d' \rangle_g \in M_k(\Gamma; A)$ as desired. $\square$

Definition 4.1.14. For a W-algebra X, we define the p-adic Hecke algebras $\mathbf{H}_k(N; X)$ (resp. $\mathbf{h}_k(N; X)$) by the subalgebra of $\mathrm{End}_X(M_k(\Gamma_1(N); X))$ (resp. $\mathrm{End}_X(S_k(\Gamma_1(N); X))$) generated by Hecke operators $T(n)$ and diamond operators $\langle z \rangle$. Similarly, for a subgroup Γ with $\Gamma_1(N) \subset \Gamma \subset \Gamma_0(N)$, assuming W is a $\mathbb{Z}[\psi]$-algebra, we define $\mathbf{H}_k(\Gamma, \psi; X)$ (resp. $\mathbf{h}_k(\Gamma, \psi; X)$) by the subalgebra of $\mathrm{End}_X(M_{k,\psi}(\Gamma; X))$ (resp. $\mathrm{End}_X(S_{k,\psi}(\Gamma; X))$) generated over X by Hecke operators $T(n)$ and diamond operators $\langle z \rangle_g$.

As is clear from the proof of Theorem 4.1.13, in the above definition, we can replace the geometric diamond operators $\langle z \rangle_g$ by diamond operators $\langle z \rangle_k$ of weight k, and the outcome is the same.

Proposition 4.1.15. *Suppose $k \geq 2$. Let A be a ring either embeddable into one of $\{\mathbb{C}, \overline{\mathbb{Q}}_p, \overline{\mathbb{F}}_p\}$ or $W/\mathfrak{m}_W^n$, and assume that A is a $\mathbb{Z}_p$-algebra if $A \subset \overline{\mathbb{Q}}_p$. Then we have $\mathbf{h}_k(N, A) \cong \mathbf{h}_k(N; \mathbb{Z}) \otimes_{\mathbb{Z}} A$ and $\mathbf{H}_k(N, A) \cong \mathbf{H}_k(N; \mathbb{Z}) \otimes_{\mathbb{Z}} A$. The isomorphism sends $T(n)$ to $T(n) \otimes 1$ and $\langle z \rangle$ to $\langle z \rangle \otimes 1$. The same assertion also holds for $\mathbf{H}_k(\Gamma, \psi; A)$ and $\mathbf{h}_k(\Gamma, \psi; A)$.*

We give an argument for $\Gamma = \Gamma_1(N)$ as the general case can be treated similarly.

Proof. As long as $S_k(\Gamma_1(N), R) \to R[[q]]$ is injective, $S_k(\Gamma_1(N); R) = S_k(\Gamma_1(N); \mathbb{Z}) \otimes_{\mathbb{Z}} R$ is free of finite rank $r = \mathrm{rank}_{\mathbb{Z}}\, S_k(\Gamma_1(N); \mathbb{Z})$ over R, its R-dual $\mathbf{h}_k(N; R)$ is free of the same rank r over R by Proposition 3.2.9. The injectivity is true if $R = A$ as in the proposition as verified above Definition 4.1.14. Since we have a natural map $\mathbf{h}_k(N; \mathbb{Z}) \otimes_{\mathbb{Z}} A \to \mathbf{h}_k(N; A)$ which is onto as the two sides are both generated by $T(n)$ and both free of the same rank $\mathrm{rank}_{\mathbb{Z}}\, S_k(\Gamma_1(N); \mathbb{Z})$ over A, we conclude $\mathbf{h}_k(N; \mathbb{Z}) \otimes_{\mathbb{Z}} A \cong h_k(N; A)$ when A is a domain. If $A = W/\mathfrak{m}_W^n$, by counting the number of elements of the two sides, we again conclude $\mathbf{h}_k(N; \mathbb{Z}) \otimes_{\mathbb{Z}} A \cong h_k(N; A)$.

The above argument for cusp forms is from the rank identity of the two Hecke algebras. Since the rank identity holds for $\mathbf{H}_k$ as

$$\mathrm{rank}\,\mathbf{H}_k = \mathrm{rank}\,\mathbf{h}_k + \#\{\text{regular cusps of } X_1(N)\}$$

if $k > 2$ and

$$\mathrm{rank}\,\mathbf{H}_2 = \mathrm{rank}\,\mathbf{h}_2 + \#\{\text{regular cusps of } X_1(N)\} - 1,$$

we still have the identity for $\mathbf{H}_k$ in characteristic 0 case, and the result follows. See [MFM, §1.5] for *regular cusps* and [MFM, Chapter 7] for the dimension formulas of the space of Eisenstein series.

More generally, for a domain A of characteristic 0, we can define $\widetilde{M}_k(\Gamma_1(N); A) := \{f \in M_k(\Gamma_1(N); \mathrm{Frac}(A)) | a(n, f) \in A \text{ for all } n > 0\}$. Then by the pairing $(f, h) = a(1, f|h)$, again $\mathbf{H}_k(N; A)$ and $\widetilde{M}_k(\Gamma_1(N); A)$ are perfect A-dual as the constant term $a(0, f)$ is determined by the data of $a(n, f)$ for all $n > 0$. Using this fact, and defining $\widetilde{M}_k(\Gamma_1(N), W/\mathfrak{m}_W^n) := \widetilde{M}_k(\Gamma_1(N); W) \otimes_W W/\mathfrak{m}_W^n$ (which can be embedded into $qW/\mathfrak{m}_n^n[[q]]$ forgetting the constant term), we can conclude the result also for $W/\mathfrak{m}_W^n$. $\square$

Corollary 4.1.16. *Let Γ be a subgroup with $\Gamma_1(N\mathbf{p}) \subset \Gamma \subset \Gamma_0(N\mathbf{p})$. Suppose $m \geq r > 0$. Then the following two diagrams are commutative for all positive integer n:*

$$
\begin{array}{ccc}
M_k(\Gamma_1(Np^r); A) & \overset{\subset}{\longrightarrow} & M_k(\Gamma_1(Np^m); A) \\
{\scriptstyle T(n)}\downarrow & & \downarrow{\scriptstyle T(n)} \\
M_k(\Gamma_1(Np^r); A) & \overset{\subset}{\longrightarrow} & M_k(\Gamma_1(Np^m); A)
\end{array}
$$

and

$$
\begin{array}{ccc}
S_k(\Gamma_1(Np^r); A) & \overset{\subset}{\longrightarrow} & S_k(\Gamma_1(Np^m); A) \\
{\scriptstyle T(n)}\downarrow & & \downarrow{\scriptstyle T(n)} \\
S_k(\Gamma_1(Np^r); A) & \overset{\subset}{\longrightarrow} & S_k(\Gamma_1(Np^m); A).
\end{array}
$$

More generally, the following two diagrams are commutative for all positive integer n:

$$
\begin{array}{ccc}
M_k(\Gamma \cap \Gamma_1(p^r); A) & \xrightarrow{\ \subset\ } & M_k(\Gamma \cap \Gamma_1(p^m); A) \\
\downarrow{\scriptstyle T(n)} & & \downarrow{\scriptstyle T(n)} \\
M_k(\Gamma \cap \Gamma_1(p^r); A) & \xrightarrow{\ \subset\ } & M_k(\Gamma \cap \Gamma_1(p^r); A)
\end{array}
$$

and

$$
\begin{array}{ccc}
S_k(\Gamma \cap \Gamma_1(p^r); A) & \xrightarrow{\ \subset\ } & S_k(\Gamma \cap \Gamma_1(p^r); A) \\
\downarrow{\scriptstyle T(n)} & & \downarrow{\scriptstyle T(n)} \\
S_k(\Gamma \cap \Gamma_1(p^r); A) & \xrightarrow{\ \subset\ } & S_k(\Gamma \cap \Gamma_1(p^r); A).
\end{array}
$$

By Corollary 4.1.16, restriction to subspace of lower level induces algebra homomorphisms:

$$\mathbf{H}_k(Np^m; A) \to \mathbf{H}_k(Np^r; A) \text{ and } \mathbf{h}_k(Np^m; A) \to \mathbf{h}_k(Np^r; A)$$

sending $T(n)$ to $T(n)$ for each n and $\langle z \rangle$ to $\langle z \rangle$.

Definition 4.1.17. We define

$$\mathbf{H}_k(Np^\infty; A) = \varprojlim_r \mathbf{H}_k(Np^r; A) \text{ and } \mathbf{h}_k(Np^\infty; A) = \varprojlim_r \mathbf{h}_k(Np^r; A).$$

More generally, taking $\Gamma = \Gamma_0(N\mathbf{p})$, we define

$$\mathbf{H}_k(Np^\infty, \psi; A) = \varprojlim_r \mathbf{H}_k(\Gamma \cap \Gamma_1(p^r), \psi; A)$$

$$\text{and } \mathbf{h}_k(Np^\infty, \psi; A) = \varprojlim_r \mathbf{h}_k(\Gamma \cap \Gamma_1(p^r), \psi; A).$$

Let $u := \Phi_1(0) = \mathcal{E}_1(0)$ which is a unit in $\mathbb{Z}_p$ by Theorem 4.1.5. Since $H = u^{-1}\mathcal{E}_1(\gamma - 1) \equiv 1 \mod p$ for $\gamma = 1 + \mathbf{p}$,

$$H^{p^{m-1}} \equiv 1 \mod p^m. \tag{4.12}$$

Thus $f(q) \mapsto H^{p^{m-1}} f(q)$ induces a Hecke equivariant inclusion

$$M_2(Np^r; W/p^mW) \hookrightarrow M_{2+p^{m-1}}(Np^s; W/p^mW) \text{ for } s \geq r.$$

Since $\mathbf{H}_{2+p^{m-1}}(Np^r; W/p^mW) \subset \operatorname{End}_W(M_k(\Gamma \cap \Gamma_1(p^r); W/p^mW))$, we have

$$\mathbf{H}_{2+p^{m-1}}(Np^r; W/p^mW) \twoheadrightarrow \mathbf{H}_2(Np^r; W/p^mW).$$

Since this $W/p^m W$-algebra homomorphism sends generators $T(n)$ to $T(n)$, it is a surjective map. By duality in §3.2.5, we have a surjective algebra homomorphism sending $T(n)$ to $T(n)$:
$$\mathbf{h}_{2+p^{m-1}}(Np^r; W/p^m W) \twoheadrightarrow \mathbf{h}_2(Np^r; W/p^m W).$$
Note the duality between Hecke algebra and cusp forms in §3.2.5 is not always valid for $\mathbf{H}_k$ and M_k (as the infomation of the constant term is missing on the side of Hecke algebra); so, the surjectivity for $\mathbf{H}_{2+p^{m-1}} \twoheadrightarrow \mathbf{H}_2$ has to be proven without duality as explained above.

Since all the Hecke algebras appearing here are p-profinite compact, passing to the projective limit with respect to r and m preserves surjection, and hence we have a unique surjective W-algebra homomorphism
$$\mathbf{h}_\infty(Np^\infty; W) := \varprojlim_m \varprojlim_r \mathbf{h}_{2+p^{m-1}}(Np^r; W/p^m W)$$
$$\twoheadrightarrow \varprojlim_r \mathbf{h}_2(Np^r; W/p^m W) =: \mathbf{h}_2(Np^\infty; W) \quad (4.13)$$
sending $T(n)$ to $T(n)$. Similarly,
$$\mathbf{H}_\infty(Np^\infty; W) := \varprojlim_m \varprojlim_r \mathbf{H}_{2+p^{m-1}}(Np^r; W/p^m W)$$
$$\twoheadrightarrow \varprojlim_r \mathbf{H}_2(Np^r; W/p^m W) =: \mathbf{h}_2(Np^\infty; W) \quad (4.14)$$
sending $T(n)$ to $T(n)$. We can restart with $k \geq 1$ in place of the above argument, we find a surjective algebra homomorphism
$$\mathbf{h}_\infty(Np^\infty; W) := \varprojlim_m \varprojlim_r \mathbf{h}_{k+p^{m-1}}(Np^r; W/p^m W)$$
$$\twoheadrightarrow \varprojlim_r \mathbf{h}_k(Np^r; W/p^m W) =: \mathbf{h}_k(Np^\infty; W) \quad (4.15)$$
sending $T(n)$ to $T(n)$.

Since these weight change morphisms preserve $T(\ell^j)$, $\langle \ell \rangle_k = \ell^{1-k}(T(\ell)^2 - T(\ell^2))$ is sent to $\ell^{1-k'}\langle \ell \rangle_{k'}$; so, it is natural to normalize the diamond operator $\langle \ell \rangle_k$ to make it independent of the weight k. There are two normalizations: The first one is to put $\langle z \rangle_g := z_p^k \langle z \rangle_k \in \mathbf{H}_k(Np^\infty; W)$ for $z = (z_p, z_N) \in \mathbb{Z}_p^\times \times (\mathbb{Z}/N\mathbb{Z})^\times$. This normalization is called *geometric*, and we have already introduced this in the proof of Theorem 4.1.13. We define another normalization by putting $\langle z \rangle_a := z_p^{k-1} \langle z \rangle_k \in \mathbf{H}_k(Np^\infty; W)$, which is called the *automorphic* normalization. The Galois representation ρ_f associated to a cusp form f of weight k satisfies $\det(\rho_f(\mathrm{Frob}_\ell)) = \langle \ell \rangle_a$ for primes ℓ outside the level. When the normalization is clear from the context, we omit subscripts "g" and "a". The *normalized* diamond operator $\langle z \rangle$ is then preserved by the weight change morphism and hence is independent of the weight.

4.1.6 *Slope of modular forms*

A modular form $f \in M_k(\Gamma_1(p^r N); K)$ $(r > 0)$ has slope α if $f|(U(p)-a)^M = 0$ (for a sufficiently large integer $M \gg 0$) and $|a|_p = p^{-\alpha}$. If $f|U(p)^M = 0$ (for $M \gg 0$), we call f to have infinite slope. A slope 0 form is called an *ordinary* form.

Lemma 4.1.18. *Let A be a p-profinite local ring. Let $X = \varinjlim_n X_n$ for profinite A-modules $X_n = \varprojlim_j X_{n,j}$ for finite $X_{n,j}$ and $T : X \to X$ be a A-linear operator preserving X_n for all n given by $\varprojlim_j T_{n,j}$ for $T_{n,j} \in \mathrm{End}_A(X_{n,j})$ for all n and j. Then the p-profinite limit $e = \lim_{n\to\infty} T^{n!}$ exists in $\mathrm{End}_A(X)$ and gives an idempotent of $\mathrm{End}_A(X)$.*

Proof. Once e is well defined on X_n, it extends to $\varinjlim_n X_n$. So we may assume that $X = \varprojlim_j X_j$ for finite A-modules X_j and $T = \varprojlim_j T_j$ for $T_j \in \mathrm{End}_A(X_j)$. By definition, X_j is a finite module. Thus the subalgebra $B_j \subset \mathrm{End}_A(X_j)$ generated over A_j by $T_j : X_j = X \to X_j$ is a finite ring with p-power order. Thus $B_j = \prod_{R_j} R_j$ for finitely many local rings R_j. Let T_{R_j} be the projection of T_j in R_j. If T_{R_j} is in the maximal ideal $\mathfrak{m}_{R_j}$ of R_j, $T_{R_j}^m = 0$ for $m \gg 0$ with $\mathfrak{m}_{R_j}^m = 0$. This $\lim_{n\to\infty} T_{R_j}^{n!} = 0$.

If $T_{R_j} \notin \mathfrak{m}_{R_j}$, then T_{R_j} is a unit. Then $T_{R_j}^{|R_j|^\times} = e_j$ is the idempotent of R_j. Therefore $\lim_{n\to\infty} T_{R_j}^{n!} = e_j$. This shows $E_j := \lim_{n\to\infty} T_n^{n!} = \sum_{j:T_{R_j}\notin\mathfrak{m}_{R_j}} e_j$ is an idempotent. Then $e = \lim_n T^{n!} = \varprojlim_j E_j$ which is the desired idempotent. $\qquad\square$

Exercise 4.1.18. Under the notation of the above proof, prove that B_j is a product of finitely many local rings.

Let $e = \lim_{n\to\infty} U(p)^{n!}$ in $\mathrm{End}_W(M_k(\Gamma_1(Np^r); A)$ (with $r > 0$) for $A = W$ or K, and define

$$M_k^{ord}(\Gamma_1(Np^r); A) = e(M_k(\Gamma_1(Np^r); A)).$$

The following lemma is easy:

Lemma 4.1.19. $f \in M_k(\Gamma_1(Np^r); W))$ *is of slope zero if and only if $f \in M_k^{ord}(\Gamma_1(Np^r); W)$ and f is an eigenform for $U(p)$.*

Exercise 4.1.19. Prove the above lemma.

Definition 4.1.20. Respectively, we define a Hecke algebra $\mathbf{H}_k^{ord}(Np^r; A)$ and $\mathbf{h}_k^{ord}(Np^r; A)$ by the A-subalgebra of the linear endomorphism algebra $\mathrm{End}_A(M_k^{ord}(\Gamma_1(Np^r); A))$ and $\mathrm{End}_A(S_k^{ord}(\Gamma_1(Np^r); A))$ generated by

Hecke operators $T(n)$ for all n. More generally, taking $\Gamma = \Gamma_0(N\mathbf{p})$, for a character $\psi : (\mathbb{Z}/N\mathbf{p}\mathbb{Z})^\times \to W^\times$ and a W-algebra A, we define $\mathbf{H}_k^{ord}(Np^r, \psi; A)$ and $\mathbf{h}_k^{ord}(Np^r, \psi; A)$ by the A-subalgebra of $\mathrm{End}_A(M_{k,\psi}^{ord}(\Gamma \cap \Gamma_1(p^r); A))$ and $\mathrm{End}_A(S_{k,\psi}^{ord}(\Gamma \cap \Gamma_1(p^r); A))$ generated by Hecke operators $T(n)$ for all n.

Recall $\Lambda = \Lambda_W = W[[\mathbf{\Gamma}]]$. We define the spaces $M_\psi^{ord}(N; \Lambda)$ of p-ordinary analytic families as follows:

Definition 4.1.21. Define

$$M_\psi^{ord}(N; \Lambda) = \left\{ F \in M_\psi(N; \Lambda) \big| F(\gamma^k - 1) \in M_{k,\psi\omega^{-k}}^{ord}(\Gamma_0(Np)) \ \forall k \gg 0 \right\}$$

and $S_\psi^{ord}(N; \Lambda) = S_\psi(N; \Lambda) \cap M_\psi^{ord}(N; \Lambda)$.

Lemma 4.1.22. *We have a projector* $e : M_\psi(N; \Lambda) \twoheadrightarrow M_\psi^{ord}(N; \Lambda)$ *given by* $\lim_{n\to\infty} U(p)^{n!}$*, which induces a projector* $e : S_\psi(N; \Lambda) \twoheadrightarrow S_\psi^{ord}(N; \Lambda)$.

Proof. We only need to prove the existence of $e : M_\psi(N; \Lambda) \twoheadrightarrow M_\psi^{ord}(N; \Lambda)$. Define $M_\psi^j(N; \Lambda)$ for $j > 0$ by

$$\{F \in M_\psi(N; \Lambda) | F(\gamma^k - 1) \in M_k(\Gamma_0(N\mathbf{p}), \psi\omega^{-k}; W) \text{ for all } k \geq j\}.$$

Plainly $M_\psi^i(N; \Lambda) \hookrightarrow M_\psi^j(N; \Lambda)$ if $j > i$. Then by definition $M_\psi(N; \Lambda) = \varinjlim_j M_\psi^j(N; \Lambda)$. Thus we need to show the existence of e on $M_\psi^j(N; \Lambda)$. Consider the map $I_j : M_\psi^j(N; \Lambda) \to \prod_{k=j}^\infty M_k(\Gamma_0(\mathbf{p}), \psi\omega^{-k}; W)$ given by $F \mapsto (F(\gamma^k - 1))_k$. If $I_j(F) = 0$, $a(n, F)(\gamma^k - 1) = 0$ for all n and all $k \geq j$. Thus by Weierstrass preparation Theorem 1.10.3, we get $a(n, F) = 0$ and hence $F = 0$. At each weight $e_k = \lim_{n\to\infty} U(p)^{n!} \in \mathrm{End}_W(M_k(\Gamma_0(\mathbf{p}), \psi\omega^{-k}; W))$, and plainly $\prod_k e_k$ preserves the image of I_j getting e on $M_\psi^j(N; \Lambda)$ as desired. $\qquad\square$

Corollary 4.1.23. *The specialization maps*

$$M_\psi^{ord}(N; \Lambda) \to M_{k,\psi\omega^{-k}}^{ord}(\Gamma_0(N\mathbf{p}); W)$$

and $S_\psi^{ord}(N; \Lambda) \to S_{k,\psi\omega^{-k}}^{ord}(\Gamma_0(N\mathbf{p}); W)$ *are surjective for all* $k \geq 0$.

By this corollary, any element in $M_{k,\psi}^{ord}(\Gamma_0(N\mathbf{p}); W)$ for $k \geq 0$ can be lifted to a p-adic analytic family.

Proof. Since the proof is the same for modular forms and cusp forms, we give an argument for modular form. Note that $u := \mathcal{E}_1(0) \in \mathbb{Z}_p^\times$ by Theorem 4.1.5. Thus for $f \in M_k^{ord}(\Gamma_0(Np), \psi\omega^{-k}; W)$, the fact $e(f * u^{-1}\mathcal{E}_1)$ specializes to $e(f) = f$ shows the desired result. $\qquad\square$

The following theorem is proven in [H86a] and also in [H86b]:

Theorem 4.1.24. $M_\psi^{ord}(N;\Lambda)$ *and* $S_\psi^{ord}(N;\Lambda)$ *are free of finite rank over* Λ. *The rank over* Λ *is given by* $\text{rank}_W M_{k,\psi\omega^{-k}}^{ord}(\Gamma_0(pN);W)$ *and* $\text{rank}_W S_{k,\psi\omega^{-k}}^{ord}(\Gamma_0(pN);W)$ *for any* $k \geq 2$, *respectively.*

We will prove Λ-freeness of this theorem assuming the following rank constancy which will be proven in Corollary 4.2.32 by cohomological method.

Lemma 4.1.25. *If* $k \geq 2$, *we have*

$$\text{rank}_W M_{k,\psi\omega^{-k}}^{ord}(\Gamma_0(pN);W) = \text{rank}_W M_{2,\psi\omega^{-2}}^{ord}(\Gamma_0(pN);W),$$

$$\text{rank}_W S_{k,\psi\omega^{-k}}^{ord}(\Gamma_0(pN);W) = \text{rank}_W S_{2,\psi\omega^{-2}}^{ord}(\Gamma_0(pN);W),$$

which are independent of $k \geq 2$.

Proof of Theorem 4.1.24. We reproduce an argument to good extent due to Wiles [W88, §1.4] given in the proof of [LFE, Theorem 7.3.1]. The proof is the same for $M^{ord} = M_\psi^{ord}(N;\Lambda)$ and $S_\psi^{ord}(N;\Lambda)$. We shall give a proof only for M_ψ^{ord}. We prove first that M^{ord} is finitely generated and is Λ-torsion-free. By definition, M^{ord} is a Λ-submodule of the power series ring $\Lambda[[q]]$. Therefore it is Λ-torsion-free.

We now prove that the rank of any finitely generated free Λ-submodule M of M^{ord} is bounded. Let $F_1, F_2, \ldots, F_r$ be a basis of M over Λ. Since $F_1, \ldots, F_r$ are linearly independent over Λ, we can find positive integers $n_1, \ldots, n_r$ such that $D(T) = \det(a(n_i, F_j)) \neq 0$ in Λ. By the Weierstrass preparation Theorem 1.10.3, we can take the weight k so that $D(\gamma^k - 1) \neq 0$ and $F_i(\gamma^k - 1)$ is classical, namely, is an element of $M_k^{ord}(\Gamma_0(p^\alpha p), \psi\omega^{-k};W)$ for all i. Write f_i for $F_i(\gamma^k - 1)$. Then $D(\gamma^k - 1) = \det(a(n_i, f_j)) \neq 0$. Thus the modular forms $f_1, \ldots, f_r$ span a W-free module of rank r in $M_k^{ord}(\Gamma_0(p^\alpha p), \psi\omega^{-k};W)$ whose rank is bounded independent of the weight k. Thus r is bounded by a positive number independent of M. This shows that if $F_1, \ldots, F_r$ is a maximal set of linearly independent elements in M^{ord}, any element in M^{ord} can be expressed as a linear combination of the F_i's if one allows coefficients in the quotient field L of Λ. We thus consider $V = M^{ord} \otimes_\Lambda L$, which is a finite dimensional space over Λ embedded in $L[[q]]$. For each $F \in M^{ord}$, write $F = \sum_i x_i F_i$ with $x_i \in L$. Then the column vector $x = (x_i)_i$ is the solution of the linear equations:

$$(a(n_i, F_j))_{i,j} x = (a(n_i, F))_i \in \Lambda^r.$$

Therefore $Dx_i \in \Lambda$, and thus DM^{ord} is contained in $\Lambda F_1 + \cdots + \Lambda F_r$. Therefore M^{ord} is finitely generated since Λ is noetherian.

To prove the freeness over Λ, we remark the following facts:

(i) Λ is a unique factorization domain Corollary 1.10.5.
(ii) Λ is a compact ring.

Since M^{ord} is finitely generated, we can find a weight k so that $F(\gamma^k - 1)$ is classical for all F in M^{ord}. If $F(\gamma^k - 1) = 0$, then $a(n, F)(\gamma^k - 1)$ is divisible by $P = P_k = X - (\gamma^{k-1})$ for all n. Thus by dividing F by P, we still have an element of M^{ord}, because $(F/P_k)(\gamma^j - 1) = F(\gamma^j - 1)/(\gamma^j - \gamma^k)$ for all $j \neq k$, which is a modular form. Namely

$$PM^{ord} = \{F \in M^{ord} | F(\gamma^k - 1) = 0\}.$$

Thus M^{ord}/PM^{ord} can be embedded into $M_k^{ord}(\Gamma_0(p^\alpha \mathbf{p}), \psi\omega^{-k}; W)$. Thus M^{ord}/PM^{ord} is W-free of finite rank.

Let us take F_i $(i = 1, \ldots, r)$ so that F_i mod PM^{ord} gives a W-basis of M^{ord}/PM^{ord}. This is possible by Corollary 4.1.23. Note that the F_i's are linearly independent over Λ. In fact, if not, we may suppose that $\lambda_1 F_1 + \cdots + \lambda_r F_r = 0$ with at least one of the λ_i's not divisible by P. Then reducing modulo P, we have a non-trivial linear relation between F_i mod P, which is a contradiction, and hence the F_i's are linearly independent. Consider $M = \Lambda F_1 + \cdots + \Lambda F_r$. Then M is a Λ-free module of rank r and M/PM coincides with M^{ord}/PM^{ord} because if F is an element of M^{ord}, then we can find a finite linear combination G_0 of the F_j's such that $F - G_0$ is divisible by P. We now apply this argument to $(F - G_0)/P$ in place of F and get another linear combination G_1 of the F_i's such that $(F - G_0)/P - G_1$ is divisible by P. Repeating this process, we can find the G_i's which are linear combinations of the F_i's such that $F \equiv G_0 + G_1 P + \cdots + G_{i-1} P^{i-1}$ mod P^i. Thus $M/P^i M = M^{ord}/P^i M^{ord}$. Note that the series $G_0 + G_1 P + \cdots + G_{i-1} P^{i-1}$ converges in M by identifying M with Λ^r by the basis F_i's. Thus $M = M^{ord}$ and hence M^{ord} is Λ-free of rank $r = \text{rank}_W(M_k^{ord}(\Gamma_0(p^\alpha \mathbf{p}), \psi\omega^{-k}; W))$. $\qquad\square$

From the above proof, we can choose any k sufficiently large such that $D(\gamma^k - 1) \neq 0$ to start our argument, and we find $M^{ord}/P_k M^{ord} \cong M_k^{ord}(\Gamma_0(p^\alpha \mathbf{p}), \psi\omega^{-k}; W)$. Thus we have proven the following theorem for k sufficiently large:

Theorem 4.1.26. *If $k \geq 2$, the specialization map induces isomorphisms for $t = 1 + T$*

$$M_\psi^{ord}(N; \Lambda) \otimes_\Lambda \Lambda/(t - \gamma^k)) \cong M_{k,\psi\omega^{-k}}^{ord}(\Gamma_0(\mathbf{p}N); W),$$

$$S_\psi^{ord}(N; \Lambda) \otimes_\Lambda \Lambda/(t - \gamma^k)) \cong S_{k,\psi\omega^{-k}}^{ord}(\Gamma_0(\mathbf{p}N); W)$$

for all $k \geq 2$.

Proof. By Corollary 4.1.23, we have surjective specialization map:

$$M_\psi^{ord}(N; \Lambda) \otimes_\Lambda \Lambda/(t - \gamma^k)) \twoheadrightarrow M_{k,\psi\omega^{-k}}^{ord}(\Gamma_0(\mathbf{p}N); W).$$

However the W-rank of the two sides are equal if $k \geq 2$ by Lemma 4.1.25. This shows the isomorphism. The proof is the same for cusp forms. □

Definition 4.1.27. Let $0 < k \in \mathbb{Z}$ be an integer divisible by the least common multiple of 2 and $p-1$ (so, $\omega^k = 1$). A weak p-adic analytic family of modular forms (centered at $0 < k \in \mathbb{Z}$) is a formal power series $F = \sum_{n=0}^\infty a(n, F)(T)q^n$ with $a(n, F)(T) \in K[[T]]$ ($K = \mathrm{Frac}(W)$) convergent at $\gamma^{k'} - 1$ for all k' in a small p-adic neighborhood U in $k \cdot \Gamma \subset \mathbb{Z}_p^\times$ of k such that $F(\gamma^{k'} - 1) \in M_{k',\psi}(\Gamma_0(Np); K)$ for all $k' \gg 0$ in U.

For a given slope $\alpha \in \mathbb{Q}$, we define $M_{k,\psi}^{(\alpha)}(\Gamma_0(pN); K)$ to be the space spanned by slope α modular forms in $M_{k,\psi}(\Gamma_0(pN); K)$ and put $S_{k,\psi}^{(\alpha)}(\Gamma_0(pN); K) = M_{k,\psi}^{(\alpha)}(\Gamma_0(pN); K) \cap S_{k,\psi}(\Gamma_0(pN); K)$. By definition, we have

$$M_{k,\psi}^{(0)}(\Gamma_0(pN); K) = M_{k,\psi}^{ord}(\Gamma_0(pN); K).$$

Moreover, we have

$$M_{k,\psi}(\Gamma_0(pN); K) = \bigoplus_\alpha M_{k,\psi}^{(\alpha)}(\Gamma_0(pN); K).$$

Exercise 4.1.20. Prove the above decomposition.

This type of weak families was introduced by Mazur and Gouvêa in 1992 and such families have been constructed by Coleman for positive slope modular forms [C97]. Then Coleman and Mazur further went on to globalize the (local) parameter space U_k of modular Hecke eigenforms to a rigid analytic curve (the so-called eigencurve) in [CM98].

Though there is some speculation in [GM92] for the size of the neighborhood U (dependent on k), but the problems of determining U precisely and to find minimal k such that $F(\gamma^k - 1)$ is classical are still open (see [BC04] and [W98]).

Exercise 4.1.21. Prove that $\mathrm{Spec}(\Lambda)(\overline{\mathbb{Q}}_p) = \mathrm{Hom}_{W\text{-alg}}(\Lambda, \overline{\mathbb{Q}}_p)$ is isomorphic to the open unit disk D in $\overline{\mathbb{Q}}_p$ (centered at the origin 0) by sending a W-algebra homomorphism $\phi : \Lambda \to \overline{\mathbb{Q}}_p$ to $\phi(T)$.

4.1.7   Control of Hecke algebra

We start with a definition:

Definition 4.1.28. We define
$$\mathbf{h}_\psi^{ord}(N; \Lambda) := \mathcal{H}(S_\psi^{ord}(N; \Lambda)) = \Lambda[T(n)|n = 1, 2, \dots] \subset \mathrm{End}_\Lambda(S_\psi^{ord}(N; \Lambda)).$$
Similarly replacing $S_\psi^{ord}(N; \Lambda)$ by $M_\psi^{ord}(N; \Lambda)$, we define
$$\mathbf{H}_\psi^{ord}(N; \Lambda) := \mathcal{H}(M_\psi^{ord}(N; \Lambda)).$$

By the argument in §3.2.5 (combined with Definition 4.1.10),
$$\begin{aligned}
\mathrm{Hom}_\Lambda(S_\psi^{ord}(N; \Lambda), \Lambda) &\cong \mathbf{h}_\psi^{ord}(N; \Lambda), \\
\mathrm{Hom}_\Lambda(\mathbf{h}_\psi^{ord}(N; \Lambda), \Lambda) &\cong S_\psi^{ord}(N; \Lambda).
\end{aligned} \tag{4.16}$$

By Theorem 4.1.24, $S_\psi^{ord}(N; \Lambda)$ is free of finite rank over Λ with $\mathrm{rank}_\Lambda \, S^{ord}(N; \Lambda) = \mathrm{rank}_W \, S_{k,\psi\omega^{-k}}^{ord}(\Gamma_0(\mathbf{p}N); W) = r$ for any weight $k \geq 2$. Thus by (4.16), $\mathbf{h}_\psi^{ord}(N; \Lambda), \Lambda)$ is free of finite rank r over Λ. By Theorem 4.1.26, we have a surjective W-algebra homomorphism $i : \mathbf{h}_\psi^{ord}(N; \Lambda), \Lambda) \otimes_\Lambda \Lambda/(t - \gamma^k) \twoheadrightarrow \mathbf{h}_{k,\psi\omega^{-k}}^{ord}(\Gamma_0(\mathbf{p}N); W)$ sending $T(n)$ to $T(n)$ for all n. By the duality between the Hecek algebra and cusp forms, we have $r = \mathrm{rank}_W \, \mathbf{h}_{k,\psi\omega^{-k}}^{ord}(\Gamma_0(\mathbf{p}N); W)$. Comparing the W-rank of the two side of the morphism i, we get an isomorphism. Thus we have proven the following fact:

Theorem 4.1.29. *The Λ-algebra $\mathbf{h}_\psi^{ord}(N; \Lambda)$ is free of finite rank over Λ, and for each weight $k \geq 2$, we have an isomorphism:*
$$\mathbf{h}_\psi^{ord}(N; \Lambda), \Lambda) \otimes_\Lambda \Lambda/(t - \gamma^k) \xrightarrow{\sim} \mathbf{h}_{k,\psi\omega^{-k}}^{ord}(\Gamma_0(\mathbf{p}N); W)$$
sending $T(n)$ to $T(n)$ for all n.

If $k \geq 3$ and $\psi\omega^{-k}$ is defined modulo N, we can lower the level to N removing the p-factor $\mathbf{p}$:

Corollary 4.1.30. *Suppose $k \geq 3$ and that $\psi\omega^{-k}$ is defined modulo N. Then we have an isomorphism:*
$$\mathbf{h}_\psi^{ord}(N; \Lambda), \Lambda) \otimes_\Lambda \Lambda/(t - \gamma^k) \xrightarrow{\sim} \mathbf{h}_{k,\psi\omega^{-k}}^{ord}(\Gamma_0(N); W)$$
sending $T(n)$ to $T(n)$ for all n prime to p. Here $\mathbf{h}_{k,\psi\omega^{-k}}^{ord}(\Gamma_0(N); W) := e_0 \mathbf{h}_{k,\psi\omega^{-k}}^{ord}(\Gamma_0(N); W)$ for $e_0 = \lim_{n\to\infty} T(p)^{n!}$.

Proof.   Write $\varphi := \psi\omega^{-k}$. By the level lowering property of $U(p)$ in §4.2.8, for $A = W, K = \text{Frac}(W)$, $\mathbf{h}_{k,\varphi}^{ord}(\Gamma_0(Np); A) = \mathbf{h}_{k,\varphi}^{ord}(\Gamma_0(Np); A)$. So we prove $\mathbf{h}_{k,\varphi}^{ord}(\Gamma_0(Np); W) \cong \mathbf{h}_{k,\varphi}^{ord}(\Gamma_0(N); W)$ under $k \geq 3$.

By Lemma 4.1.8, if $k \geq 2$,

$$a(n, f|T(p)) = a(np, f) + p^{k-1}a(\frac{n}{p}, f|\langle p\rangle_k) \equiv a(n, f|U(p)) \mod p$$

for $f \in S_k(\Gamma_0(N), \varphi)$. Here the left-hand-side is the $T(p)$ given by $[\Gamma_0(N)\text{diag}[1,p]\Gamma_0(N)]$ and the $U(p)$ is on $S_k(\Gamma_0(Np), \varphi)$. Thus the inclusion $S_k^{ord}(\Gamma_0(N), \varphi; \mathbb{F}) \hookrightarrow e_0 S_k(\Gamma_0(Np), \varphi; \mathbb{F}) = S_k^{ord}(\Gamma_0(Np), \varphi; \mathbb{F})$ is Hecke equivariant. By duality, we have a surjective algebra homomorphism $\pi_{\mathbb{F}} : \mathbf{h}_{k,\varphi}^{ord}(\Gamma_0(Np); \mathbb{F}) \twoheadrightarrow \mathbf{h}_{k,\varphi}^{ord}(\Gamma_0(N); \mathbb{F})$. We consider the composite map $e : S_k(\Gamma_0(N), \varphi; W) \to S_k(\Gamma_0(Np), \varphi; W)$ regarding f of level N as having level Np and applying the projector $e = \lim_{n\to\infty} U(p)^{n!}$. Since $e \equiv e_0 \mod p$, this is injective and by duality induces a W-linear map $\pi_A : \mathbf{h}_{k,\varphi}^{ord}(\Gamma_0(Np); A) \twoheadrightarrow \mathbf{h}_{k,\varphi}^{ord}(\Gamma_0(N); A)$ for $A = W, K = \text{Frac}(W)$. Since $\pi_W \otimes 1 = \pi_{\mathbb{F}}$, by Nakayama's Lemma 1.8.3, π_W and π_K are onto. The morphism is an algebra homomorphism on the subalgebra of the source generated by $T(n)$ for n prime to p. As before, define $f|[p] = \sum_{n=1}^{\infty} a(n, f)q^{pn}$ (a p-old form coming from $f \in S_k(\Gamma_0(N), \varphi)$). Since $U(p) \circ [p]$ is the identity map on $S_k(\Gamma_0(N), \varphi))$, in $S_k^{ord}(\Gamma_0(N), \varphi))$, there is no non-zero p-old form. Thus by the theory of new forms [MFM, Theorem 4.6.19], $\mathbf{h}_{k,\varphi}^{ord}(\Gamma_0(Np); K)$ is generated over K by $T(n)$ for n prime to p. Thus π_K is a K-algebra homomorphism. Since π_W is the restriction of π_K to the W-integral Hecke algebra, π_W is a W-algebra homomorphism.

Regarding $\Gamma_0(N)$ as a union of double cosets of $\Gamma_0(Np)$, we have the double coset action $\text{Tr} := [\Gamma_0(N)] : S_k(\Gamma_0(Np), \varphi; K) \to S_k(\Gamma_0(N), \varphi; K)$. If $k \geq 3$, cusp forms in $\text{Ker}(\text{Tr})$ are killed by e_0 by [MFM, Theorem 4.6.17]; so, π_K is an isomorphism, and we also find π_W is an isomorphism. $\square$

Exercise 4.1.22. Under the circumstance of Corollary 4.1.30, prove that $U(p)$ is sent by π_W to the unit root of $X^2 - T(p)X + p^{k-1}\langle p\rangle_k = 0$ in $\mathbf{h}_{k,\varphi}^{ord}(\Gamma_0(N); W)$ by the construction in §3.2.8

4.2   Analytic families via cohomology groups

We studied the duality theorem between Hecke algebra and the space of modular forms in Proposition 3.2.10. The duality is useful to recover the space of cusp forms $S_2(\Gamma_0(N), A)$ as the linear dual $\text{Hom}_A(h(N, A), A)$

and the set of Hecke eigenforms is in bijection with $\mathrm{Spec}(h(N; A))(A) = \mathrm{Hom}_{A\text{-alg}}(h(N, A), A)$.

In this section, we interpolate Hecke algebras varying weight and characters of p-power conductor via Betti cohomological techniques (first introduced in [H86b]). We are able to construct the ordinary part $\mathbf{H}^{ord}(N; \Lambda)$ of the interpolated Hecke algebra without referring to the space of cusp forms. The algebra $\mathbf{H}^{ord}(N; \Lambda)$ is naturally an algebra over Λ, and then we redefine the space of Λ-adic forms $M^{ord}(N; \Lambda)$ to be the Λ-linear dual of $\mathbf{H}^{ord}(N; \Lambda)$. Then we study the ring structure $\mathbf{H}^{ord}(N; \Lambda)$ which eventually leads to a new proof of Theorem 4.1.24. In this section, we use the automorphic normalization of the diamond operator as it has the clear relation to Hecke operators: $\langle l \rangle = \langle l \rangle_a = T(l)^2 - T(l^2)$.

The first six subsections until §4.2.6 describe basic cohomological tools we use with fairly detailed proofs. An important observation by Eichler and Shimura connecting cohomology and modular forms is recalled in §4.2.4. The next eight subsections until §4.2.14 introduces Hecke operators acting on cohomology groups. Taking p-adic coefficient modules, increasing the level (fixing the weight k), we construct the limit Hecke algebra of level Np^∞ of weight k. By interpreting elementary morphisms between coefficients modules into modulo p-power congruences of Hecke operators, we prove that the constructed Hecke algebra is independent of weight (as long as $k \geq 2$). The independence is stated in §4.2.9 and the proof ends in §4.2.14. We do not need ordinarity to prove independence of weight. By duality, this means that a higher weight modular form can be approximated p-adically by lower weight, say 2, modular forms (density of weight 2 forms in the space of p-adic modular forms). In the rest of this chapter, we prove control theorems depicting an exact relation of the big Hecke algebra $\mathbf{H}^{ord}(N; \Lambda)$ with each Hecke algebra of finite weight and character. At the end, we come back to ordinary Λ-adic forms from Hecke algebras via duality.

Here is an outline of the cohomological way of making $\mathbf{H}^{ord}(N; \Lambda)$. Taking the limit increasing the level by p-power, we create a big module, call it X. Recall $Y_1(Np^r) = X_1(Np^r) - \{\text{cusps}\} = \Gamma_1(Np^r)\backslash\mathfrak{H}$. The module X can be the projective limit $\varprojlim_n X_n$ for cohomology groups $X_n = H^{ord}_1(Y_1(Np^n), \mathbb{Z}_p)$ or its Pontryagin dual $\varinjlim_n H^1_{ord}(Y_1(Np^n), \mathbb{Q}_p/\mathbb{Z}_p)$ or $\varprojlim_n X_n$ for $X_n = S^{ord}_2(\Gamma_1(Np^n); \mathbb{Z}_p)$. Here the subscript or superscript "*ord*" indicate the image of the projector $e = \lim_{n\to\infty} U(p)^{n!}$ (e.g., H^q_{ord} for cohomology and H^{ord}_q for homology). Note that $(\mathbb{Z}/Np^r\mathbb{Z})^\times$ acts on X_n by the diamond operator $\langle\cdot\rangle$, and $\Gamma/\Gamma^{p^{r-1}} \hookrightarrow (\mathbb{Z}/p^r\mathbb{Z})^\times \subset (\mathbb{Z}/Np^r\mathbb{Z})^\times$.

Thus $W[[\Gamma]] := \varprojlim_r W[\Gamma/\Gamma^{p^{r-1}}]$ acts on X. Note that this algebra $W[[\Gamma]]$ is isomorphic to Λ by sending $\gamma \in \Gamma \subset W[[\Gamma]]$ to $1 + T$. Then we define $\mathbf{H}^{ord}(N; \Lambda)$ to be the Λ-subalgebra of $\mathrm{End}_\Lambda(X)$ generated by Hecke operators. An important point is that the algebra $\mathbf{H}^{ord}(N; \Lambda)$ does not depend on the choice of X_n.

Here is a review of *Pontryagin duality*. Consider a profinite group G and a continuous G-module X. Assume that X has either discrete torsion or profinite topology. For any abelian profinite compact or torsion discrete module X, we define the Pontryagin dual module $X^\vee$ by $X^\vee = \mathrm{Hom}_{cont}(X, \mathbb{Q}/\mathbb{Z})$ and give $X^\vee$ the topology of uniform convergence on every compact subgroup of X. For an obvious reason, we may replace $\mathbb{Q}/\mathbb{Z}$ by $\mathbb{Q}_p/\mathbb{Z}_p$ if X is either p-profinite or p-torsion. The G-action on $f \in X^\vee$ is given by $\sigma f(x) = f(\sigma^{-1}x)$. Then by Pontryagin duality theory (e.g., [LFE, §8.3]), we have $(X^\vee)^\vee \cong X$ canonically. By this fact, if $X^\vee$ is the dual of a profinite module $X = \varprojlim_n X_n$ for finite modules X_n with surjections $X_m \twoheadrightarrow X_n$ for $m > n$, $X^\vee = \bigcup_n X_n^\vee$ is a discrete module which is a union of finite modules $X_n^\vee$.

Exercise 4.2.1. Show that $X^\vee \cong X$ noncanonically if X is finite.

Exercise 4.2.2. Prove

(1) $X^\vee$ is a discrete module if X is p-profinite, and $X^\vee$ is compact if X is discrete (e.g., [LFE, Lemma 8.3.1]).
(2) projective limit is an exact functor from the category of projective systems of profinite modules into the category of profinite modules (i.e., projective limit of short exact sequences of profinite modules remain exact after passing to the limit).

By this fact, if $X^\vee$ is the dual of a profinite module $X = \varprojlim_n X_n$ for finite modules X_n with surjections $X_m \twoheadrightarrow X_n$ for $m > n$, $X^\vee = \bigcup_n X_n^\vee$ is a discrete module which is a union of finite modules $X_n^\vee$ and vice versa. We note the following fact:

Exercise 4.2.3. Prove the following statement. If X is a profinite A-module, then by the duality, we have, for an ideal $\mathfrak{a}$ of A,

$$X^\vee[\mathfrak{a}] := \{x \in X^\vee | ax = 0 \text{ for all } a \in \mathfrak{a}\} \cong (X/\mathfrak{a}X)^\vee$$

naturally.

4.2.1 Fundamental group, again

By the theory of fundamental group $\pi_1(X)$ in §2.3.2, for a Riemann surface X, we have a universal covering $\pi : \mathcal{Z} \twoheadrightarrow X = X \cong \pi_1(X)\backslash \mathcal{Z}$ for a simply connected space $\mathcal{Z}$. The projection is unramified, and $\mathcal{Z}$ is determined uniquely (up to isomorphisms; see, for example, [AFC, §3.5]). Since $\mathfrak{H}$ is simply connected, if a subgroup $\Delta \subset PSL_2(\mathbb{Z})$ of finite index is torsion-free, $\Delta \cong \pi_1(\Delta\backslash\mathfrak{H})$.

Fix a base point $x \in X$ and cut X by 1-cycles α_i passing through x so that the resulting space is a simply connected polygon P, α_i generates $\pi_1(X)$ (basically by definition). If X is compact, and writing the edges of P counterclockwise as $\alpha_1, \beta_1, \alpha_2, \beta_2, \ldots, \alpha_g, \beta_g$, then $\pi_1(X)$ is generated by these paths with only relation

$$\prod_{j=1}^{g}(\alpha_j, \beta_j) = (\alpha_1, \beta_1)(\alpha_2, \beta_2)\cdots(\alpha_g, \beta_g) = 1, \qquad (4.17)$$

where $(\alpha, \beta) = \alpha\beta\alpha^{-1}\beta^{-1}$ (a commutator). If $Y = X - S$ for a finite set of points (outside the fixed x), ordering $S = \{s_1, \ldots, s_c\}$ according to the orientation of P (from left to right), we can take a path π_j starting at x encircling s_j, cutting P by π_j, we get a new polygon of $2g + c$ sides. In this case, $\pi_1(Y)$ is generated by $\{\pi_i, \alpha_j, \beta_j\}_{i,j}$ with only relation

$$\prod_{i=1}^{c}\pi_i \prod_{j=1}^{g}(\alpha_j, \beta_j) = \pi_1\pi_2\cdots\pi_c(\alpha_1, \beta_1)(\alpha_2, \beta_2)\cdots(\alpha_g, \beta_g) = 1. \qquad (4.18)$$

Suppose that $Y = X$ (so $S = \emptyset$) is a compact Riemann surface or an open Riemann surface of the form $Y = X - S$ as in (4.18) for a compact Riemann surface X. Then by (4.17) and (4.18), we have

$$\pi_1^{ab}(Y) = \frac{\pi_1(Y)}{(\pi_1(Y), \pi_1(Y))} \cong \bigoplus_{j=1}^{g}(\mathbb{Z}[\alpha_j] \oplus \mathbb{Z}[\beta_j]) \cong H_1(X, \mathbb{Z}), \quad \text{if } S = \emptyset,$$

$$\pi_1^{ab}(Y) \cong \left(\bigoplus_{j=1}^{c-1}\mathbb{Z}[\pi_j]\right) \oplus \left(\bigoplus_{j=1}^{g}\mathbb{Z}[\alpha_j] \oplus \mathbb{Z}[\beta_j]\right) \cong H_1(Y, \mathbb{Z}), \quad \text{otherwise.}$$

$$(4.19)$$

Thus the genus of X is given by g.

Exercise 4.2.4. Prove (4.19).

In particular, we have

Lemma 4.2.1. *We have*

$$H^1(Y, A) = \mathrm{Hom}(\pi_1(Y), A).$$

4.2.2 Group cohomology

Take an abstract group G to introduce group cohomology. For a ring A, take a $A[G]$-module M (where $A[G]$ is the group ring of G). We define a general group cohomology $H^q(G, M)$ for $q = 0, 1, 2$. We have

$$H^0(G, M) = M^G = \{m \in M | gm = m \text{ for all } g \in G\},$$

and if M is finite, the first cohomology is defined by

$$H^1(G, M) = \frac{\{G \xrightarrow{c} M | c(\sigma\tau) = \sigma c(\tau) + c(\sigma) \text{ for all } \sigma, \tau \in G\}}{\{G \xrightarrow{b} M | b(\sigma) = (\sigma - 1)x \text{ for } x \in M \text{ independent of } \sigma\}}.$$

As for the second cohomology, 2-cocycles $c : G \times G \to M$ are functions satisfying the following relation:

$$c(\alpha, \beta) + c(\alpha\beta, \gamma) = \alpha \cdot c(\beta, \gamma) + c(\alpha, \beta\gamma)$$

for all $\alpha, \beta, \gamma \in G$. For any function $b : G \to M$, we define

$$\partial b(\alpha, \beta) = b(\alpha\beta) - \alpha b(\beta) - b(\alpha). \tag{4.20}$$

Then ∂b is easily checked to be a 2-cocycle by computation. Such 2-cocycles obtained from $b : G \to M$ are called a 2-coboundary. Then

$$H^2(G, M) = \frac{\{\text{2-cocycles with values in } M\}}{\{\text{2-coboundaries with values in } M\}}.$$

Exercise 4.2.5. Check that ∂b as above is a 2-cocycle.

Remark 4.2.2. We can define cohomology group $H^q(G, M)$ of general degree q, but we apply this theory for a torsion-free subgroup Δ of finite index in $SL_2(\mathbb{Z})$, and in this case, we know that $H^q(\Delta, M) = 0$ for $q \geq 2$ (see [CGP, Example 5, page 217] or [LFE, Proposition 6.1.1]); so, we do not touch in detail the general definition of H^q (which can be found, for example, in [CGP]).

Here is another remark of topological nature:

Remark 4.2.3. If G is a topological group and M is equipped with a topology under which the action of G is continuous, requiring cocycles and coboundaries to be continuous, we define continuous cohomology group $H^q_{ct}(G, M)$ as in [CNF, II.7].

A simple application of the vanishing of H^2 is

Proposition 4.2.4. *For a torsion-free subgroup Γ of finite index in $SL_2(\mathbb{Z})$ and a Γ-module L free of finite rank over W, if $q \geq 1$, we have*

$$H^q(\Gamma, L \otimes_W A) \cong H^q(\Gamma, L) \otimes A$$

for $A = W/p^r W, K, K/W$.

Proof. From the long exact sequence attached to $W \xrightarrow{x \mapsto p^r x} W \to W.p^r W$, we get another exact sequence

$$0 \to \mathrm{Coker}(H^q(\Gamma, L) \xrightarrow{p^r} H^q(\Gamma, L)) \to H^q(\Gamma, L/p^r L) \to H^{q+1}(\Gamma, L) = 0$$

as $q + 1 > 1$. Since $\mathrm{Coker}(H^q(\Gamma, L) \xrightarrow{p^r} H^q(\Gamma, L)) = H^q(\Gamma, L) \otimes_W W/p^r W$, we get the result for $A = W/p^r W$. The result also follows for $A = K/W = \varinjlim_n W/p^r W$ as cohomology commutes with injective limit. The result for $A = K$ follows from Künneth formula. $\square$

Exercise 4.2.6. Prove that $\Gamma_1(N)$ is torsion-free if $N \geq 4$.

Here is a flavor of the general definition: Take a resolution: $\cdots \to A_{n+1} \xrightarrow{\delta_n} A_n \to \cdots \to A_0 \to M \to 0$ made of free $A[G]$-modules A_n. Then we have a reversed complex

$$0 \to \mathrm{Hom}_{A[G]}(A_0, M) \xrightarrow{\delta_0^*} \mathrm{Hom}_{A[G]}(A_1, M) \xrightarrow{\delta_1*} \cdots$$

$$\xrightarrow{\delta_{n-1}^*} \mathrm{Hom}_{A[G]}(A_n, M) \xrightarrow{\delta_n^*} \cdots,$$

and $H^q(G, M) = \mathrm{Ker}(\delta_n^*)/\mathrm{Im}(\delta_{n-1}^*)$, which is determined (up to isomorphisms) independent of the choice of the original resolution (cf. [MFG, §4.3.1]). Here $\delta_n^*(\phi) = \phi \circ \delta_n$. If G is a finite cyclic group generated by g, we have a very simple resolution $A_n = A[G]$ such that $\delta_{2n}(x) = (g-1)x$ and $\delta_{2n-1}(x) = N(x) = \sum_{h \in G} hx$. By using this resolution, we get

Proposition 4.2.5. *If G is a finite cyclic group, then $H^{2n}(G, M) = M^G/N(M)$ and $H^{2n-1}(G, M) = \mathrm{Ker}(N : M \to M)/(g-1)M$ for all $n > 0$. If G is infinite cyclic, $H^1(G, M) = M/(g-1)M$ and $H^q(G, M) = 0$ for all $q \geq 2$.*

Applying the 1-cocycle relation $c(\gamma^m) = \gamma c(\gamma^{m-1}) + c(\gamma)$ repeatedly, $c(g^m) = (1 + g + \cdot + g^{m-1})c(g)$, the isomorphism for H^1 in the above proposition is given by $c \mapsto c(g) \bmod (g-1)M$. The vanishing of $H^q(G, M)$ for $q > 1$ when G is infinite follows from the fact: $G \cong \pi_1(\mathbb{R}/\mathbb{Z})$ and that $\mathbb{R}$ has dimension 1.

Exercise 4.2.7. Give a detailed proof of the above proposition.

Remark 4.2.6. If $0 \to M_1 \xrightarrow{\iota} M_2 \xrightarrow{\pi} M_3 \to 0$ is an exact sequence of G-modules, we have a corresponding long exact sequence:

$$0 \to M_1^G \xrightarrow{\iota} M_2^G \xrightarrow{\pi} M_3^G$$

$$\xrightarrow{\delta_0} H^1(M_1) \xrightarrow{\iota_*} H^1(M_2) \xrightarrow{\pi_*} H^1(M_3) \xrightarrow{\delta} H^2(M_1) \to \cdots,$$

writing $H^q(M_j)$ for $H^q(G, M_j)$. The maps π_* and ι_* are given by composition of π and ι with 1-cocycles. The map δ_0 is given as follows. For $x \in M_3^G$, pick $\tilde{x} \in M_2^G$ with $\pi(\tilde{x}) = x$. Then $(g-1)\tilde{x} \in M_1$ because $\pi((g-1)\tilde{x}) = (g-1)x = 0$. Then $\delta_0(x)(g) = (g-1)\tilde{x}$ as a 1-cocycle. The map δ can be defined similarly. Pick a class $[c] \in H^1(M_3)$ represented by a 1-cocycle $c : G \to M_3$. Take any function $\tilde{c} : G \to M_2$ such that $\pi(\tilde{c}(g)) = c(g)$. Then $\pi(\partial\tilde{c}) = \partial c = 0$, and hence $\partial\tilde{c}$ has values in M_1 and is a 2-cocycle. Then $\delta([c]) = [\partial\tilde{c}] \in H^2(M_1)$.

Exercise 4.2.8. Prove the exactness of the long sequence up to degree 2 cohomology groups.

4.2.3 Inflation and restriction

Let U be a normal subgroup of a group G. Let M be a G-module. For a 1-cocycle $u : U \to M$ and $g \in G$, ${}^g u : (g_1) \mapsto gu(g^{-1}g_1 g)$ is again a 1-cocycle of U, and the cohomology class of ${}^g u$ is equal to that of u if $g \in U$, as easily verified by computation. Thus the quotient group G/U acts on $H^1(U, M)$ by $[u] \mapsto [{}^g u]$. We now prove

Theorem 4.2.7. *Let U be a normal subgroup of G. Then the following sequence is exact:*

$$0 \to H^1(G/U, M^U) \xrightarrow{\text{Inf}} H^1(G, M)$$
$$\xrightarrow{\text{Res}} H^0(G/U, H^1(U, M)) \xrightarrow{\text{Trans}} H^2(G/U, M^U).$$

We shall give a definition of the *transgression* Trans, due to Hochschild and Serre, in the following proof of the theorem.

Proof. For the projection $\pi : G \to G/U$ (and a cocycle $c : G/U \to M^U$), $\text{Inf}(c) = c \circ \pi$ and $\text{Res}(c) = c|_U$. For these two maps, it is easy to show the exactness by a simple computation (see Exercise 4.2.9).

We prove the exactness at $H^0(G/U, H^1(U, M))$. Let $c : U \to M$ be a 1-cocycle representing a class $[c]$ in $H^0(G/U, H^1(U, M))$. Then $gc(g^{-1}ug) - c(u) = (u-1)a(g)$ for a function $a : G \to M$, because $g[c] = [c]$. If $g \in U$, by cocycle relation, we see

$$gc(g^{-1}ug) - c(u) = c(ug) + gc(g^{-1}) - c(u) = uc(g) - c(g) = (u-1)c(g).$$

Thus we may take the function a to be c on U and hence may assume that $a(u) = c(u)$ for all $u \in U$. Then we have

$$ga(g^{-1}ug) - a(u) = (u-1)a(g).$$

Let F be the space of functions $f : U \to M$. Then we make F into a G-module by the following G-action: $gf(u) = gf(g^{-1}ug)$. Note that $(g-1)f(u) = gf(g^{-1}ug) - f(u)$. We then consider the space of functions: $C_j(G, F) = \{G^j \to F\}$. We have the differential $\partial : C_1(G, F) \to C_2(G, F)$ defined in (4.20). But note that $\partial(g \mapsto (g-1)f) = 0$, and by applying ∂ to $ga(g^{-1}ug) - a(u) = (u-1)a(g)$, we have

$$0 = \partial(x \mapsto (x-1)a(u)) = \partial(x \mapsto (u-1)a(x))(g, h)$$
$$= g(g^{-1}ug - 1)a(h) - (u-1)a(gh) + (u-1)a(g) = (u-1)(ga(h) - a(gh) + a(g)).$$

Now we put $b(g, h) = \partial(a)(g, h) = ga(h) - a(gh) + a(g)$. Then the above equation becomes:

$$(u - 1)b(g, h) = 0.$$

Thus the 2-cocycle $b : G \times G \to M$ actually has values in M^U.

Note that

$$(u - 1)(ua(g) + a(u)) = u(ga(g^{-1}ug) - a(u)) + ua(u) - a(u)$$
$$= uga(g^{-1}ug) - a(u) = (u - 1)a(ug).$$

Thus fixing a complete representative set R for $U \backslash G$ so that $1 \in R$, we may normalize a so that $a(ug) = ua(g) + a(u)$ for all $u \in U$ and all $g \in R$. Since $a|_U$ is a 1-cocycle, by computation, we conclude that $a(ug) = ua(g) + a(u)$ for all $u \in U$ and all $g \in G$ (not just in R). Then for all $u \in U$ and $g, h \in G$, we see $b(u, g) = 0$, and 2-cocycle relation is

$$ub(g, h) - b(ug, h) + b(u, gh) - b(u, g) = 0.$$

This shows that $b(g, h) = ub(g, h) = b(ug, h)$. Similarly, we can show $b(g, uh) = b(g, h)$. Thus b factors through G/U.

If $a' : G \to M$ satisfies the same properties as a, that is, $ga'(g^{-1}ug) - a'(u) = (u-1)a'(g)$ and $a' = c$ on U, then

$$(u - 1)(a(g) - a'(g)) = ga(g^{-1}ug) - a(u) - (ga'(g^{-1}ug) - a'(u)) = 0,$$

because $a = c = a'$ on U. This shows that $d(g) = a(g) - a'(g) \in M^U$. Then $b - b' = \partial(d) \in \mathrm{Im}(C_1(G/U, M^U) \xrightarrow{\partial} C_2(G/U, M^U))$, and hence we have the identity of the cohomology classes:

$$[b] = [b'] \in H^2(G/U, M^U)$$

for $b' = \partial(a')$. We then define $\mathrm{Trans}([c])$ by the cohomology class of $[b]$ in $H^2(G/U, M^U)$.

Suppose that $\text{Trans}([c]) = 0$. Then choosing a 1-cochain $d : G/U \to M^U$ such that $\partial(d) = b$, we see that $a' = a - d$ agrees with c on U and $\partial(a') = 0$; so, a' is a 1-cocycle of G inducing c. This shows

$$\text{Ker(Trans)} \supset \text{Im(Res)}.$$

By definition, if $c \in \text{Im(Res)}$, we take a to be the 1-cocycle of G restricting c on U. Thus, $\text{Im(Res)} \supset \text{Ker(Trans)}$. This proves the desired exactness for the degree 1 cohomology group. $\qquad\qquad\square$

Exercise 4.2.9. Prove the exactness at $H^1(G, M)$ of the sequence in the above theorem.

4.2.4 Eichler–Shimura isomorphism

Choose a Dirichlet character $\psi : (\mathbb{Z}/N\mathbb{Z})^\times \to A^\times$ for a commutative ring A with identity. Let $L(n, \psi; A)$ be the space of homogeneous polynomials in (X, Y) of degree n with coefficients in A. We let $\gamma = \left(\begin{smallmatrix} a & b \\ c & d \end{smallmatrix}\right) \in \Gamma_0(N)$ act on $P(X, Y) \in L(n, \psi; A)$ by

$$\gamma \cdot P(X, Y) = \psi(d) P((X, Y)^t \gamma^\iota),$$

where $\left(\begin{smallmatrix} a & b \\ c & d \end{smallmatrix}\right)^\iota = \left(\begin{smallmatrix} d & -b \\ -c & a \end{smallmatrix}\right)$.

Exercise 4.2.10. Prove the following facts:

(1) $\gamma^\iota \delta^\iota = (\delta\gamma)^\iota$ for any $\gamma, \delta \in M_2(A)$,
(2) $\gamma^\iota + \gamma = \text{Tr}(\gamma)$ and $\gamma^\iota \gamma = \gamma\gamma^\iota = \det(\gamma)$,
(3) $L(n, \psi; A)$ is a left $\Gamma_0(N)$-module,
(4) If $\psi(-1) = (-1)^n$, the action of $\Gamma_0(N)$ on $L(n, \psi; A)$ factors through the image $\overline{\Gamma}_0(N)$ of $\Gamma_0(N)$ in $\text{PSL}_2(\mathbb{Z})$.

Let $f \in M_{k,\psi}(\Gamma_0(N))$ $(k \geq 2)$, and define an $L(k - 2, \psi; \mathbb{C})$-valued differential form $\omega(f)$ by $\omega(f) = f(z)(X - zY)^n dz$ for $n = k - 2$. For the complex conjugate $\overline{M}_{k,\psi^{-1}}(\Gamma_0(N))$ of $M_{k,\psi^{-1}}(\Gamma_0(N))$ and $f \in \overline{M}_{k,\psi^{-1}}(\Gamma_0(N))$, we similarly define $\omega(f) = f(z)(X - \bar{z}Y)^n d\bar{z}$.

Exercise 4.2.11. Prove the following facts:

(1) $\omega(f) \circ \gamma(z) = \gamma \cdot \omega(f)(z)$ for all $\gamma \in \Gamma_0(N)$,
(2) Fix a point $z \in \mathfrak{H}$. The map $\gamma \mapsto c_f(\gamma) = \int_z^{\gamma(z)} \omega(f) \in L(k - 2, \psi; \mathbb{C})$ is a 1-cocycle of $\Gamma_0(N)$,
(3) The cohomology class of c_f in $H^1(\Gamma_0(N), L(k-2, \psi; \mathbb{C}))$ is independent of the choice of z.

For a congruence subgroup Γ of $\mathrm{PGL}_2(\mathbb{Z})$ and a Γ-module M, we define the *cuspidal cohomology* group by

$$H^1_!(\Gamma, M) := \mathrm{Ker}(H^1(\Gamma, M) \xrightarrow{\Pi_s \mathrm{Res}} \prod_{s \in S} H^1(\Gamma_s, M)),$$

$$H^2_!(\Gamma, M) := \frac{\{2\text{-cocycles with values in } M\}}{\{\partial u | u|_{\Gamma_s}(\pi) = (\pi - 1)m \; \exists m \in M, \forall \pi \in \Gamma_s, \forall s \in S\}}, \qquad (4.21)$$

where S is the set of cusps of Γ and Γ_s is the stabilizer in Γ of $s \in S$ and $u : \Gamma \to M$ is any map with ∂u as in (4.20).

If we use sheaf cohomology group associated to the locally constant sheaf $\mathcal{M}$ over the complex analytic space $Y := \Gamma \backslash \mathfrak{H}$ corresponding to M (see §4.2.5 for the definition of $\mathcal{M}$), we can define, denoting H^1_c for compactly supported sheaf cohomology group,

$$H^1_!(Y, \mathcal{M}) := \mathrm{Im}(H^1_c(Y, \mathcal{M}) \to H^1(Y, \mathcal{M})), \; H^2_! = H^2_c(Y, \mathcal{M})$$

$$\text{and } H^q_!(Y, \mathcal{M}) \cong H^q_!(\Gamma, M) \text{ (canonically) if } Y \text{ is smooth.} \qquad (4.22)$$

Here Y is smooth if and only if Γ is torsion-free. The cuspidal cohomology is called parabolic cohomology and is denoted by H^q_P in [IAT, Chapter 8]. It is also called interior cohomology sometimes. The isomorphism in (4.22) is well known (see [LFE, Theorem A.1 and Propostion A.4] for a proof).

In this book, we take the following well-known theorem of Eichler and Shimura for granted (with a very brief sketch of a proof at the end of the following subsection; see [IAT, Chapter 8] or [LFE, Chapter 6] for different proofs):

Theorem 4.2.8. *For a congruence subgroup* Γ *with* $\Gamma_1(N) \subset \Gamma \subset \Gamma_0(N)$, *by* $f \mapsto [c_f]$, *we get isomorphisms:*

$$M_{k,\psi}(\Gamma) \oplus \overline{S}_{k,\psi^{-1}}(\Gamma) \cong H^1(\Gamma, L(k - 2, \psi; \mathbb{C}))$$

$$S_{k,\psi}(\Gamma) \oplus \overline{S}_{k,\psi^{-1}}(\Gamma) \cong H^1_!(\Gamma, L(k - 2, \psi; \mathbb{C})).$$

These isomorphisms are equivariant under the Hecke operators (will be recalled in §4.2.7 for cohomology groups).

Exercise 4.2.12. If $k = 2$, check that the above theorem is just a reinterpretation of Proposition 3.2.4.

4.2.5 Betti cohomology

We shall give a slight generalization of the argument given in §4.2.2 for Betti cohomology groups to facilitate tools for a sketch of a proof of Theorem 4.2.8. Let $\mathfrak{H} := \{z \in \mathbb{C} \,|\, \mathrm{Im}(z) = \frac{z-\bar{z}}{2i} > 0\}$ on which $\alpha = \left(\begin{smallmatrix} a & b \\ c & d \end{smallmatrix}\right) \in \mathrm{GL}_2^+(\mathbb{R})$ acts by $z \mapsto \alpha(z) = \frac{az+b}{cz+d}$. For a congruence subgroup Γ of $\mathrm{PSL}_2(\mathbb{Z})$, the compactified modular curve $X(\Gamma)$ whose complex points is given by $\Gamma \backslash \mathfrak{H} \sqcup \mathbf{P}^1(\mathbb{Q})$ has cusps given by $S = S_\Gamma = \Gamma \backslash \mathbf{P}^1(\mathbb{Q})/\Gamma_\infty$ as in §2.4. Then $Y(\Gamma)(\mathbb{C}) = X(\Gamma)(\mathbb{C}) - S = \Gamma \backslash \mathfrak{H}$. Thus $\pi_1(Y) \cong \Gamma$.

Write $Y = Y(\Gamma)(\mathbb{C})$. For a discrete Γ-module M, we define a manifold $M_\Gamma := \Gamma \backslash (\mathfrak{H} \times M)$ (equipped with the quotient topology of the product topology of $\mathfrak{H} \times M$) which in an obvious way gives a covering $\pi : M_\Gamma = \Gamma \backslash (\mathfrak{H} \times M) \to Y$.

Consider the sheaf $\mathcal{M}$ associated to the covering. Thus for an open subset U of Y, $\mathcal{M}(U)$ is made of continuous functions $f : U \to M_\Gamma$ such that $\pi(f(u)) = u$ for all $u \in U$. If U is simply connected, the pull-back $\pi^{-1}U$ is just the product $U \times M$; so, we have just $\mathcal{M}(U) = M$. For a closed subset C, we define $\mathcal{M}(C) = \varinjlim_{C \subset U : \text{open}} \mathcal{M}(U)$ with respect to the restriction map $\mathcal{M}(U) \ni s \mapsto s|_V \in \mathcal{M}(V)$ $(U \supset V)$.

Since the simplex Δ^n is simply connected, for a C^∞-map $\phi : \Delta^n \to Y$, $\mathcal{M}(\phi(\Delta^n)) = M$; so, we can think of n-chains which are a linear combination of ϕ with coefficients in $\mathcal{M}(\phi(\Delta^n))$. In this way, we define $C_n(Y; \mathcal{M})$ and $C_n(\overline{Y}, \partial\overline{Y}; \mathcal{M})$ as in Subsections 3.1.1 and 3.1.3. In this way, we have simplicial complexes

$$0 \to C_2(Y; \mathcal{M}) \xrightarrow{\partial} C_1(Y; \mathcal{M}) \xrightarrow{\partial} C_0(Y; \mathcal{M}) \xrightarrow{\deg} M \to 0,$$

$$0 \to C_2(\overline{Y}, \partial\overline{Y}; \mathcal{M}) \xrightarrow{\partial} C_1(\overline{Y}, \partial\overline{Y}; \mathcal{M}) \xrightarrow{\partial} C_0(\overline{Y}, \partial\overline{Y}; \mathcal{M}) \xrightarrow{\deg} M \to 0.$$

$$\tag{4.23}$$

If M is A-free of finite rank, we define $M^* = \mathrm{Hom}_A(M, A)$ on which $\gamma \in \Gamma$ acts by $\gamma\phi(m) = \phi(\gamma^\iota m)$. Then we have cochain complexes by applying A-duality to (4.23) replacing M by M^* (and $\mathcal{M}$ by $\mathcal{M}^*$ associated to M^*). Thus we have homology groups and cohomology groups: $H_q(Y, \mathcal{M})$, $H^q(Y, \mathcal{M})$, $H_q(\overline{Y}, \partial\overline{Y}, \mathcal{M})$ and $H_c^q(Y, \mathcal{M}) = H^q(\overline{Y}, \partial\overline{Y}, \mathcal{M})$ for $q = 0, 1, 2$. The cohomology group $H_c^q(Y, \mathcal{M})$ is called the cohomology group of compact support with coefficients in $\mathcal{M}$. We quote [IAT, Proposition 8.2] (see also [LFE, Proposition 6.1.1]):

Proposition 4.2.9. *For any Γ-module M, writing $DM = \sum_{\gamma \in \Gamma}(\gamma-1)M$, we have $H_c^2(Y, \mathcal{M}) \cong H_1^2(\Gamma, M) = M/DM$.*

The cohomology group $H_1^2(\Gamma, M)$ is computed by simplicial complex, and for the 2-simplicial Δ obtained from $\Gamma \backslash \mathfrak{H}$ cut by the basis (α_i, β_i) and π_j, the association sending each 2-simplicial cocycle u to the value $u(\Delta) \in M/DM$ induces the isomorphism.

Similar to Remark 3.1.5, if M is an $\mathbb{R}$ vector space, we have de Rham complexes:

$$0 \to M \to \Omega^0(Y;\mathcal{M}) \xrightarrow{d} \Omega^1(Y;\mathcal{M}) \xrightarrow{d} \Omega^2(Y;\mathcal{M}) \to 0,$$
$$0 \to M \to \Omega^0(\overline{Y}, \partial\overline{Y};\mathcal{M}) \xrightarrow{d} \Omega^1(\overline{Y}, \partial\overline{Y};\mathcal{M}) \xrightarrow{d} \Omega^2(\overline{Y}, \partial\overline{Y};\mathcal{M}) \to 0,$$

(4.24)

and we have de Rham cohomology groups:

$$H^q_{DR}(Y, \mathcal{M}) \quad \text{and} \quad H^q_{DR}(\overline{Y}, \partial\overline{Y}, \mathcal{M})$$

for $q = 0, 1, 2$. We have de Rham isomorphisms:

$$H^q_{DR}(Y, \mathcal{M}) \cong H^q(Y, \mathcal{M}), \quad \text{and} \quad H^q_{DR}(\overline{Y}, \partial\overline{Y}, \mathcal{M}) \cong H^q(\overline{Y}, \partial\overline{Y}, \mathcal{M}).$$

(4.25)

Sketch of a proof of Theorem 4.2.8. The sheaf cohomology and group cohomology of the fundamental group are equivalent (e.g., [LFE, Appendix]). Applying this, we may replace group cohomology by de Rham cohomology in Theorem 4.2.8. Then the association $f \mapsto \omega(f)$ (and its complex conjugate) gives an isomorphism via standard harmonic analysis as in [DS52]. This proves the formula between the space of cusp forms and the cuspidal cohomology $H^1_!$. This method is used in [LFE, Theorem 6.2.1]. For the space of modular forms, the identity follows from the analysis of the space of Eisenstein series and the boundary cohomology group $\bigoplus_{s \in S} H^1(\Gamma_s, L(k-2; \mathbb{C}))$. $\qquad\qquad\square$

The proof in [IAT, Chapter 8] goes as follows. The injectivity of the map comes from the compatibility of Poincaré self-duality on the cohomology side and Petersson inner product on the space of cusp forms. Then by the Euler characteristic formula, one gets a dimension formula of the cohomology, which coincides with (twice of) the dimension of the space of cusp forms (proven by Riemann-Roch theorem).

4.2.6 *Duality of cohomology groups*

We insert here a sketch of Poincaré duality in sheaf and group cohomology. Let $\Gamma \subset SL_2(\mathbb{Z})$ be a subgroup of finite index and write $Y := \Gamma\backslash\mathfrak{H}$ and $X := \Gamma\backslash(\mathfrak{H} \cup \mathbf{P}^1(\mathbb{Q}))$ for its smooth compactification. We write $\mathcal{L}(n, \psi; A)$ for the locally constant sheaf on Y associated to $L(n, \psi; A)$. We have a self $\mathbb{Z}$-linear pairing $[\cdot, \cdot] : L(1; \mathbb{Z}) \otimes_{\mathbb{Z}} L(1; \mathbb{Z}) \to \mathbb{Z}$ given by $[aX + bY, cX + d] = (ad - bc)$. Since $L(n; \mathbb{Z})$ is isomorphic to the n-th symmetric tensor power of $L(1, \mathbb{Z})$ as Γ-modules by $X^{n-j}Y^j \leftrightarrow X^{\otimes n-j} \otimes Y^{\otimes j}$, this pairing induces a pairing $[\cdot, \cdot] : L(n; \mathbb{Z}) \otimes_{\mathbb{Z}} L(n; \mathbb{Z}) \to \mathbb{Z}$ satisfying

$$[(aX + bY)^n, (cX + dY)^n] = (ad - bc)^n.$$

(4.26)

Note that $[\alpha P, Q] = [P, \alpha^\iota Q]$ for $P, Q \in L(n; \mathbb{Z})$ and $\alpha \in M_2(\mathbb{Z}) \cap \mathrm{GL}_2(\mathbb{Q})$ and $[\gamma P, \gamma Q] = [P, Q]$ for $\gamma \in \mathrm{GL}_2(\mathbb{Z})$.

Twisting by ψ, we have a pairing $L(n, \psi; \mathbb{Z}[\psi]) \times L(n, \psi^{-1}; \mathbb{Z}[\psi]) \to \mathbb{Z}[\psi]$ and sheaf pairing $\mathcal{L}(n, \psi; \mathbb{Z}[\psi]) \times \mathcal{L}(n, \psi^{-1}; \mathbb{Z}[\psi]) \to \mathbb{Z}[\psi]$ which we still denote $[\cdot, \cdot]$. We regard $L(n, \psi; \mathbb{Z}[\psi])$ as a $\Delta_0(Np^r)$-module (resp. $\Delta_0(Np^r)^\iota$-module) by $\alpha P(X, Y) = \psi^{-1}(a) P((X, Y)^t \alpha^\iota)$ and $\alpha^\iota P((X, Y)) = \psi(a) P((X, Y)^t \alpha)$ for $\alpha = \left(\begin{smallmatrix} a & * \\ * & * \end{smallmatrix} \right) \in \Delta_0(Np^r)$. In $\Gamma_0(Np^r) \subset \Delta_0(Np^r) \cap \Delta_0(Np^r)^\iota$, the two actions coincide. By extending scalars, the pairing

$$[\cdot, \cdot] : L(n, \psi; A) \otimes_A L(n, \psi^{-1}; A) \to A$$

is well defined for any $\mathbb{Z}[\psi]$-algebra (and module) A. It is easy to check that this pairing is $\Delta_0(Np^r)$-equivariant in the following sense: $[\alpha P, Q] = [P, \alpha^\iota Q]$ for $\alpha \in \Delta_0(Np^r)$ (so, $[\gamma P, \gamma Q] = [P, Q]$ for $\gamma \in \Gamma_0(Np^r)$). Then by the cup product, we have an A-bilinear morphism

$$H_c^q(Y, \mathcal{L}(n, \psi; A)) \otimes_A H^{2-q}(Y, \mathcal{L}(n, \psi^{-1}; A))$$
$$\xrightarrow{\cup} H_c^2(Y, \mathcal{L}(n, \psi; A) \otimes_A \mathcal{L}(n, \psi^{-1}; A)).$$

Composing this bilinear map with the A-linear map induced by the pairing $[\cdot, \cdot]$:

$$H_c^2(Y, \mathcal{L}(n, \psi; A) \otimes_A \mathcal{L}(n, \psi^{-1}; A)) \xrightarrow{[\cdot, \cdot]} H_c^2(Y, A) \cong A,$$

we get the Poincaré duality pairing:

$$H_c^q(Y, \mathcal{L}(n, \psi; A)) \otimes_A H^{2-q}(Y, \mathcal{L}(n, \psi^{-1}; A)) \xrightarrow{[\cdot, \cdot]} A. \qquad (4.27)$$

Similarly, writing $L^*(n, \psi; W)$ for the dual W-lattice in $L(n, \psi; K)$ of $L(n, \psi^{-1}; W)$ under $[\cdot, \cdot]$ and $\mathcal{L}^*(n, \psi; W)$ for the corresponding locally constant sheaf on Y, we define

$$\mathcal{L}^*(n, \psi; K/W) := \mathcal{L}^*(n, \psi; W) \otimes_W K/W.$$

Then for $X = W, K/W$, we have a compact-discrete pairing

$$H_c^q(Y, \mathcal{L}(n, \psi; X)) \otimes_A H^{2-q}(Y, \mathcal{L}^*(n, \psi^{-1}; X^\vee)) \xrightarrow{[\cdot, \cdot]} K/W. \qquad (4.28)$$

We still write these pairings as $[\cdot, \cdot]$. Since the natural maps

$$H_c^1(Y, \mathcal{L}(n, \psi; A)) \to H_!^1(Y, \mathcal{L}(n, \psi; A)) \quad \text{and}$$
$$H_!^1(Y, \mathcal{L}(n, \psi^{-1}; A)) \to H^1(Y, \mathcal{L}(n, \psi; A))$$

are dual to each other under $[\cdot, \cdot]$, this induces a pairing of cuspidal cohomology groups:

$$H_!^1(Y, \mathcal{L}(n, \psi; A)) \otimes_A H_!^1(Y, \mathcal{L}(n, \psi^{-1}; A)) \xrightarrow{[\cdot, \cdot]} A. \qquad (4.29)$$

Here are some known facts (see [CGP, VIII.10] and [STH, V.9]):

Proposition 4.2.10.

(1) *If* Γ *is torsion-free, (4.28) gives Pontryagin duality.*
(2) *For any coefficient ring A, if $n = 0$, the pairings (4.27) and (4.29) are perfect.*
(3) *If A is a field of characteristic 0, the pairings (4.27) and (4.29) are perfect.*

The assertion (2) is the classical Poincaré duality.

4.2.7   Hecke operator on cohomology groups

Let M be $A[\Delta_0(N)^\iota]$-module. Let $\Delta = \Gamma_0(N)$. Then in particular M is a Δ-module, because $\Delta \subset \Delta_0(N)^\iota$. For any left coset decomposition $\Delta\alpha\Delta = \bigsqcup_j \Delta\alpha_j$, we have $\alpha_j\gamma = \gamma_j\alpha_{\gamma^*(j)}$ with $\gamma_j \in \Delta$ for a permutation $j \mapsto \gamma^*(j)$ of indices. Then for a 1-cocycle c: we define a map $c|[\Delta\alpha\Delta](\gamma) = \sum_j \alpha_j^\iota c(\gamma_j)$.

Lemma 4.2.11. *If $c : \Delta \to M$ is a 1-cocycle, $c|[\Delta\alpha\Delta]$ is 1-cocycle, and the cohomology class of $c|[\Delta\alpha\Delta]$ is uniquely determined by the cohomology class of c. In other words, $[\Delta\alpha\Delta] \in \mathrm{End}_A(H^1(\Delta, M))$.*

Proof. Let $\gamma, \delta \in \Delta$, and write $c' = c|[\Delta\alpha\Delta]$. Then $\alpha_j\gamma\delta = \gamma_j\alpha_{\gamma^*(j)}\delta = \gamma_j\delta_{\gamma^*(j)}\alpha_{\delta^*\gamma^*(j)}$. Note that $\alpha_j\gamma = \gamma_j\alpha_{\gamma^*(j)} \Leftrightarrow \gamma_j^{-1}\alpha_j = \alpha_{\gamma^*(j)}\gamma^{-1}$. We do a computation to verify that c' is a cocycle:

$$c'(\gamma\delta) = \sum_j \alpha_j^\iota c(\gamma_j\delta_{\gamma^*(j)}) = \sum_j \{\alpha_j^\iota\gamma_j c(\delta_{\gamma^*(j)}) + \alpha_j^\iota c(\gamma_j)\}$$

$$= \sum_j (\gamma_j^{-1}\alpha_j)^\iota c(\delta_{\gamma^*(j)}) + c'(\gamma)$$

$$= \sum_j (\alpha_{\gamma^*(j)}\gamma^{-1})^\iota c(\delta_{\gamma^*(j)}) + c'(\gamma) \stackrel{\gamma^*(j) \mapsto j}{=} \gamma c'(\delta) + c'(\gamma).$$

If $c(\gamma) = (\gamma - 1)m$ for $m \in M$, again by $\gamma_j^{-1}\alpha_j = \alpha_{\gamma^*(j)}\gamma^{-1}$, we have

$$c'(\gamma) = \sum_j \alpha_j^\iota(\gamma_j - 1)m = \sum_j (\gamma_j^{-1}\alpha_j)^\iota m - \sum_j \alpha_j^\iota m$$

$$= \sum_j (\alpha_{\gamma^*(j)}\gamma^{-1})^\iota m - \sum_j \alpha_j^\iota m = \gamma \sum_j \alpha_{\gamma^*(j)}^\iota m - \sum_j \alpha_j^\iota m$$

$$= (\gamma - 1) \sum_j \alpha_j^\iota m.$$

Thus $c \mapsto c|[\Delta\alpha\Delta]$ preserves the subspace of 1-coboundaries. Thus it induces a unique linear endomorphism of $H^1(\Delta, M)$. $\qquad\qquad\square$

We can let $\delta = \begin{pmatrix} a & b \\ c & d \end{pmatrix} \in \Delta_0(N)^\iota$ on $L(k-2, \psi; A)$ by $\delta \cdot P(X, Y) = \psi(d)P((X,Y)^t\delta^\iota)$, and this action coincides with the original action of $\Gamma_0(N)$ on $\Gamma_0(N) \subset \Delta_0(N)^\iota$. Thus by the above lemma, we have a well defined Hecke operator $T(n)$ on $H^1(\Gamma_0(N), L(k-2, \psi; A))$. In other words, splitting $T_n = \bigsqcup_\alpha \Delta\alpha\Delta$, define $T(n) = \sum_\alpha [\Delta\alpha\Delta]$.

Exercise 4.2.13. Check that the isomorphism of Theorem 4.2.8 is equivariant under Hecke operators $T(n)$.

The above action of $\Delta_0(N)^\iota$ induces an action of $\Delta_1(N)^\iota$ on $L(k-2, \psi; A)$. Since ψ is modulo N, the action of $\Delta_1(N)^\iota$ does not depend on ψ; so, we write simply $L(k-2; A)$ for this $A[\Delta_1(N)^\iota]$-module $L(k-2, \psi; A)$. Then in the same manner as above, replacing $\Delta_0(N)$ by $\Delta_1(N)$, we have a natural action of $T(n)$ on $H^1(\Gamma_1(N), L(k-2; A))$.

We can generalize a bit the above definition. Let $\Delta, \Delta' \subset \Delta_0(N)^\iota$ be two subgroups of finite index. Decomposition $\Delta\alpha\Delta' = \bigsqcup_j \Delta\alpha_j$ and defining $\alpha_j\gamma' = \gamma_j\alpha_{\gamma'^*(j)}$ for $\gamma' \in \Delta'$ and $\gamma_j \in \Delta$, we can define $c|[\Delta\alpha\Delta'](\gamma') = \sum_j \alpha_j^\iota c(\gamma_j)$. This linear map induces, by the same proof as above,

$$[\Delta\alpha\Delta'] : H^1(\Delta, M) \to H^1(\Delta', M). \qquad (4.30)$$

If $\Delta \subset \Delta'$, the double coset $\Delta' = \Delta 1 \Delta'$ induces a map $[\Delta'] : H^1(\Delta, M) \to H^1(\Delta', M)$ which is called a trace map $\mathrm{Tr} = \mathrm{Tr}_{\Delta'/\Delta}$.

Exercise 4.2.14. Define the restriction map Res $: H^1(\Delta', M) \to H^1(\Delta, M)$ by restricting cocycle of Δ' to Δ (if $\Delta' \supset \Delta$). Prove

$$\mathrm{Tr} \circ \mathrm{Res}([c]) = [\Delta' : \Delta] \cdot [c].$$

Using this fact, prove that $H^1(G, M) = 0$ for a finite group G and a finite G-module M if $|G|$ is prime to $|M|$.

The fact $H^q(G, M) = 0$ for a finite group G and a finite G-module M with $(|G|, |M|) = 1$ is true in general (and one can give a similar proof as above using a higher degree version of Tr and Res).

We can define in the same way the double coset action on $H^q(\Delta, M)$ and $H^q(\Delta', M)$. For example, if c is a two cocycle and $\Delta\alpha\Delta' = \bigsqcup_j \Delta\alpha_j$, then defining $\alpha_j\delta = \delta_j\alpha_i$ and $\alpha_j\delta' = \delta'_j\alpha_{i'}$ for some i, i' and $\delta_j, \delta'_j \in \Delta$, we define $c'(\delta, \delta') = \sum_j \alpha_j^\iota c(\delta_j, \delta'_j)$, and this action $c \mapsto c'$ preserves coboundaries. So we have a linear map $[\Delta\alpha\Delta'] : H^2(\Delta, M) \to H^2(\Delta', M)$.

In the geometric setting of open or compact Riemann surfaces associated to these groups, writing $Y(\Delta)$ for the surface corresponding to Δ, we have the two natural projection

$$p_L : Y(\alpha^{-1}\Delta\alpha \cap \Delta') \to Y(\Delta') \quad \text{and} \quad p_R : Y(\Delta \cap \alpha\Delta'\alpha^{-1}) \to Y(\Delta).$$

Then on the sheaf cohomology level, we have $[\Delta\alpha\Delta'] = p_{L,*} \circ \alpha \circ p_R^*$. This shows, on $H^2(Y(\Delta), C)$ for a constant sheaf C, the action of $[\Delta\alpha\Delta]$ is given as the multiplication by $\deg([\Delta\alpha\Delta])$.

We insert here a definition. Write $\mathcal{L}(n, \psi; A)$ for the locally constant sheaf over the (open) modular curve $Y(\Delta)$ associated to the Γ-module $L(n, \psi; A)$ defined in §4.2.5. Since Tr and Res is well defined operation on sheaf cohomology, we can define the double coset action on sheaf cohomology $H^1(Y(\Delta), \mathcal{L}(n, \psi; A))$. Then we can define $[\Delta\alpha\Delta] \in H^1(Y(\Delta), \mathcal{L}(n, \psi; A))$ by $p_{L,*} \circ \alpha \circ p_R^*$. This action is compatible with the group cohomological operator under the natural isomorphism $H^1(Y(\Delta), \mathcal{L}(n, \psi; A)) \cong H^1(\Delta, L(n, \psi; A))$.

If A is p-profinite and $\Gamma_1(Np^r) \subset \Gamma \subset \Gamma_0(Np^r)$ with $r > 0$, by Lemma 4.1.18, the projector $e = \lim_{n\to\infty} U(p)^{n!}$ is well defined on $H^q_?(\Gamma, L(k-2, \psi; A))$ (for $? = c, !$ or nothing). We indicate by $H^q_{?,ord}$ the image of e of the cohomology group. Since $\deg(U(p)) = p$, the operator $U(p)$ is topologically nilpotent on $H^2_!(\Gamma, L(k-2, \psi; A))$ for a p-profinite ring (or module) A, we get the special case of the following fact:

Proposition 4.2.12. *Suppose* $\Gamma_1(Np^r) \subset \Gamma \subset \Gamma_0(Np^r)$ *for* $Np^r > 2$ *with* $r > 0$. *Let* $q = 0, 2$ *and* $A = W, K, K/W, W/p^rW$. *If either* $r > 0$ *or* $n > 0$ *or* $\psi \not\equiv 1 \bmod \mathfrak{m}_W$, *we have* $H^0_{ord}(\Gamma, L(n, \psi; A)) = H^2_{!,ord}(\Gamma, L(n, \psi; A)) = 0$. *If* $r = n = 0$ *and* $\psi = 1$, *we have* $H^0_{ord}(\Gamma; A)) \cong H^2_{!,ord}(\Gamma, L(n; A)) \cong A$.

Proof. First assume $r > 0$. Write $L(A) = L(n, \psi; W)$. Define an operator $U : L(W) \to L(W)$ by $P(X, Y) \mapsto \sum_{u=0}^{p-1} \alpha_u^\iota P(X, Y)$ for $\alpha_u = \left(\begin{smallmatrix} 1 & u \\ 0 & p \end{smallmatrix}\right)$. Since $U(P(X, Y)) \equiv \sum_{u=0}^{p-1}(X + uY)^n \bmod pL$ and $\sum_{u=0}^{p-1} u^j \equiv 0 \bmod p$ if $j > 0$, $E = \lim_{n\to\infty} U^{n!}$ projects down $L(W)$ into KX^n. Note that α_u acts trivially on $L(n, \psi; K)/YL(n-1, \psi; K) \cong KX^n$, $X^n|U = pX^n$ on the quotient $L(n, \psi; K)/YL(n-1, \psi; K)$, and hence $E(L(n, \psi; K)) = 0$. For any W-lattice L stable under the action of $\Delta_0(N)^\iota$ in $L(n, \psi; K)$, we have $H^0_{ord}(\Gamma, L) \subset H^0_{ord}(\Gamma, L(n, \psi; K)) = 0$. Taking $L := L(W)$, we have $L(n, \psi; A) = L \otimes_W A$ for $A = W, K, K/W, W/p^rW$, and hence we obtain the assertion for H^0_{ord} in the proposition.

The above argument works well also for another operator $V(P(X, Y)) = {}^t\alpha_u^\iota P(X, Y)$, replacing X^n by Y^n. We get $F = \lim_{n\to\infty} V^{n!}$ which kills $H^0(\Gamma, L)$.

As already remarked, the pairing in (4.26) is $\Delta_0(Np^r)$-equivariant: $[\alpha P, Q] = [P, \alpha^\iota Q]$. Taking the dual lattice L^* of L under $[\cdot, \cdot]$, $L^A :=$ $L \otimes_W A$ is the Pontryagin dual of $L_A = L^* \otimes_W A^\vee$ for the Pontryagin dual $A^\vee$ of A. By Proposition 4.2.10, this induces a Pontryagin duality between $L_A^\Gamma = H^0(\Gamma, L_A)$ and $L^A / \sum_{\gamma \in \Gamma}(\gamma - 1)L^A = H_0(\Gamma, L^A) \cong H_!^2(\Gamma, L^A)$ (see also Proposition 4.2.9). The operator F killing $H^0(\Gamma, L_A)$ is the dual operator of E on $H_!^2(\Gamma, L^A)$; so, we get $H_{!,ord}^2(\Gamma, L^A) = 0$.

We now treat the case $r = 0$ and $n > 0$. Consider for $\alpha = \left(\begin{smallmatrix} p & 0 \\ 0 & 1 \end{smallmatrix}\right)$, we consider $T(P(X,Y)) = \alpha^\iota P(X,Y) + U(P(X,Y))$. Since $\alpha\alpha_u \equiv 0$ mod p and $(\alpha_u\alpha)^2 \equiv 0$ mod p, we find $e_0 L(n, \psi; K) \subset KX^n + KY^n$ for $e_0 = \lim_{n \to \infty} T^{n!}$. However $T(X^n) \equiv \sum_u (X + uY)^n \equiv 0 \bmod p$, we find $e_0 KX^n = 0$. Similarly $e_0 Y^n = 0$. Thus $H_{ord}^0(\Gamma, L(n, \psi; A)) = 0$ in this case, and by the duality argument, we find $H_{!,ord}^2(\Gamma, L(n, \psi; A)) = 0$.

If $r = n = 0$ and $\psi = 1$, we find $T \equiv 1 \bmod p$, and hence $H_{ord}^0(\Gamma, L(n; A)) = A$, and by duality, we find $H_{!,ord}^2(\Gamma, L(n; A)) = A$. $\square$

Corollary 4.2.13. *Let the notation and the assumption be as in Proposition 4.2.12. Then $H_{ord}^1(\Gamma, L(n, \psi; W))$ and $H_{!,ord}^1(\Gamma, L(n, \psi; W))$ are W-free of finite rank.*

Proof. By cohomology sequence attached to $L(n, \psi; W) \hookrightarrow L(n, \psi; K) \twoheadrightarrow L(n, \psi; K/W)$, we have an exact sequence

$$H_{ord}^0(\Gamma, L(n, \psi; K/W)) \to H_{ord}^1(\Gamma, L(n, \psi; W)) \to H_{ord}^1(\Gamma, L(n, \psi; K)).$$

Since $H_{ord}^0(\Gamma, L(n, \psi; K/W)) = 0$ by Proposition 4.2.12, we find $H_{ord}^1(\Gamma, L(n, \psi; W))$ is embedded into the finite dimensional K-vector space $H_{ord}^0(\Gamma, L(n, \psi; K))$; so, it is W-free of finite rank. Since $H_{!,ord}^1 \subset H_{ord}^1$ by definition, the above assertion also proves W-freeness of $H_{ord}^1(\Gamma, L(n, \psi; W))$. $\square$

4.2.8 *p-Hecke operator and level lowering*

Fix N prime to p. Write $\mathbf{p} := 4$ if $p = 2$ and $\mathbf{p} = p$ if $p > 2$. Take a subgroup Γ of $SL_2(\mathbb{Z})$ with $\Gamma_1(Np) \subset \Gamma \subset \Gamma_0(Np)$. For a closed subgroup H of $\mathbb{Z}_p^\times$, we define H_n for the image of H in $(\mathbb{Z}/p^n\mathbb{Z})^\times$ for $n > 0$. If H is open, we define $\ell(H)$ to be the minimal positive integer ℓ such that $1 + p^{\ell-1}\mathbf{p}\mathbb{Z}_p \subset H$; so, $\ell(H) > 0$. In this subsection, we assume that H is open. Define for $r > 0$

$$\Gamma_H(p^r) := \left\{ \left(\begin{smallmatrix} a & b \\ c & d \end{smallmatrix}\right) \in \Gamma \cap \Gamma_0(p^r) \,|\, (a \bmod p^r) \in H_r \right\}. \tag{4.31}$$

Lemma 4.2.14. *For* $m \geq r \geq \ell(H)$ *and* $s > 0$, *we have*

$$\Gamma_H(p^r) \begin{pmatrix} 1 & 0 \\ 0 & p^s \end{pmatrix} \Gamma_H(p^r) = \Gamma_H(p^m) \begin{pmatrix} 1 & 0 \\ 0 & p^s \end{pmatrix} \Gamma_H(p^r)$$

$$= \Gamma_H(p^m) \begin{pmatrix} 1 & 0 \\ 0 & p^s \end{pmatrix} \Gamma_H(p^m) = \bigsqcup_{u=0}^{p^s-1} \Gamma_H(p^m) \begin{pmatrix} 1 & u \\ 0 & p^s \end{pmatrix}.$$

Proof. Let $\alpha = \begin{pmatrix} 1 & 0 \\ 0 & p^s \end{pmatrix}$. Since $\alpha^{-1} \begin{pmatrix} a & b \\ c & d \end{pmatrix} \alpha = \begin{pmatrix} a & bp^s \\ c/p^s & d \end{pmatrix}$, we have for $m' = \max(m - s, r)$

$$\begin{pmatrix} a & b \\ c & d \end{pmatrix} \in \alpha^{-1}\Gamma_H(p^m)\alpha \cap \Gamma_H(p^r)$$

$$\Leftrightarrow (a \bmod p^{m'}) \in H_{m'}, \ c \equiv 0 \ \bmod p^{m'} \text{ and } b \equiv 0 \ \bmod p^s$$

$$\overset{(*)}{\Longleftrightarrow} (a \bmod p^{m'}) \in H, \ c \equiv 0 \ \bmod p^{m'} \text{ and } b \equiv 0 \ \bmod p^s \quad (4.32)$$

where the second equivalence $(*)$ holds only under $r \geq \ell(H)$ (the first equivalence is true without this assumption). Thus we have

$$\Gamma_H(p^r) = \bigsqcup_{u=0}^{p^s-1} (\alpha^{-1}\Gamma_H(p^m)\alpha \cap \Gamma_H(p^r)) \begin{pmatrix} 1 & u \\ 0 & 1 \end{pmatrix}.$$

Multiplying $\alpha^{-1}\Gamma_H(p^m)\alpha$ from the left side, we have

$$\alpha^{-1}\Gamma_H(p^m)\alpha\Gamma_H(p^r) = \bigsqcup_{u=0}^{p^s-1} \alpha^{-1}\Gamma_H(p^m)\alpha \begin{pmatrix} 1 & u \\ 0 & 1 \end{pmatrix}.$$

Without $r \geq \ell(H)$, we can have more cosets in the right-hand-side coming from $(1 + \ell^r \mathbb{Z}_p)H/H$. This implies

$$\Gamma_H(p^m)\alpha\Gamma_H(p^r) = \bigsqcup_{u=0}^{p^s-1} \Gamma_H(p^m)\alpha \begin{pmatrix} 1 & u \\ 0 & 1 \end{pmatrix} \qquad \text{(U)}$$

as desired. $\square$

We write $U(p^n)$ for the operator $[\Gamma_H(p^r) \begin{pmatrix} 1 & 0 \\ 0 & p^n \end{pmatrix} \Gamma_H(p^r)]$. If we need to indicate the p-power level, we write $U(p^n) = U_r(p^n) := [\Gamma_H(p^r) \begin{pmatrix} 1 & 0 \\ 0 & p^n \end{pmatrix} \Gamma_H(p^r)]$ and $U_{m,r}(p^n) = [\Gamma_H(p^m) \begin{pmatrix} 1 & 0 \\ 0 & p^n \end{pmatrix} \Gamma_H(p^r)]$ for $m \geq r$. Restricting cocycle $c : \Delta' \to M$ to a subgroup $\Delta \subset \Delta'$, we get the restriction morphism $\text{Res} : H^1(\Delta', M) \to H^1(\Delta, M)$. Since $\begin{pmatrix} 1 & u_1 \\ 0 & p \end{pmatrix} \begin{pmatrix} 1 & u_0 \\ 0 & p \end{pmatrix} = \begin{pmatrix} 1 & u_0 + pu_1 \\ 0 & p^2 \end{pmatrix}$, iterating this process, we get

$$\overbrace{\begin{pmatrix} 1 & u_n \\ 0 & p \end{pmatrix} \cdots \begin{pmatrix} 1 & u_0 \\ 0 & p \end{pmatrix}}^{n+1} = \begin{pmatrix} 1 & u_0 + u_1 p + \cdots + u_n p^n \\ 0 & p^{n+1} \end{pmatrix}$$

and hence

$$U(p^n) = U(p)^n. \tag{4.33}$$

By Lemma 4.1.8, $M_{k,\psi}(\Gamma_0(Np^m); A)$ is sent to $M_{k,\psi}(\Gamma_0(Np^r); A)$ by $U_{m,r}(p^{m-r})$. Applying Lemma 4.2.14, we get the following commutative diagrams.

Corollary 4.2.15. *If $m \geq r \geq \ell(H)$, the following two diagrams are commutative for $q = 0, 1, 2$:*

$$
\begin{array}{ccc}
H^q(\Gamma_H(p^r), L(k-2; A)) & \xrightarrow{\mathrm{Res}} & H^q(\Gamma_H(p^m), L(k-2; A)) \\
{\scriptstyle U_r(p^{m-r})}\downarrow & \nearrow{\scriptstyle U_{m,r}(p^{m-r})} & \downarrow{\scriptstyle U_m(p^{m-r})} \\
H^q(\Gamma_H(p^r), L(k-2; A)) & \xrightarrow[\mathrm{Res}]{} & H^q(\Gamma_H(p^m), L(k-2; A))
\end{array}
$$

and assuming that A is a $\mathbb{Z}[\psi]$-algebra,

$$
\begin{array}{ccc}
M_{k,\psi}(\Gamma_H(p^r); A) & \xhookrightarrow{\quad} & M_{k,\psi}(\Gamma_H(p^m); A) \\
{\scriptstyle U_r(p^{m-r})}\downarrow & \nearrow{\scriptstyle U_{m,r}(p^{m-r})} & \downarrow{\scriptstyle U_m(p^{m-r})} \\
M_{k,\psi}(\Gamma_H(p^r); A) & \xhookrightarrow{\quad} & M_{k,\psi}(\Gamma_H(p^m); A).
\end{array}
$$

This implies, if $m \geq r \geq \ell(H)$

$$H^q_{\mathrm{ord}}(\Gamma_H(p^r), L(k-2; A)) \cong H^q_{\mathrm{ord}}(\Gamma_H(p^m), L(k-2; A)). \tag{4.34}$$

More generally, by the decomposition given in Lemma 3.2.7 and (U) (see also [MFM] Section 4.5), we have

Corollary 4.2.16. *Write $X_H(p^r)$ for one of $H^1(\Gamma_H(p^r), L(k-2; A))$, $M_k(\Gamma_H(p^r); A)$ and $S_k(\Gamma_H(p^r); A)$. If $m \geq r > 0$, the following diagram is commutative for all positive integer n:*

$$
\begin{array}{ccc}
X_H(p^r) & \xrightarrow{\mathrm{Res}} & X_H(p^m) \\
{\scriptstyle T(n)}\downarrow & & \downarrow{\scriptstyle T(n)} \\
X_H(p^r) & \xrightarrow{\mathrm{Res}} & X_H(p^m).
\end{array}
$$

Let A be a p-profinite ring. Then applying the p-profinite limit $e = \lim_{n\to\infty} U(p)^{n!}$, we define $H^1_{\mathrm{ord}} = eH^1$. Then we have the corresponding diagram for the ordinary part ($m \geq r > 0$):

$$
\begin{array}{ccc}
H^1_{\mathrm{ord}}(\Gamma_H(p^r), L(k-2; A)) & \xrightarrow{\mathrm{Res}} & H^1_{\mathrm{ord}}(\Gamma_H(p^m), L(k-2; A)) \\
{\scriptstyle T(n)}\downarrow & & \downarrow{\scriptstyle T(n)} \\
H^1_{\mathrm{ord}}(\Gamma_H(p^r), L(k-2; A)) & \xrightarrow{\mathrm{Res}} & H^1_{\mathrm{ord}}(\Gamma_H(p^m), L(k-2; A)).
\end{array}
$$

4.2.9   Weight independence of limit Hecke algebra

Recall the commutative A-algebra $\mathbf{H}_k(Np^r; A)$ (resp. $\mathbf{h}_k(Np^r; A)$) by the A-subalgebra of $\mathrm{End}_A(M_k(Np^r; A))$ (resp. $\mathrm{End}_A(S_k(Np^r; A))$) generated by Hecke operators $T(n)$ for all n. By Corollary 4.2.16, a Hecke operator $h \in \mathbf{H}_k(Np^m; A)$ restricted to $M_k(\Gamma_H(p^r); A)$ belongs to $\mathbf{H}_k(Np^r; A)$, getting a surjective A-algebra homomorphism $\mathbf{H}_k(Np^m; A) \twoheadrightarrow \mathbf{H}_k(Np^r; A)$ sending $T(n)$ to $T(n)$ as long as $m \geq r > 0$. Replacing the space of modular forms by the space of cusp forms, we get a surjective A-algebra homomorphism $\mathbf{h}_k(Np^m; A) \twoheadrightarrow \mathbf{h}_k(Np^r; A)$. Recall the projective limits in Definition 4.1.17:

$$\mathbf{H}_k(Np^\infty; A) = \varprojlim_m \mathbf{H}_k(Np^m; A) \text{ and } \mathbf{h}_k(Np^\infty; A) = \varprojlim_m \mathbf{h}_k(Np^m; A).$$

Since the transition map takes $U(p)$ to $U(p)$, the projector $e = \lim_{n\to\infty} U(p)^{n!}$ of level Np^m is sent to e of level Np^r. Thus we have the projector $e = \lim_{n\to\infty} U(p)^{n!}$ well defined in $\mathbf{H}_k(Np^\infty; A)$. We define $\mathbf{H}_k^{ord}(Np^\infty; A) = e(\mathbf{H}_k(Np^\infty; A))$ and $\mathbf{h}_k^{ord}(Np^\infty; A) = e(\mathbf{h}_k(Np^\infty; A))$. By (4.34) (and Eichler–Shimura isomorphism), $\mathbf{H}_k^{ord}(Np^r; W)$ is isomorphic to the W-subalgebra of $\mathrm{End}_W(H_{ord}^1(\Gamma_1(Np^r); L(k-2; W)))$ generated by Hecke operators.

We know the following facts [H86b, (1.7)]:

Theorem 4.2.17. *If $k \geq 2$, we have unique isomorphisms:*

$$\mathbf{h}_k^{ord}(Np^\infty; W) \cong \mathbf{h}_2^{ord}(Np^\infty; W) \text{ and } \mathbf{H}_k^{ord}(Np^\infty; W) \cong \mathbf{H}_2^{ord}(Np^\infty; W)$$

taking $T(n)$ to $T(n)$ for all n. More generally we have $\mathbf{H}_k(Np^\infty; W) \cong \mathbf{H}_2(Np^\infty; W)$ and $\mathbf{h}_k(Np^\infty; W) \cong \mathbf{h}_2(Np^\infty; W)$ sending $T(n)$ to $T(n)$ for all n.

We will prove this result in §4.2.13. Then we identify all

$$\mathbf{h}^{ord}(Np^\infty; W) := \mathbf{h}_2^{ord}(Np^\infty; W) \text{ and } \mathbf{H}^{ord}(Np^\infty; W) := \mathbf{H}_2^{ord}(Np^\infty; W).$$

Remark 4.2.18. Consider the p-adic supremum norm $|\sum_n a_n q^n|_p := \mathrm{Sup}_n |a_n|_p$ for q-expansions in $W[[q]]$. This induces a p-adic norm on $S_k(\Gamma_1(Np^\infty), W) := \varinjlim_r S_k(\Gamma_1(Np^r); W) = \bigcup_r S_k(\Gamma_1(Np^r); W) \subset W[[q]]$ and $M_k(\Gamma_1(Np^\infty), W) := \varinjlim_r M_k(\Gamma_1(Np^r); W)$. Let $\mathbf{S}_k(Np^\infty; W)$ (resp. $\mathbf{M}_k(Np^\infty; W)$) be the completion (or equivalently the closure in $W[[q]]$) under this norm of $S_k(\Gamma_1(Np^\infty); W)$ (resp. $M_k(\Gamma_1(Np^\infty); W)$). Plainly $\mathbf{h}_k(Np^\infty; W)$ is the subalgebra of $\mathrm{End}_W(\mathbf{S}_k(Np^\infty; W))$, and we can define the norm of $h \in \mathbf{h}_k(Np^\infty; W)$ by $\mathrm{Sup}_{f \neq 0} |f|h|_p/|f|_p$ for $f \in \mathbf{S}_k(Np^\infty; W)$.

Since $h_k(\Gamma_1(Np^r); W)$ is canonically the W-dual of $S_k(\Gamma_1(Np^r); W)$ for each finite level Np^r, the duality as in §3.2.4 extends to the Banach duality between $\mathbf{h}_k(Np^\infty; W)$ and $\mathbf{S}_k(Np^\infty; W)$ [H88c, Theorem 1.3]. Thus the above theorem implies $\mathbf{S}_k(Np^\infty; W) = \mathbf{S}_2(Np^\infty; W)$ inside $W[[q]]$ for all weight $k \geq 2$ (a density theorem of weight 2 cusp forms in the space of p-adic modular forms [H88b, §5]).

We can prove a similar density result for p-adic modular forms, but the proof is more involved as the duality between the space of modular forms and the Hecke algebra misses the information of the constant term of the q-expansion. Here is a sketch of how the proof goes. Defining

$$\mathbf{M}_k^0(Np^\infty; W) = \{f \in \mathbf{M}_k(Np^\infty; W) : a(0, f|\langle z \rangle) = 0 \; \forall z \in \mathbb{Z}_p^\times \times (\mathbb{Z}/N\mathbb{Z})^\times\}$$

and its Hecke algebra $\mathbf{H}_k^0 = \mathbf{H}_k^0(Np^\infty; W) \subset \mathrm{End}_\Lambda(\mathbf{M}_k^0(Np^\infty; W))$ topologically generated over Λ by Hecke operators, the duality still works between $\mathbf{M}_k^0 = \mathbf{M}_k^0(Np^\infty; W)$ and $\mathbf{H}_k^0$, and the density follows for modular forms without constant term at cusps unramified over ∞ basically from the same argument in the case of cusp forms. Note that by $\mathbf{M}_k/\mathbf{M}_k^0$ for $\mathbf{M}_k = \mathbf{M}_k(Np^\infty; W)$ is covered by $\mathbf{M}_k^{ord} \subset \mathbf{M}_k$, and the density of weight 2 ordinary forms proves the density in the entire space.

4.2.10 Hecke operator on boundary

Let Γ be a subgroup of $\mathrm{SL}_2(\mathbb{Z})$ and write its image in $\mathrm{PSL}_2(\mathbb{Z})$ as $\overline{\Gamma}$. Let M be an $A[\Delta_0(N)]$-module. We follow [H86b] to define Hecke operators on the boundary cohomology group $G^q(\Gamma, M) := \bigoplus_{s \in S} H^q(\Gamma_s, M)$, where S is a representative of cusps $\Gamma \backslash \mathbf{P}^1(\mathbb{Q}) \cong \Gamma \backslash \mathrm{PSL}_2(\mathbb{Z})/\Gamma_\infty$ of Γ containing ∞ and Γ_s is a stabilizer of $s \in S$ in Γ. Note that $\Gamma_\infty = \Gamma \cap \{\pm 1\} U(\mathbb{Q})$ for

$$U(A) := \left\{ \left(\begin{smallmatrix} 1 & u \\ 0 & 1 \end{smallmatrix} \right) \middle| u \in A \right\}.$$

We take two groups Γ and Γ' satisfying $\Gamma_1(N) \subset \Gamma, \Gamma' \subset \Gamma_0(N)$ for a positive integer N, and write S' for a choice of representative set of $\Gamma' \backslash \mathbf{P}^1(\mathbb{Q})$ containing ∞. We want to define the action of $\Gamma \alpha \Gamma'$ on $G^q(\Gamma, M)$ for $\alpha \in \Delta_0(N)^\iota$. For a given s, we decompose $\Gamma \alpha \Gamma' = \bigsqcup_\beta \Gamma \beta \Gamma_s'$. Since $|\Gamma \backslash (\Gamma \alpha \Gamma')|$ is finite, this is a finite decomposition. Then we further decompose

$$\Gamma \beta \Gamma_s' = \bigsqcup_i \Gamma \beta \pi_{\beta,i} \text{ with finitely many } \pi_{\beta,i} \in \Gamma_s'. \qquad (4.35)$$

Note that the equivalence class of $\beta(s)$ in $\Gamma \backslash \mathbf{P}^1(\mathbb{Q})$ is uniquely determined only dependent on β. Thus replacing β by an element in $\Gamma \beta$, we may assume

$\beta(s) \in S$, and we always make this choice. Once this choice is made, we get a coset decomposition

$$\Gamma_{\beta(s)}\beta\Gamma'_s = \bigsqcup_i \Gamma_{\beta(s)}\beta\pi_{\beta,i}. \tag{4.36}$$

For simplicity, write $\beta_i := \beta\pi_{\beta,i}$.

In this way, by (4.35), $\Gamma\alpha\Gamma'$ induces a correspondence on S sending $\{\beta(s)\}_\beta \subset S$ to $s \in S'$, written as $C_\alpha : S \to S$. We can thus define, for each $\gamma \in \Gamma_s$, $\beta_i\gamma = \gamma_i\beta_j$ for an index j with $\gamma_j \in \Gamma_{\beta(s)}$. We want to define $c_{\beta(s)}|[\Gamma_{\beta(s)}\beta\Gamma'_s]$ for a q-cocycle $c_{\beta(s)} : \Gamma_{\beta(s)} \to M$. Since either $|\mathrm{Ker}(\Gamma_s \to \overline{\Gamma}_s)| \leq 2$, as long as A is p-profinite with $p > 2$, we have $H^q(\Gamma_s, M) = 0$ if $q \geq 2$. Thus we need to define the action for $q = 0, 1$. For a 1-cocycle $c_{\beta(s)} : \Gamma_{\beta(s)} \to M$, $c_{\beta(s)}|[\Gamma_{\beta(s)}\alpha\Gamma'_s](\gamma) = \sum_i \beta_i^\iota c_s(\gamma_i)$, which gives a 1-cocycle $c_{\beta(s)}|[\Gamma_{\beta(s)}\alpha\Gamma'_s] : \Gamma_s \to M$. In this way, we get an A-linear map $[\Gamma_{\beta(s)}\alpha\Gamma'_s] : H^1(\Gamma_{\beta(s)}, M) \to H^1(\Gamma_s, M)$ as in §4.2.7, and we define $[\Gamma\alpha\Gamma'] : G^1(\Gamma, M) \to G^1(\Gamma, M)$ which sends $(c_s)_{s\in S}$ to $(c_{\beta(s)}|[\Gamma_{\beta(s)}\alpha\Gamma'_s])_{s\in S'}$ at the cocycle level.

Similarly, for $c_{\beta(s)} \in H^0(\Gamma_{\beta(s)}, M)$, we define $c_{\beta(s)}|[\Gamma_{\beta(s)}\alpha\Gamma'_s] = \sum_i \beta_i^\iota c_{\beta(s)}$ which can be easily verified to be in $H^0(\Gamma'_s, M)$, and we get

$$[\Gamma\alpha\Gamma'] : G^0(\Gamma, M) \to G^0(\Gamma', M)$$

given by $(c_s)_{s\in S} \mapsto (c_{\beta(s)}|[\Gamma_{\beta(s)}\alpha\Gamma_s])_{s\in S'}$. Therefore, for $H^q(\Gamma_s, M) \subset G^q(\Gamma, M)$,

if $s \in S$ is outside $\bigcup_\beta \beta(S')$, $H^q(\Gamma_s, M)$ is annihilated by $[\Gamma\alpha\Gamma']$.   (4.37)

In view of this fact, it is important to know how α acts on S. Suppose $\Gamma_1(Np^r) \subset \Gamma \subset \Gamma_0(Np^r)$. For each cusp $s \in S$, we choose $\alpha_s = \left(\begin{smallmatrix} a & b \\ c & d \end{smallmatrix}\right) \in$ $\mathrm{SL}_2(\mathbb{Z})$ with $s = \alpha_s(\infty)$. Since $s = \alpha_s(\infty) = \frac{a}{c} \in \mathbf{P}^1(\mathbb{Q})$, we can assign s the column vector $v_s = \left[\begin{smallmatrix} a \\ c \end{smallmatrix}\right] \in (\mathbb{Z}/Np^r\mathbb{Z})^2$. Plainly $a\mathbb{Z} + c\mathbb{Z} = \mathbb{Z}$, and hence v_s has order Np^r. Since $\Gamma_1(Np^r) \mod Np^r = U(\mathbb{Z}/Np^r)$, we have

$$S \cong U(\mathbb{Z}/Np^r\mathbb{Z})\backslash(\mathbb{Z}/Np^r\mathbb{Z})^2/\{\pm 1\}.$$

Decompose $S = S_\infty \sqcup \bigsqcup_{j=0}^{r-1} S_j$ so that for $v_s = \left[\begin{smallmatrix} a \\ c \end{smallmatrix}\right]$, $s \in S_\infty \Leftrightarrow p^r|c$ and $s \in S_j \Leftrightarrow p^j \| c$ (i.e., $p^j|c$ and $p \nmid c/p^r$). Note that $\alpha_u := \left(\begin{smallmatrix} 1 & u \\ 0 & p \end{smallmatrix}\right)$ sends S_j into S_{j+1} and S_{r-1} into S_∞. Thus if $r > 0$, by (4.37)

$$U(p)^r \text{ brings } G^q(\Gamma, M) \text{ into } G^q_\infty(\Gamma, M) := \bigoplus_{s\in S_\infty} H^q(\Gamma_s, M). \tag{4.38}$$

Writing $H^q(\Gamma, M) \xrightarrow{\text{Res}} G^q(\Gamma, M)$ given by the product of restriction maps, from our construction, we get a commutative diagram:

$$
\begin{array}{ccc}
H^q(\Gamma, M) & \longrightarrow & G^q(\Gamma, M) \\
{\scriptstyle [\Gamma\alpha\Gamma']}\downarrow & & \downarrow{\scriptstyle [\Gamma\alpha\Gamma']} \\
H^q(\Gamma, M) & \longrightarrow & G^q(\Gamma, M).
\end{array}
$$

Thus the Hecke operator $[\Gamma\alpha\Gamma']$ preserves $H^q_!(\Gamma, M) = \mathrm{Ker}(H^q(\Gamma, M) \to G^q(\Gamma, M))$. In particular, we can define $G^q_{ord}(\Gamma, M) := e(G^q(\Gamma, M))$ for $e = \lim_{n \to \infty} U(p)^{n!}$ if $\Gamma_1(Np^r) \subset \Gamma \subset \Gamma_0(Np^r)$.

Proposition 4.2.19. *Suppose $\Gamma_1(Np^r) \subset \Gamma \subset \Gamma_0(Np^r)$ for $Np^r > 2$. Then $G^1_{ord}(\Gamma, L(n, \psi; W))$ is W-free of finite rank.*

We give a proof for $\Gamma = \Gamma_1(Np^r)$ as the treatment is essentially the same for any Γ in the proposition.

Proof. We first assume $r > 0$. By (4.38) and (4.36), we only need to compute the effect of $U(p)$ on $G^1_\infty(\Gamma, M)$ for $M = L(n, \psi; W)$. For $s \in S_\infty$, $\alpha_s \bmod p^r$ can be chosen upper triangular by the definition of S_∞, Γ_s surjects down to $U(\mathbb{Z}/p^r\mathbb{Z})$ modulo p^r. Therefore $\Gamma\alpha\Gamma = \Gamma\alpha\Gamma_s$ for $\alpha = \left(\begin{smallmatrix} 1 & 0 \\ 0 & p \end{smallmatrix}\right)$, and $\alpha(s) = s$ for $s \in S_\infty$. Note that $\Gamma_s\alpha\Gamma_s = \bigsqcup_{u=0}^{p-1} \Gamma\alpha\gamma_u$ with $\gamma_u \in \Gamma_s$ having congruence $\gamma_u \equiv \left(\begin{smallmatrix} 1 & u \\ 0 & 1 \end{smallmatrix}\right) \bmod p^r$. Thus $U(p) = [\Gamma\alpha\Gamma]$ brings $H^1(\Gamma_s, M)$ into itself. Thus we may only deal with the single direct summand $H^1(\Gamma_s, M)$ of $G^1_\infty(\Gamma, M)$. Moreover, we can assume that $s = \infty$ since $H^1(\Gamma_s, M) \cong H^1(\Gamma_\infty, M)$ by conjugation of α_s which is equivariant under $U(p)$ by the expression of its coset decomposition.

Note that $Y \in L(1, \psi; W)$ is fixed by Γ_∞; therefore, we have a short exact sequence of Γ_∞-modules:

$$
0 \to L(n-1, \psi; A) \xrightarrow{P(X,Y) \mapsto P(X,Y)Y} L(n, \psi; A) \xrightarrow{P(X,Y) \mapsto P(1,0)} A \to 0.
$$

We have the associated cohomology exact sequence

$$
H^1(\Gamma_\infty, L(n-1, \psi; W)) \xrightarrow{i} H^1(\Gamma_\infty, L(n, \psi; W)) \to H^1(\Gamma_\infty, W) \to 0.
$$

The last map is onto as $\Gamma_\infty \cong \pi_1(S^1)$ has cohomological dimension 1. Since the action of α^i_u for $\alpha_u = \left(\begin{smallmatrix} 1 & u \\ 0 & p \end{smallmatrix}\right)$ on $Y \in L(1, \psi; A)$ is given by $Y \mapsto pY$, on the image of i, $U(p) \equiv 0 \bmod p$; so, $e(i(H^1(\Gamma_\infty, M))) = 0$ for $e = \lim_{n \to \infty} U(p)^{n!}$. Thus we get

$$
H^1_{ord}(\Gamma_\infty, L(n, \psi; W)) \cong H^1_{ord}(\Gamma_\infty, W).
$$

Since $H^0(\Gamma_\infty, A) = A$, the short exact sequence $W \hookrightarrow K \twoheadrightarrow K/W$ produces cohomology sequence:

$$K \twoheadrightarrow K/W \to H^1(\Gamma_\infty, W) \to H^1(\Gamma_\infty, K)$$

which shows W-freeness of $H^1_{ord}(\Gamma_\infty, W) \cong H^1_{ord}(\Gamma_\infty, W)$ as desired.

Now we assume that $r = 0$. In this case, for any $s \in S$, $\Gamma\alpha\Gamma = \Gamma\alpha\Gamma_s \bigsqcup \Gamma\beta\Gamma_s$, where $\beta = \sigma\alpha^\iota$ for $\sigma \in \mathrm{SL}_2(\mathbb{Z})$ with $\sigma \equiv \begin{pmatrix} * & 0 \\ 0 & p \end{pmatrix} \bmod N$. We may assume that $\alpha(s) = s$ and $\beta(s) = s$. Note that $\Gamma_s\alpha\Gamma_s = \bigsqcup_{u=0}^{p-1} \Gamma_s\alpha_u$ for $\alpha_u \cong \begin{pmatrix} 1 & u \\ 0 & p \end{pmatrix} \bmod p^2$ and $\Gamma_s\beta\Gamma_s = \Gamma_s\beta$. Note that for $\alpha_u\beta$ and $\beta\alpha_u$ are p-adically nilpotent. On the other hand, we have $\bigcap_n \beta^n(L(n, \psi; A)) = AY^n$ and $\prod_{i=1}^\infty \alpha_{u_i}(L(n, \psi; A)Y) = 0$. We may assume that $s = \infty$ and $\Gamma_\infty = U(\mathbb{Z})$. Then for $u = \begin{pmatrix} 1 & 1 \\ 0 & 1 \end{pmatrix}$, we have $H^1(\Gamma_\infty, M) = M/(u-1)M$. Since $AY^n \subset (u-1)L(n, \psi; W)$, we may forget about β. Then by the same argument as in the case of $r > 0$, we find, writing $e_0 := \lim_{n\to\infty} T(p)^{n!}$, $e_0 H^1(\Gamma_\infty, L(n, \psi; W)) \cong e_0(\Gamma_\infty, W) \cong W$ as desired. $\quad\square$

Proposition 4.2.20. *Suppose $\Gamma_1(Np^r) \subset \Gamma \subset \Gamma_0(Np^r)$ for $Np^r > 2$. We have $G^1(\Gamma, L(n, \psi; W/p^rW)) = G^1(\Gamma, L(n, \psi; W)) \otimes_W W/p^rW$ for all $r = 0, 1, \ldots, \infty$, where $W/p^\infty W = K/W$. This implies*

$$G^1_{ord}(\Gamma, L(n, \psi; W/p^rW)) = G^1_{ord}(\Gamma, L(n, \psi; W)) \otimes_W W/p^rW.$$

Proof. By the cohomology sequence attached to

$$L(n, \psi; W) \xrightarrow[\hookrightarrow]{x \mapsto px} L(n, \psi; W) \twoheadrightarrow L(n, \psi; W/p^rW),$$

we get an exact sequence

$$G^1(\Gamma, L(n, \psi; W)) \otimes_W W/p^rW \hookrightarrow G^1(\Gamma, L(n, \psi; W/p^rW))$$
$$\to G^2(\Gamma, L(n, \psi; W)) = 0,$$

since $\Gamma_s = \pi_1(S^1)$ has cohomological dimension 1. Passing to the injective limit with respect to r, we get the result for $W/p^\infty W = K/W = \varinjlim_r W/p^rW$. The assertion for the ordinary part can be obtained applying the projector e or e_0. $\quad\square$

Proposition 4.2.21. *Suppose $\Gamma_1(Np^r) \subset \Gamma \subset \Gamma_0(Np^r)$ for $Np^r > 2$. If $r > 0$, we have $G^0_{ord}(\Gamma, L(n, \psi; A)) = 0$ for $A = W, K, K/W$, and if $r = 0$, we have $G^0_{ord}(\Gamma, L(n, \psi; A)) \cong \bigoplus_{s \in S} A$ for $A = W, K, K/W, W/p^rW$. In particular, we have*

$$G^0_{ord}(\Gamma, L(n, \psi; W/p^rW)) \cong G^0_{ord}(\Gamma, L(n, \psi; W)) \otimes_W W/p^rW$$

canonically.

Proof. We have again an exact sequence, writing $M(A) = L(n, \psi; A)$,

$$0 \to G^0_{ord}(\Gamma, M(W)) \to G^0_{ord}(\Gamma, M(K)) \to G^0_{ord}(\Gamma, M(K/W))$$
$$\xrightarrow{\delta} G^1_{ord}(\Gamma, M(W)).$$

Since $G^1_{ord}(\Gamma, M(W))$ is W-free, the connection map δ vanishes. Thus we need to prove $G^0_{ord}(\Gamma, M(K)) = 0$ if $r > 0$. By the same argument in the case of G^1 in the proof of Proposition 4.2.19, we only need to prove $H^1_{ord}(\Gamma_\infty, M(K)) = 0$. It is easy to see that $H^1(\Gamma_\infty, M(K)) = KY^n$. Under the notation introduced in the proof of Proposition 4.2.19, on KY^n $A := \Gamma_\infty \alpha \Gamma_\infty$ acts by multiplication by p^n, and $B := \Gamma_\infty \beta \Gamma_\infty$ projects every element to KY^n. Thus if $r > 0$, we have $H^1_{ord}(\Gamma_\infty, M(K)) = 0$.

Suppose $r = 0$. Then $T(p) \equiv \mathrm{id} \mod p^n$ on $H^0(\Gamma_\infty, M(W)) = WY^n$, and hence $H^0_{ord}(\Gamma_\infty, M(A)) = AY^n \cong A$ for $A = W, K, K/W$.

We have a similar exact sequence

$$0 \to G^0_{ord}(\Gamma, M(W)) \xrightarrow{x \mapsto p^r x} G^0_{ord}(\Gamma, M(W)) \to G^0_{ord}(\Gamma, M(W/p^r W)) \to 0.$$

Then from the description of $G^0(\Gamma, M(W))$, we get the assertion for $A = W/p^r W$. $\qquad\square$

We quote the following result from [LFE, Corollary A.2, page 364]:

Corollary 4.2.22 (Boundary exact sequence). *Let* Γ *be a subgroup with* $\Gamma_1(Np^r) \subset \Gamma \subset \Gamma_0(Np^r)$ *with* $r > 0$*. Suppose either that image* $\overline{\Gamma}$ *in* $\mathrm{PSL}_2(\mathbb{Z})$ *is torsion-free or* $p \geq 5$*. Let* $M = L(n, \psi; A)$ *for* $A = W, K, K/W, W/p^r W$*. Write* $\mathcal{M}$ *for the locally constant sheaf associated to* M *on* $Y := \Gamma \backslash \mathfrak{H}$*. Then we have the following long exact sequence for* $Y = \Gamma \backslash \mathfrak{H}$:

$$0 \to G^0_{ord}(\Gamma, M) \to H^1_{c,ord}(Y, \mathcal{M}) \to H^1_{ord}(Y, \mathcal{M}) \to G^1_{ord}(\Gamma, M) \to 0.$$

Proof. If $\overline{\Gamma}$ has torsion, taking a subgroup Δ with torsion-free $\overline{\Delta}$ with index $(\overline{\Gamma} : \overline{\Delta})$ prime to $p \geq 5$ (such Δ exists by intersecting $\Gamma_1(q)$ with Γ for a suitable $q \geq 4$ outside Np), by restriction-trace, $H^j(\Delta, M) \cong H^j(\Gamma, M) \oplus \mathrm{Ker}(\mathrm{Tr}_{\Gamma/\Delta})$, and the result for Δ implies that of Γ. Therefore, we give a proof when $\overline{\Gamma}$ is torsion-free.

From the boundary exact sequence (e.g., [LFE, Corollary A.2, page 364]), we have the following exact sequence (for general coefficient sheaf $\mathcal{M}$)

$$0 \to H^0(Y, \mathcal{M}) \to G^0(\Gamma, M) \to H^1_c(Y, \mathcal{M})$$
$$\to H^1(Y, \mathcal{M}) \to G^1(\Gamma, M) \to H^2_!(Y, M) \to 0. \quad (4.39)$$

Here we used the facts: $H^q(\partial \overline{Y}, \mathcal{M}) \cong G^q(\Gamma, M)$, $H^2_!(Y, M) = H^2_c(Y, \mathcal{M})$ and $H^2(Y, M) = 0$ described in [LFE, Appendix] and an obvious fact $H^0_c(Y, \mathcal{M}) = 0$ as Y is an open Riemann surface.

Applying the projector $e = \lim_{n \to \infty} U(p)$, we find $eH^0(Y, \mathcal{M}) = eH^2_!(Y, M) = 0$ by Proposition 4.2.12. This shows the result. □

Exercise 4.2.15. Prove the existence of a subgroup Δ with torsion-free $\overline{\Delta}$ with index $(\overline{\Gamma} : \overline{\Delta})$ prime to $p \geq 5$.

Corollary 4.2.23. *Let the notation and the assumption be as in Corollary 4.2.22. Suppose one of the following conditions:* (i) $r > 0$, (ii) $n > 0$, (iii) $\psi \not\equiv 1 \bmod \mathfrak{m}_W$ *and* (iv) $r = n = 0$ *and* $\psi = 1$.

$$H^1_{!,ord}(\Gamma, L(n, \psi; A)) \cong H^1_{!,ord}(\Gamma, L(n, \psi; W)) \otimes_W A.$$

Proof. The result for $A = K$ is plain; so, we prove it only for $W/p^r W$ (as $K/W = \varinjlim_r W/p^r W$). Since $H^1_{!,ord}(\Gamma, M) = \mathrm{Im}(H^1_{c,ord}(Y, \mathcal{M}) \to H^1_{ord}(Y, \mathcal{M}))$, from the boundary exact sequence, we have the following exact sequence

$$0 \to H^1_{!,ord}(\Gamma, M) \to H^1_{ord}(\Gamma, M) \to G^1_{ord}(\Gamma, M) \to H^2_{ord}(\Gamma, M) \to 0.$$

Write simply $L(A) = L(n, \psi; A)$. Under one of (i), (ii) and (iii), we have $H^2_{ord}(\Gamma, M) = 0$. By W-freeness of $G^1_{ord}(\Gamma, L(W))$, tensoring $W/p^r W$ with the exact sequence:

$$H^1_{!,ord}(\Gamma, L(W)) \hookrightarrow H^1_{ord}(\Gamma, L(W)) \twoheadrightarrow G^1_{ord}(\Gamma, L(W)),$$

we still have an exact sequence

$$H^1_{!,ord}(\Gamma, L(W)) \otimes_W W/p^r W \hookrightarrow H^1_{ord}(\Gamma, L(W)) \otimes_W W/p^r W$$
$$\twoheadrightarrow G^1_{ord}(\Gamma, L(W)) \otimes_W W/p^r W.$$

Thus we get the following commutative diagram with two vertical isomorphism:

$$H^1_{ord}(\Gamma, L(W)) \otimes_W W/p^r W \xrightarrow{\twoheadrightarrow} G^1_{ord}(\Gamma, L(W)) \otimes_W W/p^r W$$

$$\wr \Big\downarrow \text{Prop. 4.2.4} \qquad\qquad \wr \Big\downarrow \text{Prop. 4.2.20}$$

$$H^1_{ord}(\Gamma, L(W/p^r W)) \xrightarrow{\twoheadrightarrow} G^1_{ord}(\Gamma, L(W/p^r W)).$$

Taking the kernel of the horizontal maps, we get

$$H^1_{!,ord}(\Gamma, L(n, \psi; A)) \cong H^1_{!,ord}(\Gamma, L(n, \psi; W)) \otimes_W A.$$

for $A = W/p^r W$.

Now assume (iv). Then $H(A) := \mathrm{Ker}(G^1_{ord}(\Gamma, L(A)) \to H^2_{1,ord}(\Gamma, L(A)))$ is A-free of finite rank for $A = W, W/p^r W$ by Proposition 4.2.9 combined with Proposition 4.2.19, and therefore $H(W/p^r W) = H(W) \otimes_W W/p^r W$. Again we get the commutative diagram with two vertical isomorphism:

$$
\begin{array}{ccc}
H^1_{ord}(\Gamma, L(W)) \otimes_W W/p^r W & \xrightarrow{\;\twoheadrightarrow\;} & H(W) \otimes_W W/p^r W \\
\wr \downarrow & & \wr \downarrow \\
H^1_{ord}(\Gamma, L(W/p^r W)) & \xrightarrow{\;\twoheadrightarrow\;} & H(W/p^r W).
\end{array}
$$

The kernel of the two horizontal maps gives the desired identity. $\qquad\square$

4.2.11 *Weight comparison*

Let A be a $\mathbb{Z}/p^r\mathbb{Z}$-algebra and ψ be a Dirichlet character modulo Np^r with values in $A^\times$. We consider $\Gamma_0(Np^r)$-modules $A(n)$ (resp. $A(n) \otimes \psi$) such that $\gamma := \left(\begin{smallmatrix} a & b \\ Np^r c & d \end{smallmatrix}\right) \in \Gamma_0(Np^r)$ acts on $x \in A(n)$ (resp. $x \in A(n) \otimes \psi$) by $\gamma x = d^n x$ (resp. $\gamma x = \psi(d)d^n x$). We consider the morphism

$$
i = i_r : L(n, \psi; A) \ni P(X, Y) \to P(1, 0) \in A(n) \otimes \psi. \tag{4.40}
$$

Since $\gamma \equiv \left(\begin{smallmatrix} d^{-1} & b \\ 0 & d \end{smallmatrix}\right)$, we find $\gamma P(X, Y)|_{(X,Y)=(1,0)} = P((1,0)^t \gamma^\iota) = d^n P(1, 0)$, and hence i is a morphism of $\Gamma_0(p^r)$-modules. Define

$$
j = j_r : A(-n) \otimes \psi \ni y \mapsto y Y^n \in L(n, \psi; A). \tag{4.41}
$$

Similarly as above, j is a morphism of $\Gamma_0(p^r)$-modules.

For a positive integer N prime to p, the normalizer of $\Gamma_0(p^r N)$ contains the following element $\tau = \tau_r \in M_2(\mathbb{Z})$ given by

$$
\det(\tau_r) = p^r, \quad \tau_r \equiv \left(\begin{smallmatrix} 0 & -1 \\ p^r & 0 \end{smallmatrix}\right) \bmod p^{2r}, \quad \tau_r \equiv \left(\begin{smallmatrix} 1 & 0 \\ 0 & p^r \end{smallmatrix}\right) \bmod N^2. \tag{4.42}
$$

This follows from the density of $\mathrm{SL}_2(\mathbb{Z})$ in $\mathrm{SL}_2(\widehat{\mathbb{Z}})$ (the strong approximation theorem [LFE, Lemma 6.1.1]).

Exercise 4.2.16. Prove the following facts:

(1) existence of τ_r as above,

(2) $\tau_r \left(\begin{smallmatrix} a & b \\ cNp^r & d \end{smallmatrix}\right) \tau_r^{-1} \equiv \left(\begin{smallmatrix} d & -Nc \\ -bp^r & a \end{smallmatrix}\right) \bmod p^{2r}$,

(3) $\tau_r \left(\begin{smallmatrix} a & b \\ cNp^r & d \end{smallmatrix}\right) \tau_r^{-1} \equiv \left(\begin{smallmatrix} a & b[p]^{-r} \\ cN[p]^r p^r & d \end{smallmatrix}\right) \bmod N^2$, where for a class $[x] \in (\mathbb{Z}/N\mathbb{Z})^\times$ of $x \in \mathbb{Z}$, $[x]^{-1}$ is the class in $(\mathbb{Z}/N\mathbb{Z})^\times$ with $[x][x]^{-1} = [1]$,

(4) τ_r normalizes $\Gamma_0(N)$ and $\Gamma_1(N)$.

We analyze $\mathrm{Ker}(L(n, \psi; A) \xrightarrow{i} A(n) \otimes \psi)$, which is isomorphic to $L(n-1; A)$ by $P(X, Y) \mapsto P(X, Y)Y \in L(n; A)$. Therefore for $\alpha = \left(\begin{smallmatrix} 1 & u \\ 0 & p \end{smallmatrix}\right)$, we have $\alpha^\iota P(X, Y)Y = P(X + uY, pY)pY \equiv 0 \bmod pA$ and hence $U(p)P(X, Y)Y \equiv 0 \bmod pA$. Thus e kills $H^1(\Gamma_H(p^r), \mathrm{Ker}(L(n, \psi; A) \xrightarrow{i} A(n) \otimes \psi))$; so, i induces an injection

$$I := i_* : H^1_{ord}(\Gamma_H(p^r), L(n, \psi; A)) \hookrightarrow H^1_{ord}(\Gamma_H(p^r), A(n) \otimes \psi).$$

The matrix τ_r acts on 1-cocycle $c : \Gamma_H(p^r) \to A(-n) \otimes \psi$ by $c|[\tau_r](\gamma) := c(\tau_r \gamma \tau_r^{-1})$ inducing an operator $[\tau] = [\tau_r] : H^1(\Gamma_H(p^r), A(-n) \otimes \psi) \to H^1(\Gamma_H(p^r), A(n) \otimes \psi_p^{-1}\psi_N)$, where $\psi_p = \psi|_{(\mathbb{Z}/p^r\mathbb{Z})^\times}$ and $\psi_N = \psi|_{(\mathbb{Z}/N\mathbb{Z})^\times}$ (by Exercise 4.2.16 (3)). Take $\delta \in \mathrm{SL}_2(\mathbb{Z})$ such that

$$\delta \equiv \left(\begin{smallmatrix} 0 & 1 \\ -1 & 0 \end{smallmatrix}\right) \bmod p^{2r} \quad \text{and} \quad \delta \equiv 1 \bmod N^2.$$

Such a δ exists by the strong approximation theorem. It is easy to see

$$\Gamma_H(p^r)\delta\Gamma_H(p^r) = \bigsqcup_{u=0}^{p^r-1} \Gamma_H(p^r)\delta_u \quad \text{with} \quad \delta_u = \delta \left(\begin{smallmatrix} 1 & u \\ 0 & 1 \end{smallmatrix}\right). \tag{4.43}$$

Consider the double coset operator Δ associated to $\Gamma_H(p^r)\delta\Gamma_H(p^r)$

$$\Delta : H^1(\Gamma_H(p^r), L(n, \psi_p^{-1}\psi_N; A)) \to H^1(\Gamma_H(p^r), L(n, \psi; A))$$

given by $c|\Delta(\gamma) = \sum_u \delta_u^\iota c(\gamma_u)$ on a 1-cocycle c, where $\delta_u\gamma = \gamma_u\delta_v$ for a suitable v and $\gamma_u \in \Gamma_H(p^r)$. Define

$$J : H^1(\Gamma_H(p^r), A(n) \otimes \psi) \to H^1(\Gamma_H(p^r), L(n, \psi; A))$$

by $J := \Delta \circ j_* \circ [\tau_r]$. The following result was first proven in [H86b, Theorem 4.4]:

Theorem 4.2.24. *Suppose $p^r A = 0$ and $r \geq \ell(H)$. Then we have*

$$J \circ I = U_r(p^r) \quad \text{on} \quad H^1(\Gamma_H(p^r), L(n, \psi; A)),$$

$$I \circ J = U_r(p^r) \quad \text{on} \quad H^1(\Gamma_H(p^r), A(n) \otimes \psi).$$

Proof. Since the proof is basically the same for the two formulas, we prove the first one. For each 1-cocycle $c : \Gamma_H(p^r) \to L(n, \psi; A)$, we have

$$J \circ I(c)(\gamma) = \sum_{u=0}^{p^r-1} \delta_u^\iota \cdot j_r(i_r(c(\tau\gamma_u\tau^{-1}))),$$

where δ_u and γ_u are as in (4.43). Since $p^r A = 0$, we note that $\tau_r^\iota P(X, Y) = j_r(i_r(P(X, Y)))$. From this, we get

$$J \circ I(c)(\gamma) = \sum_{u=0}^{p^r-1} (\tau\delta_u)^\iota c(\tau\gamma_u\tau^{-1}),$$

where $\tau\delta_u \equiv \left(\begin{smallmatrix} 1 & u \\ 0 & p^r \end{smallmatrix}\right) \bmod Np^r$, $\det(\tau\delta_u) = p^r$ and $(\tau\delta_u)\gamma = \tau\gamma_u\tau^{-1}(\tau_u\delta_v)$ for some v. This shows the desired identity. $\square$

Remark 4.2.25. Though we have proven Theorem 4.2.24 only for H^1, by the same argument, it is valid for H^q for any $q \geq 0$ as the double coset action on higher cocycles is just a many variable version in the case of 1-cocycle as described at the end of §4.2.7.

Corollary 4.2.26. *Let the assumption be as in* Theorem 4.2.24. *The map I induces Hecke equivariant isomorphisms:*

$$H^1_{ord}(\Gamma_H(p^r), L(n, \psi; A)) \cong H^1_{ord}(\Gamma_H(p^r), A(n) \otimes \psi)$$
$$H^1_{!,ord}(\Gamma_H(p^r), L(n, \psi; A)) \cong H^1_{!,ord}(\Gamma_H(p^r), A(n) \otimes \psi),$$

where we have written $H^1_{!,ord} = eH^1_!$.

The result for $H^1_{!,ord}$ follows as the identity in Theorem 4.2.24 is valid also for the boundary cohomology groups by the definition of the action of Hecke operators in §4.2.10. See also [H86b, Proposition 4.7] for details of this fact.

Definition 4.2.27. Define the following limit with respect to the restriction maps

$$H^1_{ord}(\Gamma_H(p^\infty), L(n, \psi; K/W)) := \varinjlim_r H^1_{ord}(\Gamma_H(p^r), L(n, \psi; K/W))$$
$$H^1_{!,ord}(\Gamma_H(p^\infty), L(n, \psi; K/W)) := \varinjlim_r H^1_{!,ord}(\Gamma_H(p^r), L(n, \psi; K/W)).$$

These cohomology groups are canonically isomorphic to the following sheaf cohomology groups, respectively.

$$H^1_{ord}(Y(\Gamma_H(p^\infty)), L(n, \psi; K/W)) := \varinjlim_r H^1_{ord}(Y(\Gamma_H(p^r)), L(n, \psi; K/W))$$

$$H^1_{!,ord}(Y(\Gamma_H(p^\infty)), L(n, \psi; K/W)) := \varinjlim_r H^1_{!,ord}(Y(\Gamma_H(p^r)), L(n, \psi; K/W)).$$

More specifically, we define

$$H^1_{ord}(Y_1(Np^\infty), L(n, \psi; K/W)) := \varinjlim_r H^1_{ord}(Y_1(Np^r), L(n, \psi; K/W))$$

$$H^1_{!,ord}(Y_1(Np^\infty), L(n, \psi; K/W)) := \varinjlim_r H^1_{!,ord}(Y_1(Np^r), L(n, \psi; K/W)).$$

If we take $H = \{1\}$, $\Gamma_H(p^r) = \Gamma_1(p^r) \cap \Gamma_0(N)$ for all $r > 0$ if p is odd (when $p = 2$, this holds if $r > 1$).

Corollary 4.2.28. *Let the assumption be as in* Theorem 4.2.24. *The map I induces Hecke equivariant isomorphisms:*

$$H^1_{ord}(\Gamma_H(p^\infty), L(n, \psi; A)) \cong H^1_{ord}(\Gamma_H(p^\infty), A(n) \otimes \psi),$$
$$H^1_{!,ord}(\Gamma_H(p^\infty), L(n, \psi; A)) \cong H^1_{!,ord}(\Gamma_H(p^\infty), A(n) \otimes \psi).$$

Proof. Since the proof is the same for $H^1_{!,ord}$ and H^1_{ord}, we only deal with H^1_{ord}. Taking $A = p^{-r}W/p$, we have $p^{-r}W/W(n) \otimes \psi = L(0, \psi; p^{-r}W/W)$ in the above Corollary. Thus passing to the limit of the isomorphisms I of Corollary 4.2.26, we conclude

$$H^1_{ord}(\Gamma_H(p^\infty), L(n, \psi; K/W)) = \varinjlim_r H^1_{ord}(\Gamma_H(p^\infty), L(n, \psi; p^{-r}W/W))$$

$$\cong \varinjlim_r H^1_{ord}(\Gamma_H(p^r), L(0, \psi; p^{-r}W/W)) = H^1_{ord}(\Gamma_H(p^\infty), L(n, \psi; K/W))$$

as desired. $\qquad\qquad\qquad\qquad\qquad\qquad\qquad\qquad\qquad\qquad\qquad\qquad\square$

4.2.12 Freeness and divisibility

As we have already remarked in Remark 4.2.2, second cohomology vanishes: $H^2(\Delta, M) = 0$ for all Δ-module M if Δ is a torsion-free subgroup of finite index in $\Gamma_r := \Gamma_0(Np^r)$. We consider the following short exact sequence $0 \to L(n, \psi; W) \xrightarrow{\iota} L(n, \psi; K) \xrightarrow{\pi} L(n, \psi; K/W) \to 0$ for a character ψ modulo Np^r and the corresponding long exact sequence (Remark 4.2.6). Since $H^2(\Delta, L(n, \psi; W)) = 0$, $\pi_* : H^2(\Delta, L(n, \psi; K)) \to H^1(\Delta, L(n, \psi; K/W))$ is surjective; so, $H^1(\Delta, L(n, \psi; K/W))$ is divisible (because $H^1(\Delta, L(n, \psi; K))$ is a K-vector space). Since Δ acts trivially on $L(0; A) = A$, $H^0(\Delta, L(0; K)) = K$ mapped onto $H^0(\Delta, L(K/W)) = K/W$; so, by long exact sequence for $W \hookrightarrow K \twoheadrightarrow K/W$, $H^1(\Delta, W)$ is mapped injectively into $H^1(\Delta, K)$; so, $H^1(\Delta, W)$ is W-free (which actually follows from the expression in (4.19)).

Proposition 4.2.29. *If Δ is torsion-free, the cohomology group $H^1(\Delta, L(n, \psi; K/W))$ is p-torsion and divisible, and the cohomology group $H^1(\Delta, W)$ is W-free of finite rank.*

The torsion-ness of $H^1(\Delta, L(n, \psi; K/W))$ follows from that fact that K/W is p-torsion and a 1-cocycle is determined by its value at the (finitely many) generators of Δ.

Exercise 4.2.17. Find an example of an integer $n > 0$ such that $H^1(\Delta, L(n; W))$ has non-trivial torsion. *(Hint: Find an integer n such that $L(n; K/W)^\Delta / \pi(L(n; K)^\Delta) \neq 0$, and use the long exact sequence.)*

Let $H^1(\Delta, L(n, \psi; K/W))[p^m]$ be the submodule of $H^1(\Delta, L(n, \psi; K/W))$ killed by p^m. By the above proposition and Theorem 4.2.8,

$$H^1(\Delta, L(n, \psi; K)) = \left(\varprojlim_m H^1(\Delta, L(n, \psi; K/W))[p^m] \right) \otimes_W K,$$

and hence by Theorem 4.2.8, the Hecke algebra $\mathbf{H}_k(Np^r; W) \subset$ $\mathrm{End}_W(M_k(\Gamma_1(Np^r); W))$ generated by Hecke operators $T(n)$ is identical to the subalgebra generated by Hecke operators $T(n)$ of $\mathrm{End}_W(H^1(\Gamma_1(Np^r), L(k-2, \psi; K/W)))$. Thus we get

Corollary 4.2.30. *For simplicity, write $L(X)$ for $L(k-2, \psi; M)$ for $M = W, K, K/W$. Let Γ be a subgroup with $\Gamma_1(N\mathbf{p}) \subset \Gamma \subset \Gamma_0(N\mathbf{p})$. The Hecke algebra $\mathbf{H}_k(\Gamma \cap \Gamma_1(p^r), \psi; W)$ is isomorphic to the W-subalgebra of $\mathrm{End}_W(H^1(\Gamma \cap \Gamma_1(p^r), L(K/W)))$ generated by Hecke operators $T(n)$ for all n and diamond operators $\langle z \rangle$. The same assertion also holds for $\mathbf{h}_k(\Gamma \cap \Gamma_1(p^r), \psi; W)$, $\mathbf{H}_k^{ord}(\Gamma \cap \Gamma_1(p^r), \psi; W)$ and $\mathbf{h}_k^{ord}(\Gamma \cap \Gamma_1(p^r), \psi; W)$ replacing the cohomology group $H^1(\Gamma \cap \Gamma_1(p^r), L(K/W))$ in the above statement by the following: $H^1_!(\Gamma \cap \Gamma_1(p^r), L(W)) \otimes_W K/W$, $H^1_{ord}(\Gamma \cap \Gamma_1(p^r), L(K/W))$ and $H^1_{!,ord}(\Gamma \cap \Gamma_1(p^r), L(K/W))$, respectively.*

Since the proof is the same for any choice of Γ, in the proof, we assume $\Gamma = \Gamma_1(N\mathbf{p})$; so, for example, $\mathbf{H}_k(\Gamma \cap \Gamma_1(p^r), \psi; W) = \mathbf{H}_k(Np^r; W)$.

Proof. Since the Eichler–Shimura isomorphism is compatible with operators $T(n)$ and $\langle z \rangle$, $\mathbf{H}_k(Np^r; \mathbb{Z})$ acts on $H^1(\Gamma_1(Np^r), L(K))$ and preserves the image of $H^1(\Gamma_1(Np^r), L(W))$ in $H^1(\Gamma_1(Np^r), L(K))$; so, it acts on the quotient

$$H^1(\Gamma_1(Np^r), L(W)) \otimes_W K/W$$
$$\cong \mathrm{Coker}(H^1(\Gamma_1(Np^r), L(W)) \to H^1(\Gamma_1(Np^r), L(K))).$$

We know from Proposition 4.1.15 that $\mathbf{H}_k(Np^r; \mathbb{Z}) \otimes_{\mathbb{Z}} W = \mathbf{H}_k(Np^r; W)$. Thus $\mathbf{H}_k(Np^r; W)$ acts on $H^1(\Gamma_1(Np^r), L(K/W))$. Writing $\mathbf{H}'$ for the W-subalgebra of $\mathrm{End}(H^1(\Gamma_1(Np^r), L(W)) \otimes_W K/W)$ generated by Hecke operators and diamond operators, we have a W-algebra surjection $\mathbf{H}_k(Np^r; W) \twoheadrightarrow \mathbf{H}'$. By Eichler–Shimura isomorphism, after tensoring K over W, the morphism becomes an isomorphism. Since $\mathbf{H}_k(Np^r; W) = \mathbf{H}_k(Np^r; \mathbb{Z}) \otimes_{\mathbb{Z}} W$ is W-free of finite rank, we conclude $\mathbf{H}_k(Np^r; W) \cong \mathbf{H}'$. Since $H^1(\Gamma_1(Np^r), L(W)) \otimes_W K/W \cong H^1(\Gamma_1(Np^r), L(K/W))$ by the vanishing of H^2 in Remark 4.2.2, we may replace

$$H^1(\Gamma_1(Np^r), L(W)) \otimes_W K/W$$

by $H^1(\Gamma_1(Np^r), L(K/W))$ in the definition of $\mathbf{H}'$, which proves the assertion for $\mathbf{H}_k(Np^r; W)$. By Corollaries 4.2.13 and 4.2.23, replacing H^1_{ord} by $H^1_{!,ord}$, the above argument gives the result for $\mathbf{h}_k^{ord}$.

The same proof works for $\mathbf{h}_k(Np^r; W)$ except for the step replacing $H^1_!(\Gamma_1(Np^r), L(W)) \otimes_W K/W$ by $H^1_!(\Gamma_1(Np^r), L(K/W))$ as the natural map

$$H^1_!(\Gamma_1(Np^r), L(W)) \otimes_W K/W \to H^1_!(\Gamma_1(Np^r), L(K/W))$$

may have non-trivial cokernel. $\qquad\qquad\square$

4.2.13 Proof of Theorem 4.2.17 for the ordinary part

For simplicity, we write Γ_r for $\Gamma_1(Np^r)$. By the cohomology exact sequence attached to the short one $W \hookrightarrow K \to K/W$, we have a long exact sequence for $M(?) = L(n, \psi; ?)$ by Remark 4.2.6

$$H^0(\Gamma_r, M(K/W)) \to H^1(\Gamma_r, M(W)) \to H^1(\Gamma_r, M(K))$$
$$\to H^1(\Gamma_r, M(K/W)) \to H^2(\Gamma_r, M(W)) \overset{\text{Remark } 4.2.2}{=} 0.$$

Applying e, we find

$$0 \to H^1_{ord}(\Gamma_r, M(W)) \to H^1_{ord}(\Gamma_r, M(K)) \to H^1_{ord}(\Gamma_r, M(K/W)) \to 0$$

is exact. Therefore this shows $H^1_{ord}(\Gamma_r, M(K/W))$ is p-divisible and $H^1_{ord}(\Gamma_r, M(W))$ is W-free. Thus writing $\mathbf{H}^{ord}_k(Np^r; W/p^nW)$ for the W/p^nW-subalgebra of $\mathrm{End}(H^1_{ord}(\Gamma_r, W/p^nW))$ generated by $T(n)$ for all n, we find $\mathbf{H}^{ord}_k(Np^r; W) = \varprojlim_n \mathbf{H}^{ord}_k(Np^r, W/p^nW)$ and $\mathbf{H}^{ord}_k(Np^r, W/p^rW) \cong \mathbf{H}^{ord}_2(Np^r, W/p^rW)$ by Corollary 4.2.26. Since

$$\mathbf{H}^{ord}_k(Np^\infty, W) = \varprojlim_r \mathbf{H}^{ord}_k(Np^r, W)$$
$$= \varprojlim_r \varprojlim_n \mathbf{H}^{ord}_k(Np^r, W/p^nW) = \varprojlim_r \mathbf{H}^{ord}_k(Np^r, W/p^rW),$$

we conclude $\mathbf{H}^{ord}_k(Np^\infty, W) \cong \mathbf{H}^{ord}_2(Np^\infty, W)$.

By Corollary 4.2.23, we know $H^1_{!,ord}(\Gamma_r, M(K/W))$ is p-divisible and $H^1_{!,ord}(\Gamma_r, M(W))$ is W-free. Thus by the same argument above, we get the assertion for $\mathbf{h}^{ord}_k$.

Here is an alternative argument for $\mathbf{h}^{ord}_k$ without using Corollary 4.2.23. We have the following Hecke equivariant inclusions:

$$H^1_{!,ord}(\Gamma_r, M(W)) \otimes_W K/W \hookrightarrow H^1_{!,ord}(\Gamma_r, M(K/W))$$
$$\overset{I}{\to} \varinjlim_r H^1_{!,ord}(\Gamma_r, W/p^rW(n)) = \varinjlim_r H^1_{!,ord}(\Gamma_r, K/W)$$
$$= \varinjlim_r H^1_{!,ord}(X(Np^r), K/W). \quad (4.44)$$

The first injection follows from the injective limit (with respect to n) of the long exact sequence attached to the short one $0 \to M(W) \xrightarrow{p^n} M(W) \to M(p^{-n}W/W) \to 0$. By Corollary 4.2.30, $\mathbf{h}_k^{ord}(Np^r; W)$ is the subalgebra of $\mathrm{End}_W(H^1_{!,ord}(\Gamma_r, M(W)) \otimes_W K/W)$ generated over W by $T(n)$ for all W. The subalgebra of $\mathbf{h}_2^{ord}(Np^r; W)$ acts faithfully on $H^1_{!,ord}(X(Np^r), K/W)$, and hence $\mathbf{h}_2^{ord}(Np^\infty; W)$ acts on $\varinjlim_r H^1_{!,ord}(X(Np^r), K/W)$. Restricting the action of $\mathbf{h}_2^{ord}(Np^\infty; W)$ on the extreme right end to the cohomology of the extreme left end, we get a W-algebra homomorphism $\pi_r : \mathbf{h}_2^{ord}(Np^\infty; W) \to \mathbf{h}_k^{ord}(Np^r; W)$ with $\pi_r(T(n)) = T(n)$. Since the morphism takes generators $T(n)$ to generators $T(n)$, π_r is a surjection. Since projective limit is an exact functor from the category of compact modules, passing to the limit of r, we get a surjective algebra homomorphism $\mathbf{h}_2^{ord}(Np^\infty; W) \twoheadrightarrow \mathbf{h}_k^{ord}(Np^\infty; W)$ sending $T(n)$ to $T(n)$. Since this is true for all k, passing to the limit of k as in (4.13), we get a surjective algebra homomorphism $\pi_\infty : \mathbf{h}_2^{ord}(Np^\infty; W) \twoheadrightarrow \mathbf{h}_\infty^{ord}(Np^\infty; W)$ sending $T(n)$ to $T(n)$. As we have seen already in (4.13), we have a reverse surjection $\pi_\infty' : \mathbf{h}_\infty^{ord}(Np^\infty; W) \twoheadrightarrow \mathbf{h}_2^{ord}(Np^\infty; W)$ sending $T(n)$ to $T(n)$. By Theorem 4.2.37 proven in the following section, the Pontryagin dual of $\varinjlim_r H^1_{ord}(\Gamma_r, K/W)$ is a Λ-module of finite rank, and hence $\mathbf{H}_2^{ord}(Np^\infty; W)$ is a Λ-module of finite type. Therefore, its surjective image $\mathbf{h}_2^{ord}(Np^\infty; W)$ is a Λ-module of finite type; in particular, these rings are noetherian. Hence we conclude π_∞ and π_∞' are isomorphisms inverse each other. Since by definition $\mathbf{h}_\infty^{ord}(Np^\infty; W)$ has surjection onto $\mathbf{h}_k^{ord}(Np^\infty; W)$ sending $T(n)$ to $T(n)$, we also conclude $\mathbf{h}_k^{ord}(Np^\infty; W) \twoheadrightarrow \mathbf{h}_2^{ord}(Np^\infty; W)$ for all k. $\quad\square$

We record what we have proven in the above proof:

Lemma 4.2.31. *If $\Gamma_H(p^r)$ is torsion-free, $H^1(\Gamma_H(p^r), L(n, \psi; K/W))$ and $H^1_{ord}(\Gamma_H(p^r), L(n, \psi; K/W))$ are W-divisible of finite corank and $H^1_{ord}(\Gamma_H(p^r), L(n, \psi; W))$ is W-free of finite rank. If further $n < p$, $H^1(\Gamma_1(N), L(n, \psi; W))$ (for N prime to p) is W-free of finite rank.*

The last assertion follows from the long exact sequence and the vanishing of $H^0(\Gamma_1(N), L(n, \psi; K/W))$ as $L(n, \psi; W/\mathfrak{m}_W)$ is irreducible $SL_2(\mathbb{Z}_p)$-module if $n < p$.

Corollary 4.2.32. *Let $H = \mathbb{Z}_p^\times$. Then we have W-rank equalities:*

$$\mathrm{rank}(H^1_{ord}(\Gamma_H(\mathbf{p}), L(n, \psi; W)) = \mathrm{rank}(H^1_{ord}(\Gamma_H(\mathbf{p}p^r), L(0, \psi\omega^{-n}; W)),$$
$$\mathrm{rank}(H^1_{!,ord}(\Gamma_H(\mathbf{p}), L(n, \psi; W)) = \mathrm{rank}(H^1_{!,ord}(\Gamma_H(\mathbf{p}p^r), L(0, \psi\omega^{-n}; W)).$$

This implies

$$\text{rank } M^{ord}_{k,\psi\omega-k}(\Gamma_0(\mathbf{p}N);W) = \text{rank } M^{ord}_{2,\psi\omega-2}(\Gamma_0(\mathbf{p}N);W),$$

$$\text{rank } S^{ord}_{k,\psi\omega-k}(\Gamma_0(\mathbf{p}N);W) = \text{rank } S^{ord}_{2,\psi\omega-2}(\Gamma_0(\mathbf{p}N);W),$$

which are independent of $k \geq 2$.

Proof. Taking r to be sufficiently large so that the image of $\Gamma_H(\mathbf{p}p^r)$ in $\mathrm{PSL}_2(\mathbb{Z})$ is torsion-free. Then we have

$$H^1_{ord}(\Gamma_H(\mathbf{p}p^r), L(n,\psi;W)) \otimes_W W/pW$$

$$\overset{\underset{\text{Proposition 4.2.4}}{}}{\cong} H^1_{ord}(\Gamma_H(\mathbf{p}p^r), L(n,\psi\omega^{-n};W/pW))$$

$$\overset{\underset{\text{Corollary 4.2.26}}{}}{\cong} H^1_{ord}(\Gamma_H(\mathbf{p}p^r), L(0,\psi\omega^{-n};W/pW))$$

$$\overset{\underset{\text{Proposition 4.2.4}}{}}{\cong} H^1_{ord}(\Gamma_H(\mathbf{p}p^r), L(0,\psi\omega^{-n};W)) \otimes_W W/pW. \quad (4.45)$$

By contraction property (4.34), the identity (4.45) actually holds for $r = 0$. Then W-freeness of $H^1_{ord}(\Gamma_H(\mathbf{p}), L(n,\xi;W))$ for any character ξ modulo $N\mathbf{p}$ (Corollary 4.2.13) tells us

$$\text{rank}_W(H^1_{ord}(\Gamma_H(\mathbf{p}), L(n,\psi;W)) = \text{rank}_W(H^1_{ord}(\Gamma_H(\mathbf{p}), L(0,\psi\omega^{-n};W)).$$

Replacing the reference to Proposition 4.2.4 by Corollary 4.2.23 in the above four line isomorphisms (4.45) and then making $r = 0$, we get the result for $H^1_{!,ord}$.

Decomposition of H^1 and M^{ord}_k into generalized eigenspaces of Hecke operators, if $U(p)$-eigenvalue is a p-adic unit (with respect to our fixed embedding $\overline{\mathbb{Q}} \hookrightarrow \overline{\mathbb{Q}}_p$), the idempotent e is identity on the eigenspace, and otherwise it vanishes. Thus e is well defined on $H^1_{ord}(\Gamma_0(N\mathbf{p}), L(k-2,\xi;A))$ and $M_{k,\xi}(\Gamma_0(N\mathbf{p});A)$ for $A = \overline{\mathbb{Q}}, \overline{\mathbb{Q}}_p, \mathbb{C}$ and K. The same fact holds for $H^1_{!,ord}$ and $S^{ord}_{k,\xi}$.

Taking $\Gamma = \Gamma_0(N)$, we have $\Gamma_H(\mathbf{p}) = \Gamma_0(N\mathbf{p})$. Therefore by the above identity (4.45) with $r = 0$ and Theorem 4.2.8 combined with compatibility of base-change and W-freeness for the spaces involved, we get for $n = k - 2$

$$\text{rank}_W S^{ord}_{k,\psi\omega-k}(\Gamma_0(N\mathbf{p});W) + \text{rank}_W M^{ord}_{k,\psi\omega-k}(\Gamma_0(N\mathbf{p});W)$$

$$= \text{rank}_W H^1_{ord}(\Gamma_0(N\mathbf{p}), L(n,\psi\omega^{-n-2};W)),$$

$$2\,\text{rank}_W S_{k,\psi\omega-k}(\Gamma_0(N\mathbf{p});W) = \text{rank}_W H^1_{!,ord}(\Gamma_0(N\mathbf{p}), L(n,\psi\omega^{-n-2};W)).$$

The right-hand-side of the above identities are constant independent of n, and hence we conclude the last two identities in the corollary. $\qquad\square$

4.2.14 *Proof of Theorem 4.2.17 in general*

We give a proof of Theorem 4.2.17 based on Lazard [GAN, V.2.2.4] and Whitehead lemma [LIE]. Here is the lemma:

Theorem 4.2.33. *Let $\mathfrak{L}$ be a semi-simple Lie algebra over a field k of characteristic 0 and M be an $\mathfrak{L}$-module finite dimensional over k. Then $H^q(\mathfrak{L}, M) = 0$ for $q = 1, 2$.*

For the definition of Lie algebra cohomology and modules over Lie algebra, see [LIE, III.10]. The proof of this fact can be found in [LIE, III.10, Theorem 13]. The action ρ of $\mathrm{SL}_2(\mathbb{Z}_p)$ on $L(n; \mathbb{Q}_p)$ extends to the action of the Lie algebra $\mathfrak{sl}_2(\mathbb{Z}_p)$ by the derivation action $d\rho$. Thus $L(n; \mathbb{Q}_p)$ is also a module over $\mathfrak{sl}_2(\mathbb{Q}_p)$ [GAN, V.2.4.2]. Here is a theorem which follows from [GAN, V.2.4.10]:

Theorem 4.2.34. *For a sufficiently small open subgroup H of $\mathrm{SL}_2(\mathbb{Z}_p)$, the continuous cohomology $H^1_{ct}(H, L(n; \mathbb{Q}_p))$ is canonically isomorphic to $H^1(\mathfrak{sl}_2(\mathbb{Q}_p), L(n; \mathbb{Q}_p))$. In particular, it vanishes by Theorem 4.2.33.*

See Remark 4.2.3 for continuous cohomology groups.

Corollary 4.2.35. *For any open subgroup U of $SL_2(\mathbb{Z}_p)$, the continuous cohomology group $H^q_{ct}(U, L(n; \mathbb{Z}_p))$ is finite for $q = 0, 1$, and the cohomology group $H^q(U, L(n; \mathbb{Q}_p/\mathbb{Z}_p))$ is finite for $q = 0, 1$. Moreover for $U_s = \mathrm{Ker}(U \xrightarrow{\bmod p^s} \mathrm{SL}_2(\mathbb{Z}/p^s\mathbb{Z}))$, $|H^1(U/U_s, L(n; \mathbb{Z}/p^s\mathbb{Z}))|$ is bounded independent of s.*

Note $H^1(U, L(n; \mathbb{Q}_p/\mathbb{Z}_p)) = H^1_{ct}(U, L(n; \mathbb{Q}_p/\mathbb{Z}_p))$ as $L(n; \mathbb{Q}_p/\mathbb{Z}_p)$ is equipped with the discrete topology. The same remark applies also to $H^1(U/U_s, L(n; \mathbb{Z}/p^s\mathbb{Z}))$. There is another proof of this fact due to Shimura [CPS, 68c, 5.7].

Proof. Write $L(A) := L(n; A)$ for simplicity. By the long exact sequence attacher to the short one $L(\mathbb{Z}_p) \hookrightarrow L(\mathbb{Q}_p) \twoheadrightarrow L(\mathbb{Q}_p/\mathbb{Z}_p)$, we have an exact sequence of $\mathbb{Z}_p$-modules: $H^{q-1}(U, L(\mathbb{Q}_p/\mathbb{Z}_p)) \to H^q_{ct}(U, L(\mathbb{Z}_p)) \to H^q_{ct}(U, L(\mathbb{Q}_p)) \to H^q(U, L(\mathbb{Q}_p/\mathbb{Z}_p))$ (with putting $H^{-1}(U, L(\mathbb{Q}_p/\mathbb{Z}_p)) := 0$ as a convention). Since cohomology group of p-torsion coefficients is p-torsion, $H^q(U, L(\mathbb{Q}_p/\mathbb{Z}_p)) \otimes_{\mathbb{Z}_p} \mathbb{Q}_p = 0$. Thus we have an equality:

$$H^q_{ct}(U, L(\mathbb{Z}_p)) \otimes_{\mathbb{Z}_p} \mathbb{Q}_p \cong H^q_{ct}(U, L(\mathbb{Q}_p)) \otimes_{\mathbb{Z}_p} \mathbb{Q}_p = H^q_{ct}(U, L(\mathbb{Q}_p))$$

(see [CNF, 2.7.11] for more details of this fact). Plainly $H^0(U, L(\mathbb{Z}_p))$ is a $\mathbb{Z}_p$-module of finite type, and $H^1_{ct}(U, L(\mathbb{Z}_p))$ is also a $\mathbb{Z}_p$-module of finite type by the 1-cocycle relation as U is topologically generated by finitely many elements. Thus the vanishing $H^q_{ct}(U, L(\mathbb{Q}_p)) = H^q(\mathfrak{sl}_2(\mathbb{Z}_p), L(\mathbb{Q}_p)) = 0$ by the above two theorems tells us that $H^q_{ct}(U, L(\mathbb{Z}_p))$ is finite for $q = 0, 1$.

By the same long exact sequence as above with vanishing of $H^q_{ct}(U, L(\mathbb{Q}_p))$ for $q = 0, 1, 2$, we have $H^{q-1}(U, L(\mathbb{Q}_p/\mathbb{Z}_p)) \cong H^q_{ct}(U, L(\mathbb{Z}_p))$ for $q = 1, 2$. Thus $H^q(U, L(n; \mathbb{Q}_p/\mathbb{Z}_p))$ is finite for $q = 0, 1$.

By the long exact sequence attached to the short one: $L(\mathbb{Z}/p^s\mathbb{Z}) \hookrightarrow L(\mathbb{Q}_p/\mathbb{Z}_p) \xrightarrow[\twoheadrightarrow]{x \mapsto p^s x} L(\mathbb{Q}_p/\mathbb{Z}_p)$, we get another exact sequence:

$$H^0(U, L(\mathbb{Q}_p/\mathbb{Z}_p)) \otimes \mathbb{Z}/p^s\mathbb{Z} \to H^0(U, L(\mathbb{Z}/p^s\mathbb{Z})) \to H^1(U, L(\mathbb{Q}_p/\mathbb{Z}_p))[p^s],$$

which tells us that $|H^0(U, L(\mathbb{Z}/p^s\mathbb{Z}))|$ is bounded independently of s (as $H^1(U, L(\mathbb{Q}_p/\mathbb{Z}_p))$ and $H^0(U, L(\mathbb{Q}_p/\mathbb{Z}_p))$ are finite). By inflation-restriction, $H^1(U/U_s, L(\mathbb{Z}/p^s\mathbb{Z})) \to H^1(U, L(\mathbb{Z}/p^s\mathbb{Z})$ is injective, we find that $|H^1(U/U_s, L(\mathbb{Z}/p^s\mathbb{Z}))|$ is bounded independently of s. $\qquad\square$

Proof of Theorem 4.2.17. Let $\Phi^s_r = \Gamma_1(Np^r) \cap \Gamma(p^s)$ and $U^s_r := \Gamma_1(Np^r)/\Phi^s_r \cong \widehat{U}_r/\widehat{\Gamma}(p^s)$. Note that $\widehat{U}_r := \varprojlim_s U^s_r$ is an open subgroup of $\mathrm{SL}_2(\mathbb{Z}_p)$ (the p-adic closure of $\Gamma_1(Np^r)$ in $\mathrm{SL}_2(\mathbb{Z}_p)$). We consider the inflation-restriction exact sequence (which is also valid for continuous cohomology):

$$0 \to H^1(U^s_r, L(\mathbb{Z}/p^s\mathbb{Z})) \to H^1(\Gamma_1(Np^r), L(\mathbb{Z}/p^s\mathbb{Z}))$$
$$\xrightarrow{\mathrm{Res}_s} H^1(\Phi^s_r, L(\mathbb{Z}/p^s\mathbb{Z})) \xrightarrow[(*)]{\sim} H^1(\Phi^s_r, \mathbb{Z}/p^s\mathbb{Z}) \otimes_{\mathbb{Z}} L(\mathbb{Z}/p^s\mathbb{Z}),$$

where the identity at $(*)$ follows from the fact that Φ^s_r acts trivially on $L(\mathbb{Z}/p^s\mathbb{Z})$. By the long exact sequence attached to $\mathbb{Z}_p \xrightarrow{x \mapsto p^s x} \mathbb{Z}_p \twoheadrightarrow \mathbb{Z}/p^s\mathbb{Z}$, we have a short exact sequence:

$$0 \to H^1(\Phi^s_r, \mathbb{Z}_p) \otimes_{\mathbb{Z}} \mathbb{Z}/p^s\mathbb{Z} \to H^1(\Phi^s_r, \mathbb{Z}/p^s\mathbb{Z}) \to H^2(\Phi^s_r, \mathbb{Z}_p)[p^s] = 0.$$

Thus

$$\begin{aligned}
H^1(\Phi^s_r, \mathbb{Z}_p) \otimes_{\mathbb{Z}} \mathbb{Z}/p^s\mathbb{Z} &\cong H^1(\Phi^s_r, \mathbb{Z}/p^s\mathbb{Z}), \\
H^1(\Phi^s_r, \mathbb{Z}_p) \otimes_{\mathbb{Z}} \mathbb{Q}_p/\mathbb{Z}_p &\cong H^1(\Phi^s_r, \mathbb{Q}_p/\mathbb{Z}_p).
\end{aligned} \tag{4.46}$$

Passing to the inductive limit with respect to s, we have a morphism

$$\mathrm{Res}_\infty : H^1(\Gamma_1(Np^r), L(\mathbb{Z}/p^s\mathbb{Z})) \to \varinjlim_s H^1(\Phi^s_r, \mathbb{Z}/p^s\mathbb{Z}) \otimes_{\mathbb{Z}} L(\mathbb{Z}/p^s\mathbb{Z}).$$

Write $\Phi_r^s = \Gamma_1(Np^r) \cap \Gamma_1(p^s)$. We have the commutative diagram:

$$
\begin{array}{ccc}
H^1(\Gamma_1(Np^r), L(\mathbb{Q}_p/\mathbb{Z}_p)) & \longrightarrow & \varinjlim_s H^1(\Phi_r^s, L(\mathbb{Z}/p^s\mathbb{Z})) \\
\downarrow{\scriptstyle \mathrm{Res}_\infty} & & \downarrow{\scriptstyle I} \\
\varinjlim_s H^1(\Phi_r^s, \mathbb{Z}/p^s\mathbb{Z}) \otimes_{\mathbb{Z}} L(\mathbb{Z}/p^s\mathbb{Z}) & \longrightarrow & \varinjlim_s H^1(\Phi_r^s, \mathbb{Z}/p^s\mathbb{Z}) \otimes_{\mathbb{Z}} \mathbb{Z}/p^s\mathbb{Z} \\
\| \downarrow & & \| \downarrow \\
(\varinjlim_s H^1(\Phi_r^s, \mathbb{Z}_p)) \otimes_{\mathbb{Z}} L(\mathbb{Q}_p/\mathbb{Z}_p) & \xrightarrow{\mathrm{id}_\infty \otimes i_\infty} & (\varinjlim_s H^1(\Phi_r^s, \mathbb{Z}_p)) \otimes_{\mathbb{Z}} \mathbb{Q}_p/\mathbb{Z}_p,
\end{array}
$$

where i_∞ is the limit of $i_{s,*}$ given by $i_s(P(X,Y)) = P(1,0)$, and the identity maps id_s of $H^1(\Phi_r^s, \mathbb{Z}/p^s\mathbb{Z})$ induces the identity id_∞ of $\varinjlim_s H^1(\Phi_r^s, \mathbb{Z}_p)$. The last vertical identity follows from (4.46). Since the image of Res_∞ is stable under the action of $\widehat{U}_r$ which is irreducible on $L(\mathbb{Q}_p)$; so, $\mathrm{id}_\infty \otimes i_\infty$ on the image of Res_∞ has finite kernel. By Corollary 4.2.35, Res_∞ has finite kernel.

The commutativity of the diagram tells us that I has finite kernel. Since I is Hecke equivariant, the restriction of Hecke operators in $\mathbf{H}_2(Np^\infty; \mathbb{Z}_p)$ acting on the target of I to $I(H^1(\Gamma_1(Np^r), L(\mathbb{Q}_p/\mathbb{Z}_p)))$ induces a surjection $\mathbf{H}_2(Np^\infty; \mathbb{Z}_p) \twoheadrightarrow \mathbf{H}_k(Np^r, \mathbb{Z}_p)$. Passing to the limit with respect to r, we get a surjection $\pi_k : \mathbf{H}_2(Np^\infty, \mathbb{Z}_p) \twoheadrightarrow \mathbf{H}_k(Np^\infty, \mathbb{Z}_p)$ sending $T(n)$ to $T(n)$. Passing to the limit with respect to k, we get $\pi_\infty : \mathbf{H}_2(Np^\infty, \mathbb{Z}_p) \twoheadrightarrow \mathbf{H}_\infty(Np^\infty, \mathbb{Z}_p)$ sending $T(n)$ to $T(n)$. By (4.14), we already have a surjection $\pi' : \mathbf{H}_\infty(Np^\infty, \mathbb{Z}_p) \twoheadrightarrow \mathbf{H}_2(Np^\infty, \mathbb{Z}_p)$ sending $T(n)$ to $T(n)$ for all n; so, $\pi_\infty \circ \pi'$ and $\pi' \circ \pi_\infty$ are both identities, and we get a canonical isomorphism $\mathbf{H}_\infty(Np^\infty, \mathbb{Z}_p) \cong \mathbf{H}_2(Np^\infty, \mathbb{Z}_p)$. Since $\mathbf{H}_\infty(Np^\infty, \mathbb{Z}_p) \twoheadrightarrow \mathbf{H}_k(Np^\infty, \mathbb{Z}_p) \twoheadrightarrow \mathbf{H}_2(Np^\infty, \mathbb{Z}_p)$ are all surjections and the composite is an isomorphism, we get $\mathbf{H}_k(Np^\infty, \mathbb{Z}_p) \cong \mathbf{H}_2(Np^\infty, \mathbb{Z}_p)$ sending $T(n)$ to $T(n)$. We can argue similarly using $H_!^1$ and obtain $\mathbf{h}_k(Np^\infty, \mathbb{Z}_p) \cong \mathbf{h}_2(Np^\infty, \mathbb{Z}_p)$. $\qquad\square$

Remark 4.2.36. The finiteness of $\mathrm{Ker}(\mathrm{Res}_\infty)$ was first noticed by Shimura in [CPS, 68c, 5.7], whose proof is an astute trick moving from the quaternion algebra producing the Shimura curve to its base change quaternion algebra (over a well chosen extension field without much changing the kernel) which gives a Shimura variety of higher dimension without the first cohomology H^1 of characteristic 0 coefficients. Therefore the proof is different from the one given for Corollary 4.2.35 which is based on general cohomology theory. Note here the argument proving Corollary 4.2.35 can be generalized beyond GL(2). In Shimura's paper, he constructed p-adic Galois representations out of Hecke eigenforms on each Shimura curve over a general totally real field (reducing the construction by R_∞ to his earlier work on the congruence relation of Hecke operators of weight 2 [S67]). Since we did not know at the time the multiplicity one property for Hecke eigenforms on general Shimura curves, his results are slightly short of reaching 2-dimensional Galois representations attached to Hilbert modular Hecke eigenforms which are the Jacquet–Langlands–Shimizu image of the starting eigenforms on the curve. The construction of the 2-dimensional representation is later carried out by Ohta following Shimura's idea [O82].

4.2.15   Control theorem

Fix a subgroup Γ with $\Gamma_1(N\mathbf{p}) \subset \Gamma \subset \Gamma_0(N\mathbf{p})$. We find that $\ell\langle\ell\rangle_2 = T(\ell)^2 - T(\ell^2)$ of weight 2 is sent to $\langle\ell\rangle_a = \ell^{k-1}\langle\ell\rangle_k = T(\ell)^2 - T(\ell^2)$ of weight k in $\mathbf{H}_k^{ord}(Np^\infty; W)$ (see Exercise 4.1.16). Anyway, by $\Gamma \ni \ell \mapsto \langle\ell\rangle_a = \ell^{k-1}\langle\ell\rangle_k$, $\mathbf{H}_k(Np^\infty; W)$ has a character $\iota : \Gamma \to \mathbf{H}_k(Np^\infty; W)^\times$, which induces $W[[\Gamma]]$-algebra structure on $\mathbf{H}_k(Np^\infty; W)$ and $\mathbf{H}_k^{ord}(Np^\infty; W)$ independent of $k \geq 2$. In other words, the $W[[\Gamma]]$-algebra structure of the Hecke algebras is given by the automorphic diamond operators $\langle z\rangle_a$, and the quotient $\mathbf{H}_k(Np^\infty; W)/(t^{p^r} - \gamma^{(k-1)p^r})$ for $t = 1 + T$ surjects onto $\mathbf{H}_k(Np^r\mathbf{p}; W)$ and acts naturally on $H^1(\Gamma_1(Np^r\mathbf{p}), L(k - 2; K/W))$. The natural morphism $\mathbf{H}_k(Np^\infty; W)/(t^{p^r} - \gamma^{(k-1)p^r}) \twoheadrightarrow \mathbf{H}_k(Np^r\mathbf{p}; W)$ sending $T(n)$ to $T(n)$ has a huge kernel. However, as we will see later, for the ordinary part, the morphism is an isomorphism (we call this fact *control* of the ordinary Hecke algebra).

In this section, we study control of cohomology groups, meaning

$$H_{ord}^1(\Gamma_1(Np^\infty), L(k - 2; K/W))[t^{p^r} - \gamma^{(k-1)p^r}]$$
$$\cong H_{ord}^1(\Gamma_1(Np^r\mathbf{p}), L(k - 2; K/W))$$

under the action of $\ell \mapsto \ell^{k-1}\langle\ell\rangle_k$. The idea is to use instead the action $\langle\ell\rangle_k$ which fits well with inflation restriction sequence. As a consequence, we want to prove the following result which basically implies Theorem 4.1.24:

Theorem 4.2.37. *Let $\Lambda = W[[\Gamma]]$, and regard cohomology groups as a Λ-module by automorphic diamond operators. Then, as Λ-module, the Pontryagin dual $H_{ord}^1(Y_1(Np^\infty), L(n; K/W))^*$ of the cohomology group $H_{ord}^1(Y_1(Np^\infty), K/W)$ (which is isomorphic to $H_1^{ord}(Y_1(Np^\infty), W) = \varprojlim_r H_1^{ord}(Y_1(Np^r), W)$ by Poincaré's duality) is free of finite rank over*

Λ_W. *Moreover, writing* $t = 1 + T$ *in* Λ, *the quotient*

$$H_1^{ord}(Y_1(Np^\infty), W) \otimes_\Lambda \Lambda/(t^{p^r} - (\gamma^{p^r})^{k-1})$$

is canonically isomorphic to the Pontryagin dual

$$H_{ord}^1(\Gamma_1(Np^r\mathbf{p}), L(n, K/W))^*$$

of $H_{ord}^1(\Gamma_1(Np^r), L(n, K/W))$ *for* $n = k - 2$. *This isomorphism commutes with Hecke operators* $T(n)$ *for all* n. *In the same way, for a subgroup* Γ *with* $\Gamma_1(Np) \subset \Gamma \subset \Gamma_0(Np)$, $H_{ord}^1(\Gamma_H(p^\infty), L(n, \psi; K/W))^*$ *for* $H = \{1\}$ *as the subgroup of* $\mathbb{Z}_p^\times$ *is free of finite rank over* Λ_W, *and we have*

$$H_{ord}^1(\Gamma_H(p^\infty), W) \otimes_\Lambda \Lambda/(t^{p^r} - (\gamma^{p^r})^{k-1}) \cong H_{ord}^1(\Gamma_1(Np^r\mathbf{p}), L(n, K/W))^*.$$

If $H = \mathbb{Z}_p^\times$ *and* Γ *as above, for a finite order character* $\psi : (\mathbb{Z}/N\mathbb{Z})^\times \times \mathbb{Z}_p^\times \to W^\times$, *we have*

$$H_{ord}^1(\Gamma_H(p^\infty), W) \otimes_\Lambda \Lambda/(t - \psi(\gamma)\gamma^{k-1}) \cong H_{ord}^1(\Gamma_H(p^r), L(n, \psi; K/W))^*,$$

where r *is a positive integer such that* $\mathrm{Ker}(\psi) \supset 1 + p^r\mathbb{Z}_p$.

We have a long way to go to prove this theorem. Since $\mathbf{H}_k^{ord}(Np^\infty; W) \cong \mathbf{H}^{ord}(Np^\infty; W)$ by Theorem 4.2.17, $\mathbf{H}^{ord}(Np^\infty; W)/(t - \gamma^{k-1})$ is almost isomorphic to the level Np weight k Hecke algebra $\mathbf{H}_k^{ord}(N\mathbf{p}; W)$ by an isomorphism sending $T(n)$ to $T(n)$. Then by duality between (the ψ-part of) the Hecke algebra and $M_\psi(N; \Lambda)$, we get Theorem 4.1.24. This is our strategy of proving the theorem.

4.2.16 *Control of cohomology*

Write simply $\Phi_{r,m} = \Gamma_0(p^m) \cap \Gamma_r = \Gamma_H(p^m)$ for $H = 1 + p^r\mathbb{Z}_p$ with $\Gamma = \Gamma_1(N)$ and $\Gamma_r = \Gamma_1(p^rN)$ for $m \geq r > 0$. These groups are torsion-free if $Np^r \geq 4$ [MFM, Theorem 4.2.9] (this follows from the fact that square and cubic roots of unity is separated modulo n if $n \geq 4$). Then Γ_m is a normal subgroup of $\Phi_{r,m}$. The quotient $\overline{\Phi} = \Phi_{r,m}/\Gamma_m$ is a p-abelian group of order p^{r-m}.

Let ψ be a character of $(\mathbb{Z}/Np^m\mathbb{Z})^\times$ with values in $W^\times$. We define the action of $U(p)$ on $H^q(\overline{\Phi}, L(n, \psi; K/W)^{\Gamma_m})$ in the following way: Note that $i : \overline{\Phi} \hookrightarrow (\mathbb{Z}/p^m\mathbb{Z})^\times$ by $\begin{pmatrix} a & b \\ c & d \end{pmatrix} \mapsto (a^{-1} \mod p^m) = (d \mod p^m)(\mathbb{Z}/p^m\mathbb{Z})^\times$. Take a q-cocycle $c : \overline{\Phi}^q \to L(n, \psi; K/W)^{\Gamma_m}$. We decompose $\Phi_{r,m}\alpha_0\Phi_{r,m} = \bigsqcup_{j=0}^{p-1} \Phi_{r,m}\alpha_j$ for $\alpha_j = \begin{pmatrix} 1 & j \\ 0 & p \end{pmatrix}$. Note that $\alpha_j^t \in \Delta_0(pN)^t$ acts trivially on $L(n, \psi; A)$. For $\gamma, \ldots, \in \Phi_{r,m}$, write $\alpha_j\gamma = \gamma_j\alpha_{\gamma^*(j)}$ for $\gamma_j \in \Phi_{r,m}$. Then $c|U(p)(\gamma) = \sum_{j=0}^{p-1} c(\gamma_j)$ for 1-cocycle c and $c|U(p)(\gamma, \delta) = \sum_{j=0}^{p-1} c(\gamma_j, \delta_j)$

for 2-cocycle c. Note that $i(\gamma_j) = i(\gamma)$, and therefore, we find that $c|U(p) = p \cdot c$. By the construction of Inf, Res and Trans in the proof of Theorem 4.2.7, these maps in the exact sequence of Theorem 4.2.7 can be easily checked to be $U(p)$-linear. Thus we get the following result:

Lemma 4.2.38. *The above operator $U(p)$ on $H^q(\overline{\Phi}, L(n, \psi; K/W)^{\Gamma_m})$ is well defined, and the group $H^q(\overline{\Phi}, L(n, \psi; K/W)^{\Gamma_m})$ $(q = 1, 2)$ is killed by $U(p)^M$ for a sufficiently large integer $M > 0$.*

Exercise 4.2.18. Give a detailed proof of the above lemma.

By the above lemma, we get

Proposition 4.2.39. *Suppose that $\Phi_{r,m}$ is torsion-free (for integers $m \geq r > 0$), and let ψ be a character of $(\mathbb{Z}/Np^m\mathbb{Z})^\times$ with values in $W^\times$. Then we have*

$$H^1_{ord}(\Phi_{r,m}, L(n, \psi; K/W)) \cong H^1_{ord}(\Gamma_m, L(n, \psi; K/W))^{\overline{\Phi}},$$

and

$$H^1_{ord}(\Phi_{r,m}, K/W) \cong H^1_{ord}(\Gamma_m, K/W)[\psi],$$

where

$$H^1_{ord}(\Gamma_m, K/W)[\psi]$$
$$= \left\{ c \in H^1_{ord}(\Gamma_m, K/W) \big| c|\langle z_\Gamma \rangle_k = \psi(z)c \text{ for } z \in \overline{\Phi} \hookrightarrow (\mathbb{Z}/Np^m\mathbb{Z})^\times \right\},$$

where z_Γ is the projection of z to the image of Γ in $(\mathbb{Z}/Np^m\mathbb{Z})^\times$. In particular, all these modules are divisible of finite corank.

Proof. The assertion follows from the above lemma. Indeed, by the above lemma, $H^q_{ord}(\overline{\Phi}, L(n, \psi; K/W)^{\Gamma_m}) = 0$. Then applying e to the exact sequence in Theorem 4.2.7, we get an isomorphism $H^1_{ord}(\Phi_{r,m}, L(n, \psi; K/W)) \overset{\text{Res}}{\cong} H^1_{ord}(\Phi_{r,m}, L(n, \psi; K/W))^{\overline{\Phi}}$. Though we have $L(0, \psi; A) = A$ as Γ_m-modules, by the definition of action of $\overline{\Phi}$ on the cohomology group $H^1(\Gamma_m, L(0, \psi; A))$, we find

$$H^1(\Gamma_m, L(0, \psi; A))^{\overline{\Phi}} = H^1(\Gamma_m, A)[\psi].$$

The divisibility then follows from Proposition 4.2.29. $\square$

Proposition 4.2.40. *Let $\Gamma = 1 + p\mathbb{Z}_p$ act on the cohomology group $H^1_{ord}(\Gamma_1(Np^m), L(k - 2, K/W))$ via diamond operators $\langle z \rangle_k$ of weight k, and write $\Gamma^{(p^r)} = 1 + p^r\mathbb{Z}_p$. Then for $0 < r \leq m$, the restriction map induces an isomorphism*

$$H^1_{ord}(\Gamma_1(Np^m), L(k - 2, K/W))^{\Gamma^{(p^r)}} \cong H^1_{ord}(\Gamma_1(Np^r), L(k - 2, K/W))$$

for $r = 1, 2, \ldots, \infty$.

For $r > 0$, if $p > 2$, $\Gamma^{(p^{r+1})} = \Gamma^{p^r} = \mathbf{\Gamma}^{p^r}$ and if $p = 2$, $\Gamma^{(p^{r+2})} = \mathbf{\Gamma}^{p^r}$.

Proof. By Lemma 4.2.14 and Corollary 4.2.15, we have the following commutative diagram:

$$
\begin{array}{ccc}
H^1(\Gamma_r, L(k-2; K/W)) & \xrightarrow{\text{Res}} & H^1(\Phi_{r,m}, L(k-2; K/W)) \\
{\scriptstyle U(p)}\downarrow & {\scriptstyle u}\nearrow & \downarrow{\scriptstyle U(p)} \\
H^1(\Gamma_r, L(k-2; K/W)) & \xrightarrow{\text{Res}} & H^1(\Phi_{r,m}, L(k-2; K/W))
\end{array}
$$

for $u = [\Gamma_r \alpha \Phi_{r,m}]$ with $\alpha = \left(\begin{smallmatrix} 1 & 0 \\ 0 & p \end{smallmatrix}\right)$. Since $U(p)$ is bijective on $H^1_{ord}(\Gamma_r, L(k-2; K/W))$ and $H^1_{ord}(\Phi_{r,m}, L(k-2; K/W))$, we conclude

$$
H^1_{ord}(\Gamma_r, L(k-2; K/W)) \cong H^1_{ord}(\Phi_{r,m}, L(k-2; K/W)).
$$

Then the assertion for finite r follows from Proposition 4.2.39. Then passing to the limit, we get the assertion for $r = \infty$. $\quad\square$

4.2.17 *Co-freeness over the group algebra*

We start with a lemma:

Lemma 4.2.41. *Let A be a reduced local ring free of finite rank over W. If M be a W-free module of finite rank with an action of A such that for any algebra homomorphism $\lambda : A \to \overline{\mathbb{Q}}_p$, $M \otimes_A A/\operatorname{Ker}(\lambda)$ is W-free, then M is A-free.*

Here the reducedness means that there is no nontrivial nilpotent element in A.

Proof. Let $\mathfrak{m}_A$ be the maximal ideal of A with residue field $\mathbb{F}$. Then by Nakayama's lemma [CRT, Theorem 2.2], $m = \dim_{\mathbb{F}} M \otimes_A \mathbb{F}$ (which is finite $\le \operatorname{rank}_W M$) is the least number of generators of M over W. Pick a set of generators $\{x_1, \ldots, x_m\}$ of M over A. Then $\pi : A^m \ni (a_1, \ldots, a_m) \mapsto \sum_j a_j x_j \in M$ is a surjective A-linear map. Writing $A_\lambda = A/\operatorname{Ker}(\lambda) = \operatorname{Im}(\lambda)$, we first assume that $A_\lambda = W$ for all λ. By the transitivity of the tensor products, we have $(M \otimes_A A_\lambda) \otimes_{A_\lambda} \mathbb{F} = A \otimes_A \mathbb{F}$, and hence $M_\lambda = M \otimes_A A_\lambda$ is W-torsion-free and the least number of generators of M_λ is m; so, $M_\lambda \cong W^m$. This shows $\operatorname{Ker}(\pi) \subset \operatorname{Ker}(\lambda)^m$. By the reducedness, $\cap_\lambda \operatorname{Ker}(\lambda) = 0$, we find that π is an isomorphism. If $A_\lambda \supsetneq W$, we take sufficiently large extension K'/K such that K' contain the image of all λ and replace A by $A' = A \otimes_W W'$ and M by $M \otimes_W W'$ for the p-adic integer ring W' of K'. Then by the argument as above, we find that $M \otimes_W W'$ is A'-free of finite rank. Then M is A-free of finite rank (because W' is W-free of finite rank). $\quad\square$

Exercise 4.2.19. Give a detailed proof of the above lemma. In particular the following points:

(1) There exists a finite extension K'/K in $\overline{\mathbb{Q}}_p$ which contains the image of all W-algebra homomorphism $\lambda : A \to \overline{\mathbb{Q}}_p$;
(2) $\bigcap_\lambda \operatorname{Ker}(\lambda) = 0$ if A is reduced;
(3) M is A-free if $M \otimes_W W'$ is A'-free.

Exercise 4.2.20. Let G be a finite abelian p-group. Prove that the group algebra $W[G]$ is a local ring.

Theorem 4.2.42. *Recall* $\Gamma^{(p^m)} = 1 + p^m \mathbb{Z}_p$. *Let*

$$M = M_m := \operatorname{Hom}(H^1_{ord}(\Gamma_m, K/W), \mathbb{Q}_p/\mathbb{Z}_p) \text{ (the Pontryagin dual)},$$

and assume that $\Phi_{r,m}$ *is torsion-free. Then* M *is a* $W[\Gamma/\Gamma^{(p^m)}]$-*free module of finite rank. In particular* $M_m/\mathfrak{a}^r_m M_m \cong M_r$ *for all* $m \geq r > 0$ *for the kernel* $\mathfrak{a}^r_m$ *of the natural projection map* $W[\Gamma/\Gamma^{(p^m)}] \to W[\Gamma/\Gamma^{(p^r)}]$. *If* p *is odd or* $r \geq 2$, *the ideal* $\mathfrak{a}^r_m$ *is generated by* $\gamma^{p^r \mathbf{p}^{-1}} - 1 \in W[\Gamma/\Gamma^{(p^m)}]$.

Proof. Since $\Gamma^{(p^r)}$ is (topologically) cyclic if $p > 2$ or $r \geq 2$, we get $\mathfrak{a}^r_m = (\gamma^{p^r \mathbf{p}^{-1}} - 1)$. Note that $H^1_{ord}(\Gamma_m, K/W)^{\Gamma_r}$ is the subspace of $H^1_{ord}(\Gamma_m, K/W)$ killed by $\mathfrak{a}^r_m$ (by Proposition 4.2.40); so, by duality, we get $M_m/\mathfrak{a}^r_m M_m \cong M_r$.

Since any algebra homomorphism $W[\Gamma/\Gamma^{(p^m)}]$ is induced by a character $\psi : \Gamma/\Gamma^{(p^m)} \to \mathcal{O}Q_p^\times$, by duality (and Proposition 4.2.39), we find M_λ which is then Pontryagin dual of $H^1_{ord}(\Gamma_m, K/W)[\psi]$ is W-free of finite rank. Therefore, M is $W[\Gamma/\Gamma^{(p^m)}]$-free module of finite rank. $\square$

Corollary 4.2.43. *Let the notation be as in* Theorem 4.2.42. *The limit* $M_\infty = \varprojlim_m M_m$ *is a* Λ-*free module of finite rank, and* $M_\infty/(t - \psi(\gamma))M_\infty$ *is isomorphic to the Pontryagin dual of* $H^1_{ord}(\Gamma_0(Np^m), K/W)[\psi]$ *for all characters* $\psi : (\mathbb{Z}/Np^m\mathbb{Z})^\times \to W^\times$ *factoring through* Γ_m.

Since $\Phi_{r,m}$ is torsion-free for sufficiently large r, we find that M_∞ is $W[[\Gamma^{p^{r-1}}]]$-free of finite rank. Then actually M_∞ is also Λ-free of finite rank. Thus in this corollary, we do not need to assume torsion-freeness of $\Phi_{r,m}$. We leave a detailed proof of the last assertion to the attentive reader:

Exercise 4.2.21. Prove the above remark that a torsion-free $W[[\Gamma]]$-module M (of finite type) is actually free if it is free over $W[[\Gamma^{p^r}]]$ for $r > 0$.

We have basically proven the following result for $k = 2$ (see [H86b] for the general case of $k \geq 2$). Here we note that the Λ-algebra structure on $\mathbf{H}^{ord}(Np^\infty; W)$ is the twist twice by the (identity inclusion) character $\Gamma \ni \delta \mapsto \delta \in W^\times$ of the action on the cohomology groups.

Corollary 4.2.44. *The algebra $\mathbf{H}^{ord}(Np^\infty; W)$ is a torsion free Λ-module of finite type and has a canonical surjective homomorphism*

$$\pi_k : \mathbf{H}^{ord}(Np^\infty; W) \otimes_\Lambda \Lambda/(t - \gamma^k) \to \mathbf{H}_k^{ord}(Np; W)$$

taking $T(n)$ to $T(n)$, where $\mathbf{H}_k^{ord}(Np; W)$ is the W-algebra in the endomorphism algebra $\mathrm{End}(H_{ord}^1(\Gamma_1(Np); L(k-2; K)))$ generated by $T(n)$ over W. Moreover we have

$$\pi_k : \mathbf{H}^{ord}(Np^\infty; W) \otimes_\Lambda \Lambda/(t - \gamma^k) \otimes_W K \cong \mathbf{H}_k^{ord}(Np; K)$$

for all $k \geq 2$.

By Eichler–Shimura isomorphism, $\mathbf{H}_k^{ord}(Np; W)$ is the modular W-algebra in the endomorphism algebra $\mathrm{End}(M_k^{ord}(\Gamma_1(Np); K))$ generated by $T(n)$ over W. The last assertion can be proven, for example, by counting the dimension of the left and right-hand-side. Here is a sketch how to prove the higher weight $k > 2$ case. For simplicity, we assume that $p > 2$ (and leave the case of $p = 2$ to the reader).

Note that the evaluation of polynomial $L(k-2; W_m) \ni P \mapsto P(1,0) \in W_m(k-2)$ for $W_m = W/p^m W$ is a morphism of $\Phi_{0,m}$-modules, where $\gamma = \left(\begin{smallmatrix} a & b \\ c & d \end{smallmatrix}\right) \in \Gamma_0(Np^m)$ acts on $W_m(k-2) = W_m$ by $x \mapsto d^{k-2}x$. Thus we have a morphism $\iota : H^1(\Phi_{0,m}, L(k-2, W_m)) \to H^1(\Phi_{0,m}, W_m(k-2)) \cong H^1(\Gamma_m, W_m)[k-2]$, where $H^1(\Gamma_m, W_m)[k-2]$ is the submodule of $H^1(\Gamma_m, W_m)$ on which $\overline{\Phi} = \Phi_{0,m}/\Gamma_m \cong (\mathbb{Z}/p^m\mathbb{Z})^\times$ acts by $z \mapsto z^{k-2}$. Similarly, $W_m(2-k) \ni a \mapsto aY^{k-2} \in L(k-2, W_m)$ gives a morphism of $\Phi_{0,m}$-modules, which results $\pi_0 : H^1(\Phi_{0,m}, W_m(2-k)) \to H^1(\Phi_{0,m}, L(k-2, W_m)) \cong H^1(\Gamma_m, W_m)[k-2]$. For simplicity, suppose $N = 1$. Then $\Phi_{0,m} = \Gamma_0(p^m)$. Since $\tau = \left(\begin{smallmatrix} 0 & -1 \\ p^m & 0 \end{smallmatrix}\right)$ normalizes $\Gamma_0(p^m)$, it acts on 1-cocycles $c : \Gamma_0(p^m) \to L(k-2, W_m)$ by $\tau^\iota c(\tau\sigma\tau^{-1})$. Write this action of τ on $H^1(\Gamma_0(p^m), L(k-2, W_m))$ as $[\tau]$. Similarly, we can define $[\tau] : H^1(\Gamma_0(p^m), W_m(k-2)) \to H^1(\Gamma_0(p^m), W(2-k))$, since $\tau \left(\begin{smallmatrix} a & b \\ p^m c & d \end{smallmatrix}\right) \tau^{-1} = \left(\begin{smallmatrix} d & c \\ p^m b & a \end{smallmatrix}\right)$. Put $\pi = [\Gamma_0(p^m)\delta\Gamma_0(p^m)] \circ \pi_0 \circ [\tau]$. Then by computation, we can prove $\pi \circ \iota = U(p^m)$ on $H^1(\Gamma_0(p^m), L(k-2, W_m))$ and $\iota \circ \pi = U(p^m)$ on

$H^1(\Gamma_m, W_m)[k-2]$. Thus taking limit, we find an isomorphism

$$I : H^1_{ord}(\Gamma_0(p^\infty), L(k-2, K/W)) = \varinjlim_m H^1_{ord}(\Gamma_0(p^m), L(k-2, W_m))$$

$$\cong \varinjlim_m H^1_{ord}(\Gamma_1(p^m), W_m)[k-2] = H^1_{ord}(\Gamma_1(p^\infty), K/W)[k-2], \quad (4.47)$$

where $H^1_{ord}(\Gamma_1(p^\infty), K/W)[k-2]$ is the subspace of $H^1_{ord}(\Gamma_1(p^\infty), K/W)$ on which $\mathbb{Z}_p^\times$ acts by characters $z \mapsto z^{k-2}$. This morphism I can be checked to satisfy $T(n) \circ I = I \circ T(n)$, and in this way, we get the result of higher weight (see [H86b, Theorem 4.4] for more details).

Instead of $\varinjlim_m H^1_{ord}(\Gamma_1(Np^m), K/W)$, taking the Pontryagin dual N_∞ of the limit cohomology $\varinjlim_m H^1_{ord}(X_1(Np^m), K/W)$ and its Hecke algebra

$$\mathbf{h}^{ord}(Np^\infty; W) = \Lambda[T(n) | n = 1, 2, \dots] \subset \mathrm{End}_\Lambda(N_\infty),$$

we can prove the cuspidal version of the above corollary:

Corollary 4.2.45. *The algebra $\mathbf{h}^{ord}(Np^\infty; W)$ is a torsion free Λ-module of finite type and has a canonical surjective homomorphism*

$$\pi_k : \mathbf{h}^{ord}(Np^\infty; W) \otimes_\Lambda \Lambda/(t - \gamma^k) \to \mathbf{h}^{ord}_k(N\mathbf{p}; W)$$

taking $T(n)$ to $T(n)$, where $\mathbf{h}^{ord}_k(N\mathbf{p}; W)$ is the W-subalgebra in $\mathrm{End}(S_k(\Gamma_1(N\mathbf{p}); K))$ generated by $T(n)$ over W. Moreover we have, for $t = 1 + T$,

$$\pi_k : \mathbf{h}^{ord}(Np^\infty; W) \otimes_\Lambda \Lambda/(t - \gamma^k) \otimes_W K \cong \mathbf{h}^{ord}_k(N\mathbf{p}; K)$$

for all $k \geq 2$.

4.2.18 *Ordinary p-adic analytic families*

Write $\mathbf{h} = \mathbf{h}^{ord}(Np^\infty; W)$ simply. Define $S^{ord}(N; \Lambda) \subset \Lambda[[q]]$ by

$$S^{ord}(N; \Lambda) := \{F := \sum_{n=1}^\infty \phi(T(n))q^n | \phi \in \mathrm{Hom}_\Lambda(\mathbf{h}, \Lambda)\} \cong \mathrm{Hom}_\Lambda(\mathbf{h}, \Lambda).$$

Here the last isomorphism is an obvious one sending F to ϕ. Since at weight k and character ψ, we have from Lemma 4.1.8

$$a(m, f|T(n)) = \sum_{0 < d | (m,n), (d, pN) = 1} \psi(d) d^{k-1} \cdot a(\frac{mn}{d^2}, f).$$

Since $\mathbf{h} \otimes_\Lambda \Lambda/(t - \gamma^k) \otimes_W K \cong \mathbf{h}^{ord}_k(N\mathbf{p}, K)$ and $\mathrm{Hom}_K(\mathbf{h}^{ord}_k(N\mathbf{p}, K), K) = S_k(\Gamma_1(N\mathbf{p}), K)$, we can easily conclude that $F(\gamma^k - 1)$ is the q-expansion of an element in $S_k(\Gamma_1(N\mathbf{p}), K)$.

Writing $h_{k,\psi} = W[T(n)|n = 1, 2, \ldots] \subset \mathrm{End}_W(S_{k,\psi}(\Gamma_0(N\mathbf{p}); W))$ and identifying $S_{k,\psi}(\Gamma_0(N\mathbf{p}); W))$ with $\mathrm{Hom}_W(h_{k,\psi}, W)$, we have for $\phi \in \mathrm{Hom}_W(k_{k,\psi}, W)$,

$$\phi(T(m)T(n)) = \sum_{0<d|(m,n),(d,pN)=1} \psi(d)d^{k-1}\phi(T(\frac{mn}{d^2})).$$

Since ϕ is arbitrary, we can remove ϕ from the above formula and get an identity in $h_{k,\psi}$:

$$T(m)T(n) = \sum_{0<d|(m,n),(d,pN)=1} \psi(d)d^{k-1}T(\frac{mn}{d^2}).$$

Since k and ψ are arbitrary, we have

$$T(m)T(n) = \sum_{0<d|(m,n),(d,pN)=1} \langle d\rangle T(\frac{mn}{d^2}) \tag{4.48}$$

valid in $\mathbf{h}$, where the operator $\langle d\rangle = \langle d\rangle_a \in \mathbf{h}$ acts on $f \in S_k(\Gamma_1(Np))$ by $f \mapsto d^{k-1}f|\sigma_d$ for $\sigma_d \in \Gamma_0(Np)$ such that $\sigma_d \equiv \left(\begin{smallmatrix} a & b \\ c & d \end{smallmatrix}\right) \mod Np$. Getting back to $S^{ord}(N; \Lambda)$, the action of $T(n)$ is given by

$$a(m, F|T(n)) = \sum_{0<d|(m,n),(d,pN)=1} a(\frac{mn}{d^2}, F|\langle d\rangle). \tag{4.49}$$

Thus by Proposition 3.2.9, we find $\mathrm{Hom}_\Lambda(S^{ord}(N; \Lambda), \Lambda) \cong \mathbf{h}$. We know from group cohomology theory that $\mathbf{h}$ is a torsion-free Λ-module of finite type, and by definition $\mathbf{h}$ acts on $S^{ord}(N; \Lambda)$. Since

$$S^{ord}(N; \Lambda) \otimes_\Lambda \Lambda/(t - \boldsymbol{\gamma}^{k-1}) \cong S_k^{ord}(\Gamma_1(N\mathbf{p}); W),$$

we find that $S_k^{ord}(\Gamma_1(N\mathbf{p}); W)$ is stable under $\langle d\rangle$ for all $d \in \mathbb{Z}$ prime to Np. This fact is true without taking the ordinary part, but the proof in general requires arithmetic geometry [GME, Proposition 3.2.12]. This shows that

Theorem 4.2.46. *As Λ-modules, $\mathrm{Hom}_\Lambda(\mathbf{h}, \Lambda)$ is isomorphic to $S^{ord}(N; \Lambda)$ of ordinary analytic families of cusp forms of prime-to-p level N.*

Exercise 4.2.22. Give a detailed proof of the above theorem and Λ-freeness of $S^{ord}(N; \Lambda)$ only using the fact $S^{ord}(N; \Lambda) \cong \mathrm{Hom}_\Lambda(\mathbf{h}, \Lambda)$.

Once stability under $\langle d\rangle$ of $S_k^{ord}(\Gamma_1(N); \Lambda)$ is established, the proof of Theorem 4.1.24 proving freeness of $S_\psi^{ord}(N; \Lambda)$ still works and produces

Theorem 4.2.47. *The Λ-module $S^{ord}(N; \Lambda)$ is free of finite rank over Λ and $\mathbf{h}^{ord}(Np^\infty; W) \cong \mathrm{Hom}_\Lambda(S^{ord}(N; \Lambda), \Lambda)$ is also free of equal rank over Λ. Moreover for any weight $k \geq 2$, the specialization map induces*

$$S^{ord}(N; \Lambda) \otimes_\Lambda \Lambda/(t - \boldsymbol{\gamma}^{k-1}) \cong S_k^{ord}(\Gamma_1(N\mathbf{p}); W)$$
$$and \quad \mathbf{h}^{ord}(Np^\infty; W) \otimes_\Lambda \Lambda/(t - \boldsymbol{\gamma}^{k-1}) \cong \mathbf{h}_k^{ord}(\Gamma_1(Np); W).$$

In the above theorem, we use the Λ-module structure coming from the automorphic diamond operator $\langle z \rangle_a$. If we instead use the geometric diamond operator $\langle z \rangle_g$, we need to replace $\Lambda/(t - \gamma^{k-1})$ in the formula by $\Lambda/(t - \gamma^k)$. Thus this theorem is compatible with Theorem 4.1.26.

Proof. We have already explained why the freeness follows. As for the last assertion, the natural map $\mathbf{h}^{ord}(Np^\infty; W) \otimes_\Lambda \Lambda/(t - \gamma^{k-1}) \to \mathbf{h}_k^{ord}(\Gamma_1(Np); W)$ is a surjection, and then comparing the W-rank of the two side, we get the identity from the freeness. The surjectivity of $S^{ord}(N; \Lambda) \otimes_\Lambda \Lambda/(t - \gamma^{k-1}) \to S_k^{ord}(\Gamma_1(Np); W)$ follows arguing as in the proof of Corollary 4.1.23. Then again comparing the W-rank of the two side, we conclude the identity. $\square$

Remark 4.2.48. It is also known that

$$\mathbf{h}^{ord}(Np^\infty; W) \otimes_\Lambda \Lambda/(t - \gamma^{k-1}\psi(\gamma)) \cong \mathbf{h}_k^{ord}(\Gamma_0(Np^{r-1}\mathbf{p}), \psi; W[\psi])$$

for any Dirichlet character $\psi : (\mathbb{Z}/Np^{r-1}\mathbf{p})^\times \to W[\psi]^\times$. Here $W[\psi] \subset \overline{\mathbb{Q}}_p$ is the W-subalgebra generated by the values of ψ. See [GME, Corollary 3.2.22] for a proof.

Chapter 5

Abelian deformation

For a given group G, since representations are easier to understand than the group itself, we have a natural question: *to what extent can we describe the group G by knowing all irreducible representations of G?* If G is finite, a representation embeds G into $\mathrm{GL}_n(A)$ for a suitable ring A; so, the answer is affirmative in a naive sense. However, it is difficult to describe the image. If G is huge, even understanding the representations is a difficult task.

When G is an abelian locally compact group, the (continuous) unitary character group $\widehat{G} := \mathrm{Hom}_{ct}(G, S^1)$ ($S^1 := \{z \in \mathbb{C}^\times : |z| = 1\}$) determines G (*Pontryagin duality*). Taking G to be the Galois group of the maximal abelian extension k^{ab} of a number field k, we get an exact description of $\mathrm{Gal}(k^{ab}/k)$ (Class field theory).

If G is non-abelian, we only have its character monoid and recovering a representation from a given character is not simple. However from an appropriately chosen category Tan_G with some "multiplicative structure" (called *Tannakian category*) of representations of a compact or an algebraic group G, we can recover the group G as its automorphism group basically fixing one point and preserving tensor product. See [HMS, II] for details and [T39] for the original paper by T. Tannaka. Though the idea is profound, execution is difficult in practice as a meaningful choice of the category is often too big and complicated. For example, the motivic Galois group (far bigger than $\mathrm{Gal}(\overline{\mathbb{Q}}/\mathbb{Q})$) made this way out of the category of motives (essentially containing all projective varieties) is largely conjectural (see [MTV] for the theory of motives).

One of the ideas to increase accessibility is to fix the dimension of representations, and somehow we want to know the collection of all representation reducing to a fixed small one (deformation theory), and from that information, try to squeeze out as much arithmetic information. In this chapter, we treat the case of 1-dimensional representations (i.e., characters) and see what we can do in the easiest case.

Since any homomorphism of a finite abelian group G into a multiplicative group of a commutative ring X induces an algebra homomorphism from the group algebra $\mathbb{Z}[G]$ to A; so, $\mathrm{Hom}_{\mathrm{group}}(G, X^\times) = \mathrm{Hom}_{\mathrm{alg}}(\mathbb{Z}[G], X)$ linearizing group homomorphisms into ring homomorphisms, as rings are easier to understand and directly connected to geometry (scheme theory). The commutative ring $R := \mathbb{Z}[G]$ can be understood geometrically regarding it as an affine scheme $S = \mathrm{Spec}(R)$, and the trivial homomorphism $1 : G \to \{1\} \subset \mathbb{Z}^\times$ induces an augmentation homomorphism $\pi : R \twoheadrightarrow \mathbb{Z} =: A$ which gives rise to an inclusion $Z := \mathrm{Spec}(A) \hookrightarrow S$. The scheme S is *connected* (rather miraculously, why?), and we can recover the group G from the geometry of S. A simplest observation is that $X = Z \cup Y$ is a union of irreducible Z and another component Y which is a union of irreducible components outside Z; so, writing $Y = \mathrm{Spec}(B)$, we find $Z \cap Y = \mathrm{Spec}(A \otimes_{\mathbb{Z}[G]} B)$ following the dichotomy ($\cap \leftrightarrow \otimes$) between rings and schemes. The ring $C_0 := A \otimes_{\mathbb{Z}[G]} B$ is finite, and again miraculously $|G| = |C_0|$ (as a first approximation of G) but they are often non-isomorphic as a group. However we can show that the module of (algebro-geometric) Kähler differentials $C_1 := \Omega_{\mathbb{Z}[G]/\mathbb{Z}} \otimes_{\mathbb{Z}[G]} A$ is canonically isomorphic to G (see Theorem 5.2.9). These modules C_0 and C_1 are called congruence modules (attached to the augmentation homomorphism). In this way, just by a simple algebraic geometry, we can recover the group G from the ring R representing the functor sending a commutative algebra X to $\mathrm{Hom}_{\mathrm{group}}(G, X^\times)$. We apply this simple principle to a slightly more sophisticated setting of Galois characters having values in local rings and study associated congruence modules number theoretically in this chapter.

5.1 Abelian deformation

We describe the universal deformation ring for representations (characters) into $\mathbf{G}_m = \mathrm{GL}_1$ and introduce invariants to compute it.

We fix an odd prime p (and later vary $p > 2$). Fix a finite extension $\mathbb{F}/\mathbb{F}_p$ and a local p-profinite noetherian ring B flat over $\mathbb{Z}_p$ with residue field $\mathbb{F}$. Let $\boxed{\mathcal{C} = \mathcal{C}_B}$ be either the category of Artinian local B-algebra with residue field $\mathbb{F}$ or just p-profinite local B-algebra with residue field $\mathbb{F}$ (this category is also denoted by CL_B). Morphisms of $\mathcal{C}$ are local B-algebra homomorphisms.

Let k be a base field (a finite extension of $\mathbb{Q}$) with integer ring O. We take a Galois extension K/k, and over its Galois group G, we consider deformation. Since $G = \varprojlim_E \mathrm{Gal}(E/k)$ for E running over finite Galois extensions E/k inside K, the Galois group $G = \mathrm{Gal}(K/k)$ is a profinite group [PAF, §2.3.1]. For a representation $\rho : G \to \mathrm{GL}_n(A)$, we write $F(\rho) := K^{\mathrm{Ker}(\rho)}$ (splitting field).

5.1.1 *Deformation of a character*

The smallest choice of the base ring B is the discrete valuation ring $W = W(\mathbb{F})$ unramified over $\mathbb{Z}_p$ with residue field $\mathbb{F}$ (Witt vector ring with coefficients in $\mathbb{F}$), or you can choose a bigger one $W(\mathbb{F})[\mu_{p^r}]$ adding p^r-th roots of unity or the Iwasawa algebra $\Lambda = W[[T]]$.

We fix the origin; i.e., the starting continuous character $\overline{\rho} : G \to \mathrm{GL}_1(\mathbb{F})$. A *deformation* into $\mathbf{G}_m(A) = \mathrm{GL}_1(A)$ $(A \in \mathcal{C})$ over G is a **continuous** character $\rho_A : G \to \mathrm{GL}_1(A)$ such that $\rho_A \bmod \mathfrak{m}_A = \overline{\rho}$. The (full) deformation (covariant) functor $\mathcal{D} : \mathcal{C}_B \to SETS$ is given by

$$\mathcal{D}(A) = \{\rho_A : G \to \mathrm{GL}_1(A) : \text{deformations of } \overline{\rho}\}.$$

If $\phi \in \mathrm{Hom}_{\mathcal{C}}(A, A')$, $\rho_A \mapsto \phi \circ \rho_A$ induces covariant functoriality.

A couple $(R, \boldsymbol{\rho})$ made of an object R of $\mathcal{C}$ (or pro-category CL_B of $\mathcal{C}$) and a character $\boldsymbol{\rho} : G \to R^\times$ is called a *universal couple* for $\overline{\rho}$ if *for any deformation* $\rho : G \to A^\times$ *of* $\overline{\rho}$, *we have a unique morphism* $\phi_\rho : R \to A$ *in* CL_W (*so it is a local W-algebra homomorphism*) *such that* $\phi_\rho \circ \boldsymbol{\rho} = \rho$.

Thus $\mathcal{D}(A) \cong \mathrm{Hom}_{\mathcal{C}}(R, A)$ by $\rho_A = \phi \circ \boldsymbol{\rho} \leftrightarrow \phi \in \mathrm{Hom}_{\mathcal{C}}(R, A)$, and R (pro-)represents the functor $\mathcal{D}$. By the universality, if it exists, the couple $(R, \boldsymbol{\rho})$ is determined uniquely up to isomorphisms.

5.1.2 *Group algebra is universal*

Let G_p^{ab} be the maximal p-profinite abelian quotient $G_p = \varprojlim_n (G^{ab}/p^n G^{ab})$ for $G^{ab} = G/\overline{[G,G]}$, where $\overline{[G,G]}$ is the closure of the commutator subgroup $[G,G]$ of G. Writing $G_p^{ab} = \varprojlim_n \mathcal{G}_n$ with finite $\mathcal{G}_n$, consider the group algebra $B[[G_p^{ab}]] := \varprojlim_n B[\mathcal{G}_n]$, where the projection $\mathcal{G}_m \to \mathcal{G}_n$ induces $B[\mathcal{G}_m] \to B[\mathcal{G}_n]$ for $m > n$. Since $\mathbb{F}^\times \hookrightarrow B^\times$, we may regard $\overline{\rho}$ as a character $\rho_0 : \mathcal{G} \to B^\times$ (Teichmüller lift of $\overline{\rho}$). Define $\boldsymbol{\rho} : G \to B[[G_p^{ab}]]^\times$ by $\boldsymbol{\rho}(g) = \rho_0(g)g_p$ for the image g_p of g in G_p^{ab}. Note that $B[G_n^{ab}]$ is a local ring with residue field $\mathbb{F}$; so, is $B[[G_p^{ab}]]$.

If $A = \varprojlim_n A_n$ for finite A_n with $A_n = A/\mathfrak{m}_n$, $\rho_n := \rho_A \rho_0^{-1} \bmod \mathfrak{m}_n$:

$G \to A_n^\times$ has to factor through $\mathcal{G}_{m(n)}$ for some $m(n)$ by continuity, and we get $\varphi_n \in \mathrm{Hom}(B[\mathcal{G}_{m(n)}], A_n)$ given by $\sum_g a_g g \mapsto \sum_g a_g \rho_n \chi_0^{-1}(g) \in A$. Then $\varphi_n \circ \rho = \rho_n$. Passing to the limit, we have $\varphi \circ \rho = \rho_A$ for $\varphi = \varprojlim_n \varphi_n : B[[G_p^{ab}]] \to A$. This proves

Theorem 5.1.1. *Let the notation be as above. Then $\mathcal{D}$ is pro-represented by the universal couple $(B[[G_p^{ab}]], \rho)$.*

5.1.3 Examples of group algebras

- If G_p^{ab} is a cyclic group C of order p^r, $B[G_p^{ab}] = B[T]/(t^{p^r} - 1)$ for $t = 1 + T$ by sending a generator $g \in C$ to t.
- If $G_p^{ab} = C_1 \times \cdots \times C_n$ for p-cyclic groups C_j with order p^{r_j}, then
$$B[G_p^{ab}] = \frac{B[T_1, \dots, T_n]}{(t_1^{p^{r_1}} - 1, \dots, t_n^{p^{r_n}} - 1)} = \frac{B[[T_1, \dots, T_n]]}{(t_1^{p^{r_1}} - 1, \dots, t_n^{p^{r_n}} - 1)} \quad (t_i = 1 + T_i).$$
  Note that $f_1 := t^{p^{r_1}} - 1, \dots, f_r := t^{p^{r_n}} - 1$ in $\mathfrak{m}_{B[[T_1,\dots,T_n]]}$ is a regular sequence (e.g., §6.2.8 in the text and [CRT, §16]), and $B[G_p^{ab}]$ is free of finite rank over B. A ring of the form $B[[T_1, \dots, T_n]]/(f_1, \dots, f_n)$ with a regular sequence (f_j) in $\mathfrak{m}_{B[[T_1,\dots,T_n]]}$ is called a *local complete intersection* over B if it is free of finite rank over B [CRT, §21].
- The Iwasawa algebra $\Lambda = W[[\mathbf{\Gamma}]]$ $(\mathbf{\Gamma} = 1 + \mathbf{p}\mathbb{Z}_p = (1 + \mathbf{p})^{\mathbb{Z}_p})$ is isomorphic to $W[[T]]$ by $1 + \mathbf{p} \leftrightarrow t = 1 + T$ (e.g., Lemma 1.9.3).

5.2 A way of recovering the group and its application

The deformation functor $\mathcal{D}$ is represented by $(R = B[[G_p^{ab}]], \rho)$. Does the ring $B[[G_p^{ab}]]$ determine explicitly the group G_p^{ab}? and if yes, how? Perhaps, it is more interesting to recover the group in an arithmetic setting of deformation theory of Galois character. Since the Galois group $\mathrm{Gal}(\overline{\mathbb{Q}}/\mathbb{Q})$ is huge, we want to cut the full deformation functor $\mathcal{D}$ to a smaller piece requiring deformation to satisfy given properties $\mathcal{P}$. Fix a set $\mathcal{P}$ of properties of Galois characters. A deformation ρ_A is called $\mathcal{P}$-*deformation* if ρ_A satisfies $\mathcal{P}$. A couple (R, ρ) made of an object R of $\mathcal{C}$ (or pro-category CL_B of $\mathcal{C}$) and a character $\rho : G \to R^\times$ satisfying $\mathcal{P}$ is called a *universal couple* for $\bar{\rho}$ if

for any $\mathcal{P}$-deformation $\rho : G \to A^\times$ of $\bar{\rho}$, we have a unique morphism $\phi_\rho : R \to A$ in CL_W (so it is a local W-algebra homomorphism) such that $\phi_\rho \circ \rho = \rho$.

Examples of $\mathcal{P}$.

- Unramified everywhere (full deformation for the maximal K/k unramified everywhere);
- Unramified outside p (full deformation if we take K to be the maximal p-profinite extension of $F(\overline{\rho})$ unramified outside p);
- Unramified outside S for a fixed finite set S of places of k (full deformation if we take K to be the maximal p-profinite extension of $F(\overline{\rho})$ unramified outside S);
- Suppose that $\overline{\rho}$ is ramified at S outside p with ramification index prime to p. A deformation ρ_A is *minimal* if $\rho_A(I_l) \cong \overline{\rho}(I_l)$ by restriction for all $l \neq p$, where $I_l \subset G$ is the inertia subgroup.

The minimal deformation problem is a full deformation problem if we choose K as follows: Take $K = F^{(p)}(\overline{\rho})$ to be the maximal p-profinite extension of $F(\overline{\rho})$ unramified outside p. Since ramification of a minimal deformation ρ_A is concentrated to $F(\overline{\rho})/k$, ρ_A factors through $G = \mathrm{Gal}(G/k)$; so, our choice is this K. In the same way, we can consider the minimal p-tamely ramified deformation problem of $\overline{\rho}$ (i.e., requiring ρ_A to be minimal and tamely ramified at p). Since $F(\rho_A)/F(\overline{\rho})$ can only wildly ramify at p, ρ_A factors through $G := \mathrm{Gal}(F^{(\emptyset)}(\overline{\rho})/k)$ for the maximal p-profinite extension $F^{(\emptyset)}(\overline{\rho})$ of $F(\overline{\rho})$ unramified everywhere.

5.2.1 *Ray class groups*

Fix an O-ideal $\mathfrak{c}$. Recall

$$Cl_k(\mathfrak{c}) = \frac{\{\text{fractional } O\text{-ideals prime to } \mathfrak{c}\}}{\{(\alpha)|\alpha \equiv 1 \bmod^{\times} \mathfrak{c}\}},$$

$$Cl_k^+(\mathfrak{c}) = \frac{\{\text{fractional } O\text{-ideals prime to } \mathfrak{c}\}}{\{(\alpha)|\alpha \equiv 1 \bmod^{\times} \mathfrak{c}\infty\}}.$$

Here $\alpha \equiv 1 \bmod^{\times} \mathfrak{c}$ means that $\alpha = a/b$ for $a, b \in O$ such that $(b) + \mathfrak{c} = O$ and $a \equiv b \bmod \mathfrak{c}$ or equivalently, for all primes $\mathfrak{l}|\mathfrak{c}$, $\alpha \in O_{\mathfrak{l}}^{\times}$ and $\alpha \equiv 1$ mod $\mathfrak{l}^{v_{\mathfrak{l}}(\mathfrak{c}\infty)}$ if the $\mathfrak{l}$-primary factor of $\mathfrak{c}$ has exponent $v_{\mathfrak{l}}(\mathfrak{c})$ (if $\mathfrak{l}|\infty$, it just means α is positive at $\mathfrak{l}$). The group $Cl_k(\mathfrak{c})$ (resp. $Cl_k^+(\mathfrak{c})$) is called the (resp. strict) ray class group modulo $\mathfrak{c}$.

Write $H_{\mathfrak{c}p^n}/k$ for the ray class field modulo $\mathfrak{c}p^n$. In other words, there exists a unique abelian extension $H_{\mathfrak{c}p^n}/k$ only ramified at $\mathfrak{c}p\infty$ such that we can identify $\mathrm{Gal}(H_{\mathfrak{c}p^n}/k)$ with the strict ray class group $Cl_k^+(\mathfrak{c}p^n)$ by sending a class of prime $\mathfrak{l}$ in $Cl_k^+(\mathfrak{c}p^n)$ to the Frobenius element $\mathrm{Frob}_{\mathfrak{l}} \in$

$\mathrm{Gal}(H_{\mathfrak{c}p^n}/k)$. This map is called the Artin symbol. Writing $[\mathfrak{a}]_n$ for the class of an ideal $\mathfrak{a}$ in $Cl_k(\mathfrak{c}p^n)$, we have a commutative diagram for $m > n$:

$$
\begin{array}{ccc}
Cl_k^+(\mathfrak{c}p^m) & \xrightarrow{\;\sim\;} & \mathrm{Gal}(H_{\mathfrak{c}p^m}/k) \\
{\scriptstyle [\mathfrak{a}]_m \mapsto [\mathfrak{a}]_n}\Big\downarrow & & \Big\downarrow{\scriptstyle \sigma \mapsto \sigma|_{H_{\mathfrak{c}p^n}}} \\
Cl_k^+(\mathfrak{c}p^n) & \xrightarrow{\;\sim\;} & \mathrm{Gal}(H_{\mathfrak{c}p^n}/k).
\end{array}
$$

Sending a class $[\mathfrak{a}] \in Cl_k^+(\mathfrak{c}p^m)$ to the class $[\mathfrak{a}] \in Cl_k^+(\mathfrak{c}p^n)$, we have a projective system $\{Cl_k^+(\mathfrak{c}p^n)\}_n$. Passing to the limit, we have $Cl_k^+(\mathfrak{c}p^\infty) = \varprojlim_n Cl_k^+(\mathfrak{c}p^n) \cong \mathrm{Gal}(H_{\mathfrak{c}p^\infty}/k)$ for $H_{\mathfrak{c}p^\infty} = \bigcup_n H_{\mathfrak{c}p^n}$. Then for $H_{\mathfrak{c}p^\infty} = \bigcup_n H_{\mathfrak{c}p^n}$, $Cl_k^+(\mathfrak{c}p^\infty) \cong \mathrm{Gal}(H_{\mathfrak{c}p^\infty}/k)$ by $[\mathfrak{l}] \mapsto \mathrm{Frob}_{\mathfrak{l}}$ for primes $\mathfrak{l} \nmid \mathfrak{c}p$.

The group $Cl_k^+(\mathfrak{c}p^n)$ is finite as we have an exact sequence:

$$
(O/\mathfrak{c}p^n)^\times \xrightarrow{\;\alpha \mapsto (\alpha)\;} Cl_k^+(\mathfrak{c}p^n) \to Cl_k^+ \to 1.
$$

For the strict class group Cl_k^+ (we write the usual class group without condition at ∞ as Cl_k), note that $|Cl_k^+|/|Cl_k|$ is a factor of 2^e for the number e of real embeddings of k. Passing to the limit again, we have the following exact sequence for $O_p = \varprojlim_n O/p^n O = O \otimes_{\mathbb{Z}} \mathbb{Z}_p$:

$$
(O/\mathfrak{c})^\times \times O_p^\times \xrightarrow{\;\alpha \mapsto (\alpha)\;} Cl_k^+(\mathfrak{c}p^n) \to Cl_k^+ \to 1.
$$

Similarly we can take the limit with respect to $\mathfrak{c}$: $Cl_k(\mathfrak{c}^\infty) := \varprojlim_j Cl_k(\mathfrak{c}^j)$ which fits with the exact sequence for $O_{\mathfrak{c}} = \varprojlim_j O/\mathfrak{c}^j$

$$
O_{\mathfrak{c}}^\times \to Cl_k^+(\mathfrak{c}^\infty) \to Cl_k^+ \to 1
$$

and $Cl_k^+(\mathfrak{c}^\infty) \cong \mathrm{Gal}(H_{\mathfrak{c}^\infty}/k)$ for $H_{\mathfrak{c}^\infty} = \bigcup_j H_{\mathfrak{c}^j}$. Since $O_{\mathfrak{c}}^\times/(O_{\mathfrak{c}}^\times)^p$ is finite, this shows the following fact:

Lemma 5.2.1. *Let K/k be a Galois extension with Galois group G. If the number of primes ramifying in K/k is finite, $\mathrm{Hom}_{cont}(G, \mathbb{F}_p)$ is finite.*

Proof. Let $\mathfrak{c}$ be the product of primes ramifying in K/k. Then $Cl_k^+(\mathfrak{c}^\infty) \cong \mathrm{Gal}(H_{\mathfrak{c}^\infty}/k)$ surjects down to G^{ab}. Since the natural map $O_{\mathfrak{c}}^\times/(O_{\mathfrak{c}}^\times)^p \to Cl_k^+(\mathfrak{c}^\infty)/(Cl_k^+(\mathfrak{c}^\infty))^p$ has finite cokernel, $G^{ab}/(G^{ab})^p$ is finite. Since $\mathrm{Hom}_{cont}(G, \mathbb{F}_p) = \mathrm{Hom}_{cont}(G^{ab}, \mathbb{F}_p)$, the finiteness follows. $\square$

If $k = \mathbb{Q}$ and $\mathfrak{c} = (N)$ for $0 < N \in \mathbb{Z}$, we have $H_{\mathfrak{c}p^n}$ is the cyclotomic field $\mathbb{Q}[\mu_{Np^n}]$ for the group μ_{Np^n} of Np^n-th roots of unity; so, $Cl_{\mathbb{Q}}^+(\mathfrak{c}p^n) \cong (\mathbb{Z}/Np^n\mathbb{Z})^\times$ and $Cl_{\mathbb{Q}}^+(\mathfrak{c}p^\infty) \cong (\mathbb{Z}/N\mathbb{Z})^\times \times \mathbb{Z}_p^\times$.

Universal deformation ring for a Galois character $\bar{\rho}$. Let $C_k(p^\infty)$ (resp. C_k) for the maximal p-profinite quotient of $Cl_k^+(p^\infty)$ (resp. Cl_k^+).

Suppose $\bar{\rho}$ is minimal, and let $G = \mathrm{Gal}(K/k)$ for $K = F^{(p)}(\bar{\rho})$ (resp. $K = F^{(\emptyset)}(\bar{\rho})$). We consider minimal (resp. minimal p-unramified) deformations ρ_A. Since ramification outside l has index prime to p, we conclude $G_p^{ab} = C_k(p^\infty)$ (resp. $G_p^{ab} = C_k$). Let $H \subset H_{p^\infty}$ with $\mathrm{Gal}(H/k) = C_k(p^\infty)$ (resp. $\mathrm{Gal}(H/k) = C_k$). If $k = \mathbb{Q}$, for the minimal deformation problem, $C_k(p^\infty) = 1 + \mathbf{p}\mathbb{Z}_p =: \mathbf{\Gamma}$ and $H = \mathbb{Q}_\infty \subset \mathbb{Q}[\mu_{p^\infty}]$ for the unique $\mathbb{Z}_p$-extension $\mathbb{Q}_\infty$ of $\mathbb{Q}$ as $Cl_{\mathbb{Q}}^+(p^\infty) = \mathbb{Z}_p^\times$.

For the Teichmüller lift ρ_0 of $\bar{\rho}$ and the inclusion $\kappa : G_p^{ab} \hookrightarrow W[[G_p^{ab}]]$, we define $\rho(\sigma) := \rho_0(\sigma)\kappa(\sigma)$. The universality of the group algebra tells us

Theorem 5.2.2. *The couple $(W[[C_k(p^\infty)]], \boldsymbol{\rho})$ is universal among all minimal deformations, and $(W[C_k], \boldsymbol{\rho})$ is universal among all p-unramified minimal deformations.*

The last assertion follows from the fact $C_k \cong G_p^{ab}$ for $\mathrm{Gal}(F^{(\emptyset)}(\bar{\rho})/k)$.

Remark 5.2.3.

- As long as $\bar{\rho}$ satisfies minimality, the universal deformation ring $W[[C_k(p^\infty)]]$ is essentially independent of $\bar{\rho}$.
- If k is totally real, $\mathrm{rank}_{\mathbb{Z}_p} C_k(p^\infty)$ is expected to be 1. If k has r' complex places, then $\mathrm{rank}_{\mathbb{Z}_p} C_k(p^\infty) = r' + 1$? (Leopoldt conjecture).
- If $k = \mathbb{Q}$, $C_{\mathbb{Q}}(p^\infty) = \mathbf{\Gamma}$, so

$$W[[C_{\mathbb{Q}}(p^\infty)]] = \varprojlim_n W[\mathbf{\Gamma}/\mathbf{\Gamma}^{p^n}] = \varprojlim_n W[[T]]/(t^{p^n} - 1) = W[[T]].$$

Iwasawa algebra again shows up. In general, if $C_k = \{1\}$ and $C_k(p^\infty) \cong \mathbb{Z}_p^{r'+1}$, then $W[[C_k(p^\infty)]] \cong W[[T_1, \ldots, T_{r'+1}]]$.

5.2.2 *Differentials*

We geometrically studied tangent spaces and differentials in §2.1.2. Here we study them purely algebraically. Fix $R \in \mathcal{C}$. For a continuous R-module M, define *continuous B-derivations* by

$$Der_B(R, M) := \{\delta \in \mathrm{Hom}_B(R, M) | \delta(ab) = a\delta(b) + b\delta(a) \ (a, b \in R)\}.$$

Here B-linearity of $\delta \Leftrightarrow \delta(B) = 0$. The association $M \mapsto Der_B(R, M)$ is a covariant functor from the category $MOD_{/R}$ of continuous profinite R-modules to modules MOD, which is represented by an R-module $\Omega_{R/B}$ with universal differential $d : R \to \Omega_{R/B}$, e.g.,

$$\Omega_{R/B} = \frac{\text{free module over } R \text{ with basis } dr \ (r \in R)}{\langle\langle d(ab) - bda - adb, d(\beta a + b) - \beta da - db\rangle\rangle_{a,b \in R, \beta \in B}}.$$

Here "$\langle\langle ? \rangle\rangle$" means the $\mathfrak{m}_R$-adic closure of the R-submodule generated by "$?$".

When $\Omega_{R/B}$ is an R-module of finite type. Suppose that B is noetherian and that R is a B-module of finite type. Choose $r_1, \ldots, r_n$ so that $R = Br_1 + \cdots + Br_n$. By B-linearity, $\Omega' := \bigoplus_{r \in R} R \cdot dr / \langle d(\beta a + b) - \beta da - db \rangle_{a,b \in R}$ is generated by $dr_1, \ldots, dr_n$; so, $\langle\langle d(ab) - bda - adb \rangle\rangle_{a,b \in R, \beta \in B} \subset \Omega'$ is equal to $\langle d(ab) - bda - adb \rangle_{a,b \in R, \beta \in B}$ inside Ω'. Therefore we can replace $\langle\langle ? \rangle\rangle$ by $\langle ? \rangle$ in the definition of $\Omega_{R/B}$. In this case, by B-linearity, any B-derivation $\delta : R \to M$ is continuous.

If R is generated by one non-unit θ over B, we have $R = B[T]/(f(T))$ for a monic polynomial $f(T) \in B[T]$ with $f(T) \mod \mathfrak{m}_B = X^{\deg f(T)}$. Then we have $\Omega_{B[T]/B} = B[T]dT$ and for $f'(T) = \frac{df}{dT}(T)$.
$$\Omega_{R/B} = \Omega_{(B[T]/(f))/B} = (B[T]/(f, f'))dT \cong B[\theta]/(f'(\theta)).$$

An alternative definition of differentials. The multiplication $a \otimes b \mapsto ab$ induces a B-algebra homomorphism $m : R \widehat{\otimes}_B R \to R$ taking $a \otimes b$ to ab. We put $I = \mathrm{Ker}(m)$, which is an ideal of $R \widehat{\otimes}_B R$. Here, for profinite B-modules M and N, writing $M = \varprojlim_m M_m$ and $N = \varprojlim_n N_n$ for finite B-modules M_m and N_n, $M \widehat{\otimes}_B N = \varprojlim_{m,n} M_m \otimes_B N_n$ (profinite completion of $M \otimes_B N$). We newly define $\Omega_{R/B} = I/I^2$. The map $d : R \to \Omega_{R/B}$ given by $d(a) = a \otimes 1 - 1 \otimes a \mod I^2$ is a continuous B–derivation. Indeed
$$a \cdot d(b) + b \cdot d(a) - d(ab) = ab \otimes 1 - a \otimes b - b \otimes a + ba \otimes 1 - ab \otimes 1 + 1 \otimes ab$$
$$= ab \otimes 1 - a \otimes b - b \otimes a + 1 \otimes ab = (a \otimes 1 - 1 \otimes a)(b \otimes 1 - 1 \otimes b) \equiv 0 \mod I^2.$$
We have a morphism of functors: $\mathrm{Hom}_R(\Omega_{R/B}, ?) \to \mathrm{Der}_B(R, ?)$ ($\phi \mapsto \phi \circ d$). The equivalence of the two definitions follows from

Proposition 5.2.4 (Universality). *The above morphism of two functors $M \mapsto \mathrm{Hom}_R(\Omega_{R/B}, M)$ and $M \mapsto \mathrm{Der}_B(R, M)$ is an isomorphism, where M runs over the category of continuous R-modules. In other words, for each B-derivation $\delta : R \to M$, there exists a unique R-linear homomorphism $\phi : \Omega_{R/B} \to M$ such that $\delta = \phi \circ d$.*

Proof. The ideal I is generated over R by $d(a)$. Assume $\sum_{a,b} m(a,b)ab = 0$ (i.e., $\sum_{a,b} m(a,b)a \otimes b \in I$). Then
$$\sum_{a,b} m(a,b)a \otimes b = \sum_{a,b} m(a,b)a \otimes b - \sum_{a,b} m(a,b)ab \otimes 1$$
$$= \sum_{a,b} m(a,b)a(1 \otimes b) - b \otimes 1) = -\sum_{a,b} m(a,b)d(b).$$

Define $\phi : R \times R \to M$ by $(x, y) \mapsto x\delta(y)$ for $\delta \in Der_B(R, M)$. If $a, c \in R$ and $b \in B$, $\phi(ab, c) = ab\delta(c) = a(b\delta(c)) = b\phi(a, c)$ and $\phi(a, bc) = a\delta(bc) = ab\delta(c) = b(a\delta(c)) = b\phi(a, c)$; so, ϕ is a continuous B-bilinear map.

By the universality of the tensor product, $\phi : R \times R \to M$ extends to a B-linear map $\phi : R \widehat{\otimes}_B R \to M$. Now we see that

$$\phi(a \otimes 1 - 1 \otimes a) = a\delta(1) - \delta(a) = -\delta(a)$$

and

$$\phi((a \otimes 1 - 1 \otimes a)(b \otimes 1 - 1 \otimes b)) = \phi(ab \otimes 1 - a \otimes b - b \otimes a + 1 \otimes ab)$$
$$= -a\delta(b) - b\delta(a) + \delta(ab) = 0.$$

This shows that $\phi|_I$-factors through $I/I^2 = \Omega_{R/B}$ and $\delta = \phi \circ d$, as desired. The map ϕ is unique as $d(R)$ generates $\Omega_{R/B}$. $\qquad\square$

Corollary 5.2.5 (Fundamental exact sequences). *We have*

(i) *Suppose that $R \in CL_A$ and $A \in CL_C$ for a B-algebra C. Then we have the following natural exact sequence:*

$$\Omega_{C/B} \widehat{\otimes}_B R \longrightarrow \Omega_{R/B} \longrightarrow \Omega_{R/C} \to 0.$$

(ii) *Let $\pi : R \twoheadrightarrow C$ be a surjective morphism in CL_W, and write $J = \mathrm{Ker}(\pi)$. Then we have the following natural exact sequence:*

$$J/J^2 \xrightarrow{\beta^*} \Omega_{R/B} \widehat{\otimes}_R C \longrightarrow \Omega_{C/B} \to 0.$$

Moreover if $B = C$, then $J/J^2 \cong \Omega_{R/B} \widehat{\otimes}_R C$.

Since the proof in the two cases is similar, we only prove (ii). See [MFG, Corollary 5.13] for the details of the proof of (i).

Proof. By assumption, we have algebra morphism $B \to R \twoheadrightarrow C = R/J$. By Yoneda's lemma, we only need to prove that

$$
\begin{array}{ccccc}
Der_B(C, M) & \xrightarrow[\hookrightarrow]{\alpha} & Der_B(R, M) & \xrightarrow{\beta} & \mathrm{Hom}_C(J/J^2, M) \\
\wr \downarrow & & \wr \downarrow & & \| \downarrow \\
\mathrm{Hom}_B(\Omega_{C/B}, M) & \longrightarrow & \mathrm{Hom}_B(\Omega_{R/B} \widehat{\otimes}_R C, M) & \longrightarrow & \mathrm{Hom}_C(J/J^2, M)
\end{array}
$$

is exact for all continuous C-modules M. The first α is the pull back map. Thus the injectivity of α is obvious.

The map β is defined as follows: For a given B-derivation $D : R \to M$, we regard D as a B-linear map of J into M. Since J kills the C-module M, $D(jj') = jD(j') + j'D(j) = 0$ for $j, j' \in J$. Thus D induces C-linear map:

$J/J^2 \to M$. Then for $b \in R$ and $x \in J$, $D(bx) = bD(x) + xD(b) = bD(x)$. Thus D is C-linear, and $\beta(D) = D|_J$.

We prove the exactness at the mid-term of the exact sequence. The fact $\beta \circ \alpha = 0$ is obvious. If $\beta(D) = 0$, then D kills J and is a derivation well defined on $C = R/J$. This shows that D is in the image of α.

Now suppose that $B = C$. To show injectivity of β^*, we create a surjective C-linear map: $\gamma : \Omega_{R/B} \otimes C \twoheadrightarrow J/J^2$ such that $\gamma \circ \beta^* = \mathrm{id}$.

Let $\pi : R \to C$ be the projection and $\iota : B = C \hookrightarrow R$ be the structure homomorphism giving the B-algebra structure on R. We first look at the map $\delta : R \to J/J^2$ given by $\delta(a) = a - P(a) \mod J^2$ for $P = \iota \circ \pi$. Then

$$a\delta(b) + b\delta(a) - \delta(ab) = a(b - P(b)) + b(a - P(a)) - ab + P(ab)$$

$$\overset{P(ab)=P(a)P(b)}{=} ab - aP(b) + ba - bP(a) - ab + P(a)P(b)$$

$$= (a - P(a))(b - P(b)) \equiv 0 \mod J^2.$$

Thus δ is a B-derivation. By the universality of $\Omega_{R/B}$, we have an R-linear map $\phi : \Omega_{R/B} \to J/J^2$ such that $\phi \circ d = \delta$. By definition, $\delta(J)$ generates J/J^2 over R, and hence ϕ is surjective.

Since J kills J/J^2, the surjection ϕ factors through $\Omega_{R/B} \otimes_R C$ and induces γ. Note that $\beta(d \otimes 1_C) = d \otimes 1_C|_J$ for the identity 1_C of C; so, $\gamma \circ \beta^* = \mathrm{id}$ as desired. $\qquad\square$

Corollary 5.2.6. *Let the assumption be as in* Corollary 5.2.5. *If we restrict the functor* $M \mapsto Der_B(R, M)$ *to the category* $MOD_{/C}$ *of* C-*modules,* $\Omega_{R/B} \widehat{\otimes}_R C$ *represents* $MOD_{/C} \ni M \mapsto Der_B(R, M)$.

Proof. By Proposition 5.2.4, for each $\delta \in Der_B(R, M)$, we find a unique $\phi \in \mathrm{Hom}_R(\Omega_{R/B}, M)$ such that $\phi \circ d = \delta$. If M is a C-module, ϕ factors through $\Omega_{R/B}/J\Omega_{R/B} = \Omega_{R/B} \otimes_R C$.

Conversely, if $\phi \in \mathrm{Hom}_C(\Omega_{R/B} \otimes_R C, M)$ for a C-module M, plainly $\delta = \phi \circ (d \otimes 1)$ gives $Der_B(R, M)$; so, the result follows. $\qquad\square$

5.2.3   *Algebra* $R[M] = R \oplus M$ *and derivation*

For any continuous R-module M, we write $R[M]$ for the R-algebra with square zero ideal M. Thus $R[M] = R \oplus M$ has multiplication given by

$$(r \oplus x)(r' \oplus x') = rr' \oplus (rx' + r'x).$$

It is easy to see that $R[M] \in CNL_W$, if M is of finite type, and $R[M] \in CL_W$ if M is a p-profinite R-module. By definition,

$$Der_B(R, M) \cong \left\{ \phi \in \mathrm{Hom}_{CL_B}(R, R[M]) \big| \phi \mod M = \mathrm{id} \right\},$$

where the map is given by $\delta \mapsto (a \mapsto (a \oplus \delta(a)))$.

Note that $i : R \to R \widehat{\otimes}_B R$ given by $i(a) = a \otimes 1$ is a section of $m :$ $R \widehat{\otimes}_B R \to R$. We see easily that $R \widehat{\otimes}_B R / I^2 \cong R[\Omega_{R/B}]$ by $x \mapsto m(x) \oplus (x - i(m(x)))$. Note that $d(a) = 1 \otimes a - i(a)$ for $a \in R$.

5.2.4 Congruence modules C_0 and C_1

We now introduce some ring invariants C_0 and C_1 to recover the group G_p^{ab} out of the ring $B[[G_p^{ab}]]$. Let $\phi : R \twoheadrightarrow A \in \mathrm{Hom}_\mathcal{C}(R, A)$. We define $\boxed{C_1(\phi; A) := \Omega_{R/B} \otimes_{R,\phi} A}$. To define C_0, we assume (i) A and B have trivial nilradical (i.e., reduced), (ii) R is reduced and (iii) $R \cong B^r$ as B-modules. By reducedness, the total quotient ring $\mathrm{Frac}(R)$ can be decomposed

$$\mathrm{Frac}(R) = \mathrm{Frac}(\mathrm{Im}(\phi)) \oplus X \quad \text{(unique algebra direct sum)}.$$

Write 1_ϕ for the idempotent of $\mathrm{Frac}(\mathrm{Im}(\phi))$ in $\mathrm{Frac}(R)$. Let $\mathfrak{b} = \mathrm{Ker}(R \to X) = (1_\phi R \cap R)$, $S = \mathrm{Im}(R \to X)$ and $\mathfrak{a} = \mathrm{Ker}(\phi)$. Here the intersection $1_\phi R \cap R$ is taken in $\mathrm{Frac}(R) = \mathrm{Frac}(\mathrm{Im}(\phi)) \times X$. First note that $\mathfrak{b} = R \cap (\mathrm{Frac}(\mathrm{Im}(\phi)) \times 0)$ and $\mathfrak{a} = (0 \times X) \cap R$. Put

$$C_0 = C_0(\phi; A) := (R/\mathfrak{b}) \otimes_{R,\phi} \mathrm{Im}(\phi) \text{ and } C_1 = C_1(\phi; A) := \Omega_{R/B} \otimes_R A.$$

The module C_j is called the *congruence* module (of degree j) of ϕ. Note:

$$C_0 = \mathrm{Im}(\phi)/(\phi(\mathfrak{b})) \cong B/\mathfrak{b} \cong R/(\mathfrak{a} \oplus \mathfrak{b}) \cong S/\mathfrak{b} \tag{5.1}$$

via projection to A and S (an exercise).

Meaning of congruence modules. Suppose that B is an integral domain and $A = B$. Write $K = \mathrm{Frac}(B)$. Fix an algebraic closure $\overline{K}$ of K. Since the spectrum $\mathrm{Spec}(C_0(\phi; B))$ of the congruence ring $C_0(\phi; B)$ is the scheme theoretic intersection of $\mathrm{Spec}(B/\mathfrak{b})$ and $\mathrm{Spec}(R/\mathfrak{b})$ in $\mathrm{Spec}(R)$:

$$\mathrm{Spec}(C_0(\phi; B)) = \mathrm{Spec}(B/\mathfrak{b}) \cap \mathrm{Spec}(S/\mathfrak{a}),$$

we conclude that

Proposition 5.2.7. *Let the notation and the assumption be as above. Then a prime $\mathfrak{p}$ is in the support of $C_0(\phi; B)$ if and only if there exists an B-algebra homomorphism $\phi' : R \to \overline{K}$ factoring through $R/\mathfrak{b}$ such that $\phi(a) \equiv \phi'(a) \mod \mathfrak{p}$ for all $a \in R$.*

Proposition 5.2.8. *Let the notation and the assumption be as above. If B is a noetherian domain, we have*

$$\mathrm{Supp}_B(C_0(\phi; B)) = \mathrm{Supp}_B(C_1(\phi; B)).$$

Here is a definition valid for any commutative ring B. For a B-module M, $\mathrm{Supp}_B(M)$ is defined by a Zariski closed subset $\{P \in \mathrm{Spec}(B) | M_P \neq 0\}$ of $\mathrm{Spec}(B)$. Writing $\mathrm{Ann}_B(M) = \{x \in B | xM = 0\}$ (the *annihilator* ideal of M), we find

$$\mathrm{Supp}_B(M) = \{P \supset \mathrm{Ann}_B(M) | P \in \mathrm{Spec}(B)\}$$

if M is finitely generated over B as an B-module (see [CRT, §4]). The set $\mathrm{Ass}_B(M)$ of associated primes of M is defined to be the set of prime ideals P of B such that $P = \mathrm{Ann}_B(Bx)$ for some $x \in M$. Since the subset of minimal primes of $\mathrm{Ass}_B(M)$ is equal to the subset of minimal primes in $\mathrm{Supp}_B(M)$ (see [CRT, Theorem 6.5 (iii)]), the identity $\mathrm{Supp}_B(C_0) = \mathrm{Supp}_B(C_1)$ implies the identity of minimal associated primes.

Proof. For simplicity, we write C_j for $C_j(\phi; B)$. Note that

$$C_{1,P} = C_1 \otimes_B B_P = \Omega_{R/B} \otimes_R B_P \cong \Omega_{R_P/B_P} \otimes_{R_P} B_P$$

for $P \in \mathrm{Spec}(B)$ by [CRT, Exercise 25.4]. Thus if $C_{1,P} = 0$, by Nakayama's lemma $\Omega_{R_P/B_P} = 0$; so, R_P is étale over B_P [CRT, §25]. Therefore $R_P = B_P \oplus S_P$ as $R_P \twoheadrightarrow B_P$ splits, and hence $C_{0,P} = C_0 \otimes_B B_P = S_P \otimes_{R_P,\phi} B_P = 0$. Thus $\mathrm{Supp}_B(C_0) \subset \mathrm{Supp}_B(C_1)$.

If $C_{0,P} = 0$, then $\mathrm{Spec}(B_P) \cap \mathrm{Spec}(S_P) = \emptyset$; therefore, $R_P = B_P \oplus S_P$, $\Omega_{R_P/B_P} = \Omega_{S_P/B_P}$ and $C_{1,P} = 0$. This shows the reverse inclusion $\mathrm{Supp}_B(C_0) \supset \mathrm{Supp}_B(C_1)$, and we conclude $\mathrm{Supp}_B(C_0) = \mathrm{Supp}_B(C_1)$. $\square$

Higher congruence modules. Since R is a B-algebra, $\phi : R \to B$ is onto. We know $C_0 = S/\mathfrak{b}$ and we get $C_1 = \mathfrak{b}/\mathfrak{b}^2$ by the second fundamental exact sequence in Corollary 5.2.5 (ii):

$$0 \to \mathfrak{b}/\mathfrak{b}^2 \xrightarrow{b \mapsto db} \Omega_{R/B} \otimes_R A \to \Omega_{B/B} = 0.$$

So why not we define $C_n := \mathfrak{b}^n/\mathfrak{b}^{n+1}$. Then $\mathrm{gr}(S) := \bigoplus_j C_j$ is the graded algebra [BCM, III]. Knowledge of $gr(S)$ is almost equivalent to the knowledge of S. Once we know S, we recover

$$R = B \times_{C_0} S = \{(b, s) \in B \times S | b \mod \mathfrak{a} = s \mod \mathfrak{b}\}.$$

If $C_1 = \mathfrak{b}/\mathfrak{b}^2$ is generated by one element over B, then by Nakayama's lemma, $\mathfrak{b} = (\theta)$ for a non-zero-divisor $\theta \in S$. Then $\mathrm{gr}(S) \cong C_0[x]$ by sending $\theta \mod \mathfrak{b}^2$ to the variable x. What is S if $B = W$ and $C_0 = \mathbb{F}$? *Is there any good way to compute C_n when R is the universal deformation ring? Or is the knowledge of C_0 and C_1 sufficient to know all C_n?*

Explicit form of $C_1(\pi; \mathbb{F})$ as cotangent space. Write $\pi : R \to R/\mathfrak{m}_R = \mathbb{F}$ for the projection. Let $\mathbb{F}[\varepsilon] = \mathbb{F}[x]/(x^2)$ with $x \leftrightarrow \varepsilon$. Then $\varepsilon^2 = 0$, and ε is called a *dual number*.

For $\phi \in \mathrm{Hom}_{B\text{-alg}}(R, \mathbb{F}[\varepsilon])$, write $\phi(a) = \pi(a) + \delta(a)\varepsilon$. From
$$\phi(ab) = \pi(a)\pi(b) + \pi(a)\delta(b)\varepsilon + \pi(b)\delta(a)\varepsilon,$$
we find $\mathrm{Hom}_{B\text{-alg}}(R, \mathbb{F}[\varepsilon]) = Der_B(R, \mathbb{F})$ by $\phi \leftrightarrow \delta$.

ϕ is determined by $\phi|_{\mathfrak{m}_R}$ which kills $\mathfrak{m}_R^2 + \mathfrak{m}_B$ as $\varepsilon^2 = 0$. Thus
$$\mathrm{Hom}_{\mathbb{F}}(\Omega_{R/B} \otimes_R \mathbb{F}, \mathbb{F}) \cong \mathrm{Hom}_R(\Omega_{R/B}, \mathbb{F}) \cong Der_B(R, \mathbb{F}) = \mathrm{Hom}_R(t^*_{R/B}, \mathbb{F}),$$
for $t^*_{R/B} := \mathfrak{m}_R/(\mathfrak{m}_R^2 + \mathfrak{m}_B)$. Taking $\mathbb{F}$-dual, if $t^*_{R/B}$ is finite dimensional,
$$\Omega_{R/B} \otimes_R \mathbb{F} \cong t^*_{R/B}. \tag{5.2}$$
In particular, $\Omega_{R/B}$ is an R-module of finite type (by Nakayama's lemma).

5.2.5 *Congruence modules for group algebras*

Let H be a finite p-abelian group. If $\mathfrak{m}$ is a maximal ideal of $B[H]$, then for the inclusion $\kappa : H \hookrightarrow B[H]^\times$ with $\kappa(\sigma) = \sigma$, κ mod $\mathfrak{m}$ is trivial as the finite field $B[H]/\mathfrak{m}$ has no non-trivial p-power roots of unity; so, $\mathfrak{m}$ is generated by $\{\sigma - 1\}_{h \in H}$ and $\mathfrak{m}_B$. Thus $\mathfrak{m}$ is unique and $B[H]$ is local.

We have a canonical algebra homomorphism: $B[H] \to B$ sending every $\sigma \in H$ to 1. This homomorphism is called the *augmentation* homomorphism of the group algebra, and its kernel is called the augmentation ideal. Write this map $\pi : B[H] \to B$. Then $\mathfrak{b} = \mathrm{Ker}(\pi)$ is generated by $\sigma - 1$ for $\sigma \in H$. Thus $\mathfrak{b} = \sum_{\sigma \in H} B[H](\sigma - 1)B[H]$. We now compute the congruence module and the differential module $C_j(\pi, B)$ $(j = 0, 1)$.

Theorem 5.2.9. *Suppose B is an integral domain with characteristic 0 $\mathrm{Frac}(B)$. We have $C_0(\pi; B) \cong B/|H|B$ and $C_1(\pi; B) = H \otimes_{\mathbb{Z}} B$.*

By this theorem, if H is a finite p-group, we can recover the group H by taking $B = \mathbb{Z}_p$. Without assuming that H is a p-group, we can actually recover the finite group H as $C_1(\pi; \mathbb{Z})$ taking $B = \mathbb{Z}$. If the reader is interested, try to give a detailed proof of this fact as an exercise.

Proof. *Argument for C_0:* Let $K := \mathrm{Frac}(B)$. Then π gives rise to the algebra direct factor $K\varepsilon \subset K[H]$ for the idempotent $\varepsilon = \frac{1}{|H|} \sum_{\sigma \in H} \sigma$. Thus $\mathfrak{a} = K\varepsilon \cap B[H] = (\sum_{\sigma \in H} \sigma)$ and $\pi(B(H))/\mathfrak{a} = (\varepsilon)/\mathfrak{a} \cong B/|H|B$.

Proof for $C_1(\pi; B) = H \otimes_{\mathbb{Z}} B$:
First step: Consider the functor $\mathcal{F} : CL_B \to SETS$ given by
$$\mathcal{F}(A) = \mathrm{Hom}_{\text{group}}(H, A^\times) = \mathrm{Hom}_{B\text{-alg}}(B[H], A).$$

Thus $R := B[H]$ and the character $\boldsymbol{\rho} : H \to B[H]$ (the inclusion: $H \hookrightarrow B[H]$) are universal among characters of H with values in $A \in CL_B$. Then for any R-module M, consider $R[M] = R \oplus M$ with algebra structure given by $rx = 0$ and $xy = 0$ for all $r \in R$ and $x, y \in M$. Thus M is an ideal of $R[M]$ with $M^2 = 0$. Extend the functor $\mathcal{F}$ to all local B-algebras with residue field $\mathbb{F}$ in an obvious way. Define $\Phi(M) = \{\rho \in \mathcal{F}(R[M]) | \rho \mod M = \boldsymbol{\rho}\}$. Write $\rho(\sigma) = \boldsymbol{\rho}(\sigma) \oplus u'_\rho(\sigma)$ for $u'_\rho : H \to M$.

Second step: Since

$$\boldsymbol{\rho}(\sigma\tau) \oplus u'_\rho(\sigma\tau) = \rho(\sigma\tau) = (\boldsymbol{\rho}(\sigma) \oplus u'_\rho(\sigma))(\boldsymbol{\rho}(\tau) \oplus u'_\rho(\tau))$$
$$= \boldsymbol{\rho}(\sigma\tau) \oplus (u'_\rho(\sigma)\boldsymbol{\rho}(\tau) + \boldsymbol{\rho}(\sigma)u'_\rho(\tau)),$$

we have $u'_\rho(\sigma\tau) = u'_\rho(\sigma)\boldsymbol{\rho}(\tau) + \boldsymbol{\rho}(\sigma)u'_\rho(\tau)$, and thus $u_\rho := \boldsymbol{\rho}^{-1}u'_\rho : H \to M$ is a homomorphism from H into M. This shows $\mathrm{Hom}(H, M) = \Phi(M)$.

Third step: Any B-algebra homomorphism $\xi : R \to R[M]$ with $\xi \mod M = \mathrm{id}_R$ can be written as $\xi = \mathrm{id}_R \oplus d_\xi$ with $d_\xi : R \to M$.

Since $(r \oplus x)(r' \oplus x') = rr' \oplus rx' + r'x$ for $r, r' \in R$ and $x.x' \in M$, we have $d_\xi(rr') = rd_\xi(r') + r'd_\xi(r)$; so, $d_\xi \in Der_B(R, M)$. By universality of $(R, \boldsymbol{\rho})$, we have

$$\Phi(M) \cong \{\xi \in \mathrm{Hom}_{B\text{-alg}}(R, R[M]) | \xi \mod M = \mathrm{id}\}$$
$$= Der_B(R, M) = \mathrm{Hom}_R(\Omega_{R/B}, M).$$

Final step, Yoneda's lemma: By the second and third steps combined, we have

$$\mathrm{Hom}_B(H \otimes_{\mathbb{Z}_p} \mathbb{Z}_p, M) = \mathrm{Hom}(H, M) = \mathrm{Hom}_R(\Omega_{R/B}, M)$$
$$= \mathrm{Hom}_B(\Omega_{R/B} \otimes_{R,\pi} B, M).$$

This is true for all M, we have (essentially by Yoneda's lemma)

$$H \cong \Omega_{R/B} \otimes_{R,\pi} B = C_1(\pi; B). \qquad \square$$

Exercise 5.2.1. Prove $H \cong \mathfrak{a}/\mathfrak{a}^2$ (as abelian groups) by $H \ni h \mapsto (h \mod \mathfrak{a}^2)$ for the augmentation ideal $\mathfrak{a}$ of $\mathbb{Z}[H]$.

From Theorem 5.2.9 and Theorem 5.2.2, we obtain

Corollary 5.2.10. *Let $\overline{\rho}$ be a character $\mathrm{Gal}(\overline{k}/k) \to \mathbb{F}^\times$ for a finite field $\mathbb{F}$ of characteristic p. Let $(R, \boldsymbol{\rho})$ be the universal couple for p-unramified minimal deformation problem of $\overline{\rho}$ over W, and pick such a deformation ρ_A with values in $A^\times$. Then we have $\Omega_{R/W} \otimes_{R,\varphi} A \cong C_k \otimes_{\mathbb{Z}_p} A = Cl_k \otimes_{\mathbb{Z}} A$ for $\varphi \in \mathrm{Hom}_{CL_W}(R, A)$ with $\varphi \circ \boldsymbol{\rho} = \rho_A$.*

5.3 Cohomology of induced representations

We now insert here a few subsections dealing with continuous cohomology and induction in representation theory in order to relate the class group with an arithmetic cohomology. In the last section of this chapter, we relate the class group with an arithmetic cohomology group, and we are preparing basic facts for this.

5.3.1 *Continuous group cohomology*

We start with a brief discussion of continuous Galois cohomology. Continuous cohomology imposes continuity on cocycles. Therefore the cohomology we discussed earlier in §4.2.2 can be regarded as continuous cohomology equipping with discrete topology on groups $\mathcal{G}$ and coefficients M. In this section, the group $\mathcal{G}$ often has p-profinite topology but on the other hand, coefficients often have discrete topology. Compact modules and discrete modules are dual to each other under Pontryagin duality (this is the reason for the use of Pontryagin duality here).

In this section, we denote by $H^q(\mathcal{G}, X)$ the continuous group cohomology with coefficients in X. If we need to emphasize continuity, we write $H^q_{ct}(\mathcal{G}, X)$ for the continuous cohomology group $H^q(\mathcal{G}, X)$. If X is finite, $H^q(\mathcal{G}, X)$ is as defined in [MFG, §4.3.3]. Thus we have

$$H^0(\mathcal{G}, X) = X^{\mathcal{G}} = \{x \in X | gx = x \text{ for all } g \in \mathcal{G}\},$$

and assuming all maps are continuous,

$$H^1(\mathcal{G}, X) = \frac{\{\mathcal{G} \xrightarrow{c} X | c(\sigma\tau) = \sigma c(\tau) + c(\sigma) \text{ for all } \sigma, \tau \in \mathcal{G}\}}{\{\mathcal{G} \xrightarrow{b} X | b(\sigma) = (\sigma - 1)x \text{ for } x \in X \text{ independent of } \sigma\}},$$

and $H^2(\mathcal{G}, X)$ is given by

$$\frac{\{\mathcal{G} \xrightarrow{c} X | c(\sigma, \tau) + c(\sigma\tau, \rho) = \sigma c(\tau, \rho) + c(\sigma, \tau\rho) \text{ for all } \sigma, \tau, \rho \in \mathcal{G}\}}{\{c(\sigma, \tau) = b(\sigma) + \sigma b(\tau) - b(\sigma\tau) \text{ for } b : \mathcal{G} \to X\}}.$$

Thus if $\mathcal{G}$ acts trivially on X, we have $H^1(\mathcal{G}, X) = \text{Hom}(\mathcal{G}, X)$. If $\mathcal{G} = \text{Gal}(E/K)$, we often write $H^j(E/K, X)$, and if $E = \overline{K}$ (an algebraic closure of K), we write $H^j(K, X)$ for $\mathcal{G} = \text{Gal}(\overline{K}/K)$.

Exercise 5.3.1. Verify that the proof given earlier for the inflation-restriction sequence and long exact sequence attached to a short one are still valid for continuous cohomology.

5.3.2 *Induced representation*

We prepare several facts on induced representation for later use. Let $A \in CL_W$ and $\mathcal{G}$ be a profinite group with a closed subgroup $\mathcal{H}$. Put $\Delta := \mathcal{G}/\mathcal{H}$. Let $\mathcal{H}$ be a character $\varphi : \mathcal{G} \to A$. Let $A(\varphi) \cong A$ on which $\mathcal{H}$ acts by φ. Regard the group algebra $A[\mathcal{G}]$ as a left and right $A[\mathcal{G}]$-module by multiplication. Define $\mathrm{Ind}_{\mathcal{H}}^{\mathcal{G}} \varphi_{/A} := A[\mathcal{G}] \otimes_{A[\mathcal{H}]} A(\varphi)$ (so, $\xi h \otimes a = \xi \otimes ha = \xi \otimes \varphi(h)a = \varphi(a)(\xi \otimes a)$) for $h \in \mathcal{H}$. Let $\mathcal{G}$ act on $\mathrm{Ind}_{\mathcal{H}}^{\mathcal{G}} \varphi_{/A}$ by $g(\xi \otimes a) := (g\xi) \otimes a$. The resulted $\mathcal{G}$-module $\mathrm{Ind}_{\mathcal{H}}^{\mathcal{G}} \varphi_{/A}$ is the *induced module*.

Similarly we can think of $\mathrm{ind}_{\mathcal{H}}^{\mathcal{G}} \varphi_{/A} := \mathrm{Hom}_{A[\mathcal{H}]}(A[\mathcal{G}], A(\varphi))$ (so, $\phi(h\xi) = h\phi(\xi) = \varphi(h)\phi(\xi)$) on which $g \in \mathcal{G}$ acts by $g\phi(\xi) = \phi(\xi g)$. In some books, $\mathrm{ind}_{\mathcal{H}}^{\mathcal{G}} \varphi$ is written as $\mathrm{Coind}_{\mathcal{H}}^{\mathcal{G}} \varphi$ (co-induced representation), but they are isomorphic if $\mathcal{H}$ has finite index in $\mathcal{G}$ (as we will see soon).

Matrix form of induced representations. Assuming $(\mathcal{G} : \mathcal{H}) = 2$ for simplicity, we describe the matrix form of $\mathrm{Ind}_{\mathcal{H}}^{\mathcal{G}} \varphi$. Suppose that φ has order prime to p. Then for $\varsigma \in \mathcal{G}$ generating $\mathcal{G}$ over $\mathcal{H}$, $\varphi_\varsigma(h) := \varphi(\varsigma^{-1}h\varsigma)$ is again a character of $\mathcal{H}$. The module $\mathrm{Ind}_{\mathcal{H}}^{\mathcal{G}} \varphi$ has a basis $1_{\mathcal{G}} \otimes 1$ and $\varsigma \otimes 1$ for the identity element $1_{\mathcal{G}}$ of $\mathcal{G}$ and $1 \in A \cong A(\varphi)$.

We have

$$g(1_{\mathcal{G}} \otimes 1, \varsigma \otimes 1) = (g \otimes 1, g\varsigma \otimes 1)$$

$$= \begin{cases} (1_{\mathcal{G}} \otimes g, \varsigma \otimes \varsigma^{-1}g\varsigma) = (1_{\mathcal{G}} \otimes 1, \varsigma \otimes 1)\begin{pmatrix} \varphi(g) & 0 \\ 0 & \varphi_\varsigma(g) \end{pmatrix} & \text{if } g \in \mathcal{H}, \\ (\varsigma \otimes \varsigma^{-1}g, 1_{\mathcal{G}} \otimes g\varsigma) = (1_{\mathcal{G}} \otimes 1, \varsigma \otimes 1)\begin{pmatrix} 0 & \varphi(g\varsigma) \\ \varphi(\varsigma^{-1}g) & 0 \end{pmatrix} & \text{if } g\varsigma \in \mathcal{H}. \end{cases}$$

Thus extending φ to $\mathcal{G}$ by 0 outside $\mathcal{H}$, we get

$$\boxed{\mathrm{Ind}_{\mathcal{H}}^{\mathcal{G}} \varphi(g) = \begin{pmatrix} \varphi(g) & \varphi(g\varsigma) \\ \varphi(\varsigma^{-1}g) & \varphi(\varsigma^{-1}g\varsigma) \end{pmatrix}.} \tag{5.3}$$

Lemma 5.3.1. *We have* $\mathrm{Ind}_{\mathcal{H}}^{\mathcal{G}} \varphi \cong \mathrm{ind}_{\mathcal{H}}^{\mathcal{G}} \varphi$.

Proof. The induction $\mathrm{ind}_{\mathcal{H}}^{\mathcal{G}} \varphi$ has basis (ϕ_1, ϕ_ς) over $A[\mathcal{H}]$ given by $\phi_1(\xi + \xi'\varsigma^{-1}) = \varphi(\xi) \in A = A(\varphi)$ and $\phi_\varsigma(\xi + \xi'\varsigma^{-1}) = \varphi(\xi') \in A = A(\varphi)$ for $\xi \in A[\mathcal{H}]$; so, (*) $\phi_1(\xi' + \xi\varsigma^{-1}) = \phi_\varsigma(\xi + \xi'\varsigma^{-1})$. Then we have

$$g(\phi_1(\xi + \xi'\varsigma^{-1}), \phi_\varsigma(\xi + \xi'\varsigma^{-1}))$$

$$= (\phi_1(\xi g + \xi'\varsigma^{-1}g\varsigma\varsigma^{-1}), \phi_\varsigma(\xi g + \xi'\varsigma^{-1}g\varsigma\varsigma^{-1}))$$

$$= \begin{cases} (\phi_1(\xi), \varphi_\varsigma(\xi'))\begin{pmatrix} \varphi(g) & 0 \\ 0 & \varphi_\varsigma(g) \end{pmatrix} & (g \in \mathcal{H}), \\ (\phi_1(\xi'\varsigma^{-1}g), \phi_\varsigma(\xi g\varsigma)) \overset{(*)}{=} (\phi_1(\xi), \phi_\varsigma(\xi'))\begin{pmatrix} 0 & \varphi(g\varsigma) \\ \varphi(\varsigma^{-1}g) & 0 \end{pmatrix} & (g\varsigma \in \mathcal{H}). \end{cases}$$

Thus we get $\mathrm{Ind}_{\mathcal{H}}^{\mathcal{G}} \varphi \cong \mathrm{ind}_{\mathcal{H}}^{\mathcal{G}} \varphi$. $\qquad\square$

5.3.3 *Adjunction formula for* Hom *and* $\otimes$

Let T be an S-algebra (here T and S are possibly non-commutative rings with identity). Let M be an S-module and N be a T-module. Regard T as a right S-module by right multiplication, and consider the scalar extension $T \otimes_S M$ which is an T-module by $\alpha(a \otimes m) = (\alpha a) \otimes m$ for $\alpha, a \in T$. Let $i : M \to T \otimes_S M$ be the S-linear map $i(m) = 1_T \otimes m$. Since $i(bm) = 1 \otimes_S bm = b \otimes_S m = bi(m)$ for $b \in S$, indeed, i is S-linear. We have the following universal property

- If N is a T-module, for any S-linear map $M \xrightarrow{f} N$, there is a unique T-morphism $g : T \otimes_S M \to N$ such that $g \circ i = f$.

This follows from the universality of the tensor product applied to the T-bilinear map $T \otimes_S M \to N$ given by $a \otimes m \mapsto af(m)$. Therefore, we get the adjunction formula for the tensor product:
$$\mathrm{Hom}_T(T \otimes_S M, N) \cong \mathrm{Hom}_S(M, N).$$
See [CRT, Appendix B] and [BAL, II.4] for more details.

Dual and derived category version of adjunction. By the derived category version of this, we get
$$\mathrm{Ext}_T^q(T \otimes_S M, N) \cong \mathrm{Ext}_S^q(M, N) \quad \text{for all } q \geq 0.$$
There is a dual version. Regard $\mathrm{Hom}_S(T, M)$ as T-module by $\alpha\phi(a) = \phi(a\alpha)$ for $\phi \in \mathrm{Hom}_S(T, M)$. Let $\pi : \mathrm{Hom}_S(T, M) \to M$ by $\pi(\phi) = \phi(1_T)$, which is S-linear. Then by the universality of Hom_S, we have

- If N is a T-module, for any S-linear map $N \xrightarrow{f} M$, there is a unique T-morphism $g : N \to \mathrm{Hom}_S(T, M)$ such that $\pi \circ g = f$.

From this, we get
$$\mathrm{Hom}_T(N, \mathrm{Hom}_S(T, M)) \cong \mathrm{Hom}_S(N, M),$$
and again
$$\mathrm{Ext}_T^q(N, \mathrm{Hom}_S(T, M)) \cong \mathrm{Ext}_S^q(N, M) \quad \text{for all } q \geq 0.$$

5.3.4 *Shapiro's lemma*

Let $\mathcal{G}$ be a group and $\mathcal{H}$ be a subgroup of finite index. Take a commutative ring B with identity. We apply the above argument to the group algebras $T = B[\mathcal{G}]$ and $S = B[\mathcal{H}]$. Then we write $\mathrm{Ind}_{\mathcal{H}}^{\mathcal{G}} M := T \otimes_S M$ and $\mathrm{ind}_{\mathcal{H}}^{\mathcal{G}} M := \mathrm{Hom}_S(A, M)$ as T-modules. Then we have

Lemma 5.3.2 (Shapiro).

$$\mathrm{Ext}^q_{B[\mathcal{G}]}(\mathrm{Ind}^{\mathcal{G}}_{\mathcal{H}} M, N) \cong \mathrm{Ext}^q_{B[\mathcal{H}]}(M, N) \ \text{ for all } q \geq 0,$$

$$\mathrm{Ext}^q_{B[\mathcal{G}]}(N, \mathrm{ind}^{\mathcal{G}}_{\mathcal{H}} M) \cong \mathrm{Ext}^q_{B[\mathcal{H}]}(N, M) \ \text{ for all } q \geq 0.$$

Since the cohomology group $H^q(\mathcal{G}, N)$ can be identified with $\mathrm{Ext}^n_{B[\mathcal{G}]}(B, N)$ (see [MFG, §4.3.1]), we can reformulate this as

Corollary 5.3.3. *We have* $H^q(\mathcal{G}, \mathrm{ind}^{\mathcal{G}}_{\mathcal{H}} M) \cong H^q(\mathcal{H}, M)$ *for all* $q \geq 0$.

See [MFG, Lemma 4.20], [CNF, (1.6.4)] and [CGP, III, (6.2)] for more detailed proofs of this fact.

5.4   Class group as an arithmetic cohomology group

For a number field k, we now relate the class group Cl_k and its p-primary part C_k with a certain cohomology group. In this section, for simplicity, assume $p \nmid [k : \mathbb{Q}]$ and that $k/\mathbb{Q}$ is a Galois extension. As in the remark after Corollary 1.11.4, decompose $\mathrm{Ind}^{\mathbb{Q}}_k \mathrm{id} = \mathbf{1} \oplus \rho_k$ for the identity representation $\mathbf{1}$. Then by Exercise 1.11.1, we have the identity of the Dedekind zeta function: $\zeta_k(s) = \zeta(s)L(s, \rho_k)$. Take $H = C_k$ and $k = F$ in Corollary 1.11.4. Normalizing the "transcendental" factor of the value $L(1, \rho_k) = \lim_{s=+1}(\zeta_k(s)/\zeta(s))$ given in Corollary 1.11.4 as

$$\Omega_k = (2\pi)^{r'} R_k$$

(which is essentially the regulator times a power of $(2\pi i)$), we obtain

$$|*L(1, \rho_k)/\Omega_k|_p = \big||C_k|\big|_p, \tag{5.4}$$

where $* = \frac{2^r}{w\sqrt{|D_k|}}$. We wrote the word "transcendental" in quotes as the transcendence may not be fully known.

5.4.1   *Abelian Selmer groups*

For a continuous representation $\rho_A : \mathcal{G} \to \mathrm{GL}_n(A)$ (for $A \in CL_B$), we often identify ρ_A with the $\mathcal{G}$-module A^n on which $\mathcal{G}$ acts by ρ_A. Then we define a $\mathcal{G}$-module $\rho^*_A = \rho \otimes_A A^\vee = (A^\vee)^n$ or the Pontryagin dual $A^\vee := \mathrm{Hom}_{cont}(A, \mathbb{Q}_p/\mathbb{Z}_p)$, where $\mathcal{G}$ act on ρ^*_A through the factor ρ_A; i.e., $\mathcal{G}$ acts by the matrix representation ρ_A on $\rho^*_A = (A^\vee)^n$.

Recall K which is the maximal p-profinite extension of k unramified outside p. Thus $K/\mathbb{Q}$ is a Galois extension as k is a Galois over $\mathbb{Q}$. We tentatively define an arithmetic cohomology group (a typical example of Selmer groups) of the ρ_A by

$$\mathrm{Sel}_k(\rho_A) := \mathrm{Ker}(H^1(K/k, \rho_A^*) \to \prod_{\mathfrak{p}|p} H^1(I_{\mathfrak{p}}, \rho_A^*)),$$

where $\mathfrak{p}$ runs over a prime of k over p and $I_{\mathfrak{p}}$ denotes the $\mathfrak{p}$-inertia group in $\mathrm{Gal}(K/k)$. Taking $\mathcal{H} = \mathrm{Gal}(K/k) \lhd \mathcal{G} =: \mathrm{Gal}(K/\mathbb{Q})$ and writing $\mathrm{Ind}_k^{\mathbb{Q}} \rho_A = \mathrm{Ind}_{\mathcal{H}}^{\mathcal{G}} \rho_A$, we therefore have

$$\mathrm{Sel}_{\mathbb{Q}}(\mathrm{Ind}_k^{\mathbb{Q}} \rho_A) := \mathrm{Ker}(H^1(K/\mathbb{Q}, \mathrm{Ind}_k^{\mathbb{Q}} \rho_A^*) \to H^1(I_p, \mathrm{Ind}_k^{\mathbb{Q}} \rho_A^*))$$

for the p-inertia group $I_p \subset \mathrm{Gal}(K/\mathbb{Q})$.

Applying Shapiro's lemma for $\mathcal{G} = \mathrm{Gal}(K/\mathbb{Q})$ and $\mathcal{H} = \mathrm{Gal}(K/k)$, we find

$$H^1(K/k, \mathbf{1}) \cong H^1(K/\mathbb{Q}, \mathrm{Ind}_k^{\mathbb{Q}} \mathbf{1}) = H^1(K/\mathbb{Q}, \mathbf{1}^*) \oplus H^1(K/\mathbb{Q}, \rho_k^*).$$

Since $p \nmid [k : \mathbb{Q}]$, triviality of a 1-cocycle on $I_{\mathfrak{p}}$ and on the p-inertia group $I_p \subset \mathrm{Gal}(K/\mathbb{Q})$ is equivalent. Thus we see

$$\mathrm{Sel}_k(\mathbf{1}) \cong \mathrm{Sel}_{\mathbb{Q}}(\mathbf{1}) \oplus \mathrm{Sel}_{\mathbb{Q}}(\rho_k).$$

Exercise 5.4.1. Why are the trivialities of a 1-cocycle on $I_{\mathfrak{p}}$ and on the p-inertia group $I_p \subset \mathrm{Gal}(K/\mathbb{Q})$ equivalent? (*Hint: Use the inflation-restriction sequence with respect to $I_{\mathfrak{p}} \lhd I_p$*).

We note

$$\mathrm{Sel}_k(\mathbf{1}) = \mathrm{Hom}_{cont}(\mathrm{Gal}(K/k), \mathbf{1}^*) = \mathrm{Hom}(C_k, \mathbb{Q}_p/\mathbb{Z}_p) = C_k^{\vee}$$

as any everywhere unramified homomorphism of $\mathrm{Gal}(K/k)$ into $\mathbf{1}^* = \mathbb{Q}_p/\mathbb{Z}_p$ factors through $\mathrm{Gal}(K/k)_p^{ab} = C_k$. In the same way, $\mathrm{Sel}_{\mathbb{Q}}(\mathbf{1}) = \mathrm{Hom}(C_{\mathbb{Q}}, \mathbf{1}^*) = \{0\}$. We conclude from Corollary 5.2.10 the following algebro-geometric Class number formula:

Theorem 5.4.1. *For the augmentation homomorphism* $\pi : \mathbb{Z}_p[C_k] \to \mathbb{Z}_p$,

$$\left| * \frac{L(1, \rho_k)}{\Omega_k} \right|_p = |C_1(\pi; \mathbb{Z}_p)|^{-1} = |C_0(\pi; \mathbb{Z}_p)|^{-1} = \left\| \mathrm{Sel}_{\mathbb{Q}}(\rho_k) \right\|_p$$

and $C_1(\pi; \mathbb{Z}_p) = \Omega_{\mathbb{Z}_p[C_k]/\mathbb{Z}_p} \otimes_{\mathbb{Z}_p[C_k]} \mathbb{Z}_p \cong C_k$ *and* $C_0(\pi; \mathbb{Z}_p) \cong \mathbb{Z}_p/|C_k|\mathbb{Z}_p$. *Here $*$ is a simple algebraic constant given in* (5.4).

Here is a naive open question:

Is there some way of proving the above class number formula without using the classical ideal theory of integer ring of k but the Galois deformation theory?

There are three incarnations of C_k as the p-primary part of the class group (field arithmetic), as the Galois group of the maximal abelian unramified extension (Galois theory), and as the dual of a Selmer group (homology theory).

Chapter 6

Universal ring and compatible system

We start studying deformation theory of 2-dimensional Galois representations. The base local ring B is either W or the Iwasawa algebra $\Lambda = W[[T]]$. As an initial datum, we fix a 2-dimensional continuous *odd* absolutely irreducible representation $\bar{\rho} = \rho_{\mathbb{F}} : \mathrm{Gal}(\overline{\mathbb{Q}}/\mathbb{Q}) \to \mathrm{GL}_2(\mathbb{F})$ ramified at finitely many primes. A 2-dimensional Galois representation ρ is called "odd" if $\det(\rho(c)) = -1$ for complex conjugation c. Then $W = W_{\bar{\rho}}$ is the ring of Witt vectors with coefficients in $\mathbb{F}$. Assume that $\bar{\rho}|_{\mathrm{Gal}(\overline{\mathbb{Q}}_p/\mathbb{Q}_p)} \cong \left(\begin{smallmatrix} \bar{\epsilon} & * \\ 0 & \bar{\delta} \end{smallmatrix} \right)$ with unramified quotient character $\bar{\delta}$ distinct from $\bar{\epsilon}$ (*p-distinguished ordinarity*). The sub-character $\bar{\epsilon}$ may ramify, and we sometimes call the property: $\bar{\delta} \neq \bar{\epsilon}$ *p-distinguishedness* of $\bar{\rho}$. In the rest of this book, we assume that $\boxed{p > 2}$.

Take the maximal p-profinite extension $F^{(p)}(\bar{\rho})$ over $F(\bar{\rho})$ unramified outside p, and let $G = \mathrm{Gal}(F^{(p)}(\bar{\rho})/\mathbb{Q})$ (so G is almost p-profinite). We consider the following *deformation functors* defined over CL_B:

$$\mathcal{D}(A) := \{\rho_A : G \to \mathrm{GL}_2(A) | \rho_A \mod \mathfrak{m}_A = \bar{\rho}, \ (p), \ (\mathfrak{m})\}/\Gamma(\mathfrak{m}_A),$$

$$\mathcal{D}_\chi(A) := \{\rho_A \in \mathcal{D}(A) | \rho_A \text{ satisfies the condition } (\det)\}/\Gamma(\mathfrak{m}_A).$$

Here $\Gamma(\mathfrak{m}_A) = \mathrm{Ker}(\mathrm{GL}_2(A) \to \mathrm{GL}_2(\mathbb{F}))$ acts by conjugation, and we impose a minimality condition $(\mathfrak{m})$ and an ordinarity condition (p) which will be described in the following subsections. The determinant condition is

$(\det)$ $\det(\rho_A) = \chi$ *for a continuous character* $\chi : \mathrm{Gal}(\overline{\mathbb{Q}}/\mathbb{Q}) \to B^\times$.

Strictly speaking, the identity $\det(\rho_A) = \chi$ means $\det(\rho_A) = \iota_A \circ \chi$ for the B-algebra structure $\iota_A : B \to A$. We emphasize the fact that the 2-dimensional deformation theory has good similarity to the 1-dimensional case, and from this chapter on, we study exclusively the 2-dimensional case.

A main topic in this chapter is a generalization of the class number formula: Theorem 5.4.1. In most cases, if $\chi = \nu_p^{k-1}\psi$ ($k \geq 2$), $\mathcal{D}_\chi$ for $B = W$ is represented by the (unique) local ring $\mathbb{T}_\chi$ of the Hecke algebra

$\mathbf{h}_k(\Gamma_0(Cp), \psi)$ associated to $\bar{\rho}$ acting on $S_k(\Gamma_0(Cp), \psi; W)$ for the prime-to-p conductor C of $\bar{\rho}$. If a W-algebra homomorphism $\lambda : \mathbb{T}_\chi \to W$ is given by $f|T(n) = \lambda(T(n))f$ for a cuspidal Hecke eigenform f and its p-adic member $\rho_\mathfrak{p}$ of the compatible system of Galois representation $\rho_f = \{\rho_\mathfrak{p}\}_\mathfrak{p}$ of f, we describe the identities $\boxed{C_1(\lambda; W) \cong \mathrm{Sel}(Ad(\rho_\mathfrak{p}))}$ (the adjoint Selmer group) and the adjoint class number formula (Corollary 6.2.24):

$$|\mathrm{Sel}(Ad(\rho_\mathfrak{p}))| = |C_1(\lambda; W)| = |C_0(\lambda; W)| \stackrel{(1)}{=} \left| * \frac{L(1, Ad(f))}{\Omega_{f,+}\Omega_{f,-}} \right|_p^{-1} \quad \text{(BKC)}$$

for explicit constants $\Omega_{f,\pm}$ (called the *period* of f) and $*$ independent of p if f has weight $k \geq 2$ (see Theorem 9.3.2 for the identity (1) and §9.1.1 for $L(s, Ad(f))$). This can be considered as a systematic example of proven cases of the Bloch–Kato conjecture [BK90] (see also [DFG04]). See "Notes to the reader" in the front matter for the definition of the adjoint Galois module $Ad(\rho_\mathfrak{p})$, and the Selmer group is defined later in (6.5) and in §6.1.8. We describe the first two identities here (see Theorem 6.1.11).

Other topics and questions. Fix $f_0 \in S_{k_0}(\psi_0)$ with $f|T(n) = \lambda_0(T(n))f$, and put $\chi_0 = \nu_p^{k_0-1}\psi_0$. For $\rho_{f_0} = \{\rho_\mathfrak{p}\}_\mathfrak{p}$, consider its reduction $\{\bar{\rho}_\mathfrak{p} = \rho_\mathfrak{p} \mod \mathfrak{p}\}_\mathfrak{p}$. The bigger functor $\mathcal{D} = \mathcal{D}_\mathfrak{p}$ deforming $\bar{\rho}_\mathfrak{p}$ is (often) represented by a local ring of the big "ordinary" Hecke algebra $\mathbb{T} = \mathbb{T}_\mathfrak{p}$ free of finite rank over the Iwasawa algebra $W_\mathfrak{p}[[\Gamma]] = W_\mathfrak{p}[[T]]$ (for $W_\mathfrak{p} = W_{\bar{\rho}_\mathfrak{p}}$). Another goal of this chapter is Theorem 6.2.17 which answers

- *Supposing $k_0 \geq 2$, when is $\mathbb{T}_\mathfrak{p} = W_\mathfrak{p}[[T]]$ moving ordinary $\bar{\rho}_\mathfrak{p}$?*

Then an obvious question is:

- *What happens if $k_0 = 1$?*

When $k_0 = 1$, ρ_{f_0} has finite image independent of p (an *Artin representation*) by Deligne–Serre [DS74]; so, it looks easier. However we do not know (BKC), and we need to deal with the p-adic value $L_p(Ad(f_0))$ for the p-adic L-function $L_p(Ad(f))$ interpolating $L(1, Ad(f))/\Omega_{f,+}\Omega_{f,-}$ for f of weight k varying belonging to a p-adic analytic family; so, it depends on p. We study the second question in Chapter 8 after studying the cyclicity question in Chapter 7 for the adjoint Selmer groups. We enforce absolute irreducibility of $\bar{\rho}_\mathfrak{p}$ (which is a condition on $\mathfrak{p}$ satisfied by almost all primes [R85, Theorem 3.1]).

6.1 Adjoint Selmer groups and differentials

Recall $G = \mathrm{Gal}(F^{(p)}(\overline{\rho})/\mathbb{Q})$. Here $\overline{\rho}$ is a 2-dimensional odd continuous absolutely irreducible representation $\overline{\rho} : G \to \mathrm{GL}_2(\mathbb{F})$ for a finite extension $\mathbb{F}_{/\mathbb{F}_p}$ (so, only finitely many primes ramify in $F(\overline{\rho})/\mathbb{Q}$. Continuity is with respect to the profinite topology on G and the discrete topology on $\mathrm{GL}_2(\mathbb{F})$. A prime l is *ramified* for $\overline{\rho}$ if l ramifies in $F(\overline{\rho})/\mathbb{Q}$. Let S be the set of primes made up of ramified primes for $\overline{\rho}$ outside p. We assume the following "minimality" condition in this book:

(m) *for all primes $l \neq p$, ramification index of l in $F(\overline{\rho})/\mathbb{Q}$ is prime to p.*

This is to simplify the local condition at l of the p-torsion adjoint Selmer group. Also, by (m), all minimal deformations factors through G.

In §6.1.8, we define the adjoint Selmer group $\mathrm{Sel}(Ad(\rho))$ for ordinary deformations ρ of $\overline{\rho}$ and show that $\mathrm{Sel}(Ad(\overline{\rho})) = t_{R/B}$ and $\mathrm{Sel}(Ad(\rho))^{\vee} \cong \Omega_{R/B}$ for the universal ordinary Galois representation ρ deforming $\overline{\rho}$. We fix a decomposition group D_p of G and write I_p for the inertia group of D_p.

6.1.1 *Ordinary deformation functor*

Let $\rho : G \to \mathrm{GL}_2(A)$ $(A \in CL_B)$ be a deformation of $\overline{\rho} : G \to \mathrm{GL}_2(\mathbb{F})$ acting on $V(\rho)$. Recall that we fixed the ordinary quotient character $\overline{\delta}$ of $\overline{\rho}$ so that $\overline{\rho}|_{D_p} \cong \left(\begin{smallmatrix} \overline{\epsilon} & * \\ 0 & \overline{\delta} \end{smallmatrix} \right)$. We say ρ is *p-ordinary* if

(p) $\rho_A|_{D_p} \cong \left(\begin{smallmatrix} \epsilon_A & * \\ 0 & \delta_A \end{smallmatrix} \right)$ *for two characters* $\epsilon_A, \delta_A : D_p \to A^{\times}$ **distinct** *modulo* $\mathfrak{m}_A$ *with* δ_A *unramified and* $\delta_A \mod \mathfrak{m}_A = \overline{\delta}$.

We call δ_A the *ordinarity* (quotient) character of ρ_A.

By (m), the image $\overline{\rho}(I_l)$ of the l-inertia subgroup for $l \in S$ is either a non-trivial subgroup having values in the normalizer of a subtorus of $\mathrm{GL}_2(\mathbb{F})$ or isomorphic to A_4 or S_4 modulo center by the classification of subgroups of $\mathrm{PGL}_2(\mathbb{F})$ by Dickson described in §6.2.4 (here A_5 does not appear as I_l is a nilpotent group). Let $S_{ab} \subset S$ be the set of primes such that the image $\overline{\rho}(I_l)$ is abelian; so, it falls in a subtorus of $\mathrm{GL}_2(\mathbb{F})$. By extending scalars $\mathbb{F}$, we assume that $\overline{\rho}|_{I_l} = \overline{\epsilon}_l \oplus \overline{\delta}_l$ for $l \in S_{ab}$.

If $\overline{\delta}_l$ is non-trivial, by local class field theory, $\overline{\delta}_l$ factors through $\mathrm{Gal}(\mathbb{Q}_l[\mu_{l^f}]/\mathbb{Q}_l)$ for some exponent $f > 0$. Since $\mathrm{Gal}(\mathbb{Q}_l[\mu_{l^f}]/\mathbb{Q}_l) \cong (\mathbb{Z}/l^f\mathbb{Z})^{\times} \cong \mathrm{Gal}(\mathbb{Q}[\mu_{l^f}]/\mathbb{Q})$, we can lift $\overline{\delta}_l$ to a character $\delta_l : \mathrm{Gal}(\mathbb{Q}[\mu_{l^f}]/\mathbb{Q}) \to \mu_{q-1} \subset W(\mathbb{F})^{\times} \subset B^{\times}$ for $q = |\mathbb{F}|$. Thus $\overline{\rho} \otimes \delta_l^{-1} \cong \overline{\epsilon}_l \delta_l^{-1} \oplus 1$. The deformation theory of $\overline{\rho}$ and $\overline{\rho} \otimes \delta_l^{-1}$ is equivalent, as ρ_A is a deformation of $\bar\rho$ over G if and only if $\rho_A \otimes \delta_l^{-1}$ is a deformation of $\bar\rho \otimes \delta_l^{-1}$ over G (and also $Ad(\bar\rho) \cong Ad(\bar\rho \otimes \delta_l^{-1})$). Thus without losing generality, we impose a "twist-minimality" condition (similar to (p)) for $l \in S_{ab}$ $(l \ne p)$:

(l) *We have a non-trivial character* $\epsilon_l : I_l \to B^\times$ *of order prime to p such that* $\rho|_{I_l} \cong \left(\begin{smallmatrix} \iota_A \circ \epsilon_l & 0 \\ 0 & 1 \end{smallmatrix} \right)$.

We always impose these two conditions (p) and (l) for $l \in S_{ab}$. We fix a character $\chi : G \to B^\times$, and we often impose the condition (det) stated at the beginning of this chapter.

We consider the following functors for a fixed absolutely irreducible representation $\bar\rho : G \to \mathrm{GL}_2(\mathbb{F})$ satisfying (p) and (l). Recall $\mathcal{D}^\emptyset, \mathcal{D}, \mathcal{D}_\chi :$ $\mathcal{C} \to SETS$ given by

$$\mathcal{D}^\emptyset(A) := \{\rho_A : G \to \mathrm{GL}_2(A) | \rho_A \bmod \mathfrak{m}_A = \bar\rho\}/\Gamma(\mathfrak{m}_A),$$
$$\mathcal{D}(A) = \{\rho_A \in \mathcal{D}^\emptyset(A) | (\mathrm{m}), (p) \text{ and } (l)\}, \qquad (6.1)$$
$$\mathcal{D}_\chi(A) = \{\rho_A \in \mathcal{D}(A) | \det\rho = \iota_A \circ \chi\}.$$

When $\bar\rho|_{D_p} \cong \bar\epsilon \oplus \bar\delta$ with $\bar\epsilon\bar\delta^{-1} = \overline\omega_p$ for $\overline\omega_p$ given by the Galois action on $\mu_p(\overline{\mathbb{Q}}_p)$ (so, $\overline\omega_p = (\nu_p \bmod p)$), we need to consider a smaller deformation functor than $\mathcal{D}_\chi$ imposing flatness

$$\mathcal{D}_\chi^{fl}(A) := \{\rho_A : G \to \mathrm{GL}_2(A) \in \mathcal{D}(A) | \rho_A \otimes (\iota_A \circ \xi^{-1}) \text{ is flat over } A\}. \qquad (6.2)$$

Here a deformation ρ_A of $\bar\rho$ into $GL_2(A)$ for an Artinian local W-algebra in CL_W is said to be *flat* at p if there exists a *locally free group* scheme $\mathcal{G}_\rho$ over $\mathbb{Z}_p$ (on which A acts by endomorphisms) such that ρ_A is isomorphic to the generic geometric fiber of $\mathcal{G}_\rho$. See [GME, §1.6] for locally free groups. If $A = \varprojlim_i A_i$ is pro-Artinian with Artinian A_i, we say ρ_A flat if ρ_A is a projective limit of flat ρ_{A_i}. Then

Theorem 6.1.1. *The functors $\mathcal{D}^\emptyset$, $\mathcal{D}$, $\mathcal{D}_\chi$ and $\mathcal{D}_\chi^{fl}$ are represented by universal couples $(R^\emptyset, \rho^\emptyset)$, (R^{ord}, ρ), (R_χ, ρ_χ) and $(R_\chi^{fl}, \rho_\chi^{fl})$, respectively, so that $\mathcal{D}(A) \cong \mathrm{Hom}_\mathcal{C}(R^{ord}, A)$ by $\rho \mapsto \varphi$ with $\varphi \circ \rho \sim \rho$ (resp. $\mathcal{D}_\chi(A) \cong \mathrm{Hom}_\mathcal{C}(R_\chi, A)$ by $\rho \mapsto \varphi$ with $\varphi \circ \rho_\chi \sim \rho$).*

For a proof, see [MFG, §2.3.2, §3.2.4] or Mazur's original paper [M87, §1.2] except for $\mathcal{D}_\chi^{fl}$. For $\mathcal{D}_\chi^{fl}$, see [R93, §2] in which Ramakrishna works in finite flat group scheme theory and p-adic Hodge theory (topics we do not touch). In this book, we mainly deal with the first three (more elementary) functors, and the functor $\mathcal{D}_\chi^{fl}$ rarely shows up but we need it to state some results (e.g. Theorem 7.2.3) and questions. The representations $\rho^\emptyset$, ρ and ρ_χ are called *universal representations*.

6.1.2 Tangent space of deformation functors

Let $\mathcal{F} : CL_B \to SETS$ be a covariant functor with $|\mathcal{F}(\mathbb{F})| = 1$. Let $\mathcal{C} = SETS$ or CL_B. For morphisms $\phi' : S' \to S$ and $\phi'' : S'' \to S$ in $\mathcal{C}$,

$$S' \times_S S'' = \{(a', a'') \in S' \times S'' | \phi'(a') = \phi''(a'')\}$$

gives the fiber product of S' and S'' over S in $\mathcal{C}$. We assume that

$$|\mathcal{F}(\mathbb{F})| = 1 \text{ and } \mathcal{F}(\mathbb{F}[\varepsilon] \times_\mathbb{F} \mathbb{F}[\varepsilon]) = \mathcal{F}(\mathbb{F}[\varepsilon]) \times_{\mathcal{F}(\mathbb{F})} \mathcal{F}(\mathbb{F}[\varepsilon])$$

by two projections.

It is easy to see $\mathcal{F} = \mathcal{D}^\emptyset, \mathcal{D}$ and $\mathcal{D}_\chi$ satisfies this condition. Indeed, noting that $\mathbb{F}[\varepsilon] \times_\mathbb{F} \mathbb{F}[\varepsilon] \cong \mathbb{F}[\varepsilon'] \times_\mathbb{F} \mathbb{F}[\varepsilon''] \cong \mathbb{F}[\varepsilon', \varepsilon'']$, if $\rho' \in \mathcal{F}(\mathbb{F}[\varepsilon'])$ and $\rho'' \in \mathcal{F}(\mathbb{F}[\varepsilon''])$, $\rho' \times \rho$ has values in $\mathrm{GL}_2(\mathbb{F}[\varepsilon', \varepsilon''])$ which is an element in $\mathcal{F}(\mathbb{F}[\varepsilon'] \times_\mathbb{F} \mathbb{F}[\varepsilon''])$.

Identify $\mathbb{F}[\varepsilon] \times_\mathbb{F} \mathbb{F}[\varepsilon]$ with $\mathbb{F}[\varepsilon', \varepsilon'']$ ($\varepsilon'\varepsilon'' = 0$ and $\dim_\mathbb{F} \mathbb{F}[\varepsilon] \times_\mathbb{F} \mathbb{F}[\varepsilon] = 3$ but $\dim_\mathbb{F} \mathbb{F}[\varepsilon] \otimes_\mathbb{F} \mathbb{F}[\varepsilon] = 4$). It is easy to see that $a + b\varepsilon' + c\varepsilon'' \mapsto a + (a+c)\varepsilon$ gives an onto CL_B-morphism $a : \mathbb{F}[\varepsilon] \times_\mathbb{F} \mathbb{F}[\varepsilon] \twoheadrightarrow \mathbb{F}[\varepsilon]$ which induces

$$+ : \mathcal{F}(\mathbb{F}[\varepsilon]) \times \mathcal{F}(\mathbb{F}[\varepsilon]) = \mathcal{F}(\mathbb{F}[\varepsilon] \times_\mathbb{F} \mathbb{F}[\varepsilon]) \xrightarrow{\mathcal{F}(a)} \mathcal{F}(\mathbb{F}[\varepsilon]).$$

Plainly this is associative and commutative, and for the inclusion $0 : \mathbb{F} \hookrightarrow \mathbb{F}[\varepsilon]$, we have $\mathbf{0} := \mathrm{Im}(\mathcal{F}(0)(\mathcal{F}(\mathbb{F}))) \in \mathcal{F}(\mathbb{F}[\varepsilon])$ gives the identity. Thus $\mathcal{F}(\mathbb{F}[\varepsilon])$ is an abelian group.

For $\alpha \in \mathbb{F}$, $a + b\varepsilon \mapsto a + \alpha b\varepsilon$ is an automorphism of $\mathbb{F}[\varepsilon]$ in CL_B. This induces an $\mathbb{F}$-scalar multiplication on $\mathcal{F}(\mathbb{F}[\varepsilon])$, and $\mathcal{F}(\mathbb{F}[\varepsilon])$ is an $\mathbb{F}$-vector space, and $\mathcal{F}(\mathbb{F}[\varepsilon])$ is called the **tangent space** of the functor $\mathcal{F}$.

For any $A \in CL_B$ and an A-module X, suppose $|\mathcal{F}(A)| = 1$ and $\mathcal{F}(A[X] \times_A A[X]) = \mathcal{F}(A[X]) \times_{\mathcal{F}(A)} \mathcal{F}(A[X])$. Then $A[X] \times_A A[X] = A[X \oplus X]$. The addition on X and A-linear map $\alpha : X \to X$ induce in the same way as above CL_B-morphisms $+_* : A[X \oplus X] \to A[X]$ by $a + (x \oplus y) \mapsto a + x + y$ and $\alpha_* : A[X] \to A[X]$ by $a + x \mapsto a + \alpha(x)$. Thus we have by functoriality the "addition"

$$+ : \mathcal{F}(A[X]) \times_{\mathcal{F}(A)} \mathcal{F}(A[X]) = \mathcal{F}(A[X \oplus X]) \xrightarrow{\mathcal{F}(+_*)} \mathcal{F}(A[X])$$

and the α-action $\alpha : \mathcal{F}(A[X]) \xrightarrow{\mathcal{F}(\alpha_*)} \mathcal{F}(A[X])$. With $\mathbf{0} := \mathrm{Im}(\mathcal{F}(A) \to \mathcal{F}(A[X]))$ for the inclusion $A \hookrightarrow A[X]$, this makes $\mathcal{F}(A[X])$ an A-module.

6.1.3 Tangent space of local rings and generators

The dimension d of the cotangent space of a local ring R over B gives the number of generators of the ring R over B. We describe this fact. Using this fact, we prove that $\Omega_{R/B}$ is generated by d elements as R-modules.

Lemma 6.1.2. *Suppose that the base ring $B \in CL_W$ is noetherian. If $t^*_{R/B} = \mathfrak{m}_R/(\mathfrak{m}_R^2 + \mathfrak{m}_B)$ is a finite dimensional vector space over $\mathbb{F}$, then $R \in CL_B$ is noetherian. Moreover $\dim_{\mathbb{F}} t^*_{R/B}$ gives the minimal number of topological generators of the ring R over B.*

The space $t^*_{R/B}$ is called the *cotangent space* of R at $\mathfrak{m}_R = (\varpi) \in \mathrm{Spec}(R)$ over $\mathrm{Spec}(W)$. Define t^*_R by $\mathfrak{m}_R/\mathfrak{m}_R^2$, which is called the *absolute cotangent space* of R at $\mathfrak{m}_R$.

Proof. For the moment, we assume $B = W$. Since

$$\mathbb{F} \xrightarrow[a \mapsto a\varpi]{\sim} \mathfrak{m}_W/\mathfrak{m}_W^2 \longrightarrow t^*_R \longrightarrow t^*_{R/W} \longrightarrow 0$$

is exact, we conclude $\dim_{\mathbb{F}} t^*_R < \infty$ if $\dim_{\mathbb{F}} t^*_{R/W} < \infty$.

First suppose that $\mathfrak{m}_R^N = 0$ for sufficiently large N. Let $\overline{x}_1, \ldots, \overline{x}_m$ be an $\mathbb{F}$–basis of t^*_R. Choose $x_j \in R$ so that $x_j \bmod \mathfrak{m}_R^2 = \overline{x}_j$ and consider the ideal $\mathfrak{a}$ generated by x_j. We have $\mathfrak{a} = \sum_j Rx_j \hookrightarrow \mathfrak{m}_R$ (the inclusion).

After tensoring $R/\mathfrak{m}_R$, we have the surjectivity of the induced linear map: $\mathfrak{a}/\mathfrak{m}_R\mathfrak{a} \cong \mathfrak{a} \otimes_R R/\mathfrak{m}_R \to \mathfrak{m} \otimes_R R/\mathfrak{m}_R \cong \mathfrak{m}/\mathfrak{m}_R^2$ because $\{\overline{x}_1, \ldots, \overline{x}_m\}$ is an $\mathbb{F}$–basis of t^*_R. This shows that $\mathfrak{m}_R = \mathfrak{a} = \sum_j Rx_j$ (NAK: Nakayama's Lemma 1.8.3 applied to the cokernel of $R^m \ni (a_1, \ldots, a_m) \mapsto \sum_j a_j x_j \in \mathfrak{m}_R$). Thus $\mathfrak{m}_R^k/\mathfrak{m}_R^{k+1}$ is generated over $\mathbb{F}$ by the monomials in x_j of degree k. In particular, $\mathfrak{m}_R^{N-1}$ is generated by the monomials in $(x_0 := \varpi, x_1, \ldots x_m)$ of degree $N - 1$.

Inductive step:
Define $\pi : C = W[[X_1, \ldots, X_m]] \to R$ by $\pi(f(X_1, \ldots, X_m)) = f(x_1, \ldots, x_m)$. Since any monomial of degree $> N$ vanishes after applying π, π is a well defined W-algebra homomorphism. Let $\mathfrak{m} = \mathfrak{m}_C = (\varpi, X_1, \cdots, X_m)$ be the maximal ideal of C. By definition, $\pi(\mathfrak{m}^{N-1}) = \mathfrak{m}_R^{N-1}$. Suppose now that $\pi(\mathfrak{m}^{N-j}) = \mathfrak{m}_R^{N-j}$, and try to prove the surjectivity of $\pi(\mathfrak{m}^{N-j-1}) = \mathfrak{m}_R^{N-j-1}$.

Since $\mathfrak{m}_R^{N-j-1}/\mathfrak{m}_R^{N-j}$ is generated by monomials of degree $N - j - 1$ in x_j, for each $x \in \mathfrak{m}_R^{N-j-1}$, we find a homogeneous polynomial $P \in \mathfrak{m}^{N-j-1}$ of $x_1, \ldots, x_m$ of degree $N - j - 1$ such that $x - \pi(P) \in \mathfrak{m}_R^{N-j} = \pi(\mathfrak{m}^{N-j})$. This shows $\pi(\mathfrak{m}^{N-j-1}) = \mathfrak{m}_R^{N-j-1}$. Thus by induction on j, we get the surjectivity of π.

General case:
Write $R = \varprojlim_i R_i$ for Artinian rings R_i. The projection maps are onto: $t^*_{R_{i+1}} \twoheadrightarrow t^*_{R_i}$. Since t^*_R is of finite dimensional, $t^*_{R_{i+1}} \cong t^*_{R_i}$ for sufficiently large i. Thus choosing x_j as above in R, we have its image $x_j^{(i)}$ in R_i.

Use $x_j^{(i)}$ to construct $\pi_i : W[[X_1, \ldots, X_m]] \to R_i$ in place of x_j. Then π_i is surjective as already shown, and $\pi = \varprojlim_i \pi_i : W[[X_1, \ldots, X_m]] \to R$ remains surjective, because projective limit of continuous surjections, if all sets involved are compact sets, remains surjective; so, R is noetherian as profinite sets are compact.

Over a general base B:
Since we have an exact sequence: $\mathfrak{m}_B/\mathfrak{m}_B^2 \to \mathfrak{m}_R/\mathfrak{m}_R^2 \twoheadrightarrow t_{R/W}^*$, we conclude in the same way as above that $W[[X_1, \ldots, X_r, X_{r+1}, \ldots, X_{r+s}]]$ surjects onto R sending X_i with $i > r$ to generators of $t_{R/B}^*$. Thus the number of generators over B of R is $\dim_{\mathbb{F}} t_{R/B}^*$. $\qquad\square$

The following result follows from the proof of the above lemma:

Corollary 6.1.3. *Let the assumption be as in Lemma 6.1.2. If $t_{R/B}^* = 0$, the B-algebra structure $i_R : B \to R$ is surjective.*

6.1.4 Tangent space as adjoint cohomology group

We compute the cotangent space of $R^\emptyset$ over $B = W$ via cohomology. We recall a full adjoint G-module $ad(\overline{\rho})$ given by $\mathfrak{gl}_2(\mathbb{F}) = M_2(\mathbb{F})$ on which G acts by $x \mapsto \overline{\rho}(g) x \overline{\rho}(g)^{-1}$.

Lemma 6.1.4. *Let $R = R^\emptyset$. Then*

$$t_{R/W} = \mathrm{Hom}_{\mathbb{F}}(t_{R/W}^*, \mathbb{F}) \cong H^1(G, ad(\overline{\rho})),$$

where $H^1(G, ad(\overline{\rho}))$ is the continuous first cohomology group of the profinite group G with coefficients in the discrete G–module $ad(\overline{\rho})$.

The space $t_{R/W}$ is called the tangent space of $\mathrm{Spec}(R)_{/W}$ at $\mathfrak{m}$. In the following proof of the lemma, we write $R = R_{\overline{\rho}}$.

Proof. *Step 1, dual number.* Let $A = \mathbb{F}[\varepsilon] = \mathbb{F}[X]/(X^2)$ with $X \leftrightarrow \varepsilon$. Then $\varepsilon^2 = 0$. We claim $\mathrm{Hom}_{W\text{-alg}}(R, A) \cong t_{R/W}$. For a W-algebra homomorphism $\phi : R \to A$, write $\phi(r) = \phi_0(r) + \phi_\varepsilon(r)\varepsilon$ with $\phi_0(r), \phi_\varepsilon(r) \in \mathbb{F}$. Then the map is given by $\phi \mapsto \ell_\phi = \phi_\varepsilon|_{\mathfrak{m}_R}$.
Step 2, well definedness of ℓ_ϕ. From $\phi(ab) = \phi(a)\phi(b)$, we get

$$\phi_0(ab) = \phi_0(a)\phi_0(b) \text{ and } \phi_\varepsilon(ab) = \phi_0(a)\phi_\varepsilon(b) + \phi_0(b)\phi_\varepsilon(a).$$

Thus $\phi_\varepsilon \in Der_W(R, \mathbb{F}) \cong \mathrm{Hom}_{\mathbb{F}}(\Omega_{R/W} \otimes_R \mathbb{F}, \mathbb{F})$. Since for any derivation $\delta \in Der_W(R, \mathbb{F})$, $\phi' = \phi_0 + \delta\varepsilon \in \mathrm{Hom}_{W\text{-alg}}(R, A)$, we find

$$\mathrm{Hom}_R(\Omega_{R/W} \otimes_R \mathbb{F}, \mathbb{F}) \cong Der_W(R, A) \cong \mathrm{Hom}_{W\text{-alg}}(R, A)$$

and $\text{Ker}(\phi_0) = \mathfrak{m}_R$ because R is local. Since ϕ is W-linear, $\phi_0(a) = \bar{a} = a$ mod $\mathfrak{m}_R$. Thus ϕ kills $\mathfrak{m}_R^2$ and takes $\mathfrak{m}_R$ W-linearly into $\mathfrak{m}_A = \mathbb{F}\varepsilon$; so, $\ell_\phi : t_R^* \to \mathbb{F}$. For $r \in W$, $\bar{r} = r\phi(1) = \phi(r) = \bar{r} + \phi_\varepsilon(r)\varepsilon$, and hence ϕ_ε kills W; so, $\ell_\phi \in t_{R/W}$.

Step 3, injectivity of $\phi \mapsto \ell_\phi$. Since R shares its residue field $\mathbb{F}$ with W, any element $a \in R$ can be written as $a = r + x$ with $r \in W$ and $x \in \mathfrak{m}_R$. Thus ϕ is completely determined by the restriction ℓ_ϕ of ϕ_ε to $\mathfrak{m}_R$, which factors through $t_{R/W}^*$. Thus $\phi \mapsto \ell_\phi$ induces an injective linear map $\ell :$ $\text{Hom}_{W-alg}(R, A) \hookrightarrow \text{Hom}_\mathbb{F}(t_{R/W}^*, \mathbb{F})$.

Note $R/(\mathfrak{m}_R^2 + \mathfrak{m}_W) = \mathbb{F} \oplus t_{R/W}^* = \mathbb{F}[t_{R/W}^*]$ with the projection $\pi : R \twoheadrightarrow$ $t_{R/W}^*$ to the direct summand $t_{R/W}^*$. Indeed, writing $\bar{r} = (r \mod \mathfrak{m}_R)$, for the inclusion $\iota : \mathbb{F} = W/\mathfrak{m}_W \hookrightarrow R/(\mathfrak{m}_R^2 + \mathfrak{m}_W)$, $\pi(r) = r - \iota(\bar{r})$.

Step 4, surjectivity of $\phi \mapsto \ell_\phi$. For any $\ell \in \text{Hom}_\mathbb{F}(t_{R/W}^*, \mathbb{F})$, we extend ℓ to R by putting $\ell(r) = \ell(\pi(r))$. Then we define $\phi : R \to A$ by $\phi(r) = \bar{r} + \ell(\pi(r))\varepsilon$. Since $\varepsilon^2 = 0$ and $\pi(r)\pi(s) = 0$ in $\mathbb{F}[t_{R/W}^*]$,

$$rs = (\bar{r} + \pi(r))(\bar{s} + \pi(s)) = \bar{r}\bar{s} + \bar{s}\pi(r) + \bar{r}\pi(s)$$

$$\xrightarrow{\phi} \bar{r}\bar{s} + \bar{s}\ell(\pi(r))\varepsilon + \bar{r}\ell(\pi(s))\varepsilon = \phi(r)\phi(s)$$

is a W-algebra homomorphism; so, $\ell(\phi) = \ell$, and hence ℓ is surjective.

By $\text{Hom}_R(\Omega_{R/W} \otimes_R \mathbb{F}, \mathbb{F}) \cong \text{Hom}_{W\text{-alg}}(R, A)$, we have

$$\text{Hom}_R(\Omega_{R/W} \otimes_R \mathbb{F}, \mathbb{F}) \cong \text{Hom}_\mathbb{F}(t_{R/W}^*, \mathbb{F});$$

so, if $t_{R/W}^*$ is finite dimensional, we again get

$$\boxed{\Omega_{R/W} \otimes_R \mathbb{F} \cong t_{R/W}^*}. \tag{6.3}$$

Step 5, use of universality. By the universality, we have

$$\text{Hom}_{W-alg}(R, A) \cong \{\rho : G \to GL_n(A) | \rho \mod \mathfrak{m}_A = \bar{\rho}\}/\sim.$$

Write $\rho(g) = \bar{\rho}(g) + u_\phi'(g)\varepsilon$ for ρ corresponding to $\phi : R \to A$. From the multiplicativity, we have

$$\bar{\rho}(gh) + u_\phi'(gh)\varepsilon = \rho(gh) = \rho(g)\rho(h) = \bar{\rho}(g)\bar{\rho}(h) + (\bar{\rho}(g)u_\phi'(h) + u_\phi'(g)\bar{\rho}(h))\varepsilon.$$

Thus as a function $u' : G \to M_n(\mathbb{F})$, we have

$$u_\phi'(gh) = \bar{\rho}(g)u_\phi'(h) + u_\phi'(g)\bar{\rho}(h). \tag{6.4}$$

Step 6, getting 1-cocycle. Define a map $u_\rho = u_\phi : G \to ad(\bar{\rho})$ by $u_\phi(g) = u_\phi'(g)\bar{\rho}(g)^{-1}$. Then by computation, we get $gu_\phi(h) = \bar{\rho}(g)u_\phi(h)\bar{\rho}(g)^{-1}$ from the definition of $ad(\bar{\rho})$. Then from (6.4), we conclude that

$$\boxed{u_\phi(gh) = gu_\phi(h) + u_\phi(g)}.$$

Thus $u_\phi : G \to ad(\overline{\rho})$ is a 1-cocycle, and we get an $\mathbb{F}$-linear map

$$t_{R/W} \cong \mathrm{Hom}_{W\text{-alg}}(R, A) \to H^1(G, ad(\overline{\rho})) \quad \text{by } \ell_\phi \mapsto [u_\phi].$$

Step 7, end of the proof. By computation, for $x \in ad(\overline{\rho})$

$$\rho \sim \rho' \Leftrightarrow \overline{\rho}(g) + u'_\rho(g)\varepsilon = (1 + x\varepsilon)(\overline{\rho}(g) + u'_{\rho'}(g)\varepsilon)(1 - x\varepsilon)$$

$$\Leftrightarrow u'_\rho(g) = x\overline{\rho}(g) - \overline{\rho}(g)x + u'_{\rho'}(g) \Leftrightarrow u_\rho(g) = (1 - g)x + u_{\rho'}(g).$$

Thus the cohomology classes of u_ρ and $u_{\rho'}$ are equal if and only if $\rho \sim \rho'$. This shows:

$$\mathrm{Hom}_{\mathbb{F}}(t^*_{R/W}, \mathbb{F}) \cong \mathrm{Hom}_{W-alg}(R, A)$$

$$\cong \{\rho : G \to GL_n(A) | \rho \mod \mathfrak{m}_A = \overline{\rho}\}/\sim \,\cong H^1(G, ad(\overline{\rho})),$$

and we get a bijection between $\mathrm{Hom}_{\mathbb{F}}(t^*_{R/W}, \mathbb{F})$ and $H^1(G, ad(\overline{\rho}))$. $\qquad\square$

Lemma 6.1.5. *Let $\mathcal{F} = \mathcal{D}^\emptyset, \mathcal{D}$ or $\mathcal{D}_\chi$ and $R = R^\emptyset, R^{ord}, R_\chi$ or R_χ^{fl} accordingly. Then $t_{R/B} \cong \mathcal{F}(\mathbb{F}[\varepsilon])$ as $\mathbb{F}$-vector spaces.*

Proof. Since the proofs are almost the same, we deal with $\mathcal{D}^\emptyset$. Let $R^\emptyset$ be the universal ring for $\mathcal{D}^\emptyset$. We got a canonical bijection in Lemma 6.1.5:

$$\mathcal{D}^\emptyset(\mathcal{F}[\varepsilon]) \xrightarrow[i_1]{\text{1-1 onto}} H^1(G, ad(\overline{\rho})) \xrightarrow[i]{\sim} t_{R^\emptyset/B}$$

with a vector space isomorphism i. We have constructed a cocycle u_ρ from $\rho \in \mathcal{F}(\mathbb{F}[\varepsilon])$ writing $\rho = \overline{\rho} + u_\rho \overline{\rho}\varepsilon$. Regarding $(\rho, \rho') \in \mathcal{F}(\mathbb{F}[\varepsilon]) \times \mathcal{F}(\mathbb{F}[\varepsilon]) = \mathcal{F}(\mathbb{F}[\varepsilon] \times_{\mathbb{F}} \mathbb{F}[\varepsilon])$, we see that $+(\rho, \rho') = \overline{\rho} + (u_\rho \overline{\rho} + u_{\rho'} \overline{\rho})\varepsilon \in \mathcal{F}(\mathbb{F}[\varepsilon])$; so, i_1 is a homomorphism. Similarly, one can check that it is $\mathbb{F}$-linear. $\qquad\square$

6.1.5 Mod p adjoint Selmer group

We have identified $\mathcal{D}^\emptyset(\mathbb{F}[\varepsilon])$ with a $\mathbb{F}$-vector subspace of $H^1(G, ad(\overline{\rho}))$. We want to explicitly determine the subspace $\mathcal{D}_\chi(\mathbb{F}[\varepsilon])$. For that, we recall the adjoint Galois submodule $Ad(\overline{\rho}) \subset ad(\overline{\rho})$ from "Suggestion to the reader", which consists of trace 0 matrices in $ad(\overline{\rho}) = M_2(\mathbb{F})$. Since the associated cohomology class corresponds to a strict conjugacy class, we may choose by (p) a basis of $V(\rho)$ for $\rho \in \mathcal{D}_\chi(\mathbb{F}[\varepsilon])$ so that $\rho|_{D_p}$ is upper triangular with quotient character δ congruent to $\overline{\delta}$ modulo $\mathfrak{m}_A$. Similarly by (l), for $l \in S_{ab}$, we choose a basis so that $\rho|_{I_l} = \epsilon_l \oplus 1$ in this order. The basis is dependent on $l \in S_{ab} \cup \{p\}$. So the cocycle coming from ρ has locally a matrix form at $l \in S_{ab} \cup \{p\}$.

Theorem 6.1.6. *A 1-cocycle u gives rise to a class in $\mathcal{D}_\chi(\mathbb{F}[\varepsilon])$ if and only if $u(I_l) = 0$ for all prime $l \in S$ not equal to p, $u|_{D_p}$ is upper triangular, $u|_{I_p}$ is upper nilpotent and $\mathrm{Tr}(u) = 0$ over G.*

Note that the description of cocycles u is independent of χ; so, even if one changes χ, the tangent space $t_{R_\chi/B}$ is independent as a cohomology subgroup as long as $\mathbb{F}$ does not change.

Proof. By (det), $1 = \det(\rho\overline{\rho}^{-1}) = \det(1 + u_\rho \varepsilon) = 1 + \mathrm{Tr}(u_\rho)\varepsilon$; so, (det) $\Leftrightarrow \mathrm{Tr}(u) = 0$ over G. Thus we have $t_{R_\chi/B} \subset H^1(G, Ad(\overline{\rho}))$.

Choose a generator $w \in V(\epsilon)$ over $\mathbb{F}[\varepsilon]$. Then (w, v) is a basis of $V(\rho)$ over $\mathbb{F}[\varepsilon]$. Let $(\overline{w}, \overline{v}) = (w, v)$ mod ε and identify $V(ad(\overline{\rho}))$ with $M_2(\mathbb{F})$ with this basis. Defining $\overline{\rho}$ by $(\sigma\overline{w}, \sigma\overline{v}) = (\overline{w}, \overline{v})\overline{\rho}(\sigma)$, for $\sigma \in D_p$, we have $\overline{\rho}(\sigma) = \begin{pmatrix} \overline{\epsilon}(\sigma) & * \\ 0 & \overline{\delta}(\sigma) \end{pmatrix}$ (upper triangular). If $\sigma \in I_p$, $\rho\overline{\rho}^{-1} = 1 + u_\rho$ with lower right corner of u_ρ has to vanish as $\delta = 1$ on I_p, we have $u_\rho(\sigma) \in \{\begin{pmatrix} 0 & * \\ 0 & 0 \end{pmatrix}\}$.

Since ramification at $l \neq p$ is concentrated to $\overline{\rho}$ as $\rho(I_l)$ has order prime to p, $(l) \Leftrightarrow u_\rho(I_l) = 0$. The condition (p) is equivalent to u_ρ having the form $\begin{pmatrix} * & * \\ 0 & 0 \end{pmatrix}$ but by $\mathrm{Tr}(u_\rho) = 0$, it has to be upper nilpotent. $\qquad\square$

For $l \in S \cup \{p\}$ and $\mathcal{F} = \mathcal{D}, \mathcal{D}_\chi$, we denote by $\mathcal{F}_l$ the corresponding local deformation functor, adding subscript l; so,

$$\mathcal{D}_l(A) = \{\rho_A : \mathrm{Gal}(\overline{\mathbb{Q}}_l/\mathbb{Q}_l) \to \mathrm{GL}_2(A)|\rho_A \mod \mathfrak{m}_A = \overline{\rho},$$

$$\rho_A(I_l) \cong \overline{\rho}(I_l) \text{ for } l \in S \text{ and } \rho \text{ satisfies } (l) \text{ if } l \in S_{ab} \cup \{p\}\}/\Gamma(\mathfrak{m}_A),$$

and $\mathcal{D}_{\chi,l}(A) = \{\rho \in \mathcal{D}_l(A)| \det(\rho) = \iota_A \circ \chi\}$. Thus by the proof of Theorem 6.1.6, we find

$$\mathcal{D}_\chi(A) = \{\rho_A : G \to \mathrm{GL}_2(A) \in \mathcal{D}^\emptyset(A) : \rho_A|_{D_l} \in \mathcal{D}_{\chi,l}(A) \text{ for all } l \in S \cup \{p\}\},$$

since the global character $\det(\rho_A)$ is determined by its restriction to I_l for all $l \in S \cup \{p\}$. Therefore, in appearance, we have

$$\mathrm{Sel}(Ad(\overline{\rho})) := \mathrm{Ker}(H^1(G, Ad(\overline{\rho})) \to \prod_{l \in S \cup \{p\}} \frac{H^1(\mathbb{Q}_l, Ad(\overline{\rho}))}{\mathcal{D}_{\chi,l}(\mathbb{F}[\varepsilon])}) \cong \mathcal{D}_\chi(\mathbb{F}[\varepsilon]).$$

Since $I_l \subset G$ has order prime to p for $l \neq p$; so, $H^1(I_l, M) = 0$ for $M = ad(\overline{\rho})$ or $Ad(\overline{\rho})$. In this case, $\mathcal{D}_{\chi,l}(\mathbb{F}[\varepsilon]) = \mathrm{Ker}(H^1(\mathbb{Q}_l, Ad(\overline{\rho})) \to H^1(I_l, Ad(\overline{\rho})) = H^1(\mathbb{Q}_l, Ad(\overline{\rho}))$, and hence the factor $\frac{H^1(\mathbb{Q}_l, ad(\overline{\rho}))}{\mathcal{D}_l(\mathbb{F}[\varepsilon])} = 0$ disappears for $l \neq p$. Thus we can rewrite

$$\mathrm{Sel}(Ad(\overline{\rho})) := \mathrm{Ker}(H^1(G, Ad(\overline{\rho})) \to \frac{H^1(\mathbb{Q}_p, Ad(\overline{\rho}))}{\mathcal{D}_{\chi,p}(\mathbb{F}[\varepsilon])}) \cong t_{R_\chi/B}. \qquad (6.5)$$

6.1.6   p-finiteness condition

For each open subgroup $\mathcal{H}$ of a profinite group $\mathcal{G}$, we write $\mathcal{H}_p$ for the maximal p–profinite quotient. We consider the following condition:

(Φ)   For each open subgroup $\mathcal{H}$ of $\mathcal{G}$, $\mathrm{Hom}_{cont}(\mathcal{H}, \mathbb{F}_p)$ is a finite group.

Let $\Phi(H) = D(H)H^p$ for the topological derived group $D(H)$ of H (and $H^p := \{h^p | h \in H\}$). Then $\mathrm{Hom}_{cont}(H, \mathbb{F}_p) = \mathrm{Hom}(H/\Phi(H), \mathbb{F}_p)$; so, the condition ($\Phi$) is equivalent to the finiteness of the index $[H : \Phi(H)]$.

Proposition 6.1.7 (Mazur). *The group G satisfies* (Φ), *and* $R^\emptyset, R^{ord}$, R_χ *and* R_χ^{fl} *are noetherian. In particular,* $t^*_{R/W}$ *for* $R = R^\emptyset, R^{ord}$ *and* R^χ *is finite dimensional over* $\mathbb{F}$ *and is isomorphic to* $\Omega_{R/W} \otimes_R \mathbb{F}$ (*see* (6.3)).

By this fact, the universal ring belongs to CNL_W. Therefore the deformation functor originally defined over CL_W is actually representable by an object in the smaller full subcategory CNL_W not just pro-representable over CNL_W.

Proof.   Since the tangent space of any of the functors is a subspace of $H^1(G, Ad(\overline{\rho})) \cong \mathcal{D}^\emptyset(\mathbb{F}[\varepsilon])$, we only need to prove this for $R = R^\emptyset$. Let $H = \mathrm{Ker}(\overline{\rho})$. Then the action of H on $ad(\overline{\rho})$ is trivial. By the inflation-restriction sequence for G, we have the following exact sequence:

$$0 \to H^1(G/H, H^0(H, ad(\overline{\rho}))) \to H^1(G, ad(\overline{\rho})) \to \mathrm{Hom}(\Phi(H), M_n(\mathbb{F})).$$

From this, it is clear that $\dim_{\mathbb{F}} H^1(G, ad(\overline{\rho})) < \infty$. The fact that G satisfies (Φ) follows from class field theory. Indeed, if k is the fixed field of an open subgroup $\mathcal{H}$ of G, then $\Phi(\mathcal{H})$ fixes the maximal abelian extension M/k with $\mathrm{Gal}(M/k)$ killed by p unramified outside $\{p\} \cup S$. By Lemma 5.2.1, $[M : k]$ is finite. $\square$

Corollary 6.1.8. *Let R be as in* Proposition 6.1.7. *Then $\Omega_{R/W}$ is an R-module of finite type, and its minimal number of generators over R is equal to* $\dim_{\mathbb{F}} \Omega_{R/W} \otimes_R \mathbb{F} = \dim_{\mathbb{F}} t_{R/W}$.

Proof.   For any R-module M, Lemma 1.8.3 tells us $M \otimes_R \mathbb{F} = 0 \Rightarrow M = 0$. Choose a basis $B = \{\overline{b}\}$ of $M/\mathfrak{m}_R M = M \otimes_R \mathbb{F}$ and suppose B is finite. Lift $\overline{b}$ to $b \in M$, and consider the R-linear map $\pi : \bigoplus_{g \in B} R \ni (a_{\overline{b}})_{\overline{b} \in B} \mapsto \sum_{\overline{b}} a_{\overline{b}} b \in M$. Tensoring $\mathbb{F}$ over R, we find $\mathrm{Coker}(\pi) \otimes_R \mathbb{F} = 0$; so, $\mathrm{Coker}(\pi) = 0$. This implies that $\{b | \overline{b} \in B\}$ is the minimal generators of M over R. Apply this to $M = \Omega_{R/W}$, we get the desired result by Proposition 6.1.7. $\square$

6.1.7   Reinterpretation of the functor $\mathcal{D}$

Even if we take $B = W$, we show that R^{ord} is canonically an algebra over Λ. By Corollary 1.2.2, the finite order character $\det(\overline{\rho})$ factors through $\mathrm{Gal}(\mathbb{Q}[\mu_{N_0}]/\mathbb{Q})$ for some positive integer N_0. Let N_0 be the minimal such integer (called *conductor* of $\det(\overline{\rho})$). Write $N_0 = Np^\nu$ for N prime to p; so, N is the prime-to-p conductor of $\det(\overline{\rho})$. Note that $\det(\rho)$ factors through $\mathrm{Gal}(\mathbb{Q}[\mu_{Np^\infty}]/\mathbb{Q}) \cong \mathbb{Z}_p^\times \times (\mathbb{Z}/N\mathbb{Z})^\times$. Recall $\Gamma \cong 1 + p\mathbb{Z}_p$ (the maximal p-profinite quotient of $\mathrm{Gal}(\mathbb{Q}[\mu_{Np^\infty}]/\mathbb{Q})$). This shows that the maximal p-abelian (p-profinite) quotient G_p^{ab} is the p-primary part Γ of $\mathrm{Gal}(\mathbb{Q}[\mu_{Np^\infty}]/\mathbb{Q}) \cong (\mathbb{Z}/N\mathbb{Z})^\times \times \mathbb{Z}_p^\times$. Supposing $\chi|_{I_l}$ has values in $W^\times$, consider the deformation functor

$$D(A) = \{\varphi : G \to A^\times | \varphi \bmod \mathfrak{m}_A = \det(\overline{\rho}), \varphi|_{I_l} = \iota_A \circ \chi|_{I_l} \; \forall l \neq p\}.$$

By Theorem 5.1.1, this functor is represented by $W[[\Gamma]]$ with *universal character* $\kappa(\sigma) = \chi_0(\sigma)[\sigma]$, where χ_0 is the Teichmüller lift of $\overline{\chi} := \chi \bmod \mathfrak{m}_W$ and $[\sigma]$ is the restriction of σ to $\mathbb{Q}_\infty$ with $\mathrm{Gal}(\mathbb{Q}_\infty/\mathbb{Q}) = \Gamma$ for a subfield $\mathbb{Q}_\infty \subset \mathbb{Q}[\mu_{p^\infty}]$. Since $\det \rho \in D(R^{ord})$, we have $i = \iota_{R^{ord}} : W[[\Gamma]] \to R^{ord}$ such that $\det \rho = i \circ \kappa$.

Consider the following deformation functor $\mathcal{D}_\kappa : CL_{/\Lambda} \to SETS$

$$\mathcal{D}_\kappa(A) = \{\rho \in \mathcal{D}(A) | \det(\rho) = \iota_A \circ \kappa\}/\cong,$$

where writing $\iota_A : \Lambda \to A$ for Λ-algebra structure of A.

Proposition 6.1.9. *We have $\mathcal{D}_\kappa(A) \cong \mathrm{Hom}_{\Lambda\text{-alg}}(R^{ord}, A)$ with universal representation $\rho \in \mathcal{D}(R^{ord})$; so, for any choice of $\chi \in D(B)$ (where B is either Λ or W), $\mathcal{D}_{\kappa,p}(\mathbb{F}[\varepsilon]) = \mathcal{D}_{\chi,p}(\mathbb{F}[\varepsilon])$ and*

$$\mathrm{Sel}(Ad(\overline{\rho})) := \mathrm{Ker}(H^1(G, Ad(\overline{\rho})) \to \frac{H^1(\mathbb{Q}_p, Ad(\overline{\rho}))}{\mathcal{D}_{\chi,p}(\mathbb{F}[\varepsilon])}) \cong t_{R^{ord}/\Lambda}.$$

Proof. For any $\rho \in \mathcal{D}_\kappa(A)$, regard $\rho \in \mathcal{D}(A)$. Then we have $\varphi \in \mathrm{Hom}_{W\text{-alg}}(R^{ord}, A)$ such that $\varphi \circ \rho \cong \rho$. Thus $\varphi \circ \det(\rho) = \det(\rho)$. Since $\det(\rho) = \iota_A \circ \kappa$ and $\det(\rho) = \iota_{R^{ord}} \circ \kappa$, we find $\varphi \circ \iota_{R^{ord}} = \iota_A$, and hence $\varphi \in \mathrm{Hom}_{\Lambda\text{-alg}}(R^{ord}, A)$. Thus R^{ord} also represents $\mathcal{D}_\kappa$ over Λ.

As we already remarked just after Theorem 6.1.6, $\mathcal{D}_\kappa(\mathbb{F}[\varepsilon]) = t_{R^{ord}/\Lambda} = \mathfrak{m}_{R^{ord}}/(\mathfrak{m}_{R^{ord}}^2 + \mathfrak{m}_\Lambda)$ is independent as a subgroup of $H^1(G, Ad(\overline{\rho}))$; so, we get a new expression of $\mathrm{Sel}(Ad(\overline{\rho}))$. $\square$

By the proof, $\Omega_{R^{ord}/\Lambda} \otimes_{R^{ord}} \mathbb{F} \cong \mathrm{Sel}(Ad(\overline{\rho})) \cong \Omega_{R_\chi/W} \otimes_{R_\chi} \mathbb{F}$, so the smallest number of generators of $\Omega_{R^{ord}/\Lambda}$ as R^{ord}-modules and $\Omega_{R_\chi/W}$ as

R_χ modules is equal. In the same way, the number of generators of R^{ord} as Λ-algebras and R_χ as W-algebras is equal.

If $\rho_A \in \mathcal{D}_\chi(A)$, regarding $\rho_A \in \mathcal{D}_\kappa(A)$, we have a morphism $\varphi \in \mathrm{Hom}_{CL_\Lambda}(R^{ord}, A)$ with $\varphi \circ \rho \sim \rho_A$. Since $\varphi(\kappa(\gamma)) = \det(\rho_A([\gamma, \mathbb{Q}_p])) = \chi(\gamma)$, φ factors through $R^{ord}/(t - \chi(\gamma))$, and we obtain

Corollary 6.1.10. *The morphism* $\pi : R^{ord} \to R_\chi$ *with* $\pi \circ \rho \sim \rho_\chi$ *induces an isomorphism* $R^{ord}/(t - \chi(\gamma)) \cong R_\chi$.

Exercise 6.1.1. Why does $\varphi \circ \rho \in \mathcal{D}_\chi(A)$ for $\varphi \in \mathrm{Hom}_{CL_\Lambda}(R^{ord}, A)$ factor through $R^{ord}/(t - \chi(\gamma))$?

6.1.8 *General adjoint Selmer group*

Let us describe an *optimal* deformation ρ in each class $c \in \mathcal{D}_\chi(A)$. By (l) for $l \in S_{ab} \cup \{p\}$, the universal representation ρ_χ is equipped with a basis $(\mathbf{v}_l, \mathbf{w}_l)$ so that the matrix representation with respect this basis satisfies (l). As before we allow $\chi = \kappa$. By representability, each class $c \in \mathcal{D}_\chi(A)$ has ρ such that $V(\rho) = V(\rho_\chi) \otimes_{R_\chi, \varphi} A$ for a unique $\varphi \in \mathrm{Hom}_{B\text{-alg}}(R_\chi, A)$, we can choose a unique $\rho \in c$ having matrix form over D_l with respect to the basis $\{(v_l = \mathbf{v}_l \otimes 1, w_l = \mathbf{w}_l \otimes 1\}_l$ satisfying $\{(l){:}l \in S_{ab} \cup \{p\}\}$ compatible with specialization. We always take such an optimal representative ρ for each class $c \in \mathcal{D}_\chi(A)$ hereafter.

Take a finite A-module X and consider the ring $A[X] = A \oplus X$ with $X^2 = 0$. Then $A[X]$ is still p-profinite. Pick $\rho \in \mathcal{D}_\chi(A[X])$ such that ρ mod $X \sim \rho_0$. By our choice of representative ρ and ρ_0 as above, we may (and do) assume ρ mod $X = \rho_0$.

Here we allow $\chi = \kappa$ but if $\chi = \kappa$, we assume that $A \in CL_\Lambda$. Writing $B = W$ if χ has values in $W^\times$ and Λ if $\chi = \kappa$, the functor $\mathcal{D}_\chi$ is defined over $CL_{/B}$. Let ρ_0 act on $M_2(A)$ and $\mathfrak{sl}_2(A) = \{x \in M_2(A) | \mathrm{Tr}(x) = 0\}$ by conjugation. Write this representation $ad(\rho)$ and $Ad(\rho)$ as before. Let $ad(X) = ad(A) \otimes_A X$ and $Ad(X) = Ad(A) \otimes_A X$ and regard them as G-modules by the action on $ad(A)$ and $Ad(A)$. Then we define

$$\Phi(A[X]) = \frac{\{\rho : G \to \mathrm{GL}_2(A[X]) | (\rho \text{ mod } X) = \rho_0, [\rho] \in \mathcal{D}_\chi(A[X])\}}{1 + M_2(X)},$$

where in the isomorphism class $[\rho] \in \mathcal{D}_\chi(A)$, we take an optimal representative ρ as described above.

Take X finite as above. For $\rho \in \Phi(X)$, we can write $\rho = \rho_0 \oplus u'_\rho$ letting

ρ_0 act on $M_2(X)$ by matrix multiplication from the right. Then as before

$$\rho_0(gh) \oplus u'_\rho(gh) = (\rho_0(g) \oplus u'_\rho(g))(\rho_0(h) \oplus u'_\rho(h))$$
$$= \rho_0(gh) \oplus (u'_\rho(g)\rho_0(h) + \rho_0(g)u'_\rho(h))$$

produces $u'_\rho(gh) = u'_\rho(g)\rho_0(h) + \rho_0(g)u'_\rho(h)$ and multiplying by $\rho_0(gh)^{-1}$ from the right, we get the cocycle relation for $u_\rho(g) = u'_\rho(g)\rho_0(g)^{-1}$:

$$u_\rho(gh) = u_\rho(g) + g u_\rho(h) \quad \text{for } g u_\rho(h) = \rho_0(g)u_\rho(h)\rho_0(g)^{-1},$$

getting the map $\Phi(A[X]) \to H^1(G, ad(X))$ which factors through $H^1(G, Ad(X))$. As before, this map is injective A-linear map identifying $\Phi(A[X])$ with $\mathrm{Sel}(Ad(X))$.

We see that $u_\rho : G \to Ad(X)$ is a 1-cocycle, and we get an embedding $\Phi(A[X]) \hookrightarrow H^1(\mathbb{Q}_l, Ad(X))$ for $l \in S_{ab} \cup \{p\}$ by $\rho \mapsto [u_\rho]$. Again for $l \in S_{ab} \cup \{p\}$, we consider local version of Φ replacing G by $\mathrm{Gal}(\overline{\mathbb{Q}}_l/\mathbb{Q}_l)$:

$$\Phi_l(A[X]) = \frac{\{\rho : \mathrm{Gal}(\overline{\mathbb{Q}}_l/\mathbb{Q}_l) \to \mathrm{GL}_2(A[X]) | (\rho \bmod X) = \rho_0, [\rho] \in \mathcal{D}_{\chi,l}(A)\}}{1 + M_2(X)},$$

and we define

$$\mathrm{Sel}(Ad(X)) := \mathrm{Ker}(H^1(G, Ad(X)) \to \prod_{l \in S_{ab} \cup \{p\}} \frac{H^1(\mathbb{Q}_l, Ad(X))}{\Phi_l(A[X])}).$$

Here again, since $I_l \subset G$ has order prime to p, the factor $\frac{H^1(\mathbb{Q}_l, Ad(\overline{\rho}))}{\Phi_l(A[X])}$ in the above definition disappears for $l \in S_{ab}$ different from p. Thus

$$\mathrm{Sel}(Ad(X)) := \mathrm{Ker}(H^1(G, Ad(X)) \to \frac{H^1(\mathbb{Q}_p, Ad(X))}{\Phi_p(A[X])}). \qquad (6.6)$$

If $X = \varinjlim_i X_i$ for finite A-modules X_i, we just define

$$\mathrm{Sel}(Ad(X)) = \varinjlim_i \mathrm{Sel}(Ad(X_i)).$$

Then for finite X_i,

$$\Phi(A[X_i]) = \mathrm{Sel}(Ad(X_i)) \quad \text{and} \quad \varinjlim_i \Phi(X_i) = \mathrm{Sel}(\varinjlim_i Ad(X_i)).$$

6.1.9 *Differentials and Selmer group*

For each $[\rho_0] \in \mathcal{D}_\chi(A)$, choose an optimal representative $\rho_0 = \varphi \circ \boldsymbol{\rho}$ as in §6.1.8. Then we have a map $\Phi(A[X]) \to \mathcal{D}_\chi(A[X])$ for each finite A-module X sending optimal $\rho \in \Phi(A[X])$ to the class $[\rho] \in \mathcal{D}_\chi(A[X])$. By our choice of optimal ρ as in §6.1.8, this map is injective.

Conversely pick a class $c \in \mathcal{D}_\chi(A[X])$ over $[\rho_0] \in \mathcal{D}_\chi(A)$. Then for $\rho \in c$, we have $x \in 1 + M_2(\mathfrak{m}_{A[X]})$ such that $x\rho x^{-1} \mod X = \rho_0$. By replacing ρ with $x\rho x^{-1}$ and choosing the lifted base, we conclude $\Phi(A[X]) \cong \{[\rho] \in \mathcal{D}_\chi(A[X]) | \rho \mod X \sim \rho_0\}$; so, for finite X,

$$\mathrm{Sel}(Ad(X)) = \Phi(A[X]) = \{\phi \in \mathrm{Hom}_{B\text{-alg}}(R_\chi, A[X]) : \phi \mod X = \varphi\}$$
$$= Der_B(R_\chi, X) \xrightarrow[\sim]{\text{Corollary 5.2.6}} \mathrm{Hom}_A(\Omega_{R_\chi/B} \otimes_{R_\chi, \varphi} A, X).$$

Thus

$$\boxed{\mathrm{Sel}(Ad(X)) \cong \mathrm{Hom}_A(\Omega_{R_\chi/B} \otimes_{R_\chi, \varphi} A, X)}. \tag{6.7}$$

We obtain the following theorem of Mazur:

Theorem 6.1.11. *We have a canonical isomorphism:*
$$\mathrm{Sel}(Ad(\rho_0))^\vee \cong \Omega_{R_\chi/B} \otimes_{R_\chi, \varphi} A.$$
Here if $\chi = \kappa$, $B = \Lambda$ and otherwise, $B = W$.

Proof. Taking dual, by the adjunction formula in §5.3.3,

$$A^\vee := \mathrm{Hom}(A, \mathbb{Q}_p/\mathbb{Z}_p) = \mathrm{Hom}_{\mathbb{Z}_p}(A \otimes_B B, \mathbb{Q}_p/\mathbb{Z}_p) = \mathrm{Hom}_B(A, B^\vee).$$

Since $A = \varprojlim_i A_i$ for finite i and $\mathbb{Q}_p/\mathbb{Z}_p = \varinjlim_j p^{-j}\mathbb{Z}/\mathbb{Z}$, $A^\vee = \varinjlim_i \mathrm{Hom}(A_i, \mathbb{Q}_p/\mathbb{Z}_p) = \varinjlim_i A_i^\vee$ is a union of the finite modules $A_i^\vee$. We define $\mathrm{Sel}(Ad(\rho_0)) := \varinjlim_j \mathrm{Sel}(Ad(A_i^\vee))$. Defining $\Phi(A[A^\vee]) = \varinjlim_i \Phi(A[A_i^\vee])$, since the formation of cohomology commutes with injective limit,

$$\mathrm{Sel}(Ad(\rho_0)) = \varinjlim_j \mathrm{Ker}(H^1(G, Ad(A_i^\vee)) \to \frac{H^1(\mathbb{Q}_p, Ad(A_i^\vee))}{\Phi_p(A[A_i^\vee])}).$$

By the boxed formula (6.7),

$$\mathrm{Sel}(Ad(\rho_0)) = \varinjlim_i \mathrm{Sel}(Ad(A_i^\vee)) = \varinjlim_i \mathrm{Hom}_{R_\chi}(\Omega_{R_\chi/B} \otimes_{R_\chi} A, A_i^\vee)$$

$$= \mathrm{Hom}_A(\Omega_{R_\chi/B} \otimes_{R_\chi} A, A^\vee) = \mathrm{Hom}_A(\Omega_{R_\chi/B} \otimes_{R_\chi} A, \mathrm{Hom}_{\mathbb{Z}_p}(A, \mathbb{Z}_p))$$

$$= \mathrm{Hom}_{\mathbb{Z}_p}(\Omega_{R_\chi/B} \otimes_{R_\chi} A, \mathbb{Q}_p/\mathbb{Z}_p) = (\Omega_{R_\chi/B} \otimes_{R_\chi} A)^\vee.$$

Taking Pontryagin dual back, we finally get

$$\boxed{\mathrm{Sel}(Ad(\rho_0))^\vee \cong \Omega_{R_\chi/B} \otimes_{R_\chi, \varphi} A, \ \mathrm{Sel}(Ad(\overline{\rho}))^\vee \cong \Omega_{R_\chi/B} \otimes_{R_\chi} \mathbb{F}}. \tag{6.8}$$

In particular, we have $\mathrm{Sel}(Ad(\rho_\chi))^\vee = \Omega_{R_\chi/B}$ ($\rho_\kappa = \rho$ if $\chi = \kappa$). $\square$

This is a generalization of the formula

$$Cl_F \otimes_{\mathbb{Z}} W \cong \Omega_{W[Cl_{F,p}]/W} \otimes_{W[Cl_{F,p}]} W.$$

6.1.10　*Local condition at p*

The submodule $\Phi_p(A[X])$ in $H^1(\mathbb{Q}_p, Ad(X))$ is made of classes of

(sel) 1-*cocycles u with $u|_{I_p}$ is upper nilpotent and $u|_{D_p}$ is upper triangular with respect to the compatible basis (v_p, w_p).*

Suppose we have $\sigma \in I_p$ such that $\rho_0(\sigma) = \begin{pmatrix} \alpha & 0 \\ 0 & \beta \end{pmatrix}$ such that $\alpha \not\equiv \beta$ mod $\mathfrak{m}_A$. Suppose u is upper nilpotent over I_p. Then for $\tau \in D_p$, we have $Ad(\rho_0)(\tau)u(\tau^{-1}\sigma\tau) = (Ad(\rho_0)(\sigma) - 1)u(\tau) + u(\sigma)$. Writing $u(\tau) = \begin{pmatrix} a & b \\ c & -a \end{pmatrix}$, we find $(Ad(\rho_0)(\sigma) - 1)u(\tau) = \begin{pmatrix} 0 & (\alpha\beta^{-1} - 1)b \\ (\alpha^{-1}\beta - 1)c & 0 \end{pmatrix}$. Since $\rho_0(\tau)$ is upper triangular and $u(\tau^{-1}\sigma\tau)$ is upper nilpotent, $Ad(\rho_0)(\tau)u(\tau^{-1}\sigma\tau)$ is still upper nilpotent; so, $(\alpha^{-1}\beta - 1)c = 0$ and hence $c = 0$. Therefore u is forced to be upper triangular over D_p. Thus we get

Lemma 6.1.12. *If $\overline{\rho}(\sigma)$ for at least one $\sigma \in I_p$ has two distinct eigenvalues, $\Phi_p(A[X])$ gives rise to the subgroup of $H^1(\mathbb{Q}_p, Ad(X))$ made of classes containing a 1-cocycle whose restriction to I_p is upper nilpotent.*

6.2　Deformation rings of a compatible system

The deformation space is centered at the purely characteristic p point given by $\overline{\rho}$ (so, in this sense it is local in the "moduli" space). *Is there any global moduli theory of global Galois representations?* A naive attempt to see a global structure would be to start with a compatible system $\{\rho_{\mathfrak{p}}\}_{\mathfrak{p}}$ of $\mathfrak{p}$-adic Galois representations and glue together somehow the deformation rings $\mathbb{T}_{\mathfrak{p}}$ of $\overline{\rho}_{\mathfrak{p}} = \rho_{\mathfrak{p}}$ mod $\mathfrak{p}$ to see what happens. Though this appears naive (compared to a local-local approach via p-adic Hodge theory), the proof of Serre's mod p modularity by Khare–Wintenberger [KW09] suggests that this might work to some good extent. We explore this naive approach here. Our point of view is totally global not a local-local situation (i.e., not an attempt of making a "local moduli" of a local Galois representation over the local Galois group). We see some encouraging signs that the structure of $\mathbb{T}_{\mathfrak{p}}$ when we vary $\mathfrak{p}$ is almost the same except for a small number of primes (where we would have an obstruction to study). Though we limit ourselves to the case where $\overline{\rho}_{\mathfrak{p}}$ is ordinary, ordinarity is often miraculously sufficient to produce some properties of $\mathbb{T}_{\mathfrak{p}}$ uniform in $\mathfrak{p}$; e.g., cyclicity of adjoint Selmer groups $Sel(Ad(\rho_{\mathfrak{p}}))$ which we will study in the next chapter.

A (weakly) *compatible system* of Galois representations over a number field $K \subset \overline{\mathbb{Q}}$ with coefficients in another number field T is a system of $\mathfrak{p}$-adic continuous representations $\rho = \{\rho_{\mathfrak{p}} : \mathrm{Gal}(\overline{\mathbb{Q}}/K) \to \mathrm{GL}_n(O_{T,\mathfrak{p}})\}_{\mathfrak{p}}$ satisfying the following two properties:

• There exists a finite set of primes $\mathcal{S}$ of K such that $\rho_{\mathfrak{p}}$ is unramified outside $\mathcal{S}$ and the residual characteristic p of each prime $\mathfrak{p}$;

• The characteristic polynomial of $\rho_{\mathfrak{p}}(\mathrm{Frob}_{\mathfrak{l}})$ is in $T[X]$ independent of $\mathfrak{p}$ as long as $\mathfrak{l}$ is outside $\mathcal{S} \cup \{p\}$.

For a Hecke eigenform f of weight $k \geq 2$, recall $\mathbb{Z}[f]$ which is the integer ring of the number field generated by the Hecke eigenvalues of f. Then we have its compatible system of 2-dimensional $\mathfrak{p}$-adic Galois representation $\{\rho_{f,\mathfrak{p}}\}_{\mathfrak{p}}$ for primes $\mathfrak{p}$ running over prime ideals of $\mathbb{Z}[f]$. Defining $\overline{\rho}_{\mathfrak{p}} = (\rho_{f,\mathfrak{p}} \bmod \mathfrak{p})$, we study the dependence on $\mathfrak{p}$ of the universal rings $R_{\mathfrak{p}}^{ord}$ and $R_{\chi,\mathfrak{p}}$ representing the deformation functors $\mathcal{D} = \mathcal{D}_{\mathfrak{p}}, \mathcal{D}_\chi = \mathcal{D}_{\chi,\mathfrak{p}} : \mathcal{C} \to SETS$ of $\overline{\rho}_{\mathfrak{p}}$ defined in §6.1.1 (assuming $\overline{\rho}_{\mathfrak{p}}$ satisfies (p) and (m)).

As before, we write I_l for the inertia group of the l-decomposition subgroup $D_l \subset G$ and $\chi = \nu_p^{k-1}\psi$ $(f \in S_k(\Gamma_0(N), \psi))$. We write $S_{ab} = S_{ab,\mathfrak{p}}$ for the set of ramified primes $l \neq p$ of $\overline{\rho}$ such that $\overline{\rho}|_{I_l} \cong \overline{\epsilon}_l \oplus \mathbf{1}$. The conductor of a local or Dirichlet character ψ is written as $C(\psi)$.

Old and new forms again. For a modular form $g \in S_k(\Gamma_0(M), \varphi) = S_k(\Gamma_0(M), \varphi; \mathbb{C})$, $g(mz) = g|[m](z)$ for $0 < m \in \mathbb{Z}$ is in $S_k(\Gamma_0(Mm), \varphi)$. A linear combination in $S_k(\Gamma_0(N), \psi)$ of cusp forms of the form $g|[m]$ with $m > 1$ and g of lower level is called an *old form*. The orthogonal complement under Peterson inner product of the subspace of old forms is called the space of *new forms*. These spaces are stable under Hecke operators. Recall that a cusp form $f \in S_k(\Gamma_0(N), \psi)_{/\mathbb{C}}$ is a *Hecke eigenform* if $f|T(n) = \lambda(T(n))f$ for all n with the eigenvalues $\lambda(T(n))$ and $a(1, f) = 1$. This fact is equivalent to $f = \sum_{n=1}^{\infty} \lambda(T(n))q^n$ (by duality between cusp forms and the corresponding Hecke algebra in §3.2.5). A Hecke eigenform in $S_k(\Gamma_0(N), \psi)$ and associated λ are called *primitive* if f is new. Thus a primitive form is a new form and at the same time an eigenform.

Theorem 6.2.1. *Among cusp forms of varying levels with eigenvalues for $T(l)$ given by $\lambda(T(l))$ for almost all primes l, there exists a unique Hecke eigenform of minimal level C, and that is a primitive form.*

The level C is called the *conductor* of f (and λ). For all these, see [MFM, Chapter 4]. Hereafter the fixed eigenform f is primitive of conductor C (so, $f \in S_k(\Gamma_0(C), \psi)$ and $C(\psi)|C$).

Integral modular forms. Let $\mathbb{Z}[\psi]$ be the subring of $\mathbb{C}$ generated by the values of ψ. For an algebra $\mathbb{Z}[\psi] \subset A \subset \mathbb{C}$, recall

$$S_k(\Gamma_0(N), \psi; A) := \{f \in S_k(\Gamma_0(N), \psi) | a(n, f) \in A \text{ for all } n > 0\},$$

where $f(z) = \sum_{n=1}^{\infty} a(n, f) q^n$ with $q = \exp(2\pi i z)$. Often we write $S_k(\Gamma_0(N), \psi)_{/A}$ for $S_k(\Gamma_0(N), \psi; A)$. We then write simply

$$h_k(N, \psi)_{/A} = \mathbf{h}_k(\Gamma_0(N), \psi; A)$$
$$= A[T(n)|n = 1, 2, \dots] \subset \mathrm{End}_A(S_k(\Gamma_0(N), \psi)_{/A})).$$

As seen in (4.8), these are A-modules of finite type and

$$S_k(\Gamma_0(N), \psi)_{/A} = S_k(\Gamma_0(N), \psi)_{/\mathbb{Z}[\psi]} \otimes_{\mathbb{Z}[\psi]} A,$$
$$\mathrm{Hom}_A(S_k(\Gamma_0(N), \psi)_{/A}, A) \cong h_k(N, \psi)_{/A},$$
$$\mathrm{Hom}_A(h_k(N, \psi)_{/A}, A) \xrightarrow{i}_{\sim} S_k(\Gamma_0(N), \psi)_{/A}.$$

Here the duality between h_k and S_k is given by $\langle T, f \rangle = a(1, f|T)$. By the identity of Hecke: $\langle T(n), f \rangle = a(n, f)$, we have $i(\phi) = \sum_{n=1}^{\infty} \phi(T(n)) q^n$.

Known facts for $h_k(N, \psi)$. Recall the integrality from (4.8) and Proposition 3.2.10: $S_k(\Gamma_0(N), \psi)_{/A} = S_k(\Gamma_0(N), \psi)_{/\mathbb{Z}[\psi]} \otimes_{\mathbb{Z}[\psi]} A$ and $h_k(N, \psi)_{/A} = h_k(N, \psi)_{/\mathbb{Z}[\psi]} \otimes_{\mathbb{Z}[\psi]} A$ for a general $\mathbb{Z}[\psi]$-algebra A. Then

- $h_k(N, \psi)_{/A} = A[T(n)|n = 1, \dots] \subset \mathrm{End}_A(S_k(\Gamma_0(N), \psi)_{/A}))$ and the duality statement holds without any modification.
- If $N = C(\psi)$, $h_k(N, \psi)_{/A}$ is reduced for any $\mathbb{Z}[\psi]$-domain A flat over $\mathbb{Z}[\psi]$. This follows from the theory of new/old forms (see [MFM, §4.6] (in the language of [MFM], new form is called primitive form). Actually this is still true if $N = N_0 C(\psi)$ with square-free $N_0|C(\psi)$ (why?).

Conjecture 6.2.2. *Suppose $k \geq 2$. If N is cube-free, $h_k(N, \psi)_{/A}$ is reduced for any $\mathbb{Z}[\psi]$-domain A flat over $\mathbb{Z}[\psi]$.*

This is known if $k = 2$ unconditionally and under the Tate conjecture on algebraic cycles if $k > 2$ (Coleman/Edixhoven). There is another proof for $k = 3$ assuming a function field analogue of the Birch and Swinnerton-Dyer conjecture (D. Ulmer). See Coleman/Edixhoven [CE98] and Ulmer [U95].

For a primitive form $f \in S_k(\Gamma_0(C), \psi)$ of conductor C, consider a prime $l \nmid C$. Suppose $f|T(l) = (\alpha + \beta)f$ with $\alpha\beta = \psi(l)l^{k-1}$. Writing $\{x, y\} = \{\alpha, \beta\}$ as a set, by Lemma 3.2.20, $f_x(z) = f(z) - yf|[l]$ is a Hecke eigenform in $S_k(Cl, \psi)$ with $f_x|U(l) = xf_x$. Defining $f_0(z) = f_x(z) - xf(lz) \in S_k(Cl^2, \psi)$, we find $f_0|U(l) = 0$. Writing down the action of $U(l)$ on f, f_α, f_0 in a matrix form, we can easily show that $U(l)$ is non-semi-simple if and only if $\alpha = \beta$. This gives the only reason to have non-trivial (but only one notch deeper) nilpotence in the Hecke algebra in addition to the contribution of f_0 (i.e., the relation $f_0|[l]^m|U(l)^m = f_0$; so, $f_0|[l^m]|U(l)^{m+1} = 0$). The latter nilpotence only occurs at the level Cl^3 (so, cube-freeness is required). Since a weight 1 Hecke eigenform is associated to an Artin representation [DS74], for primes l with positive density, we have $\alpha = \beta$ and hence reducedness fails often even at cube-free level if $k = 1$.

Remark 6.2.3. Suppose f is $\mathfrak{p}$-ordinary of $k \geq 2$ for a prime $\mathfrak{p}|p$. Then by definition, $\lambda(T(p)) = \alpha + \beta$ is a $\mathfrak{p}$-unit with $|\alpha\beta|_p = |\psi(p)p^{k-1}|_p < 1$. Thus we cannot have $\alpha = \beta$; so, for $\mathfrak{p}$-ordinary forms, we do not need to worry about nilpotence coming from the problem described above.

By this remark, this characteristic 0 p-indistinguishedness $\alpha = \beta$ does not happen for ordinary deformation (as long as $k \geq 2$). The problem occurs when $a_p = 0$ as in the following exercise. Indeed, the local ring of the Hecke algebra (of any level) giving a "minimal" ordinary universal deformation is reduced as we will see in Lemma 6.2.15.

Exercise 6.2.1. Find an example of a p-super-singular rational elliptic curve whose Hilbert modular form over a real quadratic field does produce non semi-simple $U(p)$-action.

Recall $h_k(C, \psi)_{/A}$ is the A-subalgebra of $\mathrm{End}_A(S_k(\Gamma_0(C), \psi; A))$ generated by all Hecke operators $T(n)$ (writing $T(l)$ for $U(l)$ for primes $l|C$). Fix a primitive form $f \in S_k(\Gamma_0(C), \psi)$, and write simply $h := h_k(C, \psi)_{\mathbb{Z}[f]}$. Thus we get an algebra homomorphism $\lambda : h \to \mathbb{Z}[f]$ such that $f|T = \lambda(T)f$ (or equivalently $f = \sum_{n=1}^{\infty} \lambda(T(n))q^n$). If $p \nmid C$, we choose an order $\alpha, \beta \in \overline{\mathbb{Q}}_p$ as above so that $|\alpha|_p \leq |\beta|_p$. Thus λ extends to $\lambda_\alpha : h_k(Cp, \psi)_{/\mathbb{Z}[f_\alpha]} \to \mathbb{Z}[f_\alpha]$ so that $\lambda_\alpha(U(p)) = \alpha$ (or equivalently $f_\alpha|T = \lambda_\alpha(T)f_\alpha$). If $p|C$, we just define α by $f|U(p) = \alpha f$ and put $\lambda_\alpha = \lambda$. Pick a prime $\mathfrak{p}|p$ of $\mathbb{Z}[f]$ and embed the $\mathfrak{p}$-adic completion $\mathbb{Z}[f]_\mathfrak{p}$ into $\overline{\mathbb{Q}}_p$ by the $\mathfrak{p}$-adic place $\mathfrak{p}$. Let $W = W_\mathfrak{p}$ denote the minimal discrete valuation ring inside $\overline{\mathbb{Q}}_p$ containing the $\mathfrak{p}$-adic completion $\mathbb{Z}[f]_\mathfrak{p}$ and α. Plainly W is independent of choices of α (as α is either a conjugate of β or $\alpha, \beta \in \mathbb{Z}[f]_\mathfrak{p}$).

$$\text{If } \alpha \not\equiv \beta \mod \mathfrak{m}_{W_\mathfrak{p}}, \text{ we have } W_\mathfrak{p} = \mathbb{Z}[f]_\mathfrak{p}. \tag{6.9}$$

This follows from Hensel's lemma [CRT, Theorem 8.3], since $\alpha \not\equiv \beta$ mod $\mathfrak{m}_{W_\mathfrak{p}}$. Note that the assumption $\alpha \not\equiv \beta \mod \mathfrak{m}_{W_\mathfrak{p}}$ holds if $k \geq 2$ and $\lambda(T(p)) \notin \mathfrak{p}$.

Let $[C, p]$ denote the least common multiple of C and p. Assume $k \geq 2$. Since $h_{\mathfrak{p}} := h_k([C, p], \psi)_{/W_{\mathfrak{p}}}$ is free of finite rank over $W_{\mathfrak{p}}$, for any maximal ideal $\mathfrak{m}$ of $h_{\mathfrak{p}}$, we have $\mathfrak{m} \cap W = \mathfrak{m}_W$. Here we emphasize that $h_{\mathfrak{p}}$ is not necessarily the $\mathfrak{p}$-adic completion of h. Since $h_{\mathfrak{p}}/\mathfrak{m}_W h_{\mathfrak{p}}$ is a finite ring; so, there is only a finite many maximal ideal of $h_{\mathfrak{p}}$; so, $h_{\mathfrak{p}} = \prod_{\mathfrak{m}} h_{\mathfrak{m}}$ for the $\mathfrak{m}$-adic completion $h_{\mathfrak{m}}$ which is a local ring. We pick the unique local ring factor $\mathbb{T}_{\chi, \mathfrak{p}}$ of $h_{\mathfrak{p}}$ through which λ_α factors. Thus we get a system of local rings $\{\mathbb{T}_{\chi, \mathfrak{p}}\}_{\mathfrak{p}}$ of residual characteristic p. Replacing $h_{\mathfrak{p}}$ by $h_{\mathfrak{p}}^{fl} := h_k(C, \psi)_{/W_{\mathfrak{p}}}$ if $p \nmid C$ and $k \geq 2$, we define $\mathbb{T}_{\chi, \mathfrak{p}}^{fl}$ in the same way as above (see [D95, Proposition 6.1]). If $p|C$, by abusing the terminology, we put $\mathbb{T}_{\chi, \mathfrak{p}}^{fl} = \mathbb{T}_{\chi, \mathfrak{p}}$ (strictly speaking, the deformation parameterized by $\mathbb{T}_{\chi, \mathfrak{p}}^{fl}$ is "potentially flat" if $k = 2$). If $k \geq 3$, $\mathbb{T}_{\chi, \mathfrak{p}}^{fl} \cong \mathbb{T}_{\chi, \mathfrak{p}}$ by Corollary 4.1.30, and the two definition coincides (i.e., flatness has real meaning when $k = 2$). This fact on the Galois side is known, as an ordinary Galois representation ρ with filtration $\nu_p^{k-1} \hookrightarrow \rho \twoheadrightarrow 1$ is crystalline if $k \geq 3$ [P94, §3.1].

First we study how $\Omega_{\mathbb{T}_{\chi, \mathfrak{p}}/W}$ depends on $\mathfrak{p}$. The idea is to relate the $\mathfrak{p}$-dependent $\Omega_{\mathbb{T}_{\chi, \mathfrak{p}}/W}$ with $\Omega_{h/\mathbb{Z}[f]}$ which is independent of $\mathfrak{p}$.

6.2.1   *Preliminary lemmas*

To show $R_{\chi, \mathfrak{p}} = W_{\mathfrak{p}}$ for most $\mathfrak{p}$, we need some basic properties of differentials. We summarize here what we need. Proving all these are good exercises for the reader. Let A, A_j, B' be B-algebras.

(d1) $\Omega_{A_1 \times A_2/B} \cong \Omega_{A_1/B} \oplus \Omega_{A_2/B}$ $(d_{A_1 \times A_2} = d_{A_1} + d_{A_2})$ as

$$Der_B(A_1 \times A_2, M) = Der_B(A_1, M) \oplus Der_B(A_2, M).$$

(d2) $\Omega_{S^{-1}A/B} \cong \Omega_{A/B} \otimes_A S^{-1}A$ $(d_{S^{-1}A} = d_A \otimes 1)$ for a multiplicative set $1 \in S \subset A$.

(d3) $\Omega_{A \otimes_B B'/B'} \cong \Omega_{A/B} \otimes_B B'$ $(d_{A \otimes B'} = d_A \otimes 1)$.

Suppose that A is a B-module of finite type.

(d4) $\Omega_{A/B} = 0$ if A is a separable extension of a field B.

Indeed, if A is a field, $A = B[X]/(f(X))$ with θ image of X in A. Then $\Omega_{A/B} = (A/f'(\theta)A)d\theta = 0$ as $f'(\theta) \neq 0$.

(d5) $\Omega_{A/B}$ is a torsion B-module if B is an integral domain of characteristic 0 and A is reduced. This follows from (d1–2) and (d4) since $A \otimes_B \text{Frac}(B) = \text{Frac}(A) = K_1 \times \cdots \times K_r$ for separable extensions K_i. What happens if B has characteristic $p > 0$?

Recall $h := h_k(C, \psi)_{/\mathbb{Z}[f]}$.

Lemma 6.2.4. *Suppose that h is reduced. Then $\Omega_{h/\mathbb{Z}[f]}$ is a finite module.*

Proof. By (d5), $\Omega_{h/\mathbb{Z}[f]}$ is a torsion $\mathbb{Z}[f]$-module of finite type; so, it is finite. $\square$

Lemma 6.2.5. *If $p \nmid C$, assume $\alpha \not\equiv \beta \mod \mathfrak{m}_{W_\mathfrak{p}}$. If $k \geq 3$, then we have $\Omega_{\mathbb{T}_{\chi,\mathfrak{p}}/W_\mathfrak{p}} = \Omega_{h/\mathbb{Z}[f]} \otimes_h \mathbb{T}_{\chi,\mathfrak{p}}$. If $k = 2$, then we have $\Omega_{\mathbb{T}_{\chi,\mathfrak{p}}^{fl}/W_\mathfrak{p}} = \Omega_{h/\mathbb{Z}[f]} \otimes_h \mathbb{T}_{\chi,\mathfrak{p}}^{fl}$.*

Since the proof is the same, we only prove the first assertion.

Proof. If $p|C$, $\mathbb{T}_{\chi,\mathfrak{p}}$ is a completion of h by the maximal ideal of h containing $\mathrm{Ker}(\lambda)$, the assertion is clear.

Assume $p \nmid C$ and $k \geq 3$. By (6.9), $W_\mathfrak{p} = \mathbb{Z}[f]_\mathfrak{p}$. Let $\mathfrak{m}$ be the unique maximal ideal of $h \otimes_{\mathbb{Z}[f]} W_\mathfrak{p}$ containing $\mathrm{Ker}(\lambda)$. Let $h_\mathfrak{m}$ be the $\mathfrak{m}$-adic completion of h. Since $\alpha \not\equiv \beta \mod \mathfrak{m}_{W_\mathfrak{p}}$, $X^2 - T(p)X + \chi(p) \in h_\mathfrak{m}[X]$ has two distinct roots $A, B \in h_\mathfrak{m}$ with $A \equiv \alpha \mod \mathfrak{m}$. Recall the extension $\lambda_\alpha : h' := h_k(Cp, \psi)_{/\mathbb{Z}[f]_\mathfrak{p}} \to \overline{\mathbb{Q}}_p$ defined just above (6.9) for the maximal ideal $\mathfrak{m}'$ containing $\mathrm{Ker}(\lambda_\alpha)$ of $h' := h_k(Cp, \psi)_{/\mathbb{Z}[f]_\mathfrak{p}}$, by Corollary 4.1.30, we have $h'_{\mathfrak{m}'} \cong h_\mathfrak{m}$ via the natural projection $h'_\mathfrak{m} \to h_\mathfrak{m}$ sending $U(p)$ to A (and $T(n)$ to $T(n)$ for n prime to p). Thus we may regard $\mathbb{T}_{\chi,\mathfrak{p}}$ as a factor of $h_\mathfrak{p}^{fl} = h \otimes_{\mathbb{Z}[f]} W_\mathfrak{p}$. Thus by (d3), $\Omega_{h_\mathfrak{p}^{fl}/W_\mathfrak{p}} = \Omega_{h/\mathbb{Z}[f]} \otimes_{\mathbb{Z}[f]} W_\mathfrak{p}$. Since $h_\mathfrak{p}^{fl} = \prod_\mathfrak{m} h_\mathfrak{m}$ for $\mathfrak{m}$ running over maximal ideals of h, $\Omega_{h_\mathfrak{p}^{fl}/W_\mathfrak{p}} = \bigoplus_\mathfrak{m} \Omega_{h_\mathfrak{m}/W_\mathfrak{p}}$. Note that $\mathbb{T}_{\chi,\mathfrak{p}}$ is one of $h_\mathfrak{m}$. Thus $\Omega_{\mathbb{T}_{\chi,\mathfrak{p}}/W_\mathfrak{p}} = \Omega_{h_\mathfrak{m}/W_\mathfrak{p}} = \Omega_{h_\mathfrak{m}^{fl}/W_\mathfrak{p}} \otimes_{h_\mathfrak{m}^{fl}} h_\mathfrak{m}$; so, we get the desired formula. $\square$

6.2.2 Consequence of vanishing of differentials

Lemma 6.2.6. *Let $A \in CL_B$. Suppose that A is B-torsion-free. Then $\Omega_{A/B} \otimes_A A/\mathfrak{a} = 0$ for a proper A-ideal $\mathfrak{a}$ if and only if $A = B$.*

Proof. We need to prove $\Omega_{A/B} \otimes_A A/\mathfrak{a} = 0 \Rightarrow A = B$. By Nakayama's lemma, we have $\Omega_{A/B} = 0 \Leftrightarrow \Omega_{A/B} \otimes_A \mathbb{F} = 0 \Leftrightarrow \Omega_{A/B} \otimes_A A/\mathfrak{a} = 0$; so, we may assume that $\mathfrak{a} = \mathfrak{m}_A$, and we have

$$t_{A/B}^* := \mathfrak{m}_A/(\mathfrak{m}_A^2 + \mathfrak{m}_B) = \Omega_{A/B} \otimes_A \mathbb{F} = 0,$$

which implies that $\iota_B(\mathfrak{m}_B) = \mathfrak{m}_A$. Therefore by Corollary 6.1.3, we have a surjective B-algebra homomorphism $\pi : B \twoheadrightarrow A$. Thus torsion-freeness tells us $\mathrm{Ker}(\pi) = 0$ and $A = B$. $\square$

Recall $\text{Ann}_A(M) := \{x \in A | xM = 0\}$ for an A-module M. We prove

Theorem 6.2.7. *Let* $h := h_k(C, \psi)_{/\mathbb{Z}[f]}$ *for a Hecke eigenform* $f \in S_k(\Gamma_0(N), \psi)$. *Let* $\text{Ann}(f) := \text{Ann}_{\mathbb{Z}[f]}(\Omega_{h/\mathbb{Z}[f]} \otimes_{h,\lambda} \mathbb{Z}[f])$ *in* $\mathbb{Z}[f]$. *Then* $\text{Ann}(f)$ *is a non-zero ideal of* $\mathbb{Z}[f]$, *and if* $\mathfrak{p} \nmid \text{Ann}(f)$, *then* $\mathbb{T}_{\chi,\mathfrak{p}} = W_{\mathfrak{p}}$ *if* $k \geq 3$ *and* $\mathbb{T}_{\chi,\mathfrak{p}}^{fl} = W_{\mathfrak{p}}$ *if* $k = 2$.

Proof. Since the proof is the same, we only prove the assertion for $k \geq 3$. By Lemma 6.2.4, $\text{Ann}(f)$ is a non-zero ideal of $\mathbb{Z}[f]$ (could be $\mathbb{Z}[f]$ itself). By Lemma 6.2.6, we need to show that $\Omega_{\mathbb{T}_{\chi,\mathfrak{p}}/W_{\mathfrak{p}}} = 0$ if $\mathfrak{p} \nmid \text{Ann}(f)$. By Lemma 6.2.5,

$$\Omega_{\mathbb{T}_{\chi,\mathfrak{p}}/W_{\mathfrak{p}}} = \Omega_{h/\mathbb{Z}[f]} \otimes_{h,\lambda} \mathbb{T}_{\chi,\mathfrak{p}}. \qquad (*)$$

If $\mathfrak{p} \nmid \text{Ann}(f)$, then $\mathbb{Z}[f]$ contains $a \notin \mathfrak{p}$ killing $\Omega_{h/\mathbb{Z}[f]} \otimes_{h,\lambda} \mathbb{Z}[f]$ which is still a unit in $\mathbb{T}_{\chi,\mathfrak{p}}$. Therefore the multiplication by a kills the right-hand-side of $(*)$ and is an automorphism of the left-hand-side; so, $\Omega_{\mathbb{T}_{\chi,\mathfrak{p}}/W_{\mathfrak{p}}} = 0$. Since $\mathbb{T}_{\chi,\mathfrak{p}}$ is $W_{\mathfrak{p}}$-free, we conclude $\mathbb{T}_{\chi,\mathfrak{p}} = W_{\mathfrak{p}}$ from Lemma 6.2.6. $\qquad \square$

6.2.3 Modular $\mathfrak{p}$-adic Galois representation

A primitive form $f = \sum_{n=1}^{\infty} a(n, f)q^n \in S_k(\Gamma_0(C), \psi)$ of conductor C has a compatible system ρ_f of $\mathfrak{p}$-adic Galois representations $\rho_{\mathfrak{p}} = \rho_{f,\mathfrak{p}}$ with values in $\text{GL}_2(W_{\mathfrak{p}})$ for each prime $\mathfrak{p}$ of $\mathbb{Z}[f]$ satisfying the following conditions (G1–6) due to Eichler, Shimura, Igusa, Deligne, Serre, Rapoport and Langlands (see [GME, §4.2] and [DS74]).

(G1) $\rho_{\mathfrak{p}}$ is unramified outside pC $(\mathfrak{p}|p)$, and $\det(\rho_{\mathfrak{p}}) = \chi$ for $\chi := \psi\nu_p^{k-1}$;

(G2) $\det(1 - \rho_{\mathfrak{p}}(\text{Frob}_l)X) = 1 - \lambda(T(l))X + \chi(l)X^2$ for $l \nmid Cp$;

(G3) If either $a(p, f) = \lambda(T(p)) \notin \mathfrak{p}$ or $k = 1$ and $p \nmid C$, $\rho_{\mathfrak{p}}|_{D_p} \cong \begin{pmatrix} \epsilon_{\mathfrak{p}} & * \\ 0 & \delta_{\mathfrak{p}} \end{pmatrix}$ with unramified $\delta_{\mathfrak{p}}$, and $\delta_{\mathfrak{p}}(\text{Frob}_p) = \alpha$ for a $\mathfrak{p}$-adic unit root α of $X^2 - \lambda(T(p))X + \psi(p)p^{k-1} = 0$ (ordinarity). Here we agree to put $\psi(p) = 0$ if $p|C$. If $k = 1$, the ordinarity datum involves the choice of the root α (which may span a quadratic extension of $\mathbb{Z}[f]_{\mathfrak{p}}$).

If $k = 1$, even if $a(p, f) = 0$, $\rho_{\mathfrak{p}}$ can satisfy (p) (see §7.3.2, particularly the conditions (U_-) and (Ds)). We insert here a conjecture (e.g., [OCS, II, Notes, 507.3]):

Conjecture 6.2.8. *If* $k \geq 2$ *and* ρ_f *is not an induced representation from a quadratic extension of* $\mathbb{Q}$, $a(p, f) \notin \mathfrak{p}$ *for density 1 primes* $\mathfrak{p}$.

Suppose $f \in S_2(\Gamma_0(C), \mathbb{Z})$. By the bound of Deligne–Shimura, $|a(p, f)| \leq 2\sqrt{p}$. If $p|a(p, f)$, we have $p \leq |a(p, f)| \leq 2\sqrt{p}$; so, $a(p, f) = 0$ if $p \geq 5$. By [S81, (c)], $\{p|a(p, f) \equiv 0 \mod p\}$ has Dirichlet density 0 but it is known to be infinite [E87]. More generally, writing $\Sigma_f = \{p \in \mathrm{Spec}(\mathbb{Z})|a(p, f) \not\equiv 0 \mod p\}$ for a primitive form $f \in S_k(C, \psi)$, it is easy to see that Σ_f has density 1 by a similar argument [H13b, §7]. If $k = 1$, since weight 1 Hecke eigenform is associated to an Artin representation [DS74], $\mathfrak{p}$ as in the conjecture has positive density < 1. To see this, we note $\rho = \rho_f(c)$ is a conjugate of $\mathrm{diag}[-1, 1]$; so, writing C for the conjugacy class of c in $\mathrm{Gal}(F(\rho)/\mathbb{Q})$, primes p of density $|C|/[F(\rho) : \mathbb{Q}] > 0$ satisfies $a(p, f) = \mathrm{Tr}(\rho(\mathrm{Frob}_p)) \equiv 0 \mod \mathfrak{p}$ by Chebotarev density. On the other hand, by irreducibility of ρ (or Dickson's classification in §6.2.4), we can find ϕ with $\mathrm{Tr}(\rho(\phi)) \neq 0$. Then for the conjugacy class C' of ϕ, primes p of density $|C'|/[F(\rho) : \mathbb{Q}] > 0$ satisfies $a(p, f) = \mathrm{Tr}(\rho(\mathrm{Frob}_p)) \not\equiv 0 \mod \mathfrak{p}$.

Remark 6.2.9. As in Conjecture 6.2.8, a Hecke eigenform whose Galois representation is induced from a quadratic field plays a singular role. If one member $\rho_{f,\mathfrak{p}}$ is induced, then every member is induced (see §7.3.1). If $k \geq 2$ and ρ_f is induced, the quadratic field is imaginary [MFM, §4.8] (see also [HMI, §2.5.4]); so, we say a Hecke eigenform f has *CM* (or a *CM form*) if ρ_f is induced from a quadratic field. As we will see later, if $k = 1$, the quadratic field can be also real.

We now return to more local information of modular Galois representations.

(G4) Writing the l-primary part of an integer $N > 0$ as N_l, if $C_l = C(\psi)_l$ for a prime $l|C$ ($l \neq p$), then $\rho_\mathfrak{p}|_{I_l} \cong \left(\begin{smallmatrix} \psi_l & 0 \\ 0 & 1 \end{smallmatrix} \right)$;

(G5) If $C_l = l$ and $C_l(\psi) = 1$ ($l \neq p$), then $\rho_\mathfrak{p}|_{I_l}$ is indecomposable, and $\rho_\mathfrak{p}|_{D_l} \cong \left(\begin{smallmatrix} \eta\nu_p & * \\ 0 & \eta \end{smallmatrix} \right)$ for a character $\eta : D_l \to W_\mathfrak{p}^\times$ such that $\eta^2\nu_p = \psi\nu_p^{k-1}$;

(G6) If $l^2|(C/C(\psi))$ ($l \neq p$), then $\lambda(T(l)) = 0$ and $\rho|_{D_l}$ is either absolutely irreducible or isomorphic to $\left(\begin{smallmatrix} \epsilon_l & 0 \\ 0 & \delta_l \end{smallmatrix} \right)$ having conductor $C(\epsilon_l)C(\delta_l) = C_l$ with $C(\epsilon_l) > 1$ and $C(\delta_l) > 1$. See [GME, §4.2] for these facts.

Remark 6.2.10. Let $\bar{\rho} = \bar{\rho}_\mathfrak{p}$ for a prime $\mathfrak{p}|p$. Suppose that ψ has order prime to p. Then for a prime $p \neq l \in S$ (ramified in $F(\bar{\rho})/\mathbb{Q})$), by (G4), $l|C$ satisfies (l) if and only if $C_l = C_l(\psi)$. Thus $S_{ab} = \{l \in S|C_l = C_l(\psi)\}$. Also by (G6), we have $l \in S - S_{ab} \Rightarrow l^2|C/C(\psi)$.

Definition 6.2.11. For a primitive form $f \in S_k(\Gamma_0(C), \psi)$, a prime l is called *multiplicative at l* if choosing a prime $\mathfrak{p} \nmid l$ of $\mathbb{Z}[f]$ $\rho_{f,\mathfrak{p}}|_{D_l}$ is of the form $\left(\begin{smallmatrix} \eta\nu_p & * \\ 0 & \eta \end{smallmatrix} \right)$. This notion is independent of the choice of $\mathfrak{p}$ as such a form is a character twist of a Hecke eigenform f described in (G5). If f does not have multiplicative prime, we say that f has *potentially good reduction everywhere* (as Shimura's abelian variety attached to f has such good reduction property if $k = 2$; e.g, [GME, Chapter 4]) or *potentially crystalline* (a terminology in p-adic Hodge theory requiring $\rho_{f,\mathfrak{p}}$ to become crystalline over an open subgroup of D_p; see [T98, Corollary 2.2.3]).

If f has potentially good reduction, then for almost all primes $\mathfrak{p}$ ordinary for f the member $\rho_{f,\mathfrak{p}} \in \rho_f$ is minimal.

6.2.4   *Modular deformation*

Fix a primitive form $f_0 \in S_{k_0}(\Gamma_0(C), \psi_0)$ of conductor C with potentially good reduction everywhere. Take a $\mathbb{Z}[f_0]$-prime $\mathfrak{p}|p > 2$. Assume that $\rho_{\mathfrak{p}} = \rho_{f_0,\mathfrak{p}}$ is minimal and satisfies ordinarity (p). Then by Lemma 3.2.20, we have a p-stabilized form $f_0^{ord} \in S_{k_0}(\Gamma_0([C, p]), \psi_0)$ for the least common multiple $[C, p]$ of C and p such that $f_0^{ord}|U(p) = \delta_{\mathfrak{p}}([p, \mathbb{Q}_p]) f_0^{ord}$ for the unramified quotient character $\delta_{\mathfrak{p}}$ of $\rho_{f_0,\mathfrak{p}}$ (as in (p)). If $\rho_{f_0,\mathfrak{p}}$ is unramified at p, we can have two choices of $\delta_{\mathfrak{p}}$ (each giving a different theory), but this does not happen if $k_0 \geq 2$ since $\det(\rho_{\mathfrak{p}}) = \psi_0 \nu_p^{k_0-1}$ is ramified at p.

Let $\bar{\rho} = \bar{\rho}_{\mathfrak{p}} := \rho_{f_0,\mathfrak{p}} \mod \mathfrak{p}$, and consider deformation functors $\mathcal{D} = \mathcal{D}_{\mathfrak{p}}, \mathcal{D}_{\chi_0} = \mathcal{D}_{\chi_0,\mathfrak{p}}, \mathcal{D}_{\chi_0}^{fl} = \mathcal{D}_{\chi_0,\mathfrak{p}}^{fl}$ for $\bar{\rho}$ and $\chi_0 = \psi_0 \nu_p^{k_0-1}$. Recall our assumption made at the beginning of this chapter that $\bar{\rho}_{\mathfrak{p}}$ is absolutely irreducible. Write $R_{\chi_0} = R_{\chi_0,\mathfrak{p}}$ (resp. $R_{\chi_0}^{fl} = R_{\chi_0,\mathfrak{p}}^{fl}$, $R^{ord} = R_{\mathfrak{p}}^{ord}$) for the universal ring representing $\mathcal{D}_{\chi_0,\mathfrak{p}}$ (resp. $\mathcal{D}_{\chi_0,\mathfrak{p}}^{fl}, \mathcal{D}_{\mathfrak{p}}$). Consider $\mathrm{Tr}(\boldsymbol{\rho}_{\chi_0,\mathfrak{p}}) = \sum_{\lambda \in \mathrm{Hom}_{W_{\mathfrak{p}}\text{-alg}}(\mathbb{T}_{\chi_0,\mathfrak{p}}, \overline{\mathbb{Q}}_p)} \mathrm{Tr}(\rho_\lambda)$. By (G2), $\mathrm{Tr}(\rho_{\mathbb{T},\chi_0,\mathfrak{p}})(\mathrm{Frob}_l) = T(l)|_{\mathbb{T}_{\chi_0,\mathfrak{p}}}$ for all primes $l \nmid Cp$. By Chebotarev density, $\mathrm{Tr}(\rho_{\mathbb{T},\chi_0,\mathfrak{p}})$ has values in $\mathbb{T}_{\chi_0,\mathfrak{p}}$. By the theory of pseudo-representation/character [MFG, §2.2.1] and [FGS, §1.2], we have a unique Galois representation $\rho_{\mathbb{T}} = \rho_{\mathbb{T},\chi_0,\mathfrak{p}} : G \to \mathrm{GL}_2(\mathbb{T}_{\chi_0,\mathfrak{p}})$ with $\mathrm{Tr}(\rho_{\mathbb{T}})(\mathrm{Frob}_l) = T(l)|_{\mathbb{T}_{\chi_0,\mathfrak{p}}}$. By absolute irreducibility of $\bar{\rho}_{\mathfrak{p}}$ (and $\bar{\rho}_{\mathfrak{p}}(c) \sim \mathrm{diag}[1, -1]$ with $p > 2$), $\rho_{\mathbb{T}}$ has values in $\mathrm{GL}_2(\mathbb{T}_{\chi_0,\mathfrak{p}})$ [MFG, Proposition 2.16]. Since trace determines representation (if residually irreducible) over a local ring [MFC, Proposition 2.13], $\rho_{\mathbb{T}}$ is unique up to conjugation by a matrix in $\mathrm{GL}_2(\mathbb{T}_{\chi_0,\mathfrak{p}})$, and we have $\rho_{\mathbb{T}} \in \mathcal{D}_\chi(\mathbb{T}_{\chi_0,\mathfrak{p}})$. Similarly we have $\rho_{\mathbb{T}_{\chi_0,\mathfrak{p}}^{fl}} \in \mathcal{D}_{\chi_0,\mathfrak{p}}^{fl}(\mathbb{T}_{\chi_0,\mathfrak{p}}^{fl})$. Thus we have a universal map $\pi : R_{\chi_0,\mathfrak{p}} \to \mathbb{T}_{\chi_0,\mathfrak{p}}$ and $\pi^{fl} : R_{\chi_0,\mathfrak{p}}^{fl} \to \mathbb{T}_{\chi_0,\mathfrak{p}}^{fl}$ such that $\pi \circ \boldsymbol{\rho}_{\chi_0} \sim \rho_{\mathbb{T}_{\chi_0}}$ and $\pi^{fl} \circ \boldsymbol{\rho}_{\chi_0} \sim \rho_{\mathbb{T}_{\chi_0}^{fl}}$.

Here is an easy lemma:

Lemma 6.2.12. *Assume absolute irreducibility of $\overline{\rho}$ in addition to* (m), (p) *and* (l). *Pick $\rho_A \in \mathcal{D}(A)$ (resp. $\rho_A \in \mathcal{D}_\chi^?(A)$) and let $A' \subset A$ be a subalgebra of A generated over B by the values of $\mathrm{Tr}(\rho_A)$, then the morphism $\pi : R^{ord} \to A$ (resp. $\pi : R_\chi^? \to A$) with $\pi \circ \rho \cong \rho_A$ (resp. $\pi \circ \rho_\chi^? \cong \rho_A$) factors through A'; in other words, if A is generated over B by $\mathrm{Tr}(\rho_A)$, then π is surjective. Here "?" indicates either nothing or fl.*

Assuming A is generated over B by $\mathrm{Tr}(\rho_A)$, we find R^{ord} surjects down to A; so, by duality, $t_{A/B}$ injects into $\mathcal{D}(\mathbb{F}[\varepsilon])$. We prove the lemma for R^{ord} and leave the proof for $\mathcal{D}_\chi^?$ to the reader.

Proof. By (l) and (p), for $l \in S_{ab} \cup \{p\}$, we may assume that we have $g \in D_l$ such that $\overline{\rho}(g)$ has two distinct eigenvalues $\overline{\delta}_l(g)$ and $\overline{\epsilon}_l(g)$ with $\overline{\delta}_l$ unramified. Let $A' \subset A$ be the Λ-subalgebra generated by trace of ρ_A. Then by the theory of pseudo representation [MFG, §2.2.1], we have a Galois representation $\rho_{A'} : G \to \mathrm{GL}_2(A')$ with $\mathrm{Tr}\,\rho_{A'} = \mathrm{Tr}\,\rho_A$. Then by Hensel's lemma [CRT, Theorem 8.3], the $\delta_{A'}(g)$-eigen quotient $\rho_A[U(l) - \delta_{A,l}(g)]$ is a A'-direct summand of the representation space of $\rho_{A'}$ on which G acts by δ_A and hence is free of rank 1 over A'; so, the values of $\delta_{A,l}$ is contained in the subring of A' generated by trace for all $l \in S_{ab} \cup \{p\}$. This implies ρ_A descends to a deformation in $\mathcal{D}(A')$ with coefficients in A'; so, π factors through A', proving the lemma. □

The $R = \mathbb{T}$ theorem. Let $p^* = (-1)^{(p-1)/2}p$; so, $\mathbb{Q}[\sqrt{p^*}]$ is the unique quadratic subfield of $\mathbb{Q}[\mu_p]$ (for odd p). Here is a theorem of Taylor–Wiles proven in 1995 (see [MFG, §3.2.4] and [HMI, §3.2.4]):

Theorem 6.2.13. *Assume that $\overline{\rho}$ restricted to $\mathrm{Gal}(\overline{\mathbb{Q}}/\mathbb{Q}[\sqrt{p^*}])$ for $p^* = (-1)^{(p-1)/2}p$ is absolutely irreducible (Taylor–Wiles condition) in addition to* (p), (l) *and* (m). *If $k_0 \geq 2$, π induces an isomorphism $R_{\chi_0,\mathfrak{p}}^? \cong \mathbb{T}_{\chi_0,\mathfrak{p}}^?$ identifying $\rho_{\chi_0}^?$ with $\rho_{\mathbb{T}^?,\chi_0,\mathfrak{p}}$. Moreover we have a presentation $\mathbb{T}_{\chi_0,\mathfrak{p}}^? \cong \frac{W_\mathfrak{p}[[T_1,\ldots,T_r]]}{(S_1,\ldots,S_r)}$ (a local complete intersection over $W_\mathfrak{p} := \mathbb{Z}[f_0]_\mathfrak{p}$) for $r = \dim_\mathbb{F} t_{\mathbb{T}_{\chi_0,\mathfrak{p}}/W_\mathfrak{p}}$. As before, "?" is either nothing or fl.*

- By Frobenius reciprocity law (§5.3.4), the Taylor–Wiles condition fails $\Leftrightarrow \overline{\rho} \cong \mathrm{Ind}_{\mathbb{Q}[\sqrt{p^*}]}^\mathbb{Q} \varphi$ for a character $\varphi : \mathrm{Gal}(\overline{\mathbb{Q}}/\mathbb{Q}[\sqrt{p^*}]) \to \mathbb{F}^\times$.
- This condition of irreducibility over $\mathbb{Q}[\sqrt{p^*}]$ is mostly removed now by Khare/Ramakrishna/Thorne/Kalyanswamy. See [T16, §6] and [K18].
- It is known that $\mathbb{T}_{\chi_0,\mathfrak{p}}$ is reduced if the prime-to-p conductor of $\overline{\rho}$ match the prime-to-p level of f_0 (e.g., under minimality). See Lemma 6.2.15.

- We will later give a sketch of the proof of this theorem, starting with §9.3.2 towards the end of this book.

Classification of $\mathrm{Im}(\bar{\rho})$ **modulo center.** We insert here a classification due to Dickson in his book [LGF, §260] of subgroups $\mathcal{G} \subset \mathrm{PGL}_2(\mathbb{F})$ given by $\mathrm{Im}(\bar{\rho})$ modulo center (see also [W95, §3] or [MFG, pages 146–147]):

(G) If $p\,|\,|\mathcal{G}|$, $\mathcal{G}$ is conjugate to $\mathrm{PGL}_2(k)$ or $\mathrm{PSL}_2(k)$ for a subfield $k \subset \mathbb{F}$ as long as $p > 3$ (when $p = 3$, $\mathcal{G}$ can be A_5).

Suppose $p \nmid |\mathcal{G}|$ (so, $p \geq 5$). Then $\mathcal{G}$ is given as follows.

(C) $\mathcal{G}$ is cyclic ($\Rightarrow \mathrm{Im}(\bar{\rho})$ is abelian; $\bar{\rho}$ is reducible).

(D) $\mathcal{G}$ is isomorphic to a dihedral group (so, $\bar{\rho} = \mathrm{Ind}_K^{\mathbb{Q}}\,\bar{\varphi}$ for a quadratic field K), and $\mathbb{F} = \mathbb{F}_p[\bar{\varphi}]$ (the field generated by the values of $\bar{\varphi}$).

(E: Exceptional cases) $\mathcal{G}$ is isomorphic to one of the following groups: A_4, S_4 ($\mathbb{F} = \mathbb{F}_p$), and A_5 ($\mathbb{F} = \mathbb{F}_p$ if $p \equiv \pm 1, 0 \mod 5$ and $\mathbb{F}_{p^2}$ otherwise).

In Cases (G), (D), (E), $\bar{\rho}$ is absolutely irreducible (e.g., (7.25)), and in Cases (C) and (D) with $K = \mathbb{Q}[\sqrt{p^*}]$, the Taylor–Wiles condition fails.

Theorem 6.2.14. *Let the notation be as above. If $k_0 \geq 3$ (resp. $k_0 = 2$), we have $R_{\chi_0,\mathfrak{p}} = W_{\mathfrak{p}}$ and $\dim t_{R_{\chi_0,\mathfrak{p}}/W} = 0$ (resp. $R_{\chi_0,\mathfrak{p}}^{fl} = W_{\mathfrak{p}}$ and $\dim t_{R_{\chi_0,\mathfrak{p}}^{fl}/W} = 0$) for almost all ordinary primes $\mathfrak{p}$.*

Here "almost all ordinary primes" means except for finitely many primes in the set of primes $\mathfrak{p}$ of $\mathbb{Z}[f_0]$ for which f_0 is ordinary. Actually this ordinarity is not necessary for the assertion as we only need $R_{\chi_0,\mathfrak{p}}^{fl} \cong \mathbb{T}_{\chi_0,\mathfrak{p}}^{fl}$ for the assertion.

Proof. Assume $k_0 \geq 2$. By a result of Ribet [R85, Theorem 3.1], if f_0 does not have CM (i.e., not a theta series of the norm form of a quadratic field), except for finitely many $\mathfrak{p}$, $\bar{\rho}_{f_0,\mathfrak{p}}$ falls in Case G; so, it satisfies Taylor–Wiles condition. Thus the assertion follows from Theorems 6.2.7 and 6.2.13.

If f_0 has CM, f_0 is associated to an imaginary quadratic field K by Remark 6.2.9. Note that the Taylor–Wiles condition depends on $\mathfrak{p}$ as p is the residual characteristic. Thus even if f_0 has CM, for almost all primes $\mathfrak{p}$, $\bar{\rho}_{\mathfrak{p}}$ is not induced from $\mathbb{Q}[\sqrt{p^*}]$ ($p^* = (-1)^{(p-1)/2}p$), and hence we conclude the same outcome. $\qquad\square$

If $k_0 = 1$, under irreducibility (and $p > 3$), we are either in Case (D) or (E) and $\rho = \rho_{f_0,\mathfrak{p}}$ is independent of $\mathfrak{p}$ (or in short, ρ is an Artin representation [DS74]). We do not know the distribution of primes $\mathfrak{p}$ with $R_{\chi_0,\mathfrak{p}} = W_{\mathfrak{p}}$ except when $\rho = \mathrm{Ind}_K^{\mathbb{Q}}\,\varphi$ for real quadratic K. We study the real and imaginary quadratic cases later in Chapter 8. Our next goal is to study this question for $B = \Lambda$. We may ask if $R^{ord} = \Lambda$ for most $\mathfrak{p}$.

6.2.5 *"Big" ordinary Hecke algebra* $\mathbb{T}_{\mathfrak{p}}$

We vary primes $\mathfrak{p}$ of $\mathbb{Z}[f_0]$ and consider $\overline{\rho}_{\mathfrak{p}}$. Then we write $S = S_{\mathfrak{p}}$ for the set of primes different from p ramified in $F(\overline{\rho}_{\mathfrak{p}})/\mathbb{Q}$. We assume

(1) $\lambda(T(p)) \notin \mathfrak{p}$; so, $f_0 \in S^{ord}_{k_0,\psi_0}(\Gamma_0(C))$ as in §4.1.6 ($\mathfrak{p}$-ordinarity),
(2) for every prime $l \neq p$, the ramification index of l in $F(\overline{\rho}_{\mathfrak{p}})/\mathbb{Q}$ is prime to p (minimality).

As we already remarked, the condition (2) is satisfied for almost all $\mathfrak{p}$ as f_0 has potentially good reduction everywhere. If $\overline{\rho}_{\mathfrak{p}}$ satisfies p-minimality and $P|C$, by the solution of Serre's modulo p modularity conjecture [KW09] (and earlier works on level lowering), we may replace f_0 by another Hecke eigenform of level C/p without changing the residual representation $\overline{\rho}_{\mathfrak{p}}$.

Start with $\overline{\rho} = \rho_{f_0,\mathfrak{p}} \mod \mathfrak{p}$ given by a primitive form $f_0 \in S_{k_0}(\Gamma_0(C), \psi_0)$. Let $\psi = \psi_0 \omega^{k_0}$. Then for the Hecke algebra $\mathbf{h}^{ord}_{\psi}(C; \Lambda)$ defined in §4.1.7, by Theorem 4.1.29, we have $\mathbf{h}^{ord}_{\psi}(C; \Lambda) \otimes_{\Lambda} \Lambda/(t - \gamma^{k_0}) \cong \mathbf{h}^{ord}_{k,\psi_0}(\Gamma_0(pC); W)$. For $\chi_0 := \psi_0 \nu_p^{k_0-1}$, the local ring $\mathbb{T}_{\chi_0,\mathfrak{p}}$ associated to $\overline{\rho}_{\mathfrak{p}}$ is a factor of $\mathbf{h}^{ord}_{k,\psi_0}(\Gamma_0(pC); W)$. The idempotent $\mathbf{1}_0$ of $\mathbb{T}_{\chi_0,\mathfrak{p}}$ in $\mathbf{h}^{ord}_{k,\psi_0}(\Gamma_0(pC); W)$ lifts to a unique idempotent $\mathbf{1}_0$ of $\mathbf{h}^{ord}_{\psi}(C; \Lambda)$ [BCM, III.4.6]. Indeed, taking $h \in \mathbf{h}^{ord}_{\psi}(C; \Lambda)$ with $h \equiv \mathbf{1}_0 \mod(t - \gamma^{k_0})$, we have a limit formula $\mathbf{1}_0 = \lim_{n \to \infty} h^{p^n}$. Let $\mathbb{T}_{\mathfrak{p}} := \mathbf{1}_0 \mathbf{h}^{ord}_{\psi}(C; \Lambda)$.

Here is a down-to-earth description of $\mathbb{T}_{\mathfrak{p}}$. By (4.12), we have a modular form H with $H \equiv 1 \mod \mathfrak{p}$ of weight 1 of level p with coefficients in $\mathbb{Z}_p$ and character ω^{-1} for the Teichmüller character ω modulo p. Then $fH^n \equiv f \mod p$, and $i : f \mapsto fH^n$ gives a q-expansion preserving $\mathbb{F}$-linear map $S_{k_0}(\Gamma_0(Cp), \psi_0; \mathbb{F}) \hookrightarrow S_k(\Gamma_0(Cp), \psi\omega^{-k}; \mathbb{F})$ ($k = k_0 + n, \psi_k = \psi_0 \omega^{-n} = \psi\omega^{-k}$). Note that $\overline{\chi} = (\nu_p^{k_0-1}\psi_0 \mod \mathfrak{p}) = (\nu_p^{k-1}\omega^{-n}\psi \mod \mathfrak{p})$, and the action of $T(n)$ on $S_k(\Gamma_0(Cp), \psi\omega^{-k}; \mathbb{F})$ is

$$a(m, f|T(n)) = \sum_{d|m, d|n, (d,pC)=1} \overline{\chi}(d) a(\frac{mn}{d^2}, f) \quad \text{(e.g., Lemma 4.1.8)}$$

and hence i is Hecke equivariant. Thus we have $\mathbb{T}_k$ as a factor of $\mathbf{h}_k(\Gamma_0(Cp), \psi_k)_{/W_{\mathfrak{p}}}$ giving the same $\overline{\rho}$. We then define $\mathbb{T} = \mathbb{T}_{\mathfrak{p}}$ to be the subalgebra of $\prod_{k \geq k_0} \mathbb{T}_k$ topologically generated by $T(n)$ for all n (here $T(n)$ has projection to $T(n)$ in $\mathbb{T}_k$ for all $k \geq k_0$).

$\mathbb{T}_\mathfrak{p}$ **is reduced.** Essentially by Remark 6.2.3, $\mathbb{T}_k$ is reduced under minimality (m), and therefore, $\mathbb{T}_\mathfrak{p}$ is reduced. Here is a proof of this fact.

Lemma 6.2.15. *Suppose that the prime-to-p level C of $\mathbb{T} = \mathbb{T}_\mathfrak{p}$ coincides with the prime-to-p conductor of $\bar{\rho}$. Then the local ring $\mathbb{T}$ is reduced.*

We use the fact that the conductor of a Hecke eigenform f is equal to the conductor of the compatible system ρ_f [GME, Theorem 5.1.10] whose prime-to-p part is equal to the *prime-to-p conductor of $\bar{\rho}_\mathfrak{p}$* (see [LFD, IV.2] and [S81, §1.2] for the prime-to-p conductor of $\bar{\rho}$).

Proof. Write $K := W_\mathfrak{p} \otimes_{\mathbb{Z}_p} \mathbb{Q}_p$ (the field of fractions of $W_\mathfrak{p}$). The definition of conductor c_l of $\bar{\rho}$ at a prime $l \neq p$ can be given as in the case of characteristic 0 (as in [LFD, IV.2] and [S81, §1.2]). Define the prime-to-p conductor $c^{(p)}(\bar{\rho})$ of $\bar{\rho}$ by $c^{(p)}(\bar{\rho} = \prod_{l \neq p} c_l)$. By (m), the prime-to-$p$ conductor $c^{(p)}(\bar{\rho})$ gives the prime-to-p level C of the Hecke algebra, as C is the prime-to-p conductor of ρ_f for any cusp-form f belonging to $\mathbb{T}_k$.

For any old Hecke eigenform g of level $N'|C$ (so, $N' < C$), the representation ρ_g has prime-to-p conductor dividing N'. Therefore the prime-to-p conductor C of each modular deformation $\rho_{f,\mathfrak{p}}$ parameterized by $\mathbb{T}$ satisfies $C = c^{(p)}(\bar{\rho})|c^{(p)}(\rho_f) = C$; so, $c^{(p)}(\rho_f) = C$ is independent of f. Thus any Hecke eigenform belonging to $\mathbb{T}$ has prime-to-p conductor C; i.e., C-new.

Pick a weight $k > 2$ and $\mathbb{T}_k = \mathbb{T}/(t - \gamma^k)\mathbb{T}$ for $t = 1 + T$ and $\gamma = 1 + p$. By Theorem 4.1.29, $\mathbb{T}_k$ is a direct factor of $\mathbf{h}_k := \mathbf{h}_{k,\psi\omega^{-k}}^{ord}(\Gamma_0(C\mathbf{p}); W)$. We can decompose $\mathbf{h}_k \otimes_W K = \mathbf{h}_k(K)^{new} \times \mathbf{h}_k(K)^{old}$ so that $\mathbf{h}_k(K)^{new}$ (resp. $\mathbf{h}_k(K)^{old}$) acts on the subspace of $S_{k,\psi\omega^{-k}}^{ord}(\Gamma_0(C\mathbf{p}); K)$ spanned by the C-new (resp. C-old) forms. If $\psi\omega^{-k}$ has conductor prime to p and $k > 2$, every new form in $S_{k,\psi\omega^{-k}}(\Gamma_0(C\mathbf{p}); K)$ is non-p-ordinary [MFM, Theorem 4.6.17 (2)]. Thus we know $\mathbf{h}_{k,\psi\omega^{-k}}^{ord}(\Gamma_0(C\mathbf{p}); W) \cong \mathbf{h}_{k,\psi\omega^{-k}}^{ord}(\Gamma_0(C); W)$ by $k > 2$, and if $\psi\omega^{-k}$ has p-conductor divisible by p, every C-new form in $S_{k,\psi\omega^{-k}}^{ord}(\Gamma_0(C\mathbf{p}))$ is a new form.

By the theory of new forms [MFM, §4.6], $\mathbf{h}_k(K)^{new}$ is a semi-simple commutative K-algebra. Let $\mathbf{h}_k^{old}$ be the projected image of $\mathbf{h}_k$ in $\mathbf{h}_k(K)^{old}$. If $\mathbb{T}_k$ projects to $\mathbf{h}_k^{old}$ non-trivially, we have $\mathbb{T}_k/\mathfrak{m}_{\mathbb{T}_k} \cong \mathbf{h}_k^{old}/\mathfrak{m}$ for a maximal ideal $\mathfrak{m}$ of $\mathbf{h}_k^{old}$. Then the $\mathfrak{m}$-residual representation has prime-to-p conductor less than C, a contradiction; so, $\mathbb{T}_k \subset \mathbf{h}_k(K)^{new}$ and hence $\mathbb{T}_k$ is reduced. Since $\mathbb{T}$ is embedded into $\prod_k \mathbb{T}_k$ by the diagonal embedding, $\mathbb{T}$ must be reduced.   □

Big Galois representation. Consider the product $\rho_{\mathbb{T}_\mathfrak{p}} = \prod_{k \geq k_0} \rho_{\mathbb{T}_k}$: $G \to \mathrm{GL}_2(\prod_{k \geq k_0} \mathbb{T}_k)$. Then $\mathrm{Tr}(\rho_{\mathbb{T}_\mathfrak{p}}(\mathrm{Frob}_l)) = T(l) \in \mathbb{T}_\mathfrak{p}$ for all primes $l \nmid Cp$. By Chebotarev, $\mathrm{Tr}(\rho_\mathfrak{p})$ has values in $\mathbb{T}$; so, by means of pseudo characters (see [MFG, §2.2.1] and [FGS, §1.2]), if $\overline{\rho}$ is irreducible, this representation descends to $\rho_{\mathbb{T}_\mathfrak{p}} : G \to \mathrm{GL}_2(\mathbb{T}_\mathfrak{p}) \in \mathcal{D}_\mathfrak{p}(\mathbb{T}_\mathfrak{p})$. Our base ring B is $W_\mathfrak{p}$ but we can descend further to the Witt vector ring $W = W(\mathbb{F})$. Since $\det(\rho_{\mathbb{T}_\mathfrak{p}})$ is a deformation of $\det(\overline{\rho})$, we have an algebra structure $\iota_{\mathbb{T}_\mathfrak{p}} : \Lambda = W[[G_p^{ab}]] \to \mathbb{T}_\mathfrak{p}$. This is the representation constructed in 1986 in [H86b], in which the representation is made only assuming that $a(p, f_0) = \lambda(T(p)) \notin \mathfrak{p}$.

Theorem 6.2.16. *Suppose that $\mathfrak{p}$ is a prime of $\mathbb{Z}[f_0]$ such that $\rho_{f_0,\mathfrak{p}}$ satisfies* (m) *and* (p) *with irreducible $\overline{\rho}_{f,\mathfrak{p}}$. Then we have a Galois representation $\rho_{\mathbb{T}_\mathfrak{p}} : G \to \mathrm{GL}_2(\mathbb{T}_\mathfrak{p})$ in $\mathcal{D}(\mathbb{T}_\mathfrak{p})$ such that $\mathrm{Tr}(\mathrm{Frob}_l) = T(l)$ for primes $l \nmid Cp$.*

Here is a Λ-adic version of Theorem 6.2.14 (e.g., [G92]):

Theorem 6.2.17. *Let the notation and assumption be as in Theorem 6.2.16. Then the Λ-algebra $\mathbb{T}_\mathfrak{p}$ is reduced Λ-free of finite rank and $\mathbb{T}_\mathfrak{p}/(t - \chi(\gamma)) \cong \mathbb{T}_{\chi,\mathfrak{p}}$ ($\chi = \nu_p^{k-1}\psi_k$) for all $k \geq 2$. If $R_{\chi_1,\mathfrak{p}} \cong \mathbb{T}_{\chi_1,\mathfrak{p}}$ for $\chi_1 = \nu_p^{k_1-1}\psi_{k_1}$ for one weight $k_1 \geq 2$, we have $R_\mathfrak{p}^{ord} \cong \mathbb{T}_\mathfrak{p}$, and $R_{\chi,\mathfrak{p}} = \mathbb{T}_\mathfrak{p}/(t-\chi(\gamma))$ for an arbitrary character $\chi : G \to W^\times$ with $\chi \equiv \chi_0 \mod \mathfrak{m}_W$. In particular, if $k_0 \geq 3$ and $\mathfrak{p} \nmid \mathrm{Ann}(f_0)$, $R_\mathfrak{p}^{ord} \cong \Lambda = W_\mathfrak{p}[[T]]$.*

Reducedness of $\mathbb{T}_\mathfrak{p}$ is already given in Lemma 6.2.15.

Proof. By definition, $\mathbb{T}_\mathfrak{p}$ is a direct summand of $\mathbf{h}_\psi^{ord}(C; \Lambda)$ which is free of finite rank over Λ by Theorem 4.1.29. Thus $\mathbb{T}_\mathfrak{p}$ is Λ-projective and Λ-flat. Since Λ is a local ring, $\mathbb{T}_\mathfrak{p}$ is Λ-free of finite rank [CRT, Theorem 2.5]. Let $r = \mathrm{rank}_\Lambda \mathbb{T}_\mathfrak{p}$. Since $\mathbb{T}_{\chi,\mathfrak{p}} = \mathbb{T}_\mathfrak{p}/(t - \chi(\gamma))$ by Theorem 4.1.29, $\mathbb{T}_{\chi,\mathfrak{p}}$ is W-free of finite rank r for any χ as in the theorem.

By assumption, $R_{\chi_1,\mathfrak{p}} \cong \mathbb{T}_{\chi_1,\mathfrak{p}}$; so, by Nakayama's Lemma 1.8.3, the morphism $\pi : R_\mathfrak{p}^{ord} \to \mathbb{T}_\mathfrak{p}$ with $\pi \circ \rho \sim \rho_{\mathbb{T}_\mathfrak{p}}$ is surjective, and $R_\mathfrak{p}^{ord}$ is generated by r elements lifting a basis of $T_{\chi,\mathfrak{p}}$. Thus we have a surjective morphism $\pi_0 : \Lambda^r \twoheadrightarrow R_\mathfrak{p}^{ord}$. Since $\pi \circ \pi_0 : \Lambda^r \to \mathbb{T}_\mathfrak{p}$ is Λ-linear onto, comparing the Λ-rank, we find $\pi \circ \pi_0$ is an isomorphism, and hence π and π_0 are both isomorphisms.

If $k_0 \geq 3$, $\mathbb{T}_{\chi_0,\mathfrak{p}} = W_\mathfrak{p}$ if $\mathfrak{p} \nmid \mathrm{Ann}(f_0)$; so, $r = 1$ and we conclude the last assertion from Theorem 6.2.7. $\qquad\square$

Remark 6.2.18. If $k_0 = 2$, by level-raising [R84], if $\overline{\epsilon}_\mathfrak{p}/\overline{\delta}_\mathfrak{p} = \overline{\omega}_p$, we have $\mathbb{T}_{\chi_0,\mathfrak{p}}^{fl} \neq \mathbb{T}_{\chi_0,\mathfrak{p}}$; so, we cannot conclude $\mathfrak{p} \nmid \mathrm{Ann}(f_0) \Rightarrow \mathbb{T}_\mathfrak{p} = W_\mathfrak{p}[[T]]$. *Are*

there infinitely many primes $\mathfrak{p}$ with $\bar{\epsilon}_\mathfrak{p}/\bar{\delta}_\mathfrak{p} = \bar{\omega}_p$ (perhaps of density 0)?
Suppose f_0 has weight 2 with trivial ψ_0. If $\det(\bar{\rho}_\mathfrak{p}) = \bar{\delta}^2\bar{\omega}_p = \bar{\omega}_p$, we
find $(a(p, f_0) \mod \mathfrak{p})^2 = \bar{\delta}_p(\mathrm{Frob}_p)^2 = (\psi(p) \mod \mathfrak{p}) = 1$. Such primes $\mathfrak{p}$
are rare. Indeed, similar to the comments between Conjecture 6.2.8 and
Remark 6.2.9, we can show that $\{p \in \mathrm{Spec}(\mathbb{Z}) | a(p, f_0) \not\equiv \pm 1 \mod p\}$ has
positive density. This fact can be generalized to the case where $\psi_0 \neq 1$
(why?). Thus there are infinitely many primes $\mathfrak{p}$ (with positive density)
such that $\bar{\epsilon}_\mathfrak{p}/\bar{\delta}_\mathfrak{p} \neq \bar{\omega}_p$.

Proposition 6.2.19. *Define* $\mathbb{T}_{\chi,\mathfrak{p}} := \mathbb{T}_\mathfrak{p}/(t - \chi(\gamma))\mathbb{T}_\mathfrak{p}$ *for any character*
$\chi : G \to W^\times$ *with* $\chi \equiv \chi_0 \mod \mathfrak{m}_A$ *(not necessarily of the form* $\nu_p^{k-1}\psi_k$*). If*
$R_{\chi,\mathfrak{p}} \cong \mathbb{T}_{\chi,\mathfrak{p}}$ *for Zariski densely populated* χ *in* $\mathrm{Spec}(\Lambda)(W)$*, then* $R_\mathfrak{p}^{ord} \cong \mathbb{T}_\mathfrak{p}$
and $R_\mathfrak{p}^{ord}$ *is* Λ-*free of finite rank.*

Zariski density of χ's is equivalent to $\bigcap_\chi(t - \chi(\gamma)) = \{0\}$. Though we can
choose a classical form g whatever close to a given p-adic cusp form f, if
χ is not of the form $\nu_p^{k-1}\psi_k$ with $k \geq 2$, we cannot insist g to have the
same character χ; particularly, if k is not a positive rational integer, the
character of g is close to χ but never equal to χ.

Proof.   Pick one χ_0 with $R_{\chi_0,\mathfrak{p}} \cong \mathbb{T}_{\chi_0,\mathfrak{p}}$. Then for $r := \mathrm{rank}_\Lambda \mathbb{T}_\mathfrak{p} =$
$\mathrm{rank}_W \mathbb{T}_{\chi,\mathfrak{p}}$, we have a Λ-linear surjection $\pi : \Lambda^r \twoheadrightarrow R_\mathfrak{p}^{ord} \twoheadrightarrow \mathbb{T}_\mathfrak{p}$ by
Nakayama's lemma. This induces $W^r = (\Lambda/(t - \chi(\gamma)))^r \xrightarrow{\pi_\chi} \mathbb{T}_{\chi,\mathfrak{p}} \cong W^r$.
Comparing W-rank, we find π_χ is an isomorphism. Thus $\mathrm{Ker}(\pi) \subset$
$\bigcap_\chi(t - \chi(\gamma))^r = \{0\}$, which proves the desired assertion. $\square$

Presentation theorem. We now note:

Theorem 6.2.20. *Assume* $R_\mathfrak{p}^{ord} \cong \mathbb{T}_\mathfrak{p}$ *and* $\mathbb{T}_{\chi,\mathfrak{p}} \cong \frac{W[[\overline{T}_1,...,\overline{T}_r]]}{(\overline{S}_1,...,\overline{S}_r)}$ *for* $r =$
$\dim_\mathbb{F} t_{\mathbb{T}_{\chi,\mathfrak{p}}/W}$ *for* $\chi = \nu_p^{k-1}\psi_k$ $(k \geq 2)$. *Then we have*
$$R_\mathfrak{p}^{ord} \cong \Lambda[[T_1, \ldots, T_r]]/(S_1, \ldots, S_r)$$
with $T_j \mod (t - \chi(\gamma)) = \overline{T}_j$ *and* $S_j \mod (t - \chi(\gamma)) = \overline{S}_j$.

Proof.   We write $\bar{t}_i$ for the image of $\overline{T}_i$ in $\mathbb{T}_\chi$. As remarked in §1.21
$t^*_{\mathbb{T}_\chi/W} = \Omega_{\mathbb{T}_\chi/W} \otimes_{\mathbb{T}_\chi} \mathbb{F} = \Omega_{\mathbb{T}_\mathfrak{p}/\Lambda} \otimes_{\mathbb{T}_\mathfrak{p}} \mathbb{F} = t^*_{\mathbb{T}_\mathfrak{p}/\Lambda}$. Any lifts $\{t_i\}_i$ of $\{\bar{t}_i\}_i$
give rise to a basis of $t^*_{\mathbb{T}_\mathfrak{p}/\Lambda}$. Then we have a surjective CL_Λ-morphism $\pi :$
$\Lambda[[T_1, \ldots, T_r]] \twoheadrightarrow \mathbb{T}_\mathfrak{p}$ with $T_i \mapsto t_i$. Then $\mathrm{Ker}(\pi)/(t - \chi(\gamma))\mathrm{Ker}(\pi)$ is generated by $\overline{S}_1, \ldots, \overline{S}_r$, and we lift $\overline{S}_i$ to $S_i \in \mathrm{Ker}(\pi)$. So $\mathrm{Ker}(\pi) \otimes_{\Lambda[[T_1,...,T_r]]} \mathbb{F}$ is
generated by the image of $\overline{S}_i$ as $\mathbb{F}$-vector space; so, by Nakayama's lemma,
we have $\mathrm{Ker}(\pi) = (S_1, \ldots, S_r)$ as desired. $\square$

Deformation functor over Λ and R^{ord}. Note that $\kappa := \det(\rho) : G \to \Lambda^\times$ is the universal character deforming $\det(\overline{\rho})$. Thus (Λ, κ) represents

$$A \mapsto \{\xi : G \to A^\times | \xi \mod \mathfrak{m}_A = \det(\overline{\rho})\}.$$

Consider a new deformation functor $\mathcal{D}_\kappa : CL_{/\Lambda} \to SETS$:

$$\mathcal{D}_\kappa(A) = \{\rho \in \mathcal{D}(A) | \det(\rho) = \iota_A \circ \kappa\}/\Gamma(\mathfrak{m}_A),$$

slightly different from the one $\mathcal{D}$, where writing $\iota_A : \Lambda \to A$ for Λ-algebra structure of A. $\mathcal{D}_\kappa$ is again represented by $(R^{ord}, \boldsymbol{\rho})$ regarding R^{ord} as a Λ-algebra by the CL_W-morphism induced by $\det(\boldsymbol{\rho}) : G \to R^{ord\times}$. Indeed, if $\rho \in \mathcal{D}_\kappa(A)$, we have $\iota_A \circ \kappa = \det(\rho)$. Regarding $\rho \in \mathcal{D}(A)$, we have a unique CL_W-morphism $R^{ord} \xrightarrow{\phi} A$ with $\phi \circ \boldsymbol{\rho} \sim \rho$. Taking determinant, we get $\phi \circ \kappa = \det(\rho)$ showing that ϕ is compatible with $\iota_{R^{ord}}$ and ι_A; so, it is a CL_Λ-morphism, showing $\mathrm{Hom}_\Lambda(R, A) \cong \mathcal{D}_\kappa(A)$ by $\phi \leftrightarrow \rho$. Thus by (6.8), $\boxed{\mathrm{Sel}(Ad(\rho))^\vee \cong \Omega_{R^{ord}/\Lambda} \otimes_{R^{ord}, \phi} A}$.

6.2.6 Fitting ideals

We recall the theory of Fitting ideals (and characteristic ideals in §1.9.3) which is useful in describing the size of a torsion B-module. Let A be a ring in the category CL_B. Let M be an A-module with presentation:

$$A^r \xrightarrow{L} A^s \to M \to 0$$

for an A-linear map given by a multiplication by a matrix L in $M_{s,r}(A)$. If $r \geq s$, the A-ideal $\mathrm{Fitt}_A(M)$ generated by $s \times s$-minors of L is called the *Fitting ideal* of M, which is independent of the choice of the presentation (and the choice of the matrix L). If $r < s$, we simply define the Fitting ideal by $\mathrm{Fitt}_A(M) = (0)$. By definition, $\mathrm{Fitt}_B(M \otimes_{A,\phi} B) = B \cdot \phi(\mathrm{Fitt}_A(M))$ for a CL_B-morphism $\phi : A \to B$. If $r = s$, $\mathrm{Fitt}_A(M)$ is principal.

See [MW84, Appendix] for a summary of the theory of Fitting ideal. Eisenbud's book [CAG, §20] has more details.

Example 6.2.1. Let M be a torsion A-module of finite type.

- If $A = B$ is a discrete valuation ring with $\mathfrak{m}_A = (\varpi)$, $M \cong A/\varpi^{e_1}A \oplus \cdots \oplus A/\varpi^{e_r}A$ by elementary divisor theory. Define the characteristic ideal by $\mathrm{char}_A(M) = (\prod_j \varpi^{e_r})A$. Choose $L = \mathrm{diag}[\varpi^{e_1}, \ldots, \varpi^{e_r}]$, we find $\mathrm{Fitt}_A(M) = \det(L) = \mathrm{char}_A(M)$. If $A = \mathbb{Z}_p$, $\mathrm{Fitt}_{\mathbb{Z}_p}(M) = (|M|)$ and if $A = W$, $\mathrm{Fitt}_W(M) = (\prod_i \varpi^{e_i}) = (\|M\|_p^{-\mathrm{rank}_{\mathbb{Z}_p} W})$.

- If $B = A = \Lambda$ and M is a Λ-torsion module, we have a Λ-linear morphism

$i : M \to \Lambda/(f_1) \oplus \cdots \oplus \Lambda/(f_r)$ for $f_j \in \mathfrak{m}_\Lambda$ with finite kernel and finite cokernel. As in §1.9.3, we define $\mathrm{char}_\Lambda(M) := (\prod_i f_i)$ (the characteristic ideal with *characteristic power series* $\prod_i f_i$).

- Let $\wp_A$ be the set of all height 1 prime ideal of A. If $A = \Lambda$, $\wp_\Lambda$ is made of all principal non-zero prime ideals. By the flatness of Λ_P over Λ, $\mathrm{Fitt}_{\Lambda_P}(M_P) = \mathrm{Fitt}_\Lambda(M)_P$ for any prime P of Λ. Hence $\bigcap_{P \in \wp_\Lambda} \mathrm{Fitt}_\Lambda(M)_P = \bigcap_{P \in \wp_\Lambda} \mathrm{char}_\Lambda(M)_P = \mathrm{char}_\Lambda(M)$. Thus $(\bigcap_{P \in \wp_\Lambda} \mathrm{Fitt}_\Lambda(M)_P)/\mathrm{Fitt}_\Lambda(M)$ is finite. By [BCM, VII.4.3], $\mathrm{Fitt}_\Lambda(M) = \mathrm{char}_\Lambda(M)$ if $\mathrm{Fitt}_\Lambda(M)$ is principal. So for any normal noetherian domain A, as in [BCM, VII.4.5], we define the *characteristic ideal* of M by $\mathrm{char}_A(M) := \bigcap_{P \in \wp_A} \mathrm{Fitt}_A(M)_P = \bigcap_{P \in \wp_A} \mathrm{char}_{A_P}(M_P)$.
- If we have a good p-adic L-function L_ρ of a Galois representation $\rho : G \to \mathrm{GL}_n(A)$, one expects $\mathrm{char}_A(\mathrm{Sel}(\rho)^\vee) = (L_\rho)$ (Main conjecture).

Theorem 6.2.21 (Tate's theorem). *Suppose that B is a domain. Let $A \in CL_B$ be a reduced B-algebra free of finite rank over B. If $A \cong \frac{B[[T_1,\ldots,T_r]]}{(S_1,\ldots,S_r)}$, then for any B-algebra homomorphism $\lambda : A \to B$,*

$$\mathrm{Fitt}_B(C_1(\lambda;B)) = \mathrm{Fitt}_B(C_0(\lambda,B)), \quad C_0(\lambda;B) = B/\mathrm{Fitt}_B(C_0(\lambda;B)).$$

We will prove this theorem in §6.2.9.

If B is normal noetherian, we still have the notion of characteristic ideal for a torsion B-module of finite type, generalizing the case in §1.9.3 when $B = \Lambda$. A height 1 prime ideal P of B is called a *prime divisor* (as $\mathrm{Spec}(B/P) \subset \mathrm{Spec}(B)$ gives a codimension 1 closed subscheme similar to prime divisor in the curve case discussed in §2.1.5). A B-module M of finite type is called *pseudo-null*, if its annihilator $\mathrm{Ann}_B(M)$ has codimension ≥ 2 (i.e., $\dim B/\mathrm{Ann}_B(M) \leq \dim B - 2$), which is equivalent to $M_P = 0$ for all prime divisors P of B. Since $\Lambda/\mathfrak{a}$ is finite if $\dim \Lambda/\mathfrak{a} = 0$ (i.e. $\mathfrak{a}$ has codimension 2), a Λ-module M is pseudo-null if and only if $|M|$ is finite. If M is a torsion B-module of finite type, we have a finite set of prime divisors S of B such that $M_P \neq 0$. Similar to the case of $B = \Lambda$, we have a B-homomorphism $i : M \to \bigoplus_{P \in S} B/P^{e_P}$ with pseudo-null kernel and cokernel [BCM, VII.4.4]. The set S and e_P is uniquely determined by M, and $\{P^{e_P}\}_{P \in S}$ is a generalization of the elementary divisors of M in the case where B is a principal ideal domain.

We define the characteristic ideal of M by $\mathrm{char}_B(M) := \prod_{P \in S} P^{e_P}$. For an ideal $\mathfrak{a}$ of B, its *reflexive closure* $\mathrm{div}(\mathfrak{a})$ is defined by $\bigcap_{P: \text{ prime divisor}} \mathfrak{a}_P$ inside $\mathrm{Frac}(B)$ for the localization $\mathfrak{a}_P$ at P of $\mathfrak{a}$. By definition, $\mathrm{div}(\mathfrak{a})/\mathfrak{a}$ is pseudo-null. It is known that $\mathrm{div}(\mathrm{Fitt}(M)) = \mathrm{char}(M)$. Indeed, by definition, the formation of Fitting ideal and characteristic ideal commutes with localization. Localization at a prime divisor P makes B_P a discrete valuation ring, and by elementary divisor theory over B_P, we see $\mathrm{Fitt}(M_P) = \mathrm{char}(M_P) = P^{e_P}$. This shows $\mathrm{div}(\mathrm{Fitt}(M)) = \mathrm{char}(M)$. Thus we have

Corollary 6.2.22. *Let the notation and the assumption be as in Theorem 6.2.21. Suppose that B is a normal noetherian domain. Then we have*

$$\mathrm{char}(C_0) = \mathrm{char}(C_1).$$

Homological dimension. For a noetherian local ring A in $CL_{/B}$, we define the *homological dimension* $\mathrm{hdim}_B M$ of a finitely generated B-module M is the minimum length h of exact sequence $0 \to F_h \to F_{h-1} \to \cdots \to F_0 \to M \to 0$ made of R-free module F_j of finite rank. If $R_{A/B}$ is a local complete intersection free of finite rank over B, we have a presentation $A = B[[T_1, \ldots, T_r]]/(f_1, \ldots, f_r)$ for a regular sequence $f_1, \ldots, f_r$. Then the fundamental exact sequence (Corollary 5.2.5) gives an exact sequence

$$(f_1, \ldots, f_r)/(f_1, \ldots, f_r)^2 \xrightarrow{i} \Omega_{B[[T_1, \ldots, T_r]]/B} \otimes_{B[[T_1, \ldots, T_r]]} A \twoheadrightarrow \Omega_{A/B}.$$

If further B is a domain of characteristic 0 and A is reduced, $\Omega_{A/B}$ is a torsion A-module (as the extension $\mathrm{Frac}(A)/\mathrm{Frac}(B)$ is a finite semi-simple extension). Since $(f_1, \ldots, f_r)/(f_1, \ldots, f_r)^2 \cong A^r$ as $(f_1, \ldots, f_r)$ is a regular sequence, torsion-property of $\Omega_{A/B}$ tells us that i is injective; so, we get from $\Omega_{B[[T_1, \ldots, T_r]]/B} \otimes_{B[[T_1, \ldots, T_r]]} A \cong \bigoplus_j Ad T_j$

$$\mathrm{hdim}_A \Omega_{A/B} = 1 \text{ if } A \text{ is a local complete intersection over } B. \qquad (6.10)$$

6.2.7 Algebraic p-adic L-function

A classical cusp form f gives a W-algebra homomorphism $\lambda : \mathbb{T} \to \overline{\mathbb{Q}}_p$ so that $f|T(n) = \lambda(T(n)f$. As we saw in §4.2.9, such points $P := \lambda \in \mathrm{Spec}(\mathbb{T})(\overline{\mathbb{Q}}_p) = \mathrm{Hom}_{W\text{-alg}}(\mathbb{T}, \overline{\mathbb{Q}}_p)$ are populated densely in $\mathrm{Spec}(\mathbb{T})$. For each P, we have congruence modules C_0, C_1 and Selmer groups $\mathrm{Sel}(Ad(\rho_P))$ for $\rho_P = \lambda \circ \rho_{\mathbb{T}}$. Thus we have a function $P \mapsto |\mathrm{Sel}(Ad(\rho_P))|$. We describe here in terms of C_0 and C_1 how to interpolate p-adically this function originally defined over points associated to classical cusp forms to an analytic function defined all over $\mathrm{Spec}(\mathbb{T})$.

We fix a prime $\mathfrak{p}$ of $\mathbb{Z}[f_0]$ and study the algebraic p-adic L-function interpolating the size of the adjoint Selmer group. Since we have fixed $\mathfrak{p}$, we write $\mathbb{T}$ for $\mathbb{T}_\mathfrak{p}$ and $\mathbb{T}_\chi$ for $\mathbb{T}_{\chi,\mathfrak{p}}$. Similarly we write the universal rings

R^{ord} and R_χ for $R^{ord}_{\mathfrak{p}}$ and $R_{\chi,\mathfrak{p}}$ hereafter. We assume that $R^{ord} \cong \mathbb{T}$ which is valid under the Taylor–Wiles condition. To treat the case of R^{ord} and R_χ uniformly, we write $R_\kappa := R^{ord}$ if $B = \Lambda$ and $\chi = \kappa$ and R_χ as usual if $B = W$ and $\chi = \nu_p^{k-1}\psi_k$. Thus $\chi = \kappa : G \to \Lambda^\times$ is allowed with $B = \Lambda$.

Let $\rho_A : G \to \mathrm{GL}_2(A) \in \mathcal{D}_\chi(A)$ be a deformation of $\bar\rho$ such that $\rho \cong P \circ \rho$ for $P \in \mathrm{Spec}(R_\chi)(A) = \mathrm{Hom}_{CLW}(R_\chi, A)$. Let $r = \dim_{\mathbb{F}} t_{R_\chi/B}$ be as in the presentation Theorem 6.2.20; so, $R_\chi = \frac{B[[T_1,\ldots,T_r]]}{(f_1,\ldots,f_r)}$. If $r = 1$, $R_\chi = B[[T_1]]/(f_1)$ and we have a commutative diagram with exact rows for $(P : R_\chi \to A) \in \mathrm{Hom}_{B\text{-alg}}(R_\chi, A)$

$$
\begin{array}{ccccc}
A = (f_1)/(f_1^2) \otimes_{R_\chi} A & \longrightarrow & A \cdot dT_1 & \overset{\twoheadrightarrow}{\longrightarrow} & \Omega_{R_\chi/B} \otimes_{R_\chi} A \\
\| \uparrow {\scriptstyle L_\rho \mapsto f_1} & & \| \uparrow {\scriptstyle 1 \mapsto dT_1} & & \wr \uparrow \\
A \cdot L_\rho & \longrightarrow & A & \overset{\twoheadrightarrow}{\longrightarrow} & \mathrm{Sel}(Ad(\rho))^\vee.
\end{array}
$$

If $B = W$, $|L_\rho|_p^{-1} = |\mathrm{Sel}(Ad(\rho))|$ and $L_\rho(P) := P(L_\rho) = L_\rho$. If $B = \Lambda$ ($\chi = \kappa$), L_ρ gives rise to a p-adic L-function with

$$\mathrm{Spec}(R_\kappa)(W) \ni P \mapsto |L_\rho(P)|_p^{-1} = |\mathrm{Sel}(Ad(P \circ \rho))|.$$

In general, when $r > 1$, we define

$$L_\rho := \det((f_1,\ldots,f_r)/(f_1,\ldots,f_r)^2 \to \bigoplus_{j=1}^{r} R_\kappa \cdot dT_j),$$

and the outcome is the same.

Lifting to an extension $\mathbb{I}$ of Λ. We say a Hecke eigenform $f \in S_{k,\psi_k}(\Gamma_0(Cp))_{/W} = S_k(Cp, \psi_k)_{/W}$ belongs to $\mathbb{T}$ if $\lambda_f : h_k(Cp, \psi_k)_{/W} \to W$ given by $f|T(n) = \lambda_f(T(n))f$ factors through $\mathbb{T}_\chi = \mathbb{T}_k$. For an irreducible component $\mathrm{Spec}(\mathbb{I}) \subset \mathrm{Spec}(\mathbb{T})$, f (or λ_f) is called "*belonging to $\mathbb{I}$*" if λ_f factors through $\mathbb{I}$. The set of all g belonging to $\mathbb{I}$ is called the *p-adic analytic family of Hecke eigenforms of $\mathbb{I}$* (or the Hida family of $\mathbb{I}$).

Let $\lambda : R_\kappa = \mathbb{T} \twoheadrightarrow \mathbb{I}$ be a Λ-algebra onto homomorphism for an integral domain $\mathbb{I}$ finite torsion-free over Λ. Let $\mathbb{T}_{\mathbb{I}} := \mathbb{T} \otimes_\Lambda \mathbb{I}$ and $\widetilde\lambda$ be the composite

$$\widetilde\lambda : \mathbb{T}_{\mathbb{I}} \twoheadrightarrow \mathbb{I} \otimes_\Lambda \mathbb{I} \overset{a \otimes b \mapsto ab}{\underset{\twoheadrightarrow}{\longrightarrow}} \mathbb{I}. \tag{6.11}$$

Since $\mathbb{T}$ is Λ-free, the map $h \mapsto h \otimes 1$ of $\mathbb{T}$ into $\mathbb{T}_{\mathbb{I}}$ is injective. For each $P \in \mathrm{Spec}(\mathbb{I})(W) = \mathrm{Hom}_{W\text{-alg}}(\mathbb{I}, W)$, $\widetilde\lambda$ induces $\mathbb{T}_{\mathbb{I}} \overset{\widetilde\lambda}{\to} \mathbb{I} \overset{P}{\to} W$ by composition.

Write $\rho_P := P \circ \lambda \circ \rho_{\mathbb{T}}$. Then regarding $\rho_{\mathbb{T}} : G \to \mathrm{GL}_2(\mathbb{T}) \subset \mathrm{GL}_2(\mathbb{T}_{\mathbb{I}})$, we have also $\rho_P = P \circ \widetilde\lambda \circ \rho_{\mathbb{T}}$. Then $\det \rho_P$ is a deformation over G of $\det \bar\rho$; so, we have a unique morphism $\iota_P : \Lambda \to W$ such that $\iota_P \circ \kappa = \det(\rho_P)$.

Let $\mathbb{T}_P = \mathbb{T}_{\mathbb{I}} \otimes_{\mathbb{I},P} W$ under the above algebra homomorphism. By associativity of tensor product,

$$\mathbb{T}_P = \mathbb{T} \otimes_\Lambda \mathbb{I} \otimes_{\mathbb{I},P} W \cong \mathbb{T} \otimes_{\Lambda, \iota_P} W$$

and we get $\lambda_P : \mathbb{T}_P \to W$

Modular adjoint p-adic L-function. By construction, we have $\lambda_P :$ $\mathbb{T}_P \to W$ induced by λ. Even if $\iota_P = \iota_{P'}$ (i.e., $\lambda_P, \lambda_{P'} \in \mathrm{Spec}(\mathbb{T})$ is sitting over one point $\iota_P \in \mathrm{Spec}(\Lambda)$), λ_P may be different from $\lambda_{P'}$ (i.e., the fiber of $\mathrm{Spec}(\mathbb{T})$ over ι_P is not a singleton). If λ_P is associated to a Hecke eigenform of weight ≥ 1, we call P a *modular* point, and if further $k \geq 2$, we call P an *arithmetic* point. If $\mathbb{T}_P \otimes_W \mathrm{Frac}(W) = \mathrm{Frac}(W) \oplus (\mathrm{Ker}(\lambda_P) \otimes_W \mathrm{Frac}(W))$ as algebra direct sum, we call P *admissible*. If P is admissible, $C_0(\lambda_P)$ is well defined. If P is arithmetic, it is admissible (see the proof of Lemma 6.2.15).

If $\rho \in \mathcal{D}_\chi(A)$ for W-valued $\chi = \det(\rho_P)$, then $\rho \in \mathcal{D}_\kappa(A)$ and hence $\rho = \phi \circ \boldsymbol{\rho}$ for $\phi : R_\kappa \to A$. By definition, ϕ factors through

$$R_\kappa / R(\det(\boldsymbol{\rho})(g) - \chi(g))_g R = R_\kappa / R(\kappa(g) - \chi(g))_g R = R \otimes_{\Lambda, \chi} W.$$

This shows that $R_\chi = R \otimes_{\Lambda, \chi} W$ for $\chi : \Lambda = W[[\Gamma]] \to W$ induced by χ. Applying this to $\mathbb{T}_P$, we get $R_{\det(\rho_P)} = \mathbb{T}_P$ by $P \circ \lambda$.

Suppose (l) in §6.1.1 for $l \in S \cup \{p\}$. Here is the *congruence number formula* I proved long ago (e.g., [MFG, §5.3.6]) for canonical periods $\Omega_{f, \pm}$ of f:

Theorem 6.2.23. *Assume $R^{ord} \cong \mathbb{T}$. Let $\lambda : \mathbb{T} \twoheadrightarrow \mathbb{I}$ be a surjective Λ-algebra homomorphism for a domain $\mathbb{I}$ containing Λ and $\widetilde{\lambda} : \mathbb{T}_{\mathbb{I}} \to \mathbb{I}$ be its scalar extension to $\mathbb{I}$ as in (6.11). Then there exists $L^{mod} \in \mathbb{I}$ such that $C_0(\widetilde{\lambda}; \mathbb{I}) = \mathbb{I}/(L^{mod})$ and for each admissible $P \in \mathrm{Spec}(\mathbb{I})$, $C_0(\lambda_P; W) = W/P(\lambda(L^{mod}))$ and if P is arithmetic (i.e., $\rho_P \in \rho_f$ for a modular form f of weight ≥ 2), we have*

$$|C_0(\lambda_P; W)| \overset{(1)}{=} |W/\lambda_P(L^{mod})| \overset{(2)}{=} \left| * \frac{L(1, Ad(\rho_f))}{\Omega_{f,+}\Omega_{f,-}} \right|_p^{-1}$$

with an explicit constant $*$ which is equal to 1 except when ρ_{D_l} has some special form of irreducible induced representation or $Ad(\overline{\rho})$ contains $\overline{\omega}_p$ with $k = 2$; see the description of the constant $*$ in Theorem 9.3.2 and a remark after the theorem.

Recalling $f = \sum_{n=1}^\infty \lambda_P(T(n)) q^n$, if f is of weight 2, the period $\Omega_{f, \pm}$ is defined as $\Omega_\pm(\lambda_P; A)$ in §3.2.7 for the discrete valuation ring $A := \mathbb{Q}(\lambda) \cap W \hookrightarrow \overline{\mathbb{Q}}$. We will give a definition of $\Omega_{f, \pm}$ for higher weight in (9.18).

We give a proof of the identity (1) in the following two steps, and we will prove the identity (2) in Chapter 9.

Proof. *Step 1: Existence of L^{mod}.* Write $X^* := \mathrm{Hom}_{\mathbb{I}}(X, \mathbb{I})$ for an $\mathbb{I}$-module X. Let S be the image of $\mathbb{T}_{\mathbb{I}}$ in $\mathfrak{B} \otimes_{\mathbb{I}} \mathrm{Frac}(\mathbb{I})$ for $\mathfrak{B} = \mathrm{Ker}(\widetilde{\lambda})$ in the decomposition $\mathbb{T} \otimes_\Lambda \mathrm{Frac}(\mathbb{I}) = \mathrm{Frac}(\mathbb{I}) \oplus (\mathfrak{B} \otimes_{\mathbb{I}} \mathrm{Frac}(\mathbb{I}))$. Let $\mu : \mathbb{T}_{\mathbb{I}} \to S$ be the projection and put $\mathfrak{A} = \mathrm{Ker}(\mu)$. So we have a split exact sequence $\mathfrak{B} \hookrightarrow \mathbb{T}_{\mathbb{I}} \twoheadrightarrow \mathbb{I}$. By Theorem 6.2.26 (1), a local complete intersection $\mathbb{T}_{\mathbb{I}}$ over $\mathbb{I}$ has a self-dual pairing $(\cdot, \cdot)$ with values in $\mathbb{I}$ such that $(xy, z) = (x, yz)$ for $x, y, z \in \mathbb{T}_{\mathbb{I}}$. Thus $\mathfrak{B}^* \cong \mathbb{T}_{\mathbb{I}}^*/\mathbb{I}^*$, and $\mathbb{I}^* \subset \mathbb{T}_{\mathbb{I}} = \mathbb{T}_{\mathbb{I}}^*$ is a maximal submodule of $\mathbb{T}_{\mathbb{I}}$ on which $\mathbb{T}_{\mathbb{I}}$ acts through $\widetilde{\lambda}$; so, $\mathbb{I}^* = \mathfrak{A}$ inside $\mathbb{T}_{\mathbb{I}}$. This implies $\mathfrak{B}^* \cong S$; so, S is $\mathbb{I}$-free. Applying $\mathbb{I}$-duality, we get a reverse exact sequence

$$\mathbb{I}^* \overset{\hookrightarrow}{\longrightarrow} \mathbb{T}_{\mathbb{I}}^* \overset{\twoheadrightarrow}{\longrightarrow} \mathfrak{B}^*$$

$$\wr\downarrow \qquad\quad \wr\downarrow \qquad\quad \wr\downarrow$$

$$? \longrightarrow \mathbb{T}_{\mathbb{I}} \longrightarrow S$$

This shows $? = \mathfrak{A} \cong \mathbb{I}^* \cong \mathbb{I}$; so, $\mathfrak{A}$ is principal to have $L^{mod} \in \mathbb{I}$ such that $\mathfrak{A} = (L^{mod})$. Note that $C_0(\widetilde{\lambda}) = C_0(\widetilde{\lambda}; \mathbb{I}) = \mathbb{I}/\mathfrak{A}$ (see §5.2.4).

Step 2: Specialization property. We have $\mathfrak{B}^* = S$ and a split exact sequence $\mathfrak{B} \to \mathbb{T}_{\mathbb{I}} \to \mathbb{I}$; so, $\mathfrak{B}$ is an $\mathbb{I}$-direct summand of $\mathbb{T}_{\mathbb{I}}$. Tensoring W over $\mathbb{I}$ via P, $\mathfrak{B} \otimes_{\mathbb{I},P} W \to \mathbb{T}_P \to W$ is exact, and we get $\mathfrak{B}_P = \mathfrak{B} \otimes_{\mathbb{I},P} W = \mathrm{Ker}(\lambda_P)$. Since $\mathbb{T}$ is Λ-free of finite rank, $\mathbb{T}_{\mathbb{I}}$ is $\mathbb{I}$-free of finite rank. Thus $\mathfrak{B}$ is $\mathbb{I}$-projective and hence $\mathbb{I}$-free; so, $S \cong \mathfrak{B}^*$ is $\mathbb{I}$-free. Tensoring W over $\mathbb{I}$ via P, $0 \to \mathfrak{A} \otimes_{\mathbb{I},P} W \to \mathbb{T}_P \to S \otimes_{\mathbb{I},P} W \to 0$ is exact. Thus if P is admissible, $S_P := S \otimes_{\mathbb{I}, \lambda_P} W$ gives rise to the decomposition: $\mathbb{T}_P \otimes_W \mathrm{Frac}(W) = \mathrm{Frac}(W) \oplus (S_P \otimes_W \mathrm{Frac}(W))$. By $\mathfrak{B}_P = \mathfrak{B}_P \otimes_{\mathbb{I},P} W = \mathrm{Ker}(\lambda_P)$, we get $C_0(\lambda_P) = S_P/\mathfrak{B}_P = (S/\mathfrak{B}) \otimes_{\mathbb{I},P} W = C_0(\widetilde{\lambda}) \otimes_{\mathbb{I},P} W = W/\lambda_P(L^{mod})$, as desired. $\qquad\square$

Tensoring $\mathbb{I}$ with the exact sequence of $\mathbb{T}$-modules:

$$(f_1, \ldots, f_r)/(f_1, \ldots, f_r)^2 \overset{f \mapsto df}{\longrightarrow} \Omega_{\Lambda[[T_1, \ldots, T_r]]/\Lambda} \otimes_{\Lambda[[T_1, \ldots, T_r]]} \mathbb{T} \twoheadrightarrow \Omega_{\mathbb{T}/\Lambda}$$

over $\mathbb{T}$, we get an exact sequence

$$\bigoplus_j \mathbb{I} df_j \overset{d \otimes 1 = \lambda(d)}{\longrightarrow} \bigoplus_j \mathbb{I} dT_j \to \Omega_{\mathbb{T}/\Lambda} \otimes_{\mathbb{T}, \lambda} \mathbb{I} \to 0.$$

Since $\mathbb{T}_{\mathbb{I}} = \mathbb{I}[[T_1, \ldots, T_r]]/(f_1, \ldots, f_r)_{\mathbb{I}}$, we have

$$\Omega_{\mathbb{T}_{\mathbb{I}}/\mathbb{I}} \otimes_{\mathbb{T}_{\mathbb{I}}, \widetilde{\lambda}} \mathbb{I} = \bigoplus_j \mathbb{I} dT_j / \bigoplus_j \mathbb{I} df_j = \Omega_{\mathbb{T}/\Lambda} \otimes_{\mathbb{T}, \lambda} \mathbb{I}.$$

They have the same characteristic ideals (and Fitting ideals) by Tate's theorem. Thus in general, we get

$$(\lambda(L_\rho)) = (\lambda(\det(d))) = \operatorname{char}(C_1(\widetilde{\lambda})) \stackrel{\text{Tate}}{=} \operatorname{char}(C_0(\widetilde{\lambda})) = (L^{mod}).$$

Thus, combining Theorems 6.1.11, 6.2.13 and 6.2.23, we obtain the *adjoint class number formula*

Corollary 6.2.24. *Let the assumption and the notation be as in Theorem 6.2.23. Then* $\lambda(L_\rho)/L^{mod} \in \mathbb{I}^\times$ *and for each arithmetic* $P \in \operatorname{Spec}(\mathbb{T})$,

$$|\operatorname{Sel}(Ad(\rho_P))| = |\lambda_P(L_\rho)|_p^{-1} = \left| * \frac{L(1, Ad(\rho_f))}{\Omega_{f,+}\Omega_{f,-}} \right|_p^{-1}.$$

The corollary tells us that $L_{mod} \in \mathbb{I}$ glues (up to units) well to L_ρ so that the image $\lambda(L_\rho)$ of L_ρ in $\mathbb{I}$ is equal to L^{mod} of $\mathbb{I}$ up to units.

As seen in Corollary 5.2.5, $C_1 = C_1(\widetilde{\lambda}) = \mathfrak{B}/\mathfrak{B}^2$ and $C_0 = C_0(\widetilde{\lambda}) = S/\mathfrak{B}$. If $r \le 1$, C_1 is cyclic, by Nakayama's lemma, $\mathfrak{B}$ is generated by an element θ of S. Since $C_1 \cong C_0$ by Tate's theorem and C_0 is $\mathbb{I}$-torsion, θ is a non-zero-divisor of S. Thus the multiplication by θ gives rise to $C_0 \cong C_1$.

6.2.8 A detailed and stronger version of Tate's theorem

Hereafter in this chapter, we allow B to be a reduced local ring in CL_W. Let $R_{/B}$ be a commutative B-algebra free of finite rank over B. Choosing a basis $(r_1, \ldots, r_m)$ of R over B, define the regular representation $\rho : R \to M_m(B)$ by $(ar_1, \ldots, ar_m) = (r_1, \ldots, r_m)\rho(a)$. We put $\operatorname{Tr}_{R/B}(a) = \operatorname{Tr}(\rho(a))$ which is called the trace of R over B. Suppose $i : R \cong \operatorname{Hom}_B(R, B)$ as R-modules (where $a\phi(x) = \phi(ax)$ for $\phi \in \operatorname{Hom}_B(R, A)$ and $a \in R$), and define $\delta_{R/B} \in R$ by $\delta_{R/B}i(1_R) = \operatorname{Tr}_{R/B}$. This element is called the *different* of R/B, whose ideal $(\delta_{R/B})$ is unique independent of i.

Let X be a commutative algebra with identity 1_X. A sequence $f = (f_1, \ldots, f_n) \in X^n$ is called X-*regular* if $x \mapsto f_j x$ is injective on $X/(f_1, \ldots, f_{j-1})$ for all $j = 1, \ldots, n$.

Let A be a reduced local complete intersection over B with presentation $B[[(X)]] := B[[X_1, \ldots, X_r]] \xrightarrow{\alpha} A$ whose kernel is generated by a $B[[(X)]]$-regular sequence $(f) = (f_1, \ldots, f_r)$. So A is free of finite rank over B with trivial nilradical. Writing the image of X_i in A as a_i, plainly $\operatorname{Ker}(A[[(X)]] = B[[(X)]] \otimes_B A \xrightarrow{\alpha \otimes 1} A)$ is generated by a sequence $(g) = (g_1, \ldots, g_r)$ given by $g_i = X_i - a_i$.

Lemma 6.2.25. *The sequence* (g) *is* $A[[(X)]]$-*regular.*

Proof. Note $A[[(X)]]/(g_1, \ldots, g_j) \cong A[[X_{j+1}, \ldots, X_r]]$ for $0 < j < r$. Thus for a minimal prime ideal $\mathfrak{p}$ of A, writing $\bar{g}_i := (g_i \mod \mathfrak{p})$, we find $(A/\mathfrak{p})[[(X)]]/(g_1, \ldots, g_j) \cong (A/\mathfrak{p})[[X_{j+1}, \ldots, X_r]]$. Since $A/\mathfrak{p}$ is an integral domain, it is plain that the multiplication by $\bar{g}_{j+1}$ is injective on $(A/\mathfrak{p})[[X_{j+1}, \ldots, X_r]]$; so, $(\bar{g}) = (\bar{g}_1, \ldots, \bar{g}_r)$ is a $(A/\mathfrak{p})$-regular sequence. Since $A[[X_{j+1}, \ldots, X_r]] \hookrightarrow \prod_{\mathfrak{p}} (A/\mathfrak{p})[[X_{j+1}, \ldots, X_r]]$ for $\mathfrak{p}$ running over minimal primes of A, we find that the multiplication by g_{j+1} is injective on $A[[X_{j+1}, \ldots, X_r]]$, as desired. $\qquad\square$

Write $f_i = \sum_i b_{ij} g_j$ in $A[[(X)]]$. The matrix (b_{ij}) is uniquely determined by (f) and (g) by Lemma 6.2.25. The determinant $\det(b_{ij})_{ij}$ plays a key role in creating an isomorphism $\mathrm{Hom}_B(A, B) \cong A$. We restate the theorem in a different way and after proving it we deduce Theorem 6.2.21 from it:

Theorem 6.2.26. *Let the notation and the assumption be as above. Then*

(1) $\mathrm{Hom}_B(A, B) \cong A$ *as* A-*modules (so, A is Gorenstein);*
(2) *Regard* $\phi' := \phi \otimes 1_{B[[(X)]]}$ *for* $\phi \in \mathrm{Hom}_B(A, B)$ *as an element of*

$$\mathrm{Hom}(A \otimes_B B[[(X)]], B \otimes_B B[[(X)]]) = \mathrm{Hom}(A[[(X)]], B[[(X)]]).$$

Then we have A-*linear* $i : A \cong \mathrm{Hom}_B(A, B)$ *with* $\alpha(i'(\det(b_{ij}))) = 1$ *for* $i' = i(1_A) \otimes_B 1_{B[[(X)]]} \in \mathrm{Hom}(A[[(X)]], B[[(X)]])$. *So we have* $\alpha(\phi'(\det(b_{ij}))) = a$ *for* $a \in A$ *and* $\phi = a \cdot i(1_A)$;
(3) *The different* $\delta_{A/B}$ *with respect to* i *in* (2) *is given by* $\delta_{A/B} = \beta(\det(b_{ij}))$ *for* $\beta = \alpha \otimes 1 : A[[(X)]] = B[[(X)]] \otimes_B A \xrightarrow{\alpha \otimes 1} A$.

The proof of the above theorem is technical though elementary; so, for the first reading, the reader may want to skip the rest of this chapter. Our proof reproduced from [MR70, Appendix] is separated into several steps.

Koszul complexes: First we introduce two B-free resolutions of A, in order to compute $\delta_{A/B}$. We start with a general setting. Let $\mathcal{X}$ be a commutative algebra with identity $1_{\mathcal{X}}$. We now define a complex $K_{\mathcal{X}}^\bullet(h)$ (called *the Koszul complex*) out of an $\mathcal{X}$-regular sequence h (see [CRT, Section 16]). Let $V = \mathcal{X}^r$ with a standard basis $e_1, \ldots, e_r$. We consider the exterior algebra $\wedge^\bullet V = \bigoplus_{j=0}^r (\wedge^j V)$. The graded piece $\wedge^j V$ has a base $e_{i_1, \ldots, i_j} = e_{i_1} \wedge e_{i_2} \wedge \cdots \wedge e_{i_j}$ indexed by sequences $(i_1, \ldots, i_j)$ satisfying $0 < i_1 < i_2 < \cdots < i_j \leq r$. We agree to put $\wedge^0 V = \mathcal{X}$ and $\wedge^j V = 0$ if $j > r$. We define

$\mathcal{X}$-linear differential $d : \bigwedge^j \mathcal{X} \to \bigwedge^{j-1} \mathcal{X}$ by

$$d(e_{i_1} \wedge e_{i_2} \wedge \cdots \wedge e_{i_j}) = \sum_{n=1}^{j} (-1)^{n-1} h_{i_n} e_{i_1} \wedge \cdots \wedge e_{i_{n-1}} \wedge e_{i_{n+1}} \wedge \cdots \wedge e_{i_j}.$$

In particular, $d(e_j) = h_j$ and hence, $\bigwedge^0 V / d(\bigwedge^1 V) = \mathcal{X}/(h)$. Thus, $(K_{\mathcal{X}}^\bullet(h), d)$ is a complex and $\mathcal{X}$-free resolution of $\mathcal{X}/(h)$. We also have

$$d(e_1 \wedge e_2 \wedge \cdots \wedge e_r) = \sum_{j=1}^{n} (-1)^{j-1} h_j e_1 \wedge \cdots \wedge e_{j-1} \wedge e_{j+1} \wedge \cdots \wedge e_r.$$

Suppose now that $\mathcal{X}$ is a $\mathcal{B}$-algebra for a commutative ring $\mathcal{B}$ with identity. For a $\mathcal{B}$-module Y, we have a reversed complex $(\mathrm{Hom}_{\mathcal{B}}(K_{\mathcal{X}}^\bullet(h), Y), d^*)$ for the pull-back differential d^*. Identifying $\bigwedge^{r-1} V$ with V $\mathcal{X}$-linearly by

$$e_1 \wedge \cdots \wedge e_{j-1} \wedge e_{j+1} \wedge \cdots \wedge e_n \mapsto e_j$$

and $\bigwedge^r V$ with $\mathcal{X}$ by $e_1 \wedge e_2 \wedge \cdots \wedge e_r \mapsto 1$, we have, for a $\mathcal{B}$-module Y,

$$\mathrm{Im}(d^* : \mathrm{Hom}_{\mathcal{B}}(\bigwedge^{r-1} V, Y) \to \mathrm{Hom}_{\mathcal{B}}(\bigwedge^r V, Y)) \cong (h) \mathrm{Hom}_{\mathcal{B}}(\mathcal{X}, Y),$$

where $(h) \mathrm{Hom}_{\mathcal{B}}(\mathcal{X}, Y) = \sum_j h_j \mathrm{Hom}_{\mathcal{B}}(\mathcal{X}, Y)$, regarding $\mathrm{Hom}_{\mathcal{B}}(\mathcal{X}, Y)$ as an $\mathcal{X}$-module by $a\phi(x) = \phi(xa)$ $(a, x \in \mathcal{X})$. If $\mathcal{X}$ is an $\mathcal{B}$-algebra free of finite rank over $\mathcal{B}$, $K_{\mathcal{X}}^\bullet(h)$ is a $\mathcal{B}$-free resolution of $\mathcal{X}/(h)$, and

$$\mathrm{Ext}_{\mathcal{B}}^r(\mathcal{X}/(h), Y) = H^r(\mathrm{Hom}_{\mathcal{B}}(K_{\mathcal{X}}^\bullet(h), Y)) \cong \frac{\mathrm{Hom}_{\mathcal{B}}(\mathcal{X}, Y)}{(h) \mathrm{Hom}_{\mathcal{B}}(\mathcal{X}, Y)} \quad (6.12)$$

for any $\mathcal{B}$-module Y (see [MFG, §4.2.4] for the extension functor).

Resolutions: We now suppose that A is free of finite rank over B and $A \cong B[[(X)]]/(f)$ for a regular sequence $(f) = (f_1, \ldots, f_r)$. Write $a_j = (X_j \bmod(f)) \in \mathfrak{m}_A$. We consider $A[[(X)]] = B[[(X)]] \otimes_B A$. Then $A = A[[(X)]]/(X_1 - a_1, \ldots, X_r - a_r)$, and $(g) = (X_1 - a_1, \ldots, X_r - a_r)$ is a regular sequence in $A[[(X)]]$ by Lemma 6.2.25. Since $A[[(X)]]$ is $B[[(X)]]$-free of finite rank, the two complexes $K_{B[[(X)]]}^\bullet(f) \twoheadrightarrow A$ and $K_{A[[(X)]]}^\bullet(g) \twoheadrightarrow A$ are $B[[(X)]]$-free resolutions of A.

We have an $B[[(X)]]$-algebra homomorphism $\Phi : B[[(X)]] \hookrightarrow A[[(X)]]$ given by $\Phi(x) = x \otimes 1$. We extend Φ to $\Phi^\bullet : K_{B[[(X)]]}^\bullet(f) \to K_{A[[(X)]]}^\bullet(g)$ in the following way. Write $f_i = \sum_{j=1}^r b_{ij} g_j$. Then we define $\Phi^1 : K_{B[[(X)]]}^1(f) \to K_{A[[(X)]]}^1(g)$ by $\Phi^1(e_i) = \sum_{j=1}^r b_{ij} e_j$. Then $\Phi^j = \bigwedge^j \Phi^1$. One can check that this map $\Phi^\bullet$ is a morphism of complexes. In particular,

$$\Phi_r(e_1 \wedge \cdots \wedge e_r) = \det(b_{ij}) e_1 \wedge \cdots \wedge e_r. \quad (6.13)$$

Since $\Phi^\bullet$ is the lift of the identity map of A to the $B[[(X)]]$-projective resolutions $K^\bullet_{B[[(X)]]}(f)$ and $K^\bullet_{A[[(X)]]}(g)$, it induces an isomorphism of extension groups computed by $K^\bullet_{A[[(X)]]}(g)$ and $K^\bullet_{B[[(X)]]}(f)$:

$$\Phi^* : H^\bullet(\mathrm{Hom}_{B[[(X)]]}(K^\bullet_{A[[(X)]]}(g), B[[(X)]])) \cong \mathrm{Ext}^j_{B[[(X)]]}(A, B[[(X)]])$$
$$\cong H^\bullet(\mathrm{Hom}_{B[[(X)]]}(K^\bullet_{B[[(X)]]}(f), B[[(X)]])).$$

In particular, identifying $\bigwedge^r B[[(X)]]^r = B[[(X)]]$, we have from (6.12) that

$$H^n(\mathrm{Hom}_{B[[(X)]]}(K^\bullet_{B[[(X)]]}(f), B[[(X)]]))$$
$$= \mathrm{Hom}_{B[[(X)]]}(B[[(X)]], B[[(X)]])/(f)\,\mathrm{Hom}_{B[[(X)]]}(B[[(X)]], B[[(X)]])$$

which is equal to $B[[(X)]]/(f) = A$, and similarly

$$H^n(\mathrm{Hom}_{B[[(X)]]}(K^\bullet_{A[[(X)]]}(g), B[[(X)]]))$$
$$= \frac{\mathrm{Hom}_{B[[(X)]]}(A[[(X)]], B[[(X)]])}{(g)\,\mathrm{Hom}_{B[[(X)]]}(A[[(X)]], B[[(X)]])}.$$

The isomorphism between A and $\frac{\mathrm{Hom}_{B[[(X)]]}(A[[(X)]], B[[(X)]])}{(g)\,\mathrm{Hom}_{B[[(X)]]}(A[[(X)]], B[[(X)]])}$ is induced by Φ_n which is a multiplication by $d = \det(b_{ij})$ (see (6.13)). Thus we have

Lemma 6.2.27. *Assume that A is a local complete intersection over B. Recall the projection $\alpha : B[[(X)]] \twoheadrightarrow A$. We have an isomorphism:*

$$h : \frac{\mathrm{Hom}_{B[[(X)]]}(A[[(X)]], B[[(X)]])}{(X_1 - a_1, \ldots, X_r - a_r)\,\mathrm{Hom}_{B[[(X)]]}(A[[(X)]], B[[(X)]])} \cong A$$

given by $h(\phi) = \alpha(\phi(d))$ for $d = \det(b_{ij}) \in A[[(X)]]$.

Proof of Theorem 6.2.26. We have a base change map:

$$\iota : \mathrm{Hom}_B(A, B) \longrightarrow \mathrm{Hom}_{B[[(X)]]}(B[[(X)]] \otimes_B A, B[[(X)]] \otimes_B B)$$
$$= \mathrm{Hom}_{B[[(X)]]}(A[[(X)]], B[[(X)]]),$$

taking ϕ to $1 \otimes \phi$. The map $\iota(\phi)$ is just applying the original ϕ to coefficients of power series in $B[[(X)]]$. We define $I = h \circ \iota : \mathrm{Hom}_B(A, B) \to A$, which gives the Gorenstein property in the theorem by the following lemma.

Lemma 6.2.28. *Suppose that A is a local complete intersection over A. Then the above map I is an A-linear isomorphism, satisfying $I(\phi) = \alpha(\iota(\phi)(d))$. Thus the ring A is Gorenstein.*

Proof. We first check that I is an A-linear map. Since $I(\phi) = \alpha(\iota(\phi)(d))$, we compute $I(\iota(\phi) \circ b))$ and $rI(\phi)$ for $b \in B[[(X)]]$ and $r = \alpha(b)$. By definition, we see $I(\alpha(b)) = \alpha(\iota(\phi)((r \otimes 1)d))$ and $rI(\phi) = \alpha(b\iota(\phi)(d))$. Thus we need to check $\alpha(\iota(\phi)((r \otimes 1 - 1 \otimes b)d)) = 0$. This follows from:

$$r \otimes 1 - 1 \otimes b \in (g) \quad \text{and} \quad \det(b_{ij})g_i = \sum_i b'_{ij}f_i,$$

where b'_{ij} are the (i, j)-cofactors of the matrix (b_{ij}). Thus I is A-linear. Since $\iota \mod \mathfrak{m}_{B[[(X)]]}$ for the maximal ideal $\mathfrak{m}_{B[[(X)]]}$ of $B[[(X)]]$ is an isomorphism from $\mathrm{Hom}_B((B/\mathfrak{m}_B)^r, B/\mathfrak{m}_B) = \mathrm{Hom}_B(A, B) \otimes_B B/\mathfrak{m}_B$ onto

$$\mathrm{Hom}_{B[[(X)]]}((B[[(X)]]/\mathfrak{m}_{B[[(X)]]})^r, B[[(X)]]/\mathfrak{m}_{B[[(X)]]})$$
$$= \mathrm{Hom}_{B[[(X)]]}(A[[(X)]], B[[(X)]]) \otimes_{B[[(X)]]} B[[(X)]]/\mathfrak{m}_{B[[(X)]]},$$

the map ι is non-trivial modulo $\mathfrak{m}_{A[[(X)]]}$. Thus $I \mod \mathfrak{m}_A$ is non-trivial. Since h in Lemma 6.2.27 is an onto isomorphism,

$$\mathrm{Hom}_{B[[(X)]]}(A[[(X)]], B[[(X)]]) \otimes_{A[[(X)]]} A[[(X)]]/\mathfrak{m}_{A[[(X)]]}$$

is 1-dimensional, and hence $I \mod \mathfrak{m}_A$ is surjective. By Nakayama's lemma, I itself is surjective. Since the target and the source of I are A-free of equal rank, the surjectivity of I tells us its injectivity. $\square$

The following corollary finishes the proof of Theorem 6.2.26:

Corollary 6.2.29. *Suppose that A is a local complete intersection over B. We have $I(\mathrm{Tr}_{A/B}) = \alpha(d)$ for $d = \det(b_{ij})$, and hence $(\delta_{A/B}) = (\alpha(d))$.*

Proof. The last assertion follows from the first by $I(\phi) = \alpha(\iota(\phi)(d))$. To show the first, we choose dual basis $x_1, \ldots, x_r$ of A/B and $\phi_1, \ldots, \phi_r$ of $\mathrm{Hom}_B(A, B)$. Thus for $x \in A$, writing $xx_i = \sum_i a_{ij}x_j$, we have $\mathrm{Tr}_{A/B}(x) = \sum_i a_{ii} = \sum_i \phi_i(xx_i) = \sum_i x_i\phi_i(x)$. Thus $\mathrm{Tr} = \sum_i x_i\phi_i$.

Since x_i is also a base of $A[[(X)]]$ over $B[[(X)]]$, we can write $d = \sum_j b_jx_i$ with $\iota(\phi_i)(d) = b_i$. Then we have

$$I(\mathrm{Tr}_{A/B}) = \sum_i x_iI(\phi_i) = \sum_i x_i\alpha(\iota(\phi_i)(d)) = \sum_i x_i\alpha(b_i) = \alpha(d).$$

This shows the desired assertion. $\square$

6.2.9 Proof of Tate's Theorem 6.2.21

We start with

Lemma 6.2.30. *Let B be a reduced local ring in CL_W and A be an B-algebra. Suppose the following three conditions:*

(1) A *is free of finite rank over* B;
(2) A *is Gorenstein; i.e., we have* $i : \mathrm{Hom}_B(A, B) \cong A$ *as* A-*modules;*
(3) $A \otimes_B \mathrm{Frac}(B) = \mathrm{Frac}(B) \oplus (\mathrm{Ker}(P) \otimes_B \mathrm{Frac}(B))$ *as algebra direct sum for a* B-*algebra homomorphism* $P : A \to B$.

Then we have $C_0 \cong B/P(i(\mathrm{Tr}_{A/B}))B$ *for an* B-*algebra homomorphism* $P : A \to B$. *In particular,* $\mathrm{length}_B C_0$ *is equal to the valuation of* $d = P(i(\mathrm{Tr}_{A/B}))$ *if* B *is a discrete valuation ring.*

Proof. Let $\phi = i^{-1}(1)$. Then $\mathrm{Tr}_{A/B} = \delta\phi$. The pairing $(x, y) \mapsto \phi(x)(y) \in B$ is a perfect pairing over B. Since A is commutative, $(xy, z) = (y, xz)$. Decomposing $A \otimes_B \mathrm{Frac}(B) = \mathrm{Frac}(B) \oplus X$, we have

$$C_0 \cong B/A \cap (\mathrm{Frac}(B) \oplus 0).$$

Then it is easy to conclude that the pairing $(\,,\,)$ induces a perfect B-duality between $A \cap (\mathrm{Frac}(B) \oplus 0)$ and $B \oplus 0$. Thus $A \cap (\mathrm{Frac}(B) \oplus 0)$ is generated by $P(\delta) = P(i(\delta)) = P(i(\mathrm{Tr}_{R/A}))$. $\qquad\square$

The theorem then follows from

Lemma 6.2.31. *Let the notation and the assumption be as in Theorem 6.2.21. Then we have* $\mathrm{Fitt}_B C_0 = \mathrm{Fitt}_B C_1$.

If B is a discrete valuation ring, actually the assertion of the proposition is equivalent to A being a local complete intersection over B [L95].

Proof. Let M be a B-module, and suppose that we have an exact sequence: $B^r \xrightarrow{L} B^r \to M \to 0$ of A-modules. Then we have $\mathrm{Fitt}_B M = \mathrm{Fitt}_B(B/\det(L)B) = (\det(L))$. Since A is reduced, $\Omega_{A/B}$ is a torsion A-module, and hence $\Omega_{A/B} \otimes_A B = C_1$ is a torsion A-module. Since A is a local complete intersection over B, we can write $A \cong B[[X_1, \ldots, X_r]]/(f_1, \ldots, f_r)$. Then by Corollary 5.2.5 (ii), we have the following exact sequence for $J = (f_1, \ldots, f_r)$:

$$J/J^2 \otimes_{B[[X_1,\ldots,X_r]]} B \to \Omega_{B[[X_1,\ldots,X_r]]/B} \otimes_{B[[X_1,\ldots,X_r]]} B \to \Omega_{A/B} \otimes_A B \to 0.$$

This gives rise to the following exact sequence:

$$\bigoplus_j B df_j \xrightarrow{L} \bigoplus_j B dX_j \to C_1 \to 0,$$

where $df_j = f_j \mod J^2$. Since C_1 is a torsion A-module, we see that $\mathrm{length}_B(B/\det(L)B) = \mathrm{length}_B C_1$. Since $g = (X_1 - a_1, \ldots, X_r - a_r)$, we see easily that $\det(L) = \alpha(\lambda(d))$. This combined with Corollary 6.2.29 and Lemma 6.2.30 shows the desired assertion. $\qquad\square$

Chapter 7

Cyclicity of adjoint Selmer groups

We assume $p \geq 3$ in this chapter unless otherwise explicitly mentioned, since many results in this chapter are not valid for $p = 2$. For a number field K, by class field theory, the maximal p-abelian extension $H_{p/K}$ unramified everywhere has Galois group canonically isomorphic to the p-class group $Cl_{K,p}$ of K. So its Pontryagin dual $\mathrm{Hom}(Cl_{K,p}, \mathbb{Q}_p/\mathbb{Z}_p)$ can be defined by

$$\mathrm{Sel}_K = \mathrm{Ker}(H^1(\overline{\mathbb{Q}}/K, \mathbb{Q}_p/\mathbb{Z}_p) \to \prod_{\mathfrak{l}} H^1(I_{\mathfrak{l}}, \mathbb{Q}_p/\mathbb{Z}_p))$$

for $\mathfrak{l}$ running over all primes of K with inertia subgroup $I_{\mathfrak{l}}$. Writing the induced representation $\mathrm{Ind}_K^{\mathbb{Q}} \mathbf{1} = \mathbf{1} \oplus \phi$, we have the celebrated class number formula giving the size $|Cl_K|$ by the integral part of the value $L(1, \phi)$ (Artin L-value) up to a canonical transcendental factor.

We have studied the following question:

When is $\mathrm{Sel}_K^{\vee} \cong Cl_{K,p}$ *cyclic?*

in Chapter 1 when K is cyclotomic, and if cyclicity holds, the structure of Sel_K is determined therefore by the value $L(1, \phi)$. As seen in Chapter 2, a similar result holds also for the value at $s = 2$, but in this case, the group is the cuspidal class group in the geometric and modular setting.

Iwasawa's theory described in Chapter 1 properly belongs to the automorphic theory for GL(1), the Iwasawa algebra (the base ring of the cyclicity in Chapter 1) can be defined as a universal deformation ring of GL(1)-representations (Chapter 5) and also as a Hecke algebra for $\mathbf{G}_m = \mathrm{GL}(1)$ as we briefly touched upon in §3.1. The theory of Kubert–Lang sketched in Chapter 2 is a bridge from the GL(1)-theory to the GL(2)-theory, and the Hecke operators acting on the cuspidal class group span the Eisenstein component ($\cong \Lambda$) of the GL(2) Hecke algebra (and this fact is essential in the proof of the main conjecture by Mazur–Wiles [MW84]). It is natural to expect a cyclicity theory for GL(2) to have $\mathbb{T}$ as its base ring.

Start with the compatible system $\rho_f = \{\rho_{\mathfrak{p}}\}_{\mathfrak{p}}$ of a primitive form f of weight k such that most members of ρ_f satisfy minimality (l). Minimality for most $\mathfrak{p}$ forces f to have potentially good reduction everywhere, and this condition is automatic if $k = 1$. Let $\mathrm{Gal}(\overline{\mathbb{Q}}/\mathbb{Q})$ act on $\mathfrak{sl}_2(W_{\mathfrak{p}})$ by the adjoint action of $\rho_{\mathfrak{p}}$, which results in a 3-dimensional compatible system $Ad(\rho_f) := \{Ad(\rho_{\mathfrak{p}})\}_{\mathfrak{p}}$. We have the formula of $|\,\mathrm{Sel}(Ad(\rho_{\mathfrak{p}}))|$ by the L-value $L(1, Ad(\rho_f))$ (a non-abelian class number formula: Theorem 6.2.23). Since the theory in Chapter 1 is based on such a formula, we want to see if cyclicity of $\mathrm{Sel}(Ad(\rho_{\mathfrak{p}}))$ holds for most of $\mathfrak{p}$ in this case.

Though our main focus in this chapter is the question when $\mathrm{Sel}(Ad(\rho_f))$ is cyclic, if ρ_f is a 2-dimensional Artin representation (i.e., $k = 1$), we obtain explicit formulas of $\big\|\,\mathrm{Sel}(Ad(\rho_f))\big\|_p$ consistent with a p-adic version of Stark's conjecture [G81], [T81] and [T84]. Since $s = 1$ is not critical for the Artin representation ρ_f, for the (weight 1) algebra homomorphism $\lambda_P :$ $\mathbb{T}_\mathfrak{p} \to W$ corresponding to ρ_f, the p-adic L-value $L_\rho(P) := \lambda_P(L_\rho)$ for the algebraic p-adic L-function L_ρ in §6.2.7 is a p-adic (perhaps transcendental) number not directly equal to the complex L-value $L(1, Ad(\rho_f))$. The p-adic L-function L_ρ interpolates $L(1, Ad(\rho_g))$ for weight ≥ 2 forms g running over the analytic family containing f, and hence it is different from the cyclotomic p-adic L-function studied classically.

Suppose $k = 1$. Then the value $L_\rho(P)$ would be factored into the product of the p-adic regulator and the size of the $Ad(\rho_f)$-isotypic component of the class group of the splitting field of $Ad(\rho_f)$. In this 2-dimensional Artin case, the regulator part would be the p-adic logarithm of the $Ad(\rho_f)$-isotypic projection ε of a Minkowski unit. Here a Minkowski unit [M00] is a generator (if it exists) over $\mathbb{Z}_{(p)}[\mathrm{Gal}(F/\mathbb{Q})]$ of the p-localized unit group of $F = F(Ad(\rho_f))$, and $|\log_p(\varepsilon)|_p^{-1}$ is essentially the index of the subgroup generated by ε in the entire $Ad(\rho_f)$-isotypic component of p-local units. Since $\big\|\,\mathrm{Sel}(Ad(\rho_f))\big\|_p = |L_\rho(P)|_p$ by Theorem 6.2.23 and Corollary 6.2.24, we expect a description of $\mathrm{Sel}(Ad(\rho_f))^\vee$ as an extension of the $Ad(\rho_f)$-part of the class group by the local unit group modulo a subgroup generated by ε. This expectation is shown to be true under mild assumptions (see Theorems 7.5.4 and 7.7.3).

In §7.1.2, we give a summary of the results we discuss here and the next chapter, and in the following section we start our discussion in depth of the topic. Gothic letters $\mathfrak{p}, \mathfrak{l}, \mathfrak{q}$ mean prime ideals of a number field or a local field and their Roman counterparts indicate their residual characteristic.

7.1 Basic set-up

Here is a slightly more detailed sketch of what we are going to do and what we have done in the earlier chapters; so, no proofs are given in this section (just short explanation of concepts).

7.1.1 Greenberg's Selmer group

For a number field $K \subset \overline{\mathbb{Q}}$, we take a continuous Galois representation $\phi : \mathrm{Gal}(\overline{\mathbb{Q}}/K) \to \mathrm{GL}_n(W)$ unramified outside a finite set of primes S (for a valuation ring W finite flat over $\mathbb{Z}_p$). We include $p \in S$. Let K^S/K be the maximal extension of K inside $\overline{\mathbb{Q}}$ unramified outside S and ∞, and we put $\mathfrak{G} = \mathfrak{G}_K^S := \mathrm{Gal}(K^S/K)$. Then ϕ factors through $\mathfrak{G}$. Let $\phi^* = \phi \otimes_{\mathbb{Z}_p} \mathbb{Q}_p/\mathbb{Z}_p$ as a discrete $\mathfrak{G}$-module. For a datum $\mathcal{L}$ of subgroup $L_{\mathfrak{q}} \subset H^1(K_{\mathfrak{q}}, \phi^*)$ for each prime $\mathfrak{q}$ of K, we define, following [G94, (4)],

$$\mathrm{Sel}_{\mathcal{L}}(\phi) = \mathrm{Ker}(H^1(\mathfrak{G}, \phi^*) \to \prod_{\mathfrak{q}} H^1(K_{\mathfrak{q}}, \phi^*)/L_{\mathfrak{q}}).$$

If we take $L_{\mathfrak{q}} := \mathrm{Ker}(H^1(K_{\mathfrak{q}}, \phi^*) \to H^1(I_{\mathfrak{q}}, \phi^*))$ for the inertia subgroup $\mathfrak{I}_{\mathfrak{q}}$ of $\mathfrak{q}$ in $\mathfrak{G}$, then $\mathrm{Sel}_{\mathcal{L}}(\phi) = \mathrm{Ker}(H^1(\mathfrak{G}, \phi^*) \to \prod_{\mathfrak{q}} H^1(\mathfrak{I}_{\mathfrak{q}}, \phi^*))$. If ϕ is given by the trivial representation $\mathbf{1}$, as we studied in Chapter 5,

$$\mathrm{Sel}_K(\mathbf{1}) := \mathrm{Sel}_{\mathcal{L}}(\phi_p) \cong Cl_K^* \otimes_{\mathbb{Z}} \mathbb{Z}_p \quad \text{for the above choice of } \mathcal{L}.$$

Supposing that ϕ is a member of a compatible system of Galois representation $\{\phi_{\mathfrak{p}}\}_{\mathfrak{p}}$ with coefficients in a number field T associated to a motive (as described in [MTV]), define the L function of ϕ by $L(s, \phi) = \prod_l \det(1 - \phi_{\mathfrak{p}}(\mathrm{Frob}_l) N_{K/\mathbb{Q}}(l)^{-s})^{-1}$ and assume analytic continuation and functional equation as predicted by Serre [HMI, §1.2.1]. If ϕ is critical (i.e., the $L(s, \phi)$ does not have a pole at $s = 0$ and the Γ-factor of $L(s, \phi)$ and its counterpart of the functional equation are finite at $s = 0$), we expect, as a weak variant of Bloch–Kato conjecture,

$$|\mathrm{Sel}_{\mathcal{L}}(\phi_{\mathfrak{p}})| = \left| \frac{L(0, \phi_{\mathfrak{p}})}{\text{period}} \right|_{\mathfrak{p}}^{-1}$$

for a suitable transcendental factor "**period**" and a suitable data $\mathcal{L}$ (depending on how to define "period"). See [HMI, §1.2.1] for functional equation and periods. Thus at least conjecturally we can compute $|\mathrm{Sel}_{\mathcal{L}}(\phi_{\mathfrak{p}})|$. Our main questions are

- *Is there any way to determine the structure of the module* $\mathrm{Sel}_{\mathcal{L}}(\phi_{\mathfrak{p}})$?

- *Or at least, is there any way to compute the number of generators of* $\mathrm{Sel}_{\mathcal{L}}(\phi_{\mathfrak{p}})$ *over* $O_{T_{\mathfrak{p}}}$?

We try to answer these questions for the compatible system $\phi = Ad(\rho_f)$.

7.1.2 *Summary*

For a given primitive form $f \in S_k(\Gamma_0(C), \psi)$ of conductor C, we have a 2-dimensional compatible system of Galois representation $\rho_f = \{\rho_{f,\mathfrak{l}}\}_{\mathfrak{l}}$ with coefficients in $\mathbb{Z}[f]$. Pick $\rho = \rho_{f,\mathfrak{p}} : \mathfrak{G}_{\mathbb{Q}}^S \to \mathrm{GL}_2(W) \in \rho_f$ for $W = W_{\mathfrak{p}} = \mathbb{Z}[f]_{\mathfrak{p}}$. Write $C^{(p)}$ for the prime-to-p part of C (for the residual characteristic p of $\mathfrak{p}$). Having ρ act on $\mathrm{SL}(2)$-Lie algebra $\mathfrak{sl}_2$ by adjoint (conjugate) action, we get a 3-dimensional representation $Ad(\rho)$. We describe the formula of the order of the p-adic arithmetic cohomology group $\mathrm{Sel}(Ad(\rho))$ (called the adjoint Selmer group) via the L-value $L(1, Ad(\rho_f)) = L(1, Ad(f))$ and explore the question when the Selmer group is cyclic (having one generator) over the coefficient ring? In this summary, we give an outline of our discussion (without details and omitting explicit description of the assumptions). A detailed description starts with the following subsection.

0. Set-up for this subsection.
- Fix a prime $\mathfrak{p}|p \geq 3$, and put $S = \{l|C\} \cup \{p\}$ and $\mathcal{S} := \{l|C^{(p)}\}$. Let $\overline{\rho} = \overline{\rho}_{\mathfrak{p}} = \rho_{f,\mathfrak{p}} \bmod \mathfrak{p} : \mathfrak{G}_{\mathbb{Q}}^S \to \mathrm{GL}_2(\mathbb{F})$: an odd representation unramified outside S ($\mathbb{F} := W/\mathfrak{m}_W$). We assume (m) and (p) in §6.1 (the prime-to-p conductor of $\overline{\rho} = C^{(p)}$).
- Recall the maximal p-profinite extension $F^{(p)}(\overline{\rho})$ of $F(\overline{\rho})$ unramified outside p and $G := \mathrm{Gal}(F^{(p)}(\overline{\rho})/\mathbb{Q})$.
- Fix a decomposition subgroup $D_l \subset G$ of l with its inertia subgroup I_l.
- Assume $\overline{\rho}|_{D_p} = \left(\begin{smallmatrix} \overline{\epsilon} & * \\ 0 & \overline{\delta} \end{smallmatrix}\right)$; $\overline{\delta} \neq \overline{\epsilon}$; $\overline{\delta}$ unramified.
- Fix a character $\chi : G \to B^\times$ with $B = W, \Lambda$ as before and write $(R_\chi, \rho_\chi : G \to \mathrm{GL}_2(R_\chi))$ for the universal pair among p-ordinary deformations with fixed determinant χ over CL_B (if $B = \Lambda$, $\chi = \kappa$). This means $\mathcal{D}_\chi(A) \cong \mathrm{Hom}_{W\text{-alg}}(R_\chi, A)$ for

$$\mathfrak{F}_\chi(A) = \{\rho : G \to \mathrm{GL}_2(A) | \rho \bmod \mathfrak{m}_A = \overline{\rho}, \ \rho|_{D_p} = \left(\begin{smallmatrix} \epsilon & * \\ 0 & \delta \end{smallmatrix}\right) \text{ and } \det \rho = \chi\},$$
$$\mathcal{D}_\chi(A) = \mathfrak{F}_\chi(A)/\Gamma(\mathfrak{m}_A) \cong \mathrm{Hom}_{CNL}(R_\chi, A) \text{ (unramified } \delta).$$

Here, strictly speaking, the condition "$\det \rho = \chi$" means that $\det \rho = \iota_A \circ \chi$ for the structure morphism $W \xrightarrow{\iota_A} A$. We often write $R = R_\chi$.

- The ramification index of $F(\bar{\rho})/\mathbb{Q}$ of any prime is prime to p.
- Define $Ad(\rho)$ by the conjugation action via ρ on $\mathfrak{sl}_2(A) \subset \mathrm{End}_A(\rho)$.

1. Dual adjoint Selmer group. Let $\rho \in \mathfrak{F}_\chi(A)$. Put $U_l := \mathrm{Ker}(H^1(\mathbb{Q}_l, Ad(\rho)^*) \to H^1(\mathfrak{I}_l, Ad(\rho)^*))$ if $l \neq p$ for the inertia subgroup $\mathfrak{I}_l$ of $\mathrm{Gal}(\overline{\mathbb{Q}}_l/\mathbb{Q}_l)$. For p, choose $\rho|_{D_p} = \left(\begin{smallmatrix} \epsilon & * \\ 0 & \delta \end{smallmatrix}\right)$ with δ unramified and $\delta \mod \mathfrak{m} = \bar{\delta}$. Define $U_p = F_-^+ H^1(?, Ad(\rho)^*)$ for $? = \overline{\mathbb{Q}}_p/\mathbb{Q}_p$ or D_p for the subgroup spanned by classes of 1-cocycles upper triangular over $\mathrm{Gal}(\overline{\mathbb{Q}}_p/\mathbb{Q}_p)$ and upper nilpotent over $\mathfrak{I}_p$. Recall

$$\mathrm{Sel}(Ad(\rho)) \overset{(*)}{:=} \mathrm{Ker}(H^1(G, Ad(\rho)^*) \xrightarrow{\mathrm{Res}} \frac{H^1(D_p, Ad(\rho)^*))}{F_-^+ H^1(D_l, Ad(\rho)^*)})$$

$$\cong \mathrm{Ker}(H^1(\mathfrak{G}_\mathbb{Q}^S, Ad(\rho)^*) \xrightarrow{\Pi_l \, \mathrm{Res}} \prod_{l|Cp} H^1(\overline{\mathbb{Q}}_l/\mathbb{Q}_l, Ad(\rho)^*))/U_l). \quad (7.1)$$

In the original definition $(*)$, by minimality (m), the order of $I_l \subset D_l$ is prime to p, and hence unramifiedness at $l \in S$ is automatic. The second isomorphism comes from the fact that any Selmer cocycle on $\mathfrak{G}_\mathbb{Q}^S$ factors through G as $Ad(\rho)^*$ is p-torsion. Since Tate duality is formulated with respect to $\mathfrak{G}_\mathbb{Q}^S$, the second definition is useful when we need Tate duality.

Define the **dual Selmer group** $\mathrm{Sel}^\perp(Ad(\rho)(1))$ replacing U_l by its orthogonal complement $U_l^\perp$ under local Tate duality:

$$\mathrm{Sel}^\perp(Ad(\rho)) := \mathrm{Ker}(H^1(\mathfrak{G}_\mathbb{Q}^S, Ad(\rho)^*) \xrightarrow{\Pi_l \, \mathrm{Res}} \prod_{l|Cp} H^1(\overline{\mathbb{Q}}_l/\mathbb{Q}_l, Ad(\rho)^*)/U_l^\perp).$$

$$(7.2)$$

2. Number of generators of R. As we have seen in §6.1.4,

$$\boxed{t_R^* := \mathfrak{m}_R/\mathfrak{m}_R^2 + \mathfrak{m}_W = \Omega_{R/W} \otimes_R \mathbb{F} \cong \mathrm{Sel}(Ad(\bar{\rho}))^\vee.}$$

Here "$\vee$" indicates Pontryagin dual. So the number of generators of $R_{/W}$ is $r_0 := \dim_\mathbb{F} \mathrm{Sel}(Ad(\bar{\rho}))$. More generally, by Theorem 6.1.11,

$$\boxed{\Omega_{R/W} \otimes_{R,\varphi} A \cong \mathrm{Sel}(Ad(\rho))^\vee} \text{ (Selmer control)}$$

for all $\rho \in \mathfrak{F}_\chi(A)$ with $\varphi \circ \boldsymbol{\rho} \cong \rho$ (so, under the identification of $R \cong \mathbb{T}$, $\varphi = \lambda$ if ρ is associated to f). By tensoring with respect to the specialization morphism φ over the deformation ring, we recover the specialized Selmer group without any error (an exact control theorem of this type is rare to find for arithmetic cohomology groups).

There is a generalization by Greenberg and Wiles of the reflection theorem of Leopoldt that appeared as Theorem 1.8.1 in Iwasawa's cyclicity theory. They proved the following fact without assuming (p) and (m):

Theorem 7.1.1. *Suppose either $p|C(\psi)$ or $k \geq 3$. Then*

$$r_0 := \dim_{\mathbb{F}} \mathrm{Sel}(Ad(\overline{\rho})) \leq \dim_{\mathbb{F}} \mathrm{Sel}^{\perp}(Ad(\overline{\rho})(1)) =: r.$$

The right-hand-side is often **computable** by Kummer theory (which played an important role in Chapters 1 and 2). Though $\mathrm{Sel}^{\perp}(Ad(\overline{\rho})(1))$ is called the dual Selmer group, it is not always the $\mathbb{F}$-dual of $\mathrm{Sel}^{\perp}(Ad(\overline{\rho}))$, as the reflection theorems are inequality. Though the inequality is strict from time to time, in the abelian case we have an optimal bound of the difference (e.g., [ANT, Theorems 2.116-117]). It is interesting to know what is an optimal bound of the failure $r_1 - r_0$ in Theorem 7.1.1. Even if $p \nmid C$ and $k = 2$, if we take the smaller functor $\mathcal{D}_{\chi}^{fl}$, the reflection theorem Theorem 7.1.1 is still valid. We prove this theorem as Theorem 7.2.3 later.

3. Presentation Theorem: $\mathbb{T} \cong \frac{W[[X_1,\ldots,X_r]]}{(S_1,\ldots,S_r)}$. On the way to prove "$R_{\chi} = \mathbb{T}_{\chi}$", Taylor–Wiles got the presentation (Theorem 6.2.13). They proved the presentation originally for $r = \dim_{\mathbb{F}} \mathrm{Sel}^{\perp}(Ad(\overline{\rho})(1))$.

On the other hand, the minimal number of generators of $R = \mathbb{T}$ is given by the dimension r_0 of its co-tangent space $\mathbb{F}$-dual to $\mathrm{Sel}(Ad(\overline{\rho}))$ (Lemma 6.1.2). By a general ring theory (for example, Matsumura's book [CRT, Theorem 21.2 (ii)]), we can reduce the number of variables to $r_0 \leq r$:

$$\mathbb{T} \cong \frac{W[[T_1,\ldots,T_{r_0}]]}{(s_1,\ldots,s_{r_0})} \quad \text{(local complete intersection over } W\text{)}.$$

This implies $|C_0| = |C_1|$ by Tate, and

$$\boxed{\mathrm{Sel}(Ad(\rho))^{\vee} \cong C_1 = \Omega_{\mathbb{T}/W} \otimes_{\mathbb{T},\varphi} A = \frac{A \cdot dT_1 + \cdots + A \cdot dT_{r_0}}{A \cdot ds_1 + \cdots + A \cdot ds_{r_0}}.}$$

This type of presentation holds in quite a general setting of adjoint Selmer groups as long as we have "$R = \mathbb{T}$" theorem (e.g. [PAM, Chapter 4]).

4. Cyclicity: When is $r \leq 1$? Let $F := F(Ad(\overline{\rho}))$ with integer ring O and $\overline{G} := \mathrm{Gal}(F/\mathbb{Q}) \cong \mathrm{Im}(Ad(\overline{\rho}))$. By Kummer theory, $\mathrm{Sel}^{\perp}(Ad(\overline{\rho})(1))$ (restricted to the stabilizer $\mathcal{H}$ of F in $\overline{G}$) is generated by Kummer cocycle $u(g) = u_{\xi}(g) = \sqrt[p]{\xi}^{(g-1)}$ for $\xi \in F^{\times}$ **very unramified**. This means that u comes from a unit $\xi \in O^{\times}$ (in particular, it is unramified outside p). Let $\widehat{O}^{\times} = O^{\times} \otimes_{\mathbb{Z}} \mathbb{Z}_p$. Assume $\boxed{\widehat{O}^{\times} = \mathbb{Z}_p[\overline{G}]\varepsilon}$ (cyclicity of $\widehat{O}^{\times}$ over $\mathbb{Z}_p[\overline{G}]$) for a unit $\varepsilon \in O^{\times}$ which is implied by $\boxed{p \nmid |\overline{G}|}$ and $\mu_p(O) = \{1\}$ (i.e., $\overline{\rho}$ is a reduction modulo $p \gg 1$ of an **Artin representation** ρ). The unit ε is called a (p-optimal) *Minkowski unit*. It is hard to know about Cl_F;

so, we assume $\boxed{p \nmid |Cl_F[Ad]|}$ for $Ad(\bar{\rho})$-isotypical component $Cl_F[Ad] :=$ $Cl_F \otimes_{\mathbb{Z}[\overline{G}]} Ad(\bar{\rho})$. As $u = u_\xi$ for $\xi \in O^\times$, **cyclicity** is implied by

$$\boxed{\dim_{\mathbb{F}} \text{Sel}^\perp(Ad(\bar{\rho})(1)) \leq \dim_{\mathbb{F}} \text{Hom}_{\mathbb{F}[\overline{G}]}(O^\times \otimes \mathbb{F}, Ad(\bar{\rho})) =: r_1}.$$

Without $p \nmid |Cl_F[Ad]|$ but assuming that $\rho = \rho_f$ is Artin, if $r_1 \leq 1$, we get an exact sequence for ρ,

$$\text{Hom}_{\overline{G}}(Cl_F, Ad(\rho)^*) \hookrightarrow \text{Sel}(Ad(\rho)) \twoheadrightarrow \text{Hom}_{\mathbb{Z}_p}(\widehat{O}_{\mathfrak{p}}^\times[\delta^{-1}\epsilon]/\langle \varepsilon_{\delta^{-1}\epsilon} \rangle, W^\vee),$$

where and $\varepsilon_{\epsilon\delta^{-1}}$ is the projection of ε in the $\epsilon\delta^{-1}$-eigenspace $\widehat{O}_{\mathfrak{p}}^\times[\epsilon\delta^{-1}] \subset \widehat{O}_{\mathfrak{p}}^\times$ for the prime $\mathfrak{p}|p$ associated to D_p.

5. Reaching cyclicity by Dirichlet's unit theorem.

Theorem 7.1.2. *Assume* $(O^\times \otimes_{\mathbb{Z}} \mathbb{Z}_p) = \mathbb{Z}_p[\overline{G}]\varepsilon$ *or* $p \nmid |\overline{G}|$. *Then we have* $\dim_{\mathbb{F}} \text{Hom}_{\mathbb{F}[\overline{G}]}(O^\times \otimes \mathbb{F}, Ad(\bar{\rho})) \leq \dim_{\mathbb{F}} Ad(\bar{\rho})^{c=1} = 1$.

By the proof of Dirichlet's unit theorem, for the subgroup C generated by a complex conjugation c, $(O^\times \otimes_{\mathbb{Z}} \mathbb{Q}) \oplus \mathbb{Q} \cong \mathbb{Q}[\overline{G}/C] = \text{Ind}_C^{\overline{G}} \mathbb{Q}$, and hence

$$\mathbb{Z}_p[\overline{G}/C] \hookrightarrow (O^\times \otimes_{\mathbb{Z}} \mathbb{Z}_p) \oplus \mathbb{Z}_p \hookrightarrow \mathbb{Z}_p[\overline{G}/C] \cong \text{Ind}_C^{\overline{G}} \mathbb{Z}_p.$$

Assuming $(O^\times \otimes_{\mathbb{Z}} \mathbb{Z}_p) = \mathbb{Z}_p[\overline{G}]\varepsilon$, the above inclusions are isomorphisms, and by Shapiro's lemma in §5.3.4,

$$\text{Hom}_{\mathbb{Z}_p[\overline{G}]}(O^\times \otimes_{\mathbb{Z}} \mathbb{Z}_p, Ad(\bar{\rho})) \hookrightarrow \text{Hom}_{\mathbb{Z}_p[\overline{G}]}(\text{Ind}_C^{\overline{G}} \mathbb{Z}_p, Ad(\bar{\rho}))$$

$$= \text{Hom}_{\mathbb{Z}_p[C]}(\mathbb{Z}_p, Ad(\bar{\rho})) \cong Ad(\bar{\rho})^{c=1} \text{ (the } c\text{-fixed subspace).}$$

Since $Ad(\bar{\rho})(c) \sim \text{diag}[-1, 1, -1]$, we get $\dim_{\mathbb{F}} \text{Sel}(Ad(\bar{\rho})(1)) \leq 1$.

6. Questions towards general cyclicity.

Starting the compatible system $\{\rho_{\mathfrak{p}}\}_{\mathfrak{p}}$ associated to a cusp form f, if $F := F(Ad(\bar{\rho}_{\mathfrak{p}}))$ for $\bar{\rho}_{\mathfrak{p}} = \rho_{\mathfrak{p}} \mod \mathfrak{p}$ is independent of p (for almost all p), $p \nmid |Cl_F|$ gives a condition for cyclicity (e.g., when ρ_f is an Artin representation). Assuming $\bar{\rho}$ comes from an Artin representation, we prove cyclicity of $\text{Sel}(Ad(\rho_{\mathfrak{p}}))^\vee$ over W, which implies cyclicity of $\text{Sel}(Ad(\rho))^\vee$ over $\mathbb{T}$ (even if $\text{Sel}(Ad(\rho_{\mathfrak{p}}))$, $\text{Sel}(Ad(\rho))$ and $\mathbb{T}$ depend on $\mathfrak{p}$).

In the general non-Artin case (of weight $k \geq 2$), we ask an open question:

Is $\mathfrak{p} \nmid |Cl_F[Ad]|$ *for most* $\mathfrak{p}$ *(even if* F *depends on* $\mathfrak{p}$*)?*

Another ingredient we need to prove cyclicity of $\text{Sel}(Ad(\rho))$ and $\text{Sel}(Ad(\rho))$ is cyclicity of $O^\times \otimes_{\mathbb{Z}} \mathbb{Z}_p$ over $\mathbb{Z}_p[\overline{G}]$; so, we also ask:

Is $O^\times \otimes_{\mathbb{Z}} \mathbb{Z}_p$ *cyclic as a* $\mathbb{Z}_p[\overline{G}]$-*module for most of* $\mathfrak{p}$*?*

The cyclicity of $O^\times \otimes_{\mathbb{Z}} \mathbb{Z}_p$ over $\mathbb{Z}_p[\overline{G}]$ implies that the dimension of the dual Selmer group to be defined by (7.3) is equal to 1. This dimension gives an upper bound of the number of the generator of $\mathrm{Sel}(Ad(\rho))$ by Greenberg–Wiles reflection theorem (Theorem 7.2.3) similar to Iwasawa's treatment in §1.8. Since this is an upper bound, one might be able to circumvent (see a different approach to cyclicity to be described in §8.1). If $\mathfrak{p}|(L(1, Ad(f))/\Omega_f^+\Omega_f^-)$ and the above questions are affirmative, $\mathrm{Sel}(Ad(\rho))^\vee \neq 0$ is cyclic over A for every $\rho \in \mathfrak{F}_\chi(A)$.

In any case, if $k \geq 2$, $|L(1, Ad(f))/\Omega_f^+\Omega_f^-|_{\mathfrak{p}} = 1$ for almost all $\mathfrak{p}$; so, $\mathrm{Sel}(Ad(\rho_f)) = 0$ for almost all ordinary $\mathfrak{p}$ (Theorem 6.2.7). Therefore cyclicity for almost all $\mathfrak{p}$ ordinary for f is trivially true if $k \geq 2$. This is because $s = 1$ is critical for $L(s, Ad(\rho_f))$ (so, the p-adic L interpolating the L-value over the slope 0 p-adic analytic family of f exactly reflects the size of the entire Selmer group not just its p-primary part). From this, we could guess that pure cyclicity holds for almost all ordinary primes even when one starts with a cusp form of weight 1 (i.e., a compatible system of an Artin representation). If $k = 1$, $s = 1$ is no longer critical for $Ad(\rho_f)$; so, the p-adic adjoint L-value has a purely p-local component (supposed to be the motivic p-adic regulator), and the adjoint p-Selmer group could be non-trivial for infinitely many primes p. Therefore we study the case of odd Artin representations (i.e., $k = 1$) carefully.

7.2 Upper bound of the number of Selmer generators

By Kummer theory, we give an upper bound of the dimension $\dim t_{R^{ord}/\Lambda} = \dim t_{R_\chi/W}$ by the dimension of the dual Selmer group, which turns out to be often optimal.

7.2.1 *Local theory*

We summarize facts from l-adic class field theory [LCF, VI-VII] (or [CFN, III]). Let $K_{/\mathbb{Q}_l}$ be a finite extension with integer ring O_K. Take an algebraic closure $\overline{K}$ of K. Let ϖ be a generator of $\mathfrak{m}_{O_K}$. Write $D := \mathrm{Gal}(\overline{K}/K)$ fixing an algebraic closure $\overline{K}/K$. Let $D \triangleright I$ be the inertia subgroup and D^{ab} be its maximal continuous abelian quotient with inertia subgroup I_{ab}.

- $x \mapsto [x, K] : K^\times \hookrightarrow D^{ab}$ (the local Artin symbol);
- $[\varpi, K]$ modulo I_{ab} is the Frobenius element Frob;
- $K^\times/(K^\times)^m = K^\times \otimes_{\mathbb{Z}} \mathbb{Z}/m\mathbb{Z} \cong D^{ab}/mD^{ab}$ by Artin symbol ($0 < m \in \mathbb{Z}$);

- $O_K^\times \cong I_{ab}$ by Artin symbol.

We summarize facts from local cohomology.
- $inv : H^2(K, \mu_m(\overline{K})) \cong \mathbb{Z}/m\mathbb{Z}$ (the invariant map);
- $H^1(K, \mu_m) \cong K^\times/(K^\times)^m$ (Kummer theory valid for any field $K \supset \mathbb{Q}$).
This follows from the long exact sequence of $H^?(M) := H^?(K, M)$ associated to $\mu_m(\overline{K}) \hookrightarrow \overline{K}^\times \xrightarrow{x \mapsto x^m} \overline{K}^\times$:

$$H^0(\overline{K}^\times) \xrightarrow{x \mapsto x^m} H^0(\overline{K}^\times) \longrightarrow H^1(\mu_m) \longrightarrow H^1(\overline{K}^\times) \stackrel{(*)}{=} 0$$

$$K^\times \xrightarrow[x \mapsto x^m]{} K^\times \longrightarrow K^\times/(K^\times)^m,$$

where the vanishing $(*)$ follows from Hilbert's theorem 90.

Let $M^*(1) := \mathrm{Hom}(M, \mu_m(\overline{K}))$ (called Tate dual) with a Galois action given by $g \cdot \phi(x) = g(\phi(g^{-1}x))$ for any finite (continuous) D-module M killed by $0 < m \in \mathbb{Z}$. Then $M^*(1) \otimes_{\mathbb{Z}/m\mathbb{Z}} M \ni \phi \otimes x \mapsto \phi(x) \in \mu_m$ is a $\mathbb{Z}[D]$-morphism inducing a cup product pairing

$$H^r(M^*(1)) \times H^{2-r}(M) \to H^2(\mu_m) \xrightarrow{inv} \mathbb{Z}/m\mathbb{Z}.$$

Theorem 7.2.1 (Tate). *Cohomological dimension of D is equal to 2 and the above pairing is perfect for $r = 0, 1, 2$.*

If $M = \mu_m(\overline{K})$, by definition $\mu_m = (\mathbb{Z}/m\mathbb{Z})^*(1)$. We know $H^1(\mu_m) = K^\times/(K^\times)^m$ and $H^1(\mathbb{Z}/m\mathbb{Z}) = \mathrm{Hom}(D^{ab}/mD^{ab}, \mathbb{Z}/m\mathbb{Z})$. By local class field theory, $D^{ab}/mD^{ab} \cong K^\times/(K^\times)^m$; so, the duality follows. One can deduce a general proof from this special case by restricting to $\mathrm{Gal}(\overline{K}/F(M))$ for the splitting field $F(M)$ of M (see [MFG, Theorem 4.43]).

7.2.2 Another example of local Tate duality

Consider $\mathrm{Hom}(\mathrm{Frob}^{\widehat{\mathbb{Z}}}, M) \subset H^1(K, M)$ for a finite $\mathbb{Z}/m\mathbb{Z}$-module M on which D acts trivially. Here Frob is the Frobenius element in D/I.

Lemma 7.2.2. *The orthogonal complement of $\mathrm{Hom}(\mathrm{Frob}^{\widehat{\mathbb{Z}}}, M) \subset H^1(K, M)$ in the dual $H^1(K, M^*(1)) = K^\times \otimes_{\mathbb{Z}} M$ is given by $H^1_{fl}(K, M^*(1)) := O_K^\times \otimes_{\mathbb{Z}} M$. In particular, the Tate duality between $H^1(K, \mu_m)$ and $H^1(K, \mathbb{Z}/m\mathbb{Z})$ gives rise to the tautological duality between $\mathrm{Frob}^{\widehat{\mathbb{Z}}}/m\,\mathrm{Frob}^{\widehat{\mathbb{Z}}}$ and $\mathrm{Hom}(\mathrm{Frob}^{\widehat{\mathbb{Z}}}, \mathbb{Z}/m\mathbb{Z})$.*

The cohomology group $H^1_{fl}(K, M^*(1))$ is called *flat cohomology* groups, as it is computed in the category of finite-flat group schemes [ADT, III].

Proof. The result for general M follows from extending scalar to M; so, we may assume $M = \mathbb{Z}/m\mathbb{Z}$. The last statement follows from the construction of pairing between $H^1(K, \mu_m)$ and $H^1(K, \mathbb{Z}/m\mathbb{Z})$ described in §7.2.1. By the inflation-restriction sequence (Theorem 4.2.7), the sequence

$$0 \to \mathrm{Hom}(D/I, \mathbb{Z}/m\mathbb{Z}) \to \mathrm{Hom}(D, \mathbb{Z}/m\mathbb{Z}) \to \mathrm{Hom}(I, \mathbb{Z}/m\mathbb{Z}) \to 0$$

is exact for the inertia group $I \lhd D$. Since $D/I = \mathrm{Frob}^{\widehat{\mathbb{Z}}}$, we have the following commutative diagram with exact rows:

$$
\begin{array}{ccccc}
(O_K^\times/(O_K^\times)^m) & \overset{\hookrightarrow}{\longrightarrow} & (K^\times/(K^\times)^m) & \overset{\twoheadrightarrow}{\longrightarrow} & \mathrm{Frob}^{\widehat{\mathbb{Z}}}/\mathrm{Frob}^{m\widehat{\mathbb{Z}}} \\
\wr \downarrow & & \wr \downarrow & & \wr \downarrow \\
H^1(I, \mathbb{Z}/m\mathbb{Z})^\vee & \underset{\hookrightarrow}{\longrightarrow} & H^1(K, \mathbb{Z}/m\mathbb{Z})^\vee & \underset{\twoheadrightarrow}{\longrightarrow} & H^1(D/I, \mathbb{Z}/m\mathbb{Z})^\vee.
\end{array}
$$

Since the image of I in D^{ab} is given by $O_K^\times$, the result follows. $\quad\square$

7.2.3   *An adjoint reflection theorem*

By trace pairing $(x, y) = \mathrm{Tr}(xy)$ the Galois modules $ad(\bar\rho)$ and $Ad(\bar\rho)$ are self dual (if $p > 2$); so, $ad(\bar\rho)^*(1) = ad(\bar\rho)(1)$ and $Ad(\bar\rho)^*(1) = Ad(\bar\rho)(1)$. The dual Selmer group of $Ad(\bar\rho)$ can be redefined as follows:

$$\mathrm{Sel}^\perp(Ad(\bar\rho)(1)) := \mathrm{Ker}(H^1(\mathfrak{G}_\mathbb{Q}^S, Ad(\bar\rho)(1)) \to \prod_{l \in S \cup \{p\}} \frac{H^1(\mathbb{Q}_l, Ad(\bar\rho)(1))}{\mathcal{D}_{\chi,l}(\mathbb{F}[\varepsilon])^\perp}).$$

$$(7.3)$$

Here "$\perp$" indicates the orthogonal complement under the Tate duality.

Assume that $\bar\rho = \rho_{f,\mathfrak{p}} \mod \mathfrak{p}$ for a primitive form $f \in S_k(\Gamma_0(C), \psi)$. To state the non-abelian reflection theorem, we need to modify the mod p adjoint Selmer group slightly in the exceptional case when $\bar\rho|_{D_p}$ satisfies $\bar\epsilon/\bar\delta = \bar\omega_p$ for the Teichimüller character $\bar\omega_p$, $k = 2$ and $p \nmid C$ (the flat case). In this case, we need to replace $\mathcal{D}_\chi$ by the smaller flat deformation functor $\mathcal{D}_\chi^{fl}$ defined in (6.2). We then define flat Selmer groups $\mathrm{Sel}_{fl}(Ad(\bar\rho)) :=$ $\mathcal{D}_\chi^{fl}(\mathbb{F}[\varepsilon])$ and $\mathrm{Sel}_{fl}^\perp(Ad(\bar\rho))$ replacing $\mathcal{D}_{\chi,l}$ by $\mathcal{D}_{\chi,l}^{fl}$ in (7.3).

We have the following bound due to R. Greenberg and A. Wiles:

Theorem 7.2.3. *If either* $k \geq 3$ *or* $\bar\epsilon/\bar\delta \neq \bar\omega_p$ *or* $p|C$,

$$\dim_\mathbb{F} \mathrm{Sel}(Ad(\bar\rho)) \leq \dim_\mathbb{F} \mathrm{Sel}^\perp(Ad(\bar\rho)(1)).$$

If $\bar\rho|_{D_p}$ *is flat (so,* $p \nmid C$ *and* $k = 2$), *we have*

$$\dim_\mathbb{F} \mathrm{Sel}_{fl}(Ad(\bar\rho)) \leq \dim_\mathbb{F} \mathrm{Sel}_{fl}^\perp(Ad(\bar\rho)(1)).$$

The proof of this theorem is more advanced than the material we presented so far; so, for the first reading, the reader who does not have good knowledge of Tate duality can skip the proof and move on to §7.2.5 after the proof.

To prove the theorem in the following subsection, we prepare

Lemma 7.2.4. *Let* $\phi : \mathrm{Gal}(\overline{\mathbb{Q}}_p/\mathbb{Q}_p) \to \mathbb{F}[\varepsilon]^\times$ *(for the dual number* ε*) be a character. Put* $\overline{\phi} = (\phi \mod (\varepsilon))$. *Then we have*

$$\dim_{\mathbb{F}}(\mathrm{Ker}(H^1(\phi) \xrightarrow{\mathrm{mod} \ (\varepsilon)} H^1(\overline{\phi}))) \leq \dim_{\mathbb{F}} H^1(\overline{\phi}),$$

where $H^q(?) = H^q(\mathbb{Q}_p, ?)$.

Proof. By the long exact sequence attached to $\overline{\phi} \to \phi \xrightarrow{\mathrm{mod} \ (\varepsilon)} \overline{\phi} \to 0$, we have an exact sequence:

$$0 \to H^1(\overline{\phi}) \to H^1(\phi) \to H^1(\overline{\phi}) \to H^2(\overline{\phi}) \to H^2(\phi) \to H^2(\overline{\phi}) \to 0. \quad (7.4)$$

From this, the estimate of the dimension follows. □

Next, we quote the following result from [W95, Proposition 1.9]:

Proposition 7.2.5. *Assume* (p). *Let* $\xi_p = \overline{\epsilon}\overline{\delta}^{-1}$ *on* D_p. *We have*

(1) $\dim_{\mathbb{F}} \mathcal{D}_{\chi,p}(\mathbb{F}[\varepsilon]) \leq 1 + \dim_{\mathbb{F}} H^1(\mathbb{Q}_p, \xi_p)$ *without assuming* $p \notin S$;
(2) *If* $p \notin S$, $\dim_{\mathbb{F}} \mathcal{D}_{\chi,p}(\mathbb{F}[\varepsilon]) \leq \dim_{\mathbb{F}} H^1(\mathbb{Q}_p, \xi_p)$. *Note that* $\xi_p = \overline{\omega}_p$ *in this case; so,* $H^1(\mathbb{Q}_p, \mathbb{F}(1)) = H^1(\mathbb{Q}_p, \xi_p)$.

Since we only use this result when p is tamely ramified in $F(\overline{\rho})$, we give a proof assuming this (see [HMI, Proposition 3.20] for the general case). We shall repeat the proof by Fujiwara in [F06]. In this proof, we work with $\mathfrak{G} = \mathfrak{G}_{\mathbb{Q}}^S$ as Tate duality is defined over $\mathfrak{G}$ (not G), and as before, we write $H^q(?) = H^q(\mathbb{Q}_p, ?)$.

Proof. The tangent space $\mathcal{D}_{\chi,p}(\mathbb{F}[\varepsilon])$ of $\mathcal{D}_{\chi,p}$ classifies extensions $\varepsilon_p \hookrightarrow \rho \twoheadrightarrow \delta_p$ with coefficients in $\mathbb{F}[\varepsilon]$ such that

- $\delta_p \mod(\varepsilon) \equiv \overline{\delta}$,
- δ_p is unramified,
- $\varepsilon_p \delta_p = \iota_{\mathbb{F}[\varepsilon]} \circ \chi$.

Let $\mathbb{Q}_p^{unr}/\mathbb{Q}_p$ be the maximal unramified extension $\overline{\mathbb{Q}}_p^{I_p}$ with integer ring O_p^{unr}, and taking a modular form f whose Galois representation ρ_f is in $\mathcal{D}_\chi(W)$, write $\rho_f|_{D_p} = \begin{pmatrix} \eta_2 & * \\ 0 & \eta_1 \end{pmatrix}$. We consider the map $\pi_{unr} : \mathcal{D}_{\chi,p}(\mathbb{F}[\varepsilon]) \to \mathrm{Hom}(\mathrm{Gal}(\mathbb{Q}_p^{unr}/\mathbb{Q}_p), \mathbb{F}) \cong \mathbb{F}$ given by $\pi_{unr}(\varepsilon_p \hookrightarrow \rho \twoheadrightarrow \delta_p) = \phi$ for ϕ with

$\delta_p \eta_1^{-1}(\sigma) = 1 + \phi(\sigma)\varepsilon \in \mathbb{F}[\varepsilon]^\times$. We need to determine the fiber of π_{unr}. The fiber at ϕ is determined by fixing δ_p to be $\delta_\phi := (1 + \phi\varepsilon)\eta_2$; so, ϵ_p has to be equal to $\chi\delta_\phi^{-1}$. Thus $\pi_{unr}^{-1}(\phi) \subset \mathrm{Ext}^1(\delta_\phi, \chi\delta_\phi^{-1}) \cong H^1(\chi\delta_\phi^{-2})$. Since the reduction mod (ε) of each element of $\pi_{unr}^{-1}(\phi)$ is a single element $\overline{\rho}$, we need to compute the dimension of the fiber of $H^1(\chi\delta_\phi^{-2})$ over $H^1(\xi_p)$. The desired estimate follows from Lemma 7.2.4 applied to $\phi = \chi\delta_\phi^{-2}$.

Now suppose $p \notin S$. We write the class of ρ in $H^1(\mathbb{F}[\varepsilon](1))$ as c_ρ. We write H_{unr}^q for cohomology group of $\mathrm{Gal}(\mathbb{Q}_p^{unr}/\mathbb{Q}_p)$. We consider the inflation and restriction sequence:

$$0 \to H_{unr}^1(H^0(I_p, \mathbb{F}[\varepsilon](1))) \to H^1(\mathbb{F}[\varepsilon](1))$$
$$\xrightarrow{\mathrm{Res}} H^0(\mathrm{Gal}(\mathbb{Q}_p^{unr}/\mathbb{Q}_p), H^1(\mathbb{Q}_p^{unr}, \mathbb{F}[\varepsilon](1))) \to H_{unr}^2(H^0(I_p, \mathbb{F}[\varepsilon](1))).$$

By the p-distinguishedness (p), we have $\mu_p \not\subset \mathbb{Q}_p$; so, $\mathbb{Q}_p[\mu_p]/\mathbb{Q}_p$ tamely ramifies. Therefore $H^0(I_p, \mathbb{F}[\varepsilon](1)) = 0$ and the restriction map Res is an isomorphism. By $p \notin S$, $\mathrm{Res}(c_\rho)$ has to land in the flat cohomology $H^0(\mathrm{Gal}(\mathbb{Q}_p^{unr}/\mathbb{Q}_p), H_{fl}^1(\mathbb{Q}_p^{unr}, \mathbb{F}[\varepsilon](1)))$, which is

$$H^0(\mathrm{Gal}(\mathbb{Q}_p^{unr}/\mathbb{Q}_p), \mathbb{F}[\varepsilon] \otimes_\mathbb{F} \varprojlim_n \frac{O_p^{unr,\times}}{(O_p^{unr,\times})^{p^n}}) = \mathbb{F}[\varepsilon] \otimes_\mathbb{F} \varprojlim_n \frac{O_p^{unr,\times}}{(O_p^{unr,\times})^{p^n}}.$$

On the other hand, $H^1(\mathbb{F}[\varepsilon](1)) = \mathbb{F}[\varepsilon] \otimes_\mathbb{F} \varprojlim_n \mathbb{Q}_p^\times/(\mathbb{Q}_p^\times)^{p^n}$ by Kummer's theory. Thus we lose one dimension for each choice of ϕ, and the desired assertion follows. $\qquad\square$

Remark 7.2.6. When $\overline{\rho}|_{D_p} \cong \overline{\omega}_p\eta \oplus \eta$ for the Teichmüller character $\overline{\omega}_p$ giving Galois action on $\mu_p(\overline{\mathbb{F}}_p)$ and a character $\eta : D_p \to \mathbb{F}^\times$, the representation $\overline{\rho} \otimes \eta^{-1}$ itself is realized on a flat group scheme $\mathbb{Z}/p\mathbb{Z} \oplus \mu_p$ defined over $\mathbb{Z}_p$, but its deformation $\rho \in \mathcal{D}(\mathbb{F}[\varepsilon])$ twisted by η^{-1} may not be flat over $\mathbb{Z}_p$ giving a non-trivial extension $1 \hookrightarrow \rho \otimes \eta^{-1} \twoheadrightarrow \mu_{p/\mathbb{F}[\varepsilon]}$. The above Proposition 7.2.5 only deals with flat deformations (up to twists) not non-flat (semi-stable) deformations.

7.2.4 *Proof of Theorem 7.2.3*

We follow the proof of [W95, Proposition 1.6] with some simplification by Fujiwara (particularly on the estimate of h_p below). In this proof, we take $\mathfrak{G} = \mathfrak{G}_\mathbb{Q}^S$ as we need the global Euler characteristic formula (and the dual Selmer groups are defined for $\mathfrak{G}_\mathbb{Q}^S$). Write $Ad := Ad(\overline{\rho})$ simply. Note that

$$H^0(\mathfrak{G}, Ad(1)) = H^0(\mathfrak{G}, Ad) = 0$$

by absolute irreducibility of $\overline{\rho}$ which implies $H^0(\mathfrak{G}, ad(\overline{\rho})) = \mathrm{End}_{\mathbb{F}[\mathfrak{G}]}(\overline{\rho}) \cong \mathbb{F}$. Let $\Sigma = S \cup \{p, \infty\} = \{l | C\} \cup \{p, \infty\}$ for the prime-to-p conductor C of $\overline{\rho}$. We have by the Poitou–Tate exact sequence (cf. [MFG, Theorem 4.50 (5)]) combined with local Tate duality (cf. [MFG, Theorem 4.43]):

$$0 \to \mathrm{Sel}(Ad) \to H^1(\mathfrak{G}, Ad) \to \prod_{l \in \Sigma} \frac{H^1(\mathbb{Q}_l, Ad)}{\mathcal{D}_{\chi,l}(\mathbb{F}[\varepsilon])} \to \mathrm{Sel}^\perp(Ad(1))^*$$

$$\to H^2(\mathfrak{G}, Ad) \to \prod_{l \in \Sigma} H^0(\mathbb{Q}_l, Ad(1)) \to H^0(\mathfrak{G}, Ad(1)) = 0. \quad (7.5)$$

This combined with the global Euler characteristic formula:

$$\dim_{\mathbb{F}} H^2(\mathfrak{G}, Ad) - \dim_{\mathbb{F}} H^1(\mathfrak{G}, Ad) = -2$$

tells us $\dim_{\mathbb{F}} \mathrm{Sel}(Ad) - \dim_{\mathbb{F}} \mathrm{Sel}^\perp(Ad(1)) = 2 + \sum_{l \in \Sigma} h_l$, where $h_l = \dim_{\mathbb{F}} H^0(\mathbb{Q}_l, Ad(1)) - \dim_{\mathbb{F}} H^1(\mathbb{Q}_l, Ad) + \dim_{\mathbb{F}} \mathcal{D}_{\chi,l}(\mathbb{F}[\varepsilon])$. If $l \in \Sigma$ is prime to p, we have

$$\mathcal{D}_{\chi,l}(\mathbb{F}[\varepsilon]) = \mathrm{Ker}(H^1(\mathbb{Q}_l, Ad) \to H^1(I_l, Ad))$$

$$\cong H^1(\mathrm{Gal}(\mathbb{Q}_l^{unr}/\mathbb{Q}_l), H^0(I_l, Ad))$$

$$\cong H^0(I_l, Ad)/(\mathrm{Frob}_l - 1)H^0(I_l, Ad) \overset{(*)}{\cong} H^0(\mathbb{Q}_l, Ad),$$

and by the local Euler characteristic formula [MFG, Theorem 4.52], $\boxed{h_l = 0}$. Here, the identity $(*)$ follows from $\dim_{\mathbb{F}} H^0(I_l, Ad) \leq 1$ (the minimality assumptions (m) and (l)).

Now we assume that $l = p$. Then we have an exact sequence:

$$0 \to \xi_p \to Ad \to \mathbb{F} \oplus \xi_p^{-1} \to 0, \quad (7.6)$$

where $\xi_p = \overline{\epsilon}\overline{\delta}^{-1}$ for $\overline{\epsilon} = (\epsilon \mod \mathfrak{m}_W)$ and $\overline{\delta} = (\delta \mod \mathfrak{m}_W)$. The sequence $\overline{\epsilon} \hookrightarrow \overline{\rho} \twoheadrightarrow \overline{\delta}$ is split if $\xi_p \neq \overline{\omega}$ by Kummer's theory, and hence the above sequence (7.6) is also split. Thus, if $\xi_p \neq \overline{\omega}$, we have $\dim_{\mathbb{F}} H^0(\mathbb{Q}_p, Ad) = 1$ because $\xi_p \neq 1$ by the condition (p). Again by the local Euler characteristic formula (cf. [MFG, Theorem 4.52]), we get the following two formulas:

$$h_p = \dim_{\mathbb{F}} H^0(\mathbb{Q}_p, Ad(1)) - \dim_{\mathbb{F}} H^1(\mathbb{Q}_p, Ad) + \dim_{\mathbb{F}} \mathcal{D}_{\chi,p}(\mathbb{F}[\varepsilon])$$

$$= -(\dim_{\mathbb{F}} Ad(1)) - \dim_{\mathbb{F}} H^0(\mathbb{Q}_p, Ad) + \dim_{\mathbb{F}} \mathcal{D}_{\chi,p}(\mathbb{F}[\varepsilon])$$

$$= -3 \dim_{\mathbb{F}} H^0(\mathbb{Q}_p, Ad) + \dim_{\mathbb{F}} \mathcal{D}_{\chi,p}(\mathbb{F}[\varepsilon]).$$

and $\dim_{\mathbb{F}} H^1(\mathbb{Q}_p, \xi_p) = \begin{cases} 1 & \text{if } \xi_p \notin \{\overline{\omega}_p, 1\}, \\ 2 & \text{if } \xi_p = \overline{\omega}_p. \end{cases}$

We have by Proposition 7.2.5,

$$\dim_{\mathbb{F}} \mathcal{D}_{\chi,p}(\mathbb{F}[\varepsilon]) \leq \begin{cases} 1 + \dim_{\mathbb{F}} H^1(\mathbb{Q}_p, \xi_p) & \text{if } \xi_p \notin \{\overline{\omega}_p, 1\}, \\ \dim H^1(\mathbb{Q}_p, \xi_p) & \text{if } \xi_p = \overline{\omega}_p. \end{cases}$$

This shows $-\dim_{\mathbb{F}} H^0(\mathbb{Q}_p, Ad) + \dim_{\mathbb{F}} \mathcal{D}_{\chi,p}(\mathbb{F}[\varepsilon]) \le 1$ and hence $\boxed{h_p \le -2}$. Then the desired assertion follows. $\qquad\square$

7.2.5 Details of $H^1(K, \mu_p) \cong K^\times \otimes_{\mathbb{Z}} \mathbb{F}_p$

Here K is any field. Recall from Remark 4.2.6 the connection map δ of the long exact sequence $H^0(K, M) \to H^0(K, N) \xrightarrow{\delta} H^1(K, L)$ of a short one $L \hookrightarrow M \twoheadrightarrow N$: Pick $n \in H^0(K, N)$ and lift it to $m \in M$. Then for $\sigma \in \mathrm{Gal}(\overline{K}/K)$, $(\sigma - 1)m$ is sent to $(\sigma - 1)n = 0$ as n is fixed by σ. Thus we may regard $u_m : \sigma \mapsto (\sigma - 1)m$ as a 1-cocycle with values in L. If we choose another lift m', then $m' - m = l \in L$ and hence $u_{m'} - u_m = (\sigma - 1)l$ which is a coboundary. Thus we get the map δ sending m to the class $[u_m]$.

Applying this to $\mu_p(\overline{K}) \hookrightarrow \overline{K}^\times \xrightarrow[\twoheadrightarrow]{\xi \mapsto \xi^p} \overline{K}^\times$, the cocycle u_ξ corresponding $\xi \in K^\times/(K^\times)^p = K^\times \otimes \mathbb{F}_p$ is given by $\boxed{u_\xi(\sigma) = {}^{\sigma-1}(\sqrt[p]{\xi})}$.

Let K be an $\mathfrak{l}$-adic field which is a finite extension of $\mathbb{Q}_l$ for a prime $l \ne p$. Write $O_{\mathfrak{l}}$ for the $\mathfrak{l}$-adic integer ring of K. If $\xi \notin (K^\times)^p$, $\xi' := l^{p^N} \xi \notin (K^\times)^p$ with $K[\sqrt[p]{\xi}] = K[\sqrt[p]{\xi'}]$ and $u_\xi = u_{\xi'}$. Replacing ξ by ξ' for a sufficiently large exponent N, we may assume that $\xi \in O_{\mathfrak{l}} \cap K^\times$.

Assume $\xi \notin (K^\times)^p$. Then the minimal equation of $\sqrt[p]{\xi}$ is $f(X) = X^p - \xi$. Since the derivative of $f(X)$ is given by $f'(X) = pX^{p-1}$, the different of $K[\sqrt[p]{\xi}]/K$ is a factor of $p\sqrt[p]{\xi}^{p-1}$ [NTH, II.7]. Thus we find

Lemma 7.2.7. *Let the notation and the assumption be as above. If $l \ne p$, then the Kummer cocycle u_ξ ($\xi \in K^\times$) is unramified if and only if $\xi \in O_{\mathfrak{l}}^\times$, choosing $\xi \in O_{\mathfrak{l}} \cap K^\times$.*

This can also be shown by noting that all conjugates of $\sqrt[p]{\xi}$ is given by $\{\zeta \sqrt[p]{\xi} | \zeta \in \mu_p\}$ which has p distinct elements modulo $\mathfrak{l}$ if and only if $\xi \in O_{\mathfrak{l}}^\times$.

7.2.6 Restriction to the splitting field of $Ad(\overline{\rho})$

Let F be the splitting field of $Ad := Ad(\overline{\rho})$; so, $F = F(Ad) = \overline{\mathbb{Q}}^{\mathrm{Ker}(Ad)}$, and $K := F[\mu_p]$ is the splitting field of $Ad(1)$. Write $\overline{G} := \mathrm{Gal}(F/\mathbb{Q})$. We realize $\mathrm{Sel}^\perp(Ad(1))$ inside $H^1(F, Ad(1)) = F^\times \otimes_{\mathbb{Z}} Ad$. Assume

$$H^j(F/\mathbb{Q}, Ad(1)^{\mathfrak{G}_F}) = 0 \quad \text{for } j = 1, 2, \tag{7.7}$$

which follows if $K = F[\mu_p] \ne F$ or $p \nmid [F : \mathbb{Q}]$. If $F[\mu_p] \ne F$, we see $Ad(1)^{\mathfrak{G}_F} = 0$ as Ad is trivial over $\mathfrak{G}_F$. If $p \nmid [F : \mathbb{Q}] = |\overline{G}|$, we note $H^q(\overline{G}, M) = 0$ for any $\mathbb{F}[\overline{G}]$-module M [MFG, Proposition 4.21]. By inflation-restriction,

$$H^1(\overline{G}, Ad(1)^{\mathfrak{G}_F}) \hookrightarrow H^1(\mathbb{Q}, Ad(1)) \to H^1(F, Ad(1))^{\overline{G}} \to H^2(\overline{G}, Ad(1)^{\mathfrak{G}_F})$$

is exact. So $\boxed{H^1(\mathbb{Q}, Ad(1)) \cong (F^\times \otimes_{\mathbb{Z}} Ad)^{\overline{G}}}$.

Kummer theory applied to $H^1(\mathbb{Q}, Ad(1))$. We analyze how $\overline{G}$ acts on $F^\times \otimes_{\mathbb{F}_p} Ad$. The action of $\tau \in \overline{G}$ is given by $^\tau u(g) = \tau u(\tau^{-1} g \tau) = Ad(\tau) u(\tau^{-1} g \tau)$ $(\tau \in \overline{G})$ for cocycle u giving rise to a class in $H^1(F, Ad(1))$. For a basis (v_1, v_2, v_3) of Ad giving an identification $Ad = \mathbb{F}^3$, and write $u = v \cdot \underline{u}$ for $\underline{u} := {}^t(u_1, u_2, u_3)$ (column vector) for $v = (v_1, v_2, v_3)$ (row vector) as a $\mathbb{F}^3$ valued cocycle; so, $\tau v = (\tau v_1, \tau v_2, \tau v_3) = v \cdot {}^t Ad(\tau)$. Since $u_j(g) = u_{\xi_j}(g) = {}^{g-1}\sqrt[p]{\xi_j}$ for $\xi_j \in F^\times \otimes_{\mathbb{Z}} \mathbb{F}$, rewriting $u_\xi := \underline{u}$, we have $\tau(v \cdot {}^\tau u_\xi(\tau^{-1} g \tau)) = v \cdot {}^t Ad(\tau) u_{\tau_\xi}(g)$. Thus τ-invariance implies

$$u_{\tau_\xi} := {}^t(u_{\tau_{\xi_1}}, u_{\tau_{\xi_2}}, u_{\tau_{\xi_3}}) = {}^t Ad(\tau)^{-1} u_\xi \Leftrightarrow v \cdot {}^t Ad(\tau) u_{\tau_\xi}(g) = v \cdot u_\xi.$$

Therefore inside $F^\times \otimes_{\mathbb{Z}} \mathbb{F}$, ξ_js span an $\mathbb{F}$-vector space on which $\overline{G}$ acts by a factor of $Ad \cong {}^t Ad^{-1}$. Thus we get

$$H^1(\mathbb{Q}, Ad(1)) \cong \operatorname{Hom}_{\mathbb{F}[\overline{G}]}(Ad, F^\times \otimes_{\mathbb{Z}} \mathbb{F}) =: (F^\times \otimes_{\mathbb{Z}} \mathbb{F})[Ad]. \qquad (7.8)$$

7.2.7 Selmer group as a subgroup of $F^\times \otimes_{\mathbb{Z}} \mathbb{F}$

Recall that F denotes the splitting field of $Ad(\overline{\rho})$. Fix a prime $\mathfrak{p}|p$ of F, we write $\overline{D} := \operatorname{Gal}(F_\mathfrak{p}/\mathbb{Q}_p) = \overline{G}$.

Theorem 7.2.8. Let O be the integer ring of F. Let $\operatorname{Sel}_?^\perp(Ad(\overline{\rho})(1)) = \operatorname{Sel}^\perp(Ad(\overline{\rho})(1))$ unless $\overline{\delta}\overline{\epsilon}^{-1} = \overline{\omega}$ over D_p and otherwise $\operatorname{Sel}_?^\perp(Ad(\overline{\rho})(1)) = \operatorname{Sel}_{fl}^\perp(Ad(\overline{\rho})(1))$. If $p \nmid h_F = |Cl_F|$, we have the following inclusion

$$\operatorname{Sel}_?^\perp(Ad(\overline{\rho})(1)) \hookrightarrow O^\times \otimes_{\mathbb{Z}} \mathbb{F}[Ad(\overline{\rho})].$$

Later we ease the condition $p \nmid h_F$ to $Cl_F \otimes_{\mathbb{Z}[\operatorname{Gal}(F/\mathbb{Q})]} Ad(\overline{\rho}) = 0$.

Proof. Let $[u] \in \operatorname{Sel}^\perp(Ad(\overline{\rho})(1))$ for a cocycle $u : \mathfrak{G}_\mathbb{Q}^S \to Ad(\overline{\rho})(1)$. Thus $u|_{\mathfrak{G}_F}$ gives rise to u_ξ for $\xi \in F^\times \otimes_{\mathbb{Z}} \mathbb{F}[Ad(\overline{\rho})]$ by Kummer theory. Consider the fractional ideal $(\xi) = \xi O[\frac{1}{p}]$. Make a prime decomposition $(\xi) = \prod_{\mathfrak{l}} \mathfrak{l}^{e(\mathfrak{l})}$ in $O[\frac{1}{p}]$. Since u_ξ gives a class in $H_{fl}^1(F_\mathfrak{l}, Ad(\overline{\rho})(1))$, at each prime $\mathfrak{l} \nmid p$, by Lemma 7.2.2, we find $\boxed{p|e(\mathfrak{l})}$; so, u_ξ is unramified at $\mathfrak{l} \nmid p$. So $(\xi) = \mathfrak{a}^p$ for $\mathfrak{a} = \prod_{\mathfrak{l}} \mathfrak{l}^{e(\mathfrak{l})/p}$

Step 1: l-integrality (l ≠ p). If $p \nmid h := h_F = |Cl_F|$, replacing ξ by ξ^h only changes the Kummer cocycle by a non-zero scalar. We do this replacement and write ξ instead of ξ^h. Then $\mathfrak{a}$ is replaced by the principal ideal $\mathfrak{a}^h = (\xi')$, and we find that $\xi = \varepsilon \xi'^p$ for $\varepsilon \in O[\frac{1}{p}]^\times$. Thus $u_\xi = u_\varepsilon$. Therefore $\mathrm{Sel}^\perp(Ad(1)) \subset (O[\frac{1}{p}]^\times \otimes_\mathbb{Z} \mathbb{F})[Ad]$.

Step 2: Case where $\overline{\rho}|_{\overline{D}}$ is indecomposable over $\overline{D} = \mathrm{Gal}(F_\mathfrak{p}/\mathbb{Q}_p)$. By indecomposability, the matrix form of $Ad(\sigma)$ if $\overline{\rho}(\sigma) = \left(\begin{smallmatrix} \bar{\varepsilon} & a \\ 0 & \bar{\delta} \end{smallmatrix}\right)$ $(a \neq 0)$ with respect to the basis $\{\left(\begin{smallmatrix} 0 & 1 \\ 0 & 0 \end{smallmatrix}\right), \left(\begin{smallmatrix} 1 & 0 \\ 0 & -1 \end{smallmatrix}\right), \left(\begin{smallmatrix} 0 & 0 \\ 1 & 0 \end{smallmatrix}\right)\}$ is

$$\begin{pmatrix} \bar{\varepsilon}\bar{\delta}^{-1} & -2\bar{\delta}^{-1}a & -(\bar{\varepsilon}\bar{\delta})^{-1}a^2 \\ 0 & 1 & \bar{\varepsilon}^{-1}a \\ 0 & 0 & \bar{\varepsilon}^{-1}\bar{\delta} \end{pmatrix},$$

in short, Ad is also an indecomposable $\overline{D}$-module without trivial quotient. Decomposing $(\varepsilon) = \prod_{\mathfrak{p}^\sigma} \mathfrak{p}^{e_\sigma \sigma}$ for the O-fractional ideal (ε) generated by $\varepsilon \in O[\frac{1}{p}]^\times$, we define $\mathrm{div}(\varepsilon) = \sum_{\sigma \in \overline{G}/\overline{D}} e_\sigma \mathfrak{p}^\sigma \in \oplus_{\sigma \in \overline{G}/\overline{D}} \mathbb{Z}\mathfrak{p}^\sigma$. We have an exact sequence of $\overline{G}$-modules: $1 \to O^\times \overset{i}{\hookrightarrow} O[\frac{1}{p}]^\times \xrightarrow{\varepsilon \mapsto \mathrm{div}(\varepsilon)} \oplus_{\sigma \in \overline{G}/\overline{D}} \mathbb{Z}\mathfrak{p}^\sigma$. The cokernel $\mathrm{Coker}(\mathrm{div})$ lands in Cl_F whose order is prime to p by our assumption. Localizing at p, we have $\mathrm{Coker}(\mathrm{div}) \otimes_\mathbb{Z} \mathbb{Z}_p = 0$; thus, the above sequence turns into a short exact sequence:

$$1 \to O^\times \otimes_\mathbb{Z} \mathbb{Z}_p \overset{i}{\hookrightarrow} O[\frac{1}{p}]^\times \otimes_\mathbb{Z} \mathbb{Z}_p \xrightarrow{\mathrm{div}} \oplus_{\sigma \in \overline{G}/\overline{D}} \mathbb{Z}_p \mathfrak{p}^\sigma \to 0.$$

Since $\mathrm{Im}(\mathrm{div}) = \oplus_{\sigma \in \overline{G}/\overline{D}} \mathbb{Z}_p \mathfrak{p}^{a\sigma}$ is a $\mathbb{Z}_p$-free module, the exactness is kept after tensoring $\mathbb{F}$ [BCM, I.2.5]. Therefore, noting $\oplus_{\sigma \in \overline{G}/\overline{D}} \mathbb{Z}_p \mathfrak{p}^\sigma \cong \mathrm{Ind}_{\overline{D}}^{\overline{G}} \mathbb{Z}_p$ as $\overline{G}$-modules, we get an exact sequence of $\overline{G}$-modules:

$$1 \to O^\times \otimes_\mathbb{Z} \mathbb{F} \xrightarrow{i \otimes 1} O[\frac{1}{p}]^\times \otimes_\mathbb{Z} \mathbb{F} \xrightarrow[\twoheadrightarrow]{\varepsilon \mapsto \mathrm{div}(\varepsilon) \otimes 1} \mathrm{Ind}_{\overline{D}}^{\overline{G}} \mathbb{F} \to 0. \qquad (7.9)$$

By Shapiro's Lemma 5.3.2,

$$(\mathrm{Ind}_{\overline{D}}^{\overline{G}} \mathbb{F})[Ad] = \mathrm{Hom}_{\mathbb{F}[\overline{G}]}(Ad, \mathrm{Ind}_{\overline{D}}^{\overline{G}} \mathbb{F}) = \mathrm{Hom}_{\overline{D}}(Ad|_{\overline{D}}, 1) = 0$$

by indecomposability; so, $\boxed{\mathrm{Sel}^\perp(Ad(1)) \subset (O^\times \otimes \mathbb{F})[Ad]}$.

Step 3: Case where $\overline{\rho}|_{\overline{D}} \cong \bar{\varepsilon} \oplus \bar{\delta}$. Since $Ad|_{\overline{D}} = \bar{\delta}\bar{\varepsilon}^{-1} \oplus 1 \oplus \bar{\delta}^{-1}\bar{\varepsilon}$, by the exact sequence (7.9), $\mathrm{Coker}(i \otimes 1) \cong \mathrm{Ind}_{\overline{D}}^{\overline{G}} 1$ corresponds to the classes of cocycles on $\overline{D}$ with values in the factor 1. In this case, we have

$$\mathrm{Ind}_{\overline{D}}^{\overline{G}} 1[Ad] = \mathrm{Hom}_{\mathbb{F}[\overline{G}]}(Ad, \mathrm{Ind}_{\overline{D}}^{\overline{G}} 1) = \mathrm{Hom}_{\overline{D}}(Ad|_{\overline{D}}, 1) = \mathbb{F}.$$

Pick a cocycle $u : D_\mathfrak{p} \to Ad(1)$ of the decomposition group $D_\mathfrak{p} = \mathrm{Gal}(\overline{\mathbb{Q}}_p/F_\mathfrak{p})$ at $\mathfrak{p}$ projecting down non-trivially to $F_\mathfrak{p}^\times \otimes \mathbb{F}[1]$ (i.e., $u \in$

$H^1(F_{\mathfrak{p}}, \mathbb{F}(1)) \subset H^1(F_{\mathfrak{p}}, Ad(1)))$. It has values in the $\overline{\omega}$-eigenspace $\mathbb{F}(1)$ of $Ad(1)$. Thus if it is a dual Selmer cocycle, it is in the orthogonal complement of $\mathrm{Hom}_{\mathbb{Z}_p}(\mathrm{Frob}_{\mathfrak{p}}^{\mathbb{Z}_p}, \mathbb{F}) \subset H^1(F_{\mathfrak{p}}, Ad)$ as $\mu_p \otimes_{\mathbb{Z}} \mathbb{F} \subset Ad(1)$ is the dual of $Ad \twoheadrightarrow Ad \otimes_{\mathbb{F}[\overline{G}]} 1$. Then by Lemma 7.2.2, if u is a dual Selmer cocycle it corresponds to an element $\xi_{\mathfrak{p}}$ in $H^1_{fl}(F_{\mathfrak{p}}, \mathbb{F}(1)) = O_{\mathfrak{p}}^\times \otimes \mathbb{F}$. Since $\mathfrak{p}|p$ is arbitrary, we may assume that $u = u_\xi$ for $\xi \in O[\frac{1}{p}]^\times$ whose image in $F_{\mathfrak{p}}^\times \otimes_{\mathbb{Z}} \mathbb{F}$ is $\xi_{\mathfrak{p}}$. Thus we conclude again $\boxed{\mathrm{Sel}_?^\perp(Ad(1)) \subset (O^\times \otimes_{\mathbb{Z}} \mathbb{F})[Ad]}$. This finishes the proof of the theorem. $\qquad\square$

In the last boxed formula, $\mathrm{Sel}_?^\perp(Ad(1)) = \mathrm{Sel}^\perp(Ad(1))$ unless $\overline{\epsilon}\overline{\delta}^{-1} = \overline{\omega}_p$.

7.2.8 *Dirichlet's unit theorem*

Fix a complex conjugation $c \in \overline{G}$ and C be the subgroup generated by c. Let ∞ be the set of complex places of F. Dirichlet's unit theorem [LFE, Theorem 1.2.3] is proven by considering

$$O^\times \xrightarrow{Log} \mathbb{R}^\infty := \prod_\infty \mathbb{R}$$

given by $Log(\varepsilon) = (\log|\varepsilon|_v)_{v \in \infty}$ and showing $\mathrm{Im}(Log) \otimes_{\mathbb{Z}} \mathbb{R} = \mathrm{Ker}(\mathbb{R}^\infty \xrightarrow{\mathrm{Tr}} \mathbb{R})$ for $\mathrm{Tr}(x_v)_v = \sum_v x_v$. The Galois group $\overline{G}$ acts by permutation on $\infty \cong \overline{G}/C$. Therefore $\mathbb{R}^\infty \cong \mathrm{Ind}_C^{\overline{G}} 1$. Thus $(O^\times \otimes \mathbb{Q}) \oplus 1 \cong \mathrm{Ind}_C^{\overline{G}} \mathbb{Q}1$.

If $p \nmid |\overline{G}|$, any $\mathbb{F}[\overline{G}]$-module over $\mathbb{F}$ is semi-simple; so, characterized by its trace [MFG, §2.1.5]. Therefore this descends to $(O^\times/\mu_p(F)) \otimes_{\mathbb{Z}} \mathbb{F}$ and

$$\boxed{\mathrm{Ind}_C^{\overline{G}} \mathbb{F}1 \cong ((O^\times/\mu_p(F)) \otimes_{\mathbb{Z}} \mathbb{F}) \oplus \mathbb{F}1} \qquad (7.10)$$

This shows that $(O^\times/\mu_p(F)) \otimes_{\mathbb{Z}} \mathbb{Z}_p$ is generated by one element $\varepsilon \in O^\times$ over $\mathbb{Z}_p[\overline{G}]$. This unit ε defined in [M00] is well defined as long as $p \nmid |\overline{G}|$ and is called the *Minkowski unit* (at p).

Consider the following conditions for $F = F(Ad(\overline{\rho}))$ and $\overline{G} = \mathrm{Gal}(F/\mathbb{Q})$:

(C0) $p \nmid h_F$;
(C1) $Cl_F \otimes_{\mathbb{Z}[\overline{G}]} Ad(\overline{\rho}) = 0$;
(C2) $Ad(\overline{\rho}) \otimes_{\mathbb{F}_p[D_p]} \mathbb{F}_p(1) = 0$ (i.e., $\overline{\epsilon}\overline{\delta}^{-1} \neq \overline{\omega}$).

Plainly (C0) implies (C1).

Theorem 7.2.9. *Assume* (C0). *If* $p \nmid |\overline{G}|$, *we have* $\dim_{\mathbb{F}} \mathrm{Sel}_?^\perp(Ad(1)) \leq 1$. *If further* (C2) *is satisfied,* $\dim_{\mathbb{F}} \mathrm{Sel}^\perp(Ad(1)) \leq 1$.

Proof. By Shapiro's lemma in §5.3.4, we have

$$((O^\times/\mu_p(F)) \otimes_\mathbb{Z} \mathbb{F})[Ad] = \mathrm{Hom}_{\overline{G}}(Ad, (O^\times/\mu_p(F)) \otimes_\mathbb{Z} \mathbb{F})$$

$$\cong \mathrm{Hom}_{\overline{G}}(Ad, \mathrm{Ind}_C^{\overline{G}} \mathbb{F}1) \cong \mathrm{Hom}_C(Ad|_C, \mathbb{F}1) \cong \mathbb{F},$$

since $Ad(c) \sim \mathrm{diag}[-1, 1, -1]$. By irreducibility, we have $\mu_p(F)[Ad] = 0$; so, $(O^\times \otimes_\mathbb{Z} \mathbb{F})[Ad] \cong \mathbb{F}$. By Theorem 7.2.8, we have

$$\mathrm{Sel}_?^\perp(Ad(1)) \hookrightarrow (O^\times \otimes_\mathbb{Z} \mathbb{F})[Ad] \cong \mathbb{F},$$

we conclude $\boxed{\dim_\mathbb{F} \mathrm{Sel}_?^\perp(Ad(1)) \leq 1}$. $\qquad\square$

By Theorem 7.2.3 combined with Proposition 6.1.9 (see also §6.1.3), we prove the following corollary under (C0) and (C2). In Corollary 8.1.2, we will ease the assumption (C0) to (C1) of the following result.

Corollary 7.2.10. *Assume* (C0) *and* (C2) *and* $R^{ord} \cong \mathbb{T}$. *If* $p \nmid |\overline{G}|$, *then for any deformation* $\rho \in \mathcal{D}(A)$, $\mathrm{Sel}(Ad(\rho))^\vee$ *is generated by at most one element over* A.

Proof. Since $\mathrm{Sel}(Ad(\overline{\rho}))^\vee \cong \Omega_{\mathbb{T}/\Lambda} \otimes_\mathbb{T} \mathbb{F}$. By Theorem 7.2.3 combined with Proposition 6.1.9, $\dim_\mathbb{F} \Omega_{\mathbb{T}/\Lambda} \otimes_\mathbb{T} \mathbb{F} \leq 1$. If $\dim_\mathbb{F} \Omega_{\mathbb{T}/\Lambda} \otimes_\mathbb{T} \mathbb{F} = 0$, by Nakayama's Lemma 1.8.3, $\Omega_{\mathbb{T}/\Lambda} = 0$. If $\dim_\mathbb{F} \Omega_{\mathbb{T}/\Lambda} \otimes_\mathbb{T} \mathbb{F} = 1$, by Corollary 1.8.4, $\omega_{\mathbb{T}/\Lambda}$ is generated by one element. Taking the Λ-algebra homomorphism $\pi : \mathbb{T} \to A$ with $\pi \circ \rho \cong \rho$, by Theorem 6.1.11, we have $\mathrm{Sel}(Ad(\rho))^\vee \cong \Omega_{\mathbb{T}/\Lambda} \otimes_{\mathbb{T}, \pi} A$ which is generated by at most one element. $\quad\square$

7.3   Induced modular Galois representation

Assuming that $\rho = \mathrm{Ind}_K^\mathbb{Q} \varphi$ for a quadratic field $K = \mathbb{Q}[\sqrt{D}]$ (with discriminant D) and a character $\varphi : \mathfrak{G}_K \to W^\times$ of finite order prime to p, after this section, we explore the meaning of the cyclicity of $\mathrm{Sel}(Ad(\rho))^\vee$ in terms of Iwasawa theory over K. The cyclicity question is not studied much for p-abelian extensions of K classically. However $\mathrm{Sel}(Ad(\mathrm{Ind}_K^\mathbb{Q} \Phi))$ for a deformation of φ has an intimate relation to (anticyclotomic) Iwasawa modules classically studied.

A cyclic p-abelian extension $F_{/K}$ is called *anticyclotomic* if it is Galois over $\mathbb{Q}$ and $\sigma \in \mathrm{Gal}(F/\mathbb{Q})$ non-trivial over K acts by -1 via conjugation on $\mathrm{Gal}(F/K)$ (i.e., $\sigma\tau\sigma^{-1} = \tau^{-1}$ for $\tau \in \mathrm{Gal}(K/F)$). For such an anticyclotomic extension F, a p-abelian extension L/F Galois over K produces an Iwasawa module $X = \mathrm{Gal}(L/F)$, and we ask if X is cyclic over

$\mathbb{Z}_p[[\mathrm{Gal}(K/F)]]$. Often assuming splitting of the prime p into $\wp\wp^\sigma$ in K (basically to have natural p-ordinarity), people studied the case when L/F is a maximal p-abelian extension unramified outside (a fixed factor) $\wp$. Especially the case where K is imaginary has been studied extensively in the framework of elliptic Iwasawa theory and complex multiplication. In this section, we allow K to be real also, and describe basics of induced representations to study cyclicity of anticyclotomic Iwasawa modules in the following sections. Particularly, we determine the Galois module structure of the unit group of F and classify cases of p-ordinary induced representations (i.e., cases of $\mathrm{Ind}_K^{\mathbb{Q}} \varphi$ satisfying (p)).

Write $\overline{\varphi} := (\varphi \bmod \mathfrak{m}_W)$ and $\overline{\rho} = \mathrm{Ind}_K^{\mathbb{Q}} \overline{\varphi}$ in the standard matrix form in (5.3). We denote now by O the integer ring of K and pick $\sigma \in \mathfrak{G}$ with $\sigma|_K \neq 1$. If K is imaginary, we choose σ to be a complex conjugation; so, $\sigma^2 = 1$. For a fractional ideal $\mathfrak{a}$ outside primes ramified for φ, define $\varphi(\mathfrak{a}) := \varphi([\mathfrak{a}, K])$ for the Artin symbol $[\mathfrak{a}, K]$. Then the starting Hecke eigenform is given by $f_0 := \sum_{0 \neq \mathfrak{a} \in \mathrm{Spec}(O)} \varphi(\mathfrak{a}) q^{N(\mathfrak{a})} \in S_1(C; \psi_1)$ where $C = N(\mathfrak{f})D$ for the conductor $\mathfrak{f}$ of φ and $\psi_1(n) = \varphi(n)\alpha(n)$ ($n \in \mathbb{Z}$) with $\alpha = \left(\frac{K/\mathbb{Q}}{\cdot} \right)$ (see [HMI, Theorem 2.71]).

7.3.1 *Induced representation and self-twist*

We return to the setting of §5.3.2; so, $\mathcal{G}$ is a profinite group and $\mathcal{H}$ is an open subgroup of index 2 of $\mathcal{G}$ with $\mathcal{G} = \mathcal{H} \sqcup \mathcal{H}\varsigma$. For each representation $\rho = \left(\begin{smallmatrix} \rho_{11} & \rho_{12} \\ \rho_{21} & \rho_{22} \end{smallmatrix} \right) : G \to \mathrm{GL}_2(A)$ and a character $\xi : G \to A^\times$, we write a matrix form of $\rho \otimes \xi$ as $\rho \cdot \xi := \left(\begin{smallmatrix} \rho_{11}\xi & \rho_{12}\xi \\ \rho_{21}\xi & \rho_{22}\xi \end{smallmatrix} \right)$. Let $J = \left(\begin{smallmatrix} 1 & 0 \\ 0 & -1 \end{smallmatrix} \right)$. Write α for the order two character: $\mathcal{G}/\mathcal{H} \cong \{\pm 1\}$ (which will be $\alpha = \left(\frac{K/\mathbb{Q}}{\cdot} \right)$ in our application). Extending φ to $\mathcal{G}$ by 0 outside $\mathcal{H}$, we find

$$(\mathrm{Ind}_{\mathcal{H}}^{\mathcal{G}} \varphi) \otimes \alpha(g) \cong (\mathrm{Ind}_{\mathcal{H}}^{\mathcal{G}} \varphi) \cdot \alpha(g)$$

$$= \begin{cases} \left(\begin{smallmatrix} \varphi(g) & 0 \\ 0 & \varphi(\varsigma^{-1}g\varsigma) \end{smallmatrix} \right) = J \left(\begin{smallmatrix} \varphi(g) & 0 \\ 0 & \varphi(\varsigma^{-1}g\varsigma) \end{smallmatrix} \right) J^{-1} & (g \in \mathcal{H}), \\ -\left(\begin{smallmatrix} 0 & \varphi(g\varsigma) \\ \varphi(\varsigma^{-1}g) & 0 \end{smallmatrix} \right) = J \left(\begin{smallmatrix} 0 & \varphi(g\varsigma) \\ \varphi(\varsigma^{-1}g) & 0 \end{smallmatrix} \right) J^{-1} & (g\varsigma \in \mathcal{H}). \end{cases}$$

Thus J induces the $\mathcal{G}$ equivariant map i_α:

$$\boxed{(\mathrm{Ind}_{\mathcal{H}}^{\mathcal{G}} \varphi) \cdot \alpha = J(\mathrm{Ind}_{\mathcal{H}}^{\mathcal{G}} \varphi)J^{-1} \xrightarrow[i_\alpha]{\sim} \mathrm{Ind}_{\mathcal{H}}^{\mathcal{G}} \varphi}. \qquad (7.11)$$

Writing $Ad(\mathrm{Ind}_{\mathcal{H}}^{\mathcal{G}} \varphi) = \{x \in \mathrm{End}_A(\mathrm{Ind}_{\mathcal{H}}^{\mathcal{G}} \varphi)) | \mathrm{Tr}(x) = 0\}$, we see $i_\alpha \in Ad(\mathrm{Ind}_{\mathcal{H}}^{\mathcal{G}} \varphi)$ as $\mathrm{Tr}(J) = 0$.

Let $\overline{\varphi} := (\varphi \mod \mathfrak{m}_A)$. Suppose $\overline{\varphi}_\varsigma \neq \overline{\varphi}$ (where $\overline{\varphi}_\varsigma(h) = \overline{\varphi}(\varsigma h \varsigma^{-1})$). Since $\mathrm{Ind}_{\mathcal{H}}^{\mathcal{G}} \overline{\varphi}(\mathcal{H})$ contains diagonal matrices with distinct eigenvalues, its normalizer is $\mathrm{Ind}_{\mathcal{H}}^{\mathcal{G}} \overline{\varphi}(\mathcal{G})$. Thus the centralizer $Z(\mathrm{Ind}_{\mathcal{H}}^{\mathcal{G}} \overline{\varphi}) = \mathbb{F}^\times$ (scalar matrices). Since $\mathrm{Ind}_{\mathcal{H}}^{\mathcal{G}} \overline{\varphi}(\varsigma)$ interchanges $\overline{\varphi}$ and $\overline{\varphi}_\varsigma$, $\mathrm{Ind}_{\mathcal{H}}^{\mathcal{G}} \overline{\varphi}$ is absolutely **irreducible**. Since $\mathrm{Aut}(\overline{\rho}) = \mathbb{F}^\times$, i_α for $\overline{\rho}$ is unique up to scalars.

We now apply the above characterization to a deformation $\rho : \mathcal{G} \to \mathrm{GL}_2(A)$ with $\rho \mod \mathfrak{m}_A = \mathrm{Ind}_{\mathcal{H}}^{\mathcal{G}} \overline{\varphi}$. Write $\alpha : \mathcal{G}/\mathcal{H} \cong \{\pm 1\} \subset W^\times$. Pick $g \in \mathcal{G}$ such that $\alpha(g) = -1$. Suppose $\rho \otimes \alpha \cong \rho$. So we have $j \in \mathrm{GL}_2(A)$ such that $j\rho j^{-1} = \rho \otimes \alpha$. Since $\alpha^2 = 1$, j^2 is scalar. We may normalize $j \equiv J \mod \mathfrak{m}_A$ as $j \mod \mathfrak{m}_A = zJ$ for a scalar $z \in A^\times$. Thus j has two eigenvalues $\epsilon_\pm$ with $\epsilon_\pm \equiv \pm z \mod \mathfrak{m}_A$. Let $A_\pm$ be $\epsilon_\pm$-eigenspace of j. Since $j\rho|_{\mathcal{H}} = \rho|_{\mathcal{H}} j$, $A_\pm \cong A$ is stable under $\mathcal{H}$. Thus we find a character $\varphi : \mathcal{H} \to A^\times$ acting on A_+. Plainly $\mathcal{H}$ acts on A_- by φ_ς. This shows $\rho \cong \mathrm{Ind}_H^G \varphi$ as $V(\rho) = A_+ \oplus \rho(\varsigma)A_+$. We record what we have proven:

Lemma 7.3.1. *Let* $\rho : \mathcal{G} \to \mathrm{GL}_2(A)$ *be a representation such that* $\rho \mod \mathfrak{m}_A = \mathrm{Ind}_{\mathcal{H}}^{\mathcal{G}} \overline{\varphi}$, *and assume that* $\mathrm{Ind}_{\mathcal{H}}^{\mathcal{G}} \overline{\varphi}$ *is absolutely irreducible. Then* $\rho \otimes \alpha \cong \rho$ *if and only if there exists a character* $\varphi : \mathcal{H} \to A^\times$ *such that* $\rho \cong \mathrm{Ind}_{\mathcal{H}}^{\mathcal{G}} \varphi$.

Decomposition of adjoint induced representation.

Proposition 7.3.2. *We have* $Ad(\mathrm{Ind}_{\mathcal{H}}^{\mathcal{G}} \varphi) \cong \alpha \oplus \mathrm{Ind}_{\mathcal{H}}^{\mathcal{G}} \varphi^-$ *as representation of* $\mathcal{G}$. *If* $Ad(\rho)$ *for a residually absolutely irreducible representation* $\rho : \mathcal{G} \to \mathrm{GL}_2(A)$ *has a proper A-direct summand stable under* $\mathcal{G}$, *then there exists a subgroup* $\mathcal{H}'$ *of index 2 of* $\mathcal{G}$ *such that* $\rho \cong \mathrm{Ind}_{\mathcal{H}'}^{\mathcal{G}} \varphi$ *for a character* $\varphi : \mathcal{H}' \to A^\times$ *(under the assumption* $p > 2$).

Here $\varphi^-(g) = \varphi(g)\varphi_\varsigma^{-1}(g) = \varphi(\varsigma^{-1}g^{-1}\varsigma g)$ and $\mathrm{Ind}_{\mathcal{H}}^{\mathcal{G}} \varphi^-$ is irreducible if $\varphi^- \neq \varphi_\varsigma^- = (\varphi^-)^{-1}$ (i.e., φ^- has order ≥ 3).

Proof. On $\mathcal{H}$, $\rho := \mathrm{Ind}_{\mathcal{H}}^{\mathcal{G}} \varphi = \left(\begin{smallmatrix} \varphi & 0 \\ 0 & \varphi_\varsigma \end{smallmatrix} \right)$. Therefore

$$Ad(\mathrm{Ind}_{\mathcal{H}}^{\mathcal{G}} \varphi)(h) \left(\begin{smallmatrix} x & y \\ z & -x \end{smallmatrix} \right) = \rho(h) \left(\begin{smallmatrix} x & y \\ z & -x \end{smallmatrix} \right) \rho^{-1}(h) = \left(\begin{smallmatrix} x & \varphi^-(h)y \\ (\varphi^-)^{-1}(h)z & -x \end{smallmatrix} \right),$$

and

$$Ad(\mathrm{Ind}_{\mathcal{H}}^{\mathcal{G}} \varphi)(\varsigma) \left(\begin{smallmatrix} x & y \\ z & -x \end{smallmatrix} \right) = \left(\begin{smallmatrix} 0 & \varphi(\varsigma^2) \\ 1 & 0 \end{smallmatrix} \right) \left(\begin{smallmatrix} x & y \\ z & -x \end{smallmatrix} \right) \left(\begin{smallmatrix} 0 & 1 \\ \varphi(\varsigma^{-2}) & 0 \end{smallmatrix} \right) = \left(\begin{smallmatrix} \alpha(\varsigma)x & \varphi(\varsigma)^2 z \\ \varphi(\varsigma)^{-2}y & -\alpha(\varsigma)x \end{smallmatrix} \right).$$

Thus α is realized on diagonal matrices, and $\mathrm{Ind}_{\mathcal{H}}^{\mathcal{G}} \varphi^-$ is realized on the anti-diagonal matrices. This proves the first assertion.

Now assume that $Ad(\rho) = V \oplus V'$ with V stable under $\mathcal{G}$, V is A-free as A is local. If $\mathrm{rank}_A V = 2$, then, by the perfect duality pairing $(x, y) \mapsto \mathrm{Tr}(xy)$ on $\mathfrak{sl}_2(A)$ under $p > 2$, the orthogonal complement $V^\perp$ has rank 1 stable under $\mathcal{G}$. Thus we may assume that $\mathrm{rank}_A V = 1$. Then $\mathcal{G}$ acts on V by a character ξ. Pick a generator $v \in V$ over A, and regard $v \in \mathrm{End}_A(\rho)$ with $\mathrm{Tr}(v) = 0$. If $\xi = 1$, $\overline{\rho} = (\rho \mod \mathfrak{m}_A)$ commutes with $\overline{v} := (v \mod \mathfrak{m}_A)$. Since $\overline{v}$ is the centralizer of an absolutely irreducible $\overline{\rho}$, $\overline{v} \neq 0$ is a scalar, which is impossible as $\mathrm{Tr}(\overline{v}) = 0$. Thus $\xi \neq 1$. Since $\rho(g)v\rho(g)^{-1} = \xi(g)v$, v induces isomorphisms $\rho \otimes \xi \cong \rho$ and $\overline{\rho} \otimes \xi \cong \overline{\rho}$. Taking the determinant, $\xi^2 = 1$ (but $\xi \neq 1$); so, for $\mathcal{H}' = \mathrm{Ker}(\xi)$, by Lemma 7.3.1, we find a character φ so that $\rho \cong \mathrm{Ind}_{\mathcal{H}'}^{\mathcal{G}} \varphi$. $\square$

Lemma 7.3.3. *The representation* $\mathrm{Ind}_{\mathcal{H}}^{\mathcal{G}} \overline{\varphi}^-$ *is irreducible if and only if* $\overline{\varphi}^- \neq \overline{\varphi}_\varsigma^- = (\overline{\varphi}^-)^{-1}$ *(i.e.,* $\overline{\varphi}^-$ *has order* ≥ 3*). If* $\overline{\varphi}^-$ *has order* ≤ 2*, then* $\overline{\varphi}^-$ *extends to a character* $\overline{\phi} : \mathcal{G} \to \mathbb{F}^\times$ *and* $\mathrm{Ind}_{\mathcal{H}}^{\mathcal{G}} \varphi^- \cong \overline{\phi} \oplus \overline{\phi}\alpha$.

Proof. Note $\varphi^-(\varsigma^2) = \varphi(\varsigma^2)\varphi(\varsigma^{-1}\varsigma^2\varsigma)^{-1} = \varphi(1) = 1$. The irreducibility of $\mathrm{Ind}_{\mathcal{H}}^{\mathcal{G}} \overline{\varphi}^-$ under $\overline{\varphi}^- \neq \overline{\varphi}_\varsigma^-$ follows from the argument just below (7.11) proving irreducibility of $\mathrm{Ind}_{\mathcal{H}}^{\mathcal{G}} \overline{\varphi}$ under $\overline{\varphi} \neq \overline{\varphi}_\varsigma$. Suppose $\overline{\varphi}^-$ has order ≤ 2 (so, $\overline{\varphi}^- = \overline{\varphi}_\varsigma^-$). Choose a root $\zeta = \pm 1$ of $X^2 - \varphi_\varsigma^-(\varsigma^2) = X^2 - 1$ in $\mathbb{F}$. Define $\overline{\phi} = \overline{\varphi}_-$ on $\mathcal{H}$ and $\overline{\phi}(\varsigma h) = \zeta\overline{\varphi}^-(h)$. For $h, h' \in \mathcal{H}$,

$$\overline{\phi}(\varsigma h \varsigma h') = \overline{\phi}(\varsigma^2 \varsigma^{-1} h \varsigma h') = \overline{\varphi}^-(\varsigma^2))\overline{\varphi}_\varsigma^-(h)\overline{\varphi}^-(h')$$
$$= \zeta^2 \overline{\varphi}^-(hh') = \overline{\varphi}^-(\varsigma h)\overline{\varphi}^-(\varsigma h').$$

Similarly $\overline{\phi}(h\varsigma h') = \overline{\phi}(\varsigma\varsigma^{-1}hch') = \zeta\overline{\varphi}_\varsigma(h)\overline{\varphi}(h') = \overline{\phi}(h)\overline{\phi}(\varsigma h')$; so, $\overline{\phi}$ is a character. Then we have $\mathbb{F}[\zeta][\mathcal{G}] \otimes_{\mathbb{F}[\mathcal{H}]} \mathbb{F}[\zeta](\overline{\varphi}^-) \cong \mathbb{F}[\zeta](\overline{\phi})$ as $\mathcal{G}$-modules by $a \otimes b \mapsto \overline{\phi}(a)b$. $\square$

7.3.2 Galois action on unit groups

Consider a character $\phi : H \to \overline{\mathbb{Q}}^\times$. We apply the results in §7.3.1 taking $\mathcal{G} = G$ and $\mathcal{H} = H$ and $\varsigma = \sigma$. We study Galois module structure of $\mathfrak{R}^\times \otimes_{\mathbb{Z}} \overline{\mathbb{Q}}$ for the integer ring $\mathfrak{R}$ of $F(\phi)$:

Proposition 7.3.4. *Let the notation be as above. Suppose that* K *is imaginary and that* $\phi_\sigma = \phi^{-1}$ *for complex conjugation* $\sigma \in G$. *Write* a *for the order of* ϕ. *Then we have, as* $\mathrm{Gal}(F(\phi)/\mathbb{Q})$*-modules,*

$$\mathfrak{R}^\times \otimes_{\mathbb{Z}} \overline{\mathbb{Q}} \cong \begin{cases} \xi \oplus \bigoplus_{j=1}^{b-1} \mathrm{Ind}_K^{\mathbb{Q}} \phi^j & \text{if } a \text{ is even with } a = 2b, \\ \bigoplus_{j=1}^{b} \mathrm{Ind}_K^{\mathbb{Q}} \phi^j & \text{if } a \text{ is odd with } a = 2b + 1, \end{cases}$$

where $\xi \ : \ \text{Gal}(F(\phi)/\mathbb{Q}) \ \to \ \{\pm 1\}$ *is a quadratic character such that* $\xi|_{\text{Gal}(F(\phi)/K)} = \phi^b$ *and* ξ *is even at the infinite place of* $\mathbb{Q}$.

Here we do not claim that the factors $\text{Ind}_K^{\mathbb{Q}} \phi^j$ are irreducible (and the same remarks apply to the proposition following this one).

Proof. If $a = 2$, we have $\text{Gal}(F(\phi)/\mathbb{Q}) \cong \{\pm 1\}^2$ as $\text{Im}(\text{Ind}_K^{\mathbb{Q}} \phi)$ is dihedral of order 4. Since complex conjugation $c \in \text{Gal}(F(\phi)/\mathbb{Q})$ fixes a unique totally real quadratic extension $k'/\mathbb{Q}$ with $\xi = \left(\frac{k'/\mathbb{Q}}{} \right)$. Then $F(\phi)$ is a CM quadratic extension of k', and the assertion is clear.

Now suppose that $a > 2$ is even. The fixed field $F(\phi^b)$ of $\text{Im}(\phi^2) \subset \text{Im}(\phi) = \text{Gal}(F(\phi)/K)$ is the composite of K and another quadratic extension k' of $\mathbb{Q}$. By the argument in the case of $a = 2$, we may assume that k' is real, and Kk' contains another imaginary quadratic extension $K'_{/\mathbb{Q}}$. Thus $\xi := \left(\frac{k'/\mathbb{Q}}{} \right)$ has multiplicity 1 in $\mathfrak{R}^{\times} \otimes_{\mathbb{Z}} \overline{\mathbb{Q}}$ as the unit group of k' has rank 1. The maximal abelian quotient of $\text{Gal}(F(\phi)/\mathbb{Q})$ is equal to $\text{Gal}(Kk'/\mathbb{Q})$. Writing $a = 2b$ with $1 < b \in \mathbb{Z}$, the action of $\text{Gal}(F(\phi)/\mathbb{Q})$ on $\mathfrak{R}^{\times} \otimes_{\mathbb{Z}} \overline{\mathbb{Q}}$ is therefore isomorphic to

$$\mathfrak{R}^{\times} \otimes_{\mathbb{Z}} \overline{\mathbb{Q}} \cong \xi \oplus \bigoplus_{j=1}^{b-1} m(j) \, \text{Ind}_K^{\mathbb{Q}} \phi^j,$$

since $\{\phi^j, \phi^{-j}\}$ $(j = 1, \ldots, b-1)$ and $\{\phi^b = \xi|_G\}$ give conjugacy classes of characters under conjugation of c. Thus we have $1 + \sum_j 2m(j) = a - 1$; so,

$$\sum_j m(j) = (b - 1).$$

Write $\Sigma(\phi)$ for the set of infinite places of $F(\phi)$. The $\text{Gal}(F(\phi)/\mathbb{Q})$-module $\mathfrak{R}^{\times} \otimes_{\mathbb{Z}} \mathbb{C}$ is embedded into the Galois module $V := \text{Im}(\text{Tr}_{\mathbb{C}/\mathbb{R}} : F(\phi) \otimes_{\mathbb{Q}} \mathbb{R} \to \mathbb{R}^{\Sigma(\phi)}) \otimes_{\mathbb{R}} \mathbb{C}$ by the $(x_v)_v \otimes z \mapsto (z \log |x_v|^2)_v$ for infinite places v of $F(\phi)$ (e.g., the proof of Dirichlet's unit theorem). The cokernel of this embedding is identified with the trivial $\text{Gal}(F(\phi)/\mathbb{Q})$-module $\mathbb{C}$ by the degree map $\deg(x_v) = \sum_v x_v$. Let $\text{Gal}(F(\phi)/\mathbb{Q})$ act on $\Sigma(\phi)$ by permutation; so, the space of $\mathbb{C}$-valued functions $\mathbb{C}[\Sigma(\phi)]$ on $\Sigma(\phi)$ is a $\text{Gal}(F(\phi)/\mathbb{Q})$-module. The $\text{Gal}(F(\phi)/\mathbb{Q})$-module V is isomorphic to $\mathbb{C}[\Sigma(\phi)]$. We claim

$$\mathbb{C}[\Sigma(\phi)] \cong \mathbf{1} \oplus \xi \oplus \bigoplus_{j=1}^{b-1} \text{Ind}_K^{\mathbb{Q}} \phi^j. \tag{7.12}$$

We prove this claim. The complex conjugation ϱ_v at v coincide with c on K, and hence $\text{Ind}_K^{\mathbb{Q}} \phi^j$ for all $j = 1, \ldots, b-1$ appears along with real characters

ξ and $\mathbf{1}$. Since c acts non-trivially on $K_\infty := K \otimes_\mathbb{Q} \mathbb{R} \cong \mathbb{C}$, this shows the desired formula from the exact sequence $\mathfrak{R}^\times \otimes_\mathbb{Z} \mathbb{C} \hookrightarrow \mathbb{C}[\Sigma(\phi)] \twoheadrightarrow \mathbb{C}$.

We now assume that $a = 2b + 1$ is odd. Then we have

$$\mathbb{C}[\Sigma(\phi)] \cong \mathbf{1} \oplus \bigoplus_{j=1}^{2b} \phi^j \tag{7.13}$$

as $\mathrm{Gal}(F(\phi)/K)$-modules, and $c \in \mathrm{Gal}(K/\mathbb{Q})$ interchanges ϕ^j and ϕ^{-j}, which implies

$$\mathbb{C}[\Sigma(\phi)] \cong \mathbf{1} \oplus \bigoplus_{j=1}^{b} \mathrm{Ind}_K^\mathbb{Q} \phi^j \tag{7.14}$$

as $\mathrm{Gal}(F(\phi)/\mathbb{Q})$-modules. Thus we conclude the desired formula. $\qquad\square$

Now we treat the case where K is real. For simplicity, we suppose $\phi_\sigma = \phi^{-1}$ and that $F(\phi)$ is totally imaginary. Since $F(\phi)/K$ is abelian, complex conjugation $c \in \mathrm{Gal}(F(\phi)/\mathbb{Q})$ is in the center of $\mathrm{Gal}(F(\phi)/\mathbb{Q})$; thus, by [IAT, Proposition 5.11], $F(\phi^2)$ is totally real and $F(\phi)$ is a totally imaginary quadratic extension of $F(\phi^2)$. Thus $F(\phi)$ is a CM field. Then by Dirichlet's unit theorem [LFE, Theorem 1.2.3], $\mathfrak{R}^\times \otimes_\mathbb{Z} \mathbb{Q} = \mathfrak{r}^\times \otimes_\mathbb{Z} \mathbb{Q}$ for the integer ring $\mathfrak{r}$ of $F(\phi^2)$.

Proposition 7.3.5. *Suppose that K is real, $\phi_\sigma = \phi^{-1}$ and that $F(\phi)$ is totally imaginary. Let $\mathfrak{R}$ be the integer ring of $F(\phi)$. Write a for the order of ϕ. Then a is even with $a = 2b$ for $0 < b \in \mathbb{Z}$, and we have*

$$\mathfrak{R}^\times \otimes_\mathbb{Z} \overline{\mathbb{Q}} \cong \alpha \oplus \bigoplus_{j=1}^{b-1} \mathrm{Ind}_K^\mathbb{Q} \phi^{2j}$$

as $\mathrm{Gal}(\overline{\mathbb{Q}}/\mathbb{Q})$-modules, where $\alpha : \mathrm{Gal}(K/\mathbb{Q}) \cong \{\pm 1\}$.

Proof. Since $F(\phi)$ is imaginary, $a = [F(\phi) : K]$ has to be even, and hence $a = 2b$. As already remarked, $F(\phi)$ is a CM field with maximal totally real field $F(\phi^2)$; so, c acts trivially on $\mathfrak{R}^\times \otimes_\mathbb{Z} \mathbb{Q}$ and as Galois modules, we have

$$\mathbf{1} \oplus (\mathfrak{R}^\times \otimes_\mathbb{Z} \mathbb{Q}) \cong \mathbf{1} \oplus (\mathfrak{r}^\times \otimes_\mathbb{Z} \mathbb{Q}) \cong \mathrm{Ind}_{F(\phi^2)}^\mathbb{Q} \mathbf{1}$$

for the identity character $\mathbf{1}$. Then the assertion is clear from this. $\qquad\square$

7.3.3   *Ordinarity for induced representation*

By minimality, for the prime-to-p conductor $\mathfrak{c}$ of φ, $\bar{\rho}$ has prime-to-p conductor $C := N_{K/\mathbb{Q}}(\mathfrak{c})D$ [GME, Theorem 5.1.9], and the Hecke algebra $\mathbb{T}$ has prime-to-p level C. For the conductor $\mathfrak{f}$ of φ, put $N = N_{K/\mathbb{Q}}(\mathfrak{f})D$; so, $C|N|pC$ (why?). Recall $\phi^- = \phi\phi_\sigma^{-1}$ for a character ϕ of H. We classify our consideration into the following five cases.

(R$_-$) If $p|N$ and $\alpha(p) = -1$, $\bar{\varphi}$ and $\bar{\varphi}_\sigma$ both ramify at p; so, no p-unramified quotient of $\bar{\rho}|_{D_p}$ (no ordinary case).

(R$_+$) If $p|N$ and $\alpha(p) = 1$ (so, $(p) = \wp\wp^\sigma$ with $\wp \neq \wp^\sigma$), to have p-unramified quotient, we assume that $\wp|\mathfrak{f}$ and $\wp^\sigma \nmid \mathfrak{f}$ for a prime $\wp|p$ of K; so, $\delta = \varphi_\sigma$ (and automatically $\bar{\rho}$ is p-distinguished).

(Ds) If $p|N$ and $\alpha(p) = 0$ (so, $(p) = \wp^2$). To have p-unramified quotient, we assume $\wp \nmid \mathfrak{f}$ for the ramified prime $\wp|p$ of K. Then $\delta(\mathrm{Frob}_p) = \varphi(\mathrm{Frob}_\wp)$ and $\rho|_{D_p} = \mathrm{diag}[\alpha\delta, \delta]$ (so $\bar{\rho}$ is p-distinguished).

(U$_\pm$) If $\alpha(p) = \pm 1$ and $p \nmid N$, ρ and $\bar{\rho}$ is unramified. We choose a prime factor $\wp|p$. In this case, $\bar{\rho}$ is ordinary but if $\alpha(p) = 1$, we need to assume p-distinguishedness.

Choice of δ and $\bar{\delta}$ in Cases (Ds) and (U$_\pm$).

- In Case U$_+$, $\rho(\mathrm{Frob}_p) = \mathrm{diag}[\varphi(\wp), \varphi_\sigma(\wp)]$. Take $\bar{\delta} := (\varphi_\sigma \bmod \mathfrak{p})$ (in this case, $D_p = D_\wp$) and $\delta = \varphi_\sigma$. The p-distinguishedness assumption $\bar{\varphi}^-(\mathrm{Frob}_p) \neq 1$ we make is equivalent to representability of $\mathcal{D}$ (and $\mathcal{D}_\chi$).

- In Case U$_-$, taking $\sigma = \mathrm{Frob}_p$ and choosing a square root $\delta = \delta(\mathrm{Frob}_p)$ of $\varphi(\mathrm{Frob}_p^2)$, we have $\rho(\mathrm{Frob}_p) = \left(\begin{smallmatrix} 0 & \delta^2 \\ 1 & 0 \end{smallmatrix}\right)$ and $\rho(\mathrm{Frob}_p) \sim \mathrm{diag}[\delta, -\delta]$ (so p-distinguished). Choose a character $\delta : D_p/I_p = \langle \mathrm{Frob}_p \rangle \to \overline{\mathbb{Q}}^\times$ such that $\delta^2 = \varphi|_{D_p}$ and $\delta(\mathrm{Frob}_p) = \delta$. Put $\bar{\delta} = (\delta \bmod \mathfrak{p})$.

- In Case Ds, φ is unramified at p, and hence $\varphi|_{D_p} = \varphi_\sigma|_{D_p}$. Then $\rho|_{D_p}$ is a direct sum of subspaces on which I_p acts trivially and by α. The action of D_p on $H^0(I_p, \mathrm{Ind}_F^\mathbb{Q} \varphi)$ gives a character $\delta : D \to \mathbb{Z}[\varphi]^\times$ and $\rho|_{D_p} = \mathrm{diag}[\delta\alpha, \delta]$ (so p-distinguished).

As is clear from the description above, the splitting condition on p in K according to the values of $\alpha(p)$ affects quite the outcome; so, we separately study the case $\alpha(p) = 1$ in §7.4 and the case $\alpha(p) \neq 1$ in §7.6.

p-Stabilization in Case U$_\pm$. We recall the following fact: If $g \in S_k(\Gamma_0(N), \psi)$ is a Hecke eigenform for the level N prime to p, writing $g|T(p) = \lambda(T(p))g$ with $\lambda(T(p)) = \alpha + \beta$ and $\alpha\beta = \chi(p)$ ($\chi = \nu_p^{k-1}\psi$), define $g_\alpha(z) = g(z) - \beta g(pz)$. As seen in §3.2.8, g is a Hecke eigenform of level Np with $g_\alpha|U(p) = \alpha \cdot g_\alpha$. The form g_α is called the p-stabilization of g with $U(p)$-eigenvalues α.

Let $\beta = \varphi(\wp)$ if $\alpha(p) = 1$ and $\beta = -\delta(\mathrm{Frob}_p)$ if $\alpha(p) = -1$. Replacing f by $f^{ord} := f_\alpha$ $(\alpha\beta = \psi(p))$, we have $f^{ord}|U(p) = \delta(\mathrm{Frob}_p)f^{ord}$. Hereafter we choose the p-stabilized form f^{ord} in the place of f (and write it f). Thus δ has values in $\mathbb{Z}[f]^\times$, and write the level of f as N (so, $N = C$ if $p|C$ and $N = Cp$ otherwise). Put $\overline{\rho}_{\mathfrak{p}} := \rho_{f,\mathfrak{p}} \mod \mathfrak{p}$. If $\alpha(p) = 1$, p-distinguishedness $\Leftrightarrow \overline{\varphi}^-|_{D_p} \neq 1$ for $\overline{\varphi}^- = \overline{\varphi\varphi_\sigma^{-1}}$.

7.3.4 Induction in three ways

We would like to prove

Proposition 7.3.6. *Let A be a p-profinite local integral domain for $p > 2$ with finite residue field. Let M and K be two distinct quadratic fields in $\overline{\mathbb{Q}}$. Suppose that we have continuous characters $\varphi : \mathrm{Gal}(\overline{\mathbb{Q}}/K) \to A^\times$ and $\phi : \mathrm{Gal}(\overline{\mathbb{Q}}/M) \to A^\times$ with absolutely irreducible $\rho := \mathrm{Ind}_K^\mathbb{Q} \varphi$ over $\mathrm{Frac}(A)$ such that $\mathrm{Ind}_K^\mathbb{Q} \varphi \cong \mathrm{Ind}_M^\mathbb{Q} \phi$. Assume that one of these representations and characters is unramified outside a finite set of primes. Write φ_σ for the character $\mathrm{Gal}(\overline{\mathbb{Q}}/K) \ni \tau \mapsto \varphi(\sigma\tau\sigma^{-1})$ for $\sigma \in \mathrm{Gal}(\overline{\mathbb{Q}}/\mathbb{Q})$ inducing the generator of $\mathrm{Gal}(K/\mathbb{Q})$. If the representations $\mathrm{Ind}_K^\mathbb{Q} \varphi \cong \mathrm{Ind}_M^\mathbb{Q} \phi$ are ordinary at p, then we have*

(1) *φ and ϕ are both of finite order,*
(2) *We have $\varphi_\sigma = \left(\dfrac{MK/K}{}\right)\varphi$.*
(3) *If p does not ramify in $KM/\mathbb{Q}$, φ and ϕ are both unramified at p.*
(4) *If φ ramifies at a prime factor of p, then p splits in K, φ is unramified at another prime factor of p, p ramifies in M and ϕ is unramified at p.*
(5) *If M is real and $\mathrm{Ind}_K^\mathbb{Q} \varphi$ is odd, K is imaginary and at infinite places, ϕ ramifies at exactly one real place.*

Conversely, if φ^- has order 2 and ρ is odd, we have two quadratic fields M, L distinct from K with $KM = LM$ and finite order characters ϕ, ϕ' such that $\mathrm{Ind}_K^\mathbb{Q} \varphi \cong \mathrm{Ind}_M^\mathbb{Q} \phi \cong \mathrm{Ind}_L^\mathbb{Q} \phi'$.

Proof. Suppose $\mathrm{Ind}_M^\mathbb{Q} \phi \cong \mathrm{Ind}_K^\mathbb{Q} \varphi$. We first prove the assertion (2). Let N be the prime-to-p Artin conductor of $\mathrm{Ind}_M^\mathbb{Q} \phi$. For any prime l outside Np inert in K and split in M (such primes have positive density by Chebotarev density theorem [CFN, (6.4)]), we have

$$0 = \mathrm{Tr}(\mathrm{Ind}_K^{\mathbb{Q}}\, \varphi(\mathrm{Frob}_l)) = \mathrm{Tr}(\mathrm{Ind}_M^{\mathbb{Q}}\, \phi(\mathrm{Frob}_l)) = \phi(\mathfrak{l}) + \phi(\mathfrak{l}^\sigma)$$

for $\sigma \in \mathrm{Gal}(\overline{\mathbb{Q}}/\mathbb{Q})$ inducing a generator of $\mathrm{Gal}(M/\mathbb{Q})$. Thus we have $\phi^-(\mathrm{Frob}_l) = -1$ if l is inert in K and split in M (note here that $-1 \neq 1$ because $p > 2$). For any other primes q outside Np inert in K and split in M, $\phi^-(\mathrm{Frob}_l) = -1 = \phi^-(\mathrm{Frob}_q)$. Since $\mathrm{Frob}_l\, \mathrm{Frob}_q^{-1}$ fix MK, by moving q, again Chebotarev density tells us that ϕ^- factors through $\mathrm{Gal}(KM/M)$. Since $\mathrm{Ind}_M^{\mathbb{Q}}\, \phi$ is absolutely irreducible, we have $\phi \neq \phi_\sigma$ (i.e., $\phi^- \neq 1$). Thus we conclude $\varphi = \left(\frac{KM/K}{\cdot}\right)\varphi_\sigma$. This proves (2).

If D_p acts trivially on K, we have $\rho|_{D_p} \cong \varphi \oplus \varphi_\sigma$. Otherwise, we are in Case U$_-$ or Case Ds. Thus $\varphi|_{D_p} = \varphi_\sigma|_{D_p}$ extends to a character δ of D_p ([MFG, §4.3.5] or [GME, §5.1.1]) and $\rho|_{D_p} \cong \delta \oplus \delta\left(\frac{K/\mathbb{Q}}{\cdot}\right)$. Then ordinarity implies that φ is at least unramified at one prime in K over p.

We now deal with the assertions (3) and (4). We may assume that φ is unramified at one prime factor $\wp^\sigma$ of p. If there is only one prime factor in K over p, this forces φ_σ to be unramified at p. If there are two factors of p in K, either φ is unramified also at $\wp$ or M ramifies at p by (2). If M ramifies at p, there is only one prime factor in M over p, this forces ϕ to be unramified at p. Thus if $KM/\mathbb{Q}$ is unramified at p, $\left(\frac{KM/M}{\cdot}\right)$ is unramified at p, and ϕ and φ are both unramified at p. This proves (3) and (4).

Since $\mathrm{ind}_K^{\mathbb{Q}}\, \varphi \cong \mathrm{Ind}_M^{\mathbb{Q}}\, \phi$, the two are unramified outside a finite set S. We regard φ and ϕ class characters by class field theory. Thus φ (resp. ϕ) factors through $Cl_K(\prod_{l \in S} l^\infty)$ (resp. $Cl_M(\prod_{l \in S} l^\infty)$). We have an exact sequence $\prod_{l \in S} O_l^\times \to Cl_K(\prod_{l \in S} l^\infty) \to Cl_K^+ \to 1$ for the integer ring O of K. Since $A^\times$ is almost p-profinite (i.e., the prime-to-p part of A is finite isomorphic to $(A/\mathfrak{m}_A)^\times$) and $O_l^\times$ is almost l-profinite, if $l \neq p$, $\varphi(O_l^\times)$ is finite. Therefore, if φ is infinite order, $\varphi(O_p^\times)$ has infinite order, and hence φ ramifies at p. The same assertion holds for ϕ for the same reason.

To show (1), first suppose that φ ramifies at a prime factor $\wp|p$. Thus p ramifies in M and splits in K. Then $\left(\frac{KM/M}{\cdot}\right)$ ramifies at two primes $\wp$ and $\wp^\sigma$, and therefore φ has to be unramified at $\wp^\sigma$. In short, φ ramifies at $\wp$ and unramified at $\wp^\sigma$. Since p ramifies in M, ordinarity of $\mathrm{Ind}_M^{\mathbb{Q}}\, \phi$ forces ϕ to be unramified at p; so, ϕ has to have finite order. Thus $\mathrm{Ind}_M^{\mathbb{Q}}\, \phi \cong \mathrm{Ind}_K^{\mathbb{Q}}\, \varphi$ has finite image; so, ϕ has finite order.

Next suppose that ϕ is unramified at p. Then ϕ factors through the finite ray class group $Cl_M(\mathfrak{f})$ of M modulo $\mathfrak{f}$ for the prime-to-p conductor $\mathfrak{f}$ of ϕ. Now $\mathrm{Ind}_M^{\mathbb{Q}}\, \phi$ has finite image, and we conclude that φ is of finite order (this proves (1)).

To prove (5), write $c \in \mathrm{Gal}(\overline{\mathbb{Q}}/\mathbb{Q})$ for complex conjugation. Since $\mathrm{Ind}_K^{\mathbb{Q}} \varphi$ is odd by assumption, we have $\mathrm{Tr}(\mathrm{Ind}_K^{\mathbb{Q}} \varphi(c)) = 0$. Regard ϕ as an idele character of $M_{\mathbb{A}}^{\times}$. Since $\mathrm{Ind}_K^{\mathbb{Q}} \varphi$ is odd,

$$0 = \mathrm{Tr}(\mathrm{Ind}_K^{\mathbb{Q}} \varphi(c)) = \mathrm{Tr}(\mathrm{Ind}_M^{\mathbb{Q}} \phi(c)) = \phi(-1_{\infty}) + \phi(-1_{\infty'})$$

where ∞ is an infinite place of M and ∞' is the other, and $1_{\infty'}$ is the identity of the ∞'-component $M_{\infty'}^{\times} = \mathbb{R} \subset M_{\mathbb{A}}^{\times}$. Thus ϕ ramifies at exactly one infinite place of M as M is real. If K is also real, $\phi_{\sigma}/\phi = \left(\frac{MK/M}{}\right)$ is unramified at the two infinite places of M; so, either ϕ ramifies at the two infinite places or unramified at the two infinite places; so, this is impossible (finishing the proof of (5)).

Suppose now that φ^{-} has order 2. Then the image of $\varrho := \mathrm{Ind}_K^{\mathbb{Q}} \varphi^{-}$ is isomorphic to a dihedral group of order 4; so, isomorphic to $\{\pm 1\} \times \{\pm 1\}$. Thus we have three distinct quadratic subfields K, M and L inside $F = F(\varrho)$. Since we have a commutative diagram with exact rows

$$
\begin{array}{ccccccc}
Z & \stackrel{\hookrightarrow}{\longrightarrow} & \mathrm{Gal}(F(\rho)/\mathbb{Q}) & \stackrel{\tau \mapsto \tau|_F}{\longrightarrow} & \mathrm{Gal}(F/\mathbb{Q}) & \to & 1 \\
& \wr \downarrow \rho & & \wr \downarrow \rho & & \wr \downarrow \varrho & \\
\rho(Z) & \stackrel{\hookrightarrow}{\longrightarrow} & \mathrm{Im}(\rho) & \stackrel{\gamma \mapsto Ad(\gamma)}{\twoheadrightarrow} & \mathrm{Im}(Ad(\rho)) & \cong & \mathrm{Im}(\varrho).
\end{array}
$$

Note $Z = \{z \in \mathrm{Gal}(F(\rho)/\mathbb{Q}) | \gamma z \gamma^{-1} = z \text{ for all } \gamma \in \mathrm{Gal}(F(\rho)/\mathbb{Q})\}$, which is the center of $\mathrm{Gal}(F(\rho)/\mathbb{Q})$ and hence is an abelian group. Pick a quadratic subfield $X \subset F$. Let $C_X := \mathrm{Ker}(\mathrm{Gal}(F(\rho)/\mathbb{Q}) \to \mathrm{Gal}(X/\mathbb{Q}))$. Then $C_X \hookrightarrow \mathrm{Gal}(F(\rho)/\mathbb{Q}) \twoheadrightarrow \mathrm{Gal}(X/\mathbb{Q})$ is an extension and C_K is cyclic isomorphic to $\mathrm{Im}(\varphi)$. Since $\xi := \varphi|_Z$ satisfies $\xi_{\varsigma} = \xi$ for $\varsigma \in C_M$ outside Z, ξ extends to a character ϕ of C_M [GME, Proposition 5.1.2], and $\phi_{\tau}/\phi = \left(\frac{F/M}{}\right)$ for $\tau \in \mathrm{Gal}(F(\rho)/\mathbb{Q})$ non-trivial on M (this implies that C_M is cyclic and $C_M \cong C_K$). Thus $\mathrm{Ind}_M^{\mathbb{Q}} \phi \cong \mathrm{Ind}_K^{\mathbb{Q}} \varphi$. Similarly we find a character ϕ' of C_L extending ξ such that $\mathrm{Ind}_K^{\mathbb{Q}} \varphi \cong \mathrm{Ind}_L^{\mathbb{Q}} \phi'$.[1] $\square$

Corollary 7.3.7. *Let the notation and the assumption be as in Proposition 7.3.6. Suppose $\rho := \mathrm{Ind}_K^{\mathbb{Q}} \varphi \cong \mathrm{Ind}_M^{\mathbb{Q}} \phi$ for two distinct quadratic extensions M, K over $\mathbb{Q}$. Put $F = MK$. Suppose ρ is p-ordinary, p-distinguished and absolutely irreducible. Then the p-decomposition subgroup $\overline{D}$ of $\mathrm{Gal}(F/\mathbb{Q})$ has order 2. Taking $K := F^{\overline{D}}$, we can decompose $(p) = \wp\wp^{\sigma}$*

---

[1]There is another proof of this converse [H15, Proposition 5.2].

in K for $\mathrm{Gal}(K/\mathbb{Q}) = \langle\sigma\rangle$ so that φ is unramified at $\wp^\sigma$ and that ρ has unramified quotient character φ_σ on the $\wp$-decomposition group $D_\wp$, where $\varphi_\sigma(g) = \varphi(\sigma g \sigma^{-1})$ for $\sigma \in \mathrm{Gal}(F(\rho)/\mathbb{Q})$ with $\sigma|_K \neq 1$.

Proof. Since ρ is p-distinguished, the p-decomposition group $\overline{D}$ of $\mathrm{Gal}(F/\mathbb{Q})$ is non-trivial. If $\overline{D} = \mathrm{Gal}(F/\mathbb{Q})$, the inertia subgroup $\overline{I}$ of $\overline{D}$ is non-trivial as $\overline{D}/\overline{I}$ is cyclic, while $\mathrm{Gal}(F/\mathbb{Q}) \cong \{\pm 1\} \times \{\pm 1\}$. Since $p > 2$, there is no single fully ramified $(2,2)$-extension of $\mathbb{Q}_p$. Suppose $\overline{I} \cong \{\pm 1\}$ and $\overline{D}/\overline{I} \cong \{\pm 1\}$. Then, taking $K = F^{\overline{I}}$, $\varphi^-|_{\overline{I}}$ is a ramified character of order 2, and hence ρ cannot have unramified quotient; so, ρ cannot be p-ordinary. Thus $\overline{D} \cong \{\pm 1\}$. Now let $K = F^{\overline{D}}$ in which p splits into $\wp\wp^\sigma$ for $\sigma \in \mathrm{Gal}(F/\mathbb{Q})$ non-trivial on K. By p-ordinarity, φ can ramify only at one factor of p, say $\wp$. Then the p-ordinary quotient unramified character is given by φ_σ on $D_\wp$. $\square$

7.4   Cyclicity when p splits in K

In this section, we assume that $p > 2$ splits as $(p) = \wp\wp^\sigma$ in K and that φ is unramified at $\wp^\sigma$ with absolutely irreducible $\overline{\rho} = \mathrm{Ind}_K^{\mathbb{Q}} \overline{\varphi}$. We study cyclicity of $\mathrm{Sel}(Ad(\rho))$ for a minimal ordinary deformation ρ of $\mathrm{Ind}_K^{\mathbb{Q}} \overline{\varphi}$. So we are either in Case U$_+$ or in Case R$_+$.

7.4.1   *Identity of two deformation functors*

Let χ be the Teichmüller lift of $\det(\overline{\rho})$. Recall the maximal p-profinite extension $F^{(p)}(\overline{\rho})$ unramified outside p of $F(\overline{\rho})$ and its Galois group $G = \mathrm{Gal}(F^{(p)}(\overline{\rho})/\mathbb{Q})$. Put $H = \mathrm{Gal}(F^{(p)}(\overline{\rho})/K)$. Consider the deformation functor $\mathcal{D}_? : CL_B \to SETS$ for χ and κ in §6.1.1. Let

$$\mathcal{F}_H(A) = \{\varphi : H \to A^\times \,|\, \varphi \bmod \mathfrak{m}_A = \overline{\varphi} \text{ unramified outside } \mathfrak{c}\wp\}$$

and $\mathcal{D}_?^{\widehat{\Delta}}(A) = \{\rho \in \mathcal{D}_?(A) | J(\rho \cdot \alpha)J^{-1} \sim \rho, \det\rho = ?\}/\Gamma(\mathfrak{m}_A)$, where "$\sim$" is conjugation by an element in $\Gamma(\mathfrak{m}_A) = \mathrm{Ker}(\mathrm{GL}_2(A) \xrightarrow{\bmod \mathfrak{m}_A} \mathrm{GL}_2(\mathbb{F}))$. Recall $\Delta = G/H$ and write $\widehat{\Delta} = \{\alpha, \mathbf{1}\}$ for its character group. By (7.11), $J(\overline{\rho} \cdot \alpha)J^{-1} = \overline{\rho}$; so, $\widehat{\Delta}$ acts on $\mathcal{D}_?$ by $\rho \mapsto J(\rho \cdot \alpha)J^{-1}$.

Theorem 7.4.1. *The map $\mathcal{F}_H(A) \ni \varphi \mapsto \mathrm{Ind}_H^G \varphi \in \mathcal{D}(A)^{\widehat{\Delta}}$ induces an isomorphism: $\mathcal{F}_H \cong \mathcal{D}_?^{\widehat{\Delta}}$ of the functors if $\overline{\varphi} \neq \overline{\varphi}_c$.*

This fact was first noticed in [CV03, §2].

Proof. If $\rho = \operatorname{Ind}_H^G \varphi$ is specified as in (5.3), we have $J(\rho \cdot \alpha)J^{-1} = \rho$ by (7.11). Conversely, by the characterization in Lemma 7.3.1, if $\rho \in \mathcal{D}_?^{\widehat{\Delta}}(A)$, we find a character $\varphi : H \to A^\times$ such that $\operatorname{Ind}_H^G \varphi \sim \rho$.

We need to show that φ is unramified outside $\mathfrak{c}\wp$. We choose $j \in \operatorname{GL}_2(A)$ with $j \equiv J \mod \mathfrak{m}_A$ as in §7.3.1. Then $A_+ = A(\varphi)$ for a character $\varphi : H \to A^\times$. Note that $\varphi \mod \mathfrak{m}_A = \overline{\varphi}$ by the construction in §7.3.1. By (l) for $l \in S$ and (p), $\overline{\varphi}_\sigma$ acting on A_- is unramified at each prime $l|\mathfrak{c}\wp$. Thus we conclude $\mathcal{F}_H \cong \mathcal{D}_?^{\widehat{\Delta}}$. $\qquad\square$

By $\rho \mapsto J(\rho \cdot \alpha)J^{-1}$, $\widehat{\Delta}$ acts on $\mathcal{D}_?$. For the universal representation $\boldsymbol{\rho}_? \in \mathcal{D}_?(R_?)$, therefore, we have an involution $[\alpha] \in \operatorname{Aut}_{B\text{-alg}}(R_?)$ such that $[\alpha] \circ \boldsymbol{\rho}_? \sim J(\boldsymbol{\rho}_? \cdot \alpha)J^{-1}$. Define $R_?^\pm := \{x \in R_?|[\alpha](x) = \pm x\}$.

For a character $\phi : H \to \mathbb{F}^\times$, let $K^{(\wp)}$ be the maximal p-abelian extension of K unramified outside $\wp$. Let $\Gamma_\wp = \operatorname{Gal}(K^{(\wp)}/K)$ which is a p-profinite abelian group.

Corollary 7.4.2. *We have a canonical isomorphism* $R_\kappa/R_\kappa([\alpha] - 1)R_\kappa \cong W[[\Gamma_\wp]]$, *where* $R_\kappa([\alpha] - 1)R_\kappa$ *is the* R_κ*-ideal generated by* $[\alpha](x) - x$ *for all* $x \in R_\kappa$.

If a finite cyclic group $\langle \gamma \rangle$ acts on $R \in CL_B$ fixing B, then

$$\boxed{\operatorname{Hom}_{B\text{-alg}}(R, A)^{\langle\gamma\rangle} = \operatorname{Hom}_{B\text{-alg}}(R/R(\gamma - 1)R, A)}$$

Indeed, $f \in \operatorname{Hom}_{B\text{-alg}}(R, A)^{\langle\gamma\rangle}$, then $f \circ \gamma = f$; so, $f(R(\gamma-1)R) = 0$. Thus $\operatorname{Hom}_{B\text{-alg}}(R, A)^{\langle\gamma\rangle} \hookrightarrow \operatorname{Hom}_{B\text{-alg}}(R/R(\gamma-1)R, A)$. Surjectivity is plain.

Proof. Since $\mathcal{F}_H = \mathcal{D}_\kappa^{\widehat{\Delta}}$, we find

$$\mathcal{F}_H(A) = \operatorname{Hom}_{\Lambda\text{-alg}}(R_\kappa, A)^{\widehat{\Delta}} = \operatorname{Hom}_{\Lambda\text{-alg}}(R_\kappa/(R_\kappa([\alpha] - 1)R_\kappa), A).$$

Thus $\mathcal{F}_H$ is represented by $R_\kappa/(R_\kappa([\alpha] - 1)R_\kappa)$.

Let $\varphi_0 : H \to W^\times$ be the Teichmüller lift of $\overline{\varphi}$. Define $\boldsymbol{\varphi} : H \to W[[\Gamma_\wp]]^\times$ by $\boldsymbol{\varphi}(h) = \varphi_0(h)h|_{K^{(\wp)}} \in W[[\Gamma_\wp]]$. We show that $(W[[\Gamma_\wp]], \boldsymbol{\varphi})$ is a universal couple for $\mathcal{F}_H$, which implies the identity of the corollary. Pick a deformation $\varphi \in \mathcal{F}_H(A)$. Then $(\iota_A \circ \varphi_0)^{-1}\varphi$ (for the structure morphism $\iota_A : W \to A$) has values in $1 + \mathfrak{m}_A$ unramified outside $\wp$ as the ramification at $l \in S$ different from p is absorbed by that of $\overline{\varphi}$ by the fact that the inertia group at l in H is isomorphic to the inertia group at l of $\operatorname{Gal}(F(\overline{\varphi})/K)$ for $F(\overline{\varphi}) = \overline{\mathbb{Q}}^{\operatorname{Ker}(\overline{\varphi})}$. Thus $(\iota_A \circ \varphi_0)^{-1}\varphi$ factors through $\Gamma_\wp$, and induces a unique W-algebra homomorphism $W[[\Gamma_\wp]] \xrightarrow{\phi} A$ such that $\varphi = \phi \circ \boldsymbol{\varphi}$. $\qquad\square$

Remark 7.4.3. For the automorphic representation π generated by the adelic form $\mathbf{f} : GL_2(\mathbb{Q})\backslash GL_2(\mathbb{A}) \to \mathbb{C}$ associated to a Hecke eigenform $f \in S_k(\Gamma_0(N), \psi)$ [MFG, §3.1] and a Dirichlet character α regarded as an idele character, the automorphic representation generated by $(\mathbf{f} \otimes \alpha)(g) = \alpha(\det(g))\mathbf{f}(g)$ produces compatible system $\rho_f \otimes \alpha$. Because of this fact, the Hecke algebra $\mathbb{T}$ also has involution $\alpha_{\mathbb{T}}$ such that $\alpha_{\mathbb{T}} \circ \rho_{\mathbb{T}} \cong \rho_{\mathbb{T}} \otimes \alpha$ and $\pi \circ [\alpha] = \alpha_{\mathbb{T}} \circ \pi$ for the projection $\pi : R_\kappa \twoheadrightarrow \mathbb{T}$ with $\pi \circ \rho \cong \rho_{\mathbb{T}}$ (even if we do not have $R^{ord} \cong \mathbb{T}$). This fact has a more elementary proof if the prime-to-p level N of $\mathbb{T}$ can be factored as $N = N'D$ with N' prime to D [MFM, (4.6.22-23)].

7.4.2 Decomposition of $\mathrm{Sel}(Ad(\mathrm{Ind}_K^{\mathbb{Q}} \varphi))$

Pick a deformation $\varphi \in \mathcal{F}_H(A)$. By Proposition 7.3.2, $Ad(\mathrm{Ind}_K^{\mathbb{Q}} \varphi) = \alpha \oplus \mathrm{Ind}_K^{\mathbb{Q}} \varphi^-$, and by Corollary 5.3.3, the cohomology is decomposed accordingly: $H^1(G, Ad(\mathrm{Ind}_K^{\mathbb{Q}} \varphi)) \cong H^1(G, \alpha) \oplus H^1(H, \varphi^-)$. Since Selmer cocycles are upper triangular over D_p and upper nilpotent over I_p, noting the fact that $\alpha \subset Ad(\mathrm{Ind}_K^{\mathbb{Q}} \varphi)$ is realized on diagonal matrices, and $\mathrm{Ind}_K^{\mathbb{Q}} \varphi^-$ is realized on anti-diagonal matrices, the Selmer conditions are compatible with the above factorization; so, we have

Theorem 7.4.4. *We have* $\mathrm{Sel}(Ad(\mathrm{Ind}_K^{\mathbb{Q}} \varphi)) = \mathrm{Sel}(\alpha) \oplus \mathrm{Sel}_K(\varphi^-)$, *where* $\mathrm{Sel}_K(\varphi^-) \subset H^1(H, \varphi^-)$ *is made of classes unramified outside $\wp$ and vanishes over $D_{\wp^\sigma}$ and $\mathrm{Sel}(\alpha)$ is made of classes in $H^1(G, \alpha)$ unramified everywhere. Thus* $\mathrm{Sel}(\alpha) = \mathrm{Hom}(Cl_K, A^\vee) = \mathrm{Hom}(Cl_K \otimes_{\mathbb{Z}} A, \mathbb{Q}_p/\mathbb{Z}_p)$ *and* $\mathrm{Sel}(\alpha)^\vee = Cl_K \otimes_{\mathbb{Z}} A$.

Proof. Pick a Selmer cocycle $u : G \to Ad(\rho_0)^*$. Projecting down to α, it has diagonal form; so, the projection u_α restricted to D_p is unramified. Therefore u_α factors through Cl_K. Start with an unramified homomorphism $u : Cl_K \to A^\vee$ and regard it as having values in diagonal matrices in $Ad(\rho_0)^*$, its class falls in $\mathrm{Sel}(Ad(\rho_0))$.

Similarly, the projection u^{Ind} of u to the factor $\mathrm{Ind}_K^{\mathbb{Q}} \varphi^-$ is anti-diagonal of the form $\left(\begin{smallmatrix} 0 & u^+ \\ u^- & 0 \end{smallmatrix} \right)$. Noting $H^j(\Delta, ((\mathrm{Ind}_K^{\mathbb{Q}} \varphi^-)^*)^H) = 0$ $(j = 1, 2)$, by inflation-restriction sequence Theorem 4.2.7,

$$H^1(G, (\mathrm{Ind}_K^{\mathbb{Q}} \varphi^-)^*) \cong (H^1(H, (\varphi^-)^*) \oplus H^1(H, (\varphi_\sigma^-)^*))^\Delta.$$

So $u^-(\sigma^{-1}g\sigma) = u^+(g)$ as $\sigma \in \Delta$ interchanges $H^1(H, (\varphi^-)^*)$ and $H^1(H, (\varphi_\sigma^-)^*)$. Moreover $u^+ : H \to (\varphi^-)^*$ is unramified outside $\wp$. Since $u^-|_{D_\wp} = 0$, u^+ vanishes on $D_{\wp^\sigma}$ by $u^-(\sigma^{-1}g\sigma) = u^+(g)$. $\qquad\square$

7.4.3 An exotic identity of Galois groups

We start with a result from class field theory:

Proposition 7.4.5. *If $p > 2$, we have an exact sequence*

$$1 \to (1 + p\mathbb{Z}_p)/\varepsilon^{(p-1)\mathbb{Z}_p} \to \Gamma_\wp \to Cl_K \otimes_\mathbb{Z} \mathbb{Z}_p \to 1,$$

where $\varepsilon = 1$ if K is imaginary, and ε is a fundamental unit of K if K is real. Thus $\Gamma_\wp$ is finite if K is real.

Proof. Since $\Gamma_\wp = Cl_K(\wp^\infty) \otimes_\mathbb{Z} \mathbb{Z}_p$, the exact sequence is the p-primary part of the exact sequence of the class field theory:

$$1 \to O_\wp^\times/\overline{O}^\times \to Cl_K(\wp^\infty) \to Cl_K \to 1.$$

Thus tensoring $\mathbb{Z}_p$ over $\mathbb{Z}$, we get the desired exact sequence, since $O_\wp \cong \mathbb{Z}_p$ canonically. Note here $\varepsilon^{p-1} \in 1 + p\mathbb{Z}_p = 1 + \wp O_\wp$. $\qquad\square$

Recall the universal character $\varphi : H \to W[[\Gamma_\wp]]^\times$ in the proof of Corollary 7.4.2. Define $K_{/K}^-$ by the maximal p-abelian anticyclotomic extension unramified outside p (so, $\sigma\gamma\sigma^{-1} = \gamma^{-1}$). The extension K^-/K is also called the anticyclotomic p-ramified p-abelian extension. The fixed subfield of $F^{(p)}(\overline{p})$ by $\mathrm{Ker}(\varphi^-)$ is given by $F(\varphi^-)K^-$. So $\Gamma^- := \mathrm{Gal}(K^-/K)$ is the maximal p-abelian quotient of $\mathrm{Im}(\varphi^-)$; i.e., $\mathrm{Gal}(K^-/K) \cong \Gamma^- \times \mathrm{Gal}(F(\overline{\wp_0^-})/K)$. Note that $\varphi^-(h) = \varphi(h)\varphi(\sigma^{-1}h\sigma)^{-1} \in \Gamma_\wp \subset W[[\Gamma_\wp]]$ if $h \in \Gamma^-$. Thus we have a group theoretic homomorphism $\Gamma^- \to \Gamma_\wp$ given by $h \mapsto \varphi^-(h)$. Though this homomorphism appears "exotic", its inverse looks more natural as it is the projection of $\Gamma_\wp$ to the "$-$" eigenspace of σ in the Galois group of the maximal p-abelian extension of K unramified outside p. We have an exact sequence for $C_K := Cl_K \otimes_\mathbb{Z} \mathbb{Z}_p$:

$$1 \to ((1 + pO_p)/\varepsilon^{(p-1)\mathbb{Z}_p})^{\sigma=-1} \to \Gamma^- \to C_K \to 1,$$

which is the "$-$"-eigenspaces of the action of σ on the exact sequence with $C_K(p^\infty) := \varprojlim_n Cl_K(p^\infty)/p^n Cl_K(p^\infty)$:

$$1 \to (1 + pO_p)/\varepsilon^{(p-1)\mathbb{Z}_p} \to C_K(p^\infty) \to C_K \to 1.$$

Therefore the above homomorphism induces an isomorphism $\Gamma^- \cong \Gamma_\wp$, and in this way, we identify $W[[\Gamma^-]]$ with $W[[\Gamma_\wp]]$.

7.4.4   Cyclicity of anticyclotomic Iwasawa modules

To allow deformations φ of order divisible by p, write φ_0 for the Teichmüller lift of $\overline{\varphi}$, and pick a deformation $\varphi \in \mathcal{F}_H(A)$. Let L (resp. L') be the maximal p-abelian extension of $K^- F(\varphi_0^-)$ (resp. $F(\varphi^-)$) unramified outside $\wp$ totally split at $\wp^\sigma$ (so $L' \subset L$). Put $\mathcal{Y} := \mathrm{Gal}(L/K^- F(\varphi_0^-))$. Write $\overline{H} := \mathrm{Gal}(F(\varphi_0^-)/K)$. By conjugation, $\overline{H} \times \Gamma^- = \mathrm{Gal}(K^- F(\varphi_0^-)/K)$ acts on $\mathcal{Y}$; so, we put $\mathcal{Y}(\varphi_0^-) = \mathcal{Y} \otimes_{\mathbb{Z}_p[\overline{H}]} \varphi_0^-$ (the maximal quotient of $\mathcal{Y}$ on which $\overline{H} \subset \mathrm{Gal}(K^- F(\varphi_0^-)/K)$ acts by φ_0^-). Then $\mathcal{Y}(\varphi_0)$ is a module over $W[[\Gamma^-]]$ (an Iwasawa module). As long as

$$F(\varphi_0^-)K^- \text{ has a unique prime above each prime of } F(\varphi^-) \text{ over } \wp^\sigma, \tag{7.15}$$

the Galois group $\mathrm{Gal}(L'/F(\varphi^-))$ (resp. $\mathrm{Gal}(L'/F(\varphi^-)) \otimes_{\mathbb{Z}_p[\overline{H}]} \varphi_0^-$) is a quotient of $\mathcal{Y}$ (resp. $\mathcal{Y}(\varphi_0)$). Indeed, since $\wp^\sigma$ totally splits in $L'/F(\varphi^-)$, if (7.15) holds, L' and $K^- F(\varphi^-)$ are linearly disjoint over $F(\varphi^-)$, and hence $\mathrm{Gal}(L/K^- F(\varphi_0^-))$ surjects onto $\mathrm{Gal}(L'/F(\varphi^-))$. In general, the cokernel of $\mathrm{Res} : \mathrm{Gal}(L/K^- F(\varphi_0^-)) \to \mathrm{Gal}(L'/F(\varphi^-))$ is isomorphic to $\mathrm{Gal}((L' \cap K^- F(\varphi^-))/F(\varphi^-))$ on which $\overline{H}$ acts trivially; so, as long as $\varphi_0^- \neq 1$, the cokernel restricted to the φ_0^--isotypical quotient vanishes. In particular, if φ induces a surjection $W[[\Gamma^-]] \twoheadrightarrow A$ and $\varphi_0^- \neq 1$, we have $\mathrm{Gal}(L'/F(\varphi^-)) \otimes_{W[[\mathrm{Gal}(F(\varphi^-)/F(\varphi_0^-))]], \varphi^-} A = \mathcal{Y}(\varphi_0^-) \otimes_{W[[\Gamma^-]], \varphi^-} A$.

We have an inflation-restriction exact sequence (Theorem 4.2.7):

$$H^1(F(\varphi^-)/K, (\varphi^-)^*) \hookrightarrow H^1(H, (\varphi^-)^*) \to$$

$$\mathrm{Hom}_{\mathrm{Gal}(F(\varphi^-)/K)}(\mathrm{Gal}(F(\overline{\rho})^{(p)}/F(\varphi^-)), (\varphi^-)^*) \to H^2(F(\varphi^-)/K, (\varphi^-)^*).$$

Lemma 7.4.6. *If either Γ^- is cyclic with $\overline{\varphi}^- \neq 1$ or φ is the universal character $\varphi : H \to W[[\Gamma_\wp]]^\times$, then $H^j(F(\varphi^-)/K, (\varphi^-)^*) = 0$ for $j > 0$.*

For a finite cyclic group C generated by γ, we have

$$H^{2n+1}(C, M) = \mathrm{Ker}(\mathrm{Tr})/\mathrm{Im}(\gamma - 1), \; H^{2n}(C, M) = \mathrm{Ker}(\gamma - 1)/\mathrm{Im}(\mathrm{Tr}),$$

where $\mathrm{Tr}(x) = \sum_{c \in C} cx$ and $(\gamma - 1)(x) = \gamma x - x$ for $x \in M$ [CNF, I.7]. If C is infinite with M discrete, $H^q(C, M) = \varinjlim_{C' \subset C} H^q(C/C', M^{C'})$.

Proof.   First suppose that Γ^- is topologically cyclic (so, $\Gamma^- \times \overline{H}$ is also cyclic). Thus if $\overline{\varphi}^-(\gamma) \neq 1$ for a generator γ of $\Gamma^- \times \overline{H}$, we find

$$H^j(F(\varphi^-)/K, (\varphi^-)^*) = 0$$

as $\gamma - 1 : (\varphi^-)^* \to (\varphi^-)^*$ is a bijection.

If $\varphi = \varphi$, φ^- is a factor of the $\Gamma^- \times \overline{H}$-module $W[[\Gamma \times \overline{H}]]$ and hence $(\varphi^-)^*$ is a cohomologically trivial $\Gamma^- \times \overline{H}$-module. $\square$

Under the assumptions of Lemma 7.4.6, we get

$$H^1(H, (\varphi^-)^*) \cong \mathrm{Hom}_{\mathrm{Gal}(F(\varphi^-)/K)}(\mathrm{Gal}(F(\overline{\rho})^{(p)}/F(\varphi^-)), (\varphi^-)^*).$$

By Theorem 7.4.4, Selmer cocycles factor through $\mathcal{Y}$. We get

$$\mathrm{Sel}(\varphi^-) = \mathrm{Hom}_{\mathbb{Z}_p[\overline{G}]}(\mathcal{Y}, (\varphi^-)^*) \cong \mathrm{Hom}_{W[[\Gamma^-]]}(\mathcal{Y}(\varphi_0^-), (\varphi^-)^*)$$

$$\cong \mathrm{Hom}_W(\mathcal{Y}(\varphi_0^-) \otimes_{W[[\Gamma^-]], \varphi^-} A, \mathbb{Q}_p/\mathbb{Z}_p),$$

where $\overline{G} := \mathrm{Gal}(F(\varphi^-)/K)$. Recall $R_\kappa/R_\kappa([\alpha]-1)R_\kappa \cong W[[\Gamma_\wp]] = W[[\Gamma^-]]$, and write this morphism as $\lambda : R_\kappa \to W[[\Gamma^-]]$.

Theorem 7.4.7. *We have*

$$\Omega_{R_\kappa/\Lambda} \otimes_{R_\kappa, \lambda} W[[\Gamma^-]] \cong Cl_K \otimes_{\mathbb{Z}} W[[\Gamma^-]] \oplus \mathcal{Y}(\varphi_0^-),$$

and under either $\varphi_0^- \neq 1$ or (7.15), $\mathrm{Sel}(\varphi^-) \cong \mathrm{Hom}_{W[[\Gamma^-]]}(\mathcal{Y}(\varphi_0^-), (\varphi^-)^*)$ all as $W[[\Gamma^-]]$-modules.

This follows from Theorems 6.1.11 and 7.4.4 combined with Lemma 7.4.6.

Since $p \nmid [F(\varphi_0^-) : K]$, the p-Hilbert class field $H_{/K}$ and $F = F(Ad(\overline{\rho})) = F(\varphi_0^-)$ are linearly disjoint over K; so, we have $[H : K] = [HF, F]$; so, $p \nmid h_F$ implies $p \nmid h_K$. Since p splits in K, $\overline{\varphi}|_{D_p} \neq \overline{\omega}_p$ implies the condition (C2): $Ad(\overline{\rho}) \otimes_{\mathbb{F}_p[\overline{G}]} \mathbb{F}_p(1) = 0$. Thus combining the above theorem with the cyclicity result in Corollary 7.2.10, we get the following fact under $p \nmid h_F$:

Corollary 7.4.8. *If $Cl_F \otimes_{\mathbb{Z}[\overline{H}]} \overline{\varphi}^- = 0$ and $\varphi_0^- \neq 1$ and $\overline{\varphi}^-|_{D_p} \neq \overline{\omega}_p$, then $\mathcal{Y}(\varphi_0^-)$ is a cyclic module over $W[[\Gamma^-]]$.*

We proved the corollary under the stronger assumption $p \nmid h_F$. The result in Corollary 7.4.8 is valid only assuming triviality of the $\overline{\varphi}^-$-isotypical component $Cl_F \otimes_{\mathbb{Z}[\overline{H}]} \overline{\varphi}^-$ of $Cl_F \otimes_{\mathbb{Z}} \mathbb{F}$ as we will prove at the end of §8.3.3.

7.5 Iwasawa theory over quadratic fields

Let $K/\mathbb{Q}$ be a quadratic extension. We keep assuming that p splits into $\wp\wp^\sigma$ in K. Assuming that $\overline{\rho} = \mathrm{Ind}_K^\mathbb{Q} \overline{\varphi}$ for a character $\overline{\varphi} : \mathfrak{G}_K \to \mathbb{F}^\times$, we describe the size of its adjoint Selmer group in terms of a Minkowski unit. Let $\overline{G} = \mathrm{Gal}(F/\mathbb{Q}) \cong \mathrm{Im}(Ad(\overline{\rho}))$. Let φ be the Teichmüller lift of $\overline{\varphi}$, and put $\rho = \mathrm{Ind}_K^\mathbb{Q} \varphi$. Then $\overline{G} \cong \mathrm{Im}(Ad(\rho)) = \mathrm{Im}(Ad(\overline{\rho}))$. We write $\mathbb{F}$ for the field generated by the values of $\overline{\varphi}$. As seen in Proposition 7.3.2, $Ad(\rho) \cong \alpha \oplus \mathrm{Ind}_K^\mathbb{Q} \varphi^-$ for $\alpha = \left(\frac{K/\mathbb{Q}}{\cdot}\right)$. So $F = F(Ad(\overline{\rho})) = F(\varphi^-)$.

Then we take W to be the unramified extension of $\mathbb{Z}_p$ with $W/\mathfrak{m}_W = \mathbb{F}$. We write O (resp. O_K) for the integer ring F (resp. K). Fix a prime $\mathcal{P}|p$ in O and a prime $\mathfrak{P}|\mathcal{P}$ in $F(\varphi) = F(\bar{\rho})$. We write $\overline{D} \subset \overline{G}$ (resp. $D' \subset G_F := \mathrm{Gal}(F^{(p)}(\bar{\rho})/F))$ for the decomposition group of $\mathcal{P}$ (resp. $\mathfrak{P}$) such that $\rho|_{D'} = \left(\begin{smallmatrix} \epsilon & 0 \\ 0 & \delta \end{smallmatrix}\right)$ with $\delta = \varphi_\sigma|_{D'}$ unramified. We need to assume $\mu_p(F) = \{1\}$. If $\mu_p(F) \neq \{1\}$, then $\overline{G}$ has cyclic quotient $\mathrm{Gal}(\mathbb{Q}[\mu_p]/\mathbb{Q})$ of order $p-1$. Since $\overline{G}$ is dihedral, its maximal cyclic quotient is of order 2; so, $p = 3$. If $p = 3$, by Proposition 7.3.6 (4), this could happen. So we just assume hereafter in this chapter

$$\mu_3(F) = \{1\} \ \text{when } p = 3. \tag{mu3}$$

7.5.1   Galois action on global units

Recall $(O^\times \otimes_\mathbb{Z} W) \oplus W \cong \mathrm{Ind}_C^{\overline{G}} W \cong W[\overline{G}/C]$ from §7.2.8, where we simply write W for the Galois module with trivial action (so, $W = \mathbf{1}$ for the trivial character $\mathbf{1}$). Here C is the subgroup of $\overline{G}$ generated by the fixed complex conjugation c. Let ε be the Minkowski unit at p; so, $O^\times \otimes_\mathbb{Z} W = W[\overline{G}]\varepsilon$. For each irreducible representation ξ of a subgroup X in $\mathrm{Gal}(F/\mathbb{Q})$ into $\mathrm{GL}_n(W)$, we write ε_ξ for the image of ε in $O^\times \otimes_{\mathbb{Z}[X]} \xi$. We often take $\alpha : \mathrm{Gal}(F/\mathbb{Q}) \to \{\pm 1\}$ and $\varphi^- : \mathrm{Gal}(F/K) \to W^\times$ for ξ.

Proposition 7.5.1. *We have*

$$\mathrm{Hom}_{W[\overline{G}]}(\mathrm{Ind}_K^\mathbb{Q} \varphi^-, O^\times \otimes_\mathbb{Z} W) = \begin{cases} 0 & \text{if } K \text{ is real,} \\ W & \text{if } K \text{ is imaginary,} \end{cases}$$

$$\mathrm{Hom}_{\mathbb{Z}_p[\overline{G}]}(\alpha, O^\times \otimes_\mathbb{Z} W) = \begin{cases} W & \text{if } K \text{ is real,} \\ 0 & \text{if } K \text{ is imaginary.} \end{cases}$$

If K is imaginary, $\varepsilon_{\varphi^-} \neq 1$ and $\varepsilon_\alpha = 1$ and if K is real, $\varepsilon_\alpha \neq 1$ and $\varepsilon_{\varphi^-} = 1$.

We have $\mathrm{Hom}_{W[\overline{G}]}(\mathrm{Ind}_K^\mathbb{Q} \varphi^-, \mathrm{Ind}_C^{\overline{G}} W) = \mathrm{Hom}_{W[C]}(\mathrm{Ind}_K^\mathbb{Q} \varphi^-|_C, W)$ and $\mathrm{Hom}_{\mathbb{Z}_p[\overline{G}]}(\alpha, \mathrm{Ind}_C^{\overline{G}} \mathbf{1}) = \mathrm{Hom}_{\mathbb{Z}_p[C]}(\alpha|_C, \mathbf{1})$ by Lemma 5.3.2. The last assertion of the proposition is clear from the second identity.

Proof. Pick $\sigma \in \overline{G}$ such that $\sigma|K$ is non-trivial. If K is imaginary, $\mathrm{Ind}_K^\mathbb{Q} \varphi^-|_C = \mathbf{1} \oplus \alpha$ as $\mathrm{Tr}(\mathrm{Ind}_K^\mathbb{Q} \varphi^-)(c)) = 0$. Therefore

$$\mathrm{Hom}_{W[C]}(\mathrm{Ind}_K^\mathbb{Q} \varphi^-|_C, W) = \mathrm{Hom}_{W[C]}(\mathbf{1} \oplus \alpha, \mathbf{1}) = W.$$

Suppose that K is real. Since
$$Ad(\bar{\rho})(c) \sim \mathrm{diag}[-1, 1, -1] \sim \mathrm{diag}[\varphi^-(c), \alpha(c), (\varphi^-)^{-1}(c)],$$
$\alpha(c) = 1$ implies $\varphi^-(c) = \varphi_\sigma^-(c) = -1$. Therefore
$$\mathrm{Hom}_{W[C]}(\mathrm{Ind}_K^{\mathbb{Q}} \varphi^-|_C, \mathbf{1}) = \mathrm{Hom}_{W[C]}(\varphi^- \oplus \varphi^-, \mathbf{1}) = 0$$
for $\varphi^- : C \cong \{\pm 1\}$. $\qquad\qquad\qquad\qquad\qquad\qquad\qquad\square$

7.5.2 Selmer group and ray class group

As we will see in Lemma 7.7.2, we have the following fact:

Lemma 7.5.2. *We have a canonical inclusion*
$$\mathrm{Sel}(Ad(\rho)) \subset \mathrm{Hom}_{\mathbb{Z}_p[\overline{G}]}(C_F(p^\infty), Ad(\rho)^*),$$
where $C_F(p^\infty) = \varprojlim_n C_F(p^n)$ *with* $C_F(p^n) = Cl_F(p^n) \otimes_{\mathbb{Z}} \mathbb{Z}_p$.

We put $\mathfrak{h} := \mathrm{Gal}(F^{(p)}/F)$ (resp. $\mathfrak{h}^{ab}$) for the maximal p-profinite extension $F^{(p)}$ of F unramified outside p (resp. the maximal p-profinite abelian quotient of $\mathfrak{h}$), and we study decomposition group in $\mathfrak{H}^{ab}$ as $\overline{D}$-modules. Recall the fixed prime factor $\mathcal{P}|p$ in O with its decomposition subgroup $\overline{D} \subset \overline{G}$. Write simply $M_{\mathcal{P}} := \widehat{F_{\mathcal{P}}^\times} \otimes_{\mathbb{Z}} W$ for $\widehat{F_{\mathcal{P}}^\times} := \varprojlim_n (F_{\mathcal{P}}^\times \otimes_{\mathbb{Z}} \mathbb{Z}/p^n\mathbb{Z})$ and $U_{\mathcal{P}} := \widehat{O_{\mathcal{P}}^\times} \otimes_{\mathbb{Z}_p} W$. Then for each character $\xi : \overline{D} \to W^\times$, $M_{\mathcal{P}}$ contains as a direct factor the ξ-eigenspace $M_{\mathcal{P}}[\xi]$. Then writing $\mu_p(F_{\mathcal{P}})_{/\mathbb{F}} = \mu_p(F_{\mathcal{P}}) \otimes_{\mathbb{Z}} \mathbb{F}$

(U) $M_{\mathcal{P}}[\xi] = U_{\mathcal{P}}[\xi] \cong \begin{cases} W & \text{if } \xi \notin \{\mathbf{1}, \omega\}, \\ W \oplus \mu_p(F_{\mathcal{P}})_{/\mathbb{F}} & \text{if } \xi = \omega. \end{cases}$

(M) We have an exact sequence $0 \to U_{\mathcal{P}}[\mathbf{1}] \to M_{\mathcal{P}}[\mathbf{1}] \xrightarrow{\mathrm{ord}_{\mathcal{P}}} W \to 0$ induced by the valuation $\mathrm{ord}_{\mathcal{P}} : F_{\mathcal{P}}^\times \twoheadrightarrow \mathbb{Z}$ at $\mathcal{P}$, and $U_{\mathcal{P}}[\mathbf{1}] \cong W$.

7.5.3 Structure of $M_p[Ad]$ as a $\overline{G}$-module in Case D

For each irreducible factor ϕ of Ad, we consider the ϕ-isotypical component $X[\phi]$, and write $\mu_p(F_{\mathcal{P}})_{\mathbb{F}} = \mu_p(F_{\mathcal{P}}) \otimes_{\mathbb{Z}} \mathbb{F}$.

Lemma 7.5.3. *Assume* $\varphi^-|_{\overline{D}} \neq 1$ *and* $p \geq 5$. *Then* $\mathrm{Hom}_{\overline{G}}(M_p, \phi^*)$ *is given by*
$$\begin{cases} \mathrm{Hom}_{\overline{D}}(U_{\mathcal{P}}[\varphi^-] \oplus U_{\mathcal{P}}[\varphi_\sigma^-], (\phi|_{\overline{D}})^*) \cong W^2 \oplus \mu_p(F_{\mathcal{P}})_{\mathbb{F}}[\xi^{\pm 1}] & \dim \phi = 2, \\ \mathrm{Hom}_{\overline{D}}(U_{\mathcal{P}}[\phi], \phi^*) \cong W & \phi \subsetneq \mathrm{Ind}_K^{\mathbb{Q}} \varphi^-, \\ \mathrm{Hom}_{\overline{D}}(M_{\mathcal{P}}[\mathbf{1}], \mathbf{1}^*) \cong W^2 & \phi = \alpha, \end{cases}$$
where $\xi = \varphi^-$ *in the first case.*

Proof. Since $M_p = \text{Ind}_{\overline{D}}^{\overline{G}} M_{\mathcal{P}}$, we have

$$\text{Hom}_{\overline{G}}(M_p, \phi^*) = \text{Hom}_{\overline{D}}(M_{\mathcal{P}}, \phi^*|_{\overline{D}})$$

by Shapiro's Lemma 5.3.2. If $\varphi^-|_{\overline{D}} \neq 1$, $\phi|_{\overline{D}}$ is

- $(\varphi^- \oplus \varphi_\sigma^-)|_{\overline{D}}$ when $\phi = \text{Ind}_K^{\mathbb{Q}} \varphi^-$ is irreducible ($\text{ord}(\varphi^-) \geq 3$),

- $\varphi^-|_{\overline{D}}$ when $\phi \subsetneq \text{Ind}_K^{\mathbb{Q}} \varphi^-$ ($\text{ord}(\varphi^-) = 2$),

- 1 when $\phi = \alpha$.

Let $\xi = \varphi^-|_{\overline{D}}$. Since $M_{\mathcal{P}}[\xi^{\pm 1}] = U_{\mathcal{P}}[\xi^{\pm 1}]$ (by $\xi \neq 1$),

$$(\text{Ind}_{\overline{D}}^{\overline{G}} U_{\mathcal{P}}[\xi^{\pm 1}])[Ad] = \begin{cases} \phi \oplus \text{Ind}_{\overline{D}}^{\overline{G}} \mu_p(F_{\mathcal{P}})[\phi] & \text{if } \xi \neq \xi^{-1} \text{ and } \dim \phi = 2, \\ \phi \oplus \phi\alpha & \text{if } \phi \subsetneq \text{Ind}_K^{\mathbb{Q}} \varphi^-, \\ 0 & \text{if } \phi = \alpha, \end{cases}$$

$$(\text{Ind}_{\overline{D}}^{\overline{G}} M_{\mathcal{P}}[1])[Ad] = \begin{cases} 0 & \text{if } \phi \subset \text{Ind}_K^{\mathbb{Q}} \varphi^-, \\ \alpha \oplus \alpha & \text{if } \phi = \alpha. \end{cases}$$

This is because $M_{\mathcal{P}}[\xi^{\pm 1}] = U_{\mathcal{P}}[\xi^{\pm 1}] \cong W \oplus \mu_p(F_{\mathcal{P}})_{\mathbb{F}}$ by (U) and by Shapiro's lemma

$$\text{Hom}_{\overline{G}}(\text{Ind}_K^{\mathbb{Q}} \varphi^-, \text{Ind}_{\overline{D}}^{\overline{G}} U_{\mathcal{P}}[\xi]) = \text{Hom}_{\overline{D}}(\text{Ind}_K^{\mathbb{Q}} \varphi^-|_{\overline{D}}, \xi \oplus (\mu_p(F_{\mathcal{P}}) \otimes_{\mathbb{Z}} \mathbb{F}))$$
$$= \text{Hom}_{\overline{D}}(\xi \oplus \xi^{-1}, \xi \oplus (\mu_p(F_{\mathcal{P}}) \otimes_{\mathbb{Z}} \mathbb{F}))$$

as $\overline{D} \subset \text{Gal}(F/K)$. The second formula follows from (M). $\square$

7.5.4 *Theorem for* $\text{Sel}(\text{Ind}_K^{\mathbb{Q}} \varphi^-)$

The representations $\Phi := \text{Ind}_K^{\mathbb{Q}} \varphi^-$ and α in $Ad(\rho)$ fit into the following exact sequence of $\overline{G}$-modules:

$$0 \to \overbrace{\Phi \oplus \text{Ind}_{\overline{D}}^{\overline{G}}(\mu_p(F_{\mathcal{P}}) \otimes_{\mathbb{Z}} \mathbb{F})[\Phi]}^{\text{inertia part}} \oplus \alpha \oplus \Phi \to M_p[Ad] \to \alpha \to 0.$$

Here Φ can be reducible.

Theorem 7.5.4. *Assume that we are in Case D with irreducible* $\text{Ind}_K^{\mathbb{Q}} \overline{\varphi}$. *Then we have an exact sequence*

$$\text{Hom}_{\mathbb{Z}_p[\overline{G}]}(C_F, \Phi^*) \hookrightarrow \text{Sel}(\text{Ind}_K^{\mathbb{Q}} \varphi^-) \twoheadrightarrow \text{Hom}_{W[\overline{D}]}(U_{\mathcal{P}}[\varphi^-]/\overline{\langle \varepsilon_{\varphi^-} \rangle}, W^\vee),$$

where ε is a Minkowski unit, ε_{φ^-} is the projection of ε in the direct summand $U_{\mathcal{P}}[\varphi^-]$ under $O^\times \to U_p \twoheadrightarrow U_{\mathcal{P}}[\varphi^-]$, and $\overline{\langle \varepsilon_{\varphi^-} \rangle}$ is the p-adic closure of the subgroup $\varepsilon_{\varphi^-}^{\mathbb{Z}}$ generated by ε_{φ^-}.

This theorem describes the structure of the Selmer group $\mathrm{Sel}(\mathrm{Ind}_K^{\mathbb{Q}}\varphi^-)$ explicitly via class group and a specific unit. There is a conjecture by Stark [St75], [T81] and [T84] about the size (not the structure) of the Selmer group via the Artin L-value at 1 or equivalently at 0 (in terms of a Minkowski unit) and also an index formula similar to Theorem 1.11.9 by Robert (when K is imaginary) in terms of elliptic units [UNE] (and [MUN, Chapters 10, 13]) completing a project started by Siegel. See also [G80] on the relation between Minkowski units and elliptic units and [MUN, Theorem 9.5.1] for a class number formula as an unit index.

Proof. *Argument for* $\mathrm{Hom}_{\mathbb{Z}_p[\overline{G}]}(C_F, \Phi^*) \hookrightarrow \mathrm{Sel}(\Phi)$. We proceed as in Case E (in Step 1 of the proof of Theorem 7.7.3) replacing Ad by Φ. Let $C_F^{(p)}$ be the subgroup of C_F generated by the projection to C_F of classes of primes above p. Since $C_F^{(p)}$ (surjective image of $\mathrm{Ind}_{\overline{D}}^{\overline{G}}\mathbf{1}$) does not contain $\Phi = \mathrm{Ind}_K^{\mathbb{Q}}\varphi^-$, we can ignore it and can work with the entire C_F. Elements in $\mathrm{Hom}_{\mathbb{Z}_p[\overline{G}]}(C_F, \Phi^*)$ are everywhere unramified and trivial at p; so, they give rise to a subgroup of $\mathrm{Sel}(\Phi)$ of unramified classes trivial at p. Indeed, by $H^1(\mathfrak{G}, \Phi^*) \cong \mathrm{Hom}_{\mathbb{Z}_p[\overline{G}]}(\mathfrak{h}^{ab}, \Phi^*)$, any $u \in \mathrm{Hom}_{\mathbb{Z}_p[\overline{G}]}(C_F, \Phi^*)$ extends uniquely the cocycle $u : \mathfrak{G} \to \Phi^*$ unramified everywhere over $\mathfrak{h}$. Since the inertia group $I_l \subset \overline{G}$ of any prime $l \in S$ has order prime to p, $u|_{I_l} = 0$, and hence $[u] \in \mathrm{Sel}(\Phi)$.

Elements of $\mathrm{Sel}(\Phi)$ modulo $\mathrm{Hom}_{\mathbb{Z}_p[\overline{G}]}(C_F, \Phi^*)$ are determined by its restriction to M_p as they are unramified outside p as they factor through $C_F(p^\infty)$ and $p \nmid |I_l|$.

Inertia part. Recall $\xi = \epsilon\delta^{-1} = \varphi^-$. A Selmer cocycle $u|_{\mathfrak{h}^{ab}}$ can be regarded as a $W[\overline{G}]$-linear homomorphism of M_p into $(\varphi^-)^*$. Since $M_p/U_p \cong \mathrm{Ind}_{\overline{D}}^{\overline{G}}\mathbf{1}$ does not contain Φ, we can ignore M_p/U_p. By its $\overline{G}$-equivariance,

$$u|_{U_p} \in \mathrm{Hom}_{W[\overline{G}]}(U_p, \Phi^*).$$

By Shapiro's Lemma 5.3.2,

$$\mathrm{Hom}_{W[\overline{G}]}(U_p, \Phi^*) \cong \mathrm{Hom}_{W[\overline{D}]}(U_{\mathcal{P}}[\varphi^-], (\varphi^-)^*).$$

Since u is trivial over $\overline{O^\times}$, u factors through $U_{\mathcal{P}}[\varphi^-]/\overline{\langle\varepsilon_{\varphi^-}\rangle}$. $\square$

Corollary 7.5.5. *If K is imaginary, we have*

$$|\mathrm{Sel}(\mathrm{Ind}_K^{\mathbb{Q}}\varphi^-)| = |C_F \otimes_{\mathbb{Z}_p[\mathrm{Gal}(F/K)]} \varphi^-||(U_{\mathcal{P}}[\varphi^-]/\overline{\langle\varepsilon_{\varphi^-}\rangle})|$$

which is finite, otherwise $\mathrm{Sel}(\mathrm{Ind}_K^{\mathbb{Q}}\varphi^-)^\vee$ *has W-rank 1 (up to finite W-torsion).*

In view of Corollary 6.2.24, this result is more in a spirit of a p-adic version [G81] of Stark's conjecture by B. Gross, though in [G81], a conventional p-adic L function different from ours is studied (see also Corollary 7.7.4).

Proof. By Proposition 7.5.1, $\varepsilon_{\varphi^-} \neq 1$ only when K is imaginary. Thus the finiteness of $\mathrm{Sel}(\mathrm{Ind}_K^{\mathbb{Q}} \varphi^-)$ follows. When K is real, $U_{\mathcal{P}}[\varphi^-] \cong W$ up to torsion, and from Theorem 7.5.4, the Selmer group has corank 1.   □

Remark 7.5.6. Suppose that K is imaginary. We choose a Minkowski unit ε. Then $[O_F^\times : \mathbb{Z}[\overline{G}]\varepsilon]$ is finite. Note $|\log_p(\varepsilon)_{\varphi^-}|_p < 1$ always. The span $\overline{\langle \varepsilon_{\varphi^-} \rangle}$ is equal to $\langle \varepsilon_{\varphi^-}^{N(\mathcal{P})-1} \rangle$ for the norm $N(\mathcal{P})$. For $p \nmid h_F[O_F^\times : \mathbb{Z}[\overline{G}]\varepsilon]$, $|\log_p \varepsilon_{\varphi^-}|_p \leq p^{-2}$ if and only if $\varepsilon_{\varphi^-}^{N(\mathcal{P})-1}$ is a p-power in the unit group $U_{\mathcal{P}}[\varphi^-]$; so, $|(U_{\mathcal{P}}[\varphi^-]/\overline{\langle \varepsilon_{\varphi^-} \rangle})|$ is non-trivial, and this implies $\dim_{\mathbb{F}}(Ad(\overline{\rho}_{\mathfrak{p}})) \geq 1$, which in turn implies $\mathbb{T}_{\mathfrak{p}} \neq \Lambda$. We therefore ask

$$if \ |\log_p \varepsilon_{\varphi^-}|_p \leq p^{-2} \ for \ infinitely \ many \ p \ splits \ in \ K?$$

If affirmative, we have infinity of $\mathfrak{p}$ with $\mathbb{T}_{\mathfrak{p}} \neq \Lambda$. It is an interesting question to find out the density of split primes with $|\log_p \varepsilon_{\varphi^-}|_p \leq p^{-2}$. See §8.3.6 for a down-to-earth discussion of a similar question when K is real.

7.6   Selmer groups in the non-split case

We study the cyclicity problem in the remaining cases (i.e., Cases Ds and U$_-$). Let $\overline{G} := \mathrm{Gal}(F(\overline{\rho})/\mathbb{Q})$, $\overline{H} := \mathrm{Gal}(F(\overline{\rho})/K)$ and $G_{\overline{\rho}} := \mathrm{Gal}(F^{(p)}(\overline{\rho})/F(\overline{\rho}))$. We can decompose $G = \overline{G} \ltimes G_{\overline{\rho}}$, $H = \overline{H} \ltimes G_{\overline{\rho}}$ and $D_p = \Delta_p \ltimes D_{\mathcal{P}}$ with $\Delta_p \subset \overline{G}$ and choose σ in Δ_p; so, $\sigma \in D_p$ normalizes I_p. In Case Ds, the inertia group in Δ_p has order 2; so, we may assume $\sigma^2 = 1$. However, in Case U$_-$, σ can have even order > 2.

7.6.1   Set-up in Cases Ds and U$_-$

Let $\rho = \mathrm{Ind}_K^{\mathbb{Q}} \varphi$ in the standard form in (5.3). Then $\rho(\sigma) = \begin{pmatrix} 0 & \varphi(\sigma^2) \\ 1 & 0 \end{pmatrix}$ and $\rho(\sigma^2) = \begin{pmatrix} \varphi(\sigma^2) & 0 \\ 0 & \varphi(\sigma^2) \end{pmatrix}$. Pick a root of unity ζ such that $\zeta^2 = \varphi(\sigma^2)$.
    Recall the set-up of the two cases:

(U$_-$)   Suppose $\alpha(p) = -1$. Put $\delta_\pm$ to be the unramified character of D_p such that $\delta_\pm(\sigma) = \pm\alpha(p)\zeta$. Then we normalize $\rho(\sigma) \sim \mathrm{diag}[\zeta, \alpha(p)\zeta]$ over $\mathbb{Z}[\varphi][\frac{1}{2}, \zeta]$ (automatically p-distinguished). Pick an odd prime

$\mathfrak{p}$ of $\mathbb{Z}[\varphi][\frac{1}{2}, \zeta]$ and put $W = W_{\mathfrak{p}} = \mathbb{Z}[\varphi, \zeta]_{\mathfrak{p}}$. We consider two ordinary deformation problem: one corresponds to the ordinary quotient character $\overline{\delta} = \overline{\delta}_+ := (\delta_+ \bmod \mathfrak{m}_W) : H \to \mathbb{F}^\times$ and another having $\overline{\delta} = \overline{\delta}_- := (\delta_- \bmod \mathfrak{m}_W)$. Note in this case, $\delta_- = \alpha\delta_+$.

(Ds) We assume $\alpha(p) = 0$. The character φ is unramified at p, and hence $\varphi|_{D_\varphi} = \varphi_\sigma|_{D_\varphi}$. Then $\rho|_{D_p}$ is a direct sum of subspaces on which I_p acts trivially and by α. The action of D_p on $H^0(I_p, \mathrm{Ind}_F^{\mathbb{Q}} \varphi)$ gives a character $\delta : D \to \mathbb{Z}[\varphi]^\times$ and $\rho|_{D_p} = \mathrm{diag}[\delta\alpha, \delta]$ (p-distinguished). Pick an odd prime $\mathfrak{p}$ of $\mathbb{Z}[\varphi]$, and we write $W = W_{\mathfrak{p}} := \mathbb{Z}[\varphi]_{\mathfrak{p}}$. Put $\overline{\delta}_+ := (\delta \bmod \mathfrak{m}_W)$ and $\overline{\delta}_- = \overline{\alpha}\overline{\delta}_+$. We take $\zeta := -\delta(\sigma)$.

In Cases U$_-$ and Ds, for $\Xi := \left(\begin{smallmatrix} \zeta & -\zeta \\ 1 & 1 \end{smallmatrix}\right)$, we get by computation

$$\begin{pmatrix} 0 & \zeta^2 \\ 1 & 0 \end{pmatrix} \Xi = \Xi \begin{pmatrix} \zeta & 0 \\ 0 & -\zeta \end{pmatrix}$$

and $\Xi^{-1}J'\Xi = \left(\begin{smallmatrix} 0 & 1 \\ 1 & 0 \end{smallmatrix}\right) =: j$ for $J' = \mathrm{diag}[-1, 1] = -J$.

Let $\beta = \mp\delta_+(\mathrm{Frob}_p)$ if $\alpha(p) = -1$ and $\beta = \varphi(\wp)$ if $\alpha(p) = 0, 1$. Replacing $f = \sum_{0 \neq \mathfrak{a}: O\text{-ideals}} \varphi(\mathfrak{a})q^{N(\mathfrak{a})}$ by $f_\pm(z) := f(z) - \beta f(pz)$, we have $f_\pm|U(p) = \pm\delta(\mathrm{Frob}_p)f_\pm$. Hereafter we choose the p-stabilized form $f_\pm$ in the place of f. Thus $\delta_\pm$ has values in $\mathbb{Z}[f_\pm]^\times$, and hereafter write the level of $f_\pm$ as N (so, $N = N(\mathfrak{f})D$ if $p|N$ and $N = pN(\mathfrak{f})D$ otherwise).

7.6.2 *Local Galois action in non-split case*

Let $\varphi^- = \varphi\varphi_\sigma^{-1} \neq 1$ be an anticyclotomic character of $\mathfrak{G}_K^S$, and suppose that the order of φ^- is prime to p. The splitting field $F := F(\varphi^-)$ is a dihedral extension of $\mathbb{Q}$, and we suppose that F is totally imaginary if K is real. Recall $\overline{G} := \mathrm{Gal}(F/\mathbb{Q})$ and $\overline{H} := \mathrm{Gal}(F/K)$. Write simply Y for the maximal p-profinite quotient C_F of $Cl_F(p^\infty)$; so, by class field theory $Y = \mathrm{Gal}(F_\infty/F)$ for the maximal p-abelian extension F_∞ of $F = F(\varphi^-)$ unramified outside p. For any profinite module X, we put $\widehat{X} := \varprojlim_n X/p^n X$ (the maximal p-profinite quotient of X).

Writing O (resp. O_p) for the integer ring of F (resp. its p-adic completion), we have an exact sequence of $\mathbb{Z}_p[\overline{G}]$-modules by class field theory:

$$1 \to \widehat{O}_p^\times/\overline{O}^\times \xrightarrow{i} Y \to C_F \to 1, \tag{7.16}$$

for $C_F := Cl_F \otimes_{\mathbb{Z}} \mathbb{Z}_p$, where $\overline{O}^\times$ is the closure of the image of $O^\times$ in $\widehat{O}_p^\times$. The $\mathbb{Z}_p$-module Y fits into the exact sequence $Y \hookrightarrow \mathrm{Gal}(F_\infty/\mathbb{Q}) \twoheadrightarrow \overline{G} := \mathrm{Gal}(F/\mathbb{Q})$. Here, $\overline{G}$ acts on Y by conjugation.

Let $\widehat{O}^{\times} := \varprojlim_n O^{\times}/(O^{\times})^{p^n}$, and regard it as a $\mathbb{Z}_p[G]$-module. As $p \nmid |\overline{H}|$, an (abelian) irreducible $\mathbb{F}_p[\overline{G}]$-module $\overline{\xi}$ lifts uniquely to a $\mathbb{Z}_p[\overline{G}]$-module, which is irreducible over $\mathbb{Q}_p$. Take an absolutely irreducible factor η of ξ; so, $\eta : \overline{H} \to \mathbb{Z}_p[\eta]^{\times}$ is a character. Then the corresponding factor $\overline{\eta}$ of $\overline{\xi}$ has splitting field $\mathbb{F}_p[\overline{\eta}]$ and $\mathbb{Z}_p[\eta] = W(\mathbb{F}[\overline{\eta}])$. The $\overline{H}$-module ξ (resp. $\overline{\xi}$) is a direct sum of Galois conjugates η^{σ} (resp. $\overline{\eta}^{\sigma}$) of η (resp. $\overline{\eta}$). Here σ runs over $\mathrm{Gal}(\mathbb{F}_p[\eta]/\mathbb{F}_p) = \mathrm{Aut}(W(\mathbb{F}[\overline{\eta}])/\mathbb{Z}_p)$. Therefore $\widehat{O}^{\times}[\xi] := \widehat{O}^{\times} \otimes_{\mathbb{Z}_p[\overline{H}]} \xi \cong \widehat{O}^{\times} \otimes_{\mathbb{Z}_p[\overline{G}]} \eta \cong \mathbb{Z}_p[\eta]$ as $\mathbb{Z}_p$-modules, though the right-hand-side is naturally a $\mathbb{Z}_p[\eta][\overline{H}]$-module.

Pick a Minkowski unit ε of $O^{\times}$ at p. Then the projection ε_{ξ} of ε in $\widehat{O}^{\times}[\xi]$ is a generator of $\widehat{O}^{\times}[\xi]$ over $\mathbb{Z}_p[\overline{H}]$. When we regard ε_{ξ} as an element of $\widehat{O}^{\times} \otimes_{\mathbb{Z}_p[\overline{H}]} \eta$ via the isomorphism $\widehat{O}^{\times}[\xi] \cong \widehat{O}^{\times} \otimes_{\mathbb{Z}_p[\overline{G}]} \eta$, we write it as ε_{η}. Under the inclusion $\widehat{O}^{\times} \to \widehat{O}_p^{\times}$, we can further bring ε_{η} into $\widehat{O}_p^{\times}$. We write $(\varepsilon_{\eta}) := \mathbb{Z}_p[\overline{H}]\varepsilon_{\eta} \subset \widehat{O}_p^{\times}[\eta]$, where $\mathbb{Z}_p[\overline{H}]\varepsilon_{\eta}$ is the sub $\mathbb{Z}_p[\eta][\overline{G}]$-module generated by ε_{η}. If $\widehat{O}_p^{\times}$ is p-torsion-free, the p-torsion part $T[\eta]$ of $\widehat{O}_p^{\times}[\eta]/\widehat{O}^{\times}[\eta]$ is isomorphic to $W/(\log_p(\varepsilon_{\eta})/\log_p(1+p))W$. In general, we have $T[\eta] \cong W/[\varepsilon]_{\eta}W \oplus \mu_{p^{\infty}}(O_p)/\mu_{p^{\infty}}(O)$ for $[\varepsilon]_{\eta} := \log_p(\varepsilon_{\eta})/\log_p(1+p)$.

Proposition 7.6.1. *Let the notation be as above. Suppose that $\alpha(p) \neq 1$ and φ^- is unramified at p and ramified at ∞ if K is real. Put $I := \mathrm{Im}(i)$ in (7.16). For any representation $\xi :? \to \mathrm{GL}_n(W)$ for $? = \overline{G}, \overline{H}$, writing ξ also for the W-module with the action given by ξ, put $I[\xi] = I \otimes_{\mathbb{Z}_p[?]} \xi$. Realize $\mathrm{Ind}_K^{\mathbb{Q}} \varphi^-$ in the standard form as in (5.3) over W. Then*

(1) *If K is real and φ^- has order ≥ 3, $I[\mathrm{Ind}_K^{\mathbb{Q}} \varphi^-] \cong (\mathrm{Ind}_K^{\mathbb{Q}} \varphi^-)^2$ and $I[\varphi^-] \cong \varphi^- \oplus \varphi^-$.*

(2) *If K is imaginary and φ^- has order ≥ 3, up to finite error, $I[\mathrm{Ind}_K^{\mathbb{Q}} \varphi^-] \cong \mathrm{Ind}_K^{\mathbb{Q}} \varphi^-$ and $I[\varphi^-] \cong \varphi^-$. Indeed, the ϕ-eigenspace for $\phi := \varphi^-$ and φ_{σ}^- of $\overline{O}^{\times}$ is free of rank one over W with a generator ε_{ϕ} given by $\sum_{h \in \overline{H}} \phi^{-1}(h)h(\varepsilon)$ for a Minkowski unit $\varepsilon \in O^{\times}$ at p, and $I[\mathrm{Ind}_K^{\mathbb{Q}} \varphi^-] \cong \mathrm{Ind}_K^{\mathbb{Q}} \varphi^- \oplus (\mathrm{Ind}_K^{\mathbb{Q}} \varphi^- \otimes_W W/[\varepsilon]_{\varphi^-}W)$.*

(3) *If K is real and φ^- has order 2, then φ^- extends to an odd character $\widetilde{\varphi}^-$ of $\overline{G}$ of order 2 and $I[\widetilde{\varphi}^-] \cong \widetilde{\varphi}^-$.*

(4) *If K is imaginary and φ^- has order 2, then φ^- extends to an even character $\varphi_{\mathbb{R}}^-$ of $\overline{G}$ of order 2 and another odd character $\varphi_{\mathbb{R}}^- \alpha$ of order 2. We have $I[\varphi_{\mathbb{R}}^- \alpha] \cong \varphi_{\mathbb{R}}^- \alpha$, and the $\varphi_{\mathbb{R}}^-$-component $I[\varphi_{\mathbb{R}}^-]$ is finite isomorphic to $W/[\varepsilon]W$. Here $[\varepsilon] = \log_p(\varepsilon)/\log_p(1+p)$ for a fundamental unit ε of $F(\varphi_{\mathbb{R}}^-)$.*

Proof. Note that $\overline{O}^{\times}$ is torsion-free as φ^- is unramified at p. Since $p \nmid |G|$, the isomorphism class of $\widehat{O}_p^{\times} \otimes_{\mathbb{Z}_p} W$ as a $W[\overline{G}]$-module is determined by the isomorphism class of $\widehat{O}_p^{\times} \otimes_{\mathbb{Z}_p} \mathrm{Frac}(W)$ as a $\mathrm{Frac}(W)[\overline{G}]$-module. By p-adic logarithm map, the $\overline{G}$-module $\widehat{O}_p^{\times} \otimes_{\mathbb{Z}_p} \mathrm{Frac}(W)$ is isomorphic to

$$O_p \otimes_{\mathbb{Z}_p} \mathrm{Frac}(W) \cong \mathrm{Frac}(W)[\overline{G}]$$

by the existence of a normal basis in Galois theory. Thus $\mathrm{Ind}_K^{\mathbb{Q}} \varphi^-$ appears in $\widehat{O}_p^{\times} \otimes_{\mathbb{Z}_p} W$ with multiplicity 2. If φ^- has order ≥ 3, by Propositions 7.3.4 and 7.3.5, in $\overline{O}^{\times} \otimes_{\mathbb{Z}_p} W$, the multiplicity of $\mathrm{Ind}_K^{\mathbb{Q}} \varphi^-$ is 1 if K is imaginary and 0 if K is real. This shows the first two assertions (except for the description of the finite error term in (2)), since $\mathrm{Ind}_K^{\mathbb{Q}} \varphi^- \cong \varphi^- \oplus (\varphi^-)^{-1}$ as $W[\overline{H}]$-modules.

Note that $\overline{O}^{\times}[\mathrm{Ind}_K^{\mathbb{Q}} \varphi^-] = \overline{O}^{\times}[\varphi^-] \oplus \overline{O}^{\times}[\varphi_\sigma^-] = (\varepsilon_{\varphi^-}) \oplus (\varepsilon_{\varphi_\sigma^-})$ (the sum of φ^- and φ_σ^--eigenspaces under $\overline{H}$). We find generators $[\varepsilon]_{\varphi^-}$ and $[\varepsilon]_{\varphi_\sigma^-}$ over W so that $\overline{O}^{\times}[\varphi^-] = W[\varepsilon]_{\varphi^-}$ and $\overline{O}^{\times}[\varphi_\sigma^-] = W[\varepsilon]_{\varphi_\sigma^-}$. Since $\mathrm{Ind}_K^{\mathbb{Q}} \overline{\varphi}^-$ is irreducible, we have $W/[\varepsilon]_{\varphi^-} \cong W/[\varepsilon]_{\varphi_\sigma^-}$ as W-modules. Since $\widehat{O}_p^{\times}[\mathrm{Ind}_K^{\mathbb{Q}} \varphi^-] \cong (\mathrm{Ind}_K^{\mathbb{Q}} \varphi^-)^2$, the W-torsion free quotient of $I[\mathrm{Ind}_K^{\mathbb{Q}} \varphi^-]$ is isomorphic to $\mathrm{Ind}_K^{\mathbb{Q}} \varphi^-$, and the W-torsion part of $I[\mathrm{Ind}_K^{\mathbb{Q}} \varphi^-]$ is given by $\mathrm{Ind}_K^{\mathbb{Q}} \varphi^- \otimes_W W/[\varepsilon]_{\varphi_\sigma^-} W$.

Suppose now that φ^- has order 2. Then $\varphi^- = \varphi_\sigma^-$, and hence φ^- extends to a character $\widetilde{\varphi}^-$ of $\overline{G}$ [GME, Proposition 5.1.1]. Since $\overline{G} \cong \{\pm 1\} \times \{\pm 1\}$, $\widetilde{\varphi}^-$ is of order 2. By Proposition 7.3.6, we find a quadratic field M which is real if K is imaginary and is imaginary if K is real such that $F = KM$. Then the $\overline{G}$-module $\overline{O}^{\times}$ contains an even quadratic character ξ such that $\xi = \alpha$ if K is real and $\xi = \left(\frac{M/\mathbb{Q}}{}\right)$ if K is imaginary. If K is imaginary, $\xi = \varphi_{\mathbb{R}}^-$ is even and extends φ^- by Proposition 7.3.6 and hence $I(\widetilde{\varphi}^-) = I(\xi)$ which is finite isomorphic to $W/[\varepsilon]W$ as in the proposition. Note that $(\varepsilon) = (\varepsilon_{\varphi_{\mathbb{R}}^-})$ for the Minkowski unit at p by definition. Another extension $\varphi_{\mathbb{R}}^- \alpha$ does not show up in $\overline{O}^{\times}$, and hence $I[\varphi_{\mathbb{R}}^- \alpha] \cong \varphi_{\mathbb{R}}^- \alpha$. If K is real, φ^- is non-trivial over $\mathfrak{G}_K$ and hence $\widetilde{\varphi}^- \neq \alpha$ for any extension $\widetilde{\varphi}^-$ of φ^-. Thus we find $I(\widetilde{\varphi}^-) \cong \widetilde{\varphi}^-$ as $\overline{G}$-modules. $\qquad\square$

7.6.3 *Generality of Selmer cocycle in non-split case*

Assume that $\alpha(p) \neq 1$; so, we are either in Case U$_-$ or in Case Ds in this subsection. In Case U$_-$, we write $\mathcal{D}_{\pm}$ for the p-ordinary deformation functor $\mathcal{D}$ with p-ordinarity quotient character $\overline{\delta}_{\pm} = (\delta_{\pm} \bmod \mathfrak{m}_W)$. In Case Ds, we write $\mathcal{D}_+$ for the p-ordinary deformation functor $\mathcal{D}$ with p-ordinarity quotient character $\bar{\delta}_+ = (\delta_+ \mod \mathfrak{m}_W)$, and define

$$\mathcal{D}_-(A) := \{\rho_A : G \to GL_2(A) | \rho_A \otimes \alpha \in \mathcal{D}_+(A)\}.$$

Thus $\rho_A \in \mathcal{D}_+(A)$ is made of classes of deformations of $\bar{\rho} = \operatorname{Ind}_K^{\mathbb{Q}} \bar{\varphi}$ over G satisfying $\rho_A|_{D_p} = \begin{pmatrix} * & * \\ 0 & \delta_A \end{pmatrix}$ with $\delta_A \alpha^{-1}$ unramified.

Recall the $\mathbb{Z}_p$-extension $\mathbb{Q}_\infty/\mathbb{Q}$ inside $\mathbb{Q}[\mu_{p^\infty}]$. Since $\alpha|_{D_p} \neq 1$, if a deformation $\rho_A \in \mathcal{D}_\pm(A)$ ramifies over $\mathbb{Q}_\infty F(\det(\bar{\rho}))$ (i.e., has non-trivial image over the unipotent part of the p-inertia group), $\rho_A \otimes \alpha$ has specified quotient character $\delta_\mp$. Thus $\rho \mapsto \rho \otimes \alpha$ does not give an involution of $\mathcal{D}_\pm$ but induces an isomorphism $\mathcal{D}_\pm \cong \mathcal{D}_\mp$. We write $\mathbb{T}_+$ for the Hecke algebra $\mathbb{T}_p$ corresponding to $\mathcal{D}_+$ and assume

$$R_p^{ord} \cong \mathbb{T}_+. \tag{RT}$$

By $\mathcal{D}_+ \cong \mathcal{D}_-$, another local ring $\mathbb{T}_-$ isomorphic to $\mathbb{T}_+$ represents $\mathcal{D}_-$. The isomorphism has an effect such that the image $a(n)$ of $T(n)$ in $\mathbb{T}_+$ is sent to $\alpha(n)a(n) \in \mathbb{T}_-$ as long as n is prime to D. Let $\rho_+ := \rho_{\mathbb{T}_+}$ and $\rho_- = j(\rho_{\mathbb{T}_+} \cdot \alpha)j^{-1}$. In other words, choosing $\bar{\rho} = \rho_+ \mod \mathfrak{m}_{\mathbb{T}_+}$ with $\bar{\rho}|_{D_\wp} = \operatorname{diag}[\bar{\delta}_-, \bar{\delta}_+]$, we may assume that

$$\rho_+|_{D_\wp} = \begin{pmatrix} \epsilon_{\mathbb{T}_+} & * \\ 0 & \delta_{\mathbb{T}_+} \end{pmatrix} \quad \text{and} \quad \rho_-|_{D_\wp} = \begin{pmatrix} \alpha\delta_{\mathbb{T}_+} & 0 \\ * & \alpha\epsilon_{\mathbb{T}_+} \end{pmatrix} = j\rho_+ \cdot \alpha j^{-1}. \tag{7.17}$$

Therefore, in Case U_-, $\mathbb{T}_\pm$ gives a different local ring of the big ordinary Hecke algebra $h_\psi^{ord}(N; \Lambda)$ lifting $f_\pm$. Since $f_+ \not\equiv f_- \mod \mathfrak{m}_W$, $\mathbb{T}_+$ and $\mathbb{T}_-$ are distinct. In Case Ds, as the character $\bar{\delta}_-$ ramifies at p, $\mathbb{T}_-$ does not show up as a local ring of $h_\psi^{ord}(N, \Lambda)$. In any case, we have two representations $\rho_+ := \rho_{\mathbb{T}_+} : G \to GL_2(\mathbb{T}_+)$ and $\rho_- := \rho_+ \otimes \alpha : G \to GL_2(\mathbb{T}_-)$. Thus $(\mathbb{T}_\pm, \rho_\pm)$ represents $\mathcal{D}_\pm$.

Pick a deformation $\phi = \phi_A : H \to A^\times$ with $\phi_A \mod \mathfrak{m}_A = \bar{\varphi}$ which factors through the universal character Φ. Choosing the matrix realization:

$$\rho_A(g) = \operatorname{Ind}_K^{\mathbb{Q}} \phi_A(g) = \begin{pmatrix} \phi(g) & \phi(g\sigma) \\ \phi(\sigma^{-1}g) & \phi(\sigma^{-1}g\sigma) \end{pmatrix},$$

we find the image of ρ_A modulo center of $\rho_A(D_p)$ is $\left\langle \begin{pmatrix} 0 & \phi(\sigma^2) \\ 1 & 0 \end{pmatrix} \right\rangle$. By Proposition 7.3.2, we have $Ad(\operatorname{Ind}_K^{\mathbb{Q}} \phi)^* \cong \alpha_A^* \oplus (\operatorname{Ind}_K^{\mathbb{Q}} \phi^-)^*$, where we write α_A for α having values in $A^\times$. Note that $\alpha_A^* \subset Ad(\operatorname{Ind}_K^{\mathbb{Q}} \phi)^*$ is realized on diagonal matrices, and $(\operatorname{Ind}_K^{\mathbb{Q}} \phi^-)^*$ is realized on anti-diagonal matrices. The cohomology is decomposed accordingly:

$$H^1(G, Ad(\operatorname{Ind}_K^{\mathbb{Q}} \phi)^*) = H^1(G, \alpha_A^*) \oplus H^1(G, (\operatorname{Ind}_K^{\mathbb{Q}} \phi^-)^*).$$

We decompose a Selmer cocycle $U : G \to Ad(\rho_A)^*$ according to the above decomposition

$$U = \begin{pmatrix} -u & u_+ \\ u_- & u \end{pmatrix} = u \cdot j + \begin{pmatrix} 0 & u_+ \\ u_- & 0 \end{pmatrix} \quad \text{with } u, u_\pm : G \to A^\vee \text{ and } j = \operatorname{diag}[-1,1]. \tag{7.18}$$

Recall $\Xi := \begin{pmatrix} \zeta & -\zeta \\ 1 & 1 \end{pmatrix}$ in §7.6.1. Since the matrix form of $\Xi^{-1}\rho_A\Xi|_{D_p}$ is upper triangular with δ at the lower right corner and is upper nilpotent over I_p, after conjugating by Ξ, Selmer cocycles $U : G \to Ad(\rho_A)^*$ are upper triangular over D_p and upper nilpotent over I_p. By computation, we have

$$\Xi^{-1}U\Xi = \begin{pmatrix} \frac{\zeta u_-}{2} + \frac{u_+}{2\zeta} & -\frac{\zeta u_-}{2} + \frac{u_+}{2\zeta} + u \\ \frac{\zeta u_-}{2} - \frac{u_+}{2\zeta} + u & -\frac{\zeta u_-}{2} - \frac{u_+}{2\zeta} \end{pmatrix}.$$

Thus we get the relation

$$\frac{\zeta u_-}{2} - \frac{u_+}{2\zeta} + u = 0 \text{ on } D_p \text{ and } u_- + \zeta^{-2}u_+ = 0 \text{ on } I_p. \tag{7.19}$$

Therefore

$$\Xi^{-1}U\Xi|_{D_p} = \begin{pmatrix} \frac{\zeta u_-}{2} + \frac{u_+}{2\zeta} & -\zeta u_- + \zeta^{-1}u_+ \\ 0 & -\frac{\zeta u_-}{2} - \frac{u_+}{2\zeta} \end{pmatrix} \Leftrightarrow U|_{D_p} = \begin{pmatrix} \frac{\zeta u_-}{2} - \frac{u_+}{2\zeta} & u_+ \\ u_- & -\frac{\zeta u_-}{2} + \frac{u_+}{2\zeta} \end{pmatrix},$$

$$\Xi^{-1}U\Xi|_{I_p} = \begin{pmatrix} 0 & \frac{2u_+}{\zeta} \\ 0 & 0 \end{pmatrix} \Leftrightarrow U|_{I_p} = \begin{pmatrix} -\zeta^{-1}u_+ & u_+ \\ -\zeta^{-2}u_+ & \zeta^{-1}u_+ \end{pmatrix} = u_+ \begin{pmatrix} -\zeta^{-1} & 1 \\ -\zeta^{-2} & \zeta^{-1} \end{pmatrix}. \tag{7.20}$$

For a subfield M of $F(\rho_+)$, writing G_M for the subgroup of G fixing M and $D_\mathcal{P} \subset G_M$ for the decomposition subgroup of a prime $\mathcal{P}|p$ of M, we define the Selmer group $\operatorname{Sel}_M(Ad(\rho_A))$ for any $\rho_A \in \mathcal{D}(A)$ by

$$\operatorname{Sel}_M(Ad(\rho_A)) := \operatorname{Ker}(H^1(G_M, Ad(\rho_A)^*) \to \prod_{\mathcal{P}|p} \frac{H^1(D_\mathcal{P}, Ad(\rho_A)^*)}{F^+_{-,\mathcal{P}}H^1(D_\mathcal{P}, Ad(\rho_A)^*)}), \tag{7.21}$$

where $\mathcal{P}$ runs over all prime factors of p in M, and choosing $a_\mathcal{P} \in \operatorname{GL}_2(A)$ so that $a_\mathcal{P}\rho_A a_\mathcal{P}^{-1}|_{D_\mathcal{P}} = \begin{pmatrix} \epsilon_\mathcal{P} & * \\ 0 & \delta_\mathcal{P} \end{pmatrix}$ with $\delta_\mathcal{P}$ unramified and $\delta_\mathcal{P} \bmod \mathfrak{m}_A = \bar{\delta}$, $a_\mathcal{P}\Xi F^+_{-,\mathcal{P}}H^1(D_\mathcal{P}, Ad(\rho_A)^*)(a_\mathcal{P}\Xi)^{-1}$ is made of cohomology classes *upper triangular over $D_\mathcal{P}$ and upper nilpotent over the inertia subgroup $I_\mathcal{P}$ of $D_\mathcal{P}$*. We just write $\operatorname{Sel}(Ad(\rho_A))$ for $\operatorname{Sel}_\mathbb{Q}(Ad(\rho_A))$ as before.

Recall $\rho = \operatorname{Ind}_K^\mathbb{Q} \varphi$. Consider the upper nilpotent part u_+ of a Selmer cocycle $U : G \to Ad(\rho)^*$ and its restriction to G_F for $F = F(Ad(\rho)) = F(\varphi^-)$. Since $\varphi^-(\operatorname{Frob}_{p^2}) = 1$, $\bar{D} \subset \bar{G}$ (which is the image of D_p such that $\Xi^{-1}U\Xi|_{D_p}$ upper triangular) is isomorphic to $\{\pm 1\}$. Let $\mathfrak{P}$ for a prime of $\bar{\mathbb{Q}}$ with $\mathfrak{P} \cap F(\rho_+)$ corresponding to D_p, and put $\mathcal{P} = \mathfrak{P} \cap F$. Taking

$\sigma = \mathrm{Frob}_p$ in Case U$_-$ and $\sigma \in I_p$ in Case Ds, for $g \in G$,

$$Ad(\rho)(\sigma) \begin{pmatrix} 0 & u_+(\sigma^{-1}g\sigma) \\ u_-(\sigma^{-1}g\sigma) & 0 \end{pmatrix}$$

$$= (Ad(\rho)(g) - 1) \begin{pmatrix} 0 & u_+(\sigma) \\ u_-(\sigma) & 0 \end{pmatrix} + \begin{pmatrix} 0 & u_+(g) \\ u_-(g) & 0 \end{pmatrix}.$$

Thus

$$\begin{pmatrix} 0 & u_-(\sigma^{-1}g\sigma) \\ u_+(\sigma^{-1}g\sigma) & 0 \end{pmatrix} = (Ad(\rho)(g) - 1) \begin{pmatrix} 0 & u_+(\sigma) \\ u_-(\sigma) & 0 \end{pmatrix} + \begin{pmatrix} 0 & u_+(g) \\ u_-(g) & 0 \end{pmatrix}.$$
$$(7.22)$$

We get $u_-(\sigma^{-1}h\sigma) = (\varphi^-(h)-1)u_+(\sigma)+u_+(h)$ for $h \in H$. Since $\varphi_-(h) = 1$ for $h \in G_F$, u_- is determined by u_+ on G_F. By inflation-restriction, $\mathrm{Res} : H^1(G, \mathrm{Ind}_K^Q \varphi^{-*}) \to H^1(G_F, \mathrm{Ind}_K^Q \varphi^{-*})$ is an injection; so,

($\pm$) *the class of u_+ in $H^1(H, \varphi^-)$ determines that of u_- and hence the class of $H^1(G, \mathrm{Ind}_K^Q \varphi^-)$ on G.*

Thus we may concentrate on u_+ to analyze the $\mathrm{Ind}_K^Q \varphi^-$ component of the adjoint Selmer group. Similarly, u_+ determines u on $D_\wp$ by (7.19).

Since $D_\wp$ acts trivially on $\mathrm{Ind}_K^Q \varphi^-$, $D_\wp \subset G_F$ and $u_+|_{D_\wp} \in \mathrm{Hom}(D_\wp, (\varphi^-)^*)$; so,

$$\zeta^2 u_-(\sigma^{-1}h\sigma) = u_+(h) \text{ for } h \in D_\wp. \qquad (7.23)$$

If $u_+|_{D_\wp}$ is non-trivial (resp. ramifies), then u_- is also non-trivial (resp. ramifies) as $\sigma^{-1}D_\wp\sigma = D_\wp$ (resp. $\sigma^{-1}I_p\sigma = I_p \subset D_\wp$). By p-ordinarity and (7.19), $(u_- + \zeta^{-2}u_+)|_{I_p} = 0$ and hence $u_\pm(\sigma^{-1}i\sigma) = -u_\pm(i)$ for $i \in I_p$. This implies u_+ factors through $Y = \mathrm{Gal}(F_\infty/F(\varphi^-))$ defined in §7.6.2.

Regard u_+ as a $W[\overline{H}]$-linear map $u_+ : Y \to \varphi^-$. Indeed, for $\delta \in \overline{H}$,

$$u_\pm(\delta^{-1}h\delta) = (\varphi^-)^{\pm 1}(\delta)u_\pm(h) \text{ for } h \in G_F. \qquad (7.24)$$

So, $u_+|_{G_F}$ factors through $Y[\varphi^-] := Y \otimes_{\mathbb{Z}_p[\overline{H}]} \varphi^-$.

7.6.4 Adjoint Selmer groups in non-split case

Let $\overline{\rho} = \mathrm{Ind}_K^Q \overline{\varphi}$ and $\rho = \mathrm{Ind}_K^Q \varphi$ for the Teichmüller lift φ of $\overline{\varphi}$.

Theorem 7.6.2. *Let the notation and the assumption be as above. Suppose that $\alpha(p) \neq 1$ and put $F = F(\varphi^-)$. Then the following assertion holds:*

(1) *If φ^- has order ≥ 3, then*

$$\mathrm{Sel}(Ad(\overline{\rho})) \cong \mathrm{Hom}(Cl_K, \mathbb{F}) \oplus \mathrm{Hom}(Cl_F[\varphi^-], \mathbb{F}),$$

$$\mathrm{Sel}(Ad(\rho)) \cong \mathrm{Hom}(Cl_K, W^\vee) \oplus \mathrm{Hom}(Cl_F[\varphi^-], W^\vee),$$

where $Cl_F[\varphi^-] = Cl_F \otimes_{\mathbb{Z}[\overline{H}]} \varphi^-$. So, if $p \nmid h_K$ and $Cl_{F(\varphi^-)} \otimes_{\mathbb{Z}[\overline{H}]} \overline{\varphi}^- = 0$, we have $\mathrm{Sel}(Ad(\overline{\rho})) = \mathrm{Sel}(Ad(\rho)) = 0$.

(2) *If φ^- has order 2, then we have a unique quadratic subfield $M \subset F$ such that p splits in M. Moreover for the character $\beta := \left(\frac{M/\mathbb{Q}}{\,}\right)$, $\alpha\beta$ is an odd character, and*

$$\mathrm{Sel}(Ad(\overline{\rho})) \cong \mathrm{Hom}(Cl_K, \mathbb{F}) \oplus \mathrm{Hom}(Y[\alpha\beta], \mathbb{F}) \oplus \mathrm{Hom}(Cl_M, \mathbb{F}),$$
$$\mathrm{Sel}(Ad(\rho)) \cong \mathrm{Hom}(Cl_K, W^\vee) \oplus \mathrm{Hom}(Y[\alpha\beta], W^\vee) \oplus \mathrm{Hom}(Cl_M, W^\vee).$$

(3) *If φ^- has order 2, for the maximal anticyclotomic p-abelian extension J^- of $J := F(\alpha\beta)$, we have $Y[\alpha\beta] \cong \mathrm{Gal}(J^-/J)$, $\mathrm{rank}_{\mathbb{Z}_p} Y[\alpha\beta] = 1$, $\dim_\mathbb{F} \mathrm{Sel}(Ad(\overline{\rho})) \geq 1$, and under $Cl_F \otimes_{\mathbb{Z}[\overline{G}]} Ad(\overline{\rho}) = 0 (\Leftrightarrow p \nmid h_K h_M h_J)$, we have $\dim_\mathbb{F} \mathrm{Sel}(Ad(\overline{\rho})) = 1$.*

We prove the theorem only for $\rho = \mathrm{Ind}_K^\mathbb{Q} \varphi$ as the proof for $\mathrm{Sel}(Ad(\overline{\rho}))$ is almost identical.

Proof. For the moment, we suppose that φ^- has order ≥ 3. Note that $Y[\mathrm{Ind}_K^\mathbb{Q} \varphi^-] := Y \otimes_{\mathbb{Z}_p[\overline{G}]} \mathrm{Ind}_K^\mathbb{Q} \varphi^-$ is the direct sum $Y[\varphi^-] \oplus Y[(\varphi^-)^{-1}]$. Consider the $\mathrm{Ind}_K^\mathbb{Q} \varphi^-$-isotypic ($p$-inertia) component $I[\mathrm{Ind}_K^\mathbb{Q} \varphi^-]$ defined in Proposition 7.6.1 of $Y[\mathrm{Ind}_K^\mathbb{Q} \varphi^-]$. By Proposition 7.6.1, we know the $W[\overline{G}]$-module structure of $I[\mathrm{Ind}_K^\mathbb{Q} \varphi^-]$

$$I[\mathrm{Ind}_K^\mathbb{Q} \varphi^-] \cong \mathrm{Ind}_K^\mathbb{Q} \varphi^- \oplus (\mathrm{Ind}_K^\mathbb{Q} \varphi^- \otimes_W W/[\varepsilon]_{\varphi^-} W)$$

as $\overline{G}$-modules, where ε_{φ^-} is the φ^- projection of the Minkowski unit at p; so, $\varepsilon_{\varphi^-} = 0$ if K is imaginary. However, in $\mathrm{Ind}_K^\mathbb{Q} \varphi^-$, σ brings the φ^--eigenspace over H to the $(\varphi^-)^{-1}$-eigenspace. Therefore the maximal quotient of $I[\mathrm{Ind}_K^\mathbb{Q} \varphi^-]$ on which σ acts by -1 and $\overline{H}$ acts by φ^- is trivial. If φ^- has order > 2, $u_+|_{G_F}$ factors through $Y[\mathrm{Ind}_K^\mathbb{Q} \varphi^-]/I[\mathrm{Ind}_K^\mathbb{Q} \varphi^-]$ which is isomorphic to $Cl_F[\mathrm{Ind}_K^\mathbb{Q} \varphi^-] = Cl_F \otimes_{\mathbb{Z}[\overline{G}]} \mathrm{Ind}_K^\mathbb{Q} \varphi^-$. Therefore the homomorphisms $u_\pm, u : G_F \to W^\vee$ are unramified everywhere. Then $u|_{G_F}$ factors through $Cl_F \otimes_{\mathbb{Z}[\overline{G}]} \alpha = Cl_F[\alpha] \cong Cl_K \otimes_\mathbb{Z} \mathbb{Z}_p$. Thus we see $u|_H \in \mathrm{Hom}(Cl_K \otimes_\mathbb{Z} \mathbb{Z}_p, W^\vee)$. Since $\wp = (p)$ is principal in O if $\alpha(p) = -1$ and $\wp$ gives an ambiguous class in Cl_K of order ≤ 2 if $\alpha(p) = 0$, u is trivial on $D_\wp$. Thus by (7.19), $\zeta u_- - \zeta^{-1} u_+ = 0$ on $D_\wp$. This is compatible with u_+ determining u_- on G (given in $(\pm)$ in §7.6.3).

Conversely, start with $u' \in \mathrm{Hom}(Cl_K \otimes_\mathbb{Z} \mathbb{Z}_p, W^\vee)$ and $u'_+ \in \mathrm{Hom}(Cl_F[\varphi^-], W^\vee)$. Extend u' (resp. u'_+) to a 1-cocycle $u : G \to \alpha^* = \alpha \otimes_\mathbb{Z} W^\vee$ (resp. $u''_+ : G \to (\varphi^-)^*$) by class field theory (up to coboundary). Then by Shapiro's lemma, u''_+ extends to $\widetilde{u} : G \to (\mathrm{Ind}_K^\mathbb{Q} \varphi^-)^*$. Writing $\widetilde{u} = {}^t(u_+, u_-)$ with $u_\pm : G \to ((\varphi^-)^{\pm 1})^*$ and normalizing $u_\pm$ so that (7.22) holds, and define $U := j \cdot u + \begin{pmatrix} 0 & u_+ \\ u_- & 0 \end{pmatrix}$. Then U gives rise to a Selmer cocycle. Thus by $U \leftrightarrow (u|_H, u_+|_{G_F})$,

$$\mathrm{Sel}(Ad(\rho)) = \mathrm{Hom}(Cl_K \otimes_{\mathbb{Z}} \mathbb{Z}_p, W^{\vee}) \oplus \mathrm{Hom}(Cl_F[\varphi^-], W^{\vee}).$$

Suppose $(\varphi^-)^2 = 1$ (and $\varphi^- \neq 1$). The existence and uniqueness of M follows from Corollary 7.3.7 (interchanging the role of K and M in the corollary). If K is real, $M = F(\beta)$ with $\beta := \left(\frac{M/\mathbb{Q}}{} \right)$ and $\alpha\beta$ both odd. If K is imaginary, the p-decomposition subgroup $D \subset \mathrm{Gal}(F/\mathbb{Q})$ restricts surjectively on $\mathrm{Gal}(K/\mathbb{Q})$ by non-splitting of p in K. Thus we have $D = \langle c \rangle$ for complex conjugation c. This implies $M = F(\beta)$ is real; so, $\alpha\beta$ is odd.

By Proposition 7.6.1, we find $\mathrm{Ind}_K^{\mathbb{Q}} \varphi^- \cong \beta \oplus \beta\alpha$. Decompose $\mathrm{Gal}(F/\mathbb{Q}) = \mathrm{Gal}(K/\mathbb{Q}) \times \mathrm{Gal}(M/\mathbb{Q})$ so that $\sigma|_M = 1$ (i.e., $\beta(\sigma) = 1$). Write $\langle \varsigma \rangle = \mathrm{Gal}(M/\mathbb{Q})$ and regard $\varsigma \in \mathrm{Gal}(F/\mathbb{Q})$ by the above decomposition. Indicating $\pm$-eigenspace of σ by superscript "$\pm$", we have $I[\varphi^-]^+ = I[\beta]$ and $I[\varphi^-]^- = I[\alpha\beta]$. Write $(p) = \wp_M \wp_M^{\varsigma}$ in M. Since $p > 2$, we have $C_F := Cl_F \otimes_{\mathbb{Z}} \mathbb{Z}_p = C_F[\beta] \oplus C_F[\alpha] \oplus C_F[\alpha\beta]$ for ?-eigenspace $C_F[?]$ under $\mathrm{Gal}(F/\mathbb{Q})$. By the norm map: $\mathfrak{a} \mapsto \mathfrak{a}\mathfrak{a}^{\varsigma}$, $C_F^+ = C_F[\beta] \cong C_M$ and $C_F^- = C_F[\alpha] \oplus C_F[\alpha\beta] \cong C_M \oplus C_{K'}$ for $J = F(\alpha\beta)$. The group $Y[\alpha\beta]$ is the Galois group of the maximal anticyclotomic extension J^-/J unramified outside p. We have an exact sequence: $0 \to I[\varphi^-]^{\pm} \to Y[\varphi^-]^{\pm} \to C_F^{\pm} \to 0$. By (7.23), u_+ factors through

$$Y[\varphi^-]/I[\varphi^-]^+ = Y[\alpha\beta] \oplus Y[\beta]/I[\beta] = Y[\alpha\beta] \oplus C_F[\beta] \cong \mathrm{Gal}(J^-/J) \oplus C_M.$$

By (7.19), $u = \zeta^{-1}u_+$ on I_p. On $u|_{G_F}$, $\overline{H}$ acts trivially; so, by (7.24) we conclude $u|_{I_p} = 0$ as $\varphi^- \not\equiv 1 \mod \mathfrak{m}_W$. Thus $u|_H$ is unramified everywhere factoring through C_K. Reversing the argument, we obtain the assertion (2).

By Proposition 7.6.1, $Y[\alpha\beta] = I[\alpha\beta] \cong \mathbb{Z}_p$ under the condition of (3), which finishes the proof. $\qquad \square$

Remark 7.6.3. As long as φ^- has order ≥ 3, for a deformation $\phi : H \to A^{\times}$ of $\overline{\varphi}$, writing $\rho_A = \mathrm{Ind}_K^{\mathbb{Q}} \phi$, we can express $\mathrm{Sel}(Ad(\rho_A))$ as in the formula of Theorem 7.6.2 (1) just replacing $W^{\vee}$ and Hom by $(\phi^-)^*$ and $\mathrm{Hom}_{\mathbb{Z}_p[[\Gamma^-]]}$. We leave the proof to the attentive reader. If φ^- has order 2, for a nontrivial deformation $\phi : H \to A^{\times}$ of $\overline{\varphi}$, we can determine the Selmer group similarly to Theorem 7.6.2 (2) but more complicated as ϕ^- has order $2p^r$ and the argument becomes a mixture of the proof of the two assertions (1) and (2). We do not give details of the structure as it is anyway treated in Theorems 7.4.4 and 7.4.7 as we will see in the following §7.6.5.

Corollary 7.6.4. *Suppose that* $\alpha(p) \neq 1$ *and* $R_{\mathfrak{p}}^{ord} = \mathbb{T}_{\mathfrak{p}}$. *Then we have the following two assertions:*

(1) *Suppose that* $\overline{\varphi}^-$ *has order* ≥ 3. *If* $p \nmid h_K$ *and* $Cl_F \otimes_{\mathbb{Z}[\overline{H}]} \overline{\varphi}^- = 0$, *then*
$R_{\mathfrak{p}}^{ord} \cong \mathbb{T}_{\mathfrak{p}} \cong \Lambda$.

(2) *Suppose that* $\overline{\varphi}^-$ *has order* 2. *Then* $R_{\mathfrak{p}}^{ord} \supsetneq \Lambda$, *and if* $Cl_F \otimes_{\mathbb{Z}[\overline{G}]} Ad(\overline{\rho}) = 0$, $\mathrm{Sel}(Ad(\rho_A))^\vee$ *is a non-trivial cyclic* A-*module for all* $\rho_A \in \mathcal{D}_\kappa(A)$.

Remark 7.6.5. As an example of the assertion (1) of Corollary 7.6.4, we can offer $p = 23$ and $K = \mathbb{Q}[\sqrt{-23}]$ with $\varphi : Cl_K \cong \mu_3(\overline{\mathbb{Q}})$. Though this example does not satisfy the Taylor–Wiles condition, we know from [S71, 3.3] that $\mathbb{T} \cong \Lambda$ with the residual representation $\mathrm{Ind}_K^{\mathbb{Q}} \overline{\varphi}$ to which Ramanujan's Δ function in $S_{12}(\mathrm{SL}_2(\mathbb{Z}))$ belongs.

7.6.5 *Comparison of Theorems 7.6.2 and 7.4.4*

Assume that φ^- has order 2. By the assertion (2) of Theorem 7.6.2, p splits in M; so, we can apply the results obtained in Theorem 7.4.4 and Theorem 7.4.7 to $\mathrm{Ind}_M^{\mathbb{Q}} \phi \cong \mathrm{Ind}_K^{\mathbb{Q}} \varphi$. We now describe the equivalence of the two results. Recall $\alpha = \left(\frac{K/\mathbb{Q}}{\cdot}\right)$, $\beta = \left(\frac{M/\mathbb{Q}}{\cdot}\right)$ and $J = F(\alpha\beta)$ (the splitting field of $\alpha\beta$). Since p splits in $M/\mathbb{Q}$, we may apply the result in split case to $\rho = \mathrm{Ind}_K^{\mathbb{Q}} \varphi = \mathrm{Ind}_M^{\mathbb{Q}} \phi$ choosing $\phi : G \to W^\times$ as in the proof of Corollary 7.3.7. Let $\varsigma \in \mathrm{Gal}(F(\rho)/\mathbb{Q})$ with $\varsigma|_M \neq 1$, and choose a prime factor $\mathcal{P}|p$ in M so that ϕ_ς is unramified at $\mathcal{P}$.

We recall the notation and the result given in Theorem 7.4.7. We have an involution $[\beta] \in \mathrm{Aut}(\mathbb{T})$ so that $[\beta] \circ \rho_{\mathbb{T}} \cong \rho_{\mathbb{T}} \otimes \beta$. Recall the maximal anticyclotomic p-abelian extension M^-/M unramified outside p and the maximal p-abelian extension L/FM^- unramified outside $\wp_M$ and totally split at $\wp_M^\varsigma$ for the Teichmüller lift ϕ_0 of $\overline{\phi}$. We have an exact sequence

$$1 \to \Gamma^- \to \mathrm{Gal}(M^-/M) \to C_M \to 1$$

for $C_M = Cl_M \otimes_{\mathbb{Z}} \mathbb{Z}_p$ and $\Gamma^- \cong \mathbb{Z}_p$ if M is imaginary and otherwise, $\Gamma^- \cong \Gamma/\Gamma^{\log_p(\varepsilon)/\log_p(1+p)}$ for a fundamental unit $\varepsilon \in M$. Let $\mathcal{Y} = \mathrm{Gal}(L/FM^-)$ and $\Gamma^- := \mathrm{Gal}(M^-/M)$; so, $\mathrm{Gal}(FM^-/M) \cong \mathrm{Gal}(F/M) \times \Gamma^-$ and $\mathrm{Gal}(FM^-/\mathbb{Q}) \cong \mathrm{Gal}(F/\mathbb{Q}) \times \Gamma^-$. Splitting $\mathrm{Gal}(L/M) = (\mathrm{Gal}(F/M) \times \Gamma^-) \ltimes \mathcal{Y}$, define $\mathcal{Y}(\phi^-) = \mathcal{Y} \otimes_{\mathbb{Z}[\mathrm{Gal}(F/M)]} \phi^-$. By Theorem 7.4.7 combined with Theorem 7.4.4, we have

$$\mathrm{Sel}(Ad(\rho)) \cong \mathrm{Hom}(Cl_M, W^\vee) \oplus \mathrm{Hom}_{\mathbb{Z}_p[[\Gamma^-]]}(\mathcal{Y}(\phi^-), (\phi^-)^*).$$

Identifying $W[[\Gamma^-]] = W[[\mathcal{T}]]$ with $(\mathcal{T})$ giving the augmentation ideal, we have

$$\mathrm{Hom}_{\mathbb{Z}_p[[\Gamma^-]]}(\mathcal{Y}(\phi^-), (\phi^-)^*) = \mathrm{Hom}(\mathcal{Y}(\phi^-)/\mathcal{T}\mathcal{Y}(\phi^-), W^\vee).$$

Therefore, our task is to prove

Proposition 7.6.6. *Let the notation and assumption be as above. We use the notation introduced in* Theorem 7.6.2 (2). *In particular, we assume φ^- has order 2 and $\rho = \mathrm{Ind}_K^{\mathbb{Q}} \varphi \cong \mathrm{Ind}_M^{\mathbb{Q}} \phi$ for M as in* Theorem 7.6.2 (2). *Recall $\langle \varsigma \rangle = \mathrm{Gal}(M/\mathbb{Q})$ and $(p) = \wp_M \wp_M^\varsigma$ with $\wp_M \neq \wp_M^\varsigma$ in O_M so that $\wp_M$ is the specified prime factor of p to define $\mathcal{Y}$ for M. Then we have*

$$\mathcal{Y}(\phi^-)/\mathcal{T}\mathcal{Y}(\phi^-) \cong Y[\alpha\beta] \oplus C_K \quad (\text{canonically}).$$

Proof. Regarding $\mathrm{Gal}(F/\mathbb{Q})$ as a subgroup of $\mathrm{Gal}(L/\mathbb{Q})$ by the identity $\mathrm{Gal}(L/\mathbb{Q}) = (\mathrm{Gal}(F/\mathbb{Q}) \times \Gamma^-) \ltimes \mathcal{Y}$, we regard $\mathcal{Y}(\phi^-)$ as a $\mathrm{Gal}(F/\mathbb{Q})$-module. Since $\phi^-|_{\mathrm{Gal}(K/\mathbb{Q})} = \alpha$ under the decomposition $\mathrm{Gal}(F/\mathbb{Q}) = \mathrm{Gal}(M/\mathbb{Q}) \times \mathrm{Gal}(K/\mathbb{Q})$, we find $\mathcal{Y}(\phi^-) = \mathcal{Y}[\alpha] \oplus \mathcal{Y}[\alpha\beta]$. Since K has only one prime over p, unramifiedness at $\wp_M^\varsigma$ of $\mathcal{Y}[\alpha]$ implies $C_K \twoheadrightarrow \mathcal{Y}[\alpha]/\mathcal{T}\mathcal{Y}[\alpha]$. Since we are in Case U$_-$ or Case Ds, if inert, $\wp$ is principal, and if ramified, $\wp$ is in an ambiguous class having order ≤ 2 in Cl_K. Thus we find $\wp_M^\varsigma$-splitting condition is automatic, and hence $C_K \cong \mathcal{Y}[\alpha]/\mathcal{T}\mathcal{Y}[\alpha]$.

Now we would like to show $Y[\alpha\beta] \cong \mathcal{Y}[\alpha\beta]/\mathcal{T}\mathcal{Y}[\alpha\beta]$. For simplicity, we write $\mathcal{Y}' := \mathcal{Y}[\alpha\beta]/\mathcal{T}\mathcal{Y}[\alpha\beta]$ (the right-hand-side). Let X_∞/X be the maximal p-abelian extension of a number field X unramified outside p. Since F is a CM field abelian over $\mathbb{Q}$, $\mathrm{rank}_{\mathbb{Z}_p} \mathrm{Gal}(F_\infty/F) = 3$ with $\mathrm{rank}\, \mathrm{Gal}(F_\infty/F)[\alpha\beta] = \mathrm{rank}_{\mathbb{Z}_p} \mathrm{Gal}(F_\infty/F)[\gamma] = 1$ for the odd character $\gamma \neq \alpha\beta$ different from $\alpha\beta$. Take $F_\infty^{\alpha\beta} \subset F_\infty$ such that $\mathrm{Gal}(F_\infty^{\alpha\beta}) = \mathrm{Gal}(F_\infty/F)[\alpha\beta]$. We want to show $\mathcal{Y}' \cong \mathrm{Gal}(F_\infty^{\alpha\beta} M^-/M^-) \cong \mathrm{Gal}(F_\infty^{\alpha\beta}/F) \cong Y[\alpha\beta]$.

Note that F has two primes $\mathcal{P}$ and $\mathcal{P}^\varsigma$ over $\wp_M$ and $\wp_M^\varsigma$, respectively. Since F is a CM field of degree 4, $O^\times = \varepsilon^{\mathbb{Z}} \times \mu$ for a finite cyclic group μ and a fundamental unit ε. Thus $\widehat{O}_\mathcal{P}^\times/\overline{O}^\times$ is isomorphic to the inertia group $I \subset \mathrm{Gal}(F_\infty/F)$ of $\mathcal{P}$ which has rank 1 over $\mathbb{Z}_p$. Since $\mathrm{Gal}(M^-/M)$ has rank 1 $\mathcal{P}$-inertia subgroup, $\mathrm{Gal}(F_\infty/FM^-)$ can only have finite inertia subgroup at $\mathcal{P}$, and the outcome is the same for $\mathcal{P}^\varsigma$. Here is an exercise:

Exercise 7.6.1. Prove that the class of $\mathcal{P}^\varsigma$ in F_∞/F has infinite order.

Then for the maximal $\mathcal{P}^\varsigma$-splitting subextension L' of F_∞/FM^-, $\mathrm{Gal}(L'/FM^-)$ has $\mathbb{Z}_p$-rank 1. The action of $\mathrm{Gal}(F/\mathbb{Q})$ on the $\mathbb{Z}_p$-free quotient of $\mathrm{Gal}(L'/F)$ is given by a character γ different from β (since ς brings $\mathcal{P}$ to $\mathcal{P}^\varsigma$). It cannot be α as the α-eigenspace is exhausted by $\mathrm{Gal}(K^- M^-/M^-)$ which has $\mathbb{Z}_p$-rank 1. Thus the restriction map $\mathrm{Res}: \mathcal{Y}' \to Y[\alpha\beta]$ has finite kernel and finite cokernel.

Recall $J = F(\alpha\beta)$. Since p is unramified in J^-M^-/M^- and $\mathrm{Gal}(J^-M^-/M^-)$ is the quotient of $\alpha\beta$-eigenspace, $L' \supset J^-$ and $Y[\alpha\beta] = \mathrm{Gal}(J^-F/F) \cong \mathrm{Gal}(J^-/J)$, Res is onto. Thus the torsion-free quotient of $\mathcal{Y}'$ and $Y[\alpha\beta]$ are identical. Moreover the p-inertia subgroup of $Y[\alpha\beta]$ is $\mathbb{Z}_p$-free of rank 1 and isomorphic to each other (because it is the $\alpha\beta$ eigenspace of $\widehat{O}_{J,p}^\times \cong \mathbb{Z}_p^2$). Thus $\mathrm{Ker}(\mathrm{Res})$ is in the torsion part $\mathcal{Y}_t'$ of $\mathcal{Y}'$. Realizing $\mathcal{Y}_t'$ as a quotient of $\mathcal{Y}'$, we have an extension $\mathcal{H}/M^-$ inside L' giving $\mathrm{Gal}(\mathcal{H}/M^-) = \mathcal{Y}_t'$. By the action of $\mathrm{Gal}(F/\mathbb{Q})$, $\mathcal{H} = H'F_\infty$ for a finite p-extension of H' unramified everywhere (as Res is onto). In the p-Hilbert class field H_p of J, p totally splits, as $\mathcal{P} \cap H'$ has order ≤ 2 in $Cl_{H'}$. This implies $H' = H_p \subset J^-$, and hence Res is an isomorphism. $\qquad\square$

7.7 Selmer group of exceptional Artin representation

In this section, assuming that $\bar\rho$ comes from an exceptional Artin representation $\rho : \mathfrak{G}_\mathbb{Q} \to \mathrm{GL}_2(W)$ with $\mathrm{Im}(Ad(\bar\rho)) = \mathrm{Im}(Ad(\rho))$ and supposing $\boxed{p \geq 5}$, we explore a way to describe the structure of its adjoint Selmer group in terms of a global unit of the splitting field F, similar to Theorem 7.5.4 (so the remark about Stark's conjecture after Theorem 7.5.4 also applies to this case). We say a two-dimensional Artin representation is *exceptional* if its image falls in Case E in Dickson's classification. In this section, we assume that ρ is exceptional. Our main result is Theorem 7.7.3. We often write Ad for $Ad(\rho)$. Let $\overline{G} := \mathrm{Gal}(F/\mathbb{Q}) \cong \mathrm{Im}(Ad(\bar\rho))$ for $F := F(Ad(\bar\rho))$. Assume also $\boxed{p \nmid |\overline{G}|}$ and irreducibility of $\bar\rho$ throughout this section. Then $\overline{G} \cong \mathrm{Im}(Ad(\rho)) = \mathrm{Im}(Ad(\bar\rho))$. We write $\mathbb{F}$ for the minimal field of rationality of $Ad(\bar\rho)$. Noting that $Ad(\bar\rho)$ factors through $\mathrm{PGL}_2(\mathbb{F})$, $\mathbb{F}$ is the minimal subfield of $\overline{\mathbb{F}}_p$ with $\mathrm{Im}(Ad(\bar\rho)) \subset \mathrm{PGL}_2(\mathbb{F})$. We take W to be the unramified extension of $\mathbb{Z}_p$ with $W/\mathfrak{m}_W = \mathbb{F}$; so, $W = W(\mathbb{F})$ (the ring of Witt vectors with coefficients in $\mathbb{F}$). We write O for the integer ring F. Fix a prime $\wp|p$ in O. We write $\overline{D} \subset \overline{G}$ for the decomposition group of $\wp$ and assume $\rho|_{\overline{D}} = \left(\begin{smallmatrix} \epsilon & 0 \\ 0 & \delta \end{smallmatrix}\right)$ with δ unramified.

7.7.1 *Classification of subgroups of* $\mathrm{PGL}_2(\mathbb{F})$

Identify $\overline{G}$ with the subgroup $\mathrm{Im}(Ad(\bar\rho))$ of $\mathrm{PGL}_2(\mathbb{F})$. Recall Dickson's classification of $\overline{G} \subset \mathrm{PGL}_2(\mathbb{F})$ in §6.2.4:

(G) If p is a factor of $|\overline{G}|$, $\overline{G}$ is conjugate to $\mathrm{PGL}_2(k)$ or $\mathrm{PSL}_2(k)$ for a subfield $k \subset \mathbb{F}$ as long as $p \geq 5$ (when $p = 3$, $\overline{G}$ can be A_5).

One of the reasons we suppose $p \geq 5$ is this fact: $p \| |\overline{G}| \Rightarrow$ Case G. Suppose $p \nmid |\overline{G}|$ (so, $p \geq 5$). Then $\overline{G}$ is given as follows.

(C) $\overline{G}$ is cyclic ($\Rightarrow \mathrm{Im}(\overline{\rho})$ is abelian).

(D) $\overline{G}$ is a dihedral group D_a of order $2a$ (so, $\overline{\rho} = \mathrm{Ind}_K^{\mathbb{Q}} \overline{\varphi}$ for a quadratic field K), and $\mathbb{F} = \mathbb{F}_p[\overline{\varphi}]$ (the field spanned by the values of $\overline{\varphi}$).

(E) $\overline{G}$ is isomorphic to one of the following groups: A_4, S_4 ($\mathbb{F} = \mathbb{F}_p$ and $W = \mathbb{Z}_p$), and A_5 ($\mathbb{F} \cong \mathbb{Z}_p[\sqrt{5}]/\wp$ for a prime $\wp|p$ by the character table of A_5; so, $\mathbb{F} = \mathbb{F}_p$ or $\mathbb{F}_{p^2}$).

By the assumption $p \geq 5$ and $p \nmid |\overline{G}|$, when $\overline{G} \cong A_4$ and S_4, $p = 5$ is allowed, but if $\overline{G} \cong A_5$, we actually assume $p \geq 7$. Here is a remark about cyclic subgroups and cyclic quotients of $\overline{G}$ in Case E:

Lemma 7.7.1. *The groups A_4, S_4 and A_5 do not have quotient isomorphic to the cyclic group of order $p - 1$ for $p \geq 5$, and the order of non-trivial cyclic subgroups has the following orders:*

$$A_4 : 2, 3, S_4 : 2, 3, 4, A_5 : 2, 3, 5.$$

For this list, see [LRF, §5.7-8 and §18.6].

We prove that $Ad(\overline{\rho})$ is absolutely irreducible in Case E. If $Ad(\overline{\rho})$ is reducible, it contains a 1-dimensional subspace or quotient stable under $\overline{G}$-action. We regard $\overline{\rho}$ as having values in the algebraic closure $\overline{\mathbb{F}}_p$. Since $Ad(\overline{\rho})$ is self dual, the dual of the quotient is a subspace; so, it always contains subspace of dimension 1 spanned by $0 \neq i \in \mathrm{End}_{\overline{\mathbb{F}}_p}(\overline{\rho})$ with $\mathrm{Tr}(i) = 0$. Thus $\overline{G}$ acts on i by a character α: $\overline{\rho}(g) \circ i \circ \overline{\rho}(g)^{-1} = \alpha(g)i$ ($\Leftrightarrow \overline{\rho} \circ i = i \circ (\overline{\rho} \otimes \alpha)$). This implies that i gives an isomorphism $\overline{\rho} \cong \overline{\rho} \otimes \alpha$ as $\overline{\rho}$ is irreducible. Taking the determinant of this identity, $\det(\overline{\rho}) = \det(\overline{\rho})\alpha^2$; so, $\alpha^2 = 1$. If $\alpha = 1$, i commutes with absolutely irreducible $\overline{\rho}$; so, by Schur's lemma, $i \neq 0$ is a scalar, contradicting $\mathrm{Tr}(i) = 0$ (by $p > 2$). Thus α is quadratic, and as seen in Lemma 7.3.1, $\overline{\rho} = \mathrm{Ind}_K^{\mathbb{Q}} \varphi$ for a quadratic extension $K/\mathbb{Q}$ fixed by $\mathrm{Ker}(\alpha)$. This means we are in Case D or Case C. Thus

$$Ad(\overline{\rho}) \text{ is absolutely irreducible in Case E.} \tag{7.25}$$

In this section, we study Case E but until §7.7.3, we do not suppose that we are in case E.

Under $p \nmid |\overline{G}|$, by lifting $\overline{\rho}$ to characteristic 0 representation, often we can go forward and back between residual representations and their deformations without changing the image of the representation as a group. We describe this fact here. The group $\mathbf{G} := \mathrm{Gal}(F(\overline{\rho})/\mathbb{Q}) \cong \mathrm{Im}(\overline{\rho})$ fits into an exact sequence for the center Z (scalar matrices) of GL_2:

$$1 \to Z(\mathbb{F}) \cap \mathbf{G} \to \mathbf{G} \to \overline{G} \to 1.$$

Since $|Z(\mathbb{F})| = |\mathbb{F}^\times|$ is prime to p, we find $p \nmid |\mathbf{G}|$. Under this circumstance, the set of irreducible representations of $\mathbf{G}$ with coefficients in $\mathbb{F}$ is in bijection to representations with coefficients in W irreducible over $\mathrm{Frac}(W)$ by reduction modulo $\mathfrak{m}_W$ (cf. [MFG, Corollary 2.7]).

Writing $\rho : \mathrm{Gal}(\overline{\mathbb{Q}}/\mathbb{Q}) \to \mathrm{GL}_2(W)$ (factoring through $\mathbf{G}$) for the lifted representation, we have $\mathrm{Im}(Ad(\rho)) = \mathrm{Im}(Ad(\overline{\rho})) \cong \overline{G}$. Recall the splitting field F of $Ad(\overline{\rho})$; so, $\overline{G} = \mathrm{Gal}(F/\mathbb{Q})$. In Case E, $\overline{G}$ has no abelian cyclic quotient of order $p - 1$; so, $\mu_p(F) = \{1\}$ by Galois theory.

7.7.2 Minkowski unit

We recall Minkowski unit in our setting. Assume $p \nmid |\overline{G}|$. Let $O_f^\times := O^\times/\mu_p(F)$. We have shown in (7.10) that $(O_f^\times \otimes_{\mathbb{Z}} \mathbb{F}) \oplus \mathbb{F}1 \cong \mathrm{Ind}_C^{\overline{G}} 1 \cong \mathbb{F}[\overline{G}/C]$ by (the proof of) Dirichlet's unit theorem. Here C is the subgroup of $\overline{G}$ generated by the fixed complex conjugation c. By the same argument, we find $(O_f^\times \otimes_{\mathbb{Z}} \mathfrak{m}_W^n/\mathfrak{m}_W^{n+1}) \oplus \mathfrak{m}_W^n/\mathfrak{m}_W^{n+1}1 \cong \mathfrak{m}_W^n/\mathfrak{m}^{n+1}[\overline{G}/C]$; so,

$$(O_f^\times \otimes_{\mathbb{Z}} W/\mathfrak{m}_W^n) \oplus W/\mathfrak{m}_W^n 1 \cong W/\mathfrak{m}_W^n[\overline{G}/C].$$

Passing to the (projective) limit, we get

$$(O_f^\times \otimes_{\mathbb{Z}} W) \oplus W \cong W[\overline{G}/C]$$

as $\overline{G}$-module. Take $W = \mathbb{Z}_p$. Since $\mathbb{Z}_p[\overline{G}/C]/\mathbb{Z}_p1$ is a cyclic $\mathbb{Z}_p[\overline{G}]$-module, there is a generator $\varepsilon \otimes 1 \in O_f^\times \otimes_{\mathbb{Z}} \mathbb{Z}_p$ ($\varepsilon \in O_f^\times$) over $\mathbb{Z}_p[\overline{G}]$. This unit ε is called a Minkowski unit (at p), and we fix one ε. By our choice, the conjugates $\{\varepsilon^\sigma | \sigma \in \overline{G}/C\}$ have a unique relation $\prod_{\sigma \in \overline{G}/C} \varepsilon^\sigma = 1$ modulo torsion and generates a subgroup of $O_f^\times$ of finite index prime to p. For each general $W_{/\mathbb{Z}_p}$, $\varepsilon \otimes 1$ is a generator of $O_f^\times \otimes_{\mathbb{Z}} W$ over $W[\overline{G}]$.

Recall $Cl_F(p^\infty) = \varprojlim_n Cl_F(p^n)$, and we have an exact sequence

$$O^\times \to (O/p^n O)^\times \to Cl_F(p^n) \to Cl_F \to 1.$$

Passing to the limit, we get

$$1 \to \overline{O^\times} \to O_p^\times \to Cl_F(p^\infty) \to Cl_F \to 1,$$

where $O_p = \varprojlim_n O/p^n O$ and $\overline{O^\times} = \varprojlim_n \mathrm{Im}(O^\times \to (O/p^n O)^\times)$.

Adding "$\widehat{}$", we denote the p-profinite part of each group in the sequence, getting another exact sequence

$$1 \to \widehat{O^\times} \to \widehat{O_p^\times} \to C_F(p^\infty) \to C_F \to 1,$$

where we have written simply $\widehat{O^\times}$ for $\widehat{\overline{O}^\times}$. Except for Case E, we could have p-torsion in $\widehat{O^\times}$ (i.e., $\mu_p(F) \neq 1$) and in $\widehat{O_p^\times}$ (i.e., $\epsilon/\delta = \omega_p$ is the Teichmüller character).

7.7.3 Selmer group revisited

We often write simply Ad for $Ad(\rho)$ and recall the set S of ramified primes outside p of $F(Ad(\rho))/\mathbb{Q}$. Let $k^{(p)}$ be the maximal p-profinite extension of a number field k unramified outside p, put $\mathcal{G} = \mathrm{Gal}(F^{(p)}/\mathbb{Q})$, $\mathcal{G}_F = \mathrm{Gal}(F^{(p)}/F)$, and recall $G = \mathrm{Gal}(F(\overline{\rho})^{(p)}/\mathbb{Q})$, $G_F = \mathrm{Gal}(F(\overline{\rho})^{(p)}/F)$. Recall also

$$\mathrm{Sel}(Ad(\rho)) := \mathrm{Ker}(H^1(G, Ad^*) \to \frac{H^1(\mathbb{Q}_p, Ad^*)}{F_-^+ Ad^*}),$$

where $F_-^+ Ad^*$ is a subgroup of $H^1(\mathbb{Q}_p, Ad^*)$ made of classes of cocycles upper triangular over the p-decomposition group and upper nilpotent over the p-inertia group. In the same manner as above, we define the Selmer group over $F = F(Ad(\overline{\rho}))$ by

$$\mathrm{Sel}_F(Ad(\rho)) = \mathrm{Ker}(H^1(\mathcal{G}_F, Ad^*) \to \prod_{\wp|p} \frac{H^1(F_\wp, Ad^*)}{F_{-,\wp}^+ Ad^*}),$$

where $F_{-,\wp}^+ Ad^*$ for primes $\wp|p$ of F is a subgroup of $H^1(F_\wp, Ad^*)$ made of classes of cocycles upper triangular over $D_\wp$ and upper nilpotent over $I_\wp$ (after suitable conjugation as in (7.21)).

Lemma 7.7.2. *Assume $p \nmid |\overline{G}| = [F : \mathbb{Q}]$. Then we have a canonical isomorphism induced by the restriction map:*

$$\mathrm{Sel}(Ad(\rho)) \cong \mathrm{Sel}_F(Ad(\rho))^{\overline{G}} \subset \mathrm{Hom}_{\mathbb{Z}_p[\overline{G}]}(C_F(p^\infty), Ad(\rho)^*),$$

where $\mathrm{Sel}_F(Ad(\rho))^{\overline{G}}$ is the subgroup of $\mathrm{Sel}_F(Ad(\rho))$ fixed by $\overline{G}$.

Proof. For a topological group X, write X^{ab} for the maximal continuous abelian quotient of X. Let $u : G \to Ad^*$ be a Selmer cocycle. Let $u' = u|_{G_F} : G_F \to Ad^*$, which is a $\overline{G}$-equivariant homomorphism. By inflation-restriction, through $u \mapsto u'$,

$$\mathrm{Sel}(Ad(\rho)) \hookrightarrow H^1(G, Ad^*) \cong \mathrm{Hom}_{\mathbb{Z}_p[\overline{G}]}(G_F^{ab}, Ad^*),$$

since $H^q(\overline{G}, Ad(\rho)^*) = 0$ for $q > 0$ by $p \nmid [F : \mathbb{Q}]$.

Since the ramification of a prime l outside p is concentrated in $\mathrm{Gal}(F(\overline{p})/F)$, the inertia group $I_{\mathfrak{l}} \subset G_F$ for a prime ideal $\mathfrak{l}$ of O above is finite of order prime to p. This implies $u'(I_{\mathfrak{l}}) = 0$ as Ad^* is p-torsion. Thus u' factors through $G_F^{ab} \twoheadrightarrow \mathcal{G}_F^{ab}$ as $\mathcal{G}_F$ is the Galois group over F of the maximal p-profinite extension $F^{(p)}$ of F unramified outside p. By class field theory, we know $\mathcal{G}_F^{ab} \cong C_F(p^\infty)$.

By definition, the image of $\mathrm{Sel}(Ad(\rho))$ in $\mathrm{Hom}_{\mathbb{Z}_p[\overline{G}]}(C_F(p^\infty), Ad(\rho)^*)$ is contained in $\mathrm{Sel}_F(Ad(\rho))^{\overline{G}}$.

We now start with a Selmer homomorphism $u' \in \mathrm{Sel}_F^{\overline{G}}(Ad(\rho))$ and pick a 1-cocycle $u : \mathcal{G} \to Ad(\rho)^*$ in the inflated image of u' in $H^1(\mathcal{G}, Ad(\rho^*))$. Since $\overline{G}$ has order prime to p, we have $\mathcal{G} \cong \overline{G} \ltimes \mathcal{G}_F$. We can decompose compatibly $D_p = \overline{D} \ltimes D_\wp$ for the decomposition groups $D_p \subset \mathcal{G}$ at p and $\overline{D} \subset \overline{G}$ with $D_\wp \subset D_p$. Regarding $\overline{G}$ as a subgroup of $\mathcal{G}$, we consider $u|_{\overline{G}}$. Since $p \nmid |\overline{G}|$, we may choose u so that $u|_{\overline{G}} = 0$ regarding $\overline{G} \subset \mathcal{G}$; in particular, $u|_{\overline{D}} = 0$. Since $u|_{\mathcal{G}_F} = u'$, we find that u is a Selmer 1-cocycle. Thus the restriction map $\mathrm{Res} : \mathrm{Sel}(Ad(\rho)) \to \mathrm{Sel}_F(Ad(\rho))^{\overline{G}}$ is an isomorphism. $\qquad\square$

7.7.4 *Decomposition group as Galois module*

We describe Galois module structure of p-decomposition groups. An essential part of $C_F(p^\infty)$ comes from $\widehat{O}_p^\times$ which is the product of p-inertia subgroup of $\mathcal{G}_F^{ab}$; so, we study decomposition group in $\mathcal{G}_F^{ab}$ as $\overline{D}$-modules. We fix a prime factor $\wp|p$ in O with its decomposition subgroup $\overline{D} \subset \overline{G}$. Write simply $M_\wp := \widehat{F_\wp^\times} \otimes_{\mathbb{Z}_p} W$ and $U_\wp := \widehat{O_\wp^\times} \otimes_{\mathbb{Z}_p} W$. Then for each character $\xi : \overline{D} \to W^\times$, $M_\wp$ contains as a direct factor the ξ-eigenspace $M_\wp[\xi] = 1_\xi M_\wp$ for $1_\xi = |\overline{D}|^{-1} \sum_{g \in \overline{D}} \xi^{-1}(g)g \in W[\overline{D}]$. Then

- A canonical exact sequence $U_\wp[1] \hookrightarrow M_\wp[1] \xrightarrow{\mathrm{ord}_\wp} W$ induced by the valuation $\mathrm{ord}_\wp : F_\wp^\times \twoheadrightarrow \mathbb{Z}$ at $\wp$, and $U_\wp[1] \cong W$ as $\mu_p(F_\wp)[1] = 0$.
- $M_\wp[\xi]$ is a direct summand of $U_\wp$ if $\xi \neq 1$. Since all other prime factors of p are of the form $\sigma(\wp)$ for $\sigma \in \overline{G}/\overline{D}$, we have $M_p := \widehat{F_p^\times} \otimes_{\mathbb{Z}_p} W \cong \mathrm{Ind}_{\overline{D}}^{\overline{G}} M_\wp$ as $\overline{G}$-modules (for $F_p = F \otimes_{\mathbb{Q}} \mathbb{Q}_p$). Put $U_p := \widehat{O_p^\times} \otimes_{\mathbb{Z}_p} W$.

Since we are in Case E, $F^\times$ is p-torsion-free. Indeed, if $\mu_p(F) \neq 1$, $\overline{G}$ has a quotient $\mathrm{Gal}(\mathbb{Q}[\mu_p]/\mathbb{Q})$ which is a cyclic group of order $p - 1 \geq 4$. Thus by Lemma 7.7.1 (as $p \geq 5$), this is impossible. Write $Ad = Ad(\rho)$. We want to know the structure of $M_p[Ad]$ as a $\overline{G}$-module. If M_p has p-torsion, $\mu_p(F_\wp) \neq 1$. Thus $\overline{G}$ has the decomposition subgroup $\overline{D}$ at $\wp$ which surjects down to a cyclic group $\mathrm{Gal}(\mathbb{Q}_p[\mu_p]/\mathbb{Q}_p)$ of order $p-1$. Since $\mathbb{Q}_p[\mu_p]/\mathbb{Q}_p$ is fully ramified, the inertia group $\overline{I}$ of $\overline{D}$ has tame cyclic quotient of order $p-1$, which lifts to a subgroup of I as the determination of the tame inertia group [MFG, §3.2.5]. Thus $\mu_p(F_\wp) = \{1\}$ again by Lemma 7.7.1 if $p \geq 7$, and the $\mathbb{Z}_p$-module M_p is free if $p \geq 7$.

For the idempotent 1_{Ad} of $W[\overline{G}]$ corresponding to $Ad(\rho)$ and a W-free $W[\overline{G}]$-module X, we consider the Ad-isotypical component $X[Ad] = 1_{Ad}X \cong X \otimes_{W[\overline{G}]} Ad$ of X. Since $M_p = \mathrm{Ind}_{\overline{D}}^{\overline{G}} M_\wp$, by Shapiro's lemma Lemma 5.3.2, we have for $\xi = \epsilon\delta^{-1}$

$$\mathrm{Hom}_{\overline{G}}(M_p, Ad^*) = \mathrm{Hom}_{\overline{D}}(M_\wp, Ad^*|_{\overline{D}}) = \mathrm{Hom}_{\overline{D}}(M_\wp, \xi^* \oplus \mathbf{1}^* \oplus (\xi^{-1})^*).$$

Since $M_\wp[\xi^{\pm1}] = U_\wp[\xi^{\pm1}]$ (by $\xi \neq \mathbf{1}$),

$$(\mathrm{Ind}_{\overline{D}}^{\overline{G}} U_\wp[\xi] \oplus \mathrm{Ind}_{\overline{D}}^{\overline{G}} U_\wp[\mathbf{1}] \oplus \mathrm{Ind}_{\overline{D}}^{\overline{G}} U_\wp[\xi^{-1}])[Ad] = Ad_\xi \oplus Ad_{\mathbf{1}} \oplus Ad_{\xi^{-1}},$$

where $Ad_? = \mathrm{Ind}_{\overline{D}}^{\overline{G}} ?[Ad]$. This fits into the following exact sequence of $\overline{G}$-modules:

$$0 \to \overbrace{Ad_\xi \oplus Ad_{\mathbf{1}} \oplus Ad_{\xi^{-1}}}^{\text{inertia part}} \to M_p[Ad] \xrightarrow{\prod_{\sigma \in \overline{G}/\overline{D}} \mathrm{ord}_{\sigma(\wp)}} \overbrace{(\mathrm{Ind}_{\overline{D}}^{\overline{G}} \mathbf{1})[Ad]}^{\text{Frobenius part}} \to 0.$$

7.7.5   *Structure of Selmer groups*

Let $C_F^{(p)}$ be the subgroup of $C_F = Cl_F \otimes_{\mathbb{Z}} \mathbb{Z}_p$ generated by the image of $\sigma(\wp)$ for $\sigma \in \overline{G}$ under the projection $Cl_F \twoheadrightarrow C_F$. We then define $\widehat{C}_F := C_F/C_F^{(p)}$.

Theorem 7.7.3. *Assume that we are in Case E and assume $p \geq 5$ with $p \nmid |\overline{G}| = [F : \mathbb{Q}]$. Then we have an inclusion $i : \mathrm{Hom}_{\mathbb{Z}_p[\overline{G}]}(\widehat{C}_F, Ad(\rho)^*) \hookrightarrow \mathrm{Sel}(Ad(\rho))$ whose cokernel fits into the following short exact sequence*

$$\mathrm{Hom}_{\mathbb{Z}_p[\overline{G}]}(C_F^{(p)}, Ad(\rho)^*) \hookrightarrow \mathrm{Coker}(i) \twoheadrightarrow \mathrm{Hom}_{W[\overline{D}]}(U_\wp[\epsilon\delta^{-1}]/\overline{\langle \varepsilon_{\epsilon\delta^{-1}} \rangle}, W^\vee),$$

where ε is the fixed Minkowski unit with $\widehat{O}^\times = \mathbb{Z}_p[\overline{G}]\varepsilon$, $\varepsilon_{\epsilon\delta^{-1}}$ is the projection of ε in the direct summand $U_\wp[\epsilon_\wp\delta_\wp^{-1}]$ under $O^\times \to U_p \twoheadrightarrow U_\wp[\epsilon\delta^{-1}]$, and $\overline{\langle \varepsilon_{\epsilon\delta^{-1}} \rangle}$ is the W-submodule of $U_\wp[\epsilon\delta^{-1}]$ generated by $\varepsilon_{\epsilon\delta^{-1}}$.

Corollary 7.7.4. *Let the assumption be as in Theorem 7.7.3. Then*

$$|\mathrm{Sel}(Ad(\rho))| = |C_F \otimes_{\mathbb{Z}_p[\overline{G}]} Ad(\rho)||(U_\wp[\epsilon\delta^{-1}]/\overline{\langle \varepsilon_{\epsilon\delta^{-1}} \rangle})|,$$

which is finite.

Remark 7.7.5. Write $\rho = \rho_f$ for a weight 1 form f. The conjugacy class of $\overline{D}$ in $\overline{G}$ and characters ϵ and δ depends on primes $\mathfrak{p}$ of $\mathbb{Z}[f]$. By Chebotarev density, for a given conjugacy class of a cyclic subgroup $\overline{D} \neq 1$, we have infinitely many primes $\mathfrak{p}$ giving rise to $\overline{D}$ and for such $\mathfrak{p}$, we may assume the unordered set $\{\epsilon, \delta\}$ are independent of $\mathfrak{p}$. Fix such a choice, we ask if $|(U_\wp[\epsilon\delta^{-1}]/\langle\varepsilon_{\epsilon\delta^{-1}}\rangle)| > 1$ for infinitely many primes $\mathfrak{p}$. Similar to Wall–Sun–Sun primes in §8.3.6, there could be infinitely many of them for each choice of $\overline{D}$? If this is the case, then $\mathbb{T}_\mathfrak{p} \neq \Lambda$ for infinitely many $\mathfrak{p}$'s.

7.7.6 Proof of Theorem 7.7.3

We prove the theorem separating it into several steps. After finishing the proof of the theorem, we prove the corollary.

Proof. Step 0: Preliminary.

By Lemma 7.7.2, we have $\mathrm{Sel}(Ad(\rho)) \cong \mathrm{Sel}_F(Ad(\rho))^{\overline{G}}$; so, we can replace $\mathrm{Sel}(Ad(\rho))$ in the assertion of the theorem by $\mathrm{Sel}_F(Ad(\rho))^{\overline{G}}$. Writing the closure of the image of $O^\times$ in U_p as $\overline{O}^\times$, we have the commutative diagram with exact rows:

$$
\begin{array}{ccccc}
U_p/\overline{O}^\times & \hookrightarrow & M_p/\overline{O}^\times & \twoheadrightarrow & C_F^{(p)} \\
\| \downarrow & & \cap \downarrow & & i \downarrow \cap \\
U_p/\overline{O}^\times & \hookrightarrow & C_F(p^\infty) & \twoheadrightarrow & C_F.
\end{array}
\qquad (7.26)
$$

This produces an short exact sequence:

$$1 \to M_p/\overline{O}^\times \to C_F(p^\infty) \to \widehat{C}_F \to 1.$$

Taking the Ad-eigenspace keeps exactness as $p \nmid |\overline{G}|$, and we get another short exact sequence:

$$1 \to (M_p/\overline{O}^\times)[Ad] \to C_F(p^\infty)[Ad] \to \widehat{C}_F[Ad] \to 1. \qquad (7.27)$$

Step 1: $\mathrm{Hom}_{\mathbb{Z}_p[\overline{G}]}(\widehat{C}_F, Ad^*) \xrightarrow{i}{\hookrightarrow} \mathrm{Sel}_F(Ad(\rho))^{\overline{G}}$.

Elements in $\mathrm{Hom}_{\mathbb{Z}_p[\overline{G}]}(\widehat{C}_F, Ad^*)$ are everywhere unramified and trivial at p; so, they gives rise to a subgroup of $\mathrm{Sel}_F(Ad(\rho))^{\overline{G}}$ of classes everywhere unramified and trivial at p. Indeed, by $H^1(\mathcal{G}, Ad^*) \cong \mathrm{Hom}_{\mathbb{Z}_p[\overline{G}]}(\mathcal{G}_F^{ab}, Ad^*)$, any $u \in \mathrm{Hom}_{\mathbb{Z}_p[\overline{G}]}(\widehat{C}_F, Ad^*)$ extends uniquely the cocycle $u : \mathcal{G} \to Ad^*$ trivial at $D_p = \overline{D} \ltimes D_\wp$ unramified everywhere over $\mathcal{G}_F$. Since the inertia group $I_l \subset G$ of any prime $l \neq p$ has order prime to p, $u|_{I_l} = 0$, and hence $[u] \in \mathrm{Sel}(Ad(\rho))$ by Lemma 7.7.2.

Let $D_\wp$ be the decomposition group at $\wp$ of $\mathcal{G}_F^{ab}$ with inertia subgroup $I_\wp$. Then by local class field theory described in §7.2.1

$$\prod_{\sigma \in \overline{G}/\overline{D}} \sigma D_\wp \sigma^{-1} \cong M_p \quad \text{and} \quad \prod_{\sigma \in \overline{G}/\overline{D}} \sigma I_\wp \sigma^{-1} \cong U_p.$$

Elements of $\mathrm{Sel}_F(Ad(\rho))^{\overline{G}}$ modulo $\mathrm{Hom}_{\mathbb{Z}_p[\overline{G}]}(\widehat{C}_F, Ad^*)$ are determined by its restriction to M_p as they are unramified outside p as they factor through $C_F(p^\infty)$ and $p \nmid |I_l|$.

Step 2: Restriction to $D_\wp$.
By applying the functor $? \mapsto \mathrm{Hom}_{\mathbb{Z}_p[\overline{G}]}(?, Ad(\rho)^*)$ to the top row of (7.26) produces the exact sequence in the theorem for $\mathrm{Coker}(i)$. Here the divisibility of $Ad(\rho)^*$ preserves exactness, as the functor is essentially the Pontryagin dual functor (see (4.27)). Recall $\xi = \epsilon \delta^{-1}$. We study

$$u_\wp = u|_{D_\wp} \in \mathrm{Hom}_{\mathbb{Z}_p[\overline{D}]}(D_\wp, Ad^*) = \mathrm{Hom}_{\mathbb{Z}_p[\overline{D}]}(M_\wp, Ad^*)$$

for a 1-cocycle $u : \mathcal{G} \to Ad^*$. Since $Ad = Ad[\xi] \oplus Ad[1] \oplus Ad[\xi^{-1}]$, we have a decomposition:

$$
\mathrm{Hom}_{\mathbb{Z}_p[\overline{D}]}(M_\wp, Ad^*) = \overbrace{\mathrm{Hom}_{\mathbb{Z}_p[\overline{D}]}(U_\wp[\xi], Ad[\xi]^*)}^{\text{upper nilpotent}}
$$

$$
\oplus \underbrace{\mathrm{Hom}_{\mathbb{Z}_p[\overline{D}]}(M_\wp[1], Ad[1]^*)}_{\text{diagonal}} \oplus \overbrace{\mathrm{Hom}_{\mathbb{Z}_p[\overline{D}]}(U_\wp[\xi^{-1}], Ad[\xi^{-1}]^*)}^{\text{lower nilpotent}}.
$$

Thus a Selmer cocycle u projects down to the first two factors:

$$
\overbrace{\mathrm{Hom}_{\mathbb{Z}_p[\overline{D}]}(U_\wp[\xi], Ad[\xi]^*)}^{\text{upper nilpotent}} \oplus \overbrace{\mathrm{Hom}_{\mathbb{Z}_p[\overline{D}]}(M_\wp[1], Ad[1]^*)}^{\text{diagonal}}.
$$

Write $u_\wp^+$ (resp. $u_\wp^0$) for the upper nilpotent projection (resp. the diagonal projection) of u.

Step 3: Inertia part u_+.
We have $u_\wp^+ : I_\wp[\xi] = U_\wp[\xi] \to Ad[\xi]^*$ and $u_{\sigma(\wp)}^+ : U_{\sigma(\wp)}[\xi_\sigma] \dashrightarrow Ad[\xi_\sigma]^*$ for $D_{\sigma(\wp)} = \sigma D_\wp \sigma^{-1} \xrightarrow{\xi_\sigma} A^\times$ given by $\xi_\sigma(h) = \xi(\sigma^{-1} h \sigma)$. Note $Ad[\xi_\sigma]^* = \sigma(Ad[\xi]^*)$ and $U_{\sigma(\wp)}[\xi_\sigma] = \sigma(U_\wp[\xi])$ and $u_{\sigma(\wp)}(h) = u_\wp(\sigma^{-1} h \sigma)$. Since u is a cocycle over $\mathcal{G}$, out of each restriction $u_{\sigma(\wp)}^+$, we create the map

$$u_+ := (u_{\sigma(\wp)}^+)_\sigma : \prod_{\sigma \in \overline{G}/\overline{D}} \sigma(U_\wp[\xi]) \to \prod_{\sigma \in \overline{G}/\overline{D}} \sigma(Ad[\xi])^*.$$

Note $\prod_\sigma \sigma(U_\wp[\xi]) = \mathrm{Ind}_{\overline{D}}^{\overline{G}} U_\wp[\xi]$ and $\prod_\sigma \sigma(Ad[\xi])^* \cong \mathrm{Ind}_{\overline{D}}^{\overline{G}} Ad[\xi]^*$ as $\overline{G}$-modules. Since u is a cocycle defined over $\mathcal{G}$, we get a $\overline{G}$-equivariant commutative diagram:

$$
\begin{array}{ccc}
\mathrm{Ind}_{\overline{D}}^{\overline{G}} U_\wp[\xi] & \xrightarrow{\;u_+\;} & \mathrm{Ind}_{\overline{D}}^{\overline{G}} Ad[\xi]^* \\
\downarrow & & \downarrow \\
U_p[Ad] & \xrightarrow{\;u|_{U_p}\;} & Ad^*.
\end{array}
$$

Step 4: Determination of inertia part $u|_{U_p}$.

By the above argument, the restriction $u|_{U_p}$ falls into $\mathrm{Hom}_{\mathbb{Z}_p[\overline{G}]}(Ad_\xi, Ad^*)$ induced from u_+. Though $U_p[Ad] \cong Ad^m$ for $m = 3$ if ξ has order 3 and $m = 2$ if ξ has order 2, as Shapiro's isomorphism (Lemma 5.3.2)

$$
S : \mathrm{Hom}_{\mathbb{Z}_p[\overline{G}]}(\mathrm{Ind}_{\overline{D}}^{\overline{G}} U_\wp[\xi], Ad) \cong \mathrm{Hom}_{\overline{D}}(\xi, \xi_+ \oplus \mathbf{1} \oplus \xi_-^{-1})
$$

with $\xi_+ = \xi$ realized on upper nilpotent matrices and $\xi_- = \xi$ realized on lower nilpotent matrices. The restriction $u|_{U_p}$ only has values in ξ_+; so, ξ_-^{-1}-component does not show up as $u|_{U_\wp}$ is upper nilpotent, we have $S(u|_{U_p}) \in \mathrm{Hom}_{W[\overline{D}]}(U_\wp[\xi], \xi_+^*)$. Since u factors through $\widehat{O}_p^\times/\overline{O^\times}$,

$$
S(u|_{U_p}) \text{ factors through } U_\wp[\xi]/\langle \overline{\varepsilon_\xi} \rangle.
$$

Starting from $u_\wp \in \mathrm{Hom}(U_\wp[\xi]/\langle \overline{\varepsilon_\xi} \rangle, \xi^*)$, we can recreate $u = S^{-1}(u_\wp) : U_p/\widehat{O}^\times[Ad] \to Ad^*$; so, we have $\mathrm{Sel}(Ad) \twoheadrightarrow \mathrm{Hom}(U_\wp[\xi]/\langle \overline{\varepsilon_\xi} \rangle, \xi_+^*)$.

Step 5: Frobenius part.

Writing $\mathbf{1}$ (resp. $W/p^h W$) for the $\overline{D}$-module W (resp. $W/p^h W$) with trivial action, note $M_p/U_p = \mathrm{Ind}_{\overline{D}}^{\overline{G}} \mathbf{1} \cong \bigoplus_{\sigma(\wp):\sigma \in \overline{G}/\overline{D}} W\sigma(\wp)$ as $W[\overline{G}]$-modules. We have a projection $\pi : \mathrm{Ind}_{\overline{D}}^{\overline{G}} \mathbf{1} \twoheadrightarrow C_F^{(p)} \otimes_{\mathbb{Z}_p} W$ sending $\sigma(\wp)$ to its class in $C_F^{(p)}$. If $\wp$ has order p^h in $C_F^{(p)}$, this induces a surjection $\mathrm{Ind}_{\overline{D}}^{\overline{G}} W/p^h W \to C_F^{(p)} \otimes_{\mathbb{Z}_p} W$, which gives rise to an isomorphism:

$$
(\mathrm{Ind}_{\overline{D}}^{\overline{G}} W/p^h W)[Ad] \cong (C_F^{(p)} \otimes_{\mathbb{Z}_p} W)[Ad] =: C_F^{(p)}[Ad] \tag{$*$}
$$

by the irreducibility of $Ad(\overline{\rho})$. Therefore

$$
u_0 \in \mathrm{Hom}_{W[\overline{G}]}(C_F^{(p)}[Ad], Ad^*) \overset{(*)}{=} \mathrm{Hom}_{W[\overline{G}]}(\mathrm{Ind}_{\overline{D}}^{\overline{G}} W/p^h W, Ad^*)
$$

$$
\overset{\text{Lemma 5.3.2}}{=} \mathrm{Hom}_{\overline{D}}(W/p^h W, Ad^*|_{\overline{D}}) = W/p^h W.
$$

Reversing the argument, the Frobenius part is given by

$$
\mathrm{Hom}_{W[\overline{G}]}(C_F^{(p)}[Ad], Ad^*) \cong \mathrm{Hom}_{\mathbb{Z}_p[\overline{G}]}(C_F^{(p)}, Ad^*).
$$

This finishes the proof of the theorem. $\qquad\square$

7.7.7 Proof of Corollary 7.7.4

By Lemma 7.7.2, we only need to prove the assertion for $|\operatorname{Sel}_F(Ad(\rho))^{\overline{G}}|$: We shall do this in two steps.

Step 1: Proof of the formula.
Since $Ad^* = Ad(\rho) \otimes_{\mathbb{Z}_p} \mathbb{Q}_p/\mathbb{Z}_p \cong \operatorname{Hom}_{\mathbb{Z}_p}(Ad, \mathbb{Q}_p/\mathbb{Z}_p)$ and $\otimes$-Hom adjunction formula [BAL, II.4.1], we have

$$\operatorname{Hom}_{\mathbb{Z}_p[\overline{G}]}(C_F^{(p)}, Ad^*) \cong \operatorname{Hom}_{\mathbb{Z}_p[\overline{G}]}(C_F^{(p)}, \operatorname{Hom}_{\mathbb{Z}_p}(Ad, \mathbb{Q}_p/\mathbb{Z}_p))$$
$$\cong \operatorname{Hom}_{\mathbb{Z}_p}(C_F^{(p)} \otimes_{\mathbb{Z}_p[\overline{G}]} Ad, \mathbb{Q}_p/\mathbb{Z}_p). \quad (7.28)$$

Similarly we have

$$\operatorname{Hom}_{\mathbb{Z}_p[\overline{G}]}(\widehat{C}_F, Ad^*)) \cong \operatorname{Hom}_{\mathbb{Z}_p[\overline{G}]}(\widehat{C}_F, \operatorname{Hom}_{\mathbb{Z}_p}(Ad, \mathbb{Q}_p/\mathbb{Z}_p))$$
$$\cong \operatorname{Hom}_{\mathbb{Z}_p}(\widehat{C}_F \otimes_{\mathbb{Z}_p[\overline{G}]} Ad, \mathbb{Q}_p/\mathbb{Z}_p).$$

The W-corank of the Selmer group is positive when $\varepsilon_{\epsilon\delta^{-1}} = 1$. If this happens, it is equal to $\operatorname{rank}_W U_\wp[\epsilon\delta^{-1}]$. Since $U_\wp[\epsilon\delta^{-1}] = \widehat{O_\wp^\times} \otimes_{\mathbb{Z}_p} W[\epsilon\delta^{-1}]$ has the same rank with $O_\wp \otimes_{\mathbb{Z}_p} W[\epsilon\delta^{-1}]$ by $\overline{D}$-equivariance of logarithm, we get $\operatorname{rank}_W(O_\wp \otimes_{\mathbb{Z}_p} W)[\epsilon\delta^{-1}] = 1$, since $F_\wp$ has normal basis over $\mathbb{Q}_p$.

Step 2: Galois action on global units.
Recall $(O^\times \otimes_{\mathbb{Z}} W) \oplus 1 \cong \operatorname{Ind}_C^{\overline{G}} 1 \cong W[\overline{G}/C]$ from (7.10) (as $\mu_p(F) = \{1\}$). Here C is the subgroup of $\overline{G}$ generated by the fixed complex conjugation c. The following lemma finishes the proof.

Lemma 7.7.6. *We have a $W[\overline{G}]$-linear surjective homomorphism ϕ : $O^\times \otimes_{\mathbb{Z}} W \twoheadrightarrow Ad$ and $\varepsilon_{\epsilon\delta^{-1}} \neq 1$.*

Since $Ad(\overline{\rho})$ is irreducible over $\mathbb{F}$, if a $W[\overline{G}]$-linear map $M \to Ad$ for a $W[\overline{G}]$-module M is non-trivial modulo $\mathfrak{m}_W$, the map is surjective modulo $\mathfrak{m}_W$, and by Nakayama's lemma, the original map is surjective.

Proof. Since Ad is irreducible of dimension 3 over $\operatorname{Frac}(W)$, a non-zero homomorphism $\phi \in \operatorname{Hom}_{W[\overline{G}]}(O^\times \otimes_{\mathbb{Z}} W \oplus 1, Ad)$ has to factor through $O^\times \otimes_{\mathbb{Z}} W = W[\overline{G}]\varepsilon$. By Shapiro's lemma, we have, for $\alpha : C \cong \{\pm 1\}$,

$$\operatorname{Hom}_{W[\overline{G}]}(O^\times \otimes_{\mathbb{Z}} W, Ad) = \operatorname{Hom}_{W[\overline{G}]}(\operatorname{Ind}_C^{\overline{G}} 1, Ad)$$
$$= \operatorname{Hom}_{W[C]}(1, Ad|_C) = \operatorname{Hom}_{W[C]}(1, \alpha \oplus 1 \oplus \alpha) = W.$$

Thus we have a $W[\overline{G}]$-linear homomorphism $\phi : O^\times \otimes_{\mathbb{Z}} W \to Ad$ non-zero modulo $\mathfrak{m}_W$. Therefore, the $W[\overline{G}]$-linear homomorphism $\phi : O^\times \otimes_{\mathbb{Z}} W \to Ad$ is onto, and Ad is generated over $W[\overline{G}]$ by the image of ε. Since $Ad|_{\overline{D}} = \xi \oplus 1 \oplus \xi^{-1}$, the composed ξ-projection $O^\times \otimes_{\mathbb{Z}} W \xrightarrow{\phi} Ad \twoheadrightarrow Ad[\xi] = W\xi$ is onto producing a non-zero multiple of $\varepsilon_{\epsilon\delta^{-1}}$ as its image. $\qquad \square$

7.7.8 *Concluding remarks and questions on cyclicity*

In this chapter, we studied cyclicity for the adjoint Selmer group of an Artin compatible system. Our conclusion is that the cyclicity holds except for finitely many explicitly determined primes. If we start with a compatible system of higher weigh $k_0 \geq 3$, since $\mathbb{T}_{\mathfrak{p}} = \Lambda$ for almost all $\mathfrak{p}$, the cyclicity is trivially true. The case of $k_0 = 2$ is still mysterious. We expect that

$$\mathrm{Sel}(Ad(\rho_{\mathbb{T}_{\mathfrak{p}}}) \otimes \chi)^{\vee} \text{ is cyclic over } \mathbb{T}_{\mathfrak{p}} \text{ for almost all } \mathfrak{p}.$$

for a fixed even Dirichlet character χ. The case of odd Dirichlet character χ is more mysterious (but perhaps doable by the reflection theorem).

If we twist $Ad(\rho_{\mathbb{T}_{\mathfrak{p}}})$ by an irreducible Artin representation ρ, plainly we need to ask the cyclicity question over the group algebra $\mathbb{T}_{\mathfrak{p}}[\mathrm{Gal}(F(\rho)/\mathbb{Q})]$. When ρ is totally even (i.e., $\rho(c) = 1$ for complex conjugation c), there might be a good chance to know if the question has affirmative answer for almost all $\mathfrak{p}$ with a well specified set of exceptional primes.

Of course, we can ask beyond cyclicity:

$$\text{Is } \mathrm{Sel}(Ad(\rho_{\mathbb{T}_{\mathfrak{p}}}) \otimes \chi)^{\vee} \otimes_{\mathbb{Z}_p} \mathbb{Q}_p \text{ a semi-simple } \mathbb{T}_{\mathfrak{p}}\text{-module?}$$

Further, we can ask if there is any series of systematic examples outside representations of adjoint type for which one can expect cyclicity for "most" primes.

Chapter 8

Local indecomposability of modular Galois representation

The universal ring R^{ord} is generated over the Iwasawa algebra Λ by (at most) one element if and only if the adjoint Selmer group is cyclic over R^{ord}. Under this circumstance, assuming $R^{ord} \neq \Lambda$, we have a generator Θ of R^{ord} over Λ. If $R^{ord} = \Lambda$, R^{ord} is generated over W by $t = \det \rho(\gamma)$ for the cyclotomic p-inertia generator $\gamma = [1 + \mathbf{p}, \mathbb{Q}_p]$. Since the influence of the p-decomposition group to the arithmetic of $F(\rho)$ is strong (as is clear from Iwasawa's theory and p-adic Hodge theory), we expect that the elements related to the Frobenius element Frob_p and the inertia group I_p would generate a sizable subring of R^{ord} and $\mathbb{T}$. Thus when we have one generator over Λ, it is natural to try to find a relation of the generator Θ and the non-cyclotomic part of the p-inertia group (and also to Frob_p).

Having this principle in mind, we find, even if R^{ord} may not be generated by one element over Λ, an element θ which satisfies the following characterization. For any deformation $\rho_A \in \mathcal{D}(A)$ with $\rho_A \sim \lambda \circ \rho$ (for the deformation functor $\mathcal{D}$ defined at the beginning of Chapter 6), $\lambda(\theta) \neq 0$ if and only if $\rho_A|_{I_p}$ is a non-trivial extension of ϵ_A by δ_A. In other words, p-local indecomposability of ρ_A is described completely by θ, and in fact, θ generates the unipotent part of the p-inertia subgroup of $\mathrm{Gal}(F(\rho)/F(\overline{\rho}))$ over the subalgebra generated by Λ and $\delta_{R^{ord}}(\mathrm{Frob}_p)$.

Long ago, Greenberg asked if an ordinary $\rho_{f,\mathfrak{p}}$ is indecomposable over I_p as long as f has weight $k \geq 2$ and does not have CM (see [GV04, (1.2)]). The first progress towards the solution of this question is made by Ghate–Vatsal [GV04, Theorem 2], and they proved indecomposability for almost all such forms of fixed prime-to-p level under the Taylor–Wiles condition and p-distinguishedness ("almost all" means here "except for finitely many"). Later for a fixed primitive cusp form of level 1 with coefficients in $\mathbb{Z}$, they proved indecomposability for all ordinary primes except for primes at which

341

$\bar{\rho}_p$ is p-tame with image containing $SL_2(\mathbb{F}_p)$ [GV11]. In [Z14] (based on a result in [H13b]), the indecomposability is proved for all weight 2 non-CM forms (ordinary at p) without exception. More recently, assuming $\bar{\rho} = \text{Ind}_K^{\mathbb{Q}} \bar{\varphi}$ for imaginary K in which p splits, Castella and Wang-Erickson [CW19] have proved indecomposability under $Cl_{F(\bar{\varphi}^-)} \otimes_{\mathbb{Z}[\text{Gal}(\bar{\varphi}^-)/\mathbb{Q})]} \bar{\varphi}^- = 0$, $(\bar{\varphi}^-)^2 \neq 1$ and $\mathfrak{f} + \mathfrak{f}^\sigma = O_K$ for the conductor $\mathfrak{f}$ of $\bar{\varphi}$.

In this chapter, we explicitly compute the ideal $(\theta) \cap \Lambda$ if $\bar{\rho} = \text{Ind}_K^{\mathbb{Q}} \varphi$ for K real in which p splits, and by doing this, we prove in Theorem 8.4.8 indecomposability for weight $k \geq 2$ under p-distinguishedness of $\bar{\rho}$. We will also remove the assumption $\mathfrak{f} + \mathfrak{f}^\sigma = O_K$ and ease the assumption $(\bar{\varphi}^-)^2 \neq 1$ to $\bar{\varphi}^- \neq 1$ in §8.5 if K is imaginary.

We continue to assume $p > 2$ and $\mu_3(F(Ad(\bar{\rho}))) = \{1\}$ if $p = 3$.

8.1 One generator theorem of $\mathbb{T}$ over Λ

Start with a primitive form $f \in S_k(C, \psi)$ and its compatible system $\rho_f = \{\rho_{\mathfrak{p}}\}_{\mathfrak{p}}$. Choose a prime $\mathfrak{p}$ with p-ordinary $\rho_{\mathfrak{p}}$, and put $\bar{\rho} = \bar{\rho}_{\mathfrak{p}} := \rho_{\mathfrak{p}} \mod \mathfrak{p}$. In this section, we prove the following perhaps a most general one-generator theorem which does not require $R^{ord} \cong \mathbb{T}$ (but only p-tameness for $\bar{\rho}$):

Theorem 8.1.1. *Suppose minimality* (m), $\bar{\rho}|_{D_p} = \bar{\epsilon} \oplus \bar{\delta}$ *with* $\bar{\epsilon} \neq \bar{\delta}$ *and either* $t_{\mathbb{T}/\Lambda} \hookrightarrow \mathcal{D}(\mathbb{F}[\varepsilon])$ *and* $p \nmid |\overline{G}|$ *or absolute irreducibility of* $\bar{\rho}$. *Assume that* $Cl_F \otimes_{\mathbb{Z}[G]} Ad(\bar{\rho}) = 0$ *for* $F = F(Ad(\bar{\rho}))$ *and that the Galois module* $Ad(\bar{\rho})|_{D_p}$ *does not contain* $\mathbb{F}_p(1)$ *as a Galois subquotient. Then we have* $\mathbb{T} \cong \Lambda[X]/(D(X))$ *for a distinguished polynomial* $D(X)$ *with respect to* $\mathfrak{m}_\Lambda$ *(i.e.,* $\dim_{\mathbb{F}} t_{\mathbb{T}/\Lambda} \leq 1$*).*

Here are some remarks and open questions:

- The Galois module $\mathbb{F}_p(1)$ is isomorphic $\mu_p(\overline{\mathbb{Q}})$ (so, the Galois action is given by the Teichmüller character ω_p if $p > 2$).
- If $\bar{\rho}|_{I_p}$ is already indecomposable, $F(\bar{\rho})/\mathbb{Q}$ is wildly ramified, and all deformations of $\bar{\rho}$ are indecomposable; so, for the purpose of studying indecomposability, p-tameness assumption is harmless.
- A polynomial $D(X) \in A[X]$ is *distinguished with respect to a prime* $P \in \text{Spec}(A)$ if $D(X)$ is monic and $D(X) \equiv X^{\deg(D)} \mod P$.
- If $k = 1$, $p \nmid |\text{Im}(\rho_f)| \Rightarrow p$ is tamely ramified in $F(\bar{\rho})/\mathbb{Q}$.
- Suppose $p \nmid C$. For $2 \leq k \leq p$, p is tamely ramified if and only if there exists $g \in S_{p+1-k}(C, \psi)_{/\mathbb{F}}$ with $na(n, f) \equiv n^k a(n, g) \mod p$ for all n.

This is conjectured by Serre and a theorem of B. Gross [G90] and R. Coleman–J. F. Voloch [CV92] (but once modular lifting is known, the outcome is almost obvious; so, this result is superseded by [KW09]). The form g is called a *companion form* of f. A weight 1 form is a companion of a weight p form. See [R18] (and Remark 7.6.5) for examples of companion forms.

- *Is the p-tameness necessary for the assertion of the theorem?*
- *What is the density of primes with p-tamely ramified $\bar{\rho}_{\mathfrak{p}}$?* As already mentioned, $\bar{\rho}_{\mathfrak{p}}$ is p-tame for almost all $\mathfrak{p}$ if $k = 1$; and this also holds if f has CM. Thus the question is about a primitive form f without CM of weight ≥ 2. See [OCS, III, Notes, 16.1].

8.1.1 *Proof of one generator theorem*

Step 0: Since $\mathbb{T}$ is free of finite rank over Λ, if $\mathbb{T}$ is generated by one element $\Theta \in \mathfrak{m}_{\mathbb{T}}$ over Λ, the multiplication by Θ on $\mathbb{T}$ has its characteristic polynomial $D(X)$ of degree $e = \mathrm{rank}_{\Lambda} \mathbb{T}$ which is a distinguished polynomial with respect to $\mathfrak{m}_{\Lambda}$ satisfying $\mathbb{T} = \Lambda[[X]]/(D(X))$. Indeed, $D(X) \bmod \mathfrak{m}_{\Lambda}$ is the characteristic polynomial of the multiplication by Θ on $\mathbb{T}/\mathfrak{m}_{\Lambda}\mathbb{T} \cong \mathbb{F}^e$, which is nilpotent; so, $D(X) \equiv X^e \bmod \mathfrak{m}_{\Lambda}$.

Either by Lemma 6.2.12 under absolute irreducibility of $\bar{\rho}$ or by the assumption $t_{\mathbb{T}/\Lambda} \hookrightarrow \mathcal{D}(\mathbb{F}[\varepsilon])$, $t_{\mathbb{T}/\Lambda}$ injects into $\mathcal{D}(\mathbb{F}[\varepsilon]) \cong \mathrm{Sel}(Ad(\bar{\rho}))$; so, $\dim_{\mathbb{F}} t_{\mathbb{T}/\Lambda} \leq \dim_{\mathbb{F}} \mathrm{Sel}(Ad(\bar{\rho}))$, we need to prove that $\mathrm{Sel}(Ad(\bar{\rho}))$ is generated by (at most) one element over $\mathbb{F}$; i.e., we prove that $\mathrm{Sel}(Ad(\bar{\rho}))$ has dimension ≤ 1 over $\mathbb{F}$.

Step 1, Restriction: Write $\overline{G} = \mathrm{Gal}(F(Ad(\bar{\rho}))/\mathbb{Q})$. If $p \nmid |\overline{G}|$, plainly $H^1(\overline{G}, Ad(\bar{\rho})^*) = 0$. Otherwise, by Dickson's classification §7.7.1, $\overline{G}$ is isomorphic to either $\mathrm{PSL}_2(\mathbb{F}')$, $\mathrm{PGL}_2(\mathbb{F}')$ for a subfield $\mathbb{F}'$ of $\mathbb{F}$ or A_5 (when $p = 3$), and we know $H^1(\overline{G}, Ad(\bar{\rho})^*) = 0$ (e.g., [CPS75, Theorem 4.2] and [W95, Proposition 1.11]). By restriction, for $F = F(Ad(\bar{\rho}))$, we find

$$\mathrm{Sel}(Ad(\bar{\rho})) \hookrightarrow \mathrm{Hom}_{\mathbb{Z}[G]}(G_F, Ad(\bar{\rho})^*),$$

where $G_F = \mathrm{Gal}(F^{(p)}(\bar{\rho})/F)$. The image of this map falls in $\mathrm{Sel}_F(Ad(\bar{\rho}))^{\overline{G}}$:

$$\mathrm{Sel}(Ad(\bar{\rho})) \hookrightarrow \mathrm{Sel}_F(Ad(\bar{\rho}))^{\overline{G}}. \tag{8.1}$$

Thus we need to show $\dim_{\mathbb{F}} \mathrm{Sel}_F(Ad(\bar{\rho}))^{\overline{G}} \leq 1$ under our assumptions.

Some notation. Let O be the integer ring of F, $O_p = O \otimes_{\mathbb{Z}} \mathbb{Z}_p$ and $\widehat{O_p^{\times}} = \varprojlim_n O_p^{\times}/(O_p^{\times})^{p^n}$ (the maximal p-profinite quotient of $O_p^{\times}$). Similarly set $\widehat{O_{\wp}^{\times}} = \varprojlim_n O_{\wp}^{\times}/(O_{\wp}^{\times})^{p^n}$ for each prime factor $\wp|p$.

Step 2, Selmer sequence: We fix a prime $\wp_0|p$ of F and choose the inertia group I_0 at $\wp_0$ of G_F so that $\rho_T|_{I_0}$ has values in upper triangular subgroup with the trivial quotient. For each $\wp|p$, we pick $g_\wp \in G$ and put $I_\wp := g_\wp I_0 g_\wp^{-1} \subset G_F$ as an inertia subgroup of $\wp$. By class field theory, writing $G_{F,p}^{ab}$ for the maximal p-abelian quotient of G_F, $\widehat{O_p^\times} \to G_{F,p}^{ab} \twoheadrightarrow C_F$ for $C_F := \widehat{Cl}_F \otimes_{\mathbb{Z}} \mathbb{Z}_p$ is exact, and applying $\mathrm{Hom}_{\mathbb{Z}[\overline{G}]}(?, Ad(\overline{\rho}))$,

$$\mathrm{Hom}_{\mathbb{Z}[\overline{G}]}(C_F, Ad(\overline{\rho})) \hookrightarrow \mathrm{Sel}_F(Ad(\overline{\rho}))^{\overline{G}} \xrightarrow{\pi} \mathrm{Hom}_{\mathbb{Z}_p[\overline{G}]}(\widehat{O_p^\times}, Ad(\overline{\rho})) \quad (8.2)$$

with $\mathrm{Im}(\pi)$ made of ramified Selmer cocycles at p. To validate this exact sequence, we need to show $\mathrm{Hom}_{\mathbb{Z}[\overline{G}]}(C_F, Ad(\overline{\rho})) \hookrightarrow \mathrm{Sel}_F(Ad(\overline{\rho}))^{\overline{G}}$. We proceed similarly to the proof of Theorem 7.7.3 (particularly, Steps 0–2). Let $C_F^{(p)} \subset C_F \otimes_{\mathbb{Z}} \mathbb{F}$ be generated by the classes of prime factors of p. Write $\overline{D} \subset \overline{G}$ for its decomposition group. Then we have a $\mathbb{Z}[\overline{G}]$-equivariant surjection $\mathrm{Ind}_{\overline{D}}^{\overline{G}} \mathbb{F} \cong \bigoplus_{\sigma \in \overline{G}/\overline{D}} \mathbb{F}\sigma(\wp_0) \twoheadrightarrow C_F^{(p)}$, and hence $\mathrm{Hom}_{\mathbb{Z}[\overline{G}]}(C_F, Ad(\overline{\rho})) \hookrightarrow \mathrm{Hom}_{\mathbb{Z}[\overline{G}]}(\mathrm{Ind}_{\overline{D}}^{\overline{G}} \mathbb{F}, Ad(\overline{\rho}))$.

By Lemma 5.3.2, $\mathrm{Hom}_{\mathbb{Z}[\overline{G}]}(\mathrm{Ind}_{\overline{D}}^{\overline{G}} \mathbb{F}, Ad(\overline{\rho})) = \mathrm{Hom}_{\mathbb{F}}(\mathbb{F}, Ad(\overline{\rho})|_{\overline{D}})$. Since $Ad(\overline{\rho}) = \overline{\epsilon}\overline{\delta}^{-1} \oplus 1 \oplus \overline{\epsilon}^{-1}\overline{\delta}$ with $\overline{\epsilon}\overline{\delta}^{-1} \neq 1$, we find the image of $\phi \in \mathrm{Hom}_{\mathbb{F}}(\mathbb{F}, Ad(\overline{\rho})|_{\overline{D}})$ is in $Ad(\overline{\rho})^{\overline{D}} = \{\mathrm{diag}[a, -a] \in \mathfrak{sl}_2(\mathbb{F}) | a \in \mathbb{F}\}$ (diagonal). This shows $\mathrm{Hom}_{\mathbb{Z}[\overline{G}]}(C_F, Ad(\overline{\rho})) \hookrightarrow \mathrm{Sel}_F(Ad(\overline{\rho}))^{\overline{G}}$.

By (8.2), identifying the image of I_0 in $G_{F,p}^{ab}$ with $\widehat{O_{\wp_0}^\times}$ by class field theory, $\phi \in \pi(\mathrm{Sel}_F(Ad(\overline{\rho}))^{\overline{G}})$ has values over $\widehat{O_{\wp_0}^\times}$ in the upper nilpotent subalgebra $\mathfrak{n} \subset \mathfrak{sl}_2(\mathbb{F})$ and in $Ad(\overline{\rho}(g_\wp))(\mathfrak{n}) = g_\wp \mathfrak{n} g_\wp^{-1}$ over $\widehat{O_\wp^\times}$.

Step 3, Use of Shapiro's lemma: Let $\overline{D}$ be the p-decomposition subgroup $\overline{D} \subset \overline{G}$ of $\wp_0$. Note $p \nmid |\overline{D}|$ by assumption. The isomorphism class of a p-torsion-free $\mathbb{Z}_p[\overline{D}]$-module L of finite type is determined by the isomorphism class of $\mathbb{Q}_p[\overline{D}]$-modules $L \otimes_{\mathbb{Z}_p} \mathbb{Q}_p$. By p-adic logarithm and the normal basis theorem, $\widehat{O_{\wp_0}^\times} \otimes_{\mathbb{Z}_p} \mathbb{Q}_p \cong \mathrm{Ind}_1^{\overline{D}} \mathbb{Q}_p = \mathbb{Q}_p[\overline{D}]$ as $\mathbb{Q}_p[\overline{D}]$-modules. We conclude $\widehat{O_{\wp_0}^\times} \cong \mu_p(F_{\wp_0}) \oplus \mathrm{Ind}_1^{\overline{D}} \mathbb{Z}_p$. Since $\widehat{O_p^\times} = \prod_{g \in \overline{D}\backslash\overline{G}} \widehat{O_{\wp_0^g}^\times} \cong \mathrm{Ind}_{\overline{D}}^{\overline{G}} \widehat{O_{\wp_0}^\times}$, the p-profinite completion $\widehat{O_p^\times}$ is isomorphic to $\mathbb{Z}_p[\overline{G}] = \mathrm{Ind}_1^{\overline{G}} \mathbb{Z}_p$ up to p-torsion. If $\mu_p(F_{\wp_0}) = \{1\}$, from Shapiro's lemma

$$\mathrm{Hom}_{\mathbb{Z}_p[\overline{G}]}(\widehat{O_p^\times}, Ad(\overline{\rho})) = \mathrm{Hom}_{\mathbb{Z}_p}(\mathbb{Z}_p, Ad(\overline{\rho})) \cong Ad(\overline{\rho})$$

in which $\pi(\mathrm{Sel}_F(Ad(\overline{\rho}))^{\overline{G}})$ is sent into $\mathfrak{n}$ having dimension 1 over $\mathbb{F}$. Thus the theorem follows from $\mathrm{Hom}_{\mathbb{Z}[\overline{G}]}(Cl_F, Ad(\overline{\rho})) = Cl_F \otimes_{\mathbb{Z}[\overline{G}]} Ad(\overline{\rho}) = 0$; so, $\dim_{\mathbb{F}} \mathrm{Sel}_F(Ad(\overline{\rho}))^{\overline{G}} = \dim_{\mathbb{F}} \pi(\mathrm{Sel}_F(Ad(\overline{\rho}))^{\overline{G}}) \leq 1$. This finishes the proof when $\mu_p(F_{\wp_0}) = \{1\}$.

Final step, $\mu_p(F_{\wp_0}) \neq \{1\}$: Now assume that $\mu_p(F_{\wp_0})$ has order p. By our assumption, $Ad(\overline{\rho})|_{\overline{D}}$ does not contain $\overline{\omega}_p = (\omega_p \mod p\mathbb{Z}_p)$. We have $\widehat{O_p^\times} \cong \text{Ind}_{\overline{D}}^{\overline{G}} \mu_p(\overline{\mathbb{Q}}) \oplus \text{Ind}_1^{\overline{G}} \mathbb{Z}_p$ as $\widehat{O_{\wp_0}^\times} \cong \mu_p(F_{\wp_0}) \oplus \text{Ind}_1^{\overline{D}} \mathbb{Z}_p$. Since $Ad(\overline{\rho})|_{\overline{D}}$ does not contain $\overline{\omega}_p$, by Shapiro's lemma, $\text{Ind}_{\overline{D}}^{\overline{G}} \mu_p(F_{\wp_0}) \otimes_{\mathbb{Z}[\overline{G}]} Ad(\overline{\rho}) = 0$, and we find

$$\text{Hom}_{\mathbb{Z}_p[\overline{G}]}(\widehat{O_p^\times}, Ad(\overline{\rho})) \cong \text{Hom}_{\mathbb{Z}_p[\overline{G}]}(\text{Ind}_1^{\overline{G}} \mathbb{Z}_p, Ad(\overline{\rho})).$$

Then by the same argument as above, we conclude

$$\dim_{\mathbb{F}} \text{Sel}_F(Ad(\overline{\rho}))^{\overline{G}} = \dim_{\mathbb{F}} \pi(\text{Sel}_F(Ad(\overline{\rho}))^{\overline{G}}) \leq 1$$

as desired. $\qquad\qquad\square$

8.1.2 Cyclicity of the adjoint Selmer group again

If $\mathbb{T} = \Lambda[[X]]/(D(X))$ for a distinguished polynomial $D(X)$, we have an exact sequence: $0 \to (D(X))/(D(X))^2 \xrightarrow{d} \Omega_{\Lambda[[X]]/\Lambda} \otimes_{\Lambda[[X]]} \mathbb{T} \to \Omega_{\mathbb{T}/\Lambda} \to 0$. Let $L_\rho = d(D(X))/dX \in \mathbb{T}$. Then in the same manner as in the proof of Theorem 6.1.11, Theorem 6.2.23 and Corollary 6.2.24, we get

Corollary 8.1.2. *Suppose*

- *either f has weight 1 with $p \nmid |\text{Im}(\rho_f)|$ or p is tamely ramified in $F(\overline{\rho})$,*
- *$\overline{\rho}$ satisfies (p), (m) and $\mathbb{T} = R^{ord}$,*
- *$Ad(\overline{\rho}) \otimes_{\mathbb{Z}[D_p]} \mu_p(F_{\wp_0}) = 0$,*
- *$Cl_F \otimes_{\mathbb{Z}[\overline{G}]} Ad(\overline{\rho}) = 0$ for $F := F(Ad(\overline{\rho}))$.*

Then if $A \in CL_\Lambda$, $\text{Sel}(Ad(\rho_A))^\vee \cong A/L_\rho(\varphi)A$ $(L_\rho(\varphi) := \varphi(L_\rho))$ as A-modules for each $\varphi \in \text{Hom}_{CL_\Lambda}(R^{ord}, A)$.

Proof. By the second fundamental exact sequence in Corollary 5.2.5

$$\mathbb{T} \cong (D)/(D^2) \xrightarrow{x \mapsto L_\rho x} \mathbb{T} \cong \mathbb{T} dX \to \Omega_{\mathbb{T}/\Lambda} \to 0.$$

Tensoring A, we get $\text{Sel}(Ad(\rho_A))^\vee \cong \Omega_{\mathbb{T}/\Lambda} \otimes_{\mathbb{T}} A \cong A/L_\rho(\varphi)A$ by Theorem 6.1.11 as desired. $\qquad\qquad\square$

Here are some remarks comparing this corollary and the cyclicity results covered in Chapter 7:

- For an exceptional Artin representation ρ_f and an induced Artin representation in the non-split case, for primes p outside $|\text{Im}(\rho_f)|$ for which ρ_f satisfies p-distinguishedness, this corollary applies, and it also shows that the annihilator of the Selmer group is a principal ideal.

- For induced Artin representation in the split case, Corollary 7.4.8 deals with the Iwasawa module $\mathcal{Y}(\varphi_0)$ which is a factor of $\text{Sel}(Ad(\rho))$ under the milder assumption $Cl_F \otimes_{\mathbb{Z}[G]} \overline{\varphi}^- = 0$ than $Cl_F \otimes_{\mathbb{Z}[G]} Ad(\overline{\rho}) = 0$.
- Perhaps, the structure theorems of adjoint Selmer groups of Artin representations in Chapter 7 would possibly imply the above corollary, though the proof of this corollary does not require a detailed analysis of p-local units on which structure theorems are based on.

8.1.3   *Open questions*

If a Hecke eigenform f has weight $k \geq 3$, we have $R_{\mathfrak{p}}^{ord} = \mathbb{T}_{\mathfrak{p}} = \Lambda$ for almost all ordinary primes $\mathfrak{p}$ of $\mathbb{Z}[f]$ by Theorem 6.2.17; so, $\text{Sel}(Ad(\overline{\rho}_{\mathfrak{p}})) = 0$ for almost all ordinary $\mathfrak{p}$. If $k = 2$, this case is still mysterious, and $R_{\mathfrak{p}}^{ord} = \mathbb{T}_{\mathfrak{p}} = \Lambda$ ($\Leftrightarrow \text{Sel}(Ad(\overline{\rho}_{\mathfrak{p}})) = 0$) perhaps for $\mathfrak{p}$ of density 1 as discussed in Remark 6.2.18. Suppose that f does not have CM (i.e., not a binary theta series of an imaginary quadratic field). Then for almost all $\mathfrak{p}$, $\text{Im}(\overline{\rho}_{\mathfrak{p}})$ modulo center is isomorphic to $\text{PGL}_2(\mathbb{F}')$ or $\text{PSL}_2(\mathbb{F}')$ (for $\mathbb{F}' \subset \mathbb{F}$) by a result of Ribet [R85]. For these classical groups, group theorists proved $H^1(\overline{G}, Ad(\overline{\rho})) = 0$ [CPS75, Theorem 4.2], but $H^2(\overline{G}, Ad(\overline{\rho}))$ is 1-dimensional [U12, Theorem 1.2.5] (see also [H81b, Proposition 2.3]). Therefore for $F = F(Ad(\overline{\rho}_{\mathfrak{p}}))$, the map $\text{Res} : \text{Sel}(Ad(\overline{\rho}_{\mathfrak{p}})) \hookrightarrow \text{Sel}_F(Ad(\overline{\rho}_{\mathfrak{p}}))^{\overline{G}}$ in (8.1) may not be surjective.

> *Is this restriction map* Res *an onto isomorphism?*

If so, $\text{Hom}_{\mathbb{Z}[\overline{G}]}(Cl_F, Ad(\overline{\rho}_{\mathfrak{p}})) = 0$ (if $k \geq 3$) by the exact sequence in (8.2).

> *Is this vanishing of* $\text{Hom}_{\mathbb{Z}[\overline{G}]}(Cl_F, Ad(\overline{\rho}_{\mathfrak{p}}))$ ($\Leftrightarrow Cl_F \otimes_{\mathbb{Z}[\overline{G}]} Ad(\overline{\rho}) = 0$)
> *true for almost all* $\mathfrak{p}$?

In the 1-dimensional abelian case, for $\overline{\omega}_p$, if we fix an integer $k > 0$,

$$Cl_{\mathbb{Q}[\mu_p]} \otimes_{Z[\overline{G}]} \overline{\omega}_p^{1-2k} = 0 \Leftrightarrow p \nmid \zeta(1 - 2k)$$

by Herbrand–Ribet theorem [R76]. Kummer–Vandiver conjecture (Conjecture 1.6.1) tells us $Cl_{\mathbb{Q}[\mu_p]} \otimes_{Z[\overline{G}]} \overline{\omega}_p^{2k} = 0$ for all p.

8.2   Local structure at p of modular deformation

We study when a modular p-ordinary Galois representation is indecomposable over the p-inertia subgroup. In this section, $\overline{\rho}$ is a general absolutely irreducible odd 2-dimensional p-ordinary p-distinguished representation.

8.2.1 *Local indecomposability conjecture*

Conjecture 8.2.1 (Greenberg). *For a p-ordinary Hecke eigenform f of weight $k \geq 2$, if f has no CM (not a theta series of a norm form of an imaginary quadratic field), then $\rho_{f,\mathfrak{p}}|_{I_p}$ is indecomposable.*

For a cusp form $f = \sum_{n=1}^{\infty} a_n q^n \in S_k(N, \psi)_{/W}$, we define $d := q\frac{d}{dq}$ as a differential operator on $W[[q]]$. It is well known that $d^m f$ is a p-adic limit of classical cusp forms (why?). Assume $p \nmid N$. If f is a $\mathfrak{p}$-ordinary Hecke eigenform with $f|T(n) = \lambda(T(n))f$, then we can distinguish two roots α, β of $X^2 - \lambda(T(p))X + \chi(p) = 0$ so that $|\alpha|_p = 1$ and $|\beta|_p = p^{1-k}$ (i.e., $p^{k-1}\|\beta$). By Lemma 3.2.20, we have two p-stabilizations $f^{ord}|U(p) = \alpha f^{ord}$ and $f^{crt}|U(p) = \beta f^{crt}$.

Conjecture 8.2.2 (Coleman). $f^{crt} = d^{k-1}g$ *for a p-adic limit $g \in W[[q]]$ of cusp forms if and only if f has CM.*

It is known that Conjecture 8.2.1 $\Leftrightarrow$ Conjecture 8.2.2 by Breuil–Emerton [BE10, Theorems 4.3.3 and 4.4.8].

Exercise 8.2.1. Prove that $f^{crt} = d^{k-1}g$ if f has CM (i.e., "$\Leftarrow$" of Conjecture 8.2.2).

8.2.2 *Inertia theorem*

We start studying an exact form of $\rho_{\mathbb{T}}(I_p)$ via local Iwasawa theory described in §1.12.2. Pick $\phi_0 \in D_p$ so that $\overline{\rho}(\phi_0) = \left(\begin{smallmatrix} \overline{a} & 0 \\ 0 & \overline{b} \end{smallmatrix}\right)$ with $\overline{a} \neq \overline{b}$. Fix a prime $\wp|p$ of $M = F(\overline{\rho})$, and choose the $\wp$-decomposition group of $\mathrm{Gal}(F(\rho_{\mathbb{T}})/M)$ so that $\rho_{\mathbb{T}}(D_\wp)$ is upper triangular with quotient unramified character congruent to $\overline{\delta}$ modulo $\mathfrak{m}_{\mathbb{T}}$. Identify $\Gamma = \mathrm{Gal}(\mathbb{Q}_{p,\infty}/\mathbb{Q}_p)$ for the cyclotomic $\mathbb{Z}_p$-extension $\mathbb{Q}_{p,\infty}/\mathbb{Q}_p$, and for the (unique) unramified $\mathbb{Z}_p$-extension $k_\infty/\mathbb{Q}_p$, put $\Upsilon := \mathrm{Gal}(k_\infty/\mathbb{Q}_p)$. Pick a generator $\upsilon \in \Upsilon$. Define $\phi = \lim_{n\to\infty} \phi_0^{q^n}$ ($q = |\mathbb{F}|$). Let $I = \overline{I}_\wp$ (resp. $D = \overline{D}_\wp$) be the wild $\wp$-inertia (resp. $\wp$-decomposition) subgroup of $\mathrm{Gal}(F(\rho_{\mathbb{T}})/F(\overline{\rho}))$ for $\wp$.

Exercise 8.2.2. Prove $\nu_p([p, \mathbb{Q}_p]) = 1$.

By this exercise, $\kappa([p, \mathbb{Q}_p]) = \det(\rho_\mathbb{T}([p, \mathbb{Q}_p])) = 1$ since $\kappa(g) = t^{\log_p \nu_p(g)/\log_p(\gamma)}$ for $g \in \mathrm{Gal}(F(\rho_\mathbb{T})/M)$. Thus $\rho_\mathbb{T}([p, \mathbb{Q}_p]) = \begin{pmatrix} u^{-1} & * \\ 0 & u \end{pmatrix}$ for $u \in \mathbb{T}^\times$. Choose $\upsilon := [p, \mathbb{Q}_p]^h \in D$ for the residual degree h of $\wp$, and write $\rho_\mathbb{T}([p, \mathbb{Q}_p]^h) = \begin{pmatrix} u^{-h} & * \\ 0 & u^h \end{pmatrix}$. Write $\overline{\rho}|_{D_p} = \begin{pmatrix} \overline{\epsilon} & * \\ 0 & \overline{\delta} \end{pmatrix}$ for the choice of p-ordinarity quotient character $\overline{\delta}$. Let $\varphi : D_p \to W^\times$ be the Teichmüller lift of $\overline{\epsilon}\,\overline{\delta}^{-1}$. Let W_1 be the $\mathbb{Z}_p$-subalgebra of W generated by the values of φ over D_p. Put $\Lambda_0 := \mathbb{Z}_p[[T]] \subset \Lambda_1 := W_1[[T, a]] \subset \mathbb{T}$ for $a = u^{2h} - 1 \in \mathfrak{m}_{\Lambda_1}$, which is the image of $W_1[[\Gamma \times \Upsilon]]$ in $\mathbb{T}$. Here Λ_1 is a Λ_0-module of finite type; so, a and T have a non-trivial relation. Therefore Λ_1 is a closed subalgebra generated topologically by a over $W_1[[T]]$. Note $\Upsilon = \upsilon^{\mathbb{Z}_p}$.

Theorem 8.2.3. *Let the notation be as above. Suppose that $\overline{\rho}$ is p-distinguished ordinary, absolutely irreducible and minimal. Then,*

(1) *after choosing I suitably in its conjugacy class, we have an exact sequence $\mathcal{U} \hookrightarrow I \twoheadrightarrow t^{\mathbb{Z}_p}$ with $\rho_\mathbb{T}(\mathcal{U})$ made of unipotent matrices,*
(2) *there exists an element $\theta \in \mathbb{T}$ satisfying $\mathcal{U} = \Lambda_1 \theta$; in other words, we have $\rho_\mathbb{T}(I) = \left\{ \begin{pmatrix} a & b \\ 0 & 1 \end{pmatrix} \big| a \in t^{\mathbb{Z}_p}, b \in \theta\Lambda_1 \right\}.$*

By conjugation, $\rho_\mathbb{T}([p, \mathbb{Q}_p]^f)$ acts on $\mathcal{U}$ giving the action of $\mathbb{Z}_p[[\Upsilon]]$. Similarly $\rho_\mathbb{T}([1 + \mathbf{p}, \mathbb{Q}_p])$ acts on $\mathcal{U}$; so, $\mathcal{U}$ is a module over $\mathbb{Z}_p[[\Gamma]]$. Finally conjugation by $\begin{pmatrix} \epsilon & * \\ 0 & \delta \end{pmatrix}$ for Teichmüller lifts ϵ and δ of $\overline{\epsilon}$ and $\overline{\delta}$ induces the action of W_1, making $\mathcal{U}$ a $W_1[[\Gamma \times \Upsilon]]$-module.

***Proof of* (1).** From p-ordinarity, we know $\rho_\mathbb{T}(I) \subset M(\mathbb{T})$ for the mirabolic subgroup $\left\{ \begin{pmatrix} a & b \\ 0 & 1 \end{pmatrix} \big| a \in \mathbb{T}^\times, b \in \mathbb{T} \right\}$. Since $\mathrm{Gal}(\mathbb{Q}_p^{ab}/\mathbb{Q}_p) = [p, \mathbb{Q}_p]^{\widehat{\mathbb{Z}}} \ltimes \mathbb{Z}_p^\times$ for the maximal abelian extension $\mathbb{Q}_p^{ab}/\mathbb{Q}_p$ and the local Artin symbol $[p, \mathbb{Q}_p]$ (see §7.2.1), we find

$$\rho_\mathbb{T}(I) \subset \left\{ \begin{pmatrix} a & b \\ 0 & 1 \end{pmatrix} \big| a \in t^{\mathbb{Z}_p}, b \in \mathbb{T} \right\},$$

and $\det(\rho_\mathbb{T}(I)) = \mathcal{T} := t^{\mathbb{Z}_p} \subset \Lambda^\times$. Thus we have an extension

$$1 \to \mathcal{U} \to \rho_\mathbb{T}(I) \to \mathcal{T} \to 1.$$

Recall $\phi_0 \in D_\wp$ with $\overline{\rho}(\phi_0) = \begin{pmatrix} a & \upsilon \\ 0 & b \end{pmatrix}$ $(\overline{a} \neq \overline{b})$ and $\phi = \lim_{n \to \infty} \phi_0^{q^n}$ inside $\mathrm{Gal}(F(\rho_\mathbb{T})/K)$. This extension is split by the conjugation action of ϕ_0 with $\mathcal{U}$ characterized to be an eigenspace on which ϕ_0 acts by ab^{-1} for the Teichmüller lift a, b of $\overline{a}, \overline{b}$; so, we may assume to have a section $s : \mathcal{T} \hookrightarrow \rho_\mathbb{T}(I)$ identifying $\mathcal{T}$ with $\left\{ \begin{pmatrix} a & 0 \\ 0 & 1 \end{pmatrix} \big| a \in t^{\mathbb{Z}_p} \right\}$. Thus $\mathcal{U}$ is made of unipotent matrices. $\square$

Proof of (2). We have $\mathcal{U} \subset \mathbb{T}$ and regard φ as an abelian irreducible $\mathbb{Z}_p$-representation acting on W_1 regarded as a $\mathbb{Z}_p$-module. Apply Theorem 1.12.4 to the splitting field $k = F(\varphi)$ of φ. Then, under the notation of Corollary 1.12.3, the Galois group $X'[\varphi]$ is cyclic over $W_1[[\Gamma \times \Upsilon]]$ ($\Gamma = \gamma^{\mathbb{Z}_p} \cong t^{\mathbb{Z}_p}$) and surjects onto $\mathcal{U}$. Since the action of $W_1[[\Gamma \times \Upsilon]]$ factors through Λ_1, $\mathcal{U}$ is cyclic over Λ_1; so, we have $\mathcal{U} \cong \Lambda_1$. Thus we conclude $\mathcal{U} = \left\{ \left(\begin{smallmatrix} 1 & a \\ 0 & 1 \end{smallmatrix} \right) \big| a \in \theta \Lambda_1 \right\}$ inside $\operatorname{Im}(\rho_{\mathbb{T}})$ (for a generator $\theta \in \mathbb{T}$). $\qquad\square$

8.2.3 *Generic non-triviality of* $\mathcal{U}$

We describe non-triviality of θ of Theorem 8.2.3 without giving a detailed proof. For details, one may refer to [Z14, Theorem 5.7]. Let $\operatorname{Spec}(\mathbb{I}) \subset \operatorname{Spec}(\mathbb{T})$ be an irreducible component with the projection $\pi : \mathbb{T} \twoheadrightarrow \mathbb{I}$.

Theorem 8.2.4. *Let the notation and assumption be as above and as in Theorem 8.2.3. Then $\pi(\theta) = 0$ if and only if $F(\rho_{\mathbb{I}})$ for $\rho_{\mathbb{I}} := \pi \circ \rho_{\mathbb{T}}$ is abelian over an imaginary quadratic field.*

By Shapiro's lemma, $F(\rho_{\mathbb{I}})$ is abelian over an imaginary quadratic field K if and only if $\rho_{\mathbb{I}} \cong \operatorname{Ind}_K^{\mathbb{Q}} \Phi$ for a character $\Phi : \mathfrak{G}_K \to \mathbb{I}^\times$. An irreducible component $\operatorname{Spec}(\mathbb{I})$ or its ring $\mathbb{I}$ is called a *CM component* of $\operatorname{Spec}(\mathbb{T})$ or $\mathbb{T}$ if $F(\rho_{\mathbb{I}})$ is abelian over an imaginary quadratic field K. Since $\mathbb{T}$ is reduced by Theorem 6.2.17 (or Lemma 6.2.15), we can decompose $\operatorname{Spec}(\mathbb{T}) = \operatorname{Spec}(\mathbb{T}^{cm}) \cup \operatorname{Spec}(\mathbb{T}^{ncm})$, where $\operatorname{Spec}(\mathbb{T}^{cm})$ and $\operatorname{Spec}(\mathbb{T}^{ncm})$ are a union of irreducible components of $\operatorname{Spec}(\mathbb{T})$ and $\mathbb{I}$ is a CM component $\Leftrightarrow \operatorname{Spec}(\mathbb{I}) \subset \operatorname{Spec}(\mathbb{T}^{cm})$.

This theorem was first proven by Ghate–Vatsal [GV04, Theorem 3] under the Taylor–Wiles condition (and p-distinguishedness), whose proof is based on an argument of Buzzard [B03] (and Buzzard–Taylor [BT99]) proving modularity of odd Artin two dimensional Galois representations. For general $\bar{\rho}$, this follows from the following two facts:

(a) By Theorems 4.1.29 and 4.2.47 combined with Remark 4.2.48, for a W-algebra homomorphism $\lambda : \mathbb{T}^{ncm} \to \overline{\mathbb{Q}}_p$, if $\lambda|_{\Gamma} : \Gamma \to \overline{\mathbb{Q}}_p^\times$ coincides with ν_p up to a finite order character ϵ, $f := \sum_{n=1}^{\infty} \lambda(T(n)) q^n$ is a weight 2 cusp form in $S_2(Cp^r, \psi_2 \epsilon)$ (see also [LFE, §7.3]);

(b) The Galois representation $\rho_{\lambda, \mathfrak{p}} = \rho_{f, \mathfrak{p}}$ for f of weight 2 as in (a) is locally indecomposable. The assumptions may be precisely referred to [Z14, Introduction]. The result in [Z14] is based on the fact that ρ_f for a non-CM cusp form f of weight 2 (restricted to an open subgroup) is realized on the Tate module of a non-CM abelian variety A of Hilbert modular type (i.e. $\text{End}(A)$ is the integer ring of a totally real field with $\dim A = \text{rank}_{\mathbb{Z}} \text{End}(A)$) [GME, §4.2]. The Tate module of a non-CM abelian variety of Hilbert modular type is p-indecomposable by a geometric argument in [H13b] via the Serre–Tate coordinate (which is closely related to θ) around the point of the Hilbert modular Shimura variety carrying the abelian variety.

Theorem 4.2.17 tells us $\bigcap_\lambda \text{Ker}(\lambda) = 0$ (the intersection over λ's as in (a)). By (b), the restriction of θ to any irreducible component of $\text{Spec}(\mathbb{T}^{ncm})$ is non-zero. Since $\mathbb{T}$ is reduced by Lemma 6.2.15, $\mathbb{T}^{ncm} \hookrightarrow \prod_{\mathbb{I}} \mathbb{I}$ for $\text{Spec}(\mathbb{I})$ running over all irreducible components of $\text{Spec}(\mathbb{T}^{ncm})$. This shows θ is a non-zero divisor of $\mathbb{T}^{ncm}$, and $\mathcal{U}$ contains a non-zero divisor of $\mathbb{T}^{ncm}$. Thus $\mathcal{U}$ is "highly" non-zero.

8.3 Universal ring for $\text{Ind}_K^{\mathbb{Q}} \varphi$ for K real at split p

We now study a very specific case of weight 1 which produces $\mathbb{T}_{\mathfrak{p}} \neq \Lambda$ for ordinary $\mathfrak{p}$ of positive density. Fix a quadratic field $K = \mathbb{Q}[\sqrt{D}]$ with discriminant D and integer ring O_K. In this section, mostly we suppose that K is real. However for our later use, in some subsections, K can be imaginary. We assume that p splits in K as $(p) = \wp\wp^\sigma$ for $\sigma \in \mathfrak{G}_{\mathbb{Q}}$ non-trivial on K.

Fix a character $\varphi : \text{Gal}(\overline{\mathbb{Q}}/K) \to \mathbb{Z}[\varphi]^\times$ of finite order a. As before, let $\overline{\varphi} = \overline{\varphi}_{\mathfrak{p}} = (\varphi \bmod \mathfrak{p})$ for each prime $\mathfrak{p}$ of $\mathbb{Z}[f]$ (which is a subring of $\mathbb{Z}[\mu_a]$). If $\mathfrak{p} \nmid a$, the order of $\overline{\varphi}_{\mathfrak{p}}$ is equal to a and $F(\overline{\rho}) = F(\rho)$. Replacing φ by the Teichmüller lift of $\overline{\varphi}$, we always assume that $p \nmid a$. To have a well defined universal ring $R_{\mathfrak{p}}^{ord}$, we always assume that $\boxed{\overline{\varphi}^-|_{D_p} \neq 1}$ (p-distinguishedness). If K is real, $R_{\mathfrak{p}}^{ord} \cong \mathbb{T}_{\mathfrak{p}}$ is valid for all ordinary $\mathfrak{p}$ with $\overline{\varphi}^-|_{D_p} \neq 1$ by [T16, §6], including the assertion of the presentation theorem (Theorem 6.2.20).

Write $\mathfrak{f}$ for the conductor of φ, and suppose $\wp^\sigma \nmid \mathfrak{f}$ (so, we are in Cases U_+ or R_+ in §7.3.3). By class field theory, we may regard $\varphi : Cl_K^+(\mathfrak{f}) \to \mathbb{Z}[\varphi]^\times$. When K is real, assume that $\varphi(\xi) = -1$ for any totally negative $\xi \in O_K$ with $\xi \equiv 1 \bmod \mathfrak{f}$ (i.e., φ ramifies at one infinite place ∞ of K). Under this assumption, by Hecke [HMI, Proposition 2.71] or [MFM, §4.8],

$$f = \sum_{0 \neq \mathfrak{a} \,:\, O_K\text{-ideals}} \varphi(\mathfrak{a}) q^{N(\mathfrak{a})}.$$

is in $S_1(\Gamma_0(C), \psi)$ with $C = D \cdot N(\mathfrak{f})$ for $\psi := \varphi|_{\mathbb{Z}}\alpha$ with $\alpha := \left(\frac{K/\mathbb{Q}}{}\right)$ and is a primitive form.

The associated Galois representation is given by the Artin representation $\rho_f = \mathrm{Ind}_K^{\mathbb{Q}} \varphi$. Indeed, by the explicit form of induced representation, $\mathrm{Tr}(\rho_f(\mathrm{Frob}_l)) = 0 = a(l, f)$ if a prime l is inert in K and $\mathrm{Tr}(\rho_f(\mathrm{Frob}_l)) = \varphi(\mathfrak{l}) + \varphi(\mathfrak{l}^\sigma) = a(l, f)$ if $(l) = \mathfrak{l}\mathfrak{l}^\sigma$ with $\mathfrak{l} \neq \mathfrak{l}^\sigma$ for the non-trivial automorphism σ of K.

8.3.1 R^{ord} is a non-trivial extension of Λ if K is real

Recall the automorphism $[\alpha]$ of R^{ord} and R_χ induced by $\rho_A \mapsto J\rho_A \cdot \alpha J^{-1}$ described in §7.4.1. The subring of R_χ fixed by $[\alpha]$ is denoted by R_χ^+. We recall Corollary 7.4.2 in a slightly different way:

Proposition 8.3.1. *Assume* $\mathrm{Ind}_F^{\mathbb{Q}} \varphi$ *is minimal (i.e., the order of* φ *is prime to* p*) and put* $\mathfrak{d} := R^{ord}([\alpha] - 1)R^{ord}$. *We have* $R^{ord}/\mathfrak{d} \cong W[\Gamma_\varphi]$ *for* Γ_φ *in* §7.4.1 *so that* $\rho \mod \mathfrak{d} \cong \mathrm{Ind}_K^{\mathbb{Q}} \varphi$ *with* $\varphi(h) = \varphi(h)h|_{H_\varphi}$ *for the modulo* $\wp^\infty$ *ray class field* H_φ *with* $\mathrm{Gal}(H_\varphi/F) \cong \Gamma_\varphi$.

Corollary 8.3.2. *If* K *is real,* $R^{ord} \neq R_+^{ord}$, $R_\chi \neq R_\chi^+$ *and* $\Omega_{R^{ord}/\Lambda} \neq 0$.

Proof. Since K is real, Γ_φ is finite. If $R_+^{ord} = R^{ord}$, then $\mathfrak{d} = 0$ and hence $R^{ord} = W[\Gamma_\varphi]$. This is impossible as R^{ord} surjects down to $\mathbb{T}_\mathfrak{p}$ which has infinite rank over W. As we have seen in Corollary 6.1.3, if $\Omega_{A/B} = 0$, $B \twoheadrightarrow A$. Thus $R_+^{ord} \neq R^{ord}$ implies $\Omega_{R^{ord}/R_+^{ord}} \neq 0$. By the fundamental exact sequence, $\Omega_{R^{ord}/\Lambda}$ surjects down to $\Omega_{R^{ord}/R_+^{ord}}$; so, $\Omega_{R^{ord}/\Lambda} \neq 0$.

Recall $R_\chi = R^{ord} \otimes_\Lambda \Lambda/(t - \chi(\gamma))$ from Corollary 6.1.10. Taking "+"-eigenspace of this identity, we get $R_\chi^+ = R_+^{ord} \otimes_\Lambda \Lambda/(t - \chi(\gamma))$. Tensoring $\Lambda/(t - \chi(\gamma))$ over Λ with the exact sequence $R_+^{ord} \to R^{ord} \to C \to 0$ with $C \neq 0$, we find $R_\chi^+ \to R_\chi \to C/(t - \chi(\gamma))C \to 0$. By Nakayama's lemma, $C = 0 \Leftrightarrow C/(t - \chi(\gamma))C = 0$. Thus $R_\chi^+ \neq R_\chi$. $\square$

In contrast to Theorem 6.2.17, as $\mathbb{T}_\mathfrak{p} = R^{ord} \supset R_+^{ord} \supset \Lambda$, this corollary shows $R^{ord} \neq \Lambda$ for all p split in K under p-distinguishedness.

8.3.2 Questions on structure of the universal ring

Assuming that $\overline{\rho} = \mathrm{Ind}_K^{\mathbb{Q}} \overline{\varphi}$ is minimal p-ordinary, the next goal is to specify the single generator Θ of $\mathbb{T}_\mathfrak{p}$ and study the ring structure of $\mathbb{T}_\mathfrak{p}$. For example, we ask

when is $\mathbb{T}_{\mathfrak{p}}$ a regular local ring?

In this setting, writing Θ for a well chosen generator, $\mathbb{T}_{\mathfrak{p}}$ is a regular local ring if and only if $\mathbb{T}_{\mathfrak{p}} \cong W_{\mathfrak{p}}[[\Theta]] \cong W_{\mathfrak{p}}[[X]]$ (the one variable power series ring) by $\Theta \mapsto X$, but still $\Lambda = W_{\mathfrak{p}}[[T]] \subsetneq \mathbb{T}_{\mathfrak{p}}$. The power series ring $W[[X]]$ has a property that for any given $\phi \in \mathrm{Hom}_{CL_W}(W[[X]], A)$ and a CL_W-morphism $\pi : A' \to A$, we can lift ϕ to $\phi' \in \mathrm{Hom}_{CL_W}(W[[X]], A')$ so that $\pi \circ \phi' = \phi$. In other words, the deformation $\rho_A = \phi \circ \rho_{\mathbb{T}} \in \mathcal{D}(A)$ can be lifted to $\rho_{A'} = \phi' \circ \rho_{\mathbb{T}}$ (so, no obstruction of lifting any given deformation $\rho_A \in \mathcal{D}(A)$ along π if $\mathbb{T}_{\mathfrak{p}} = R^{ord}$ is regular).

The distribution of such primes is another question we want to ask:

are such primes infinitely many? Or even of positive density?

We explore these questions. In this weight 1 case, though one expects abundance of $\mathfrak{p}$ with regular $\mathbb{T}_{\mathfrak{p}}$, even the existence of such primes is a subtle question, as we discuss how to find them numerically in §8.3.6 when K is real.

8.3.3   *Action of $[\alpha]$ on $\mathrm{Sel}(Ad(\rho))$*

Since $\widehat{\Delta} := \mathrm{Hom}(\Delta, \{\pm 1\})$ for $\Delta = \mathrm{Gal}(K/\mathbb{Q})$ acts on R^{ord} by Theorem 7.4.1, it acts on $\Omega_{R^{ord}/\Lambda} \cong \mathrm{Sel}(Ad(\rho))^{\vee}$. We make explicit the action on Selmer cocycles. In this subsection, K can be real and imaginary.

Pick $\rho_A \in \mathcal{D}_\kappa(A)$. Let $\pi : R^{ord} \to A$ with $\rho_A \sim \pi \circ \rho$. Suppose we have $\alpha_A \in \mathrm{Aut}(A)$ such that $\alpha_A \circ \pi = \pi \circ [\alpha]$. Recall $j(\rho \cdot \alpha)j^{-1} = (\rho)^{[\alpha]}$ for $j := \left(\begin{smallmatrix} -1 & 0 \\ 0 & 1 \end{smallmatrix} \right)$. For each 1-cocycle $u : G \to Ad(\rho_A)^*$, we define $u^\alpha(g) = ju(g)^{\alpha_A}j^{-1}$. From $u(gh) = Ad(\rho_A)(g)u(h) + u(g)$, we find

$$u^\alpha(gh) = j\rho_A^{\alpha_A}(g)jju(h)^{\alpha_A}jj\rho_A(g^{-1})^{\alpha_A}j + ju(g)^{\alpha_A}j$$
$$= Ad(j\rho_A^{\alpha_A}j)(g)u^\alpha(h) + u^\alpha(g) = Ad(\rho_A \cdot \chi)(g)u^\alpha(h) + u^\alpha(g)$$
$$= Ad(\rho_A)(g)u^\alpha(h) + u^\alpha(g).$$

Since the conjugation of j preserves the upper triangular p-decomposition subgroup and p-inertia subgroup of $\mathrm{Gal}(F(\rho)/\mathbb{Q})$, in this way, $[\alpha]$ acts on $\mathrm{Sel}(Ad(\rho_A))$. In particular,

if α_A is trivial (i.e., $\rho_A = \mathrm{Ind}_K^{\mathbb{Q}} \phi$), $u \mapsto u^\alpha$ is just a conjugate action of j.
(8.3)

Decomposition of $\mathrm{Sel}(Ad(\mathrm{Ind}_K^{\mathbb{Q}} \phi))$. Recall $G = \mathrm{Gal}(F^{(p)}(\overline{\rho})/\mathbb{Q})$ and write $H = \mathrm{Gal}(F^{(p)}(\overline{\rho})/K)$. Let $\phi : H \to A^\times$ be a deformation of $\overline{\varphi}$ (so,

ϕ^- mod $\mathfrak{m}_A = \overline{\varphi}^- \neq 1$). Recall the splitting into indecomposable factors in §7.3.1:

$$Ad(\mathrm{Ind}_K^{\mathbb{Q}} \phi) = \begin{cases} \alpha_A \oplus \mathrm{Ind}_K^{\mathbb{Q}} \phi^- & \text{if } (\phi^-)^2 \neq 1, \\ \alpha_A \oplus \widetilde{\phi}^- \oplus \alpha\widetilde{\phi}^- & \text{if } (\phi^-)^2 = 1. \end{cases}$$

Here $\widetilde{\phi}^-$ is an extension of ϕ^- to G (which exists if $(\phi^-)^2 = 1$) and we have written $\alpha_A : G \to A^\times$ for α with values in A.

Here is a restatement of Theorem 7.4.7 in a slightly different form:

$$\mathrm{Sel}(\alpha_A) := \mathrm{Hom}(Cl_K \otimes_{\mathbb{Z}} A, A^\vee) \cong (Cl_K \otimes_{\mathbb{Z}} A)^\vee,$$

$$\mathrm{Sel}(\mathrm{Ind}_K^{\mathbb{Q}} \phi^-) := \mathrm{Hom}(\mathcal{Y}(\varphi) \otimes_{\mathbb{Z}[\Gamma^-]} \phi^-, A^\vee).$$

In Theorem 7.4.7, φ is denoted by φ_0 and ϕ is denoted by φ (as we did not replace φ by the Teichmüller lift of $\overline{\varphi}$). Perhaps some explanation on the second formula is due. In Theorem 7.4.7, $\mathrm{Sel}(\mathrm{ind}_K^{\mathbb{Q}} \phi^-)$ is identified with $\mathrm{Hom}_{W[[\Gamma^-]]}(\mathcal{Y}(\varphi), (\phi^-)^*)$ which is isomorphic to $\mathrm{Hom}(\mathcal{Y}(\varphi) \otimes_{\mathbb{Z}[\Gamma^-]} \phi^-, A^\vee)$ by the adjunction formula in §5.3.3.

Theorem 8.3.3. *Write* $\rho_A^- := \mathrm{Ind}_K^{\mathbb{Q}} \phi^-$. *Then we have* $\mathrm{Sel}(Ad(\mathrm{Ind}_K^{\mathbb{Q}} \phi)) \cong \mathrm{Sel}(\alpha_A) \oplus \mathrm{Sel}(\rho_A^-)$.

Proof for the α-factor. Write $\rho_A := \mathrm{Ind}_K^{\mathbb{Q}} \phi$. Then $H^1(G, Ad(\rho_A))$ is isomorphic to

$$H^1(G, \alpha_A^*) \oplus H^1(G, (\rho_A^-)^*) = H^1(G, \alpha_A^*) \oplus H^1(H, (\phi^-)^*).$$

The identity of the second factors is by Shapiro's lemma. Since α_A is realized on the diagonal matrix, by the definition of $\mathrm{Sel}(Ad(\rho_A))$, it is unramified everywhere; so, it factors through $Cl_K \otimes_{\mathbb{Z}} A$ over H. Since G/H has order 2, this map is realized by the restriction map to H, and the result follows. $\square$

Proof for the ρ_A^--factor. The isomorphism of Shapiro's lemma is realized by the restriction map

$$H^1(G, (\rho_A^-)^*) \xrightarrow[\sim]{\mathrm{Res}} H^1(H, (\rho_A^-)^*)^G = (H^1(H, (\phi^-)^*) \oplus H^1(H, (\phi_\sigma^-)^*))^G,$$

which is an isomorphism. In the last factor, G acts on cocycles $u(g) \mapsto u_\sigma(g) = \sigma u(\sigma^{-1} g \sigma)$; so, interchanges the two factors. Therefore $H^1(G, (\rho_A^-)^*) \cong H^1(H, (\phi^-)^*)$. If $U : G \to (\rho_A^-)^*$ is a Selmer cocycle, we have $U(h) = \begin{pmatrix} 0 & u \\ u_\sigma & 0 \end{pmatrix}$ for a cocycle $u : H \to (\phi^-)^*$. By Selmer condition that $U|_{D_{\wp^\sigma}}$ is lower triangular, we have $u|_{G_F}$ factors through

$\mathcal{Y}$. Note that $H^q(H/G_F, \phi^-) = 0$ for $q > 0$ as $\phi^- \not\equiv 1 \mod \mathfrak{m}_A$ (e.g. Lemma 7.4.6). Since it is a $\mathbb{Z}[H]$-morphism into ϕ^-, it factors through $\mathcal{Y} \otimes_{\mathbb{Z}[H]} \phi^- = \mathcal{Y}(\varphi) \otimes_{W[[\Gamma^-]]} \phi^-$. Reversing the argument, it is an isomorphism. $\qquad\square$

The $\pm$-eigenspace of $[\alpha]$ in $\mathrm{Sel}(Ad(\rho_A))$. We show that the above decomposition of the Selmer group is actually an eigenspace decomposition under the automorphism $[\alpha]$.

Lemma 8.3.4. *The involution $[\alpha]$ acts on $\mathrm{Sel}(\mathrm{Ind}_K^{\mathbb{Q}} \phi^-)$ (resp. $\mathrm{Sel}(\alpha_A)$) by -1 (resp. $+1$).*

Proof. In the decomposition of Theorem 8.3.3, α_A is realized on the subspace $\mathfrak{t}$ of diagonal matrices in $Ad(\bar{\rho}) = \mathfrak{sl}_2(\mathbb{F})$, and $\mathrm{Ind}_K^{\mathbb{Q}} \phi^-$ is realized on anti-diagonal matrices $\mathfrak{a} \subset Ad(\bar{\rho})$. Since j acts by $+1$ on $\mathfrak{t}$ and -1 on $\mathfrak{a}$ and the action of $[\alpha]$ on cocycle is conjugation by j as seen in (8.3), the action of $[\alpha]$ on $\mathrm{Sel}(\alpha_A)$ is by $+1$ and on $\mathrm{Sel}(\mathrm{Ind}_K^{\mathbb{Q}} \phi^-)$ is by -1. $\qquad\square$

Final touch on the proof of Corollary 7.4.8. We prove the cyclicity asserted in Corollary 7.4.8 only assuming $Cl_F \otimes_{\mathbb{Z}[H]} \overline{\varphi}^- = 0$. We use the notation introduced in Theorem 7.4.7 and Corollary 7.4.8. For a W-module M on which $[\alpha]$ acts, we write $M^\pm$ for the "$\pm$" eigenspace of M. By Theorem 7.4.7, we have $Cl_K \otimes_{\mathbb{Z}} W[[\Gamma^-]] \oplus \mathcal{Y}(\varphi_0) \cong \Omega_{R_\kappa/\Lambda} \otimes_{R_\kappa, \lambda} W[[\Gamma^-]]$. Taking the universal character $\varphi : H \to W[[\Gamma_\varphi]]^\times$ defined in the proof of Corollary 7.4.2, again by Theorem 7.4.7, we have $\mathrm{Sel}(\mathrm{Ind}_K^{\mathbb{Q}} \varphi^-) = \mathrm{Sel}_K(\varphi^-) = \mathrm{Hom}_{W[[\Gamma^-]]}(\mathcal{Y}(\varphi_0), (\varphi^-)^*)$. By Lemma 8.3.4, $\mathcal{Y}(\varphi_0) = \mathrm{Sel}(\mathrm{Ind}_K^{\mathbb{Q}} \varphi^-)^\vee = M^-$ for $M = \Omega_{R_\kappa/\Lambda} \otimes_{R_\kappa, \lambda} W[[\Gamma^-]]$. This shows $M^- \otimes_{W[[\Gamma^-]]} \mathbb{F} = (\Omega_{R_\kappa/\Lambda} \otimes_{R_\kappa, \lambda} \mathbb{F})^- \cong \mathrm{Sel}(Ad(\bar{\rho}))^-$ which has dimension at most 1 over $\mathbb{F}$ if $Cl_F \otimes_{\mathbb{Z}[H]} \overline{\varphi}^- = 0$ (cf. Theorem 8.3.5). Thus M^- is cyclic over $W[[\Gamma^-]]$ as desired. $\qquad\square$

8.3.4   *Set-up for a structure theorem*

Hereafter, we write $F := F(Ad(\bar{\rho})) = F(\overline{\varphi}^-)$. Recall the fixed prime $\wp | p$ in K. We give here the list of the four conditions which repeatedly appear in the description of our results for deformation of an induced representation (in the following conditions, $K/\mathbb{Q}$ can be real or imaginary):

(H0) *the local character $\overline{\varphi}^-|_{D_p}$ is non-trivial* ($\Rightarrow$ irreducibility of $\bar{\rho}$).

(H1) $\wp^\sigma \nmid \mathfrak{f}$ *for the conductor $\mathfrak{f}$ of φ (ordinarity in Cases R_+ and U_+; $N = N_{K/\mathbb{Q}}(\mathfrak{f})D$ with its prime-to-p part C).*

(H2) $Cl_F \otimes_{\mathbb{Z}[H]} \overline{\varphi}^- = 0$, and the local character $\overline{\varphi}^-|_{\mathrm{Gal}(\overline{\mathbb{Q}}_p/\mathbb{Q}_p)}$ *is different from the reduction* $\overline{\omega}_p$ *modulo* p *of the Teichmüller character* $\omega = \omega_p$ *acting on* $\mu_p(\overline{\mathbb{Q}}_p)$.

(H3) $h_K = |Cl_K|$ *is prime to* p $(Cl_F \otimes_{\mathbb{Z}[G]} Ad(\overline{\rho}) = 0 \Leftrightarrow$ (H2–3)$)$.

Replacing φ by the Teichmüller lift of $\overline{\varphi}$, we assume *the order of* φ *is prime to* p. Throughout this section, we assume (H0) and K is real in which p splits. As already remarked, by J. Thorne, Taylor–Wiles condition is removed; so, $R_{\mathfrak{p}}^{ord} \cong \mathbb{T}_{\mathfrak{p}}$ under (H0). We write $\mathbb{T}_{\mathfrak{p}}^{\pm} := \{x \in \mathbb{T}_{\mathfrak{p}} | [\alpha](x) = \pm x\}$.

8.3.5 Structure theorems

We first state the presentation theorem in a slightly different way. Let $r_+ := \dim_{\mathbb{F}} \mathrm{Sel}(\overline{\alpha})$ for $\overline{\alpha} = \alpha_{\mathbb{F}}$ and $r_- = \mathrm{Sel}(\mathrm{Ind}_K^{\mathbb{Q}} \overline{\varphi}^-)$.

Theorem 8.3.5. *Assume* $R_{\mathfrak{p}}^{ord} = \mathbb{T}_{\mathfrak{p}}$. *Then we have*

$$\mathbb{T}_{\mathfrak{p}} \cong \Lambda[[X_1^+, \ldots, X_{r_+}^+, X_1^-, \cdots, X_{r_-}^-]]/(S_1, \ldots, S_r)$$

for $r = r_+ + r_-$ *so that* $[\alpha]$ *fixes the image of* X_j^+ *in* $\mathbb{T}_{\mathfrak{p}}$ *and acts by* -1 *on the image in* $\mathbb{T}_{\mathfrak{p}}$ *of* X_i^-. *If* $r_+ + r_- \leq 1$, *without assuming* $R_{\mathfrak{p}}^{ord} = \mathbb{T}_{\mathfrak{p}}$, *the same assertion holds.*

If K is real, $R_{\mathfrak{p}}^{ord} = \mathbb{T}_{\mathfrak{p}}$ holds under (H0–1) [T16, §6]. If K is imaginary, it holds under the Taylor–Wiles condition, which can be verified under one of the following two conditions:

- $\overline{\varphi}^-$ has order ≥ 3,
- $\overline{\varphi}^-$ has order 2 and $F(\overline{\varphi}^-)$ does not contain $\mathbb{Q}[\sqrt{p^*}]$.

By Shapiro's lemma, if Taylor–Wiles condition fails, $\mathrm{Ind}_K^{\mathbb{Q}} \overline{\varphi} \cong \mathrm{Ind}_{\mathbb{Q}[\sqrt{p^*}]}^{\mathbb{Q}} \overline{\phi}$. By Proposition 7.3.6, under the above conditions, this cannot happen.

Proof. The last assertion is a restatement of Theorem 8.1.1. We prove the first assertion. From the presentation Theorem 6.2.20, we have

$$\mathbb{T} = \Lambda[[X_1, \ldots, X_r]]/(S_1, \ldots, S_r).$$

Let $x_j \in \mathbb{T}$ denote the image of X_j. For a matrix $A = (a_{ij}) \in M_r(\Lambda)$, we can replace the generators $(x_1, \ldots, x_r)$ by $(x_1, \ldots, x_r)A$ in the presentation as long as A induces an $\mathbb{F}$-linear automorphism of $t_{\mathbb{T}/\Lambda}^*$. In particular, we can arrange $(x_1, \ldots, x_r) = (x_1^+, \ldots, x_{r_+}^+, x_1^-, \cdots, x_{r_-}^-)$ with $[\alpha](x_j^{\pm}) = \pm x_j^{\pm}$. Then we get the presentation as in the theorem.

Assuming $r_+ = 0$ ($\Leftrightarrow p \nmid h_K = |Cl_K|$) and $r_- = 1$, we now prove the presentation theorem itself. So $[\alpha]$ acts on $t^*_{\mathbb{T}/\mathbb{T}^+}$ by -1. If $r_- = 1$, we can choose a generator Θ so that $[\alpha](\Theta) = -\Theta$. Then $x \mapsto \Theta x$ is a Λ-linear map of $\mathbb{T} \cong \Lambda^e$. Writing this map as a $d \times d$ matrix form L and define $D(X) = \det(X1_e - L)$. Then $\mathbb{T} = \Lambda[X]/(D(X))$ and $\mathbb{T}$ is a local complete intersection with $2|e$ as $[\alpha]$ acts non-trivially on $\mathbb{T}_{/\Lambda}$. $\square$

As remarked in the above proof, we have the following fact:

Corollary 8.3.6. *Assuming* (H0–2) *and* $r_- = 1$, *we have* $\Theta \in \mathbb{T}^-$ *generating* $\mathbb{T}$ *over* $\mathbb{T}^+$. *The principal ideal* (Θ) *is equal to the* $\mathbb{T}$*-ideal* $\mathfrak{d}_{\mathbb{T}} = \mathbb{T}([\alpha] - 1)\mathbb{T}$ *generated by* $[\alpha](x) - x$ *for* $x \in \mathbb{T}$.

Proof. Under $R^{ord}_{\mathfrak{p}} \cong \mathbb{T}_{\mathfrak{p}}$, this follows directly from Theorem 8.3.5. Otherwise, from the exact sequence $t^*_{\mathbb{T}^+/\Lambda} \to t^*_{\mathbb{T}/\Lambda} \to t^*_{\mathbb{T}/\mathbb{T}^+} \to 0$, taking the "$-$" eigenspace of $[\alpha]$, we find $(t^*_{\mathbb{T}/\Lambda})^- = t^*_{\mathbb{T}/\mathbb{T}^+}$. By Lemma 8.3.4, we find $t^*_{\mathbb{T}/\mathbb{T}^+} \cong \mathrm{Sel}(\mathrm{Ind}^{\mathbb{Q}}_K \overline{\varphi}^-)$ which has dimension 1 by our assumption. Therefore $\mathbb{T}$ is generated by one element Θ over $\mathbb{T}^+$. Since we can choose $\Theta \in \mathbb{T}^-$, we have $(\Theta) = \mathfrak{d}_{\mathbb{T}}$. $\square$

In the theory of dualizing modules, there is a general notion of "different" for a finite flat extension A/B of rings (see [MR70, Appendix]). In our case where $A/B = \mathbb{T}/\mathbb{T}_+$, following the computation as in Corollary 6.2.29 by Tate, the different equals $\mathfrak{d} = \mathfrak{d}_{\mathbb{T}/\mathbb{T}_+} := \mathbb{T}([\alpha] - 1)\mathbb{T}$ which is well defined without assuming the assumption (H2) but just flatness of $\mathbb{T}/\mathbb{T}_+$, though it may not be principal without assuming (H2). The number of generators of $\mathfrak{d}$ is equal to $r_- = \dim_{\mathbb{F}} \mathrm{Sel}(\mathrm{Ind}^{\mathbb{Q}}_K \overline{\varphi}^-)$) by Theorem 8.3.5. By Proposition 7.3.5, $\varepsilon_{\varphi^-} = 1$ for ε_{φ^-} in Theorem 7.5.4. This shows $r_- \geq 2$ if the exact sequence of Theorem 7.5.4 is split and (H2) fails.

Hereafter in this section, we suppose that K is real. Identify $\widehat{O}^{\times}_{K,\wp}$ with Γ by the isomorphism induced by the inclusion $\mathbb{Z} \hookrightarrow O_K$, and write the image of the fundamental unit ε of K in $\widehat{O}^{\times}_{K,\wp} = \Gamma$ as $\langle \varepsilon \rangle$. If we identify $\Gamma = t^{\mathbb{Z}_p}$ for $t = 1 + T$ with $T \in \Lambda$, we have $\langle \varepsilon \rangle = t^{\log_p(\varepsilon)/\log_p(\gamma)}$ for $\gamma = 1 + p$. In this way, we consider $\langle \varepsilon \rangle \in \Lambda = W_{\mathfrak{p}}[[T]]$. We put $\Lambda_\varepsilon := \Lambda/(\langle \varepsilon \rangle - 1)$. Here is a structure theorem for $\mathbb{T} = \mathbb{T}_{\mathfrak{p}}$ and $\mathbb{T}^+ = \mathbb{T}^+_{\mathfrak{p}}$:

Theorem 8.3.7. *Assume* (H0–3) *and that* K *is real. Write* $A = \mathbb{T}$ *or* $\mathbb{T}^+$ *and* $W = W_{\mathfrak{p}}$. *Let* $e = \mathrm{rank}_\Lambda \mathbb{T}$. *Then the following five assertions hold:*

(1) *If* $\langle \varepsilon \rangle - 1$ *is a prime in* Λ, *then the ring* A *is isomorphic to a power series ring* $W[[x]]$ *of one variable over* W; *hence,* A *is a regular local domain and is factorial;*

(2) *The ring A is an integral domain fully ramified at $(\langle \varepsilon \rangle - 1)$ generated by one element over Λ;*

(3) *If p is prime to e, the ramification locus of $A_{/\Lambda}$ is given by $\mathrm{Spec}(\Lambda_\varepsilon)$ for $\Lambda_\varepsilon := \Lambda/(\langle \varepsilon \rangle - 1))$, the different for A/Λ is principal and generated by Θ^{a-1} for $a = \mathrm{rank}_\Lambda A$, and A is a normal integral domain of dimension 2 unramified outside $(\langle \varepsilon \rangle - 1)$ over Λ;*

(4) *If $p|e$, $\mathbb{T}[\frac{1}{p}] := \mathbb{T} \otimes_{\mathbb{Z}} \mathbb{Q}$ is a Dedekind domain unramified outside $(\langle \varepsilon \rangle - 1)$ over $\Lambda \otimes_{\mathbb{Z}} \mathbb{Q}$, and the relative different for $\mathbb{T}[\frac{1}{p}]/\Lambda[\frac{1}{p}]$ is principal and generated by Θ^{e-1};*

(5) *If $e = 2$, $\mathbb{T}^+ = \Lambda$ and $\mathbb{T} = \Lambda[\sqrt{1 - \langle \varepsilon \rangle}]$.*

We will prove this theorem in §8.3.7.

Conjecture 8.3.8 (Semi-simplicity). *Assume that K is real. Then* $\boxed{e = 2}$ *under* (H0–3).

Note that $\mathrm{Sel}(Ad(\rho_{\mathbb{T}}))^\vee \cong \Lambda/(\langle \varepsilon \rangle - 1)$ if and only if the above conjecture holds. Since $(\langle \varepsilon \rangle - 1) = (t^{p^{r-1}} - 1)$ for $r = \mathrm{ord}_\wp(\varepsilon^{p-1} - 1) \geq 1$, this ideal is a product of primitive cyclotomic polynomials and hence square-free. Thus this conjecture is equivalent to the semi-simplicity of the Λ-module $\mathrm{Sel}(Ad(\rho_{\mathbb{T}}))$ and hence is an analogue of Iwasawa's conjecture in the cyclotomic theory [CPI, II, U3]. We record the following fact:

Corollary 8.3.9. *Suppose* (H0–2) *and that K is real. We have $\mathrm{Sel}(\mathrm{Ind}_K^{\mathbb{Q}} \Phi^-)^\vee \cong W[\Gamma_\wp]$ and hence $\mathrm{Sel}(\mathrm{Ind}_K^{\mathbb{Q}} \Phi^-)^\vee \otimes_{\mathbb{Z}_p} \mathbb{Q}_p \cong \mathrm{Frac}(W)[\Gamma_\wp]$ is a semi-simple Λ-module finite dimensional over $\mathrm{Frac}(W)$.*

Proof. We have $\mathrm{Sel}(Ad(\mathrm{Ind}_K^{\mathbb{Q}} \Phi))^\vee \cong \Omega_{\mathbb{T}/\Lambda} \otimes_{\mathbb{T}} W[\Gamma_\wp]$ by Theorem 7.4.1 combined with Theorem 6.1.11. From the first fundamental sequence for $\Lambda \to \mathbb{T}^+ \to \mathbb{T}$ in Corollary 5.2.5,

$$\Omega_{\mathbb{T}^+/\Lambda} \otimes_{\mathbb{T}^+/\Lambda} W[\Gamma_\wp] \to \Omega_{\mathbb{T}/\Lambda} \otimes_{\mathbb{T}} W[\Gamma_\wp] \to \Omega_{\mathbb{T}/\mathbb{T}^+} \otimes_{\mathbb{T}} W[\Gamma_\wp] \to 0$$

is exact. Taking the "$-$"-eigenspace of $[\alpha]$, we find $\mathrm{Sel}(\mathrm{Ind}_K^{\mathbb{Q}} \Phi^-)^\vee = \Omega_{\mathbb{T}/\Lambda}^- \otimes_{\mathbb{T}} W[\Gamma_\wp] \cong \Omega_{\mathbb{T}/\mathbb{T}^+} \otimes_{\mathbb{T}} W[\Gamma_\wp]$. Since $\mathbb{T} = \mathbb{T}^+[X]/(X^2 - \Theta^2)$, we have $\Omega_{\mathbb{T}/\mathbb{T}^+} = (\mathbb{T}/(2\Theta))dX \cong \mathbb{T}/(\Theta) \cong W[\Gamma_\wp]$; so, $\Omega_{\mathbb{T}/\mathbb{T}^+} \otimes_{\mathbb{T}} W[\Gamma_\wp] \cong W[\Gamma_\wp]$, and we obtain the desired identity. $\square$

Thus $\mathrm{Sel}(\mathrm{Ind}_K^{\mathbb{Q}} \Phi^-)^\vee \otimes_{\mathbb{Z}_p} \mathbb{Q}_p \cong \mathrm{Frac}(W[\Gamma_\wp])$ is a semi-simple Λ-module, and Conjecture 8.3.8 also claims $\mathrm{Sel}(Ad(\rho_{\mathbb{T}})) \cong \mathrm{Sel}(\mathrm{Ind}_K^{\mathbb{Q}} \Phi^-)$ under (H0–3).

8.3.6 Wall–Sun–Sun primes

If $\langle \varepsilon \rangle - 1$ is not a prime in Λ ($\Leftrightarrow \varepsilon^{p-1} \equiv 1 \mod \wp^2$), $\langle \varepsilon \rangle - 1$ is a product of at least two primes. The ambiguous classes of primes of $\mathbb{T}$ ramifying over $\mathbb{T}^+$ (i.e., those dividing $\mathfrak{d}_{\mathbb{T}/\mathbb{T}^+} = \mathbb{T}([\alpha] - 1)\mathbb{T}$) cannot be principal. Hence $\mathbb{T}$ cannot be factorial. Indeed, every height 1 prime $P|(\langle \varepsilon \rangle - 1)$ in $\mathbb{T} = \mathbb{T}_\mathfrak{p}$ ramifies over $\mathbb{T}_\mathfrak{p}^+$ with ramification index 2. In the class group $\mathrm{Pic}(\mathbb{T}[\frac{1}{p}])$ of the normal integral domain $\mathbb{T}[\frac{1}{p}]$, the class of P lives in the ambiguous class, and has order divisible by 2 or infinite; so, P cannot be principal.

There is no example known of a prime $p \geq 5$ split in $K = \mathbb{Q}[\sqrt{5}]$ such that $((\langle \varepsilon \rangle - 1) \neq (T)$. There are no such primes less than 6.7×10^{15} [DK11] and this bound is extended to 2.6×10^{17} by the PrimeGrid project.

Taking $K = \mathbb{Q}[\sqrt{d}]$ with square-free $0 < d \in \mathbb{Z}$, we describe how to decide if $\wp^2|\varepsilon^{k-1} - 1$. Since $p > 2$, $\wp^2|(\varepsilon^{p-1} - 1) \Leftrightarrow \wp^2|(\varepsilon^{2(p-1)} - 1)$. On the other hand, $\varepsilon^{2(p-1)} - 1 = \varepsilon^{2(p-1)} - \varepsilon^{p-1}\varepsilon^{\sigma(p-1)} = \varepsilon^{p-1}(\varepsilon^{p-1} - \varepsilon^{\sigma(p-1)})$. Define $\alpha \in \mathbb{Z}$ so that $\varepsilon^2 - \alpha\varepsilon \pm 1 = 0$. Consider the corresponding Fibonacci type recurrence relation $f_n = \alpha f_{n-1} \mp f_{n-2}$. For the solution f_n with initial values $f_0 = 0$ and $f_1 = 1$, we have $f_n = \frac{\varepsilon^n - \varepsilon^{n\sigma}}{\varepsilon - \varepsilon^\sigma}$. Thus we have $\frac{\varepsilon^{p-1} - \varepsilon^{\sigma(p-1)}}{\sqrt{d}} = f_{p-1}C$ for $C = \frac{\varepsilon - \varepsilon^\sigma}{\sqrt{d}}$. If $d = 5$, we have $C = 1$.

In any case, $\langle \varepsilon \rangle - 1$ is not a prime in $\Lambda \Leftrightarrow p^2|f_{p-1}C$. The primes satisfying this condition are called Wall–Sun–Sun primes (which was studied in [SS92] for $d = 5$). For $K = \mathbb{Q}[\sqrt{10}]$, $p = 191,643$ are such primes. It is conjectured that Wall–Sun–Sun primes are infinitely many [K07].

8.3.7 Proof of Theorem 8.3.7

We give a proof for $A = \mathbb{T}$, as the argument is the same for $\mathbb{T}^+$.

Proof of (1). Note $\mathfrak{d} = \Theta\mathbb{T}$ and put $\mathfrak{d}^0 = \mathbb{T}$. For all $0 \neq u \in \mathbb{T}$, $[u]$: $x \mapsto ux$ induces a linear endomorphism $\mathrm{gr}(u)$ of the corresponding graded algebra $\mathrm{gr}_\mathfrak{d}(\mathbb{T}) := \bigoplus_{n=0}^\infty \mathfrak{d}^n/\mathfrak{d}^{n+1}$. Then $[u]$ is injective if $\mathrm{gr}(u)$ is injective [BCM, III.2.8, Corollary 1]. We have $\mathrm{gr}_\mathfrak{d}(\mathbb{T}) \cong \Lambda_\varepsilon[x]$ for the polynomial ring $\Lambda_\varepsilon[x]$ where the variable x corresponds to the image $\overline{\Theta}$ of Θ in the first graded piece $\mathfrak{d}/\mathfrak{d}^2$. Take n so that $u \in \mathfrak{d}^n$ but $u \notin \mathfrak{d}^{n+1}$. Then $\mathrm{gr}(u) : \mathrm{gr}_\mathfrak{d}(\mathbb{T}) \to \mathrm{gr}_\mathfrak{d}(\mathbb{T})$ is multiplication by a polynomial of degree n. Assume that $\langle \varepsilon \rangle - 1$ is a prime; so, $((\langle \varepsilon \rangle - 1) = (T)$ in Λ and $\Lambda_\varepsilon = W$. Then $\mathrm{gr}_\mathfrak{d}(\mathbb{T})$ is an integral domain isomorphic to the polynomial ring $W[x]$; so, if $u \neq 0$, $\mathrm{gr}(u)$ is injective, and hence, $[u]$ is injective; so, u is not a zero divisor. We conclude that $\mathbb{T}$ is an integral domain and $\mathbb{T} = \varprojlim_n \mathbb{T}/\mathfrak{d}^n \cong W[[x]]$ by sending Θ to x. By Corollary 1.10.5, a power series ring over a discrete valuation ring is a unique factorization domain and is regular; so, we get the assertion (1). $\qquad\square$

Proof of (2). By class field theory, recalling $\Gamma_\wp := Cl_K(\wp^\infty) \otimes_{\mathbb{Z}} \mathbb{Z}_p$, we have an exact sequence: $1 \to \widehat{O}_{K,\wp}^\times / \overline{O_K^\times} \to \Gamma_\wp \to C_K \to 1$. By (H3), $C_K = \{1\}$, and hence $t^{\mathbb{Z}_p}/\langle\varepsilon\rangle^{\mathbb{Z}_p} \cong \widehat{O}_{K,\wp}^\times / \overline{O_K^\times} \cong \Gamma_\wp$. Thus we conclude $\Lambda_\varepsilon = W[\Gamma_\wp]$. By Corollary 7.4.2, $\mathbb{T}/\mathbb{T}([\alpha]-1)\mathbb{T} = \mathbb{T}/(\Theta) = W[\Gamma_\wp]$. Then $\Lambda/(\langle\varepsilon\rangle - 1) \cong \mathbb{T}/(\Theta) = \Lambda[[X]]/(X, D(X)) = \Lambda/(D(0))$, and we find the identity of ideals

$$(D(0)) = \mathrm{Ann}_\Lambda(\Lambda/(D(0)) = \mathrm{Ann}_\Lambda(\Lambda/(\langle\varepsilon\rangle - 1)) = (\langle\varepsilon\rangle - 1).$$

Thus $(D(0))$ is square-free. Let $P|(\langle\varepsilon\rangle - 1)$ be a prime factor; so, the localization Λ_P and its completion $\widehat{\Lambda}_P = \varprojlim_n \Lambda_P/P^n\Lambda_P$ are discrete valuation rings.

Consider the localization $\mathbb{T}_P = \mathbb{T} \otimes_\Lambda \Lambda_P$ and $\mathfrak{d}_P = \mathfrak{d}\mathbb{T}_P$. Then $\mathfrak{d}_P^n/\mathfrak{d}_P^{n+1} \cong \Lambda_\varepsilon \otimes_\Lambda \Lambda_P = P^n\mathbb{T}_P/P^{n+1}\mathbb{T}_P \cong \kappa(P)$ for $\kappa(P) = \Lambda_P/P\Lambda_P$. Therefore $\mathfrak{d}\mathbb{T}_P = P\mathbb{T}_P = \Theta\mathbb{T}_P$ by Nakayama's lemma. Thus $\mathrm{gr}_{\mathfrak{d}\mathbb{T}_P}(\mathbb{T}_P)$ is isomorphic to the polynomial ring $\kappa(P)[x]$ with $x = \overline{\Theta}$ over the residue field $\kappa(P) = \Lambda_P/P\Lambda_P$. Therefore, by [CRT, Theorem 28.3], the P-adic completion $\widehat{\mathbb{T}}_P$ of $\mathbb{T}_P$ is isomorphic to the power series ring $\kappa(P)[[x]]$ by sending Θ to x, and hence $\mathbb{T}_P$ is a discrete valuation ring with a prime element Θ. Since $\mathbb{T}_P/P\mathbb{T}_P = \kappa(P)$, $\mathbb{T}_P$ is fully ramified over Λ_P. Since $\mathbb{T}$ is free of finite rank e over Λ, so is $\mathbb{T}_P$ over Λ_P, and $\mathbb{T}$ injects into $\mathbb{T}_P$. In particular $\mathbb{T}$ is an integral domain. This proves the assertion (2).

As remarked in §8.1.1, $D(X)$ is the characteristic polynomial of $x \mapsto \Theta x$ acting on $\widehat{\mathbb{T}}_P$ over $\widehat{\Lambda}_P$. Since $\widehat{\mathbb{T}}_P$ is a discrete valuation ring fully ramified over $\widehat{\Lambda}_P$, the polynomial $D(X)$ is a distinguished polynomial with respect to P. Since $D(0)$ is square free, $D(X)$ is an Eisenstein polynomial with respect to P for every prime divisor P of $\langle\varepsilon\rangle - 1$. $\qquad\square$

Proof of (3). By (2), $\mathbb{T}/\Lambda$ is fully ramified at each prime factor of $\langle\varepsilon\rangle - 1$ with ramification index $e = \mathrm{rank}_\Lambda \mathbb{T}$. Then $\mathbb{T} = \Lambda[X]/(D(X))$ is a local domain by [BCM, VIII.5.4]. The polynomial $D(X) = X^e + a_{e-1}X^{e-1} + \cdots + a_0$ satisfies $(\langle\varepsilon\rangle - 1)|a_i$ and $(a_0) = (\langle\varepsilon\rangle - 1)$. Write $u = \frac{\langle\varepsilon\rangle - 1}{\Theta^e} \in \mathbb{T}^\times$ and $a_j = (\langle\varepsilon\rangle - 1)^{\alpha_j}u_j$ for $\alpha_j \geq 1$ and $\alpha_0 = 1$ ($0 \leq j < e$) such that $u_j \in \Lambda$ is either 0 or $(\langle\varepsilon\rangle - 1) \nmid u_j$ for $j = 1, \ldots, e-1$ and $u_0 \in \Lambda^\times$. Then

$$0 = D(\Theta) = \sum_{j=0}^{e} a_j\Theta^j = (\langle\varepsilon\rangle - 1)(u^{-1} + u_0 + \sum_{j=1}^{e-1}(\langle\varepsilon\rangle - 1)^{\alpha_j - 1}u_j\Theta^j).$$

Since $\langle \varepsilon \rangle - 1$ is not a zero-divisor of $\mathbb{T}$, we find

$$u^{-1} = -u_0 - \sum_{j=1}^{e-1} (\langle \varepsilon \rangle - 1)^{\alpha_j - 1} u_j \Theta^j \in \mathbb{T},$$

and hence $u \in \mathbb{T}^\times$. Therefore

$$\frac{dD(X)}{dX}(\Theta) = \sum_{j=1}^{e} j a_j \Theta^{j-1} = \Theta^{e-1}(e + \sum_{j=1}^{e-1} u^{\alpha_j} u_j j \Theta^{\alpha_j e - e + j}).$$

If $p \nmid e$, $(e + \sum_{j=1}^{e-1} u^{\alpha_j} u_j j \Theta^{\alpha_j e - e + j})$ is a unit; so, $(\frac{dD(X)}{dX}(\Theta)) = (\Theta^{e-1})$, and the ramification of $\mathbb{T}/\Lambda$ is limited to prime factors of (Θ^{e-1}). Since $(\Theta^{e-1}) \cap \Lambda = (\langle \varepsilon \rangle - 1)$, the ramification locus of $\mathbb{T}_{/\Lambda}$ is $\mathrm{Spec}(\Lambda/((\langle \varepsilon \rangle - 1))$ if $p \nmid e$. Therefore if $p \nmid e$, $\mathbb{T}_P$ is a discrete valuation ring for all height 1 prime $P | \langle \varepsilon \rangle - 1$. Since $\mathbb{T}$ is Λ-free, we have $\mathbb{T} = \bigcap_P \mathbb{T}_P$ [BCM, VII.4.3] which is a normal local domain (proving (3)).   □

Proof of (4). Suppose $p | e$. Write $e = p^r e'$ with $p \nmid e'$. Since

$$0 = D(\Theta) = \Theta^e + a_{e-1}\Theta^{e-1} + \cdots + a_0 = \Theta^e(1 + u_0 u + \sum_{j=1}^{e-1} u^{\alpha_j} u_j \Theta^{\alpha_j e - e + j}),$$

we have $-uu_0 \equiv 1 \mod \Theta$; so, $v = (-uu_0)^{1/e'} \in \mathbb{T}$. Then $\mathbb{T}[v^{1/p^r}]/\mathbb{T}$ can ramify only at p as v is a unit. By replacing Θ by $\Theta' = v^{1/p^r}\Theta$, we find $\Theta'^e = -a_0$, and $\mathbb{T}[v^{1/p^r}] = \Lambda[v^{1/p^r}, \Theta']$, which can ramify over Λ only at p and prime factors of $\langle \varepsilon \rangle - 1$ (as $(\langle \varepsilon \rangle - 1) = (a_0)$). In particular $\mathbb{T}[\frac{1}{p}] = \bigcap_P \mathbb{T}_P$ for P running all prime divisors outside p, and $\mathbb{T}_P$ with $P \nmid (p)$ is a discrete valuation ring. This shows (4).   □

Proof of (5). If $e = 2$, $\sigma(\Theta) = -\Theta$ implies $\Theta^2 \in \Lambda$. Thus $D(X) = X^2 - \Theta^2$ with $\Theta^2 = -a_0 = u_0(1 - \langle \varepsilon \rangle) \in \Lambda$ with $u = -u_0 \in \Lambda^\times$. Thus $u_0^2 = -uu_0 \equiv 1 \mod (\Theta)$. Since $(\Theta) \cap \Lambda = (\langle \varepsilon \rangle - 1)$ and $u_0, uu_0 \in \Lambda$, we find $u_0^2 = -uu_0 \equiv 1 \mod ((\langle \varepsilon \rangle - 1)$, and $\sqrt{u_0} \in \Lambda^\times$. Thus replacing Θ by $\sqrt{u_0}\Theta$, we get $\Theta = \sqrt{1 - \langle \varepsilon \rangle}$ and $\mathbb{T} = \Lambda[\sqrt{1 - \langle \varepsilon \rangle}]$, proving (5).   □

Remark 8.3.10. By the proof of the assertion (2) of Theorem 8.3.7 and $[\alpha](\Theta) = -\Theta$, we see $D(X) = D^+(X^2)$ for an Eisenstein polynomial $D^+(X) \in \Lambda[X]$ with respect to each prime factor of $(\langle \varepsilon \rangle - 1)$ such that $\mathbb{T}^+ = \Lambda[X]/(D^+(X))$.

8.4 Indecomposability of deformations of $\mathrm{Ind}_K^{\mathbb{Q}}\,\varphi$ for K real

In this section, until §8.4.3, we allow imaginary K. In the study of induced representations in §7.4.1 and Theorem 8.3.5, we have a characterization of the generator $\Theta \in \mathbb{T}$ over $\mathbb{T}^+$: for $\pi \in \mathrm{Hom}_{CL_W}(\mathbb{T}, A)$, $\pi \circ \rho_{\mathbb{T}}$ is an induced representation from K if and only if $\pi(\Theta) = 0$. On the other hand, if ρ_A is indecomposable over the inertia group I_p, it cannot be an induced representation. Thus one would expect that Θ would span the unipotent part of $\rho_{\mathbb{T}}(I_p)$. Our next goal is to prove $\theta\mathbb{T} = \Theta\mathbb{T}$ for the generator θ of the unipotent part of $\rho_{\mathbb{T}}(I_p)$ in Theorem 8.2.3. Assuming K is real, we further show the indecomposability $\pi(\theta) \neq 0$ if π is associated to a Hecke eigenform of weight ≥ 2 (even if $\mathbb{T}$ has more than one generator over $\mathbb{T}^+$). If $\mathbb{T}$ has more then one generator over Λ, this point of the non-vanishing of the specialization $\pi(\theta)$ is not known in full generality in the imaginary quadratic case (cf. §8.5.4).

We assume that p split in K unless explicitly mentioned otherwise.

8.4.1 *Generalized matrix algebras*

Recall $H = \mathrm{Gal}(F^{(p)}(\overline{p})/K)$. The representation $\overline{\rho}|_H = \overline{\varphi} \oplus \overline{\varphi}_\sigma$ is no longer absolutely irreducible; so, a deformation $\rho_A \in \mathcal{D}(A)$ over G restricted to H generally spans a proper A-subalgebra of $M_2(A)$ called a *generalized matrix A-algebra* (abbreviated as GMA).

We describe briefly representations of an algebra R with values in a GMA $E_{/A}$ following [FGS]. For a representation of a group $\mathcal{G} \to E^\times$ over a base algebra $B \in CL_W$, we take R to be the group algebra $B[\mathcal{G}]$ and the linear extension of the original group representation to the algebra representation of $B[\mathcal{G}]$. We follow [FGS, §1.3] to define a GMA A-algebra E. Let A be a commutative B-algebra and suppose that E is an A-algebra (so, E becomes also a B-algebra). We say that E is a *generalized matrix algebra* (GMA) of type $(d_1, \ldots, d_r)$ if R is equipped with:

- a family orthogonal idempotents $\mathcal{E} = \{e_1, \ldots, e_r\}$ with $\sum_i e_i = 1$,
- for each i, an A-algebra isomorphism $\psi_i : e_i E e_i \xrightarrow{\sim} M_{d_i}(A)$, such that the trace map $T : R \to A$, defined by $T(x) := \sum_i \mathrm{Tr}(\psi_i(e_i x e_i))$ satisfies $T(xy) = T(yx)$ for all $x, y \in E$. We call $\mathcal{E} = \{e_i, \psi_i, i = 1, \ldots, r\}$ the data of idempotents of E.

In this book, we assume that $r = 2$ and $d_1 = d_2 = 1$; so, we can forget about ψ_i as an A-algebra automorphism of A is unique. Once we have $\mathcal{E}$, we identify $e_i E e_i = A$ and put $B = e_1 E e_2$ and $C = e_2 E e_1$. Then a generalized matrix algebra over A is a pair of an associative A-algebra E and $\mathcal{E}$. It is isomorphic to $A \oplus B \oplus C \oplus A$ as A-modules; so, we write instead $(E, \mathcal{E}) = \left(\begin{smallmatrix} A & B \\ C & A \end{smallmatrix} \right)$ which we call a GMA structure. There are A-linear maps $\psi : B \otimes_A C \to A$ and $\psi' : C \otimes_A B \to A$ (called the product law of E) such that the multiplication in E is given by the usual 2×2 matrix product (and of course they need to satisfy rules to assure associativity; see [FGS, §1.3]). However, we only need the case where we have a commutative A-algebra $\widetilde{A}$ with injective A-algebra structure: $A \hookrightarrow \widetilde{A}$ and A-linear embeddings $B \hookrightarrow \widetilde{A}$ and $C \hookrightarrow \widetilde{A}$ so that

$$E = \left\{ \left(\begin{smallmatrix} a & b \\ c & d \end{smallmatrix} \right) \, \middle| \, a, d \in A, b \in B, c \in C \right\} \subset M_2(\widetilde{A})$$

is an A-subalgebra of $M_2(\widetilde{A})$. Thus the products ψ and ψ' are just induced by the product of $\widetilde{A}$. This means $BC \subset \widetilde{A}$ and $CB \subset \widetilde{A}$ (and we do not worry about the rules giving associativity). For $b \in B$ and $c \in C$, we often write simply $bc := \psi(b \otimes c)$ and $cb := \psi'(c \otimes b)$ if confusion is unlikely. We call A the *scalar subring* of $(E, \mathcal{E})$, and $(E, \mathcal{E})$ is called an A-GMA. The A-GMA's form a category over the category of A-algebras. Here, writing $E = \left(\begin{smallmatrix} A & B \\ C & A \end{smallmatrix} \right)$ and $E' = \left(\begin{smallmatrix} A' & B' \\ C' & A' \end{smallmatrix} \right)$, we say that $\phi_E : E \to E'$ is an A-*GMA morphism* over an algebra homomorphism $\phi_A : A \to A'$ if

(1) ϕ_E is an algebra homomorphism,
(2) ϕ_E sends each matrix entry of E to the corresponding entry of E',
(3) on the entries A, ϕ coincides with ϕ_A,
(4) $\phi_E|_B : B \to B'$ and $\phi_E|_C : C \to C'$ are morphisms of A-modules.

8.4.2 *Inertia embedding into* $\mathbb{T}^-$

Recall $\overline{H} \subset \overline{G} \subset G$ with $\overline{G} \cong \mathrm{Gal}(F(\overline{\rho})/\mathbb{Q})$, $\overline{H} \cong \mathrm{Gal}(F(\overline{\rho})/K)$, $G = \mathrm{Gal}(F^{(p)}(\overline{\rho})/\mathbb{Q}) = \overline{G} \ltimes \mathrm{Gal}(F^{(p)}(\overline{\rho})/F(\overline{\rho}))$ and $H = \mathrm{Gal}(F^{(p)}(\overline{\rho})/K) = \overline{H} \ltimes \mathrm{Gal}(F^{(p)}(\overline{\rho})/F(\overline{\rho}))$. Let $\overline{D} \subset \overline{H}$ be the $\wp$-decomposition subgroup identifying $\overline{H} = \mathrm{Gal}(F(\overline{\rho})/K)$. We may and do assume

(D0) $\rho_{\mathbb{T}}(\sigma) = \left(\begin{smallmatrix} 0 & \varphi(\sigma^2) \\ 1 & 0 \end{smallmatrix} \right)$ for $\sigma \in \overline{G}$,
(D1) $\rho_{\mathbb{T}}(\delta) = \mathrm{diag}[\varphi(\delta), \varphi_\sigma(\delta)]$ for $\delta \in \overline{D}$ (where $\mathrm{diag}[a, b] = \left(\begin{smallmatrix} a & 0 \\ 0 & b \end{smallmatrix} \right)$),
(D2) If K is real, $\rho_{\mathbb{T}}(c) = \mathrm{diag}[-1, 1] \in \mathrm{GL}_2(W) \subset \mathrm{GL}_2(\mathbb{T})$ for a complex conjugation c,
(D3) $D_{\wp^\sigma} = \sigma^{-1} D_\wp \sigma$ for the fixed decomposition group $D_\wp$ of $\wp$.

Recall $\alpha_{\mathbb{T}}$ in Remark 7.4.3 so that $\alpha_{\mathbb{T}} \circ \rho_{\mathbb{T}} \cong \rho_{\mathbb{T}} \otimes \alpha$. We simply write $[\alpha]$ for $\alpha_{\mathbb{T}}$ even if we do not know $R_{\kappa} = \mathbb{T}$. Let $\mathbb{T}^{\pm} := \{x \in \mathbb{T} | [\alpha](x) = \pm x\}$ (which is a $\mathbb{T}^{+}$-module) and $\mathfrak{d}_{\mathbb{T}} = \mathbb{T}([\alpha] - 1)\mathbb{T}$. We prove

Proposition 8.4.1. *Assume* (H0–1). *After conjugating* $\rho_{\mathbb{T}}$ *by a suitable element in* $\mathrm{GL}_2(\mathbb{T})$, *we can arrange* $\rho_{\mathbb{T}}|_H$ *to have values in the* $\mathbb{T}^{+}$-*GMA* $E_+ := \left(\begin{smallmatrix} \mathbb{T}^+ & \mathbb{T}^- \\ \mathbb{T}^- & \mathbb{T}^+ \end{smallmatrix} \right)$ *so that*

(i) *if K is real,* $\rho_{\mathbb{T}}(c) = \mathrm{diag}[-1, 1]$ *and* $\rho_{\mathbb{T}}(\delta) = \mathrm{diag}[\varphi(\delta), \varphi_\sigma(\delta)]$ *for* $\delta \in \overline{D}$ *and the complex conjugation c chosen as in* (D2),
(ii) *if K is imaginary,* $\rho_{\mathbb{T}}(\delta) = \mathrm{diag}[\varphi(\delta), \varphi_\sigma(\delta)]$ *for* $\delta \in \overline{D}$.

Here the product law: $\mathbb{T}^- \otimes_{\mathbb{T}^+} \mathbb{T}^- \to \mathbb{T}^+$ *is given by* $b \otimes c = bc$ *with the product bc in* $\mathbb{T}$.

$\mathbb{T}^{+}$-*GMA* E_+ *is a* $\mathbb{T}^{+}$-*subalgebra of* $M_2(\mathbb{T})$ *with* $B = C = \mathbb{T}^-$ (so, $A = \mathbb{T}^+$ *and* $\widetilde{A} = \mathbb{T}$ *in the above definition*).

Proof. Recall the deformation functor giving rise to R^{ord}:

$$\mathcal{D}(A) := \{\rho : G \to \mathrm{GL}_2(A) : p\text{-ordinary} | (\rho \bmod \mathfrak{m}_A) = \overline{\rho}\}/\Gamma(\mathfrak{m}_A). \quad (8.4)$$

We have the involution $[\alpha]$ on R^{ord} given in §7.4.1. By our normalization of the action of $[\alpha]$ on deformations,

$$J(\rho_{\mathbb{T}} \otimes \alpha)J^{-1} \sim \rho_{\mathbb{T}}^{[\alpha]} \quad \text{for} \quad J = \left(\begin{smallmatrix} 1 & 0 \\ 0 & -1 \end{smallmatrix} \right).$$

Since $J \left(\begin{smallmatrix} a & b \\ c & d \end{smallmatrix} \right) J^{-1} = \left(\begin{smallmatrix} a & -b \\ -c & d \end{smallmatrix} \right)$ and $(\rho_{\mathbb{T}}|_H \bmod \mathfrak{d}_{\mathbb{T}}) = \Phi \oplus \Phi_{\varsigma}$ is diagonal by Theorem 7.4.1, we have $uJ(\rho_{\mathbb{T}} \otimes \alpha)(uJ)^{-1} = \rho_{\mathbb{T}}^{[\alpha]}$ with $u \in 1 + \Theta M_2(\mathbb{T})$. Write $U = uJ$. Applying $[\alpha]$, we get $U^{[\alpha]}(\rho_{\mathbb{T}}^{[\alpha]} \otimes \alpha)U^{-[\alpha]} = \rho_{\mathbb{T}}$; so, we have

$$U\rho_{\mathbb{T}}U^{-1} = U(\rho_{\mathbb{T}} \otimes \alpha)U^{-1} \otimes \alpha = \rho_{\mathbb{T}}^{[\alpha]} \otimes \alpha = U^{-[\alpha]}\rho_{\mathbb{T}}U^{[\alpha]}.$$

Thus $U^{[\alpha]}U = Ju^{[\alpha]}Ju = z \in Z := 1 + \Theta\mathbb{T}$. Since $1 + \Theta M_2(\mathbb{T})$ is p-profinite, letting $[\alpha]$ act on $1 + \Theta M_2(\mathbb{T})$ by $x \mapsto x^{[\alpha]} := Jx^{[\alpha]}J$, we can write $u = v^{[\alpha]-1} \in (1 + \Theta M_2(\mathbb{T}))$ for $v \in 1 + \Theta M_2(\mathbb{T})$. Replacing $\rho_{\mathbb{T}}|_H$ by $v^{-1}J\rho_{\mathbb{T}}Jv|_H$, we find $J\rho_{\mathbb{T}}|_H J^{-1} = \rho_{\mathbb{T}}^{[\alpha]}|_H$. In other words, $\rho_{\mathbb{T}}|_H$ has values in $E_+ = \left(\begin{smallmatrix} \mathbb{T}^+ & \mathbb{T}^- \\ \mathbb{T}^- & \mathbb{T}^+ \end{smallmatrix} \right)$. $\square$

Corollary 8.4.2. *Let the notation be as in Theorem 8.2.3. Under* (H0–1), *we have a* Λ_1-*linear embedding* $\mathcal{U} \subset \mathbb{T}^- = \Theta\mathbb{T}^+$ *for the unipotent part of the $\wp$-inertia subgroup of* $\mathrm{Gal}(F(\rho_{\mathbb{T}})/\mathbb{Q})$. *In particular, under* (H0–2), *we have* $\Theta|\theta$ *in* $\mathbb{T}$.

Proof. Since the GMA E_+ is constructed under (H0–1) in a manner compatible to the upper triangular form of $\rho_{\mathbb{T}}(D_\wp)$, we find $\rho_{\mathbb{T}}(\mathcal{U}) \subset \{ \left(\begin{smallmatrix} 1 & a \\ 0 & 1 \end{smallmatrix} \right) \big| a \in \mathbb{T}^- \}$. Since the conjugation action of $\mathcal{T}$ and $\rho_{\mathbb{T}}([p, \mathbb{Q}_p]^f)$ gives the Λ_1-module structure on $\mathcal{U}$, the map $\mathcal{U} \ni \varsigma \mapsto \rho_{\mathbb{T}}(\varsigma) = \left(\begin{smallmatrix} 1 & a(\varsigma) \\ 0 & 1 \end{smallmatrix} \right) \mapsto a(\varsigma) \in \mathbb{T}^-$ is Λ_1-linear embedding.

Since (Θ) generates $\mathbb{T}([\alpha] - 1)\mathbb{T} = \mathbb{T}\mathbb{T}^-$ under (H0–2), we find $\mathbb{T}^- = \Theta\mathbb{T}^+$. Thus $a(\mathcal{U}) = \theta\Lambda_1 \subset \mathbb{T}^- = \Theta\mathbb{T}^+$, which implies $\Theta|\theta$. □

By (H0–2), $\mathbb{T}^- = \Theta\mathbb{T}^+$. Since $\theta \in \mathbb{T}^-$, we can write $\theta = u\Theta$ ($u \in \mathbb{T}$).

Theorem 8.4.3. *Assume that K is real. Then θ/Θ is a unit under* (H0–2).

Even when K is imaginary, the outcome is the same. However we need to modify the proof slightly, which we deal with later in the following section.

Proof. Recall the relative different $\mathfrak{d} := \mathbb{T}([\alpha] - 1)\mathbb{T}$. We have an exact sequence $\mathfrak{d} \hookrightarrow \mathbb{T} \twoheadrightarrow W[[\Gamma_\wp]]$ by Corollary 7.4.2. In this real case, we have $W[[\Gamma_\wp]] = W[\Gamma_\wp]$ as $\Gamma_\wp$ is finite, but we keep writing $W[[\Gamma_\wp]]$ as we use some argument here in the proof of the following proposition also valid in the imaginary case.

Taking the $[\alpha]$-invariant subspaces (indicated superscript "+"), $\mathbb{T}^+/\mathfrak{d}^+ \cong W[[\Gamma_\wp]]$ for $\Gamma_\wp = Cl_K(\wp^\infty) \otimes_{\mathbb{Z}} \mathbb{Z}_p$. Recall the universal character $\varphi : H \to W[[\Gamma_\wp]]^\times = (\mathbb{T}^+/\mathfrak{d}^+)^\times$ unramified outside $\wp\mathfrak{c}$ deforming $\overline{\varphi}$ (defined in the proof of Corollary 7.4.2). Write $\rho_H := \rho_{\mathbb{T}}|_H = \left(\begin{smallmatrix} \rho_{11} & \rho_{12} \\ \rho_{21} & \rho_{22} \end{smallmatrix} \right) \subset E_+$ for the $\mathbb{T}^+$-GMA E_+ in Proposition 8.4.1 and put $a = \rho_{11} \mod \mathfrak{d}^+ = \varphi$, $d = \rho_{22} \mod \mathfrak{d}^+ = \varphi_\sigma$, $b = \rho_{12} \mod \mathfrak{d}^+ : H \to \mathbb{T}^-/\mathfrak{d}^+\mathbb{T}^-$ and $c = \rho_{21} \mod \mathfrak{d}^+ : H \to \mathbb{T}^-/\mathfrak{d}^+\mathbb{T}^-$.

First we show that $b \mod \mathfrak{m}_{\mathbb{T}^+}\mathbb{T}^-$ is non-trivial. Suppose against the desired outcome that b has image in $\mathfrak{m}_{\mathbb{T}^+}(\mathbb{T}^-/\mathfrak{d}^+\mathbb{T}^-)$. By $c(g) = \varphi(\sigma^2)b(\sigma^{-1}g\sigma)$, c has values also in $\mathfrak{m}_{\mathbb{T}^+}(\mathbb{T}^-/\mathfrak{d}^+\mathbb{T}^-)$. This implies $\rho_H \mod \mathfrak{m}_{\mathbb{T}^+}\mathfrak{d}^+$ is diagonal; so, $\rho_H^{[\alpha]} = j\rho_H j^{-1}$ which implies $\rho_{\mathbb{T}}^{[\alpha]} \mod \mathfrak{m}_{\mathbb{T}}\mathfrak{d} \cong \rho_{\mathbb{T}} \otimes \alpha$, a contradiction as $\mathfrak{d}$ is the maximal ideal for which the identity holds. Thus b is onto.

Replacing ρ_H by $\rho' := \xi^{-1}\rho_H\xi$ for $\xi := \left(\begin{smallmatrix} \Theta & 0 \\ 0 & 1 \end{smallmatrix} \right)$, ρ' has values in $GL_2(\mathbb{T}^+)$ and $\rho' \mod \mathfrak{m}_{\mathbb{T}^+} = \left(\begin{smallmatrix} \overline{\varphi} & \overline{b} \\ 0 & \overline{\varphi}_\sigma \end{smallmatrix} \right)$ with $\overline{b} = b/\Theta \mod \mathfrak{m}_{\mathbb{T}^+} \neq 0$. If u is a non-unit, $\overline{b}$ is unramified at $\wp$. Since $\rho(I_{\wp^\sigma})$ is lower triangular, $\overline{b}$ is unramified at $\wp$ and $\wp^\sigma$; so, everywhere unramified over $F(\overline{\varphi}^-)$. Thus the fixed field $M = F(\overline{b})$ by $\text{Ker}(\overline{b})$ is a non-trivial abelian p-extension over $F(\overline{\varphi}^-)$ everywhere unramified (Galois over K) on which $\overline{H}$ acts by $\overline{\varphi}^-$, contradicting $Cl_{F(\overline{\varphi}^-)} \otimes_{\mathbb{Z}[H]} \overline{\varphi}^- = 0$. □

We keep using the notation introduced in the above proof of Theorem 8.4.3, but we only assume (H0–1), and K can be either real or imaginary. As we have written in the proof of Theorem 8.4.3, $\rho_H \bmod \mathfrak{d}^+ = \left(\begin{smallmatrix} a & b \\ c & d \end{smallmatrix}\right)$:

$$H \to E_+ \otimes_{\mathbb{T}^+} \mathbb{T}^+/\mathfrak{d}^+ = \begin{pmatrix} W[[\Gamma_\wp]] & \overline{\mathbb{T}}^- \\ \overline{\mathbb{T}}^- & W[[\Gamma_\wp]] \end{pmatrix} \text{ for } \overline{\mathbb{T}}^- := \mathbb{T}^-/\mathfrak{d}^+\mathbb{T}^-. \text{ Since}$$

$\mathbb{T}^-$ generates $\mathfrak{d}$, the product of E_+ brings $\mathbb{T}^- \otimes \mathbb{T}^-$ into $(\mathbb{T}^-)^2 \subset \mathfrak{d}^+$. Extend ρ_H to a representation $\Lambda[H] \to E_+$ Λ-linearly. Let A be an Artinian quotient of $\overline{\mathbb{T}}^+ := W[[\Gamma_\wp]]$, and put $\overline{\mathbb{T}}_A^- = \overline{\mathbb{T}}^- \otimes_{\overline{\mathbb{T}}^+} A$ and $E_A := E_+ \otimes_{\mathbb{T}^+} A$ with the specialized representation $\rho_A := \rho_H \otimes 1_A : H \to E_A$. For the universal character $\varphi : H \to W[[\Gamma_\wp]]^\times$, we write the A-specialization as $\varphi_A = \varphi \otimes 1_A : H \to A^\times$ and $\varphi_{\sigma,A} = \varphi_\sigma \otimes 1_A : H \to A^\times$. We quote the following result from [FGS, Theorem 1.5.5]:

Proposition 8.4.4. *Let the notation be as above. In particular, K is either real or imaginary. Assume (H0–1). Then there is a natural injective morphism of $W[[\Gamma_\wp]]$-modules:*

$$\iota_A : \mathrm{Hom}_{W[[\Gamma_\wp]]}(\overline{\mathbb{T}}_A^-, A^\vee) \hookrightarrow \mathrm{Ext}^1_{W[H]}(\varphi_A, \varphi_{\sigma,A}^*).$$

Passing to the limit, we have an injection

$$\iota : \mathrm{Hom}_{W[[\Gamma_\wp]]}(\overline{\mathbb{T}}^-, W[[\Gamma_\wp]] \otimes_W W^\vee) \hookrightarrow \mathrm{Ext}^1_{W[H]}(\varphi, \varphi_\sigma^*).$$

See [MFG, §4.2] for extension groups Ext^1. We only need to prove the first assertion. Since A is Artinian finite, $A^\vee \cong A$, so we remove the superscript "$*$" and "$\vee$" in the proof.

Proof. The map ι_A is made as follows. Pick an $f \in \mathrm{Hom}_A(\overline{\mathbb{T}}^-, A)$ and regard it as residing in $\mathrm{Hom}_{\mathbb{T}^+}(\mathbb{T}^-, A)$. We consider the following A-linear map $\Lambda[H] \to M_2(A)$,

$$x \mapsto \begin{pmatrix} \varphi_A(x) & f(b(x)) \\ 0 & \varphi_{\sigma,A}(x) \end{pmatrix}. \tag{8.5}$$

We claim that the map (8.5) is an $W[[\Gamma_\wp]]$-algebra homomorphism which gives rise to an extension of $\varphi_{\sigma,A}$ by φ_A. This map is multiplicative from $bc = 0$ by computation, where we have written $\rho_H \bmod \mathfrak{d}^+ = \left(\begin{smallmatrix} a & b \\ c & d \end{smallmatrix}\right)$ as before. As a consequence, the map (8.5) produces an element $\iota_A(f)$ in the extension group. The map ι_A is linear by the Yoneda interpretation of the addition of the extension group [MFG, §4.2.1]. It is injective because $\iota_A(f) = 0 \Leftrightarrow$ the extension is split $\Leftrightarrow f = 0$ in $\mathrm{Hom}_{W[[\Gamma_\wp]]}(\overline{\mathbb{T}}^-, A)$. $\square$

Recall that $\mathbb{F}$ is generated over $\mathbb{F}_p$ by the values of $\overline{\varphi}$ and therefore $W \supset \mathbb{Z}_p[\varphi^-]$.

Corollary 8.4.5. *Let the notation and assumption be as in Proposition 8.4.4. Then as $W[[\Gamma_\wp]]$-modules, $\mathcal{Y}(\varphi^-) \otimes_{\mathbb{Z}_p[\varphi^-]} W \cong \overline{\mathbb{T}}^-$.*

Proof. The extension class of $\iota_A(f)$ for $f \in \operatorname{Hom}_{W[[\Gamma_\wp]]}(\overline{\mathbb{T}}^-, A^\vee)$ is given by the upper triangular representation in (8.5). Since $\rho_{\mathbb{T}}(D_{\wp^\sigma})$ is upper triangular, the extension class of $\iota_A(f)$ is unramified outside $\wp$. By definition, $\operatorname{Ext}^1(\varphi, \varphi_\sigma^*) \cong H^1(H, (\varphi^-)^*)$, and hence $\iota(f)$ for $f \in \operatorname{Hom}_{W[[\Gamma_\wp]]}(\overline{\mathbb{T}}^-, W[[\Gamma_\wp]]^\vee)$ lands in $\operatorname{Sel}_K((\varphi^-)^*) = \operatorname{Sel}_{\mathbb{Q}}((\operatorname{Ind}_K^{\mathbb{Q}} \varphi^-)^*) = \operatorname{Hom}_{W[[\Gamma^-]]}(\mathcal{Y}(\varphi^-) \otimes_{\mathbb{Z}_p[\varphi^-]} W, (\varphi^-)^*)$. By the universality $R^{ord} = \mathbb{T}$, we find $\iota : \operatorname{Hom}_{W[[\Gamma_\wp]]}(\overline{\mathbb{T}}^-, W[[\Gamma_\wp]]^\vee) \to \operatorname{Hom}_{W[[\Gamma^-]]}(\mathcal{Y}(\varphi^-) \otimes_{\mathbb{Z}_p[\varphi^-]} W, (\varphi^-)^*)$ is an isomorphism. Note the exotic isomorphism $\Gamma_\wp = \Gamma^-$ in §7.4.3. Taking the Pontryagin dual back, we find $\overline{\mathbb{T}}^- \cong \mathcal{Y}(\varphi^-) \otimes_{\mathbb{Z}_p[\varphi^-]} W$.   □

8.4.3   *Local indecomposability in the real case*

Hereafter in this section, we assume that K is real. We start with

Proposition 8.4.6. *Assume* (H0–1), *and write* $\mathfrak{d} = \mathbb{T}([\alpha] - 1)\mathbb{T}$ *with* $\mathfrak{d}^+ = \mathfrak{d} \cap \mathbb{T}^+$. *Recall the universal character* $\varphi : H \to W[\Gamma_\wp]^\times$. *Let* $L := F(\varphi^-)$ *and* $W(\varphi^-)$ *be the rank one W-free module on which $\overline{H}$ acts by φ^-. Then we have a $W[\Gamma_\wp]$-linear surjection:* $Cl_L \otimes_{\mathbb{Z}[\overline{H}]} W(\varphi^-) \twoheadrightarrow \mathbb{T}^-/(\theta\mathbb{T}^+ + \mathfrak{d}^+\mathbb{T}^-)$; *so, $\mathbb{T}^-/(\theta\mathbb{T}^+ + \mathfrak{d}^+\mathbb{T}^-)$ is finite.*

Proof. Write $\overline{\theta}$ for the image of $\theta \in \mathbb{T}^-$ in $\mathbb{T}^-/\mathfrak{d}^+\mathbb{T}^-$. Since $\mathbb{T}^-$ is a $\mathbb{T}^+$-module and $\mathbb{T}^+/\mathfrak{d}^+ \cong W[\Gamma_\wp]$ (Corollary 7.4.2), we can think of $W[\Gamma_\wp]\overline{\theta} \subset \mathbb{T}^-/\mathfrak{d}^+\mathbb{T}^-$. The pull back by $\mathbb{T}^- \twoheadrightarrow \mathbb{T}^-/\mathfrak{d}^+\mathbb{T}^-$ of $W[\Gamma_\wp]\overline{\theta}$ is equal to $\theta\mathbb{T}^+ + \mathfrak{d}^+\mathbb{T}^-$. By conjugation, $\mathcal{Y}(\varphi^-) \otimes_{\mathbb{Z}_p[\varphi^-]} W$ is a $W[\Gamma_\wp]$-module, which gives the upper nilpotent part $\overline{\mathbb{T}}^-$ of the $\mathbb{T}^+$-GMA E_+ in which ρ_H has values. By Corollary 8.4.5, $\overline{\mathbb{T}}^- = \mathcal{Y}(\varphi^-) \otimes_{\mathbb{Z}_p[\varphi^-]} W$.

Let I be the wild inertia subgroup at $\wp$ in $\operatorname{Gal}(F(\rho_H)/F(\varphi))$. As in the proof of Theorem 8.2.3, we have an exact sequence $0 \to \mathcal{U} \to I \twoheadrightarrow \Gamma \to 1$ for the Galois group $\Gamma \cong t^{\mathbb{Z}_p}$ of the cyclotomic $\mathbb{Z}_p$-extension of $\mathbb{Q}_p$. Lift $\gamma \in \Gamma_\wp = \operatorname{Gal}(F(\varphi)/F(\varphi))$ to $\widetilde{\gamma} \in \operatorname{Gal}(F(\rho_H)/F(\varphi))$ so that $\rho(\widetilde{\gamma})$ is upper triangular. This is possible by the form of the $\mathbb{T}^+$-GMA E_+.

We let $\widetilde{\gamma}$ act on I and $\mathcal{U}$ by conjugation. The $W[\Gamma_\wp]$-module structure (through the identification $W[\Gamma_\wp] = \mathbb{T}^+/\mathfrak{d}^+$) of $\mathbb{T}^-/\mathfrak{d}^+\mathbb{T}^-$ is induced by the conjugation action of $\widetilde{\gamma}$ on the inertia groups, $\widetilde{\gamma}I\widetilde{\gamma}^{-1}$ and $\widetilde{\gamma}\mathcal{U}\widetilde{\gamma}^{-1}$ only depends on $\gamma \in \Gamma_\wp$; so, we write them as $X \mapsto \gamma X$. Note that $\gamma\overline{\theta}$ is the image in $\mathbb{T}^-/\mathfrak{d}^+\mathbb{T}^-$ of a generator (over Λ_1) of the unipotent part $\gamma\mathcal{U}$.

Write $\overline{\mathcal{U}}$ for the image of $\mathcal{U}$ in $\mathcal{Y}(\varphi^-)$. Consider the quotient

$$\mathcal{C} := (\mathcal{Y}(\varphi^-) \otimes_{\mathbb{Z}_p[\varphi_-]} W)/W[\Gamma_\wp]\overline{\mathcal{U}}$$

for $W[\Gamma_\wp]$-span $W[\Gamma_\wp]\overline{\mathcal{U}}$ of $\overline{\mathcal{U}}$. Then $\mathcal{C}$ is the scalar extension to W of the Galois group of the maximal p-abelian extension unramified everywhere of $L = F(\varphi^-)$ on which $\overline{H}$ acts by φ^-, that is, $Cl_L \otimes_{\mathbb{Z}[\overline{H}]} W(\varphi^-)$. Therefore we conclude $\mathcal{C} = Cl_L \otimes_{\mathbb{Z}[\overline{H}]} W(\varphi^-)$ which is finite. Since $W[\Gamma_\wp]\overline{\mathcal{U}}$ has image $(W[\Gamma_\wp]\overline{\theta} + \mathfrak{d}^+\mathbb{T}^-)/\mathfrak{d}^+\mathbb{T}^- = (\theta\mathbb{T}^+ + \mathfrak{d}^+\mathbb{T}^-)/\mathfrak{d}^+\mathbb{T}^-$ in $\mathbb{T}^-/\mathfrak{d}^+\mathbb{T}^-$, $\mathcal{C}$ surjects down to $\mathbb{T}^-/(\theta\mathbb{T}^+ + \mathfrak{d}^+\mathbb{T}^-)$, and therefore $\mathbb{T}^-/(\theta\mathbb{T}^+ + \mathfrak{d}^+\mathbb{T}^-)$ is finite. $\square$

Corollary 8.4.7. *Assume* (H0–1). *Then we have* $\theta\mathbb{T} \cap \Lambda = (\langle\varepsilon\rangle - 1)$.

Since $\theta \in \mathbb{T}^-$, $(\theta) \subset \mathfrak{d}$; so, $(\theta) \cap \Lambda \subset (\langle\varepsilon\rangle - 1)$ as $\mathbb{T}/\mathfrak{d} \cong \Lambda/(\langle\varepsilon\rangle - 1)$. Therefore we prove the vanishing: $(\langle\varepsilon\rangle - 1)/((\theta) \cap \Lambda) = 0$.

Proof. Since $\mathbb{T}^+/\mathfrak{d}^+ \cong W[\Gamma_\wp]$, $\theta\mathbb{T}^+$ has image $W[\Gamma_\wp]\overline{\theta}$ in $\mathbb{T}^-/\mathfrak{d}^+\mathbb{T}^-$, and $\mathbb{T}^-/(\theta\mathbb{T}^+ + \mathfrak{d}^+\mathbb{T}^-)$ is finite by Proposition 8.4.6.

Pick $\lambda \in \Lambda$ such that $\lambda \notin (\langle\varepsilon\rangle - 1)$ and $\Lambda/(\lambda) = W$. Let $X_\lambda := X[\frac{1}{\lambda}] = X \otimes_\Lambda \Lambda[\frac{1}{\lambda}]$ for $X = \mathbb{T}^\pm, \mathfrak{d}, \mathfrak{d}^\pm$. By $|\mathbb{T}^-/(\theta\mathbb{T}^+ + \mathfrak{d}^+\mathbb{T}^-)| < \infty$, we have $\mathbb{T}_\lambda^- = \theta\mathbb{T}_\lambda^+ + \mathfrak{d}^+\mathbb{T}_\lambda^-$. We want to prove $\mathbb{T}_\lambda^- = \theta\mathbb{T}_\lambda^+ + (\mathfrak{d}^+)^m\mathbb{T}_\lambda^-$ for all $m > 0$ by induction on m. Suppose $\mathbb{T}_\lambda^- = \theta\mathbb{T}_\lambda^+ + (\mathfrak{d}^+)^n\mathbb{T}_\lambda^-$. Then

$$\mathbb{T}_\lambda^- = \theta\mathbb{T}_\lambda^+ + (\mathfrak{d}^+)^n(\theta\mathbb{T}_\lambda^+ + \mathfrak{d}^+\mathbb{T}_\lambda^-) = \theta\mathbb{T}_\lambda^+ + (\mathfrak{d}^+)^{n+1}\mathbb{T}_\lambda^-.$$

Thus $\mathbb{T}_\lambda^- = \theta\mathbb{T}_\lambda^+ + (\mathfrak{d}^+)^n\mathbb{T}_\lambda^-$ for any $n > 0$, and we conclude

$$\mathbb{T}_\lambda^- = \bigcap_n(\theta\mathbb{T}_\lambda^+ + (\mathfrak{d}^+)^n\mathbb{T}_\lambda^-).$$

Consider $M := \mathbb{T}^-/\theta\mathbb{T}^+$ which is a $\mathbb{T}^+$-module of finite type. Then by Krull's intersection theorem [CRT, Theorem 8.9], $N := \bigcap_{n>0}(\mathfrak{d}^+)^n M$ has an element $a \equiv 1 \mod \mathfrak{d}^+$ such that $aN = 0$. Since $\mathfrak{d}^+$ is in the radical $\mathfrak{m}_{\mathbb{T}^+}$, a is a unit in $\mathbb{T}^+$; so, $N = 0$. This shows

$$\bigcap_n(\theta\mathbb{T}^+ + (\mathfrak{d}^+)^n\mathbb{T}^-) = \theta\mathbb{T}^+.$$

Since $\Lambda[\frac{1}{\lambda}]$ is Λ-flat, tensoring $\Lambda[\frac{1}{\lambda}]$ over Λ (i.e., localization by inverting λ) commutes with intersection [BCM, I.2.6]; so, we conclude the identity $(*)$ in the following equation:

$$\mathbb{T}_\lambda^- = \bigcap_n(\theta\mathbb{T}_\lambda^+ + (\mathfrak{d}^+)^n\mathbb{T}_\lambda^-) \overset{(*)}{=} \theta\mathbb{T}_\lambda^+. \tag{8.6}$$

Thus $\mathfrak{d}_\lambda = \mathfrak{d}\mathbb{T}_\lambda = \mathfrak{d}\mathbb{T}_\lambda^- = (\theta)$. This shows $\theta\mathbb{T}_\lambda \cap \Lambda_\lambda = \mathfrak{d}_\lambda \cap \Lambda_\lambda = (\langle\varepsilon\rangle - 1)\Lambda_\lambda$ as $\Lambda_\lambda = \Lambda[\frac{1}{\lambda}]$ is flat over Λ. Therefore $(\langle\varepsilon\rangle - 1)/((\theta) \cap \Lambda)$ is a λ-torsion module. Choosing a couple of λ's as above, say λ_1, λ_2, so that $(\lambda_1, \lambda_2) = \mathfrak{m}_\Lambda$, the Λ-module $(\langle \varepsilon \rangle - 1)/((\theta) \cap \Lambda)$ is λ_j-torsion for $j = 1, 2$ and hence is killed by $\mathfrak{m}_\Lambda^M$ for some $0 \ll M \in \mathbb{Z}$. Since $(\theta) \cap \Lambda$ is a reflexive Λ-module, the Λ-module $(\langle \varepsilon \rangle - 1)/((\theta) \cap \Lambda)$ does not have $\mathfrak{m}_\Lambda$ as an associated prime [BCM, VII.4.3]; so, we conclude $(\langle \varepsilon \rangle - 1)/((\theta) \cap \Lambda) = 0$. $\qquad\square$

We are now ready to prove

Theorem 8.4.8. *Assume* (H0–1) *and that K is real. If f is a Hecke eigenform belonging to $\mathbb{T}$ of weight $k \geq 2$, $\rho = \rho_{f,\mathfrak{p}}$ is indecomposable over I_p.*

Under (H0–2), this follows from the fact that $(\Theta) \cap \Lambda = (\langle \varepsilon \rangle - 1)$, and for any height 1 prime P outside $(\langle \varepsilon \rangle - 1)$, $\Theta \mod P \neq 0$, and hence $\theta \mod P \neq 0$ by Theorem 8.4.3. We remove now the assumption (H2):

Proof. By Theorem 8.2.3, the restriction $\rho|_{I_p}$ is decomposable if and only if $P|(\theta)$, which is equivalent to $P \nmid (\langle \varepsilon \rangle - 1)$ by Corollary 8.4.7. This implies that $\theta \not\equiv 0 \mod P$ if P is outside $V((\langle \varepsilon \rangle - 1)) = \mathrm{Spec}(\mathbb{T}/(\langle \varepsilon \rangle - 1))$. Thus local decomposable points are limited to weight 1 points associated to an induced representation from K. $\qquad\square$

Remark 8.4.9.

- Actually we can prove $\Lambda[\theta] \subset \mathbb{T}$ is an integral domain fully ramified over $(\langle \varepsilon \rangle - 1)$ similar to the structure theorem in Theorem 8.3.7 under (H0–1) [H21, Corollary 10.4].
- Indecomposability in Cases U$_-$ and Ds is still an open question.
- The ring $\mathbb{T}^+$ is a universal trace ring deforming $\mathrm{Tr}(\overline{\rho})$ under (H0-2). The details can be found in [H21, Theorem C].

8.5   Local indecomposability for imaginary K

In this last section of this chapter, we describe a proof of the indecomposability theorem for non-CM modular deformations of an induction from an imaginary quadratic field $K = \mathbb{Q}[\sqrt{-D}]$ with discriminant $D > 0$. Under some additional hypothesis, the result is proven in [CW19] after some earlier work [GV04]. Since this requires the solution of the anticyclotomic main conjecture via the theory of Hecke algebras ([T89], [H06] and [H09]) whose proof we do not touch upon, this section is rather sketchy. Though the anticyclotomic main conjecture was proven by Rubin earlier (e.g., [R91]) by an Euler system argument, what is relevant in this section is its relation to the adjoint Selmer group discussed in the above three papers. The proof proceeds in the same way as in the real induced case and therefore is a bit different from the one given in the main text of [CW19].

As in the real case, in addition to p-distinguishedness (H0), we assume to be in Case U_+ or R_+ that the prime $p > 2$ splits into $\wp\wp^\sigma$ in K and that φ is a character of $Cl_K(\mathfrak{f})$ of conductor $\mathfrak{f}$ with $\wp^\sigma \nmid \mathfrak{f}$ and of order prime to p (i.e., (H1)). If $\rho = \mathrm{Ind}_K^{\mathbb{Q}} \varphi$ is not induced from $\mathbb{Q}[\sqrt{p^*}]$, the Taylor–Wiles condition is satisfied, and we have $R_{\mathfrak{p}}^{ord} = \mathbb{T}_{\mathfrak{p}}$ for the universal ordinary deformation ring $R_{\mathfrak{p}}^{ord}$ of $\bar\rho := \rho \mod \mathfrak{p}$ (for a prime $\mathfrak{p}|p$ of $\mathbb{Z}[\varphi]$).

8.5.1 *CM and non-CM components*

Recall $\mathfrak{d} := \mathbb{T}([\alpha]-1)\mathbb{T}$. Then by Theorem 7.4.1, $\mathbb{T}/\mathfrak{d} \cong W[[\Gamma_\wp]] \cong W[[\Gamma^-]]$. In this imaginary case, $\mathrm{rank}_{\mathbb{Z}_p} \Gamma_\wp = 1$, and hence $\dim W[[\Gamma_\wp]] = 2$. Because of this, $\mathrm{Spec}(W[[\Gamma_\wp]]) \subset \mathrm{Spec}(\mathbb{T})$ is a union of CM irreducible components.

Though in the proof of Theorem 7.4.1, we used the identity $R^{ord} = \mathbb{T}$, we do not need $R^{ord} = \mathbb{T}$ to have ring $W[[\Gamma_\wp]]$ as a quotient of $\mathbb{T}$. Here is the proof of this fact: We have Hecke's theta series $f_\phi = \sum_{\mathfrak{a}} \phi(\mathfrak{a})q^{N(\mathfrak{a})}$ associated to $\mathrm{Ind}_F^{\mathbb{Q}} \phi$ for any p-adic algebraic Hecke character ϕ of $\Gamma_\wp$. By this fact, we have a CL_W morphism $\pi : \mathbb{T}/\mathfrak{d} \to W[[\Gamma_\wp]]$ such that $\pi \circ \rho_{\mathbb{T}} \cong \mathrm{Ind}_K^{\mathbb{Q}} \varphi$ for the universal character $\varphi : \Gamma_\wp \hookrightarrow W[[\Gamma_\wp]]^\times$. It is an easy exercise to prove from absolute irreducibility of $\bar\rho$ (e.g. [MFG, §2.1.7]) that $W[[\Gamma_\wp]]$ is generated over Λ by $\mathrm{Tr}(\mathrm{Ind}_K^{\mathbb{Q}} \varphi)(g)$ for $g \in G$, and hence π is a surjection. Therefore $\mathrm{Tr}(\rho)(g) \mapsto \mathrm{Tr}(\mathrm{Ind}_K^{\mathbb{Q}} \varphi)(g)$ factors through the universal map $R^{ord} \twoheadrightarrow \mathbb{T}$ inducing a surjection

$$W[[\Gamma_\wp]] \cong R^{ord}/R^{ord}([\alpha] - 1)R^{ord} \twoheadrightarrow \mathbb{T}/\mathfrak{d} \twoheadrightarrow W[[\Gamma_\wp]],$$

which is an isomorphism as $W[[\Gamma_\wp]]$ is noetherian. $\qquad\square$

Since $\mathbb{T}$ is a reduced algebra free of finite rank over Λ (by Theorem 6.2.17), we have a complementary quotient $\mathbb{T}^{ncm} = \mathbb{T}_{\mathfrak{p}}^{ncm}$ of $\mathbb{T}$ such that $\mathbb{T} \hookrightarrow \mathbb{T}^{cm} \times \mathbb{T}^{ncm}$ with Λ-torsion congruence module $C_0(\varphi^-) := (\mathbb{T}^{cm} \oplus \mathbb{T}^{ncm})/\mathbb{T}$. Since $[\alpha]$ acts trivially on $\mathbb{T}^{cm} = \mathbb{T}/\mathfrak{d} \cong W[[\Gamma_\wp]]$, defining $\mathbb{T}_\pm^{ncm} := \{x \in \mathbb{T}^{ncm}|[\alpha](x) = \pm x\}$, we have $\mathbb{T}^- = (0 \oplus \mathbb{T}_-^{ncm}) \cap \mathbb{T}$; so, $\mathfrak{d} = \mathfrak{d}_{\mathbb{T}} = \mathbb{T}([\alpha] - 1])\mathbb{T} = \mathbb{T}\mathbb{T}^- = (0 \oplus \mathbb{T}^{ncm}) \cap \mathbb{T}$. Put $\mathfrak{a} := (\mathbb{T}^{cm} \oplus 0) \cap \mathbb{T}$. By (5.1), we have

$$C_0(\varphi^-) = \mathbb{T}^{cm}/\mathfrak{a} \cong \mathbb{T}^{ncm}/\mathfrak{d}. \tag{8.7}$$

If the conditions (H0–2) hold, by Corollary 8.3.6, we have $\mathfrak{d} = (\Theta)$ with $[\alpha](\Theta) = -\Theta$. The decomposition: $\mathbb{T} \hookrightarrow \mathbb{T}^{cm} \times \mathbb{T}^{ncm}$ with Λ-torsion quotient $C_0(\varphi^-) := (\mathbb{T}^{cm} \times \mathbb{T}^{ncm})/\mathbb{T}$ implies that $\Theta \in \mathbb{T}^{ncm}_- = \mathbb{T}^-$. By (8.7), we have a canonical isomorphism

$$C_0(\varphi^-) \cong \mathbb{T}^{ncm}/(\Theta). \tag{8.8}$$

Theorem 8.5.1. *Assume* (H0–2) *and that* K *is an imaginary quadratic field. Then, if* $\mathbb{T}^{ncm} = 0$, *we have* $\Theta = 0$, *and otherwise,* Θ *is a zero-divisor of* $\mathbb{T}$ *but a non-zero-divisor of* $\mathbb{T}^{ncm}$ *and for each* W*-algebra homomorphism* $\lambda : \mathbb{T}^{ncm} \to \overline{\mathbb{Q}}_p$ *associated to a Hecke eigenform of weight* $k \geq 2$, $\lambda(\Theta) \neq 0$.

In [CW19], this theorem is proven under some extra hypothesis.

Proof. We only need to prove $\lambda(\Theta) \neq 0$ (assuming $\mathbb{T}^{ncm} \neq 0$) as all the other assertions are clear from the explanation given before the proposition. The isomorphism $\mathbb{Z}_p \cong O_{K,\wp}$ induces an inclusion $\Gamma \hookrightarrow \Gamma_\wp$. In this way $W[[\Gamma_\wp]]$ is a Λ-algebra. From the exact sequence: $\mathfrak{d} \hookrightarrow \mathbb{T} \twoheadrightarrow W[[\Gamma_\wp]]$ combined with Λ-freeness of $W[[\Gamma_\wp]]$, $\mathfrak{d}$ is a Λ-direct summand of $\mathbb{T}$, and hence Λ-free. By Tate's Theorem 6.2.26 (1), $\mathbb{T}$ is self Λ-dual under the trace pairing $(x, y) = \mathrm{Tr}_{\mathbb{T}/\Lambda}(xy)$. Note that $\mathrm{Hom}_\Lambda(\mathfrak{d}, \Lambda) = \mathbb{T}^{ncm}$. In particular,

$$\mathbb{T}^{ncm} \text{ is } \Lambda\text{-free of finite rank.} \tag{8.9}$$

Let $\chi := \det(\lambda \circ \rho_{\mathbb{T}})$. Since $\mathbb{T}_\chi$ is reduced, $\mathbb{T}_\chi \twoheadrightarrow W[[\Gamma_\wp]]/(t - \chi(\gamma)) = \mathbb{T}^{cm}/(t - \chi(\gamma))\mathbb{T}^{cm}$ is split after tensoring $\mathrm{Frac}(W)$, and we have an exact sequence $i_\chi : \mathbb{T}_\chi \hookrightarrow \mathbb{T}^{cm}_\chi \times \mathbb{T}^{ncm}_\chi$ for $\mathbb{T}^{ncm}_\chi := \mathbb{T}^{ncm}/(t - \chi(\gamma))\mathbb{T}^{ncm}$. By the Λ-freeness of $\mathbb{T}^{ncm}$, $\mathbb{T}^{ncm}/(t - \chi(\gamma))\mathbb{T}^{ncm}$ is W-free of rank equal to $\mathrm{rank}_\Lambda \mathbb{T}^{ncm}$. Then the natural onto map $\mathbb{T}^{ncm}/(t - \chi(\gamma))\mathbb{T}^{ncm} \twoheadrightarrow \mathbb{T}^{ncm}_\chi$ has the domain and target both W-free of equal rank; so, we have an isomorphism $\mathbb{T}^{ncm}_\chi \cong \mathbb{T}^{ncm}/(t - \chi(\gamma))\mathbb{T}^{ncm}$. Then the source and the target of i_χ are both W-free of equal rank; so,

$$\mathrm{Coker}(i_\chi) \cong C_0(\varphi^-) \otimes_\Lambda \Lambda/(t - \chi(\gamma)) = \mathbb{T}^{ncm}/(\Theta) \otimes_\Lambda \Lambda/(t - \chi(\gamma))$$

is finite. Since we have a natural surjection: $C_0(\varphi^-) \otimes_\Lambda \Lambda/(t - \chi(\gamma)) \twoheadrightarrow W/(\lambda(\Theta))$, $W/\lambda(\Theta))$ is finite, and hence $\lambda(\Theta) \neq 0$. $\square$

The short exact sequence $\mathfrak{a} \hookrightarrow \mathbb{T} \twoheadrightarrow \mathbb{T}^{ncm}$ for $\mathfrak{a}$ in (8.7) is the Λ-dual of $\mathfrak{d} \hookrightarrow \mathbb{T} \twoheadrightarrow W[[\Gamma_\wp]]$ by the self-dual pairing of Theorem 6.2.26 (1), we find $\mathfrak{a} \cong \mathrm{Hom}_\Lambda(W[[\Gamma_\wp]], \Lambda)$. Since $\Gamma_\wp \cong \mathbb{Z}_p \times \overline{\Gamma}_\wp$ with $\Gamma \hookrightarrow \mathbb{Z}_p$ for a finite p-group $\overline{\Gamma}_\wp$, identifying $W[[\mathbb{Z}_p]]$ with the regular Λ-algebra $A := W[[X]]$, $W[[\Gamma_\wp]] = A[\overline{\Gamma}_\wp]$ is a local complete intersection over A and hence over Λ (e.g., §5.1.3). Again by Theorem 6.2.26 (1), $\mathfrak{a} \cong W[[\Gamma_\wp]]$ as $\mathbb{T}$-modules. If $p \nmid h_K$, $\Gamma_\wp = \Gamma$ and $W[[\Gamma_\wp]] = \Lambda = W[[T]]$; so, one has the characteristic power series $\mathrm{char}_\Lambda(C_0(\varphi^-))$. Thus we get

Lemma 8.5.2. *Writing* $C_0(\varphi^-) = \mathbb{T}^{cm}/\mathfrak{a}$, *the ideal* $\mathfrak{a}$ *is principal, and if* $p \nmid h_K$, $\mathfrak{a}$ *is generated by* $\mathrm{Fitt}_\Lambda(C_0(\varphi^-)) = \mathrm{char}_\Lambda(C_0(\varphi^-))$.

8.5.2 Local indecomposability in the CM case

As before, we write $\overline{G} := \mathrm{Gal}(F(\overline{\rho})/\mathbb{Q})$ and $\overline{D}$ for p-decomposition subgroup of $\overline{G}$. Let $\rho_{\mathbb{T}^{ncm}} = \pi^{ncm} \circ \rho_{\mathbb{T}}$ and $\rho_H :== \rho_{\mathbb{T}^{ncm}}|_H$ for the projection $\pi^{ncm} : \mathbb{T} \twoheadrightarrow \mathbb{T}^{ncm}$ and $\mathbb{T}^{ncm}_{\pm} := \{x \in \mathbb{T}^{ncm} | [\alpha](x) = \pm x\}$ (which is a $\mathbb{T}^{ncm}_+$-module). Here is a non-CM version of Proposition 8.4.1:

Proposition 8.5.3. *Assume* (H0–1). *After conjugating* $\rho_{\mathbb{T}^{ncm}}$ *by an element in* $\mathrm{GL}_2(\mathbb{T}^{ncm})$, *we can arrange* ρ_H *to have values in the* $\mathbb{T}^{ncm}_+$-*GMA*

$$E^{ncm}_+ := \begin{pmatrix} \mathbb{T}^{ncm}_+ & \mathbb{T}^{ncm}_- \\ \mathbb{T}^{ncm}_- & \mathbb{T}^{ncm}_+ \end{pmatrix}$$

so that $\rho_{\mathbb{T}}(\delta) = \mathrm{diag}[\varphi(\delta), \varphi_\sigma(\delta)]$ *for* $\delta \in \overline{D}$, *where the product law:* $\mathbb{T}^{ncm}_- \otimes_{\mathbb{T}_+} \mathbb{T}^{ncm}_- \to \mathbb{T}^{ncm}_+$ *is given by* $b \otimes c = bc$ *with the product* bc *in* $\mathbb{T}^{ncm}$.

Proof. We follow the proof given for Proposition 8.4.1. The involution $[\alpha]$ on $\mathbb{T}$ induces an involution α^{ncm} on $\mathbb{T}^{ncm}$ as $\mathrm{Ker}(\pi^{ncm}) \subset \mathbb{T}^{cm}$ is fixed by $[\alpha]$. Then as seen in §8.3.3, the involution α^{ncm} is induced by the action on $\rho \in \mathcal{D}(\mathbb{T}^{ncm})$ given by $J(\rho_{\mathbb{T}^{ncm}} \otimes \alpha)J^{-1} \sim \rho^{\alpha^{ncm}}_{\mathbb{T}^{ncm}}$ for $J = \begin{pmatrix} 1 & 0 \\ 0 & -1 \end{pmatrix}$. Recall $\partial_{\mathbb{T}} = \mathbb{T}([\alpha] - 1)\mathbb{T}$. Since $[\alpha]$ acts trivially on $\mathbb{T}^{cm}$, we have $\mathbb{T}^{ncm}_- = \mathbb{T}^-$ and $\partial_{\mathbb{T}} \subset \mathbb{T}^{ncm}$. Since $J \begin{pmatrix} a & b \\ c & d \end{pmatrix} J^{-1} = \begin{pmatrix} a & -b \\ -c & d \end{pmatrix}$ and $(\rho_{\mathbb{T}^{ncm}}|_H \mod \partial_{\mathbb{T}}) = \varphi \oplus \varphi_\varsigma$ is diagonal by Theorem 7.4.1, we have $uJ(\rho_{\mathbb{T}^{ncm}} \otimes \alpha)(uJ)^{-1} = \rho^{\alpha^{ncm}}_{\mathbb{T}^{ncm}}$ with $u \in 1 + \partial_{\mathbb{T}} M_2(\mathbb{T}^{ncm})$. Writing $U = uJ$ and applying α^{ncm}, we get $U^{\alpha^{ncm}}(\rho^{\alpha^{ncm}}_{\mathbb{T}^{ncm}} \otimes \alpha)U^{-\alpha^{ncm}} = \rho_{\mathbb{T}^{ncm}}$; so, we have

$$U\rho_{\mathbb{T}^{ncm}}U^{-1} = U(\rho_{\mathbb{T}^{ncm}} \otimes \alpha)U^{-1} \otimes \alpha = \rho^{\alpha^{ncm}}_{\mathbb{T}^{ncm}} \otimes \alpha = U^{-\alpha^{ncm}}\rho_{\mathbb{T}^{ncm}}U^{\alpha^{ncm}}.$$

Thus $U^{\alpha^{ncm}}U = Ju^{\alpha^{ncm}}Ju = z \in Z := 1 + \partial_{\mathbb{T}}\mathbb{T}^{ncm}$. Since $1 + \partial_{\mathbb{T}} M_2(\mathbb{T}^{ncm})$ is p-profinite, letting α^{ncm} act on $1 + \partial_{\mathbb{T}} M_2(\mathbb{T}^{ncm})$ by $x \mapsto x^{\overline{\alpha^{ncm}}} := Jx^{\alpha^{ncm}}J$, we can write $u = v^{\overline{\alpha^{ncm}}-1} \in (1 + \partial_{\mathbb{T}} M_2(\mathbb{T}^{ncm}))$ for $v \in 1 + \partial_{\mathbb{T}} M_2(\mathbb{T}^{ncm})$. Replacing $\rho_H = \rho_{\mathbb{T}^{ncm}}|_H$ by $v^{-1}J\rho_H Jv$, we find $J\rho_H J^{-1} = \rho^{\alpha^{ncm}}_H$; so, ρ_H has values in $E^{ncm}_+ = \begin{pmatrix} \mathbb{T}^{ncm}_+ & \mathbb{T}^{ncm}_- \\ \mathbb{T}^{ncm}_- & \mathbb{T}^{ncm}_+ \end{pmatrix}$. $\square$

For the generator θ of the unipotent part $\mathcal{U}$ in Theorem 8.2.3, here is a version of Theorem 8.4.3:

Corollary 8.5.4. *Under* (H0–2), *the quotient* θ/Θ *is a unit in* $\mathbb{T}^{ncm}$.

We need to modify a bit the argument proving Theorem 8.4.3.

Proof. Note $\mathfrak{d} = \mathfrak{d}_{\mathbb{T}} = (\Theta) \subset \mathbb{T}^{ncm}$ and $\mathfrak{d}^+ = (\Theta^2) \subset \mathbb{T}^{ncm}_+$. We have an exact sequence $(\Theta) \hookrightarrow \mathbb{T} \twoheadrightarrow W[[\Gamma_\wp]] = \mathbb{T}^{cm}$ by Corollary 7.4.2. Taking the $[\alpha]$-invariant subspace (indicated superscript "$+$"), $\mathbb{T}^+/\mathfrak{d}^+ \cong W[[\Gamma_\wp]]$ for $\Gamma_\wp = Cl_K(\wp^\infty) \otimes_{\mathbb{Z}} \mathbb{Z}_p$ with $F = F(\overline{\varphi}^-)$. Recall the universal character $\varphi : H \to W[\Gamma_\wp] = \mathbb{T}^+/\mathfrak{d}^+$ unramified outside $\wp\mathfrak{f}$ deforming $\overline{\varphi}$. Let $C_0 := \mathbb{T}^{cm} \otimes_{\mathbb{T}} \mathbb{T}^{ncm} \cong \mathbb{T}^{ncm}/\mathfrak{d} \cong \mathbb{T}^{cm}/\mathfrak{a}$ for the congruence module for the projection $\mathbb{T} \twoheadrightarrow \mathbb{T}^{cm}$ (e.g. (5.1)). Since $[\alpha]$ acts trivially on C_0, taking the fixed part of $[\alpha]$, we also get $C_0 := \mathbb{T}^{cm} \otimes_{\mathbb{T}^+} \mathbb{T}^{ncm}_+ \cong \mathbb{T}^{ncm}_+/\mathfrak{d}^+ \cong \mathbb{T}^{cm}/\mathfrak{a}$.

Write $\rho_H = \rho_{\mathbb{T}^{ncm}}|_H = \left(\begin{smallmatrix} A & B \\ C & D \end{smallmatrix} \right) \subset E^{ncm}_+$ for the $\mathbb{T}^+$-GMA E^{ncm}_+ in Proposition 8.5.3 and put $a = A \bmod \mathfrak{d}^+ = \varphi \bmod \mathfrak{a} : H \to C_0^\times$, $d = D \bmod \mathfrak{d}^+ = \varphi_\sigma \bmod \mathfrak{a} : H \to C_0^\times$, $b = B \bmod \mathfrak{d}^+ : H \to \mathbb{T}^-/\mathfrak{d}^+\mathbb{T}^-$ and $c = C \bmod \mathfrak{d}^+ : H \to \mathbb{T}^-/\mathfrak{d}^+\mathbb{T}^-$. As we already remarked, we have $\mathbb{T}^- = \mathbb{T}^{ncm}_-$. Thus b and c have values in $\mathbb{T}^{ncm}_-/\mathfrak{d}^+\mathbb{T}^{ncm}_-$.

If b has image in $\mathfrak{m}_{\mathbb{T}^{ncm}_+}(\mathbb{T}^{ncm}_-/\mathfrak{d}^+\mathbb{T}^{ncm}_-)$, by $c(g) = \varphi(\sigma^2)b(\sigma^{-1}g\sigma)$, c has values also in $\mathfrak{m}_{\mathbb{T}^{ncm}_+}(\mathbb{T}^{ncm}_-/\mathfrak{d}^+\mathbb{T}^{ncm}_-)$. This implies $\rho_H \bmod \mathfrak{m}_{\mathbb{T}^{ncm}_+}\mathfrak{d}^+$ is diagonal; so, $\rho_H^{[\alpha]} = J\rho_H J^{-1}$ which implies $\rho_{\mathbb{T}^{ncm}}^{[\alpha]} \bmod \mathfrak{m}_{\mathbb{T}^{ncm}}\mathfrak{d} \cong \rho_{\mathbb{T}^{ncm}} \otimes \alpha$, a contradiction as $\mathfrak{d}$ is the ideal maximal among the ideals for which the identity holds. Thus b is onto.

Note that $\rho_H = \rho_{\mathbb{T}^{ncm}}|_H$ has values in $GL_2(\mathbb{T}^{ncm})$ and Θ is a non-zero-divisor of $\mathbb{T}^{ncm}$ by Theorem 8.5.1. Replacing ρ_H by $\rho' := \xi^{-1}\rho_H\xi$ for $\xi := \left(\begin{smallmatrix} \Theta & 0 \\ 0 & 1 \end{smallmatrix} \right)$, ρ' has values in $GL_2(\mathbb{T}^{ncm}_+)$ and $\rho' \bmod \mathfrak{m}_{\mathbb{T}^{ncm}_+} = \left(\begin{smallmatrix} \overline{\varphi} & \overline{b} \\ 0 & \overline{\varphi}_\sigma \end{smallmatrix} \right)$ with $\overline{b} = b/\Theta \bmod \mathfrak{m}_{\mathbb{T}^{ncm}_+} \neq 0$. If u is a non-unit, b is unramified at $\wp$ (which is unramified also at $\wp^\sigma$); so, everywhere unramified over $F(\overline{\varphi}^-)$. Since $b(hgh^{-1}) = \varphi^-(h)b(g)$ for $h \in H$ and $g \in G_F$, $\overline{b}$ factors through $Cl_{F(\overline{\varphi}^-)} \otimes_{\mathbb{Z}[H]} \overline{\varphi}^-$. This is a contradiction against (H2): $Cl_{F(\overline{\varphi}^-)} \otimes_{\mathbb{Z}[H]} \overline{\varphi}^- = 0$ and finishes the proof. $\qquad\square$

This combined with Theorem 8.5.1 proves

Corollary 8.5.5. *Assume* (H0–2) *and that K is an imaginary quadratic field. Then for each deformation ρ of $\mathrm{Ind}_K^{\mathbb{Q}} \overline{\varphi}$ associated to a non-CM Hecke eigenform of weight $k \geq 2$, $\rho|_{I_p}$ is indecomposable.*

When K is real, we succeeded in removing the assumption (H2) for the indecomposability and proved $\lambda(\theta) \neq 0$ for λ as in Theorem 8.5.1. To see if the techniques for K real works in the imaginary case, let $\mathcal{L}/F(\varphi^-)$ for the universal character $\varphi : H \to W[[\Gamma_\wp]]^\times$ denote the maximal p-abelian extension unramified everywhere, and write $\mathbb{Y} := \mathrm{Gal}(\mathcal{L}/F(\varphi^-))$.

The key ingredient of the proof of the removal of (H2) in the real case is the finiteness of $\mathbb{Y}[\varphi^-] = \mathbb{Y} \otimes_{\mathbb{Z}[\overline{H}]} \varphi^-$ in Proposition 8.4.6 (regarding $H = \overline{H} \ltimes \mathrm{Gal}(F^{(p)}(\overline{p})/F(\overline{\varphi})))$.

When K is imaginary, as seen in §7.6.5, under $(\varphi^-)^2 = 1$, the finiteness of $\mathbb{Y}[\varphi^-] = \mathbb{Y} \otimes_{\mathbb{Z}[\overline{H}]} \varphi^-$ fails, but for a special reason we describe now. In §7.6.5, we started with a real quadratic field K and under the assumption $(\varphi^-)^2 = 1$, we find an imaginary quadratic field M with a character $\phi :$ $\mathrm{Gal}(\overline{\mathbb{Q}}/M) \to W^\times$ such that $\mathrm{Ind}_M^\mathbb{Q} \phi \cong \mathrm{Ind}_K^\mathbb{Q} \varphi$. The setting is the reverse here; so, we interchange the role of K and M there. Our set-up is:

- $K/\mathbb{Q}$ is imaginary with splitting $p = \wp \wp^\sigma$ and quadratic φ^-;
- M is a unique real quadratic field such that $F(\varphi^-) = KM$ in which p is either inert or ramified.

Under this circumstance, we have proven in §7.6.5 that $\mathbb{Y}[\varphi^-]$ has a quotient isomorphic to $\mathbb{Z}_p$ on which $\Gamma_\wp$ acts trivially. Thus $T|(\Lambda \cap (\Theta))$. However this is a weight 1 prime, therefore, the "weight ≥ 2" requirement is necessary in Conjecture 8.2.1. We discuss what to expect about the index $[\mathcal{L} : F(\varphi^-)]$ after a brief introduction of the Katz–Yager p-adic L function.

8.5.3 *Katz–Yager p-adic L-function*

To introduce the Katz p-adic L-function (as in [EAI, Chapter 9]), we choose an algebraic closure $\overline{\mathbb{F}}_p$ of $\mathbb{F}_p$ and regard $\mathbb{F} \subset \overline{\mathbb{F}}_p$. Then we have the ring of Witt vectors $W(\overline{\mathbb{F}}_p) \subset \mathbb{C}_p$. We fix an embedding $i_p : \overline{\mathbb{Q}} \hookrightarrow \mathbb{C}_p$ so that i_p is continuous on K under $\wp$-adic topology. Put $\mathcal{W} := i_p(\overline{\mathbb{Q}}) \cap W(\overline{\mathbb{F}}_p)$. We fix an embedding $i_\infty : \overline{\mathbb{Q}} \hookrightarrow \mathbb{C}$ and regard $K \subset \mathbb{C}$ by i_∞. Choose an elliptic curve $E_{/\mathcal{W}}$ with complex multiplication by O_K such that $E(\mathbb{C}) = \mathbb{C}/i_\infty(O_K)$ under $i_\infty \circ i_p^{-1} : \mathcal{W} \hookrightarrow \mathbb{C}$. Since E is defined over $\mathcal{W}$, we have $H^0(E, \Omega_{E/\mathcal{W}}) \cong \mathcal{W}\omega$ for a differential ω. By the identification, $E(\mathbb{C}) = \mathbb{C}/i_\infty(i_p^{-1}(O_K))$, writing the variable of $\mathbb{C}$ as u, we have $\Omega i_{\infty,*}\omega = du$ with $\Omega_\infty \in \mathbb{C}^\times$. We fix an isomorphism of the formal group $\widehat{E}_{/W(\overline{\mathbb{F}}_p)} \cong \widehat{\mathbb{G}}_m = \mathrm{Spf}(W(\overline{\mathbb{F}}_p)[[T]])$ and define $\Omega_p \in W(\overline{\mathbb{F}}_p)^\times$ by $\frac{dt}{t} = \Omega_p i_{p,*}\omega$ $(t = 1 + T)$.

The interpolation property of the anticyclotomic Katz p-adic L-function $L_{\varphi^-} \in W(\overline{\mathbb{F}}_p)[[\Gamma^-]]$ of branch character φ^- is given as follows: Take a Hecke ideal character λ with values in $i_\infty(i_p^{-1}(\mathcal{W}))$ of weight $0 < k \in \mathbb{Z}$ such that $\lambda(\mathfrak{a}) \equiv \varphi(\mathfrak{a}) \mod \mathfrak{m}_W$ for each K-fractional ideal $\mathfrak{a}$ prime to $\mathfrak{f}p$ and $\lambda((\alpha)) = \alpha^k$ for $\alpha \in K$ with $\alpha \equiv 1 \mod^\times p$. Since $\mathcal{I} := \{[\mathfrak{a}] \in Cl_K(\mathfrak{f}p^\infty)\}$ for $\mathfrak{a}$ as above is a dense subgroup of $Cl_K(\mathfrak{f}p^\infty)$, $\lambda^-(\mathfrak{a}) = \lambda(\mathfrak{a})\lambda(\mathfrak{a}^\sigma)^{-1}$

is continuous on $\mathcal{I}$ under the topology induced from the profinite group $Cl_K(\mathfrak{f}p^\infty)$; so, it extends to $\widehat{\lambda}^-: Cl_K(\mathfrak{F}p^\infty) \to W(\overline{\mathbb{F}}_p)^\times$ for $\mathfrak{F} = \mathfrak{f} \cap \mathfrak{f}^\sigma$. This $\widehat{\lambda}^-|_{\Gamma^-}$ induces a $W(\overline{\mathbb{F}}_p)$-algebra homomorphism $\widehat{\lambda}^-: W(\overline{\mathbb{F}}_p)[[\Gamma^-]] \to \overline{\mathbb{Q}}_p$. For any $W(\overline{\mathbb{F}}_p)$-algebra homomorphism $\phi: W(\overline{\mathbb{F}}_p)[[\Gamma^-]] \to \overline{\mathbb{Q}}_p$, we define $L_{\varphi^-}(\phi) := \phi(L_{\varphi^-})$. Then for any $k > 0$

$$i_p^{-1}\left(\frac{L_{\varphi^-}(\widehat{\lambda}^-)}{W_p(\lambda^-)\Omega_p^{2k}}\right) = i_\infty^{-1}\left(\frac{|O_K^\times|\pi^k\Gamma(k)}{2\sqrt{D}^k\Omega_\infty^{2k}}(1 - \lambda^-(\wp^\sigma))(1 - \frac{1}{\lambda^-(\wp^\sigma)p})L(0,\lambda^-)\right).$$
(8.10)

Here $W_p(\lambda^-)$ is the Gauss sum defined in [EAI, (9.1.9)], and if $\mathfrak{a}$ has a common factor with $N(\mathfrak{F})$, we agree to put $\lambda^-(\mathfrak{a}) = 0$.

Actually, Katz in [K78] (and [HT93]) constructs a measure associated to each CM type Σ chosen and the measure in (8.10) is associated to the identity embedding $i_\infty: K \hookrightarrow \mathbb{C}$. If we choose the other CM type $i_\infty \circ \sigma$, the corresponding measure is the push forward of the action of the complex conjugation on $Cl_K(\mathfrak{F}p^\infty)$. Note that integration of a character ϕ under the measure associated to i_∞ is equal to that of ϕ^{-1} under the measure relative to $i_\infty \circ \sigma$. Since $L(0,\lambda^-) = L(0,\lambda^- \circ \sigma) = L(0,(\lambda^-)^{-1})$, the choice does not matter much for anticyclotomic characters. Yager [Y82] uses Katz measure associated to $i_\infty \circ \sigma$.

As described in [EAI, page 374], the power series L_{φ^-} is well defined independent of choices of all the identifications: $\widehat{E} \cong \widehat{\mathbb{G}}_m$, ω and $E(\mathbb{C}) = \mathbb{C}/i_\infty(O_K)$. What we need here is that this power series L_{φ^-} plays the role of "$\langle\varepsilon\rangle - 1$" in the case of real quadratic field K.

Write $\pi: \mathbb{T} \twoheadrightarrow \mathbb{T}^{cm}$ for the projection. Here is a theorem which is a combination of the results of [T89], [H06] and [H09].

Theorem 8.5.6. *Assume* $R^{ord} = \mathbb{T}$, *(H0–1) and* $p \nmid h_K$. *Let*

$$C_1(\varphi^-) := \Omega_{R^{ord}/\Lambda} \otimes_{R^{ord}} \mathbb{T}^{cm} \quad and \quad C_0(\varphi^-) := C_0(\pi; \mathbb{T}^{cm}).$$

Then $\mathrm{Fitt}_\Lambda(C_0(\varphi^-)) = \mathrm{Fitt}_\Lambda(C_1(\varphi^-)) = (L_{\varphi^-})$, *where we have taken* $W = W(\overline{\mathbb{F}}_p)$ *and we identify* $\Lambda = W[[\Gamma^-]]$.

The assumption: $p \nmid h_K$ is not necessary. But to get a correct statement, we need to replace Λ by the Λ-algebra $W[[\Gamma^-]] = \mathbb{T}^{cm}$ which is a reduced complete intersection but not a domain; so, we need to use the stronger version of Tate's theorem in §6.2.8 and the theory of Fitting ideals both valid for non-integral domains. In this book, we just assume $p \nmid h_K$.

Each irreducible component $\mathrm{Spec}(\mathbb{I})$ of $\mathrm{Spec}(W[[\Gamma^-]])$ is associated to ϕ^- for a finite order character ϕ of $Cl_K(\mathfrak{f}\wp^\infty)/\Gamma$ with $\phi \bmod \mathfrak{m}_W = \overline{\varphi}$ for the natural inclusion $\Gamma \subset Cl_K(\mathfrak{f}\wp^\infty)$ (so, if $p|h_K$, the order of ϕ^- can be divisible by p). In other words, the character $\varphi_{\mathbb{I}} : H \to \mathbb{I}^\times$ induced by the universal character $\varphi : H \to W[[\Gamma_\wp]]^\times \cong W[[\Gamma^-]]^\times$ has the branch character ϕ. Write L_{ϕ^-} for the projection of L_{φ^-} to $\mathbb{I}$. The assumption $R^{ord} = \mathbb{T}$ is not necessary for the identity of the characteristic ideal $\text{char}_\Lambda(\text{Sel}_K(\varphi_{\mathbb{I}}^-)) = (L_{\phi^-})$ by a result of Rubin [R91, Theorem 4.1] or [T89] (though [T89] assumes mild extra assumptions including $p \nmid h_K$).

Note $\text{Sel}(Ad(\text{Ind}_K^{\mathbb{Q}} \varphi)) = \text{Sel}(\alpha_{W[[\Gamma_\wp]]}) \oplus \text{Sel}_K(\varphi^-)$ by Theorem 7.4.4. Here we have written $\alpha_{W[[\Gamma_\wp]]}$ for α having values in $W[[\Gamma_\wp]]^\times$. Assuming $p \nmid h_K$ (which implies cyclicity of Γ^-), we have $\text{Sel}(\alpha_{W[[\Gamma_\wp]]}) = 0$ and therefore $C_1(\varphi^-) = \text{Sel}(\text{Ind}_K^{\mathbb{Q}} \varphi^-)^\vee$ by Theorem 7.4.7. We have defined the algebraic p-adic L-function L_ρ for $\rho = \rho_{\mathbb{T}}$ in §6.2.7. Then for the projection $\pi : \mathbb{T} \twoheadrightarrow \mathbb{T}^{cm} = W[[\Gamma_\wp]]$, $L^{cm} := \pi(L_\rho)$ generates $\text{char}_\Lambda(C_1(\varphi^-)) = \text{Fitt}_\Lambda(C_1(\varphi^-)) = \text{Fitt}_\Lambda(C_0(\varphi^-)) = \text{char}_\Lambda(C_0(\varphi^-))$ (by Tate's Theorem 6.2.21). Thus we obtain the following corollary:

Corollary 8.5.7. *Assume $R^{ord} = \mathbb{T}$, (H0–1) and $p \nmid h_K$. For the projection $\pi : \mathbb{T} \twoheadrightarrow \mathbb{T}^{cm} = W[[\Gamma_\wp]]$, we define $L^{cm} := \pi(L_\rho)$. Then $L^{cm}/L_{\varphi^-} \in W[[\Gamma^-]]$ is a unit under the identification $\Gamma_\wp \cong \Gamma^-$ in §7.4.3.*

When $p|h_K$, since h_K is a factor of L^{cm}, to get the corresponding identity, we need to divide L^{cm} by h_K.

Suppose $p \nmid \mathfrak{F}$. Then we can replace $\wp$ by $\wp^\sigma$ in the above discussion. Defining $\mathcal{Y}_\sigma(\varphi^-)$ just replacing $\wp$ by $\wp^\sigma$ in §7.4.4. Conjugating everything by complex conjugation σ, we have $\text{Fitt}_\Lambda(\mathcal{Y}_\sigma(\varphi^-)) = (L_{\varphi_\sigma^-}^*)$, where $*$ is the involution of Λ induced by $\Gamma^- \ni \gamma \mapsto \gamma^{-1} = \sigma\gamma\sigma^{-1} \in \Gamma^-$. Writing $\mathbb{L}/F(\varphi^-)$ (resp. $\mathbb{L}_\sigma/F(\varphi^-)$) for the $\wp$-ramified (resp. $\wp^\sigma$-ramified) maximal p-abelian extension with Galois group $\mathcal{Y}(\varphi^-)$ (resp. $\mathcal{Y}_\sigma(\varphi^-)$). Note that $\mathcal{L} := \mathbb{L} \cap \mathbb{L}_\sigma$ is unramified everywhere over $F(\varphi^-)$; so, $\mathcal{L}/F(\varphi^-)$ should be a small extension as specified in the following subsection.

8.5.4 *Local indecomposability again*

Recall $\mathbb{Y} = \text{Gal}(\mathcal{L}/F(\varphi^-))$ defined in §8.5.2 after Corollary 8.5.5. Similar to the pseudo-nullity conjecture of Greenberg [G98, Conjecture 3.5], we ask

(N) Is $\mathbb{Y}[\varphi^-]$ *finite if* $(\varphi^-)^2 \neq 1$? *and otherwise, is* $\mathbb{Y}[\varphi^-]$ *killed by a power of the augmentation ideal of Λ?*

Since $\mathcal{L}/F(\varphi^-)$ is unramified everywhere, $\mathbb{Y}[\varphi^-]$ is killed by L_{φ^-} and $L^*_{\varphi_\sigma^-}$ by Theorem 8.5.6. If $\varphi^- \neq \varphi_\sigma$ (i.e., φ^- has order ≥ 3), we expect that L_{φ^-} and $L^*_{\varphi_\sigma^-}$ are mutually prime in $W(\overline{\mathbb{F}}_p)[[\Gamma^-]]$, and if this is the case, pseudo-nullity (i.e., finiteness) of $\mathbb{Y}[\varphi^-]$ follows. If the pseudo-nullity holds, there seems to be several interesting consequences, for example, [BCG20, (1.3)]. If $(\varphi^-)^2 = 1$, our expectation is that the common factor of L_{φ^-} and $L^*_{\varphi_\sigma^-}$ would be primes fixed by the involution $*$ (i.e., a power of augmentation ideal (T)). Indeed, we know that T exactly divides once L_{φ^-} and $L^*_{\varphi_\sigma^-}$ if $(\varphi^-)^2 = 1$ by the non-vanishing of $\mathcal{L}$-invariant for Dirichlet L functions (proven by Ferrero–Greenberg [FG78] for characters appearing in $Ad(\mathrm{ind}_K^{\mathbb{Q}}\varphi)$) combined with Gross' limit formula [G80] relating p-adic Dirichlet L-function and the Katz–Yager p-adic L-function. However, the GCD of L_{φ^-} and $L^*_{\varphi_\sigma^-}$ could have some primes different from (T), and the question (N) asks if the GCD is indeed (T).

Corollary 8.5.8. *If the question* (N) *is affirmative, then* $\rho|_{I_p}$ *is indecomposable under* (H0–1) *for* ρ *as in Corollary 8.5.5.*

As explained just after stating Corollary 8.5.5, the necessary ingredient to prove this is just the finiteness of $[\mathcal{L} : F(\varphi^-)]$ when φ^- has order ≥ 3, which is true if (N) is affirmative. If φ^- has order 2, the assertion is reduced to the case of the companion real quadratic field inside $F(\varphi^-)$ as explained after Corollary 8.5.5.

8.5.5 *Semi-simplicity and a structure theorem*

Iwasawa asked if the p-adic "Riemann" zeta function Φ_{ω^a} (a even) in Theorem 4.1.2 is square-free in Λ ([I79, §2]). If yes, the $\mathbb{Z}_p[[\Gamma]]$-module $X^{(j)} \otimes_{\mathbb{Z}_p} \mathbb{Q}_p$ (for odd j) in Theorem 1.6.2 is a semi-simple Γ-module. Thus it is natural to ask

(SS) *Assuming* (H0–1), *is* $C_1(\varphi^-) \otimes_{\mathbb{Z}_p} \mathbb{Q}_p \cong \mathrm{Sel}(\mathrm{Ind}_K^{\mathbb{Q}} \varphi^-)^\vee \otimes_{\mathbb{Z}_p} \mathbb{Q}_p$ *semi-simple as a* $\Gamma_\wp$-*module?*

By Corollary 8.3.9, this question has an affirmative answer when K is real. It is known that $p \nmid L_{\varphi^-}$ [EAI, Theorem 3.37] and hence $p \nmid L^{cm}$ by Corollary 8.5.7. Though $C_1(\varphi^-) \cong \mathrm{Sel}(Ad(\mathrm{Ind}_K^{\mathbb{Q}} \varphi))^\vee$ has the factor $\mathrm{Sel}(\alpha_{W[[\Gamma_\wp]]})^\vee \cong Cl_K \otimes_{\mathbb{Z}} W[[\Gamma_\wp]]$, after inverting p (i.e., tensoring $\mathbb{Q}_p$), this factor disappears, and we get the identity $C_1(\varphi^-) \otimes_{\mathbb{Z}_p} \mathbb{Q}_p \cong \mathrm{Sel}(\mathrm{Ind}_K^{\mathbb{Q}} \varphi^-)^\vee \otimes_{\mathbb{Z}_p} \mathbb{Q}_p$ in the conjecture.

We now return to $W = W(\mathbb{F})$. Since $\mathbb{T}^{cm} = W[[\Gamma^-]]$, if we can describe $\mathbb{T}^{ncm}$, we would be able to describe $\mathbb{T}$ precisely. Note here that $\text{Fitt}_\Lambda(C_0(\varphi^-)) = \text{Fitt}_\Lambda(C_1(\varphi^-)) = (L^{cm})$ under $p \nmid h_K$; so, by Lemma 8.5.2, $C_0(\varphi^-) \cong \mathbb{T}^{ncm}/(L^{cm}) \cong \mathbb{T}^{ncm}/\mathfrak{d}$. Then, under (H0-2), we have $\mathfrak{d} = (\Theta)$ with $\Theta \in \mathbb{T}^{ncm}_-$ and

$$\mathbb{T} \cong \{(x, y) \in W[[\Gamma^-]] \times \mathbb{T}^{ncm} | x \bmod (L^{cm}) = y \bmod(\Theta)\}$$

as Λ-algebras. Here is a structure theorem for $\mathbb{T}^{ncm}$:

Theorem 8.5.9. *Assume* (H0-3), $R^{ord} = \mathbb{T}$ *and* $\mathbb{T}^{ncm} \neq 0$. *Write* $A = \mathbb{T}^{ncm}$ *or* $\mathbb{T}^{ncm}_+$. *Let* $e = \text{rank}_\Lambda \mathbb{T}^{ncm}$. *Then we have*

(0) *A is an integral domain with isomorphisms:* $\mathbb{T}^{ncm} \cong \Lambda[X]/(D(X))$ *and* $\mathbb{T}^{ncm}_+ \cong \Lambda[X]/(D^+(X))$ *for distinguished polynomials* $D(X)$ *and* $D^+(X)$ *in* $\Lambda[X]$ *with respect to each prime factor* $P|L^{cm}$ *in* Λ *such that* $D(X) = D^+(X^2)$. *The localization* $\widetilde{A}_P$ *of the normalization* $\widetilde{A}$ *at a prime* $P|L^{cm}$ *of* A *is a discrete valuation ring fully ramified over* Λ_P.

Further, suppose L^{cm} *is square-free in* Λ (*i.e.,* (SS) *is affirmative). Then the following five assertions hold:*

(1) *If* L^{cm} *is a prime in* Λ *with* $\Lambda/(L_{\varphi^-}) = W$ *(a linear prime), then the ring* A *is isomorphic to a power series ring* $W[[x]]$ *of one variable over* W; *hence,* A *is a regular local domain and is factorial;*

(2) *The ring* A *is an integral domain fully ramified at* (L^{cm});

(3) *If* p *is prime to* e, *the ramification locus of* $A_{/\Lambda}$ *is given by* $\text{Spec}(\Lambda_{cm})$ *for* $\Lambda_{cm} := \Lambda/(L^{cm})$, *the different for* A/Λ *is principal and generated by* Θ^{a-1} *for* $a = \text{rank}_\Lambda A$, *and* A *is a normal integral domain of dimension* 2 *unramified outside* (L^{cm}) *over* Λ;

(4) *If* $p|e$, $\mathbb{T}^{ncm}[\frac{1}{p}] := \mathbb{T}^{ncm} \otimes_{\mathbb{Z}} \mathbb{Q}$ *is a Dedekind domain unramified outside* (L^{cm}) *over* $\Lambda \otimes_{\mathbb{Z}} \mathbb{Q}$, *and the relative different for* $\mathbb{T}^{ncm}[\frac{1}{p}]/\Lambda[\frac{1}{p}]$ *is principal and generated by* Θ^{e-1};

(5) *If* $e = 2$, $\mathbb{T}^{ncm}_+ = \Lambda$ *and* $\mathbb{T}^{ncm} = \Lambda[\sqrt{L^{cm}}]$ *after replacing* W *by its finite extension.*

Again we can conjecture $e = 2$ analogous to Conjecture 8.3.8 (assuming (SS) has an affirmative answer). The polynomial $D(X)$ has the form $D^+(X^2)$ since its root Θ satisfies $[\alpha](\Theta) = -\Theta$ (see Remark 8.3.10). Though the proof is similar to the one given for Theorem 8.3.7 in the real case, we give a fairly detailed argument proving the assertion (0) for $A = \mathbb{T}^{ncm}$ here.

Proof of (0). Recall $\alpha \in \widehat{\Delta} = \text{Hom}(\text{Gal}(K/\mathbb{Q}), \{\pm 1\})$. As before, we write $M^{\pm} := \{m \in M | [\alpha](m) = \pm m\}$ for a $\mathbb{Z}_p[\widehat{\Delta}]$-module M. By Theorem 8.1.1,

$$\dim_{\mathbb{F}} t^-_{\mathbb{T}/\Lambda} = \dim_{\mathbb{F}} \text{Sel}(\text{Ind}_K^{\mathbb{Q}}(\overline{\varphi}^-)) \leq 1.$$

By the assumption: $\mathbb{T}^{ncm} \neq 0$ and Lemma 8.3.4, we have $\dim_{\mathbb{F}} t^-_{\mathbb{T}/\Lambda} > 0$. Thus we conclude $\dim_{\mathbb{F}} t^-_{\mathbb{T}/\Lambda} = \dim_{\mathbb{F}} t_{\mathbb{T}^{ncm}/\Lambda} = 1$, and hence $\mathbb{T} = \Lambda[\Theta]$ for $\Theta \in \mathbb{T}$ with $[\alpha](\Theta) = -\Theta$, which implies $\mathbb{T}^{ncm} = \Lambda[\Theta]$. Identify Λ and $W[[\Gamma_\wp]]$ by the structure morphism $\Lambda \hookrightarrow \mathbb{T}$ composed with the projection $\pi : \mathbb{T} \twoheadrightarrow \mathbb{T}^{cm} = W[[\Gamma_\wp]]$. Thus $L^{cm} \in \Lambda$. By (8.9), $\mathbb{T}^{ncm}$ is free of finite rank over Λ, and hence $\mathbb{T}^{ncm}$ generated by one element $\Theta \in \mathfrak{m}_{\mathbb{T}^{ncm}}$ is a local complete intersection with $\mathbb{T}^{ncm} \cong \Lambda[X]/(D(X))$ for the characteristic polynomial $D(X)$ over Λ of $x \mapsto \Theta x$ on $\mathbb{T}^{ncm}$. Recall $\mathfrak{d} = \mathbb{T}([\alpha] - 1)\mathbb{T} = \mathbb{T}^{ncm}([\alpha] - 1)\mathbb{T}^{ncm} = (\Theta)$. Hereafter we write $A = \mathbb{T}^{ncm}$ for simplicity.

Decompose $(L^{cm}) = \prod_P P^{a(P)} = \bigcap_P \mathcal{P}_P$ for primes divisors P in Λ, where $\mathcal{P}_P = P^{a(P)}$ as the primary ideals associated to P. Since $A/\mathfrak{d} \cong \Lambda/(L^{cm})$, we can pull back $P \in \text{Spec}(\Lambda/(L^{cm}))$ to a unique prime divisor $\widetilde{P} \in \text{Spec}(A)$ of $\mathfrak{d}$, and we get $\mathfrak{d} = \bigcap_P \widetilde{\mathcal{P}}_P$ for the (unique) primary ideal $\widetilde{\mathcal{P}}_P$ associated to $\widetilde{P}$ [CRT, Theorem 6.8]. Note here $\widetilde{\mathcal{P}}_P \supset \widetilde{P}^{a(P)}$ (but they may not be equal). Consider the localization at P: $A_P = A \otimes_\Lambda \Lambda_P$. Then any prime ideal Q of A_P containing $(L^{cm})A_P$ gives rise to a prime ideal $Q \cap A$ containing $\mathfrak{d}$. Thus Q has to be $\widetilde{P}$, as $A_P/\mathfrak{d}_P = \Lambda_P/(L^{cm})_P$. In other words, A_P is a local ring with maximal ideal $\widetilde{P}A_P$.

For a ring R and an R-ideal J, we define the J-graded algebra $\text{gr}_J(R) := \bigoplus_{n=0}^\infty J^n/J^{n+1}$ with $J^0 = R/J$ [BCM, III.2.3]. Let $C_0 := A/\mathfrak{d} \cong \Lambda/(L^{cm})$. We have $\text{gr}_\mathfrak{d}(A) \cong C_0[x]$ (the polynomial ring with variable x) by $(\Theta \mod \mathfrak{d}^2) \mapsto x$. Writing $\mathfrak{d}_P := \mathfrak{d} \otimes_\Lambda \Lambda_P$, we have $\text{gr}_{\mathfrak{d}_P}(A_P) \cong \Lambda_P/P^{a(P)}[x]$. Since Λ_P is a discrete valuation ring, $\text{Spec}(\text{gr}_{\mathfrak{d}_P}(A_P))$ is irreducible (but may not be reduced). The ring A is reduced, so is A_P. Since A_P is free of finite rank over the discrete valuation ring Λ_P, it is equi-dimensional of dimension 1. If $\text{Spec}(A_P) = \text{Spec}(I) \cup \text{Spec}(I^\perp)$ for an irreducible component $\text{Spec}(I)$ with complement $\text{Spec}(I^\perp)$, I and $I^\perp$ are Λ_P-free of finite rank, and $\mathfrak{d}_I := \mathfrak{d}_P I$ and $\mathfrak{d}_I^\perp := \mathfrak{d}_P I^\perp$ are proper ideals of I and $I^\perp$, respectively. They are principal generated by the non-zero image Θ_I and $\Theta_{I^\perp}$ of Θ in I and $I^\perp$ respectively. We have a short exact sequence $A_P \hookrightarrow I \oplus I^\perp \twoheadrightarrow C$ for a torsion Λ_P-module C of finite type (killed by Θ). Note that $(\mathfrak{d}_I \oplus \mathfrak{d}_I^\perp)^n \cap A_P = \mathfrak{d}_P^n = \Theta^n A_P$ as all the ideals are principal. Take the corresponding exact sequence of graded algebras [BCM, III.2.4]:

$$0 \to \text{gr}_{\mathfrak{d}_P}(A_P) \to \text{gr}_{\mathfrak{d}_I}(I) \oplus \text{gr}_{\mathfrak{d}_I^\perp}(I^\perp) \to \text{gr}(C)(\cong C) \to 0,$$

where $\mathrm{gr}(C)$ is the graded algebra with respect to the quotient filtration $\{\Theta^n C\}_{n \geq 0} = \{C, (0)\}$. Plainly $\mathrm{gr}(C)$ is an Artinian algebra. Thus if $\mathrm{gr}_{\mathfrak{d}_{\bar{I}}^{\perp}}(I^{\perp}) \neq 0$, $\mathrm{Spec}(\mathrm{gr}_{\mathfrak{d}_P}(A_P))$ cannot be irreducible. Thus A_P is an integral local domain, and $A \subset A_P$ is an integral domain.

Since A_P is a integral local domain finite flat over a discrete valuation ring Λ_P, A_P is an excellent local domain [CRT, §32]. Then by [EAI, Corollary 4.37], $\widehat{A}_P$ is reduced. The normalization $\widetilde{A}_P$ is a noetherian Dedekind semi-local domain of dimension 1. The proof of irreducibility of $\mathrm{Spec}(A_P)$ via graded ring carries over to $\mathrm{Spec}(\widehat{A}_P)$ for the P-adic completion $\widehat{A}_P$, because the formation of the graded algebra is insensitive to the completion process. Thus $\widehat{A}_P$ is a complete integral local domain finite flat over the complete discrete valuation ring $\widehat{\Lambda}_P$. Hence its normalization $\widetilde{\widehat{A}}_P$ is a discrete valuation ring, as extension of a valuation of a complete discrete valuation ring is unique [BCM, IV.8.7]. Since $\mathrm{rank}_{\Lambda_P} \widetilde{A}_P = \mathrm{rank}_{\widehat{\Lambda}_P} \widetilde{\widehat{A}}_P$, we have $\widetilde{A}_P = \widehat{\widetilde{A}}_P$,[1] which is a discrete valuation ring [CRT, Theorem 11.2].

Since $\widetilde{A}_P/A_P$ is a torsion A_P-module of finite type, we have $A_P \supset \Lambda_P + \mathfrak{m}_{A_P}^a \widetilde{A}_P$ for some $a \geq 0$. This shows $\widetilde{A}_P$ and A_P share the same residue field Λ_P/P; so, $\widetilde{A}_P$ is fully ramified over Λ_P. $\qquad \square$

Proof of (1). As seen in the above discussion, we have $\mathrm{gr}_{\mathfrak{d}}(A) \cong C_0[x]$ (the polynomial ring $C_0[x]$) by $x \leftrightarrow \overline{\Theta} = (\Theta \mod \mathfrak{d}^2)$.

Take n so that $u \in \mathfrak{d}^n$ but $u \notin \mathfrak{d}^{n+1}$. Then $\mathrm{gr}(u) : \mathrm{gr}_{\mathfrak{d}}(A) \to \mathrm{gr}_{\mathfrak{d}}(A)$ is multiplication by a polynomial of degree n. Assume that L^{cm} is a prime with $C_0 = \Lambda/(L^{cm}) = W$. Then $\mathrm{gr}_{\mathfrak{d}}(A)$ is an integral domain isomorphic to the polynomial ring $W[x]$. Then in exactly the same manner as in the proof of Theorem 8.3.7 (1), we conclude that A is an integral domain and $A = \varprojlim_n A/\mathfrak{d}^n \cong W[[x]]$ by sending Θ to x. $\qquad \square$

The assertion (2) is just a restatement of (0). Since $p \nmid L^{cm}$ as already mentioned, the assertions (3)–(4) can be proven, replacing $\langle \varepsilon \rangle - 1$ by L^{cm} in the proof of Theorem 8.3.7; so, we leave it to the reader.

Proof of (5). If $e = 2$, $[\alpha](\Theta) = -\Theta$ implies $\Theta^2 \in \Lambda$. Thus $D(X) = X^2 - \Theta^2$ with $\Theta^2 = uL^{cm} \in \Lambda$ with $u \in \Lambda^{\times}$. Take a Teichmüller lift ζ of $u \mod \mathfrak{m}_{\Lambda}$. Then $\zeta^{-1}u$ is in the p-profinite group $1 + \mathfrak{m}_{\Lambda}$, and hence we have $\sqrt{\zeta^{-1}u} \in \Lambda$ as $p > 2$. Extending scalars to $W[\sqrt{\zeta}]$, we can modify the equation to have the form $X^2 - L^{cm}$, and the result follows. $\qquad \square$

---

[1] Actually for an excellent integral domain, completion and normalization are interchangeable operations [EGA, IV.7.8.3.1 (vii)], and we do not need to argue as above to show $\widetilde{A}_P = \widehat{\widetilde{A}}_P$ by rank comparison.

8.5.6 *Open questions*

We have studied the indecomposability question when p splits in the quadratic field K. The case where p is not split is still open, but we hope to deal with this case similar to the way we described in this chapter, including the case of exceptional Artin representations. We expect that p-locally decomposable representations are concentrated to weight 1 classical modular forms (except for CM cases).

The general case of companion forms is mysterious, and we might have a p-locally decomposable non-geometric representation of a transcendental p-adic weight. The determination of such weight if any is an interesting question. In other words, if θ in Theorem 8.2.3 is a non-unit, θ can still be a power of p up to units.[2] Related to this, for a given Hecke eigenform f of weight ≥ 2 and its compatible system ρ_f, we can think of the subset $T_{sp} := \{\mathfrak{p} \in \mathrm{Spec}(\mathbb{Z}[f]) | \overline{\rho}_\mathfrak{p}$ is tamely ramified at $p\}$.

Is this set T_{sp} infinite? If yes, what is the density of T_{sp}?

Serre seems to believe T_{sp} has density 0 with infinite cardinality [OCS, III, Notes, 16.1]. A list of companion pairs with $\mathrm{Im}(Ad(\overline{\rho})) \cong \mathrm{PGL}_2(\mathbb{F}_p)$ can be found in [R18], which seems to indicate that T_{sp} is not very large.

Another interesting question is the determination of the structure of $\mathbb{T}_\mathfrak{p}$ if we start with an Artin representation. For example, we can ask

- *Does the set $T_{reg} = \{\mathfrak{p} | \mathbb{T}_\mathfrak{p}$ is a regular local ring\} have density 1?*
- *Is the complement of T_{reg} infinite or not?*

Suppose $\overline{\rho} = \mathrm{Ind}_K^\mathbb{Q} \overline{\varphi}$ for K real. By Theorem 7.6.2 (1), if $\overline{\varphi}^-$ has order ≥ 3 and p does not split in K, $\mathrm{Sel}(Ad(\overline{\rho})) = 0$ for almost all non-split primes and hence $\mathbb{T}_\mathfrak{p} = \Lambda$. However if either p splits in K or $\overline{\varphi}^-$ has order 2, as discussed in §8.3.6, the situation is quite subtle. Other cases (i.e., Case E or Case D with K imaginary) are mysterious.

In the cases where $\overline{\rho}$ is induced from K, θ is not divisible by $\mathfrak{m}_W \mathbb{T}^{ncm}$ (i.e., vanishing of the μ-invariant of $N_{\mathbb{T}^{ncm}/\Lambda}(\theta)$).

- *In general, when is $\rho_{\mathbb{T}^{ncm}}$ mod $\mathfrak{m}_W$ indecomposable over I_p?*

---

[2]In that case, there is no characteristic 0 p-locally decomposable deformation of p-locally splitting $\overline{\rho}$.

Chapter 9

Analytic and topological methods

We compute the size $|C_0(\lambda; W)|$ of the congruence module for an algebra homomorphism $\lambda : \mathbb{T}_\chi \to W$ associated to a Hecke eigenform f in terms of the adjoint L-value at $s = 1$.

Here is a technical heuristic explaining some reason why at the very beginning of his work [H81a] the author speculated that the adjoint L-value would be most accessible to a non-abelian generalization of the class number formula in Corollary 1.11.4 and Theorem 1.3.1. The proof of the formula by Dirichlet–Kummer–Dedekind proceeds as follows: first, one relates the class number $h = |\operatorname{Sel}_\mathbb{Q}(\rho_F)|$ to the residue of the Dedekind zeta function ζ_F of a number field F (Theorem 1.11.2), that is, the L-function of the self dual Galois representation

$$1 \oplus \rho_F = \operatorname{Ind}_F^\mathbb{Q} 1;$$

second, one uses the fact $\zeta_F(s) = L(s, \rho_F)\zeta(s)$ following the above decomposition with the Riemann zeta function $\zeta(s)$ and the residue formula $\operatorname{Res}_{s=1} \zeta(s) = 1$ to finish the proof of the identity (see Corollary 1.11.4).

If one has good experience of calculating the value or the residue of a well defined complex meromorphic function, it seems that a residue tends to be more accessible than the value of the function (a proto-typical example is the residue of the Riemann zeta function at $s = 1$).

Unless F is either $\mathbb{Q}$ or an imaginary quadratic field, the value $L(1, \rho_F)$ is not critical. Thus the transcendental factor in Corollary 1.11.4 involves the regulator in addition to the period (a power of $2\pi i$).

Perhaps, a simple (and natural) way to create a self dual representation containing the trivial representation $\mathbf{1}$ is to form the tensor product of a given n-dimensional Galois representation ρ (of $\operatorname{Gal}(\overline{\mathbb{Q}}/\mathbb{Q})$) with its contragredient $\widetilde{\rho}$: $\rho \otimes \widetilde{\rho}$. We define an $n^2 - 1$ dimensional representation $Ad(\rho)$ so that

$$\rho \otimes \tilde{\rho} \cong 1 \oplus Ad(\rho).$$

When $n = 2$, $s = 1$ is critical with respect to $Ad(\rho)$ if $\det \rho(c) = -1$ for complex conjugation c. Then we expect that $\frac{L(1, Ad(\rho))}{\text{a period}}$ should be somehow related to the size of $\mathrm{Sel}(Ad(\rho))$ in favorable cases. Even if $s = 1$ is not critical for $Ad(\rho)$, there seems to be a good way to define a natural transcendental factor of $L(1, Ad(\rho))$ only using the data from the Hecke side (see [H99] and [BR17]) if ρ is automorphic. Therefore, the transcendental factor automorphically defined in these papers via Whittaker model should contain to a good extent a period associated with $Ad(\rho)$ and a Beilinson regulator geometrically defined for $Ad(\rho)$ (as long as ρ is geometric in a reasonable sense) if one believes in the standard conjectures. It would be a challenging problem for us to factor the automorphic transcendental factor (given in these articles) in an automorphically natural way into the product of a period and the regulator of $Ad(M)$ when we know that ρ is associated to a motive M (see [TU20] for an interesting attempt).

In this chapter, the reader will see this heuristic is actually realized by a simple classical computation at least when ρ is associated to an elliptic Hecke eigenform of weight $k \geq 2$.

Since our purpose is to give a concise description of the adjoint class number formula, to show the basic principle in a readily accessible way, we assume in this chapter

- for each prime $l \neq p$, for the p-adic Galois representation ρ_λ of f, $\rho_\lambda(I_l)$ is finite and $\rho|_{I_l}$ satisfies the minimality condition (l) in §6.1.1,

and once we start dealing with arithmetic properties of the adjoint L-values (so, after §9.2), we also assume

- $p \geq 5$;
- the conductor of Hecke eigenform C is prime to p.

The treatment of the primes $p = 2, 3$ is more technical and complicated. If we incorporate in the formula the p-adic valuation of the root number (times C) for $L(s, \lambda)$, the case where $p|C$ can be done. After stating Theorem 9.2.3, we will give a brief description of the reason why we need to modify by the root number in the case where $p|C$.

Since our period is defined up to p-adic unit, starting from §9.2, we ignore p-adic unit constants appearing in front of integral expressions of the adjoint L function. For the reader who cares more about an exact expression, we refer to [H88a] and [DFG04] for exact formulas (though [DFG04] ignores primes $\leq k$). In an equality, if we ignore p-adic unit factors, we use "$\doteq$" in place of "$=$".

9.1   Analyticity of adjoint L-functions

We summarize here known facts on analyticity and arithmeticity of the adjoint L-function $L(s, Ad(\lambda)) = L(s, Ad(\rho_\lambda))$ for a $\mathbb{Z}[\psi]$-algebra homomorphism λ of $h_k(\Gamma_0(C), \psi; \mathbb{Z}[\psi])$ into $\overline{\mathbb{Q}}$ and the compatible system ρ_λ of Galois representations attached to λ. We allow $k = 1$ for analytic statements, but for rationality statements, we need to assume $k \geq 2$ as the adjoint L-value is critical at $s = 1$ only when $k \geq 2$. Recall here that h_k denotes the Hecke algebra of the space S_k of cusp forms.

9.1.1   *L-function of modular Galois representations*

By Theorem 6.2.1, we may assume that λ is primitive of exact level C. Recall that $\chi = \{\chi_\mathfrak{p}\}_\mathfrak{p} = \det \rho_f$ is the compatible system of Galois characters given by $\det(\rho_f)$; so, $\chi_\mathfrak{p}(\mathrm{Frob}_l) = l^{k-1}\psi(l)$ for l outside $C\mathfrak{p}$. Here Frob_l is the arithmetic Frobenius element as before.

Define the (reciprocal) Hecke polynomial at a prime l by
$$H_l(X) = \det(1 - \rho_{f,\mathfrak{p}}(\mathrm{Frob}_l)|_{V_{I_l}}).$$
Write it as $H_l(X) = 1 - \lambda(T(\ell))X + \chi(\ell)X^2 = (1 - \alpha_\ell X)(1 - \beta_\ell X)$. Here $V = V_\mathfrak{p}$ is the representation space of $\rho_{f,\mathfrak{p}}$ with $\mathfrak{p}$ chosen outside l, and V_{I_l} is the maximal quotient fixed by the l-inertia subgroup I_l; i.e., $V_{I_l} = H_0(I_l, V)$. From the construction of ρ_f (started from Eichler, Shimura and continued by Deligne and Langlands; [GME, §4.2]), we know that $H_l(X)$ is independent of the choice of $\mathfrak{p}$ as long as $\mathfrak{p} \nmid l$.

Exercise 9.1.1. Prove that we can replace V_{I_l} by the maximal I_l-fixed subspace V^{I_l} in the definition of $H_l(X)$ and obtain the same Hecke polynomial if we use the geometric Frobenius element Frob_l^{-1} in place of Frob_l.

Here is an explicit shape of $H_l(X)$:

$$H_l(X) = \begin{cases} 1 & \text{if } \rho_{f,\mathfrak{p}}|_{D_p} \text{ is irreducible or } V_{I_l} = 0, \\ (1 - \alpha_\ell X)(1 - \beta_\ell X) & \text{if } \rho_{f,\mathfrak{p}}|_{D_p} \text{ is unramified}, \\ (1 - \alpha_l X) & \text{if } \dim V_{I_l} = 1. \end{cases}$$

$$\tag{9.1}$$

For any polynomial representation $r : \mathrm{GL}(2) \to \mathrm{GL}(n)$, $r \circ \rho_f := \{r \circ \rho_{f,\mathfrak{p}}\}_{\mathfrak{p}}$ is another compatible system with $r \circ \rho_f$ acting on n-dimensional space $rV = rV_{\mathfrak{p}}$. Define $L(s, \lambda, r) := \prod_l E_l^r(l^{-s})^{-1}$ for

$$E_l^r(X) := \det(1 - r(\rho_{f,\mathfrak{p}}(\mathrm{Frob}_l))|_{(rV)_{I_l}}) = \det(1 - r(\rho_{f,\mathfrak{p}}(\mathrm{Frob}_l))|_{(rV)_{I_l}}).$$

When $r = Ad : \mathrm{GL}(2) \to \mathrm{GL}(3)$ given by the conjugation action of $\mathrm{GL}(2)$ on $\mathfrak{sl}(2)$, we write $L(s, Ad(\lambda)) = L(s, Ad(\rho_f)) = L(s, Ad(f))$ for $L(s, \lambda, Ad)$.

Similarly, for another Hecke eigenform g with $g|T(n)) = \mu(T(n))g$, we have its compatible system ρ_g with coefficients in $\mathbb{Z}[g] = \mathbb{Z}[\mu]$. Let $\tilde{\rho}_g := {}^t\rho_g^{-1}$ (the contragredient of ρ_g). Then the tensor product $\rho_f \otimes \tilde{\rho}_g$ has coefficients in the integer ring $\mathbb{Z}[f, g]$ of the composite field $\mathbb{Q}[f]$ and $\mathbb{Q}[g]$. Writing $V' = V'_{\mathfrak{p}}$ for the space on which $\tilde{\rho}_{g,\mathfrak{p}}$ act, $(\rho_f \otimes \tilde{\rho}_g)_{\mathfrak{P}}$ acts on $V \otimes V' := (V_{\mathfrak{P} \cap \mathbb{Z}[f]} \otimes_{\mathbb{Z}[f]} \mathbb{Z}[f, g]) \otimes_{\mathbb{Z}[f,g]} V'_{\mathfrak{P} \cap \mathbb{Z}[g]}$ for prime ideals $\mathfrak{P}$ of $\mathbb{Z}[f, g]$. Define

$$L(s, \lambda \otimes \tilde{\mu}) = L(s, \rho_f \otimes \tilde{\rho}_g) = L(s, f \otimes \tilde{g}) := \prod_l E_l^{\lambda \otimes \mu}(l^{-s})^{-1}$$

for $E^{\lambda \otimes \mu}(X) = \det(1 - (\rho_f \otimes \tilde{\rho}_g)_{\mathfrak{P}}(\mathrm{Frob}_l)|_{(V \otimes V')_{I_l}} X)$ choosing $\mathfrak{P} \nmid l$.

9.1.2 Explicit form of adjoint Euler factors

For a local representation $\rho := \rho_{f,\mathfrak{p}}|_{D_l}$ ($\mathfrak{p} \nmid l$), if ρ is reducible split, we have two characters $\rho = \mathrm{diag}[\alpha, \beta]$. If ρ is reducible indecomposable, it is known by Grothendieck that ρ is an extension of a character α by $\nu_p\alpha$ [GME, Theorem 4.2.4]. If ρ is irreducible, either $\rho \cong \mathrm{Ind}_K^{\mathbb{Q}_l} \varphi$ for a quadratic extension $K_{/\mathbb{Q}_l}$ (with $\varphi_\varsigma \neq \varphi$ for $\varsigma \in \mathrm{Gal}(\overline{\mathbb{Q}}_l/\mathbb{Q}_l)$ non-trivial on K) or $l = 2$ and ρ is irreducible possibly non-induced [W74] (this non-induced case is called "extraordinary"; see also [AFR, Propositions 4.9.2–3]). We split our consideration into the following five cases:

(U) ρ *is unramified (so, reducible split)*;
(R) $\rho = \mathrm{diag}[\alpha, \beta]$ *with α unramified and β ramified*;
(I) *irreducible $\rho \cong \mathrm{Ind}_K^{\mathbb{Q}_l} \varphi$ with K unramified and φ ramified*;
(I') *irreducible $\rho \cong \mathrm{Ind}_K^{\mathbb{Q}_l} \varphi$ with K ramified and φ ramified*;
(E) ρ *is extraordinary (so, $l = 2$ and $\mathrm{Im}(Ad(\rho)) \cong A_4$ or S_4)*.

By the symbol ? of the above cases, ? also indicates the set of primes in Case ?; so, for example, $l \in R$ means l is a prime in case R. Thus $S = R \sqcup I \sqcup I' \sqcup E$ with $S_{ab} = R$ by (m).

By the minimality (l) in §6.1.1, the case where α, β both ramified is avoided. The case where ρ is irreducible such that $\rho \cong \mathrm{Ind}_K^{\mathbb{Q}_l} \varphi$ with K

ramified and φ unramified is included in Case R. Indeed, in this case, $\varphi_\varsigma = \varphi$ for unramified φ, and φ extends to an unramified character ϕ of $\mathrm{Gal}(\overline{\mathbb{Q}}_l/\mathbb{Q}_l)$ and $\mathrm{Ind}_{\mathbb{Q}_l}^K \varphi \cong \phi \oplus \phi\left(\frac{K/\mathbb{Q}_l}{\cdot}\right)$. As a convention, in Case U, we put $\alpha_l := \alpha(l)$ and $\beta_l = \beta(l)$, in Case (R), we put $\alpha_l = \alpha(l)$ and $\beta_l = 0$ and in Cases (I), (I') and (E), we put $\alpha_l = \beta_l = 0$.

Using this convention, we first compute $E_l^{\lambda \otimes \widetilde{\lambda}}(X)$.

Proposition 9.1.1. *Let the notation be as above. Then we have*

$$E_l^{\lambda \otimes \widetilde{\lambda}}(X) = \begin{cases} (1-X)^2(1 - \frac{\alpha_l}{\beta_l}X)(1 - \frac{\beta_l}{\alpha_l}X) & \text{Case U,} \\ (1-X)^2 & \text{Case R,} \\ (1-X)(1+X) & \text{Case I,} \\ (1-X) & \text{Cases I' and E.} \end{cases}$$

Proof. If $\rho|_{D_l} \cong \alpha \oplus \beta$, then $\widetilde{\rho}|_{D_l} \cong \alpha^{-1} \oplus \beta^{-1}$, from which Cases U and R follow.

We can realize $\rho \otimes \widetilde{\rho}$ as a conjugation action of ρ on 2×2 matrices; so, $\rho \otimes \widetilde{\rho} = \mathbf{1} \oplus Ad(\rho)$. Suppose $\rho = \mathrm{Ind}_K^{\mathbb{Q}_l} \varphi$. Lifting ς to D_l, we define $\varphi_\varsigma(g) = \varphi(\varsigma^{-1}g\varsigma)$ as before. If φ is unramified, $\varphi_\varsigma = \varphi$ and hence, ρ has form $\mathrm{diag}[\alpha, \beta]$; so, we may assume that φ is ramified. Again regarding $\rho \otimes \widetilde{\rho}$ as ρ acting on 2×2 matrices by conjugation, by Proposition 7.3.2, we have $\rho \otimes \widetilde{\rho} = \mathbf{1} \oplus Ad(\rho) = \mathbf{1} \oplus \left(\frac{K/\mathbb{Q}_l}{\cdot}\right) \oplus \mathrm{Ind}_K^{\mathbb{Q}_l} \varphi^-$ with $\varphi^- = \varphi\varphi_\varsigma^{-1}$.

Suppose $(\varphi^-)^2 \neq 1$. Write I_l' for the inertia group at l over K. If $(\varphi^-)^2|_{I_l'} \neq 1$, $\mathrm{Ind}_K^{\mathbb{Q}_l} \varphi^-$ is irreducible on I_l, and we get the desired result in Cases (I) and (I').

Suppose $(\varphi^-)^2|_{I_l'} = 1$. If $\varphi^-|_{I_l'} = 1$, φ^- factors through the abelian quotient $\mathrm{Gal}(\overline{\mathbb{Q}}_l/K)/I_l'$; so, we find $\varphi = \varphi_\varsigma$. This implies that $\mathrm{Ind}_K^{\mathbb{Q}_l} \varphi$ is reducible (as seen in §7.3.1). This cannot happen as $\mathrm{Ind}_K^{\mathbb{Q}_l} \varphi$ is irreducible. Thus $\varphi^-|_{I_l'} \neq 1$. Since $\varphi_\varsigma^-|_{I_l'} = (\varphi^-)^{-1}|_{I_l'} = \varphi^-|_{I_l'}$, we have $\mathrm{Ind}_K^{\mathbb{Q}_l} \varphi^- = \mathrm{diag}[\phi, \left(\frac{K/\mathbb{Q}_l}{\cdot}\right)\phi]$ for an extension ϕ of φ^-. Since the two characters ϕ and $\left(\frac{K/\mathbb{Q}_l}{\cdot}\right)\phi$ both ramify, we again get the formula in Cases (I) and (I').

Suppose we are in Case E; so, $l = 2$. For a finite extension $M/\mathbb{Q}_l$, we write $D_M := \mathrm{Gal}(\overline{\mathbb{Q}}_l/M)$ with the inertia subgroup I_M. Note that $D_M = \mathrm{Frob}^{\widehat{\mathbb{Z}}} \ltimes I_M \subset D_l$. When $M = \mathbb{Q}_l$, we just write $D_l = D_{\mathbb{Q}_l}$ and $I_l = I_{\mathbb{Q}_l}$ as before. By [W74, §36] (or [K80, §5.1]), we find a finite Galois extension $M/\mathbb{Q}_l$ with $\mathrm{Gal}(M/\mathbb{Q}_l)$ isomorphic either to A_4 or S_4 such that $M := \overline{\mathbb{Q}}_l^{\mathrm{Ker}(Ad(\rho))} \subset F(\rho) = \overline{\mathbb{Q}}_l^{\mathrm{Ker}(\rho)}$ and that $\rho(D_M)$ is in the center. Then $\rho|_{I_l}$ is irreducible non-induced; so, the result follows. $\square$

Since $\rho_f \otimes \tilde{\rho}_f = \mathbf{1} \oplus Ad(\rho_f)$, we record the following fact shown in the above proof of Proposition 9.1.1:

Corollary 9.1.2. *We have*

$$
E_l^{Ad}(X) = \begin{cases}
(1-X)(1-\frac{\alpha_l}{\beta_l}X)(1-\frac{\beta_l}{\alpha_l}X) & \text{Case U,} \\
(1-X) & \text{Case R,} \\
(1+X) & \text{Case I,} \\
1 & \text{Cases I' and E.}
\end{cases}
$$

In particular,

$$
L(s, \lambda \otimes \tilde{\lambda}) = L(s, \rho_f \otimes \tilde{\rho}_f) = \zeta(s)L(s, Ad(\rho_f)) = \zeta(s)L(s, Ad(\lambda)),
$$

and $L(s, Ad(\lambda)$ converges locally uniformly and absolutely if $\mathrm{Re}(s) > 1$, and $\lim_{s \to +1} L(s, Ad(\lambda)) \neq 0$.

The convergence for $\mathrm{Re}(s) > 1$ follows from the fact $|\alpha_l| = |\beta_l| = l^{k-1}$ for l in Case U (due to Shimura if $k = 2$ [IAT, §7.4], to Deligne if $k > 1$ [D71] and Deligne and Serre if $k = 1$ [DS74]). The non-vanishing follows from the fact that $L(s, Ad(\lambda))$ is a real Dirichlet series which is finite at $s = 1$ (as $s = 1$ is the abscissa of convergence of $L(s, Ad(\lambda))$). We will see later in this section the finiteness of $L(s, Ad(\lambda))$ around $s = 1$ (after its analytic continuation).

9.1.3 *Analytic continuation*

The meromorphic continuation and a functional equation (with possibly some missing Euler factors) of this L-function is proven by Shimura in 1975 [S75]. The method of Shimura is generalized to show that $Ad(\rho_f)$ is associated to an automorphic representation of GL(3), using the language of Langlands' theory, by Gelbart and Jacquet [GJ78]. In [GJ78], an exact functional equation of $L(s, Ad(\lambda))$ is proven. Taking a primitive cusp form f such that $f|T(n) = \lambda(T(n))f$ for all n, let π be the automorphic representation of $GL_2(\mathbb{A})$ spanned by f and its right translations. We write $L(s, Ad(\pi))$ for the L-function of the adjoint lift $Ad(\pi)$ to GL(3). This L-function coincides with $L(s, Ad(\lambda))$ and has a meromorphic continuation to the whole complex s-plane and satisfies a functional equation of the form $s \leftrightarrow 1 - s$ whose Γ-factor is given by

$$
\Gamma(s, Ad(\lambda)) = \Gamma_{\mathbb{C}}(s + k - 1)\Gamma_{\mathbb{R}}(s),
$$

where $\Gamma_{\mathbb{C}}(s) = 2(2\pi)^{-s}\Gamma(s)$ and $\Gamma_{\mathbb{R}}(s) = \pi^{-s/2}\Gamma(\frac{s}{2})$.

The L-function is known to be entire, and the adjoint lift of Gelbart–Jacquet is a cusp form if ρ_f is not an induced representation of a Galois character (note that $L(s, Ad(\lambda) \otimes \alpha)$ for $\alpha = \left(\frac{K/\mathbb{Q}}{}\right)$ has a pole at $s = 1$ if $\rho_f = \mathrm{Ind}_K^{\mathbb{Q}} \varphi$ for a quadratic field).

To see this, suppose that ρ_f is an induced representation $\mathrm{Ind}_K^{\mathbb{Q}} \varphi$ for a Galois character $\varphi : \mathrm{Gal}(\overline{\mathbb{Q}}/K) \to \overline{\mathbb{Q}}_p^\times$ (associated to a Hecke character). Then we have $Ad(\rho_f) \cong \alpha \oplus \mathrm{Ind}_K^{\mathbb{Q}}(\varphi^-)$ by Proposition 7.3.2, where $\alpha = \left(\frac{K/\mathbb{Q}}{}\right)$ is the Legendre symbol. Since λ is cuspidal, ρ_f is irreducible, and hence $\varphi^- \neq 1$. Thus $L(s, Ad(\lambda)) = L(s, \alpha)L(s, \varphi^-)$ is still an entire function, but $L(s, Ad(\lambda) \otimes \alpha)$ has a simple pole at $s = 1$.

We now give a sketch of a proof of the meromorphic continuation of $L(s, Ad(\lambda))$ and its analyticity around $s = 1$ following [LFE, Chapter 9]:

Theorem 9.1.3 (Shimura). *Let the notation and the assumptions be as above. Let $\lambda : h_k(\Gamma_0(C), \psi; \mathbb{Z}[\psi]) \to \mathbb{C}$ be a $\mathbb{Z}[\psi]$-algebra homomorphism for $k \geq 2$. Then*

$$\Gamma(s, Ad(\lambda))L(s, Ad(\lambda))$$

has an analytic continuation to the whole complex s-plane and

$$\Gamma(1, Ad(\lambda))L(1, Ad(\lambda)) = 2^k C^{-1} \prod_{l \in I}(1 + l^{-1})^{-1} \int_{\Gamma_0(C) \backslash \mathfrak{H}} |f|^2 y^{k-2} dx dy,$$

where $f = \sum_{n=1}^\infty \lambda(T(n))q^n \in S_k(\Gamma_0(C), \psi)$ and $z = x + iy \in \mathfrak{H}$. If $C = 1$, we have the following functional equation:

$$\Gamma(s, Ad(\lambda))L(s, Ad(\lambda)) = \Gamma(1 - s, Ad(\lambda))L(1 - s, Ad(\lambda)).$$

The above exact form of $\Gamma(1, Ad(\lambda))L(1, Ad(\lambda))$ generalizes [H09, (3.5)].

Proof. *Step 1: Tensor product L.* We consider $L(s - k + 1, \rho_f \otimes \widetilde{\rho}_f)$ for the Galois representation associated to λ. Since $\rho_f \otimes \widetilde{\rho}_f = 1 \oplus Ad(\rho_f)$, we have

$$L(s, \rho_f \otimes \widetilde{\rho}_f) = L(s, Ad(\lambda))\zeta(s) \qquad (9.2)$$

for the Riemann zeta function $\zeta(s)$. Our method is the Rankin convolution. To perform convolution, we define

$$D(s, \lambda, \overline{\lambda}) := \sum_{n=1}^\infty \lambda(T(n))\overline{\lambda}(T(n))n^{-s} = \prod_l E^{\lambda, \overline{\lambda}}(l^{-s})^{-1},$$

where $\overline{\lambda}$ is the complex conjugate of λ. This L-function converges locally uniformly and absolutely if $\mathrm{Re}(s) > k$, as $|\lambda(T(n))| \leq Cn^{(k-1)/2+\epsilon}$ for $\epsilon > 0$ by Ramanujan–Petersson conjecture proven by Deligne via Weil's bound.

Step 2: Euler factorization. Let us compute the Euler factor $E_l^{\lambda,\overline{\lambda}}(X)$. In Cases I, I' and E, $\lambda(T(p^n)) = 0$; so, plainly $E_l^{\lambda,\overline{\lambda}}(X) = 1$. We first assume to be in Case U. Then writing $\alpha = \alpha_l$ and $\beta = \beta_l$, we have $\alpha\beta = l^{k-1}\psi(l)$, $|\alpha| = |\beta| = l^{(k-1)/2}$ and $\lambda(T(p^n)) = \frac{\alpha^{n+1}-\beta^{n+1}}{\alpha-\beta}$ by Exercise 3.2.17. Then

$$(\alpha - \beta)(\overline{\alpha} - \overline{\beta})XE_l^{\lambda,\overline{\lambda}}(X)^{-1} = \sum_{n=1}^{\infty}(\alpha^{n+1} - \beta^{n+1})(\overline{\alpha}^{n+1} - \overline{\beta}^{n+1})X^{n+1}$$

$$= \sum_{n=0}^{\infty}(\alpha^n\overline{\alpha}^n - \alpha^n\overline{\beta}^n - \overline{\alpha}^n\beta^n + \beta^n\overline{\beta}^n)X^n$$

$$= (1 - \alpha\overline{\alpha}X)^{-1} - (1 - \alpha\overline{\beta}X)^{-1} - (1 - \overline{\alpha}\beta X)^{-1} + (1 - \beta\overline{\beta}X)^{-1}$$

$$= \frac{(\alpha - \beta)(\overline{\alpha} - \overline{\beta})X(1 - \alpha\overline{\alpha}\beta\overline{\beta}X^2)}{(1 - \alpha\overline{\alpha}X)(1 - \alpha\overline{\beta}X)(1 - \overline{\alpha}\beta X)(1 - \beta\overline{\beta}X)}.$$

Thus we get

$$E_l^{\lambda,\overline{\lambda}}(X)^{-1} = \frac{(1 - \alpha\overline{\alpha}\beta\overline{\beta}X^2)}{(1 - \alpha\overline{\alpha}X)(1 - \alpha\overline{\beta}X)(1 - \overline{\alpha}\beta X)(1 - \beta\overline{\beta}X)}$$

$$= \frac{(1 - l^{2k-2}X^2)}{(1 - \alpha\overline{\alpha}X)(1 - \alpha\overline{\beta}X)(1 - \overline{\alpha}\beta X)(1 - \beta\overline{\beta}X)}. \quad (9.3)$$

Much easier computation in Case R produces

$$E_l^{\lambda,\overline{\lambda}}(X)^{-1} = \frac{1}{1 - \alpha\overline{\alpha}X} = \frac{1}{1 - l^{k-1}X}. \quad (9.4)$$

Then by Proposition 9.1.1, we get

$$E_l^{\lambda,\overline{\lambda}}(l^{1-k}X) = \begin{cases} \frac{E^{\lambda\otimes\overline{\lambda}}(X)}{(1-X^2)} & \text{in Case U,} \\ \frac{E^{\lambda\otimes\overline{\lambda}}(X)}{(1-X)} & \text{in Cases R, I' and E,} \\ \frac{E^{\lambda\otimes\overline{\lambda}}(X)}{(1-X)(1+X)} & \text{in Case I,} \end{cases} \quad (9.5)$$

and hence, noting $S - I = R \cup I' \cup E$,

$$\zeta(2s)D(s + k - 1, \lambda, \overline{\lambda}) = L(s, \lambda \otimes \widetilde{\lambda}) \prod_{l \in S-I}(1 + l^{-s})^{-1}$$

$$= \zeta(s)L(s, Ad(\lambda)) \prod_{l \in S-I}(1 + l^{-s})^{-1}. \quad (9.6)$$

Step 3: Eisenstein series. Define as in [LFE, Theorem 9.4.1] an Eisenstein series

$$E_C(z;s) = E'_{0,C}(z;s,1) := \sum_{\substack{(m,n)\in\mathbb{Z}^2-\{(0,0)\},\ n:\ \text{prime to } C}} |Cmz+n|^{-2s}.$$

As is well known (e.g., [LFE, §9.3], [MFM, Theorem 7.2.9]), this Eisenstein series has a meromorphic continuation as a function of s and real analytic with respect to z. It is holomorphic if $\mathrm{Re}(s) > 1$ and has a simple pole at $s = 1$. We have [MFM, Corollary 7.2.10] (or [LFE, §9.4])

$$\mathrm{Res}_{s=1} E_C(z;s) = \frac{\pi}{Cy} \prod_{p|C} (1 - \frac{1}{p}). \tag{9.7}$$

For each (Cm, n) in the sum of $E_C(z;s)$, we can write $(Cm, n) = d(Cm', n')$ with mutually prime Cm', n' and the GCD $d > 0$ prime to C of m, n uniquely. Note that for $\Gamma_\infty := \{(\begin{smallmatrix}1 & n \\ 0 & 1\end{smallmatrix}) \,|\, n \in \mathbb{Z}\}$, we have $\Gamma_\infty \backslash \Gamma_0(C) \cong \{(Cm', n') | Cm' \text{ prime to } n'\}/\{\pm 1\}$ by sending $(\begin{smallmatrix}* & * \\ c & d\end{smallmatrix})$ to (c, d). Thus writing $j((\begin{smallmatrix}* & * \\ c & d\end{smallmatrix}), z) := cz + d$, we find

$$E_C(z;s) = 2 \prod_{p|C} (1 - p^{-2s})\zeta(2s)E_C^*(z;s) \tag{9.8}$$

for $E_C^*(z;s) = \sum_{\gamma\in\Gamma_\infty\backslash\Gamma_0(C)} |j(\gamma,z)|^{-2s} \overset{(*)}{=} y^{-s}\sum_{\gamma\in\Gamma_\infty\backslash\Gamma_0(C)} (y\circ\gamma)^s$. The identity $(*)$ follows from $y\circ\gamma = y|j(\gamma,z)|^{-2}$.

Step 4: Rankin convolution. Since

$$\Gamma_\infty\backslash\mathfrak{H} \cong F = \{x + iy | x \in (0,1], y \in (0,\infty)\},$$

writing the multiplicative Haar measure $d^\times y = y^{-1}dy$ we have

$$\int_0^\infty \int_0^1 |f|^2 y^s dx d^\times y$$

$$= \int_0^\infty \sum_{m,n=1}^\infty \lambda(T(n))\overline{\lambda}(T(m)) \int_0^1 \exp(2\pi ix)dx \exp(-2\pi(n-m)y)y^s d^\times y$$

$$= \sum_{n=1}^\infty \int_0^\infty |\lambda(T(n))|^2 \exp(-4\pi ny)y^s d^\times y = (4\pi)^{-s}\Gamma(s)D(s,\lambda,\overline{\lambda})$$

if $\mathrm{Re}(s) > 1$. On the other hand, since $d\mu(z) = y^{-2}dxdy$ is $\Gamma_0(C)$-invariant and $y\circ\gamma = y|j(\gamma,z)|^{-2}$, taking the fundamental domain Φ of $\Gamma_0(C)\backslash\mathfrak{H}$,

$$\int_0^\infty \int_0^1 |f|^2 y^s dx d^\times y = \int_F |f|^2 y^{s+1} d\mu(z)$$

$$= \sum_{\gamma\in\Gamma_\infty\backslash\Gamma_0(C)} \int_\Phi |f(\gamma(z))|^2 (y\circ\gamma)^{s+1} d\mu(\gamma(z))$$

$$= \int_{\Gamma_0(C)\backslash\mathfrak{H}} |f|^2 y^{s+1} E_C^*(z;s+1-k)d\mu(z).$$

By (9.8), we get

$$2\prod_{p|C}(1 - p^{-2(s+1-k)})(4\pi)^{-s}\Gamma(s)\zeta(2(s + 1 - k))D(s, \lambda, \overline{\lambda})$$

$$= \int_{\Gamma_0(C)\backslash \mathfrak{H}} |f|^2 y^{s+1} E_C(z; s + 1 - k) d\mu(z).$$

Or equivalently,

$$2\prod_{p|C}(1 - p^{-2s})(4\pi)^{-s-k+1}\Gamma(s + k - 1)\zeta(2s)D(s + k - 1, \lambda, \overline{\lambda})$$

$$= \int_{\Gamma_0(C)\backslash \mathfrak{H}} |f|^2 y^{s+k} E_C(z; s) d\mu(z).$$

Incorporating (9.6) (and noting $S - I = R \cup I' \cup E$), we get

$$2\prod_{p|C}(1 - p^{-2s}) \prod_{l \in S-I} (1 + l^{-s})^{-1}(4\pi)^{-s-k+1}\Gamma(s + k - 1)L(s, \lambda \otimes \widetilde{\lambda})$$

$$= 2\prod_{p|C}(1 - p^{-2s}) \prod_{l \in S-I} (1 + l^{-s})^{-1}(4\pi)^{-s-k+1}\Gamma(s + k - 1)\zeta(s)L(s, Ad(\lambda))$$

$$= \int_{\Gamma_0(C)\backslash \mathfrak{H}} |f|^2 y^{s+k} E_C(z; s) d\mu(z).$$

Since f is rapidly decreasing towards cusps and E_C is slowly increasing (outside $s = 1$), the integral is convergent outside the singularity of $E_C(z; s)$, and we conclude meromorphy of $L(s, \lambda \otimes \widetilde{\lambda})$ over $\mathbb{C}$ and holomorphy if $\mathrm{Re}(s) > 1$.

Step 5: Residue formula. Since $E_C(z; s)$ and $\zeta(s)$ has a simple pole at $s = 1$, we find that $L(s, \lambda \otimes \widetilde{\lambda})$ has a simple pole at $s = 1$. By (9.7) combined with $\zeta(2) = \frac{\pi^2}{6}$ and $\mathrm{Res}_{s=1} \zeta(s) = 1$, we can compare the residue of the two sides and find

$$(f, f) := \int_{X_0(C)} |f|^2 y^{k-2} dx dy = 2^{-k} C \prod_{l \in I}(1 + l^{-1})\Gamma(1, Ad(\lambda))L(1, Ad(\lambda)).$$

$$(9.9)$$

Step 6: Functional equation. When $C = 1$, the functional equation of the Eisenstein series is particularly simple:

$$\Gamma_\mathbb{C}(s)E_1(z; s) = 2^{1-2s}\Gamma_\mathbb{C}(1 - s)E_1(z; 1 - s),$$

which combined with the functional equation of the Riemann zeta function (e.g. [LFE, Theorem 2.3.2 and Corollary 8.6.1]) yields the functional equation of the adjoint L-function $L(s, Ad(\lambda))$. $\square$

Exercise 9.1.2. In the above proof, justify all interchanges of the integral and the sum and verify absolute and locally uniform convergence if $\mathrm{Re}(s) \gg 0$.

9.2 Integrality of adjoint L-values

Hereafter we assume $p \nmid C$ and $p \geq 5$. By the explicit form of the Gamma factor, $\Gamma(s, Ad(\lambda))$ is finite at $s = 0, 1$ (as long as $k \geq 2$), and hence $L(1, Ad(\lambda))$ is a critical value in the sense of Deligne and Shimura, as $L(s, Ad(\lambda))$ is finite at these points by Theorem 9.1.3. Thus we expect algebraicity of the L-value divided by a period $\Omega(\pm, \lambda)$ of the Hecke eigenform $f = \sum_{n=1}^{\infty} \lambda(T(n))q^n$. This was first shown by Sturm (see [S80] and [S89]) by using Shimura's integral expression (in [S75]). Here we describe the integrality of the value, following [H81a] and [H88a]. Then we shall relate in the following section the size of the congruence module $C_0(\lambda; W)$ with the p-primary part of the critical value $\frac{\Gamma(1, Ad(\lambda))L(1, Ad(\lambda))}{\Omega(+, \lambda)\Omega(-, \lambda)}$. Since our argument can be substantially simplified if we ignore p-adic unit constants, as we mentioned, we use "$\doteq$" in place of "$=$" to indicate omission of p-adic unit constants. In particular, we often ignore a power of C, i and 2 as we suppose $p \geq 5$ and $p \nmid C$.

9.2.1 *Eichler–Shimura isomorphism again*

Consider the defining inclusion $I : SL_2(\mathbb{Z}) \hookrightarrow \mathrm{Aut}_{\mathbb{C}}(\mathbb{C}^2) = GL_2(\mathbb{C})$. Let us take the n-th symmetric tensor representation $I^{sym\otimes n}$ whose module twisted by the action of ψ, we write as $L(n, \psi; \mathbb{C})$. Recall the Eichler–Shimura isomorphism in §4.2.4,

$$\delta : S_k(\Gamma_0(C), \psi) \oplus \overline{S}_k(\Gamma_0(C), \psi) \cong H^1_!(\Gamma_0(C), L(n, \psi; \mathbb{C})), \qquad (9.10)$$

where $k = n + 2$, $\overline{S}_k(\Gamma_0(C), \psi)$ is the space of anti-holomorphic cusp forms of weight k of "Neben" type character ψ, and

$$H^1_!(\Gamma_0(C), L(n, \psi; \mathbb{C})) \subset H^1(\Gamma_0(C), L(n, \psi; \mathbb{C}))$$

is the *cuspidal* cohomology groups defined in (4.21) and (4.22).

Let A be a $\mathbb{Z}[\psi]$-algebra. The periods $\Omega(\pm, \lambda; A)$ measure the difference of two A-integral structures coming from algebro-geometric space $S_{k,\psi/A}$ and topologically defined

$$H^1_!(\Gamma_0(C), L(n, \psi; A)) \cong H^1_!(X_0(C), \mathcal{L}(n, \psi; A))$$

(for the A-rational symmetric tensors $L(n, \psi; A)$ and the associated sheaf $\mathcal{L}(n, \psi; A)$ on the modular curve $Y_0(C)$) connected by the Eichler–Shimura comparison map.

Since the isomorphism classes over $\mathbb{Q}$ of the symmetric n-th tensor $I^{sym \otimes n}$ of I can have several classes over $\mathbb{Z}$, we need to have an explicit construction of the $\Gamma_0(C)$-module $L(n, \psi; \mathbb{Z}[\psi])$. Here is a more concrete definition of $SL_2(\mathbb{Z})$-module as the space of homogeneous polynomial in (X, Y) of degree n with coefficients in A. We let $\gamma = \left(\begin{smallmatrix} a & b \\ c & d \end{smallmatrix}\right) \in M_2(\mathbb{Z}) \cap GL_2(\mathbb{Q})$ act on $P(X, Y) \in L(n, \psi; A)$ by

$$(\gamma P)(X, Y) = \psi(d) P((X, Y)^t \gamma^\iota) \quad \text{with } \gamma^\iota = (\det \gamma) \gamma^{-1} = \left(\begin{smallmatrix} d & -b \\ -c & a \end{smallmatrix}\right).$$

The interior cohomology group $H^1_!(\Gamma_0(C), L(n, \psi; A))$ is defined in (4.21) as the image of compactly supported cohomology group of the sheaf associated to $L(n, \psi; A)$.

The Eichler–Shimura map δ in (9.10) is specified as follows [LFE, §6.2]:

$$\omega(f) = \begin{cases} f(z)(X - zY)^n dz & \text{if } f \in S_k(\Gamma_0(C), \psi), \\ f(z)(X - \overline{z}Y)^n d\overline{z} & \text{if } f \in \overline{S}_k(\Gamma_0(C), \psi). \end{cases}$$

Then we associate to f the cohomology class of the 1-cocycle $\gamma \mapsto \int_z^{\gamma(z)} \omega(f)$ of $\Gamma_0(C)$ for a fixed point z on the upper half complex plane. The map δ does not depend on the choice of z.

Let us prepare preliminary facts. Let $\Gamma = \Gamma_C = \Gamma(3) \cap \Gamma_0(C)$ for $\Gamma(3) = \text{Ker}(SL_2(\mathbb{Z}) \to SL_2(\mathbb{Z}/3\mathbb{Z}))$. A key feature of Γ_C is that it acts on $\mathfrak{H}$ freely without fixed point. To see this, let Γ_z be the stabilizer of $z \in \mathfrak{H}$ in Γ. Since the stabilizer of z in $SL_2(\mathbb{R})$ is a maximal compact subgroup C_z of $SL_2(\mathbb{R})$, $\Gamma_z = \Gamma \cap C_z$ is compact-discrete and hence is finite. Thus if Γ is torsion-free, it acts freely on $\mathfrak{H}$. Pick a torsion-element $\gamma \in \Gamma$. Then two eigenvalues ζ and $\overline{\zeta}$ of γ are roots of unity complex conjugate each other. Since Γ cannot contain -1, $\zeta \notin \mathbb{R}$. Thus if $\gamma \neq 1$, we have $-2 < \text{Tr}(\gamma) = \zeta + \overline{\zeta} < 2$. Since $\gamma \equiv 1 \mod 3$, $\text{Tr}(\gamma) \equiv 2 \mod 3$, which implies $\text{Tr}(\gamma) = -1$. Thus γ satisfies $\gamma^2 + \gamma + 1 = 0$ and hence $\gamma^3 = 1$. Therefore $\mathbb{Z}[\gamma] \cong \mathbb{Z}[\omega]$ for a primitive cubic root ω. Since 3 ramifies in $\mathbb{Z}[\omega]$, $\mathbb{Z}[\omega]/3\mathbb{Z}[\omega]$ has a unique maximal ideal $\mathfrak{m}$ with $\mathfrak{m}^2 = 0$. The ideal $\mathfrak{m}$ is principal and is generated by ω. Thus the matrix $(\gamma - 1 \mod 3)$ corresponds $(\omega - 1 \mod 3)$, which is non-zero nilpotent. This $\gamma - 1 \mod 3$ is non-zero nilpotent, showing $\gamma \notin \Gamma(3)$, a contradiction.

By the above argument, the fundamental group of $Y = \Gamma_C \backslash \mathfrak{H}$ is isomorphic to Γ_C. Then we may consider the locally constant sheaf $\mathcal{L}(n, \psi; A)$ of sections associated to the following covering:

$$\mathcal{X} = \Gamma_C \backslash (\mathfrak{H} \times L(n, \psi; A)) \twoheadrightarrow Y \quad \text{via } (z, P) \mapsto z.$$

Since Γ_C acts on $\mathfrak{H}$ without fixed point, the space $\mathcal{X}$ is locally isomorphic to Y, and hence $\mathcal{L}(n, \psi; A)$ is a well defined locally constant sheaf. In this setting, there is a canonical isomorphism in (4.22):

$$H^1(\Gamma_C, L(n, \psi; A)) \cong H^1(Y, \mathcal{L}(n, \psi; A)).$$

Note that $\Gamma_0(C)/\Gamma_C$ is a finite group whose order is a factor of 24. Thus as long as 6 is invertible in A, we have by the restriction map

$$H^0(\Gamma_0(C)/\Gamma_C, H^1(\Gamma_C, L(n, \psi; A))) = H^1(\Gamma_0(C), L(n, \psi; A)). \tag{9.11}$$

As long as 6 is invertible in A, all perfectness of Poincaré duality for smooth quotient $\Gamma_C \backslash \mathfrak{H}$ descends over A to $H^1(\Gamma_0(C), L(n, \psi; A))$; so, we pretend as if $X_0(C)$ is smooth hereafter, as we assume that 6 is invertible in A.

For simplicity, we write Γ for $\Gamma_0(C)$ and $Y = Y_0(C) := \Gamma_0(C) \backslash \mathfrak{H}$. Let $\mathfrak{S} = \Gamma \backslash \mathbf{P}^1(\mathbb{Q}) \cong \Gamma \backslash SL_2(\mathbb{Z})/\Gamma_\infty$ for $\Gamma_\infty = \{\gamma \in SL_2(\mathbb{Z}) | \gamma(\infty) = \infty\}$. Thus $\mathfrak{S}$ is the set of cusps of Y. We can take a neighborhood of ∞ in Y isomorphic to the cylinder $\mathbb{C}/\mathbb{Z}$. Since we have a neighborhood of each cusp isomorphic to a given neighborhood of ∞, we can take an open neighborhood of each cusp of Y isomorphic to the cylinder. We compactify Y adding the circle $S^1 = \mathbb{R}/\mathbb{Z}$ at every cusp. We write $\overline{Y}$ for the compactified space. Then

$$\partial\overline{Y} = \bigsqcup_{\mathfrak{S}} S^1,$$

and

$$H^q(\partial\overline{Y}, \mathcal{L}(n, \psi; A)) \cong \bigoplus_{s \in \mathfrak{S}} H^q(\Gamma_s, L(n, \psi; A)),$$

where Γ_s is the stabilizer in Γ of a cusp $s \in \mathbf{P}^1(\mathbb{Q})$ representing an element in $\mathfrak{S}$. Since $\Gamma_s \cong \mathbb{Z}$, $H^q(\partial\overline{Y}, \mathcal{L}(n, \psi; A)) = 0$ if $q > 1$.

We have a commutative diagram whose horizontal arrows are given by the restriction maps:

$$
\begin{array}{ccc}
H^1(Y, \mathcal{L}(n, \psi; A)) & \xrightarrow{\text{Res}} & H^1(\partial\overline{Y}, \mathcal{L}(n, \psi; A)) \\
\wr\downarrow & & \wr\downarrow \\
H^1(\Gamma, L(n, \psi; A)) & \xrightarrow{\text{Res}} & \bigoplus_{s \in \mathfrak{S}} H^1(\Gamma_s, L(n, \psi; A)).
\end{array}
$$

We then define $H^1_!$ by the kernel of the restriction map.

We have the boundary exact sequence in (4.39):

$$0 \to H^0(Y, \mathcal{L}(n, \psi; A)) \to H^0(\partial\overline{Y}, \mathcal{L}(n, \psi; A)) \to H^1_c(Y, \mathcal{L}(n, \psi; A))$$
$$\xrightarrow{\pi} H^1(Y, \mathcal{L}(n, \psi; A)) \to H^1(\partial\overline{Y}, \mathcal{L}(n, \psi; A)) \to H^2_c(Y, \mathcal{L}(n, \psi; A)) \to 0.$$

Here H_c^1 is the sheaf cohomology group with compact support, and the map π sends each compactly supported cohomology class to its usual cohomology class. Thus $H_!^1$ is equal to the image of π, made of cohomology classes rapidly decreasing towards cusps (when $A = \mathbb{C}$). We also have (cf. Proposition 4.2.9)

$$H_c^2(Y, \mathcal{L}(n, \psi; A)) \cong L(n, \psi; A)/\sum_{\gamma \in \Gamma}(\gamma - 1)L(n, \psi; A) \text{ (so, } H_c^2(Y, A) = A),$$

$$H_c^0(Y, \mathcal{L}(n, \psi; A)) = 0 \quad \text{and} \quad H^0(Y, \mathcal{L}(n, \psi; A)) = H^0(\Gamma, L(n, \psi; A)).$$
$$\text{(9.12)}$$

When $A = \mathbb{C}$, the isomorphism $H_c^2(Y, \mathbb{C}) \cong \mathbb{C}$ is given by $[\omega] \mapsto \int_Y \omega$, where ω is a compactly supported 1-form representing the cohomology class $[\omega]$ (de Rham theory; cf. [LFE, Proposition A.6]).

9.2.2 *Modified duality pairing*

We slightly modify the cup product pairing we studied in §4.2.6. Suppose that $n!$ is invertible in A. Then $\binom{n}{j}^{-1} \in A$ for binomial symbols $\binom{n}{j}$. We can then define a pairing $[\ ,\ ] : L(n, \psi; A) \times L(n, \psi^{-1}; A) \to A$ by

$$[\sum_j a_j X^{n-j}Y^j, \sum_j b_j X^{n-j}Y^j] = \sum_{j=0}^n (-1)^j \binom{n}{j}^{-1} a_j b_{n-j}. \qquad (9.13)$$

By definition,

$$[(X - zY)^n, (X - \bar{z}Y)^n] = (z - \bar{z})^n. \qquad (9.14)$$

If we take the basis $\{X^{n-j}Y^j\}_j$ of $L(n; A)$ to identify $L(n; A)$ with A^{n+1} and write the pairing in §4.2.6 as $(x, y) = {}^t x S y$ for $x, y \in A^{n+1}$. Then the new pairing corresponds to S^{-1}. In other words, our pairing is the dual pairing of the one in §4.2.6 and is more convenient for our computation because of (9.14). This shows $[\gamma P, \gamma Q] = \det \gamma^n [P, Q]$ for $\gamma \in GL_2(A)$ as $(\gamma P, \gamma Q) = \det \gamma^n (P, Q)$. Thus we have a Γ-homomorphism $L(n, \psi; A) \otimes_A L(n, \psi^{-1}; A) \to A$, and we get the cup product pairing

$$[\cdot, \cdot] : H_c^1(Y, \mathcal{L}(n, \psi; A)) \times H^1(Y, \mathcal{L}(n, \psi^{-1}; A)) \longrightarrow H_c^2(Y, A) \cong A.$$

This pairing induces the cuspidal pairing

$$[\cdot, \cdot] : H_!^1(Y, \mathcal{L}(n, \psi; A)) \times H_!^1(Y, \mathcal{L}(n, \psi^{-1}; A)) \longrightarrow A. \qquad (9.15)$$

By (9.11), we identify the cohomology group $H_!^1(\Gamma_0(C), L(n, \psi; A))$ as a subspace of $H_!^1(Y, \mathcal{L}(n, \psi; A))$ and write $[\cdot, \cdot]$ for the pairing induced on $H_!^1(\Gamma_0(C), \mathcal{L}(n, \psi; A))$ by the above pairing of $H_!^1(Y, \mathcal{L}(n, \psi; A))$.

9.2.3 *Hecke Hermitian duality*

There are three natural operators acting on the cohomology group: one is the action of Hecke operators $T(n)$ on $H^1_!(\Gamma_0(C), L(n, \psi; A))$ defined in §4.2.7, the second is an involution τ induced by the action of $\tau = \left(\begin{smallmatrix} 0 & -1 \\ C & 0 \end{smallmatrix}\right)$, and the third is an action of complex conjugation c given by $c^* \omega(z) := \omega(-\bar{z})$ for a differential form ω. The Eichler–Shimura map δ and complex conjugation c commute with $T(n)$. We write $H^1_!(\Gamma_0(C), L(n, \psi; A))[\pm]$ for the $\pm$-eigenspace of c. Then it is known (e.g., [LFE, 6.3, (11)]) that

$$H^1_!(\Gamma_0(C), L(n, \psi; \mathbb{Q}(\lambda)))[\pm] \text{ is } h_k(C, \psi; \mathbb{Q}(\lambda))\text{-free of rank 1.} \quad (9.16)$$

Supposing that A contains the eigenvalues $\lambda(T(n))$ for all n, we write $H^1_!(\Gamma_0(C), L(n, \psi; A))[\lambda, \pm]$ for the λ-eigenspace under $T(n)$.

The action of $\tau = \left(\begin{smallmatrix} 0 & -1 \\ C & 0 \end{smallmatrix}\right)$ defines a quasi-involution on the cohomology

$$\tau : H^1_!(\Gamma_0(C), L(n, \psi; A)) \to H^1_!(\Gamma_0(C), L(n, \psi^{-1}; A)),$$

which is given by $u \mapsto \{\gamma \mapsto \tau u(\tau \gamma \tau^{-1})\}$ for each homogeneous 1-cocycle u. The cocycle $u|\tau$ has values in $L(n, \psi^{-1}; A)$ because conjugation by τ interchanges the diagonal entries of γ. We have $\tau^2 = (-C)^n$ and $[x|\tau, y] = [x, y|\tau]$. Then we modify the above pairing $[\cdot, \cdot]$ by τ and define $\langle x, y \rangle := [x, y|\tau]$ ([LFE, 6.3 (6)]). As described in §4.2.7, we have a natural action of Hecke operators $T(n)$ on $H^1_!(\Gamma_0(C), L(n, \psi; A))$. The operator $T(n)$ is symmetric with respect to this pairing (e.g., [MFM, Theorem 4.5.5]):

$$\langle x|T(n), y \rangle = \langle x, y|T(n) \rangle. \quad (9.17)$$

Theorem 9.2.1. *Assume $p \geq 5$, and let $W \subset \overline{\mathbb{Q}}_p$ be a complete discrete p-adic valuation ring containing $\mathbb{Z}_p[\psi]$. Then*

$$\langle \cdot, \cdot \rangle : H^1_{!,ord}(\Gamma_0(C), L(n, \psi; W)) \times H^1_{!,ord}(\Gamma_0(C), L(n, \psi; W)) \to W$$

is a perfect pairing.

We give a sketch of the proof. See [H81b, Proposition 3.4] and [H88a, Theorem 3.1] for more details.

Proof. If $n = 0$, $H^1_!(Y, ?)$ is identical to $H^1(X, ?)$ for the compactification X of Y. We can reduce the pairing modulo p and need to prove perfectness on $H^1_{!,ord}(\Gamma_0(C), L(n, \psi; \mathbb{F})) = H^1_{!,ord}(\Gamma_0(C), L(n, \psi; W)) \otimes_W \mathbb{F}$ by Corollaries 4.2.13 and 4.2.23. If $n = 0$, this is the well known Poincaré duality as the action of τ is an automorphism of the modular curve. When ψ is non-trivial, the duality is not often stated in textbooks on Topology, but for example, using the comparison isomorphism between Betti cohomology and étale cohomology, one can use the duality theorem in étale cohomology [ECH, Proposition V.2.2].

Assume that $n > 0$. We prove the perfectness for $(x, y) := [x, y|\tau]$ for the pairing in §4.2.6. This is enough as $\langle x, y \rangle$ we study here is the dual pairing, if $(\cdot, \cdot)$ is perfect, $\langle \cdot, \cdot \rangle$ is also perfect. We first prove $(\cdot, \cdot)$: $H^1_{!,ord}(\Gamma_0(pC), L(n, \psi; \mathbb{F})) \times H^1_{!,ord}(\Gamma_0(pC), L(n, \psi; \mathbb{F})) \to \mathbb{F}$ is perfect. One can check the maps I and J in Theorem 4.2.24 are adjoint to each other under $\langle \cdot \cdot \rangle$ up to sign. Thus the result follows for level pC as Corollary 4.2.26 reduces perfectness to the case of $n = 0$. As is well known, if $n > 0$, the trace map and restriction map induces

$$H^1_{!,ord}(\Gamma_0(pC), L(n, \psi; ?)) \cong H^1_{!,ord}(\Gamma_0(C), L(n, \psi; ?))$$

for $? = W, \mathbb{F}, W/\mathfrak{m}^n_W W$ just because $\mathrm{Tr}_{\Gamma_0(C)/\Gamma_0(pC)} \circ \alpha \circ \mathrm{Res}_{\Gamma_0(C)/\Gamma_0(pC)} = T(p)$ for $\alpha = \begin{pmatrix} p & 0 \\ 0 & 1 \end{pmatrix}$ (by computation). Thus the perfectness of level pC implies that of C. □

Exercise 9.2.1. In the same manner as in the above proof, prove that $H^1_{!,ord}(\Gamma, L(n, \psi; W))$ and $H^1_{!,ord}(\Gamma, L(n, \psi^{-1}; K/W))$ are Pontryagin dual to each other by the pairing induced by $\langle \cdot, \cdot \rangle$.

We note here

Corollary 9.2.2. *Let* $\Gamma_1(N) \subset \Gamma \subset \Gamma_0(N)$ *for* $N = C$ *or* Cp *be a subgroup with p-power index* $(\Gamma_0(N) : \Gamma)$. *Then* $H^1_{!,ord}(\Gamma, L(n, \psi; W))$ *is* $W[\Gamma_0(N)/\Gamma]$-*free of finite rank.*

Writing $P := \Gamma_0(N)/\Gamma$, we shall give a sketch of a proof.

Proof. By Lemma 4.2.31, $H^1_{!,ord}(\Gamma, L(n, \psi^{-1}; K/W))$ is W-divisible. Thus $H^1_{!,ord}(\Gamma, L(n, \psi^{-1}; W))$ is W-free of finite rank by Exercise 9.2.1.

Write simply $M = L(n, \psi^{-1}; K/W)$. The inflation restriction sequence

$$H^1(P, M^\Gamma) \hookrightarrow H^1(\Gamma_0(N), M) \to H^1(\Gamma, M) \to H^2(P, M^\Gamma)$$

is exact. As with Lemma 4.2.38, we can check that a power of $U(p)$ kills $H^q(P, M^\Gamma)$ for $q = 1, 2$. Thus $H^1_{ord}(\Gamma_0(N), M) \cong H^1_{ord}(\Gamma, M)$. The same assertion also holds for the boundary cohomology group; so, $G^1_{ord}(\Gamma_0(N), M) \cong G^1_{ord}(\Gamma, M)$ under the notation defined just above Proposition 4.2.19, and hence we get $H^1_{!,ord}(\Gamma_0(N), M) \cong H^1_{!,ord}(\Gamma, M)$.

First we suppose that W contains a primitive p^f-th root for $p^f := |P|$. The p-group P acts on M^Γ naturally. We twist the action of $\Gamma_0(N)$ on M by a character $\phi : P \to W^\times$. In other words, for a cocycle $u : \Gamma_0(N) \to M$, writing the original action $\delta u(g) = \delta u(\delta^{-1}g\delta)$ for $\delta \in \Gamma_0(N)$, the new action is given by $\delta \cdot u = \phi(\delta)(\delta u)$. Then the inflation-restriction sequence as above for this new action produces an isomorphism:

$$H^1(\Gamma, L(n, \psi^{-1}; K/W))[\phi^{-1}] \cong H(\Gamma_0(N), L(n, \psi^{-1}\phi^{-1}; K/W),$$

where "$[\phi^{-1}]$" indicates ϕ^{-1}-eigenspace under the action of P. By Lemma 4.2.31, the right-hand-side is W-divisible, so is the left-hand-side. Apply Pontryagin dual. For the kernel $\mathfrak{a}_\phi := \text{Ker}(W[P] \xrightarrow{\phi} \overline{\mathbb{Q}})$, we find $H^1_{!,ord}(\Gamma, L(n, \psi; W)) \otimes W[P]/\mathfrak{a}_\phi \cong H^1_{!,ord}(\Gamma_0(N), L(n, \psi; W))$ which is W-free of finite rank. Then by Lemma 4.2.41 (applied to the group algebra $A = W[P]$), we get the freeness of the cohomology group.

If W does not contain μ_{p^f}, we find $H^1_{!,ord}(\Gamma, L(n, \psi; W)) \otimes_W W[\mu_{p^f}] \cong H^1_{!,ord}(\Gamma, L(n, \psi; W[\mu_{p^f}]))$. Since $W[\mu_{p^f}]$ is faithfully flat over W, we still have W-freeness of

$$H^1_{!,ord}(\Gamma, L(n, \psi; W)) \otimes W[P]/\mathfrak{a}_\phi \cong H^1_{!,ord}(\Gamma_0(N), L(n, \psi; W)).$$

Thus again by Lemma 4.2.41, we get the desired freeness. $\square$

9.2.4 *Integrality theorem*

We now regard $\lambda : h_{k,\psi/\mathbb{Z}[\psi]} \to \mathbb{C}$ as actually having values in $W \cap \overline{\mathbb{Q}}$ (via the fixed embedding: $\overline{\mathbb{Q}} \hookrightarrow \overline{\mathbb{Q}}_p$). Put $A = W \cap \mathbb{Q}(\lambda)$. Then A is a valuation ring of $\mathbb{Q}(\lambda)$ of residual characteristic p. By (9.16), for the image L of $H^1_!(\Gamma_0(C), L(n, \psi; A))$ in $H^1_!(\Gamma_0(C), L(n, \psi; \mathbb{Q}(\lambda)))$,

$$H^1_!(\Gamma_0(C), L(n, \psi; \mathbb{Q}(\lambda)))[\lambda, \pm] \cap L = A\xi_\pm$$

for a generator $\xi_\pm$. Then for the normalized eigenform $f \in S_k(\Gamma_0(C), \psi)$ with $T(n)f = \lambda(T(n))f$, we define $\Omega(\pm, \lambda; A) \in \mathbb{C}^\times$ by

$$\delta(f) \pm c(\delta(f)) = \Omega(\pm, \lambda; A)\xi_\pm \tag{9.18}$$

denoting by c the action of complex conjugation defined in §9.2.3.[1]

We now compute

$$\langle \Omega(+, \lambda; A)\xi_+, \Omega(-, \lambda; A)\xi_- \rangle = \Omega(+, \lambda; A)\Omega(-, \lambda; A)\langle \xi_+, \xi_- \rangle.$$

Since $\Omega(\pm, \lambda; A)$ is determined up to A-units (and A is a valuation ring of residual characteristic $p > 0$, in our computation, we ignore p-adic unit constants (as they can be absorbed in the period $\Omega(\pm, \lambda; A)$). If the reader has interest in the exact constant factor (e.g., to compare the factor with

---

[1]The above definition of the period $\Omega(\pm, \lambda; A)$ can be generalized to the Hilbert modular case and beyond [H94, §8, §13].

local Tamagawa number factors), then take a look at an exact computation in [H88a] and [DFG04], though the result described here is sharper than [DFG04] for ordinary primes as it does not require $p \geq k$. Since p is supposed to be ≥ 5 in this section, we therefore ignore a power of 2 and a power of $i = \sqrt{-1}$.

Let $f_c = \sum_{m=1}^{\infty} \overline{\lambda(T(m))} q^m$ with $\overline{\lambda(T(m))}$ meaning complex conjugation. Then $f|\tau = W(\lambda) f_c$ for and $W(\lambda) \in \mathbb{C}$ with $|W(\lambda)| = 1$ and $\delta(f)|\tau = W(\lambda)(-1)^n C^{(n/2)} \delta(f_c)$. It is known that $|W(\lambda)|_p = 1$ if $p \nmid C$. Indeed, by the (proven) local Langlands correspondence [K80], the root number is a product of local ϵ-factors (see [MFM, Corollary 4.6.18] for an elementary form and [AFR, §4.9] for the local Langlands correspondence), and the factor is essentially a local Gauss sum which is non-p-adic unit only when $p|C$. Therefore, we also ignore $W(\lambda)$ often.

By definition, we have

$$2\Omega(+, \lambda; A)\Omega(-, \lambda; A)\langle \xi_+, \xi_- \rangle = [\delta(f) + c^* \delta(f), (\delta(f) - c^* \delta(f))|\tau],$$

which is equal to, up to p-adic unit, by (9.9)

$$\int_{Y_0(C)} [\delta(f)|\tau, c^* \delta(f)] dx \wedge dy \doteq \int_{Y_0(C)} |f_c|^2 y^{k-2} dx dy$$

$$\doteq \int_{Y_0(C)} |f|^2 y^{k-2} dx dy \doteq \prod_{l \in I} (1 + \frac{1}{l}) \Gamma(1, Ad(\lambda)) L(1, Ad(\lambda)), \quad (9.19)$$

where $Y_0(C) = \Gamma_0(C) \backslash \mathfrak{H}$. This shows

Theorem 9.2.3. *Let ψ be a character of conductor C, and assume $k \geq 2$. Let $\lambda : h_k(\Gamma_0(C), \psi; \mathbb{Z}[\psi]) \to \overline{\mathbb{Q}}$ be a $\mathbb{Z}[\psi]$-algebra homomorphism. Then for a valuation ring A of $\mathbb{Q}(\lambda)$ of residual characteristic $p > 3$, we have, up to p-adic units,*

$$\prod_{l \in I} (1 + \frac{1}{l}) \frac{\Gamma(1, Ad(\lambda)) L(1, Ad(\lambda))}{\Omega(+, \lambda; A)\Omega(-, \lambda; A)} = \langle \xi_+, \xi_- \rangle \in \mathbb{Q}(\lambda),$$

and $\langle \xi_+, \xi_- \rangle \in n!^{-1} \cdot A$. Moreover, if $\lambda(T(p)) \in A^\times$, we have $\langle \xi_+, \xi_- \rangle \in A$.

The last assertion follows from Theorem 9.2.1, as the λ-eigen cohomology class belongs to $H^1_{!,ord}$ on which perfectness of the duality holds without assuming $n! \in A^\times$.

The proof of rationality of the adjoint L-values as above can be generalized even to non-critical values $L(1, Ad(\lambda) \otimes \alpha)$ for quadratic Dirichlet characters α (see [H99]).

To have the identity (9.19) up to p-adic units, we assumed $p \nmid C$. Here is a brief description of what happens if $p|C$. In the way to reach the above integrality result, we replaced Petersson inner product $(\cdot, \cdot)$ by $(\cdot, \cdot|\tau)$. Writing $f|\tau = \varepsilon(f)f_c$ for the root number $\varepsilon(f)$ for $f_c = \sum_{n=1}^{\infty} \overline{\lambda}(T(n))q^n$. The adjoint L-value is essentially equal to $(f, f) = (f_c, f_c)$ by Theorem 9.1.3, therefore the ratio $(f|\tau, f_c)/(f_c, f_c)$ appears as an error factor. If $p \nmid C$, $\varepsilon(f)$ is a p-adic unit; so, we do not have too much trouble. However it could be a non-unit if $p|C$, and this is the main reason for our assumption $p \nmid C$. The description of $\varepsilon(f)$ is rather involved (see [H88c, (5.5b)] for an exact formula).

Let W be the completion of the valuation ring A, and write $r(W) = \mathrm{rank}_{\mathbb{Z}_p} W$. As defined in §6.1.10, $\mathbb{T}_\chi^{fl} \cong \mathbb{T}_\chi$ unless $k = 2$ and $Ad(\overline{\rho})|_{D_p}$ contains $\overline{\omega}_p$ as a subquotient. Similarly, we simply put $R_\chi^{fl} = R_\chi$ unless $k = 2$ and $Ad(\overline{\rho})|_{D_p}$ contains $\overline{\omega}_p$ as a subquotient. Recall that $\mathbb{T}_\chi^{fl}$ (resp. $\mathbb{T}_\chi$) is defined as a factor of $\mathbf{h}_k(\Gamma_0(C), \psi; W)$ (resp. $\mathbf{h}_k^{ord}(\Gamma_0(Cp), \psi; W)$), and by p-stabilization, we have a canonical surjective W-algebra homomorphism $\mathbb{T}_\chi \twoheadrightarrow \mathbb{T}_\chi^{fl}$ always. By Theorem 4.1.29, $\mathbb{T}/(t - \chi(\gamma))\mathbb{T} \cong \mathbb{T}_\chi$ canonically. Thus $\mathbb{T}_\chi^{fl}$ ($\chi = \nu_p^{k-1}\psi$) is the local ring of $h_k(\Gamma_0(C), \psi; W)$ through which λ factor through. Let $1_{\mathbb{T}_\chi^{fl}}$ be the idempotent of $\mathbb{T}_\chi^{fl}$ in the Hecke algebra. By Lemma 6.2.15, under the minimality (p), (l), $\mathbb{T}_\chi$ and $\mathbb{T}_\chi^{fl}$ are reduced. Thus for the quotient field K of W, the unique local ring $\mathbb{I}_K$ of $h_k(\Gamma_0(C), \psi; K)$ through which λ factors is isomorphic to K. Let 1_λ be the idempotent of $\mathbb{I}_K$ in $h_k(\Gamma_0(C), \psi; K)$. Then we have the following important corollary.

Corollary 9.2.4. *Let the assumption be as in* Theorem 9.2.3. *Let A be a valuation ring of $\mathbb{Q}(\lambda)$ of residual characteristic $p > 3$. Then*

$$\left| \prod_{l \in I}(1 + l^{-1}) \frac{\Gamma(1, Ad(\lambda))L(1, Ad(\lambda))}{\Omega(+, \lambda; A)\Omega(-, \lambda; A)} \right|_p^{-r(W)} = |L^\lambda/L_\lambda|,$$

where $L^\lambda = 1_\lambda L$ for the image L of $H_!^1(\Gamma_0(C), L(n, \psi; W))[+]$ in the cohomology $H_!^1(\Gamma_0(C), L(n, \psi; K))[+]$, and L_λ is given by the intersection $L^\lambda \cap L$ in $H_!^1(\Gamma_0(C), L(n, \psi; K))[+]$.

Writing $1_{\mathbb{T}_\chi^{fl}} = 1_\lambda + 1'_\lambda$ and defining $^\perp L_\lambda = 1'_\lambda L$ with $^\perp L^\lambda = {}^\perp L_\lambda \cap L$, we have $^\perp L_\lambda/^\perp L^\lambda \cong 1_{\mathbb{T}_\chi^{fl}} L/(L_\lambda \oplus {}^\perp L^\lambda) \cong L^\lambda/L_\lambda$ as modules over $\mathbb{T}_\chi^{fl}$. If $L^\lambda/L_\lambda \neq 0$ (i.e., the L-value is divisible by $\mathfrak{m}_W$), by the argument proving Proposition 5.2.7 applied to $(L_\lambda, {}^\perp L^\lambda, 1_{\mathbb{T}_\chi^{fl}} L)$ in place of $(\mathfrak{a}, \mathfrak{b}, R)$, we conclude the existence of an algebra homomorphism $\lambda' : \mathbb{T}_\chi^{fl} \to \overline{\mathbb{Q}}_p$

factoring through the complementary factor $1'_\chi \mathbb{T}^{fl}_\chi$ such that $\lambda \equiv \lambda'$ mod $\mathcal{P}$ for the maximal ideal $\mathcal{P}$ above $\mathfrak{m}_W$ in the integral closure of W in $\overline{\mathbb{Q}}_p$. In this way, the congruence criterion of [H81a] was proven.

Proof. By our choice, ξ_+ is the generator of L_λ. Similarly we define $M^\lambda = 1_\lambda M$ for the image M of $H^1_!(\Gamma_0(C), L(n, \psi; W))[-]$ in $H^1_!(\Gamma_0(C), L(n, \psi; K))[-]$, and $M_\lambda = M^\lambda \cap M$ in $H^1_!(\Gamma_0(C), L(n, \psi; K))[-]$. Then ξ_- is a generator of M_λ. Since the pairing is perfect, $L_\lambda \cong \mathrm{Hom}_W(M^\lambda, W)$ and $L^\lambda \cong \mathrm{Hom}_W(M_\lambda, W)$ under $\langle \, , \, \rangle$. Then it is an easy exercise to see that $|\langle \xi_+, \xi_- \rangle|_p^{-r(W)} = |L^\lambda / L_\lambda|$. $\qquad \square$

9.3 Congruence and adjoint L-values

In this final section, we study a non-abelian adjoint version Theorem 9.3.2 of the analytic class number formula, which follows from the way Taylor–Wiles' Theorem 6.2.13 is proven and some earlier work of the author (presented in the previous section). Actually, long before the formula was established, Doi and the author had found an intricate relation between congruence of modular forms and the adjoint L-value (see the introduction of [DHI98] for the history), and later via the work of Taylor–Wiles, it was formulated in a more precise form we discuss here. In this section, we assume that (p) and (l) for $\rho_{f,\mathfrak{p}}$ and write the Hecke eigenvalues of f as an algebra homomorphism $\lambda : \mathbf{h}_k(C, \psi; W) \to \overline{\mathbb{Q}}_p$.

9.3.1 $\mathbb{T}$-*freeness and the congruence number formula*

Recall (by definition) that $\mathbb{T}^{fl}_\chi = \mathbb{T}_\chi$ and $R^{fl}_\chi = R_\chi$ if either $k \geq 3$ or $\overline{\epsilon}/\overline{\delta} \neq \overline{\omega}_p$. Our goal in this section is to give a sketch of a proof of

Theorem 9.3.1. *Let the notation be as in* Theorems 6.2.13 *and* 9.2.3 *(except that we write $k = k_0$, $\psi = \psi_0$ and $\chi = \chi_0$ here). In particular, let $1_{\mathbb{T}^{fl}_\chi}$ $(\chi = \nu^{k-1}_p \psi)$ denote the idempotent of the factor $\mathbb{T}^{fl}_\chi$ of $\mathbf{h}_k(C, \psi; W)$. Assume $k \geq 2$ and that $\overline{\rho}$ is absolutely irreducible p-distinguished. Then if $\mathbb{T}^{fl}_\chi$ is Gorenstein, $1_{\mathbb{T}^{fl}_\chi} H^1(\Gamma_0(C), L(n, \psi; W))[\pm] \cong \mathbb{T}^{fl}_\chi$ as $\mathbb{T}^{fl}_\chi$-modules.*

We do not need the Taylor–Wiles condition for this theorem. The assumption necessary is Gorenstein-ness which follows either from the local complete intersection property for $\mathbb{T}^{fl}_{\chi/W}$ (as in Theorem 6.2.13) or the absolute irreducibility of $\overline{\rho} = \rho_f$ mod $\mathfrak{m}_W$ with $\overline{\epsilon} \neq \overline{\delta}$ and $\lambda(T(p)) \in A^\times$

(see [M77, Lemma 15.1] and [H13a, Remark 4.1]). However at the end, we eventually need the local complete intersection property in order to relate the L-value with $|C_0(\lambda; W)|$ and the size of the Selmer group via Theorem 6.2.21. So far, all known proofs of $R_\chi^{fl} \cong \mathbb{T}_\chi^{fl}$ somehow prove the complete intersection property of R_χ^{fl} and produce the presentation as in Theorem 6.2.13, though we have non-Gorenstein examples of $\mathbb{T}_\chi^{fl}$ [KW08] (when p-distinguishedness fails) and hence $\mathbb{T}_\chi^{fl}$ is not a complete intersection in these examples.

To relate the size $|L^\lambda/L_\lambda|$ to the size of the congruence module $|C_0(\lambda; W)|$ for $\lambda : \mathbb{T}_\chi^{fl} \to W$, we apply the theory in §5.2.4. To compare with the notation in §5.2.4, we write $R = \mathbb{T}_\chi^{fl}$, $\lambda = \phi : R \to W$, $K := \mathrm{Frac}(W)$ and S for the image of R in X, decomposing $R \otimes_W K = K \oplus X$ as algebra direct sum. Under the notation of Corollary 9.2.4, we get $L^\lambda \otimes_A W/L_\lambda \otimes_A W \cong L^\lambda/L_\lambda$ as $h_k(\Gamma_0(C), \psi; A)$-modules. Since on $L^\lambda \otimes_A W/L_\lambda \otimes_A W$, the Hecke algebra $h_k(\Gamma_0(C), \psi; A)$ acts through λ, it acts through R. Thus multiplying 1_R does not alter the identity $L^\lambda \otimes_A W/L_\lambda \otimes_A W \cong L^\lambda/L_\lambda$; so,

$$1_R(L^\lambda \otimes_A W)/1_R(L_\lambda \otimes_A W) \cong L^\lambda/L_\lambda.$$

Fix an isomorphism of R-modules: $1_R L \cong R$ by Theorem 9.3.1. Then we have $1_R(L_\lambda \otimes_A W) \cong R \cap (W \oplus 0) = \mathfrak{a}$ in $R \otimes_A K$ and $1_R(L^\lambda \otimes_A W) \cong R$. Thus $1_R(L^\lambda \otimes_A W)/1_R(L_\lambda \otimes_A W) \cong W/\mathfrak{a} \cong C_0(\lambda; W)$. We conclude

Theorem 9.3.2. *Let the assumption be as in* Theorem 9.3.1 *and the notation be as in* Corollary 9.2.4. *If* $R_\chi^{fl} \cong \mathbb{T}_\chi$ *with the presentation as in* Theorem 6.2.13 *holds, we have*

$$\left| \prod_{l \in I} (1 + l^{-1}) \frac{\Gamma(1, Ad(\lambda)) L(1, Ad(\lambda))}{\Omega(+, \lambda; A) \Omega(-, \lambda; A)} \right|_p^{-r(W)} = ||C_0(\lambda; W)||_p^{-1}.$$

Here I is the set of primes satisfying the condition (I) *in* §9.1.2.

The module $C_0 := C_0(\lambda; W)$ in the theorem is given by $C_0(\lambda : \mathbb{T}_\chi^{fl} \to W)$, which is equal to $C_0' := C_0(\lambda : \mathbb{T}_\chi \to W)$ we studied earlier unless $k = 2$ and $Ad(\bar\rho)|_{D_p}$ contains $\bar\omega_p$ (i.e., $\mathbb{T}_\chi \neq \mathbb{T}_\chi^{fl}$). When $\mathbb{T}_\chi \neq \mathbb{T}_\chi^{fl}$, the p-adic L-value $L_\rho(\lambda)$ for L_ρ in §6.2.7 (regarding $\lambda : \mathbb{T} \to W$) gives $|C_0'|$ slightly bigger than $|C_0|$. The ratio $|C_0'|/|C_0|$ can be computed by the level raising result of Ribet [R84] (and a later improvement by Taylor–Wiles) and is given by $|E^{Ad}(1)|_p^{-1}$ (the Euler p-factor $E^{Ad}(1)$ of $L(0, Ad(\lambda))$); see [HMI, Exercise 3.41] for this value). In summary, the constant $*$ in Theorem 6.2.23 is given by $\prod_{l \in I}(1 + l^{-1})$ as long as $\mathbb{T}_\chi = \mathbb{T}_\chi^{fl}$, and otherwise, we need to multiply it by $|E^{Ad}(1)|_p^{-1}$.

Let $h_k = \mathbf{h}_k(1; \mathbb{Z})$. As already described, primes appearing in the discriminant of the Hecke algebra gives congruence among algebra homomorphisms of the Hecke algebra into $\overline{\mathbb{Q}}$, which are points in $\mathrm{Spec}(h_k)(\overline{\mathbb{Q}})$. For the even weights $k = 26, 22, 20, 18, 16, 12$, we have $\dim_{\mathbb{C}} S_k(SL_2(\mathbb{Z})) = 1$, and the Hecke algebra h_k is just $\mathbb{Z}$ and hence the discriminant is 1. As is well known from the time of Hecke that $h_{24} \otimes_{\mathbb{Z}} \mathbb{Q} = \mathbb{Q}[\sqrt{144169}]$. The square root of the value in the following table is practically the adjoint L-value $L(1, Ad(f))$ for a Hecke eigenform $f \in S_k(SL_2(\mathbb{Z}))$ for the weight k in the table. Here is a table by Y. Maeda of the discriminant of the Hecke algebra of weight k for $S_k(SL_2(\mathbb{Z}))$ when $\dim S_k(SL_2(\mathbb{Z})) = 2$:

Discriminant of Hecke algebras.

weight	dim	Discriminant of $h_{k/\mathbb{Z}}$
24	2	$2^6 \cdot 3^2 \cdot 144169$
28	2	$2^6 \cdot 3^6 \cdot 131 \cdot 139$
30	2	$2^{12} \cdot 3^2 \cdot 51349$
32	2	$2^6 \cdot 3^2 \cdot 67 \cdot 273067$
34	2	$2^8 \cdot 3^4 \cdot 479 \cdot 4919$
38	2	$2^{10} \cdot 3^2 \cdot 181 \cdot 349 \cdot 1009$

A bigger table (computed by Maeda) can be found in [MFG, §5.3.3] and in [M15] with other conjectures of Maeda. The author believes that by computing Hecke fields in the mid 1970's, Maeda somehow reached the now famous conjecture asserting simplicity of the Hecke algebra of $S_{2k}(SL_2(\mathbb{Z}))$ (see [M15, Conjecture 1.1] and [HM97]).

9.3.2 *Taylor–Wiles system*

We quote here a ring theoretic result due to Diamond and Fujiwara and describe briefly the limiting process of Taylor–Wiles which has been exposed by many authors. We only deal with the case where $\mathbb{T}_\chi \cong \mathbb{T}_\chi^{fl}$. General cases and more details can be found in [HMI, §3.2].

Let Σ be the set of all rational primes $l > 0$. For each $l \in \Sigma$, we write Δ_l for the p-Sylow subgroup of $(\mathbb{Z}/l\mathbb{Z})^\times$ and split $(\mathbb{Z}/l\mathbb{Z})^\times = \Delta_l \times \Delta^{(l)}$. Thus $\Delta^{(l)}$ is the prime-to-p part of $(\mathbb{Z}/l\mathbb{Z})^\times$. For a finite subset $Q \subset \Sigma$, we put $\Delta_Q = \prod_{q \in Q} \Delta_q$. Let $W[\Delta_Q]$ be the group algebra of Δ_Q with augmentation ideal $\mathfrak{a}_Q$ generated by $\delta - 1$ for all $\delta \in \Delta_Q$. We write $\Delta_\emptyset$ for the trivial group with one element. By Exercise 4.2.20, $W[\Delta_Q]$ and $\Lambda[\Delta_Q]$ are local with residue field $\mathbb{F}$. Let $\mathcal{Q}$ be a set of finite subsets of Σ containing an empty set $\emptyset$ as a member. A Taylor–Wiles system $\{R_Q, M_Q\}_{Q \in \mathcal{Q}}$ consists of the following data:

(1) We have $q \equiv 1 \mod p$ for all $q \in Q$, and R_Q is a complete local $W[\Delta_Q]$-algebra.
(2) We have a surjective W-algebra homomorphism $R_Q/\mathfrak{a}_Q R_Q \twoheadrightarrow R_\emptyset$ for each $Q \in \mathcal{Q}$.
(3) M_Q is an R_Q-module for each $Q \in \mathcal{Q}$ such that
    (a) the R_Q-action on $M_Q/\mathfrak{a}_Q M_Q$ factors through $R_\emptyset$;
    (b) M_Q is free of a fixed rank d over $W[\Delta_Q]$.

As the notation suggests, in our application, R_Q is given by the universal deformation ring for the functor $\mathcal{D}_{\chi,Q}$ made of (isomorphism classes of) deformations satisfying (det), (m), (p) and (l) for $l \in S$ allowing ramification at primes in Q, assuming the following q-distinguishedness condition for each member of $Q \in \mathcal{Q}$:

- $\bar\rho|_{\mathrm{Gal}(\overline{\mathbb{Q}}_q/\mathbb{Q}_q)} \cong \begin{pmatrix} \bar\epsilon_q & * \\ 0 & \bar\delta_q \end{pmatrix}$ with $\bar\epsilon_q \neq \bar\delta_q$.

Define G_Q to be the absolute Galois group of the maximal p-profinite extension of $F(\bar\rho)$ unramified outside Q and p. Then $\rho_A : G_Q \to \mathrm{GL}_2(A) \in \mathcal{D}_{\chi,Q}(A)$ for $\chi = \nu_p^{k-1}\psi$ satisfies $\det(\rho_A) = \iota_A \circ \chi$, (m), (p) and (l) in §6.1 and defined by (6.1) replacing G there by G_Q. By q-distinguishedness, we can prove that $\mathcal{D}_{\chi,Q}$ is represented by $R_Q \in CL_W$. In particular, $R_\emptyset = R_\chi$.

We followed the definition of the Taylor–Wiles system in [HMI, §3.2.3] which is a reformulation of Fujiwara's system given in [F06, 1.1.1] (F. Diamond also made such a system in [D98]), although a weaker version of the system was invented earlier by A. Wiles and R. Taylor (cf. [TW95]).

We quote the isomorphism and freeness criterion from [HMI, Theorem 3.23]:

Theorem 9.3.3. *Let $\{R_Q, M_Q\}_{Q \in \mathcal{Q}}$ be a Taylor–Wiles system. Suppose the following four conditions:*

(1) *For any given positive integer m, there exist infinitely many disjoint sets $Q \in \mathcal{Q}$ such that $q \equiv 1 \mod p^m$ for all $q \in Q$;*
(2) *The number r of elements q of $Q \in \mathcal{Q}$ is independent of Q;*
(3) *R_Q is generated by at most r elements as a complete local W-algebra for all $Q \in \mathcal{Q}$;*
(4) *The annihilator $\mathfrak{A}$ of $M_Q/\mathfrak{a}_Q M_Q$ in $R_\emptyset$ is independent of $Q \in \mathcal{Q}$.*

Then $R_\emptyset$ is W-free of finite rank with a presentation $R_\emptyset \cong \frac{W[[T_1,\ldots,T_r]]}{(S_1,\ldots,S_r)}$ (i.e., a local complete intersection with r generators), and we have $\mathfrak{A} = 0$. If we further assume

(5) $M_Q/\mathfrak{a}_Q M_Q$ is isomorphic to a unique $R_\emptyset$-module M independent of Q,

then M is an $R_\emptyset$-free module of finite rank.

9.3.3   Level Q Hecke algebra

Hereafter we assume $k \geq 2$, and we fix a primitive form $f \in S_k(\Gamma_0(C), \psi)$ and a prime $\mathfrak{p}$ of $\mathbb{Z}[f]$ for which f satisfies (p) in §6.1.1. We suppose that $\bar\rho = \bar\rho_{f,\mathfrak{p}} \mod \mathfrak{p}$ satisfies (m) and (l). As before, we put $\chi := \nu_p^{k-1}\psi$. We write $\mathbb{T} = \mathbb{T}_\mathfrak{p}$ and $\mathbb{T}_\chi$ for the local ring of $\mathbf{h}_k(C', \psi; W)$ associated to $\bar\rho$, where C' is the level of the p-stabilized ordinary Hecke eigenform associated to f in Lemma 3.2.20.

We define

$$\Gamma_0(Q) := \{\gamma = \left(\begin{smallmatrix} a & b \\ c & d \end{smallmatrix}\right) \in SL_2(\mathbb{Z})|c \equiv 0 \mod q \text{ for all } q \in Q\},$$
$$\Gamma_1(Q) := \{\gamma = \left(\begin{smallmatrix} a & b \\ c & d \end{smallmatrix}\right) \in \Gamma_0(Q))|d \equiv 1 \mod q \text{ for all } q \in Q\}. \tag{9.20}$$

Let $\Gamma_Q^{(p)}$ be the subgroup of $\Gamma_0(Q)$ containing $\Gamma_1(Q)$ such that $\Gamma_0(Q)/\Gamma_Q^{(p)}$ is the maximal p-abelian quotient of $\Gamma_0(Q)/\Gamma_1(Q) \cong \prod_{q \in Q}(\mathbb{Z}/q\mathbb{Z})^\times$. Put

$$\Gamma_{Q,s} := \Gamma_Q^{(p)} \cap \Gamma_0(Cp^s). \tag{9.21}$$

We often write Γ_Q for $\Gamma_{Q,s}$ when s is well understood (mostly when $s = 0, 1$). Then we have

$$\Delta_Q = (\Gamma_0(Cp^s) \cap \Gamma_0(Q))/\Gamma_{Q,s}, \tag{9.22}$$

which is canonically isomorphic to the maximal p-abelian quotient of $\Gamma_0(Q)/\Gamma_1(Q)$ independent of the exponent s. If $Q = \emptyset$, we have $\Gamma_{Q,s} = \Gamma_0(Cp^s)$, and if $q \not\equiv 1 \mod p$ for all $q \in Q$, we have $\Gamma_1(C_Q p^s) \subset \Gamma_{Q,s} = \Gamma_0(C_Q p^s)$ for $C_Q := C\prod_{q \in Q} q$.

Recall the rings $\mathbb{Z}[\psi] \subset \mathbb{C}$ and $\mathbb{Z}_p[\psi] \subset \overline{\mathbb{Q}}_p$ generated over $\mathbb{Z}$ and $\mathbb{Z}_p$ by the values ψ, respectively. Recall the Hecke algebra $h = h_k(\Gamma_{Q,s}, \psi; \mathbb{Z}[\psi])$:

$$h = \mathbb{Z}[\psi][T(n)|n = 1, 2, \cdots] \subset \text{End}_\mathbb{C}(S_k(\Gamma_{Q,s}, \psi)),$$

where $T(n)$ is the Hecke operator as in §4.1.4. Define

$$h_{\chi, Q/W} = h_k(\Gamma_{Q,s}, \psi; W) := h \otimes_{\mathbb{Z}[\psi]} W \text{ for } \chi = \nu_p^{k-1}\psi.$$

Here $h_k(\Gamma_{Q,s}, \psi; W)$ acts on $S_k(\Gamma_{Q,s}, \psi; W)$ by Theorem 4.1.13. More generally for a congruence subgroup Γ containing $\Gamma_1(Cp^s)$, we write $h_k(\Gamma, \psi; W)$ for the Hecke algebra on Γ with coefficients in W acting on $S_k(\Gamma, \psi; W)$. The algebra $h_k(\Gamma, \psi; W)$ can be also realized as

$$W[T(n)|n = 1, 2, \cdots] \subset \mathrm{End}_W(S_k(\Gamma, \psi; W)).$$

When we need to indicate that our $T(l)$ is the Hecke operator of a prime factor l of Cp^s, we write it as $U(l)$ as before. The ordinary part $\mathbf{h}_Q \subset h_{\chi, Q/W}$ is the maximal ring direct summand on which $U(p)$ is invertible. We write e for the idempotent of $\mathbf{h}_Q$, and hence $e = \lim_{n \to \infty} U(p)^{n!}$ under the p-adic topology of $h_{\chi, Q/W}$. The idempotent e not only acts on the space of modular forms with coefficients in W but also on the classical space $S_k(\Gamma_{Q,s}, \psi)$ (as e descends from $S_k(\Gamma_{Q,s}, \psi, \overline{\mathbb{Q}}_p)$ to $S_k(\Gamma_{Q,s}, \psi, \overline{\mathbb{Q}})$ and ascends to $S_k(\Gamma_{Q,s}, \psi)$). We write the image $M^{\mathrm{ord}} := e(M)$ of the idempotent attaching the superscript "ord" (e.g., S_k^{ord}).

We quote

Lemma 9.3.4. *The Hecke algebra $\mathbf{h}_Q$ is flat over $\Lambda[\Delta_Q]$ with a canonical isomorphism:* $\mathbf{h}_Q / \mathfrak{A}_{\Delta_Q} \mathbf{h}_Q \cong \mathbf{h}_\emptyset$ *for the augmentation ideal $\mathfrak{A}_{\Delta_Q} \subset \Lambda[\Delta_Q]$.*

This follows from Lemma 4.2.41 applied to $\mathbf{h}_Q$ and $A = \Lambda[\Delta_Q]$. See [MFG, Corollary 3.20] for details of the proof.

Note that $\mathbb{T}_\chi$ in §6.2.5 is a local ring of $\mathbf{h}_\emptyset$. As for primes in $q \in Q$, if $q \equiv 1 \mod p$ and $\overline{\rho}(\mathrm{Frob}_q)$ has two distinct eigenvalues, we have

$$\rho_{\mathbb{T}_\chi}([z, \mathbb{Q}_q]) \sim \begin{pmatrix} \alpha_q(z) & 0 \\ 0 & \beta_q(z) \end{pmatrix} \tag{Gal$_q$}$$

with characters α_q and β_q of $\mathbb{Q}_q^\times$ for $z \in \mathbb{Q}_q^\times$, where one of α_q and β_q is unramified (e.g., [MFG, Theorem 3.32 (2)] or [HMI, Theorem 3.75]).

For each $q \in Q$, assuming that $\overline{\rho}(\mathrm{Frob}_q)$ has two distinct eigenvalues, we choose one eigenvalue $\overline{\alpha}_q$. The inclusion $S_k(\Gamma_0(Cp), \psi; W) \hookrightarrow S_k(\Gamma_{Q,s}, \psi; W)$ has the dual map $\mathbf{h}_Q \twoheadrightarrow \mathbf{h}_\emptyset$. The Hecke algebra $\mathbf{h}_Q$ has $U(q)$, and there are local rings $\mathbb{T}_Q$ of $\mathbf{h}_Q$ indexed by one choice of characters $? \in \{\alpha_q, \beta_q\}$ so that $U(q) = ?([q, \mathbb{Q}_p])$. We fix a choice $A_Q := \{\overline{\alpha}_q | q \in Q\}$ and define $\mathbb{T}_Q$ so that $U(q) \mod \mathfrak{m}_{\mathbb{T}_Q} = \overline{\alpha}_q$ according to this choice. The choice will not matter much as these local rings are all isomorphic.

9.3.4 Q-ramified deformation

Identify the image of the inertia group I_q for $q \in Q$ in the Galois group of the maximal abelian extension over $\mathbb{Q}_q$ with $\mathbb{Z}_q^\times$ by the q-adic cyclotomic character. Since we choose $q \equiv 1 \mod p^m$ for $m > 0$ for all $q \in Q$, $\Delta_q / \Delta_q^{p^n}$ for $0 < n \leq m$ is a cyclic group of order p^n.

We put $\Delta_n = \Delta_{n,Q} := \prod_{q \in Q} \Delta_q / \Delta_q^{p^n}$. By Lemma 9.3.4, the inertia action $I_q \twoheadrightarrow \mathbb{Z}_q^\times \to R_Q \twoheadrightarrow \mathbb{T}_Q$ makes $\mathbb{T}_Q$ free of finite rank over $W[\Delta_Q]$. Then Taylor and Wiles found an infinite sequence $\mathcal{Q} = \{Q_m | m = 1, 2, \dots\}$ of ordered finite sets $Q = Q_m$ of primes q (with $q \equiv 1 \mod p^m$) which produces a projective system:

$$\{((R_{n,m(n)}, \alpha = \alpha_n), \widetilde{R}_{n,m(n)}, M_m = M_{Q_m}, (f_1 = f_1^{(n)}, \dots, f_r = f_r^{(n)}))\}_n$$
(9.23)

made of the following objects:

(o1) There are several choices of M_Q.

    (a) $M_Q = \mathbb{T}_Q$,

    (b) $M_Q = 1_{\mathbb{T}_Q} H^1_{ord}(\Gamma_{Q,s}, L(n, \psi; W))$ for the idempotent $1_{\mathbb{T}_Q}$ of $\mathbb{T}_Q$ in $\mathbf{h}^Q$. Here s can be chosen to be 1. If $\bar{\rho}$ is unramified at p and $n > 0$, we can choose $s = 0$ as $H^1_{ord}(\Gamma_{Q,1}, L(n, \psi; W)) \cong H^1_{ord}(\Gamma_{Q,0}, L(n, \psi; W))$ by the restriction map.

(o2) $R_{n,m}$ is defined to be the image of $\mathbb{T}_{Q_m}$ in $\mathrm{End}_{W[\Delta_n]}(M_{n,m})$ for $M_{n,m} := M_{Q_m}/(p^n, \delta_q^{p^n} - 1)_{q \in Q_m} M_{Q_m}$. If we make the choice $M_Q = \mathbb{T}_Q$, the image is identical to $\mathbb{T}_{Q_m}/(p^n, \delta_q^{p^n} - 1)_{q \in Q_m} \mathbb{T}_{Q_m}$. An important point is that $R_{n,m}$ is a finite ring whose order is bounded independent of m (by (Q0) below).

(o3) $\widetilde{R}_{n,m} := R_{n,m}/(\delta_q - 1)_{q \in Q_m}$,

(o4) $\alpha_n : W_n[\Delta_n] \to R_{n,m}$ for $W_n := W/p^n W$ is a $W[\Delta_n]$-algebra homomorphism for $\Delta_n = \Delta_{n,Q_m}$ induced by the $W[\Delta_{Q_m}]$-algebra structure of $\mathbb{T}_{Q_m}$ (making $R_{n,m}$ finite $W[\Delta_n]$-algebras).

(o5) $(f_1 = f_1^{(n)}, \dots, f_r = f_r^{(n)})$ is an ordered subset of the maximal ideal of $R_{n,m}$.

Thus for each $n > 0$, the projection $\pi_n^{n+1} : R_{n+1,m(n+1)} \to R_{n,m(n)}$ is compatible with all the data in the system (9.23) (the meaning of this compatibility is specified below) and induces the projection $\widetilde{\pi}_n^{n+1} : \widetilde{R}_{n+1,m(n+1)} \to \widetilde{R}_{n,m(n)}$.

Remark 9.3.5. There is one more datum of an algebra homomorphism $\beta : R_{n,m} \to \mathrm{End}_{\mathbb{T}_{Q_m}}(M_{n,m}) \subset \mathrm{End}_{W[\Delta_n]}(M_{n,m})$ given in [HMI, page 191]. However to make notation simple, we do not explicitly mention β. If we choose M_Q to be $\mathbb{T}_Q$, $M_{n,m}$ is by definition $R_{n,m}$; so, β is just the identity map (and hence we can forget about it).

The infinite set $\mathcal{Q}$ satisfies the following conditions (Q0–8):

(Q0) M_{Q_m} is free of finite rank d over $W[\Delta_{Q_m}]$ with d independent of m. Lemma 9.3.4 if $M_Q := \mathbb{T}_Q$ and Corollary 9.2.2 if $M_Q = 1_{\mathbb{T}_Q} H^1_{ord}(\Gamma_{Q,s}, L(n, \psi; W))$ proves $W[\Delta_Q]$-freeness. By the proof of Lemma 9.3.4 and Corollary 9.2.2, the augmentation quotient of M_Q is isomorphic to $M_\emptyset$, and Nakayama's lemma tells us that $d = \mathrm{rank}_W M_\emptyset$ (independent of Q).

(Q1) $|Q_m| = r \geq \dim_{\mathbb{F}} \mathcal{D}_{\chi,Q_m}(\mathbb{F}[\epsilon])$ for r independent of m [HMI, Propositions 3.29 and 3.33], where ϵ is the dual number with $\epsilon^2 = 0$. The number r is given by $\dim_{\mathbb{F}} \mathrm{Sel}^\perp(Ad(\overline{\rho}))$, and a Q-ramified version of the reflection theorem [HMI, Proposition 3.29] proves $r \geq \dim_{\mathbb{F}} \mathcal{D}_{\chi,Q_m}(\mathbb{F}[\epsilon])$ always. (Note that $\dim_{\mathbb{F}} \mathcal{D}_{\chi,Q_m}(\mathbb{F}[\epsilon])$ is the minimal number of generators of R_{Q_m} over W.)

(Q2) $q \equiv 1 \mod p^m$ and $\overline{\rho}(\mathrm{Frob}_q) \sim \begin{pmatrix} \overline{\alpha}_q & 0 \\ 0 & \overline{\beta}_q \end{pmatrix}$ with $\overline{\alpha}_q \neq \overline{\beta}_q \in \mathbb{F}$ if $q \in Q_m$ (so, $|\Delta_q| =: p^{e_q} \geq p^m$).

(Q3) The set $Q_m = \{q_1, \dots, q_r\}$ is ordered so that

- $\Delta_{q_j} \subset \Delta_{Q_m}$ is identified with $\mathbb{Z}/p^{e_{q_j}}\mathbb{Z}$ by $\delta_{q_j} \mapsto 1$; so, $\Delta_n = \Delta_{n,Q_{m(n)}} = (\mathbb{Z}/p^n\mathbb{Z})^{Q_{m(n)}}$,
- $\Delta_n = (\mathbb{Z}/p^n\mathbb{Z})^{Q_{m(n)}}$ is identified with $\Delta_{n+1}/\Delta_{n+1}^{p^n}$ which is $((\mathbb{Z}/p^{n+1}\mathbb{Z})/p^n(\mathbb{Z}/p^{n+1}\mathbb{Z}))^{Q_{m(n)}}$,
- the diagram

$$
\begin{array}{ccc}
W_{n+1}[\Delta_{n+1}] & \xrightarrow{\alpha_{n+1}} & R_{n+1,m(n+1)} \\
\downarrow & & \downarrow{\scriptstyle \pi_n^{n+1}} \\
W_n[\Delta_n] & \xrightarrow{\alpha_n} & R_{n,m(n)}
\end{array}
$$

is commutative for all $n > 0$ (and by (Q0), α_n is injective for all n).

(Q4) There exists an ordered set of generators $\{f_1^{(n)}, \dots, f_r^{(n)}\} \subset \mathfrak{m}_{R_{n,m(n)}}$ of $R_{n,m(n)}$ over W for the integer r in (Q1) such that $\pi_n^{n+1}(f_j^{(n+1)}) = f_j^{(n)}$ for each $j = 1, 2, \dots, r$.

(Q5) $R_\infty := \varprojlim_n R_{n,m(n)}$ is isomorphic to $W[[T_1, \dots, T_r]]$ by sending T_j to $f_j^{(\infty)} := \varprojlim_n f_j^{(n)}$ for each j (e.g., [HMI, page 193]).

(Q6) Inside R_∞, $\varprojlim_n W_n[\Delta_n]$ is isomorphic to $W[[S_1, \dots, S_r]]$ so that $s_j := (1 + S_j)$ is sent to the generator $\delta_{q_j} \Delta_{q_j}^{p^n}$ of $\Delta_{q_j}/\Delta_{q_j}^{p^n}$ for the ordering $q_1, \dots, q_r$ of primes in Q_m in (Q3).

(Q7) $R_\infty/(S_1, \ldots, S_r) \cong \varprojlim_n \widetilde{R}_{n,m(n)} \cong R_\emptyset \cong \mathbb{T}_\emptyset$, where $R_\emptyset$ is the universal deformation ring for the deformation functor $\mathcal{D}_{\emptyset,k,\psi_k}$ and $\mathbb{T}_\emptyset$ is the local factor of the Hecke algebra $\mathbf{h}_{\emptyset,k,\psi_k}$ whose residual representation is isomorphic to $\bar{\rho}$.

(Q8) We have $R_{Q_m} \cong \mathbb{T}_{Q_m}$ by the canonical morphism, and $R_{Q_m} \cong R_\infty/\mathfrak{A}_{Q_m} R_\infty$ for the ideal $\mathfrak{A}_{Q_m} := ((1 + S_j)^{|\Delta_{q_j}|} - 1)_{j=1,2,\ldots r}$ of $W[[S_1, \ldots, S_r]]$ is a local complete intersection.

9.3.5   Taylor–Wiles primes

We recall briefly the way Taylor and Wiles chose the sets Q as we prove by this way $\mathbb{T}$-freeness of the $\pm$ eigenspace of the cohomology group. The case of an induced representation $\bar{\rho} = \operatorname{Ind}_K^\mathbb{Q} \bar{\varphi}$ for a quadratic field is more difficult than non-induced cases, and this is the place where we need the Taylor–Wiles condition:

(W) $\bar{\rho}$ *is irreducible over* $\operatorname{Gal}(\overline{\mathbb{Q}}/\mathbb{Q}[\sqrt{p^*}])$.

Recall the set S of primes outside p ramifying in $F(\bar{\rho})/\mathbb{Q}$. Let $k^{S \cup Q}$ be the maximal extension of a field $k \subset \overline{\mathbb{Q}}$ unramified outside $S \cup Q \cup \{p\} \cup \{\infty\}$, and recall $\mathfrak{G}_k^{S \cup Q} = \operatorname{Gal}(k^{S \cup Q}/k)$.

Write Ad for the adjoint representation of $\bar{\rho}$ acting on $\mathfrak{sl}_2(\mathbb{F})$ by conjugation, and put Ad^* for the $\mathbb{F}$-contragredient. Then $Ad^*(1)$ is one time Tate twist of Ad^*. Note that $Ad^* \cong Ad$ by the trace pairing as p is odd. Let Q be a finite set of primes, and consider

$$\beta_Q : H^1(\mathfrak{G}_\mathbb{Q}^{S \cup Q}, Ad^*(1)) \to \prod_{q \in Q} H^1(\mathbb{Q}_q, Ad^*(1)).$$

Here is a lemma due to A. Wiles [W95, Lemma 1.12] on which the existence proof of the sets Q_m is based. We state the lemma slightly different from [W95, Lemma 1.12], and for that, we write $F = \overline{\mathbb{Q}}^{\operatorname{Ker} Ad}$ (the splitting field of $Ad = Ad(\bar{\rho})$). Since $Ad \cong \bar{\alpha} \oplus \operatorname{Ind}_K^\mathbb{Q} \bar{\varphi}^-$ by Proposition 7.3.2, we have $F = F(\varphi^-)$.

Lemma 9.3.6. *Assume* (W). *Let $\tilde{\rho}$ be the composite of $\bar{\rho}$ with the projection* $GL_2(\mathbb{F}) \twoheadrightarrow PGL_2(\mathbb{F})$. *Pick* $0 < m \in \mathbb{Z}$. *If $\bar{\rho}$ is absolutely irreducible, for any absolutely irreducible subspace $X \subset Ad(\bar{\rho})$, there exists $\sigma \in \mathfrak{G}_\mathbb{Q}$ such that (i)* $\tilde{\rho}(\sigma) \neq 1$, *(ii) σ fixes* $\mathbb{Q}(\mu_{p^m})$, *and (iii) σ has eigenvalue 1 on X.*

Proof. Since $\bar{\rho}$ is absolutely irreducible, we have $\sigma \in \mathfrak{G}_\mathbb{Q}$ such that $\bar{\rho}(\sigma) \sim \left(\begin{smallmatrix} a & 0 \\ 0 & b \end{smallmatrix}\right)$ with $a \neq b$. Then $Ad(\bar{\rho})$ has three distinct eigenvalues

$(ab^{-1}, 1, a^{-1}b)$. If $Ad(\bar{\rho})$ is irreducible, nothing to prove. Thus we may assume that $\bar{\rho} = \mathrm{Ind}_K^{\mathbb{Q}} \bar{\varphi}$ for a quadratic field K. Then $Ad(\bar{\rho}) = \bar{\alpha} \oplus \mathrm{Ind}_K^{\mathbb{Q}} \bar{\varphi}^-$ by Proposition 7.3.2.

Now suppose that $\mathrm{Ind}_K^{\mathbb{Q}} \bar{\varphi}^-$ is irreducible (i.e., $\mathrm{ord}(\bar{\varphi}^-) \geq 3$). If $X = \bar{\alpha}$, any $1 \neq \sigma \in \mathrm{Gal}(F(\mu_{p^m})/K(\mu_{p^m}))$ satisfies the required properties. Indeed, by irreducibility of $\mathrm{Ind}_K^{\mathbb{Q}} \bar{\varphi}^-$, $F = F(\bar{\varphi}^-)$ and $\mathbb{Q}(\mu_{p^m})$ is linearly disjoint. Thus we can choose σ fixing $K(\mu_{p^m})$ with $\varphi^-(\sigma) \neq 1$. Then $\bar{\alpha}(\sigma) = 1$, as desired.

If $X = \mathrm{Ind}_K^{\mathbb{Q}} \bar{\varphi}^-$. Then by the standard form of the induced representation in (5.3), if σ non-trivial on K, $\bar{\rho}(\sigma) = \left(\begin{smallmatrix} 0 & \varsigma \\ 1 & 0 \end{smallmatrix} \right)$. The eigenvalue of $Ad(\bar{\rho}(\sigma))$ is $(-1, -1, 1)$. Since σ acts on the space $\bar{\alpha}$ by -1, we have a 1-dimensional subspace of X on which σ acts trivially.

Now suppose that φ^- has order 2. Then $Ad(\bar{\rho}) = \bar{\alpha} \oplus \bar{\beta} \oplus \bar{\gamma}$ for three distinct quadratic characters. Then X is one of them. So writing $X = \bar{\xi}$ for $\bar{\xi} \in \{\bar{\alpha}, \bar{\beta}, \bar{\gamma}\}$, we can write $\bar{\rho} = \mathrm{Ind}_{F(\bar{\xi})}^{\mathbb{Q}} \bar{\phi}$. Then, in the same manner as in the case where $X = \bar{\alpha}$ in the above argument for irreducible $\mathrm{Ind}_K^{\mathbb{Q}} \bar{\varphi}^-$ (so, in our case, we argue with $(\bar{\xi}, F(\bar{\xi}))$ in place of $(\bar{\alpha}, K)$), we just pick $\sigma \neq 1$ in $\mathrm{Gal}(F(\bar{\varphi}^-)[\mu_{p^m}]/F(\bar{\xi})[\mu_{p^m}])$ non-trivial in $\mathrm{Ind}_{F(\bar{\xi})}^{\mathbb{Q}} \bar{\phi} \subset Ad(\bar{\rho})$ and trivial on $F(\bar{\xi})$. The condition (W) guarantees that $F(\bar{\xi})[\mu_{p^m}] \subsetneq F(\bar{\varphi}^-)[\mu_{p^m}]$. $\square$

Now we state the lemma we really need.

Lemma 9.3.7. *Let the notation be as in* Lemma 9.3.6. *Assume* (W). *Pick* $0 \neq x \in \mathrm{Ker}(\beta_{\mathbb{Q}})$, *and write*

$$f_x : \mathfrak{G}_{F(\mu_p)}^{S \cup Q} \to Ad^*(1)$$

as an element of $\mathrm{Hom}_{\mathrm{Gal}(F(\mu_p)/\mathbb{Q})}(\mathfrak{G}_{F(\mu_p)}^{S \cup Q}, Ad^*(1))$ *for the restriction of the cocycle representing x to* $\mathfrak{G}_{F(\mu_p)}^{S \cup Q}$. *Then, f_x factors through* $\mathrm{Gal}(\mathbb{Q}^S/F(\mu_p))$, *and there exists $\sigma_x \in \mathfrak{G}_{\mathbb{Q}}^S$ such that*

(1) $\tilde{\rho}(\sigma_x) \neq 1$ *(so, $Ad(\sigma_x) \neq 1$)*,
(2) σ_x *fixes $\mathbb{Q}(\mu_{p^m})$ for an integer $m > 0$*,
(3) $f_?(\sigma_x^a) \neq 0$ *for the order $a := \mathrm{ord}(\tilde{\rho}(\sigma_x)) = \mathrm{ord}(Ad(\sigma_x))$ of $\tilde{\rho}(\sigma_x)$.*

Strictly speaking, [W95, Lemma 1.12] gives the above statement replacing F by the splitting field $K_0 := F(\bar{\rho})$ of $\bar{\rho}$. Since the statement is about the cohomology group of $Ad^*(1)$, we can replace K_0 in his argument by F. We note also $\mathrm{Ker}(Ad(\bar{\rho})) = \mathrm{Ker}(\tilde{\rho})$ as the kernel of the adjoint representation: $GL(2) \to GL(3)$ is the center of GL_2 (so it factors through PGL_2).

Proof. Since $x \in \mathrm{Ker}(\beta_Q)$, f_x is unramified at $q \in Q$; so, f_x factors through $\mathrm{Gal}(\mathbb{Q}^S/F(\mu_p))$.

Since $f_x(Y) \supset X[1] = \{v \in X | Ad(\sigma)(v) = v\} \neq 0$ by Lemma 9.3.6, we can find $1 \neq \tau \in Y$ such that $f_x(\tau) \in X[1]$; so, $f_x(\tau) \neq 0$. Thus τ commutes with $\sigma \in \mathrm{Gal}(M_x/\mathbb{Q})$. This shows $(\sigma\tau)^a = \sigma^a\tau^a$, and $f_x((\sigma\tau)^a) = f(\sigma^a\tau^a) = af_x(\tau) + f(\sigma^a)$. Since $af_x(\tau) \neq 0$, at least one of $f(\sigma^a\tau^a)$ and $f(\sigma^a)$ is non-zero. Then $\sigma_x = \sigma$ or $\sigma_x := \sigma\tau$ satisfies the condition (3) in addition to (1–2). $\qquad\square$

How to find $\mathcal{Q} = \{Q_m\}_m$. Choose a basis $\{x\}_x$ over $\mathbb{F}$ of the "dual" Selmer group $\mathrm{Sel}^\perp(Ad^*(1))$ defined in §7.2.3 inside $H^1(\mathfrak{G}_\mathbb{Q}^S, Ad^*(1))$. Then Wiles' choice of Q_m is a set of primes q so that $\mathrm{Frob}_q = \sigma_x$ on M_x as in the above lemma. By Chebotarev density, we have infinitely many sets Q_m with this property.

Then by a Q-ramified version of the reflection theorem [HMI, Propositions 3.29 and 3.33], Taylor–Wiles proves (Q1) for this choice. Once $\mathcal{Q}$ is chosen, for a fixed n, we have infinitely many $R_{n,m}$ varying $m > n$. Since $|R_{n,m}| < B_n$ for a bound only dependent on n, we still have infinitely many $W[\Delta_n]$-algebras $R_{m,n}$ in a $W[\Delta_n]$-isomorphism class C_n with extra information of f's in (Q4). Now by moving n one notch to $n+1$, by Pigeon-hole principle, we still have infinitely many $R_{m,n+1}$ with $R_{m,n+1} \otimes_{W[\Delta_{n+1}]} W[\Delta_n] \twoheadrightarrow R_{m,n} \in C_n$ with compatible f's. Out of these infinitely many $R_{m,n+1}$, we choose an $W[\Delta_{n+1}]$-algebra isomorphism class containing infinitely many $\{R_{m,n+1}\}_m$ with extra information of f's. Repeating this process, we get the projective system described in (Q5–8). Then we apply Theorem 9.3.3 to this Taylor–Wiles system, and we get Theorem 6.2.13. This is a brief outline of the proof of $R = \mathbb{T}$ theorem, and more details can be found in [HMI, §3.2] and [MFG, §3.2].

9.3.6 *Proof of $\mathbb{T}$-freeness of cohomology groups*

Recall Theorem 9.3.1 in a slightly different fashion.

Proposition 9.3.8. *Assume $k \geq 2$. If $\overline{\rho}$ is absolutely irreducible p-distinguished, then $1_{\mathbb{T}_\chi^{fl}} H^1_{ord}(\Gamma_0(C), L(n, \psi; W))$ is free of rank 2 over $\mathbb{T}_\chi^{fl}$, and hence $1_{\mathbb{T}_\chi^{fl}} H^1_{ord}(\Gamma_0(C), L(n, \psi; W))[\pm] \cong \mathbb{T}_\chi^{fl}$ as $\mathbb{T}_\chi^{fl}$-modules.*

Taking $M_Q = 1_{\mathbb{T}_Q} H^1_{ord}(\Gamma_{Q,s}, L(n, \psi; W))$, as a corollary of the construction of the Taylor–Wiles system, Theorem 9.3.3 proves this result under the

Taylor–Wiles condition (W). We give some details assuming (W) and $\mathbb{T}_\chi = \mathbb{T}_\chi^{fl}$.

Proof. Freeness of $M_\emptyset$ over $\mathbb{T}_\chi$ follows from Theorem 9.3.3. After extending scalar to $\mathbb{C}$, we know that $H^1_{ord}(\Gamma_0(C), L(n, \psi; \mathbb{C}))$ is isomorphic to the product of two copies of $S_k(\Gamma_0(C), \psi)$ which is rank 1 over the Hecke algebra over $\mathbb{C}$ as it is self-dual. By descent, we find the rank of $1_{\mathbb{T}_\chi} H^1_{ord}(\Gamma_0(C), L(n, \psi; W))$ over $\mathbb{T}_\chi$ is equal to 2. Since $1_{\mathbb{T}_\chi} H^1_{ord}(\Gamma_0(C), L(n, \psi; W))$ is the direct sum of $\pm$-eigenspaces both nontrivial, Krull–Schmidt theorem proves the last assertion. $\square$

Under (p) and absolute irreducibility of $\bar{\rho}$ (with some extra assumptions), the freeness $1_{\mathbb{T}_\chi} H^1_{ord}(\Gamma_0(C), \mathbb{Z}_p))[\pm] \cong \mathbb{T}_\chi$ (taking $W = \mathbb{Z}_p$) was proven much earlier by Mazur in [M77] and [MW86]. The earlier argument proves that $\mathbb{T}_\chi$ is a Gorenstein algebra over Λ but does not reach the local complete intersection property of $\mathbb{T}_\chi$ (so, slightly weaker than Theorem 6.2.13). See [H13a, §4] for a short description of these earlier results.

Bibliography

Books

[ADT]   J. S. Milne, *Arithmetic duality theorems*, Second edition. BookSurge, LLC, Charleston, SC, 2006.

[AFC]   K. Iwasawa, *Algebraic functions*, Translations of Math. Monographs, **118**. American Mathematical Society, Providence, RI, 1993.

[AFR]   D. Bump, *Automorphic Forms and Representations*, Cambridge Studies in Advance Mathematics **55**, Cambridge Univ. Press, 1996.

[ALG]   R. Hartshorne, *Algebraic Geometry*, Graduate Texts in Mathematics **52**, Springer, New York, 1977.

[ANT]   H. Koch, *Algebraic Number Theory*, Springer, Berlin-Heidelberg, 1997.

[BAL]   N. Bourbaki, *Algébre*, Hermann, Paris, 1958.

[BCM]   N. Bourbaki, *Algébre Commutative*, Masson, Paris, 1985–1998.

[BNT]   A. Weil, *Basic Number Theory*, Springer, New York, 1974.

[CAG]   D. Eisenbud, *Commutative algebra, with a view toward algebraic geometry*, Graduate Texts in Mathematics **150**, Springer, 1995.

[CFN]   J. Neukirch, *Class Field Theory*, Springer, 1986.

[CFZ]   J. Coates and R. Sujatha, *Cyclotomic Field and Zeta Values*, Springer Monographs in Mathematics, Springer, 2006.

[CGP]   K. S. Brown, *Cohomology of Groups*, Graduate Texts in Mathematics **87**, Springer, New York, 1982.

[CNF]   J. Neukirch, A. Schmidt, and K. Wingberg, Cohomology of number fields, Second Edition, Springer, 2008.

[CPI]   K. Iwasawa, *Collected Papers*, I, II, Springer, New York, 2001.

[CPS]   G. Shimura, *Collected Papers*, I–V Springer, New York, 2002, 2016.

[CRT]   H. Matsumura, *Commutative Ring Theory*, Cambridge Studies in Advanced Mathematics **8**, Cambridge Univ. Press, New York, 1986.

[EAI]   H. Hida, *Elliptic Curves and Arithmetic Invariant*, Springer Monographs in Mathematics, 2013, Springer, New York.

[ECH]   J. S. Milne, *Étale Cohomology*, Princeton University Press, 1980.

[EDM]   G. Shimura, *Elementary Dirichlet Series and Modular Forms*, Springer Monographs in Mathematics, 2007.

[EGA]　A. Grothendieck and J. Dieudonné, *Eléments de Géométrie Algébrique*, Publ. IHES **24** (1965).

[FGS]　J. Bellaïche and G. Chenevier, *Families of Galois representations and Selmer groups*. Astérisque, **324** 2009.

[GAN]　M. Lazard, *Groupes Analytiques p-Adiques*, Publ. IHES **26**, 1965.

[GME]　H. Hida, *Geometric Modular Forms and Elliptic Curves*, Second edition, World Scientific, Singapore, 2012.

[HMI]　H. Hida, *Hilbert modular forms and Iwasawa theory*, Oxford University Press, 2006.

[HMS]　P. Deligne, J. Milne, K.-y. Shih, *Hodge Cycles, Motives and Shimura Varieties*, Lecture notes in Math. **900**, Springer, 1989.

[IAT]　G. Shimura, *Introduction to the Arithmetic Theory of Automorphic Functions*, Princeton University Press, Princeton, 1971.

[ICF]　L. C. Washington, *Introduction to Cyclotomic Fields*, Graduate Text in Mathematics, **83**, 2nd edition, Springer, New York, 1997.

[LCF]　K. Iwasawa, *Local class field theory*, Oxford University Press, New York, 1986.

[LFD]　J.-P. Serre, *Local fields*, Graduate Texts in Mathematics, **67**, Springer, New York-Berlin, 1979.

[LFE]　H. Hida, *Elementary Theory of L-Functions and Eisenstein Series*, LMSST **26**, Cambridge University Press, Cambridge, 1993.

[LGF]　L. E. Dickson, *Linear Groups with an Exposition of the Galois Field Theory*, Teubner, 1901.

[LIE]　N. Jacobson, *Lie Algebras*, Dover, New York 1979.

[LRF]　J.-P. Serre, *Linear Representations of Finite Groups*, GTM **42**, Springer, 1977.

[MFG]　H. Hida, *Modular Forms and Galois Cohomology*, Cambridge St. in Adv. Math. **69**, Cambridge Univ. Press, Cambridge, 2000.

[MFM]　T. Miyake, *Modular Forms*, Springer Monographs in Mathematics, Springer, 2006.

[MTV]　U. Jannsen, S. Kleiman and J.-P. Serre, *Motives*, Proc. Symp. Pure Math. **55** Part 1 and 2, AMS, Providence, RI, 1994.

[MUN]　D. Kubert and S. Lang, *Modular units*, Grundlehren der Mathematischen Wissenschaften, **244**, 1981, Berlin-New York, Springer.

[NTH]　S. Iyanaga, ed., *Theory of Numbers*, North-Holland mathematical library, North Holland, Amsterdam, 1975.

[OCS]　J.-P. Serre, *Œuvres-Collected Papers* I–IV, Springer, New York, 2000.

[OSW]　A. Weil, *Œuvres Scientifiques*, I–III, Springer, New York, 1979.

[PAF]　H. Hida, *p-Adic Automorphic Forms on Shimura Varieties*, Springer Monographs in Mathematics, 2004, Springer.

[PAG]　P. Griffiths and J. Harris, *Principles of algebraic geometry*, Wiley Classics Library, New York: John Wiley & Sons, 1994.

[PAM]　B. Balasubramanyam, H. Hida, A. Raghuram, J. Tilouine, edited, *p-Adic Aspects of Modular Forms*, World Scientific, 2016.

[PAN]　F. Q. Gouvêa, *p-adic Numbers, An Introduction*, Third Edition, Universitext, Springer 2020.

[STH] Glen E. Bredon, *Sheaf Theory*, Second Edition, Graduate Text in Math. **170**, 1997, Springer.

[TTF] J. Igusa, *Theta functions,* Die Grundlehren der mathematischen Wissenschaften, Band **194**, Springer, New York-Heidelberg, 1972.

[UNE] G. Robert, *Unités elliptiques et formules pour le nombre de classes des extensions abéliennes d'un corps quadratique imaginaire,* Mémoires de la S. M. F., tome **36**, 1973.

Articles

[B03] K. Buzzard, Analytic continuation of overconvergent eigenforms. J. Amer. Math. Soc. **16** (2003), 29–55.

[BC04] K. Buzzard and F. Calegari, A counterexample to the Gouvêa-Mazur conjecture, C. R. Acad. Sci. Paris, Ser. I **338** (2004) 751–753.

[BCG20] F. M. Bleher, T. Chinburg, R. Greenberg, M. Kakde, G. Pappas, R. Sharifi, M. J. Taylor, Higher Chern classes in Iwasawa theory, American Journal of Math., **142** (2020), 627–682.

[BE10] C. Breuil and M. Emerton, Représentations p-adiques ordinaires de $GL_2(\mathbb{Q}_p)$ et compatibilité local-global. Astérisque **331** (2010), 255–315.

[BK90] S. Bloch and K. Kato, L-functions and Tamagawa numbers of motives. The Grothendieck Festschrift, Vol. I, 333–400, Progr. Math., **86**, Birkhäuser Boston, Boston, MA, 1990.

[BR17] B. Balasubramanyam and A. Raghuram, Special values of adjoint L-functions and congruences for automorphic forms on $GL(n)$ over a number field, Amer. J. of Math., **139** (2017), 641–679.

[BT99] K. Buzzard and R. Taylor, Companion forms and weight one forms. Ann. of Math. (2) **149** (1999), 905–919.

[C77] J. Coates, p-adic L-functions and Iwasawa's theory, Algebraic number fields: L-functions and Galois properties (Proc. Sympos., Univ. Durham, Durham, 1975), pp. 269–353.

[C97] R. F. Coleman, p-adic Banach spaces and families of modular forms, Inventiones Math. **127** (1997), 417–479.

[CM98] R. F. Coleman and B. Mazur, The eigencurve. London Math. Soc. Lecture Note **254** (1998), 1–113.

[CE98] R. F. Coleman and B. Edixhoven, On the semi-simplicity of the U_p-operator on modular forms. Math. Ann. **310** (1998), 119–127.

[CPS75] E. Cline, B. Parshall and L. Scott, Cohomology of finite groups of Lie type I, Inst. Hautes Études Sci. Publ.Math. **45** (1975), 169–191.

[CS74] J. Coates and W. Sinnott, On p-Adic L-Functions over Real Quadratic Fields, Inventiones Math. **25** (1974), 253–279.

[CV03] S. Cho and V. Vatsal, Deformations of Induced Galois Representations, J. reine angew. Math. **556** (2003), 79–97.

[CV92] R. F. Coleman and J. F. Voloch, Companion forms and Kodaira-Spencer theory. Inventiones Math. **110** (1992), 263–281.

[CW19] F. Castella and C. Wang-Erickson, Class groups and local indecomposability for non-CM forms, published online, J. of the European Mathematical Society, DOI 10.4171/JEMS/1107.

[D52] G. F. D. Duff, Differential forms in manifolds with boundary. Ann. of Math. **56**, (1952) 115–127.

[D71] P. Deligne, Formes modulaires et représentations l-adiques. Sém. Bourbaki. Vol. 1968/69: Exp. No. 355, Lecture Notes in Math., **175**, (1971), 139–172.

[D73] V. G. Drinfeld, Two theorems on modular curves, Akademija Nauk SSSR. Funkcionalnyi Analiz i ego Priloženija, **7** (1973), 83–84.

[D95] F. Diamond, The refined conjecture of Serre. Elliptic curves, modular forms, & Fermat's last theorem (Hong Kong, 1993), 22–37, Int. Press, Cambridge, MA, 1995.

[D98] F. Diamond, On deformation rings and Hecke rings. Ann. of Math. **144** (1996), 137–166.

[DFG04] F. Diamond, M. Flach and L. Guo, The Tamagawa number conjecture of adjoint motives of modular forms. Ann. Sci. École Norm. Sup. (4) **37** (2004), 663–727.

[DHI98] K. Doi, H. Hida and H. Ishii, Discriminants of Hecke fields and the twisted adjoint L-values for $GL(2)$, Inventiones Math. **134** (1998), 547–577.

[DK11] F. G. Dorais and D. Klyve, A Wieferich Prime Search up to 6.7×10^{15}, Journal of Integer Sequences, Vol. 14 (2011), Article 11.9.2.

[DS52] G. F. D. Duff and D. C. Spencer, Harmonic tensors on Riemannian manifolds with boundary, Ann. of Math. **56**, (1952), 128–156.

[DS74] P. Deligne and J.-P. Serre, Formes modulaires de poids 1, Ann. Sci. Ec. Norm. Sup. 4-th series **7** (1974), 507–530.

[E87] N. D. Elkies, The existence of infinitely many supersingular primes for every elliptic curve over Q. Invent. Math. **89** (1987), 561–567.

[F06] K. Fujiwara, Deformation rings and Hecke algebras in totally real case, 2006 arXiv:math/0602606 [math.NT].

[FG78] B. Ferrero and R. Greenberg, On the behavior of p-adic L-functions at $s = 0$, Inventiones Math. **50** (1978), 91–102.

[FW79] B. Ferrero and L. Washington, The Iwasawa invariant μ_p vanishes for abelian number fields, Ann. of Math. **109** (1979), 377–395.

[G80] B. H. Gross, On the Factorization of p-adic L-series, Inventiones Math. **57** (1980), 83–95.

[G81] B. H. Gross, p-adic L-series at $s = 0$. J. Fac. Sci. Univ. Tokyo Sect. IA Math. **28** (1981), 979–994.

[G90] B. H. Gross, A tameness criterion for Galois representations associated to modular forms (mod p), Duke Math. J. **61** (1990), 445–517.

[G92] F. Q. Gouvêa, On the ordinary Hecke algebra, Journal of Number Theory **41** (1992), 178–198.

[G94] R. Greenberg, Iwasawa theory and p-adic deformations of motives. Motives (Seattle, WA, 1991), 193–223, Proc. Sympos. Pure Math., **55**, Part 2, Amer. Math. Soc., Providence, RI, 1994.

[G98]   R. Greenberg, Iwasawa theory–past and present, Adv. Stud. Pure Math., **30** (2001), 335–385.

[GJ78]   S. Gelbart and H. Jacquet, A relation between automorphic representations of GL(2) and GL(3), Ann. Sci. École Norm. Sup. (4) **11** (1978), 471–542.

[GM92]   F. Gouvêa and B. Mazur, Families of modular eigenforms, Math. Comp. **58** (1992), 793–805.

[GV04]   E. Ghate and V. Vatsal, On the local behaviour of ordinary Λ-adic representations, Ann. Inst. Fourier **54** (2004), 2143–2162.

[GV11]   E. Ghate and V. Vatsal, Locally indecomposable Galois representations. Canad. J. Math. **63** (2011), 277–297.

[H06]   H. Hida, Anticyclotomic main conjectures, Documenta Math. Extra volume Coates (2006), 465–532.

[H09]   H. Hida, Quadratic exercises in Iwasawa theory, IMRN, Vol. **2009**, Article ID rnn151, 41 pages.

[H13a]   H. Hida, Image of Λ-adic Galois representations modulo p, Inventiones Math. **194** (2013), 1–40.

[H13b]   H. Hida, Local indecomposability of Tate modules of non CM abelian varieties with real multiplication, J. AMS. **26** (2013), 853–877.

[H15]   H. Hida, Big Galois representations and p-adic L-functions, Compositio Math. **151** (2015), 603–664.

[H20]   H. Hida, Memorial article for Goro Shimura, Notice AMS **67** (2020), 680–682.

[H21]   H. Hida, The universal ordinary deformation ring associated to a real quadratic field, preprint, 37 pages, 2020, to appear in Proc. Indian Academy of Science, Math. Science.

[H81a]   H. Hida, Congruences of cusp forms and special values of their zeta functions, Inventiones Math. **63** (1981), 225–261.

[H81b]   H. Hida, On congruence divisors of cusp forms as factors of the special values of their zeta functions, Invent. Math. **64** (1981), 221–262.

[H86a]   H. Hida, Iwasawa modules attached to congruences of cusp forms, Ann. Sci. Ec. Norm. Sup. 4th series **19** (1986), 231–273.

[H86b]   H. Hida, Galois representations into $GL_2(\mathbb{Z}_p[[X]])$ attached to ordinary cusp forms, Inventiones Math. **85** (1986), 545–613.

[H88a]   H. Hida, Modules of congruence of Hecke algebras and L-functions associated with cusp forms, Amer. J. Math. **110** (1988), 323–382.

[H88b]   H. Hida, On p-adic Hecke algebras for GL_2 over totally real fields, Ann. of Math. **128** (1988), 295–384.

[H88c]   H. Hida, A p-adic measure attached to the zeta functions associated with two elliptic modular forms, Ann. l'Institut Fourier, **38** (1988), 1–83.

[H94]   H. Hida, On the critical values of L-functions of $GL(2)$ and $GL(2) \times GL(2)$, Duke Math. J. **74** (1994), 431–529.

[H99]   H. Hida, Non-critical values of adjoint L-functions for $SL(2)$, Proc. Symp. Pure Math. **66** (1999), Part I, 123–175.

[HHO17]   W. Hart, D. Harvey and W. Ong, Irregular primes to two billion, Math. Comp. **86** (2017), 3031–3049.

[HM97]   H. Hida and Y. Maeda, Non-abelian base-change for totally real fields, Special Issue of Pacific J. Math. in memory of Olga Taussky Todd, 189–217, 1997.

[HT93]   H. Hida and J. Tilouine, Anticyclotomic Katz p-adic L-functions and congruence modules, Ann. Sci. Ec. Norm. Sup. 4th series **26** (1993), 189–259.

[I69]   K. Iwasawa, On p-adic L-functions, Ann. of Math. **89** (1969), 198–205 ([CPI, II 48]).

[I73]   K. Iwasawa, On $\mathbf{Z}_\ell$-extensions of algebraic number fields, Ann. of Math. **98** (1973), 246–326 ([CPI, II 52]).

[I79]   K. Iwasawa, Some problems on cyclotomic fields and $\mathbb{Z}_p$-extensions, (unpublished work no. 3 in [CPI, II, U3, 853–861]).

[J79]   N. Jochnowitz, The index of the Hecke ring, $\mathbf{T}_k$, in the ring of integers of $\mathbb{T}_k \otimes \mathbf{Q}$, Duke Math. J. **46** (1979), 861–869.

[K07]   J. Klaška, Short remark on Fibonacci-Wieferich primes, Acta Mathematica Universitatis Ostraviensis, **15** (2007), 21–25.

[K18]   S. Kalyanswamy, Remarks on automorphy of residually dihedral representations. Math. Res. Lett. **25** (2018), 1285–1304.

[K47]   E. Kummer, Über die Zerlegung der aus Wurzeln der Einheit gebildeten complexen Zahlen in ihre Primfactoren, J. Reine Angew. Math. **35** (1847), 327–367 (Collected Papers Vol. I No. 27).

[K78]   N. M. Katz, p-adic L-functions for CM fields, Inventiones Math. **49** (1978), 199–297.

[K80]   P. Kutsko, The Langlands conjecture for Gl$_2$ of a local field, Ann. of Math. **112** (1980), 381–412.

[KL78]   S. Lang and D. Kubert, The index of Stickelberger ideals of order 2 and cuspidal class numbers, Math. Annalen **237** (1978), 213–232.

[KW08]   L. J. P. Kilford and G. Wiese, On the failure of the Gorenstein property for Hecke algebras of prime weight. Experiment. Math. **17** (2008), 37–52.

[KW09]   C. Khare and J.-P. Wintenberger, Serre's modularity conjecture. I, II. I: Invent. Math. **178** (2009), 485–504; II. **178** (2009), 505–586.

[L95]   H. W. Lenstra, Complete intersections and Gorenstein rings, in "Conference on Elliptic Curves and Modular Forms", Hong Kong, International Press (1995), pp. 99–109.

[M00]   H. Minkowski, Zur Theorie der Einheiten in den algebraischen Zahlkörpern, Nachr. Gesellschaft Wiss. Göttingen, Math.-Phys. Kl., **1** (1900), 90–93 (Gesammelte Abh. Vol. 1, XV).

[M15]   Y. Maeda, Maeda's conjecture and related topics, RIMS Kôkyûroku Bessatsu **B53** (2015), 305–324.

[M72]   I. Manin, Parabolic points and zeta functions of modular curves, Izvestiya **36** (1972), 19–66.

[M77]   B. Mazur, Modular curves and the Eisenstein ideal, Publ. IHES **47** (1977), 33–186.

[M87]   B. Mazur, Deforming Galois Representations, MSRI Publ. **16** (1987), 385–437.

[MR70]   B. Mazur and L. Roberts, Local Euler characteristic, Inventiones Math. **9** (1970), 201–234.

[MW84]   B. Mazur and A. Wiles, Class fields of abelian extensions of **Q**. Inventiones Math. **76** (1984), 179–330.

[MW86]   B. Mazur and A. Wiles, On p-adic analytic families of Galois representations, Compositio Math. **59** (1986), 231–264.

[O82]   M. Ohta, On l-adic representations attached to automorphic forms, Japanese J. Math. **8** (1982), 1–47.

[O99]   M. Ohta, Ordinary p-adic étale cohomology groups attached to towers of elliptic modular curves, Compositio Math. **115** (1999) 241–301.

[P94]   B. Perrin-Riou, Représentations p-adiques ordinaires, In "Périodes p-adiques," Astérisque **223** (1994), 185–220.

[R18]   D. P. Roberts, $PGL_2(\mathbb{F}_l)$ number fields with rational companion forms, Int. J. Number Theory **14** (2018), 825–845.

[R76]   K. A. Ribet, A modular construction of unramified p-extensions of $\mathbb{Q}(\mu_p)$. Inventiones Math. **34** (1976), 151–162.

[R84]   K. Ribet, Congruence relations between modular forms. Proceedings of ICM, **1** (1984), 503–514, PWN, Warsaw, 1984.

[R85]   K. A. Ribet, On l-adic representations attached to modular forms. II. Glasgow Math. J. **27** (1985), 185–194.

[R91]   K. Rubin, The "main conjectures" of Iwasawa theory for imaginary quadratic fields, Inventiones Math. **103** (1991), 25–68.

[R93]   R. Ramakrishna, On a variation of Mazur's deformation functor. Compositio Math. **87** (1993), 269–286.

[S11]   R. Sharifi, A reciprocity map and the two-variable p-adic L-function, Ann. of Math. **173** (2011), 251–300.

[S67]   G. Shimura, Construction of class fields and zeta functions of algebraic curves, Ann. of Math. **85** (1967), 58–159, [CPS, II 67b].

[S70]   J.-P. Serre, Probleme des groupes de congruence pour SL_2, Ann. of Math. **92** (1970), 489–527 ([OCS, II 86]).

[S71]   J.-P. Serre, Congruences et formes modulaires (d'après H.P.F. Swinnerton-Dyer), Séminaire Bourbaki, 1971/72. ([OCS, III 95]).

[S73]   J.-P. Serre, Formes modulaires et fonctions zeta p-adiques, Lecture Notes in Math. **350** (1973), 191–268 ([OCS, III 97]).

[S75]   G. Shimura, On the holomorphy of certain Dirichlet series. Proc. London Math. Soc. (3) **31** (1975), 79–98 [CPS, II 75a].

[St75]   H. M. Stark, L-Functions at $s = 1$. II. Artin L-Functions with Rational Characters, Advances in Math. **17** (1975), 60–92.

[S80]   J. Sturm, Special values of zeta functions, and Eisenstein series of half integral weight, Amer. J. Math. **102** (1980), 219–240.

[S81]   J.-P. Serre, Quelques applications du théorème de densité de Chebotarev. Publ. IHES **54** (1981), 323–401 ([OCS, III 125]).

[S84]	W. Sinnott, On the μ-invariant of the Γ-transform of a rational function, Inventiones Math. **75** (1984), 273–282.

[S89]	J. Sturm, Evaluation of the symmetric square at the near center point, Amer. J. Math. **111** (1989), 585–598.

[S90]	L. Stickelberger, Ueber eine Vcrallgemeinerung der Kreistheilung, Mathematische Annalen, **37** (1890), 321–367.

[SS92]	Zhi-Hong Sun and Zhi-Wei Sun, Fibonacci numbers and Fermat's last theorem, Acta Arithmetica **60** (1992), 371–388.

[T16]	J. Thorne, Automorphy of some residually dihedral Galois representations. Math. Ann. **364** (2016), 589–648.

[T39]	T. Tannaka, Über den Dualitatssatz der nichtkommutativen topologischen Gruppen, Töhoku Math. J. (1) **45** (1939), 1–12.

[T89]	J. Tilouine, Sur la conjecture principale anticyclotomique. Duke Math. J. **59** (1989), 629–673.

[TU20]	J. Tilouine and E. Urban, Integral period relations and congruences, preprint, to appear in Algebra and Number Theory, posted on arXiv:1811.11166 [math.NT].

[T81]	J. Tate, On Stark's conjectures on the behavior of $L(s, \chi)$ at $s = 0$, J. Fac. Sci. Univ. Tokyo IA Math. **28** (1981), 963–978.

[T84]	J. Tate, Les conjectures de Stark sur les fonctions L d'Artin en $s = 0$, Progress in Math., Birkhäuser Boston, **47** (1984), 143–153.

[T98]	T. Tsuji, p-adic Hodge theory in the semi-stable reduction case. Proceedings of ICM, II, Doc. Math. 1998, Extra Vol. II, 207–216.

[TW95]	R. Taylor and A. Wiles, Ring theoretic properties of certain Hecke algebras, Ann. of Math. **141** (1995), 553–572.

[U95]	D. L. Ulmer, A Construction of Local Points on Elliptic Curves over Modular Curves, IMRN (1995), No. 7.

[U12]	University of Georgia VIGRE Algebra Group, Second cohomology for finite groups of Lie type, J. of Algebra **360** (2012), 21–52.

[W49]	A. Weil, Numbers of solutions of equations in finite fields, Bull. AMS **55** (1949), 497–508 ([OSW, I 1949]).

[W52]	A. Weil, Jacobi Sums as "Grössencharaktere" Transactions of the AMS, **73** (1952), 487–495 ([OSW, I 1952d]).

[W74]	A. Weil, Exercices dyadiques, Inventiones Math. **27** (1974), 1–22 ([OSW, I 1974e]).

[W88]	A. Wiles, On ordinary Λ-adic representations associated to modular forms, Inventiones Math. **94** (1988), 529–573.

[W95]	A. Wiles, Modular elliptic curves and Fermat's last theorem, Ann. of Math. **141** (1995), 443–551.

[W98]	D. Wan, Dimension variation of classical and p-adic modular forms. Inventiones Math. **133** (1998), 449–463.

[Y82]	R. I. Yager, On Two Variable p-adic L-functions, Annals of Math., **115** (1982), 411–449.

[Z14]	B. Zhao, Local indecomposability of Hilbert modular Galois representations. Ann. Inst. Fourier (Grenoble) **64** (2014), 1521–1560.

List of symbols and statements

421

List of Statements

Index